Toxicology of Marine Mammals

New Perspectives: Toxicology and the Environment
Series Editors
Donald E. Gardner, A. Wallace Hayes and John A. Thomas

Target Organ Toxicology in Marine and Freshwater Teleosts Volume 1 – Organs
Daniel Schlenk & William H. Benson (eds), 2001

Target Organ Toxicology in Marine and Freshwater Teleosts Volume 2 – Systems
Daniel Schlenk & William H. Benson (eds), 2001

Toxicology of Marine Mammals
*Joseph G. Vos, Gregory D. Bossart, Michel Fournier,
Tom O'Shea (eds), 2003*

Forthcoming

Avian Toxicology
Christian E. Grue and Susan C. Gardner (eds)

New Perspectives: Toxicology and the Environment

Toxicology of Marine Mammals

Edited by

Joseph G. Vos
National Institute of Public Health and the Environment
Bilthoven, RIVM
The Netherlands

Gregory D. Bossart
Division of Marine Mammal Research and Conservation
Harbor Branch Oceanographic Institution
Fort Pierce, Florida
USA

Michel Fournier
Centre de Recherche en Sante
INRS-Institut Armand-Frappier
Pointe-Claire, Quebec
Canada

Thomas J. O'Shea
Department of the Interior
US Geological Survey
Midcontinent Ecological Sciences Center
Ft. Collins, Colorado
USA

CRC Press
Taylor & Francis Group
Boca Raton London New York

CRC Press is an imprint of the
Taylor & Francis Group, an **informa** business

CRC Press
Taylor & Francis Group
6000 Broken Sound Parkway NW, Suite 300
Boca Raton, FL 33487-2742

First issued in paperback 2019

© 2003 by Taylor & Francis Group, LLC
CRC Press is an imprint of Taylor & Francis Group, an Informa business

Typeset in 10/12pt Times by Graphicraft Limited, Hong Kong

No claim to original U.S. Government works

ISBN-13: 978-0-415-23914-1 (hbk)
ISBN-13: 978-0-367-39563-6 (pbk)

British Library Cataloguing in Publication Data
A catalogue record for this book is available from the British
Library

Library of Congress Cataloging in Publication Data
A catalog record has been requested

**Visit the Taylor & Francis Web site at
http://www.taylorandfrancis.com**

**and the CRC Press Web site at
http://www.crcpress.com**

Contents

Contributors

Paul R. Becker National Institute of Standards and Technology, Hollings Marine Laboratory, 331 Fort Johnson Road, Charleston, SC 29412, USA

Kimberlee B. Beckmen Alaska Department of Fish and Game, 1300 College Road, Fairbanks, Alaska 99701, USA

A. Bergman Contaminant Research Group, Swedish Museum of Natural History, Box 50007, 104 05 Stockholm, and Department of Pathology, Faculty of Veterinary Medicine, Swedish University of Agricultural Sciences, Uppsala, Sweden

A. Bignert Contaminant Research Group, Swedish Museum of Natural History, Box 50007, 104 05 Stockholm, Sweden

Gregory D. Bossart Division of Marine Mammal Research and Conservation, Harbor Branch Oceanographic Institution, 5600 US 1 North, Fort Pierce, FL 34946, USA

Jean-Marie Bouquegneau Laboratory for Oceanology, University of Liège, B6 Sart-Tilman, B-4000 Liège, Belgium

Gerald Bratton Department of Veterinary Anatomy and Public Health, Texas A & M University, Texas, USA

Pauline Brousseau Biophage, 6100 Royalmount Avenue, Montreal, Québec, H4P 2R2, Canada

David L. Busbee Department of Veterinary Anatomy and Public Health, Texas A & M University, College Station, Texas 77843, USA

Michael J. Carvan III Great Lakes WATER Institute, University of Wisconsin-Milwaukee, Milwaukee, Wisconsin 53204, USA

Theo Colborn World Wildlife Fund, 1250-24th street, N.W., suite 6019, Washington DC 20037, USA

Daniel G. Cyr Centre de Recherches en Santé Humaine, INRS-Institut Armand Frappier, Université du Québec, 245 Boul. Hymus, Pointe Claire, Québec, H9R 1G6, Canada

Krishna Das Laboratory for Oceanology, University of Liège, B6 Sart-Tilman, B-4000 Liège, Belgium

Virginie Debacker Laboratory for Oceanology, University of Liège, B6 Sart-Tilman, B-4000 Liège, Belgium

Sylvain De Guise Department of Pathobiology, University of Connecticut, 61 North Eagleville Road, U-89, Storrs, Connecticut 06269, USA

Rik L. de Swart Institute of Virology, Erasmus University, Rotterdam, The Netherlands

Gregory J. Doucette Marine Biotoxins Program, NOAA National Ocean Service, Center for Coastal Environmental Health and Biomolecular Research, 219 Fort Johnson Rd, Charleston, SC 29412, USA

H. Dubeau Département des Sciences Biologiques, Université du Québec à Montréal, Montréal, Québec, H3C 3P8, Canada

Michel Fournier Santé Humaine Department, INRS-Institut Armand-Frappier, Pointe-Claire, Québec, Canada

J.M. Gauthier Département des Sciences Biologiques, Université du Québec à Montréal, Montréal, Québec, H3C 3P8, Canada

Jay Gorzelany Mote Marine Laboratory, 1600 Ken Thompson Parkway, Sarasota, Florida 34236, USA

Mary Gregory Centre de Recherches en Santé Humaine, INRS-Institut Armand Frappier, Université du Québec, 245 Boul. Hymus, Pointe Claire, Québec, H9R 1G6, Canada

Frances M.D. Gulland The Marine Mammal Center, Marin Headlands, Saulsalito, CA 94965, USA

Robert Higgins Département de Pathologie et Microbiologie, Faculté de Médecine Vétérinaire, Université de Montréal, Saint-Hyacinthe, J2S 7C5, PQ, Canada

Steven D. Holladay Virginia Polytechnic Institute, Blacksburg, Virginia 24061, USA

Margaret M. Krahn Environmental Conservation Division, Northwest Fisheries Science Center, National Marine Fisheries Service, National Oceanic and Atmospheric Administration, 2725 Montlake Boulevard East, Seattle, Washington 98112-2097, USA

Philippe Labelle Département de Pathologie et Microbiologie, Faculté de Médecine Vétérinaire, Université de Montréal, Saint-Hyacinthe, J2S 7C5, PQ, Canada

Jean-Martin Lapointe Département de Pathologie et Microbiologie, Faculté de Médecine Vétérinaire, Université de Montréal, Saint-Hyacinthe, J2S 7C5, PQ, Canada

Daniel Martineau Département de Pathologie et Microbiologie, Faculté de Médecine Vétérinaire, Université de Montréal, Saint-Hyacinthe, J2S 7C5, PQ, Canada

James P. Meador Environmental Conservation Division, Northwest Fisheries Science Center, National Marine Fisheries Service, National Oceanic and Atmospheric Administration, 2725 Montlake Boulevard East, Seattle, Washington 98112-2097, USA

Igor Mikaelian Département de Pathologie et Microbiologie, Faculté de Médecine Vétérinaire, Université de Montréal, Saint-Hyacinthe, J2S 7C5, PQ, Canada

Robert B. Moeller, Jr California Animal Health and Food Safety Laboratory, University of California, Tulare, California 93274, USA

Todd M. O'Hara Department of Wildlife Management, North Slope Burrough, Box 69, Barrow, Alaska 99723, USA

M. Olsson Contaminant Research Group, Swedish Museum of Natural History, Box 50007, 104 05 Stockholm, Sweden

Thomas J. O'Shea US Geological Survey, Midcontinent Ecological Science Center, 4512 McMurry Avenue, Fort Collins, CO 80525-3400, USA

Albert D.M.E. Osterhaus Institute of Virology, Erasmus University, 3015 GE Rotterdam, and Seal Rehabilitation and Research Center, 9968 AG Pieterburen, The Netherlands

Stéphane Pillet Laboratory for Oceanology, University of Liège, B6 Sart-Tilman, B-4000 Liège, Belgium and INRS-Santé, 245 Hymus, Pointe Claire, Québec, H9R 1G6, Canada

E. Rassart Département des Sciences Biologiques, Université du Québec à Montréal, Montréal, Québec, H3C 3P8, Canada

Michelle L. Reddy Saic Maritime Services Division, 3990 Old Town Ave Ste 105A, San Diego, California 92110, USA

Peter J.H. Reijnders Alterra, Marine and Coastal Zone Research, PO Box 167, 1790 AD Den Burg, The Netherlands

Sam H. Ridgway Saic Maritime Services Division, 3990 Old Town Ave, Ste 208A, San Diego, California 92110, USA

Peter S. Ross Institute of Ocean Sciences, Fisheries and Oceans Canada, PO Box 6000, Sidney BC V8L 4B2, Canada

Teri. L. Rowles NOAA National Marine Fisheries Service, Office of Protected Species, 1315 East-West Highway, Silver Spring, MD 20910, USA

S. Ruby Biology Department, Concordia University, Montreal, Québec, Canada

Mark P. Simmonds Natural Resources Institute, The University of Greenwich, Chatham Maritime, Kent ME4 4TB, UK

Michael J. Smolen World Wildlife Fund, 1250-24th street, N.W., suite 6019, Washington DC 20037, USA

John E. Stein Environmental Conservation Division, Northwest Fisheries Science Center, National Marine Fisheries Service, National Oceanic and Atmospheric Administration, 2725 Montlake Boulevard East, Seattle, Washington 98112-2097, USA

Shinsuke Tanabe Department of Environment Conservation, Ehime University, Tarumi 3-5-7, Matsuyama 790, Japan

Karen L. Tilbury Environmental Conservation Division, Northwest Fisheries Science Center, National Marine Fisheries Service, National Oceanic and Atmospheric Administration, 2725 Montlake Boulevard East, Seattle, Washington 98112-2097, USA

Frances M. Van Dolah Marine Biotoxins Program, NOAA National Ocean Service, Center for Coastal Environmental Health and Biomolecular Research, 219 Fort Johnson Rd, Charleston, SC 29412, USA

Henk van Loveren National Institute of Public Health and the Environment, 3720 BA Bilthoven, The Netherlands

Isabelle Voccia Santé Humaine, INRS-Institut Armand-Frappier, Pointe-Claire, Québec, Canada

Joseph G. Vos National Institute of Public Health and the Environment, 3720 BA Bilthoven, and Faculty of Veterinary Medicine, Utrecht University, Utrecht, The Netherlands

Graham A.J. Worthy Texas Marine Mammal Stranding Network, Texas A & M University, Galveston, Texas 77551-5962, USA

Victoria Woshner Department of Veterinary Biosciences, University of Illinois, Urbana, Illinois, USA

Part I

Implications of contaminants for marine mammal health

1 Pathology of marine mammals with special reference to infectious diseases

Robert B. Moeller, Jr

Introduction

Global pollution of our environment is a serious concern. Pollution of the aquatic environment, particularly our oceans, has raised many questions about the condition of our terrestrial environment. The marine environment is polluted with a variety of chemical compounds and heavy metals. Many of these compounds are known to be a threat to the health of most animal species. Since most marine mammals are carnivores, they represent animal groups that are at the top of the food chain. The health risk to these animals from exposure to various xenobiotics may be very high due to bioaccumulation of these compounds in various tissues. This is due to the high body fat of these animals and the lipophilic nature of many xenobiotics. Like humans, these animals have an extended life span that allows them to accumulate heavy concentrations of various xenobiotics in their tissues.

Like terrestrial mammals, marine mammals nurse their young. Lactating marine mammals have a high fat content in their milk. This exposes the neonate to high concentrations of lipophilic xenobiotics at critical developmental stages in their lives (Martineau *et al.*, 1994; Wade *et al.*, 1997). The excessive accumulation of these xenobiotics may predispose these young animals, as well as the older animals, to immunodeficiency leading to poor health, increased susceptibility to infectious agents, and tumor development (Mossner and Ballschmiter, 1997).

The decline of many marine mammal species and massive die-offs or strandings has heightened our concern about environmental pollution and its causal role in these events. Few studies exist regarding the health-related effects of xenobiotics that bioaccumulate in marine mammals (Reijnders, 1986). Most papers on this subject only summarize the possible health effects of bioaccumulation of xenobiotics on marine mammals (Miller, 1982; Kannan *et al.*, 1989; Rawson *et al.*, 1995; Becker *et al.*, 1997; Mossner and Ballschmiter, 1997; Tilbury *et al.*, 1997; Parson, 1998).

Xenobiotics of major concern are organochlorines such as polychlorinated biphenyls (PCBs), dibenzofurans (PCDFs), dibenzodioxins (PCDDs), organochlorine insecticides [dichlorodiphenyltrichloroethane (DDT), toxaphene, and others], and heavy metals (mercury, lead, and cadmium). Many of these compounds are known to target the immune system. Toxic levels of these substances may cause thymic atrophy, pancytopenia, and immunosuppression, which can result in reduced resistance to infectious diseases. Immunosuppression by PCBs involves primarily cell-mediated immunity through a mechanism of T-cell suppression (Burns *et al.*, 1996; see also Chapters 20–22 in this volume).

Following a discussion of the toxic effects of oil spills on marine mammals, this chapter will deal primarily with infectious agents and the consequences on the health of marine mammals.

Oil spills

Crude oil and other hydrocarbons from petroleum products are known to cause serious problems in many marine animals. These compounds can cause both short- and long-term health problems in animals that come in contact with these substances. These substances are lighter than water, resulting in the compounds floating at the water–air interface. Consequently, any animal surfacing in the sheen created by the oil spill will be exposed to the toxic compounds in the oil. Since most marine mammals live near land or ice sheets, the release of crude oil or petroleum distillates into this environment causes these compounds to be concentrated in narrow bodies of water (bays, inlets, and ice flows) due to water currents or wind action. This leads to marine mammals becoming entrapped in the oil sheen.

Crude oil and petroleum distillates are complex mixtures composed of numerous volatile and nonvolatile compounds. Crude oil varies greatly in hydrocarbon composition throughout the world (Rahimtula *et al.*, 1984; Lipscomb *et al.*, 1993). Consequently, it is almost impossible to compare the short- and long-term toxicity of one petroleum product with that of another. Most species of seals, polar bears (*Ursus maritimus*), and cetaceans do not avoid oil spills and have been noted feeding and swimming in oiled areas (Geraci and St. Aubin, 1982; Engelhardt, 1983). Consequently, repeated and prolonged exposure to these complex mixtures can lead to serious medical problems for these animals.

Bivalves, crustaceans, and zooplankton also have been recognized to bioaccumulate numerous hydrocarbons from oils (Engelhardt, 1983; Thomas *et al.*, 1999). The persistence of these compounds is possibly due to the lack of the microsomal enzyme aryl hydrolase that assists in the breakdown of complex hydrocarbons (Vandermeulen and Penrose, 1978; Geraci and St. Aubin, 1980). Because numerous marine mammals (walrus (*Odobenus rosmarus*), bearded seals (*Erignathus barbatus*), sea otters (*Enhydra lutris*), and baleen whales) utilize these marine species for all or part of their diet, the potential exists for long-term health and reproductive problems (Geraci and St. Aubin, 1982).

Contact with crude oil or petroleum distillates can cause adherence of the oil compound to the skin (particularly in furred species) and oral and nasal orifices, leading to serious complications. Contact with oil has led to oil coating the fur and hindering the seal's swimming ability (Warner, 1969; Davis and Anderson, 1976). Some species of pinnipeds tend to clear their coats of oil rapidly (ringed seals (*Phoca hispida*) in 1 day) while others may take weeks (northern elephant seal pups (*Mirounga leonine*) in 1 month) (Smith and Geraci, 1975; Engelhardt, 1983). Oil attaching to baleen hinders the filtration ability of baleen whales, leading to poor efficiency in obtaining food (Engelhardt, 1983). Oil on the pillage of sea otters and polar bear leads to hypothermia and possibly death (Oritsland *et al.*, 1981; Costa and Kooman, 1980, 1982; Engelhardt, 1983; Lipscomb *et al.*, 1993).

The absorption of various by-products of oil can lead to serious problems in marine mammals. Limited clinical data suggest that pinnipeds, cetaceans, and polar bears have great differences in their susceptibility to crude oil and petroleum distillates (Engelhardt, 1983). In polar bears, the ingestion of oil resulted in high levels of hydrocarbons (benzene and naphthalene) in the kidney, brain, and bone marrow, with resultant peripheral hemolysis of red blood cells, erythropoietic dysfunction in the bone marrow, degenerative changes in the liver, atrophy of lymphoid tissue, renal dysfunction, and gastric ulcers (Oritsland *et al.*, 1981; Engelhardt, 1983). In the sea otter, crude-oil exposure leads to mediastinal and pulmonary emphysema, gastric erosions, hepatic and renal lipidosis, and centrolobular hepatic necrosis (Lipscomb *et al.*, 1993). In seals and cetaceans, the liver and blubber tend to accumulate the highest concentrations of hydrocarbons (Engelhardt *et al.*, 1977; Geraci and St. Aubin, 1982; Engelhardt, 1983). Elevated levels of these hydrocarbons in the blubber may lead to the release of potentially dangerous compounds during lactation, which may affect the young at critical developmental stages. Little is known about the clinical or pathological effects of oil on these species. Most have not died after exposure to these substances. However, ringed seals are noted to develop a severe conjunctivitis with corneal edema, erosions, and ulcers that resolve after removal from exposure (Geraci and Smith, 1976). These animals also developed elevated liver enzymes without clinical liver disease. In toothed whales, mild skin damage characterized by cellular necrosis of the epidermis has been observed after exposure to crude oils (Geraci and St. Aubin, 1982; Engelhardt, 1983).

In summary, the exposure of marine mammals to crude oils and petroleum distillates will vary due to species variability and type of hydrocarbon compound. One needs to remember that crude oils vary in their composition and consequently the damage they will do. New crude and refined distillates will be more toxic than aged or weathered products due to the loss of the more volatile components of the oil.

Viral diseases

Viral infections can cause explosive disease outbreaks in both marine and terrestrial mammals. Usually, large numbers of animals will be presented for examination. Presenting signs may be nonspecific with serious concern that the event may be due to a toxin. Numerous major die-offs or mass strandings have been due to viral diseases. Examples of this are: influenza in harbor seals (*Phoca vitulina*) (Geraci and St. Aubin, 1982); phocine distemper in harbor seals (Kennedy *et al.*, 1988; Osterhaus 1988); and morbilliviral disease in bottlenose dolphins (*Tursiops truncatus*), and striped dolphins (*Stenella coeruleoalba*) (Domingo *et al.*, 1992; Lipscomb *et al.*, 1996; Schulman *et al.*, 1997).

Animals that present with severe bacterial pneumonia should always be suspected as having an underlying viral disease. Careful examination of the tissue should always occur in order to ensure that the initial viral infection is identified. Animals that are seriously immunocompromised and have secondary bacterial, fungal, or other viral (i.e. poxviruses or papillomavirus) diseases should also be considered potentially to have an underlying viral infection causing the immunosuppression of the animal. The utilization of the few established seal cell lines or the use of established terrestrial mammal cell lines should always be attempted when trying to culture tissue for viral diseases. New reverse transcriptase–polymerase chain reactions (RT–PCR) and *in situ* hybridization techniques and immunohistochemistry are assisting in the identification of viral agents (Neill and Seal, 1995; Osterhaus *et al.*, 1997).

When dealing with a potential viral outbreak, remember that the virus may have mutated to be more pathogenic to the susceptible animal population. This new mutation may make the virus more compatible to terrestrial mammals or humans. Some viruses, particularly caliciviruses, do pose a threat to humans and terrestrial mammals. San Miguel sea lion virus is indistinguishable from vesicular exanthema virus of swine and can cause disease in humans (Smith *et al.*, 1981, 1983a; Seal *et al.*, 1995). Numerous other caliciviruses are observed in other mammals and should be handled with caution.

Phocine morbillivirus (phocine distemper virus, PDV-1, PDV-2)

Phocine morbillivirus has caused several disease outbreaks that killed thousands of seals in north-western Europe and Lake Baikal in Siberia (Osterhaus, 1988; Osterhaus and Vedder, 1988; Grachev *et al.*, 1989; Kennedy, 1998). The morbillivirus affecting harbor seals and gray seals (*Halichoerus grypus*) in northern Europe and North America is caused by phocine distemper virus-1 (PDV-1). This virus is similar to, yet antigenically distinct from, canine distemper virus. The genetic differences between these two viruses are such that they should be considered separate viruses (Bostock *et al.*, 1990). Serologic studies have demonstrated that harbor seals, hooded seals

(*Cystophora cristata*), and ringed seals have an immune response by viral neutralization to PDV-1 (Visser *et al.*, 1993b). Phocid distemper virus-2 (PDV-2), isolated from Siberian seals (*Phoca sibirica*) in Lake Baikal, is felt to be a field strain of the canine distemper virus found in Europe (Mamaev *et al.*, 1996). A morbillivirus isolated from sick Mediterranean monk seals (*Monachus monachus*) from the coast of West Africa most likely resembles the morbillivirus seen in cetaceans (Osterhaus *et al.*, 1997; Van de Bildt *et al.*, 1999, 2000). This suggests that these morbillivirus may not be host specific and cross species lines.

Clinically, affected seals are depressed and weak with severe respiratory distress. A mucopurulent to serous oculonasal discharge is often observed. Many animals develop subcutaneous emphysema around the neck and thorax. On necropsy, these animals have diffusely edematous lungs with sharply demarcated areas of red consolidation. Emphysematous bullae are observed involving the interlobular septa and pleura of the caudal lung lobes. Congestion and a thick mucopurulent exudate are observed in the upper respiratory tract. Hydropericardium, hydrothorax, and hepatic congestion are common. Pulmonary lymph nodes are edematous. Histologically, the lung lesion consists of a bronchointerstitial pneumonia with syncytial cells and Type II pneumocyte proliferation. Eosinophilic intracytoplasmic inclusions are present in bronchial epithelium and syncytial cells. The brain has a nonsuppurative encephalitis characterized by necrosis of neurons (primarily in the cerebral cortex), gliosis, and lymphocytic perivascular cuffs. Many affected neurons contain intranuclear and intracytoplasmic inclusions. Demyelination of the subependymal white matter is observed. Prominent depletion and necrosis of lymphocytes in lymphoid tissue is present. Some seals develop a necrotizing lymphocytic myocarditis. Like canine distemper, intranuclear and intracytoplasmic inclusions are observed in the gastric mucosa and transitional epithelium of the urinary bladder and renal pelvis (Kennedy *et al.*, 1988, 1989; Bergman *et al.*, 1990; Kennedy, 1990; Osterhaus *et al.*, 1990a; Hall, 1995).

Morbillivirus in cetaceans

Several serious disease outbreaks with numerous cetacean deaths have been caused by a morbillivirus. This viral infection was first identified in harbor porpoises (*Phocoena phocoena*) from the Irish Sea during the 1988 European phocine morbillivirus outbreak (Kennedy *et al.*, 1988, 1991). This virus has killed numerous striped dolphins along the Spanish Mediterranean coast in 1990, has been identified in several outbreaks affecting bottlenose dolphins along the Eastern Atlantic and Gulf Coast of the United States, and has been seen in die-offs of common dolphins from the Black Sea (Domingo *et al.*, 1992; Lipscomb *et al.*, 1996; Schulman *et al.*, 1997, Birkun *et al.*, 1999). Pilot whales (*Globicephala* sp.) are also infected with this or a similar morbillivirus (Duignan *et al.*, 1995a).

The viruses affecting harbor porpoises and striped dolphins are closely related but antigenically distinct (Barrett *et al.*, 1993). These viruses are antigenically distinct from phocine morbillivirus and other mammalian morbilliviruses. However, the cetacean morbilliviruses appear to be related to the ruminant morbillivirus, peste-des-petits ruminants virus (Barrett *et al.*, 1993, 1995; Visser *et al.*, 1993a). Serologic evidence of morbillivirus infection has been identified in numerous odontocete cetaceans in the western Atlantic and may have a potential impact on these species. Affected cetaceans develop pulmonary and central nervous system lesions. Grossly, these animals develop severe pneumonia with multiple foci of atelectasis and consolidation. Pulmonary lymph nodes are often enlarged and edematous.

Histologically, there is a bronchointerstitial pneumonia with necrosis of bronchial and bronchiolar epithelium and a prominent mucopurulent exudate. Acidophilic intracytoplasmic inclusions are observed frequently (occasionally intranuclear inclusions are also observed) in the bronchiolar epithelium. Type II pneumocyte hyperplasia and prominent mononuclear inflammation are common in alveoli. Syncytial cells are seen in both the alveoli and bronchiolar epithelium. A nonsuppurative meningoencephalitis is observed involving the cerebral gray matter, with occasional eosinophilic intranuclear inclusions present in neurons. In some animals, there is necrosis of the bile duct epithelium and the transitional epithelium of the urinary bladder with occasional eosinophilic intracytoplasmic inclusions. The lymph nodes have prominent lymphoid depletion with scattered multinucleated syncytial cells. Multinucleated syncytial cells and inflammation have been observed in the ductal epithelium of the mammary gland. All affected areas demonstrate intense immunohistochemical staining for morbillivirus. Many affected dolphins are co-infected with serious secondary infections of *Toxoplasma*, *Aspergillus*, and other fungi (Kennedy *et al.*, 1988; Domingo *et al.*, 1992; Lipscomb *et al.*, 1996).

It should also be noted that free-ranging Florida manatees (*Trichechus manatus latirostris*) have high titers to porpoise and dolphin morbillivirus but have not developed clinical disease (Duignan *et al.*, 1995b).

Dolphin pox

Dolphin pox, also known as 'tattoo', is characterized by prominent well-delineated lines of hyperpigmentation of the epidermis with various geographic design patterns. These design patterns have been described as targets, circles, and pinhole lesions. The lesions are usually smooth and flat, but may occasionally be raised. They are primarily located on the dorsal body near the head and blowhole, flippers, dorsal fins, and fluke. Although this virus does not appear to cause serious illness in cetaceans, the development of these lesions usually coincides with periods of poor health and stress (Flom and Houk, 1979; Geraci *et al.*, 1979; Migaki, 1987).

Histologically, the lesions consist of ballooning (hydropic) degeneration of the deep layers of the stratum intermedium. Irregularly shaped or round, variably sized, eosinophilic, intracytoplasmic inclusions are present in the cells undergoing ballooning degeneration. The stratum externum may become thickened. Minimal inflammation is observed; this may be the reason for persistence of the lesion (Migaki, 1987). The cause of the hyperpigmentation is unknown, but one theory is that it is secondary to stimulation of the dermal melanocytes by the viral infection. Another theory postulates that damage to the stratum externum and intermedium leads to filling in of these defects with debris and bacteria which then cause discoloration of the epidermis (Britt and Howard, 1983).

This is an unusual pox lesion since it is *not* a proliferative lesion. The lesion persists and slowly spreads in the affected animal. Affected animals do not routinely develop antibodies to the virus; however, if antibodies are developed, the lesion regresses, with the affected skin undergoing necrosis and sloughing. A skin biopsy, or scraping the lesion, can cause regression in a zonal pattern around the biopsy site (Smith *et al.*, 1983b).

Seal pox

Seal pox is a parapoxvirus that is known to affect numerous species of pinnipeds (Wilson *et al.*, 1969, 1972; Sweeney, 1974). The disease is most prevalent in California sea lions (*Zalophus californianus*), South American sea lions (*Otaria flavescens*) and harbor seals, but other species of pinnipeds can also develop lesions (Hicks and Worthy, 1987; Osterhaus *et al.*, 1990b; Stack *et al.*, 1993, Nettleton *et al.*, 1995). An orthopox has also been isolated from a gray seal (Osterhaus *et al.*, 1990b). Cutaneous spread of this disease is mostly by head and neck rubbing, a common social behavior of pinnipeds. Infections are rarely fatal, but can cause a high morbidity with prolonged convalescence lasting up to 15 weeks until recovery (Wilson *et al.*, 1972; Sweeney, 1974).

Seal pox is a proliferative lesion characterized by the formation of numerous, 2–3 cm, cutaneous nodules, primarily over the head and neck. These dermal nodules eventually ulcerate and heal slowly. Areas of alopecia develop over the healed areas.

Microscopically, there is ballooning degeneration of the stratum spinosum with pustule formation. Affected epithelial cells have one or two eosinophilic intracytoplasmic inclusions 2–15 µm in diameter. There is marked acanthosis of the affected epidermis with variable orthokeratotic and parakeratotic hyperkeratosis. Inflammation can be prominent, especially during regression of the lesion (Wilson *et al.*, 1972; Migaki, 1987).

The histologic features of pox virus in South American sea lions are unique, with a downward proliferation of the epidermis followed by pustule formation and ulceration. Only one large eosinophilic or basophilic intracytoplasmic inclusion body is observed in affected epithelial cells. These inclusions are usually surrounded by a thin halo with compression of the nucleus. This lesion can occasionally resemble the human disease, molluscum contagiosum (Wilson and Poglayen-Neuwall, 1971; Migaki, 1987).

San Miguel sea lion virus

San Miguel sea lion virus (SMSV), is a calicivirus that is known to affect sea lions and seals. This calicivirus and the virus that causes vesicular exanthema of swine are felt to be similar viral agents. Most viral strains of San Miguel sea lion virus are antigenically related to vesicular exanthema of swine (Smith and Skilling, 1979; Neill and Seal, 1995; Seal *et al.*, 1995). This viral infection is characterized by the formation of vesicles on the flippers. These vesicles usually rupture and form prominent slow-healing ulcers. This virus has also been implicated in causing ulcerative lesions on the lips, nose, chin, and gums of infected sea lions (Gage *et al.*, 1990). Microscopically, the lesion consists of spongiosis of the stratum spinosum, followed by necrosis and microvesicle formation which later progresses to subcorneal vesicles. Intracytoplasmic or intranuclear inclusions are not present in infected cells (Sayer, 1976; Smith and Skilling, 1979; Smith *et al.*, 1983a; Migaki, 1987; Gage *et al.*, 1990).

Several serotypes of San Miguel sea lion virus have been isolated from aborting sea lions and aborted fetuses. These serotypes of San Miguel sea lion virus are indistinguishable from serotypes of vesicular exanthema virus that cause abortions in swine (Smith and Boyt, 1990). The relationship of San Miguel sea lion virus infection and abortions in affected marine mammals is suspected but not proven. The virus has been isolated from the opal-eye fish (*Girella nigricans*), which is believed to function as a reservoir for the spread of this disease (Smith *et al.*, 1980; Smith and Boyt, 1990). This fish is known to develop an active infection with viral replication in the spleen. The fish remains infected for about 31 days. Clinical disease has not been observed in infected fish.

A similar calicivirus has been isolated from Atlantic bottlenose dolphins. Infected animals develop vesicular skin lesions that erode and leave shallow ulcers. It is felt that this virus is infective for both sea lions and dolphins (Smith *et al.*, 1983a). Numerous other marine caliciviruses have also been identified in both pinnipeds and cetaceans (Smith *et al.*, 1981; Skilling *et al.*, 1987; Smith and Boyt, 1990; O'Hara *et al.*, 1998).

Sea lion hepatitis virus

Sea lion hepatitis virus is caused by an adenovirus (Britt *et al.*, 1979). This virus is not highly virulent, with only a few animals developing clinical disease. However, serological surveys demonstrate that large numbers of animals are exposed to the virus. Animals that die, usually die acutely, with little clinical evidence of infection. Affected animals usually present at necropsy with icterus, splenomegaly, mesenteric lymphadenopathy, and discoloration of the liver. Hepatic lesions consist of coagulative and lytic necrosis, most severe in the centrolobular region of hepatic lobules. Large pale eosinophilic to dark basophilic intranuclear inclusions are present in hepatocytes and occasionally in Kupffer cells. The inflammatory response is usually minimal, with a few macrophages, lymphocytes, and neutrophils at the periphery of the necrotic lesion (Britt *et al.*, 1979; Dierauf *et al.*, 1981; Britt and Howard, 1983).

Influenza virus

The influenza virus isolates found in seals have been identified as Type A, subtype H3 virus (Type A/Seal/MA/1/80, Type A/Seal/MA/133/82, Type A/Seal/MA/3807/91, Type A/Seal/MA/3810/91, Type A/seal/M/3911/92) which are closely related to the H3 avian influenza viruses (Geraci *et al.*, 1982; Callan *et al.*, 1995). This subtype of virus is most frequently detected in birds, pigs, horses, and humans. Some feel that seals, like swine, may play a role in genetic reassortment of these influenza viruses, thus causing a potential for interspecies transmission. A Type B influenza found in seals from The Netherlands (Type B/seal/Netherlands/1) is closely related to the human type B influenza that circulated in humans in 1995 and 1996. Seals after 1995 had a 2 per cent incidence of this virus and may have the potential to infect people (Osterhaus *et al.*, 2000).

These viruses have been associated with explosive outbreaks with a high mortality in harbor seals. Affected animals are usually presented weak with serious respiratory difficulties. A white mucinous to bloody discharge is observed in the trachea and bronchi. Pulmonary, mediastinal, and subcutaneous emphysema is commonly observed. Microscopically, there is a severe hemorrhagic bronchopneumonia with hemorrhage in alveoli and prominent necrosis of bronchioles and bronchi. Mycoplasma (*Mycoplasma phocidae*) and other bacteria have been isolated from affected animals. It is felt that there is a synergic effect between the virus and bacterial agents, since seals experimentally challenged with the virus alone develop a mild respiratory disease (Geraci *et al.*, 1982).

Seal herpesvirus

Herpesviruses have been isolated from harbor seals and a California sea lion (Osterhaus *et al.*, 1985; Borst *et al.*, 1986; Kennedy-Stoskopf *et al.*, 1986). Two distinct types have been isolated from affected animals. These have been identified as phocid herpesvirus type 1 and type 2 (PHV-1 and PHV-2) (Osterhaus *et al.*, 1985; Frey *et al.*, 1989; Harder *et al.*, 1996). PHV-1 is characterized as a member of the alpha-herpesvirus subfamily. PHV-2 has been classified as a gamma-herpesvirus. Numerous pinniped species have antibodies to PHV-1 and PHV-2 (ringed seals, spotted seals (*Phoca largha*), harbor seals, bearded seals, ribbon seals (*Histriophoca fasciata*), Steller's sea lions (*Eumetopias jubatus*), northern fur seals (*Callorhinus ursinus*), and walrus) (Zarnke *et al.*, 1997). Most animals with seroconversion to these viruses do not demonstrate clinical disease.

PHV-1, which affects mostly young harbor seals, has been shown to cause a serious systemic infection. Affected animals demonstrated an acute pneumonia, necrotizing hepatitis, and necrotizing adrenalitis (Borst *et al.*, 1986; Gulland *et al.*, 1997). Occasionally a nonsuppurative encephalitis is observed with rare neuronal necrosis. The pneumonia is characterized as a diffuse interstitial pneumonia with multifocal fibrinous exudation and emphysema. Within the liver and adrenal gland, there is multifocal necrosis of the hepatic and adrenocortical parenchyma with a minimal mononuclear cell infiltrate. Acidophilic intranuclear inclusion bodies are observed in the areas of necrosis. Electron microscopy demonstrates nonenveloped hexagonal viral particles 90–100 nm in diameter with a central dense core, and cytoplasmic enveloped particles 150–160 nm in diameter (Gulland *et al.*, 1997). The nature of this virus and its potential to cause disease in wild and captive populations of seals is still unknown. This virus has been inoculated into young seals, with the animals developing only a minimal nasal discharge (Horvat *et al.*, 1989). PHV-1 most likely acts similarly to other mammalian herpesviruses, with the infections most often fatal in the young and/or seriously stressed animal. Most sick animals have concurrent bacterial and/or protozoal infections that may mask the herpes viral infection. Like most herpesviruses, this virus probably expresses itself in times of stress, as either subclinical oral or genital lesions.

The phocine herpesvirus-2 was isolated from a California sea lion with a severe bacterial pneumonia (Kennedy-Stoskopf *et al.*, 1986), free-ranging harbor seals, and from seals with abortions during the early epizootic of morbillivirus infection in northern Europe (Zarnke *et al.*, 1997). The importance of this virus in the pathogenesis of this pneumonia is unknown (a nononcogenic retrovirus was also isolated from the skin of the affected sea lion). This virus is highly cell associated and causes little or no disease in pinnipeds.

A gamma-herpesvirus has been identified in a metastatic carcinoma of the lower genital tract from California sea lions, where it was associated with the neoplastic cells. The association of the virus and the development of neoplasia of the vagina and cervix are unknown (Lipscomb *et al.*, 1998).

Herpesvirus of bottlenose dolphins

A disseminated herpes viral infection was identified in an immature female bottlenose dolphin. On necropsy, the animal had an enlarged thymus, pericardial hemorrhage, and hydrothorax. Histologically there was a necrotizing interstitial pneumonia, lymphocytic myocarditis, splenic and lymphoid necrosis, and a necrotizing adrenalitis. Intranuclear inclusions were observed in numerous cells of the thymus, spleen, adrenal gland, heart, lungs, and glomeruli. Sequencing of DNA products indicated that this virus was an alpha-herpesvirus (Blanchard *et al.*, 1998).

Herpesvirus in beluga whales

A herpesvirus has been observed to cause a focal dermatitis in both captive and wild beluga whales (*Delphinapterus leucas*) (Martineau *et al.*, 1988; Barr *et al.*, 1989). The lesions seemed to occur at times when the animals were stressed or in animals that were visibly thin due to other diseases. These epidermal lesions consisted of random, variably sized, multiple discrete raised pale gray areas that eventually ulcerated and were slow to heal. Histologically the epithelial lesion involved the superficial epidermis with the epithelial cells undergoing intercellular edema, necrosis, and microvesicle formation. The infected epithelial cells contained prominent eosinophilic intranuclear inclusion bodies.

Herpesvirus of sea otters

A herpesvirus has been implicated in causing extensive oral lesions in sea otters. These lesions are commonly found on the gingiva and under the tongue. They consist of variably sized, irregular, white plaques on the gingiva, and/or deep, often bilaterally symmetrical, ulcers under the tongue. In severely affected animals, the ulcers tend to coalesce to cover extensive areas of the buccal, labial, gingival, and glossal mucosa. Infected animals rarely show a reluctance to eat. Histologically, the ulcerative lesions reveal extensive chronic ulcers with associated mixed bacterial colonies and separate foci of epithelial necrosis and intracellular edema. Numerous eosinophilic intranuclear inclusion bodies are observed in the degenerating and necrotic cells. The white plaques have epithelial hyperplasia with intracellular and intercellular edema and eosinophilic intranuclear inclusions. Rare inflammatory cells are observed in areas with an intact mucosa (Haebler and Moeller, 1993).

Hepatitis B-like virus in dolphins

A hepatitis B-like infection has been identified in a Pacific white-sided dolphin (*Lagenorhynchus obliquidens*). The animal presented with cyclic periods of inactivity, anorexia, and icterus. During the times of inactivity, blood values demonstrated a leukocytosis with neutrophilia, lymphopenia, and eosinopenia. Biochemical values showed markedly elevated alanine transaminase (ALT), aspartate transaminase (AST), gamma glutamyltransferase (GGT), lactic acid dehydrogenase, total bilirubin, direct bilirubin, and indirect bilirubin, suggesting chronic liver disease. Supportive measures were instituted and the dolphin eventually recovered. Serum from the animal was found to be positive for antihepatitis B virus core (anti-HBc) activity, hepatitis B virus DNA (HBV-DNA), and hepatitis B surface antibodies (anti-HBs). Other cetaceans and humans who had contact with this animal were examined for hepatitis B antigens. Only one killer whale (*Orcinus orca*) was positive, all other cetaceans and humans were negative (Bossart *et al.*, 1990).

Papillomavirus in cetaceans

Papillomas have been reported on the skin, penis, tongue, pharynx, and first gastric compartment of cetaceans (Lambertsen *et al.*, 1987; Martineau *et al.*, 1988). Although a papillomavirus has not always been implicated as the cause of these lesions, they should be suspected. In the beluga whale, the gastric papillomas identified in the first gastric compartment are white (like the surrounding normal mucosa), well-defined, exophytic masses with a central wart-like core composed of small filamentous papillae. Histologically, the papillomas develop into an exophytic cup-shaped mass with marked epithelial proliferation, supported by a thin fibrovascular proliferation forming numerous arborizing projections from the submucosa. The proliferating epithelium consists of a flattened basal cell layer 3–15 cells thick with a mature epithelium overlying the basal cells. Scattered amongst the hyperplastic epithelium, individual and small groups of epithelial cells undergo hydropic degeneration (cells become swollen and globular with a pale granular cytoplasm). Ultrastructurally, the cells undergoing hydropic degeneration have aggregates of small (40 nm) hexagonal viral particles observed in the cytoplasm. It is unclear if any of these papillomas cause physical problems for the affected belugas (Martineau *et al.*, 1988). Papillomas on the penis are usually raised plaques on the mucosal surface (Lambertsen *et al.*, 1987). As in humans, papillomavirus in cetaceans may possibly infect the cervical mucosa and cause neoplasia from this region.

Bacterial diseases

Most marine mammals die of bacterial diseases. Bacteria are easily isolated and commonly blamed for the ailments afflicting the marine mammal. However, one must realize that most bacteria are usually opportunistic invaders and may be part of the normal flora of a marine mammal. Usually, there is an underlying cause for a bacterial infection. Viral disease can lead to immunosuppression and/or loss of integrity of various host barriers in the skin, respiratory tract, or intestines. Parasites can cause malnutrition (which leads to immunosuppression) or damage to mucosal surfaces. Traumatic injuries can cause easy access to the vascular system, resulting in rapid hematogenous spread to other organ systems. Toxins (particularly xenobiotics) can lead to immunosuppression and increased susceptibility to bacterial infections.

Since these animals are mammals, one needs to remember that they can be infected by bacteria commonly found in their terrestrial counterparts. Finding pure cultures of a bacteria in an animal should be viewed as indication of a possible etiologic agent. This section will focus on bacteria that have been considered pathogens in marine mammals.

Remember, many of the bacterial agents *(Erysipelothrix, Leptospira, Edwardsiella, Salmonella, Mycobacterium, Klebsiella, Pseudomonas pseudomallei* and *P. aeruginosa, Nocardia,* and *Brucella)* are potential pathogens in humans. Thus, when handling a sick marine mammal or performing a necropsy, one should always follow proper laboratory procedures to insure that the examiner does not become infected by bacterial agents.

Erysipelothrix rhusiopathiae

Erysipelothrix rhusiopathiae is a small, pleomorphic, Gram-positive rod. This bacterium causes two distinct forms of disease in dolphins: a dermatological disease and a septicemic disease (Seibold and Neal, 1956; Simpson *et al.*, 1958; Geraci *et al.*, 1966; Kinsel *et al.*, 1997). The dermatological disease is characterized by dermal infarction that results in sloughing of the epidermis. The development of the skin lesion is usually due to a septicemia resulting in micro-infarcts of the dermis and the formation of the characteristic rhomboid areas of cutaneous necrosis (Simpson *et al.*, 1958; Geraci *et al.*, 1966). If untreated, these animals will usually die. The septicemic disease is usually peracute, with the animal found moribund or dead.

At necropsy, septicemic animals may demonstrate multifocal areas of necrosis and inflammation involving numerous organs. An embolic interstitial pneumonia, hepatic and splenic necrosis and inflammation, and petechial hemorrhage of serosal surfaces are common findings. Culturing the agent from affected tissues (particularly lungs, liver, spleen, and kidneys) and/or blood is the only method of diagnosis (Medway, 1980).

A killed bacterin is available for vaccination (Geraci *et al.*, 1966; Migaki, 1987). The disease has been effectively controlled by use of this vaccine. However, this vaccine has been known to cause anaphylactic reactions in some animals and should be used with caution. *Erysipelothrix rhusiopathiae* is commonly found on the slime of fish and ingestion is the probable route of infection in cetaceans (Seibold and Neal, 1956).

Pseudomonas

Pseudomonas aeruginosa is a Gram-negative, motile, slender bacillus, 0.5×2.5 μm. This bacterium is commonly found in the water and has been incriminated in causing disease in both pinnipeds and cetaceans (Rand, 1975; Diamond *et al.*, 1979). It is believed that this bacterium is opportunistic and colonizes wounds, leading to septicemia. *Pseudomonas aeruginosa* has been associated with multiple cutaneous ulcers and bronchopneumonia in

Atlantic bottlenose dolphins (Diamond *et al.*, 1979; Migaki, 1987). Infections with this organism lead to deep cutaneous ulcers, causing serious damage to the blubber and dermis. If affected animals develop a septicemia, cutaneous lesions similar to those observed with erysipelas may develop. Histologically, *Pseudomonas* septicemia causes a characteristic proliferation of Gram-negative bacilli into the wall of affected blood vessels. Culturing the lesions is the only method of diagnosis.

Pseudomonas (*Burkholderia*) *pseudomallei* is a pathogen found primarily in Southeast Asia and Australia. The organism is commonly found in water. The route of introduction into the animal is unknown, but is believed to be through wound contamination. Animals die of an acute septicemia, usually with no clinical signs. Grossly, animals may demonstrate congested edematous lungs with petechial hemorrhage and necrosis of the liver, spleen, kidney, and lymph nodes. Histologically the lungs develop an embolic pneumonia with necrosis of septal walls and acute inflammation. The liver and spleen develop multifocal areas of coagulative and lytic necrosis with acute inflammation associated with the areas of necrosis (Medway, 1980). Gram stains reveal numerous, small, Gram-negative rods in the cellular debris and inflammatory cells. This agent is a serious human pathogen, and transmission to humans is usually via cuts, abrasions, or by aerosol transmission. Caution should always be used when performing a necropsy on an animal suspected of dying of this bacterium.

Edwardsiella

Edwardsiella species are Gram-negative bacilli that are common inhabitants of water. These bacteria are noted to cause septicemia in fish and can be pathogens in both pinnipeds and cetaceans (Howard *et al.*, 1983a). Animals present with a serious necrotizing enterocolitis and/or septicemia. Animals with septicemia can develop a severe embolic interstitial or bronchointerstitial pneumonia, a necrotizing hepatitis, and a necrotizing splenitis. The necrotizing and hemorrhagic enteritis/colitis is similar to that found in *Salmonella* infections. Isolation of *Edwardsiella* from the lungs and liver is usually suggestive of a septicemia. Animals that develop this disease are usually debilitated or stressed and probably get these organisms from ingestion of contaminated fish.

Salmonella

Salmonella species are Gram-negative, non-lactose fermenting bacilli. *Salmonella* infections have been observed in both cetaceans and pinnipeds. *Salmonella typhimurium*, *S. enteritidis* and *S. newport* are the most prevalent *Salmonella* species that cause disease (Sweeney and Gilmartin, 1974; Stroud and Roelke, 1980; Minette, 1986; Baker *et al.*, 1995). Other *Salmonella* species have been isolated from marine mammals without apparent disease

(Gilmartin *et al.*, 1979; Baker *et al.*, 1995). These bacteria are of particular concern for animals housed in rehabilitation centers. Thus, recently stranded animals should be quarantined and monitored for *Salmonella* infection prior to placing with other animals. *Salmonella* infections usually occur in animals that are debilitated and/or stressed. Animals usually present with hemorrhagic diarrhea and/or septicemia. Animals with hemorrhagic diarrhea usually have a necrotizing enterocolitis. Culturing the feces is recommended for isolation of the organism. Animals that develop septicemia can die acutely with no clinical signs. On necropsy, the animal may have a broncho-pneumonia and/or a diffuse embolic interstitial pneumonia. A necrotizing hepatitis and splenitis are common; meningoencephalitis may occur (Sweeney and Gilmartin, 1974; Gilmartin *et al.*, 1979). Isolation of the organism from the lungs and liver is usually consistent with a septicemia.

Leptospira

Severe epizootics of leptospirosis have been observed in California sea lions and northern fur seals. These epizootics have been caused by *Leptospira pomona*. This spirochete causes renal disease and abortions in infected animals. Outbreaks are most prevalent in the fall, with infections seen prim-arily in juvenile and young adult males. Clinically, animals are depressed, anorectic, pyretic, and reluctant to move due to posterior limb paresis. Animals are usually icteric, have oral ulcerations and excessive thirst. Clinical pathology results show a marked leukocytosis and elevated creatinine, phosphorus and blood urea nitrogen (BUN) levels indicative of renal dis-ease. Grossly, the animals may present with generalized icterus, petechial hemorrhage of serosal surfaces, congested lungs, swollen kidneys, and swollen friable livers. On cut surfaces of the kidney, there may be hemorrhage at the corticomedullary junction and subcapsular region, with poor differentiation of the renal cortex and medulla. Histologically, there is a tubulointerstitial nephritis with lymphoplasmacytic interstitial infiltrates and tubular necrosis. Spirochetes are identified with silver stains in the renal tubular epithelium and free in the lumina (Vedros *et al.*, 1971; Smith *et al.*, 1974; Sweeney and Gilmartin, 1974; Smith *et al.*, 1977; Dierauf *et al.*, 1985; Gulland *et al.*, 1996).

In newborn and aborted fetuses, the disease is characterized by sub-cutaneous hemorrhage and hyphema (hemorrhage into the anterior cham-ber of the eye 'red eye disease'). This disease may have zoonotic significance since affected sea lions can shed the bacteria in the urine for up to 154 days (Smith *et al.*, 1974). Rising titers or titers above 1 : 3200 for *L. pomona* are indicative of an active infection.

Dermatophilus congolensis

Dermatophilus congolensis is known to cause a disfiguring cutaneous disease affecting the South American sea lion (Medway, 1980; Migaki and Jones,

1983; Migaki, 1987). This disease usually involves the entire body and is characterized by prominent layered scabs involving haired areas. Mortality is low, but morbidity is high.

Histologically, the epidermis demonstrates the characteristic multiple layers of coagulative necrosis of the epidermis with a peripheral line of degenerating neutrophils separating each necrotic layer. The organisms are easily observed histologically in the necrotic epidermis as numerous Gram-positive, branching, filamentous organisms that divide in a characteristic multidimensional fashion, forming parallel double rows of cocci (zoospores).

Mycobacterium

Mycobacterium species are Gram-positive, acid-fast bacilli. These bacteria are common in the soil and water (except for *Mycobacterium tuberculosis* complex, *M. bovis*). Several species of *Mycobacteria* have been isolated from pinnipeds. In seals, *M. tuberculosis* complex (*M. bovis*), *M. fortuitum*, and *M. chelonei* have been isolated. *Mycobacterium smegmatis* was isolated in a California sea lion and several manatees (*Trichechus inungius*) developed infections with *M. chelonei* and *M. marinum*. These infections usually present as nonhealing, chronic, cutaneous lesions. However, generalized infections with caseonecrotic granulomas in the lungs, liver, kidney, and lymph nodes are reported. Numerous acid-fast bacilli are usually observed in the cutaneous infections. Microorganisms may be more difficult to find in granulomas caused by *M. tuberculosis* complex (*M. bovis*). Persistent nonhealing cutaneous lesions should be biopsied to check for a granulomatous dermatitis with characteristic acid-fast bacilli present in the wound. It is also recommended to culture the lesion for *Mycobacterium* (Boever *et al.*, 1976; Howard *et al.*, 1983a; Morales *et al.*, 1985; Cousins *et al.*, 1990, 1993; Wells *et al.*, 1990; Forshaw and Phelps, 1991; Woods *et al.*, 1995).

Staphylococcus

Staphylococcus aureus has been implicated as one of the causes of pneumonia in dolphins maintained in captivity (Ketterer and Rosenfeld, 1974; Palmer *et al.*, 1991). *Staphylococcus aureus* has been cultured from animals that died from a septicemia with an embolic nephritis and cerebral abscesses as well as cutaneous lesions. It is a part of the normal flora of the blowhole of many normal dolphins. However, some investigators feel that if it is present in the upper respiratory tract, it should be considered a potential pathogen (Palmer *et al.*, 1991). *Staphylococcus delphini* has been isolated from dolphins with a purulent dermatitis (Varaldo *et al.*, 1998). In seals, *Staphylococcus* has been identified in animals with cutaneous abscesses or pneumonia (Van Pelt and Dieterich, 1973; Medway, 1980). A case of staphylococcal valvular endocarditis has been reported in a sea otter (Joseph *et al.*, 1990).

Clostridium perfringens

Clostridium perfringens has been reported to cause enterotoxemia in young pinnipeds (Sweeney and Gilmartin, 1974; Medway, 1980; Buck *et al.*, 1987), gas-producing myositis in dolphins and cutaneous abscesses in fur seals (Greenwood and Taylor, 1978; Medway, 1980; Baker and McCann, 1989). The necrotizing myositis observed in dolphins and the subcutaneous abscesses in pinnipeds have been attributed to injection-site contamination by this organism (Greenwood and Taylor, 1978; Medway, 1980). Since *Clostridium perfringens* and other clostridial organisms are normal inhabitants of the gut and common in dead animals, it is best to make impression smears from the lesion (gut or muscle) on clean glass slides for bacteriologic evaluation. A definitive diagnosis is usually rendered with the findings of a necrotizing myositis or enteritis with numerous Gram-positive bacilli present in the inflammation during histolopathological examination of the affected tissues.

Klebsiella

Klebsiella species are Gram-negative bacilli. These bacteria have been associated with pneumonia in marine mammals (Sweeney and Gilmartin, 1974; Medway, 1980). Pneumonia caused by these organism is consistent with most bacterial pneumonias, except that the Gram-negative bacilli can be identified with a clear capsule surrounding the organism amongst the inflammatory cells. As usual, bacterial cultures are necessary for a definitive diagnosis.

Nocardia asteroides

Nocardia asteroides has been isolated from numerous cetaceans (Pier *et al.*, 1970; Medway, 1980). The organism is a thin, filamentous, Gram-positive organism, that occasionally is weakly acid fast. Infected animals present with a necrotizing and pyogranulomatous lymphadenitis, pleuritis, encephalitis, and/or mastitis containing numerous Gram-positive filamentous bacteria.

Brucella

Brucella has been isolated from several species of cetaceans (Atlantic white-sided dolphins (*Lagenorhynchus acutus*), striped dolphins, and bottlenose dolphin), pinnipeds (hooded seal, gray seal, Pacific harbor seals (*Phoca vitulina richardii*)), and a European otter (*Lutra lutra*) (Ewalt *et al.*, 1994; Foster *et al.*, 1996; Ross *et al.*, 1996). These organisms have been cultured from an aborted bottlenose dolphin fetus with lesions in the subcutis, lymph nodes, liver, and lungs. The placenta developed a necrotizing placentitis with Gram-negative coccobacilli within the trophoblast. Lesions in lymph nodes, liver, and lungs were characterized by a multifocal granulomatous inflammation. Biotyping of the bacteria indicates that these bacteria are closely related to *Brucella abortus* or *Brucella melitensis*. However, it is felt that these may represent a new species of *Brucella* (Jahans *et al.*, 1997). It is currently unclear as to what impact this organism has on marine mammals. *Brucella* organisms have been identified in the uterus and intestines of *Parafilaroides* sp. lungworms (Garner *et al.*, 1997). The importance of this lungworm in the transmission of *Brucella* to animals is unknown. These parasites may represent an important transport host for the dissemination of this bacterium in the wild.

Streptococcus

Streptococcus spp. have been isolated from seals with pneumonia and septicemia and cetaceans with septicemia, metritis, pneumonia, and skin lesions (Medway, 1980; Baker and McCann, 1989; Skaar *et al.*, 1994). These are primarily β-hemolytic streptococcal species (Gram-positive diplococci). Most are normal inhabitants of the skin and upper respiratory tract of these animals and are probably opportunistic pathogens in times of stress and/ or debilitation. A β-hemolytic *Streptococcus*, *Streptococcus phocae*, was isolated from many of the seals that died of pneumonia during the phocine morbillivirus outbreak in the North Atlantic (Skaar *et al.*, 1994). Most animals presented clinically with dyspnea, coughing, and a nasal discharge. Pathological examination demonstrated a severe pneumonia with areas of lung consolidation, purulent exudate in the bronchi and bronchioles, interlobular edema and emphysema.

Mycotic diseases

Numerous mycotic diseases have been reported in pinnipeds and cetaceans. Most have been reports of single cases (Williamson *et al.*, 1959; Wilson *et al.*, 1974; Sweeney *et al.*, 1976; Migaki *et al.*, 1978a, b). Only a few fungal organisms are known as major pathogens of marine mammals. Since fungal organisms are usually opportunistic or secondary invaders, these organisms pose a serious health risk to the animal that is immunocompromised due to malnutrition, stress or concurrent viral infection, or on long-term antibiotic therapy.

Candida

Candida species are common commensals of the upper respiratory, intestinal, and genitourinary tracts. *Candida albicans* is the most common species of *Candida* to cause clinical disease in stressed pinnipeds and cetaceans (Sweeney *et al.*, 1976; Nakeeb *et al.*, 1977; Dunn *et al.*, 1982, 1984; Migaki and Jones, 1983). Diseased animals usually develop lesions occurring primarily at the mucocutaneous junctions of the mouth, blowhole, and vagina (Medway, 1980; Migaki and Jones, 1983; Dunn *et al.*, 1984). Mucosal lesions in the oropharynx, trachea, esophagus, and stomach are also common. Lesions appear as white or yellow creamy plaques on mucosal surfaces. In internal organs, prominent focal areas of necrosis are visible. Histologically, large colonies of septate hyphae, pseudohyphae (3–7 μm wide) and blastospores (3–5 μm diameter) are observed in the necrotic lesions. The presence of *Candida* in the epithelium of the esophagus or vagina is very common and its histological detection, without associated lesions, is considered to be an incidental finding.

Loboa loboi

Lobomycosis is a fungal disease that affects the skin of the Atlantic bottlenose dolphin and people of the tropical rainforests of northern South America (Migaki *et al.*, 1971a; Caldwell *et al.*, 1975; Migaki and Jones, 1983; Migaki, 1987). Grossly, these lesions are located anywhere on the animal's body, with the head, fin, and flukes being the most common sites. These lesions are white, multiple, and nodular with a cobblestone appearance to the skin. Histologically, there is a superficial granulomatous dermatitis involving the papillary dermis. This granulomatous dermatitis is composed almost entirely of macrophages and multinucleated giant cells containing numerous chains of round yeast forms (5–10 μm in diameter). Some yeast forms contain a 1–2 μm central body. Larger yeast forms may have a rough and spiny surface. The epidermis over these areas of inflammation is often acanthotic with downward growths of the rete pegs. Clinically, the animals are not seriously affected by this organism; however, if lesions become large, the animal may become debilitated and die (usually due to secondary bacterial infections). Therapeutic treatment has not proven successful; however, removal of the affected area has shown positive results (Migaki, 1987).

Fusarium

Outbreaks of *Fusarium*-induced dermatitis have been observed in a group of captive California sea lions, gray seals, harbor seals, Atlantic white-sided dolphins, and a pygmy sperm whale (*Kogia breviceps*) (Montali *et al.*, 1981; Migaki, 1987; Frasca *et al.*, 1996). The lesions consisted of papules and nodules on the face, trunk, flippers, and the caudal portions of the body. Histologically, there is hyperplasia of the follicular and epidermal epithelium with associated chronic active inflammation and numerous fungal hyphae (septate, branching hyphae, 2–5 μm in width with parallel sides). Treatment of the animals with ketoconazole caused the dermatitis to resolve in 3–4 weeks (Frasca *et al.*, 1996). In marine mammals, *Fusarium* spp. are most likely opportunistic invaders of the skin. Animals that are immunocompromised due to stress or illness may be most susceptible. Damage to the integument due to excessive chlorination of the water and large fluctuations in pool temperatures may also play an important role in this disease.

Aspergillus *and* Zygomycetes

Aspergillus and *Zygomycetes* infections cause serious disease in individual animals (Carrol *et al.*, 1968; Sweeney *et al.*, 1976; Migaki and Jones, 1983). As with infection with *Candida*, the affected animals are usually stressed and/or under prolonged antibiotic therapy. Infection can involve focal areas, usually the esophagus or lungs; however, systemic spread of the fungus is common. Lesions involving the esophagus and trachea usually appear as ulcers with yellow- to cream-colored plaques over the affected area. Systemic spread leads to necrosis of multiple tissues, with the liver and kidney commonly involved. Diagnosis is by observing the characteristic hyphae (*Aspergillus*: long, branched, septate hyphae, 3–4 µm wide with parallel walls. *Zygomycetes*: long, nonseptate, nonparallel walled hyphae with irregular branching).

Microsporum, Trichophyton, *and* Epidermophyton *spp.*

Cutaneous infections with *Microsporum canis* and *Trichophyton* spp. have been reported in both cetaceans and pinnipeds. *Microsporum canis* has been isolated from harbor seals; *Trichophyton* spp. in northern fur seals, Steller's sea lions, and bottlenose dolphins, and *Epidermophyton* from an unknown species of manatee (Dilbone 1965; Farnsworth *et al.*, 1975; Migaki and Jones, 1983; Migaki, 1987; Tanaka *et al.*, 1994). Infection in seals is characterized by round, depilated areas, 2–3 cm in diameter, on the face and back. These depilated areas spread over the entire body. Histologically, there is epidermal hyperplasia with hyperkeratosis, parakeratosis, necrosis, and microvesiculation and microabscess formation. Numerous neutrophils are present in the affected epidermis. Numerous branched, septate hyphae, 2–7 µm in diameter, are observed in the parakeratotic and necrotic regions.

Coccidioides immitis

Coccidioides immitis has been identified as an endemic disease in the California sea lion and as an isolated case in the California sea otter (*Enhydra lutris nereis*) and bottlenose dolphin (Reed *et al.*, 1976; Fauquier *et al.*, 1996; Reidarson *et al.*, 1998). This organism appears to be endemic in animals that inhabit the southern and central part of their range from Baja California, Mexico in the south to Monterey County, California in the north (Fauquier *et al.*, 1996). *Coccidioides immitis* is an infectious agent to numerous animal species and is a serious health risk to humans (Valley Fever) when living in endemic areas. Endemic areas are Arizona, southern and central California, Mexico, New Mexico, Nevada, Utah, and Texas. This organism probably infects sea lions by inhalation of spores and, as in most animals, usually results only in mild respiratory disease. Recovery usually occurs after a short illness. Some animals develop a serious disseminated disease. Sick animals usually beach and are presented ill. Gross necropsy findings are focal to disseminated granulomas involving the lungs, liver, pancreas, numerous lymph nodes (retropharyngeal, submandibular, mesenteric, and tracheobronchial lymph nodes), and occasionally a purulent pleuritis and peritonitis. Histologically, the affected organs develop pyogranulomatous inflammation with variable numbers of multinucleated giant cells and large, round, double contoured wall spherules, 10–70 μm in diameter. Filling the spherules are numerous small (2–5 μm) endospores. *Coccidioides immitis* is able to survive in sea water for several weeks, however it is unknown whether sea water could be a means of infecting animals (Fauquier *et al.*, 1996).

Parasitic diseases

Parasites can cause serious health problems in marine mammals. These parasites can cause a loss of valuable nutrients that are required for proper growth and maintenance of the animal. Parasitic damage to the animal's hair coat, respiratory tract, and digestive system can cause a breakdown of these protective barriers, leading to increased susceptibility of these animals to bacterial and viral infections. Environmental contamination such as heavy metals, PCBs, and other pollutants may further damage the animal's immune system. This can lead to an increase in the animal's susceptibility to parasitic infections and exacerbation of the effects of the parasite on the animal host.

Since parasites are common in marine mammals, it is often difficult to determine if these organisms caused the debilitation and/or death of the animal. Evaluation of the parasitic burden, type of parasites involved, and nature of the illness need to be determined if the parasites are to be implicated as a major contributor to the animal's poor condition or death.

Parasitic disease of pinnipeds and sea otters

Protozoal parasites of pinnipeds

Numerous protozoal diseases have been identified in pinnipeds. *Sarcocystis* species have been reported in northern fur seals (Brown *et al.*, 1974a), ringed seals (Migaki and Albert, 1980), bearded seals (Bishop, 1979), Hawaiian monk seals, *Monachus schauinslandi* (Yantis *et al.*, 1998), and California sea lion (Mense *et al.*, 1992). Most reports of *Sarcocystis* have been incidental findings with the presence of the parasite in large protozoal cysts in the skeletal muscle. These infections most likely represent host-adapted *Sarcocystis* species that cause little damage in the animal. In the California sea lion and Hawaiian monk seal, *Sarcocystis* species have caused serious disease leading to a necrotizing hepatitis in the affected animal.

Toxoplasma gondii has been identified in the California sea lion (Migaki *et al.*, 1977), northern fur seal (Holshuh *et al.*, 1985), bearded seal (Bishop,

1979) and harbor seal (Van Pelt and Dieterich, 1973). Affected animals develop disseminated disease with necrosis of many organs and numerous *Toxoplasma* tachyzoites present in the lesion. Differentiation of the tachyzoites of *Toxoplasma* from those of *Sarcocystis* can be accomplished with ultrastructural evaluation of the protozoal organisms or by immuno-histochemical techniques.

Giardia oocysts have been identified in fecal samples from ringed seals in the arctic region of Canada (Olson *et al.*, 1997). Infected animals failed to demonstrate clinical disease. However, the potential of malabsorption and diarrhea does exist, particularly in debilitated animals. The zoonotic potential of infected seals to act as a reservoir for human infection is unknown.

External parasites of pinnipeds and sea otters

Lice are a common finding on young pinnipeds (Sweeney, 1974; Conlogue *et al.*, 1980). Anopluran (sucking) lice are the major louse identified on these animals. These insects usually cause little damage. However, in some debilitated animals, severe alopecia may occur. As an animal ages, the louse burden usually decreases to the point of rarely finding these insects without a thorough evaluation of the skin. The finding of a severe infestation in an adult animal should suggest an immunocompromised animal. These lice are believed to be the intermediate host for filariad nematodes of pinnipeds (Conlogue *et al.*, 1980).

Demodectic mange mites have been observed in the California sea lion. These mites have been identified as *Demodex zalophi* (Sweeney, 1974). Many animals are infected with the mite, but only some animals develop alopecia and thickening of the skin over the flippers, axillary, ventral abdominal and inguinal regions, and genitalia. Demodectic mites have also been observed in the hair follicles of the heads of sea otters, but no severe dermatologic conditions have been observed (Moeller, personal observations).

Internal parasites of pinnipeds and sea otters

Lung mites are common parasites of seals and sea lions (Keyes, 1965; Kim and Haas, 1980). These small white mites (0.5–0.8 mm in length) are found in the nasal passages, trachea, bronchi, and bronchioles. *Orthohalarachne attenuata* inhabit the nasal passages and *Orthohalarachne diminuata* are found in the lungs. *Halarachne miroungae* are observed in the nasal cavity of sea otters. These mites are commonly found as incidental findings; however, large numbers of these parasite may cause increased mucus secretions of the respiratory tract, with nasal discharge, dyspnea, and coughing.

Parafilaroides decorus are the most common lung worm of the California sea lion (Dailey and Brownell, 1972; Sweeney, 1974). The opal-eye fish (*Girella nigricans*) is the intermediate host for this metastrongilid parasite. These parasites migrate to the lungs of infected sea lions and release larvae in the alveoli. These parasites are often noted with minimal inflammation present. However, in severe infections, they are believed to be associated with a severe bacterial bronchopneumonia. *Parafilaroides* species have been identified with *Brucella* bacteria present in their uterus and gut (Garner *et al.*, 1997). *Brucella*-infected lungworms may cause brucellosis infection in pinnipeds.

Otostrongylus circumlitus are metastrongylids that inhabit the primary and secondary bronchi of harbor seals and northern elephant seals (Wilson and Stockdale, 1970; Migaki *et al.*, 1971b; Dailey and Stroud, 1978). These parasites may be found with little damage to the bronchi or causing serious bronchiectasis. In severe infections, prominent bronchiectatic abscesses may be observed, containing large numbers of these parasites.

Contracaecum and *Anisakis* species are common nematodes of the stomach of pinnipeds (Bishop, 1979). These parasites are usually incidental findings; however, they may cause severe gastric hemorrhage and melena. Occasionally nodules with larval forms of these parasites may be found in the gastric mucosa and submucosa. Fish are the intermediate host for these parasites.

Hookworms are common in sea lions and northern fur seals (Olsen and Lyons, 1965; Brown *et al.*, 1974b; Lyons *et al.*, 1997). These parasites are a serious problem in young pups. Infected animals can be seriously debilitated by large worm burdens resulting in anemia and exsanguination. The most common hookworms are *Uncinaria lucase* and *Uncinaria hamiltoni*. Pups are infected by the ingestion of larva-laden milk from the dam. After about 3 months, the pups shed the parasite infection and are reinfected with the third-stage larvae that migrate to the blubber and mammary glands. In these tissues, the parasitic larvae remain dormant until the animals become adults and shed the larva in their milk to their young during lactation.

Acanthocephalans are common intestinal parasites of pinnipeds and sea otters (Rausch, 1953; Dailey and Walker, 1978). These parasites are often observed as incidental findings in the jejunum, ileum, and colon, where they bury their proboscis into the mucosa and submucosa, causing a mild granulomatous reaction. Most acanthocephalans are of the genus *Corynosoma*. In most pinnipeds, these rarely penetrate deep in the intestinal wall causing problems. However, in the sea otter, these parasites are known to penetrate through the intestinal wall, resulting in peritonitis and death (Rausch, 1953).

Zalophotrema hepaticum is the liver fluke of sea lions found in the intrahepatic bile ducts, gall bladder, and common bile ducts (Johnston and Ridgeway, 1969; Howard *et al.*, 1983a). This fluke causes little damage but will occasionally cause cystic cavitation in the hepatic parenchyma.

Sea otters and the bearded seals have gall-bladder flukes (*Orthosplanchnus fraterculus*) that cause cystic and nodular hyperplasia of the gall-bladder mucosa. It is felt that these parasites are responsible for a chronic fibrosing cholangitis and cholecystitis of the bile duct and gall bladder.

Parasitic diseases of cetaceans and manatees

Protozoal diseases of cetaceans and manatees

Sarcocystis spp. parasites have been seen in sperm whales (*Physeter macro-cephalus*) and striped dolphins (Cowen, 1966; Owen and Kakulas, 1967; Dailey and Stroud, 1978). These parasites appear to be incidental findings, with typical *Sarcocystis*-type cysts in striated muscle.

Toxoplasma gondii has been associated with disseminated disease in the Atlantic bottlenose dolphin, spinner dolphin (*Stenella longirostris*), and Florida manatee (*Trichechus manatus latirostris*) (Buergelt and Bonde, 1983; Inskeep *et al.*, 1990). Active infections with *T. gondii* have resulted in necrosis of multiple organs with tachyzoites present. Like infections in pinnipeds, infection with this protozoon is usually associated with immunocompromised animals.

A large ciliated protozoon is often found in necrotic cutaneous lesions in the Atlantic bottlenosed dolphin (Schulman and Lipscomb, 1999). This organism is up to 60–80 μm in diameter, with a large 20 μm diameter macronucleus. It is felt that it is an opportunistic invader since the protozoon is commonly found around the blowhole of dolphins without inflammation.

External parasites of cetaceans

Several species of lice have been isolated from the skin of cetaceans (Howard *et al.*, 1983a). These parasites appear to be incidental findings and do not cause clinical disease in these animals. Sessile and pedunculated barnacles are also commonly seen on cetaceans (Howard *et al.*, 1983b). These organisms appear not to cause damage to the skin of affected animals.

Internal parasites of cetaceans

Nasitrema flukes are common parasites of the head sinuses and occasionally the middle ear of many species of dolphins, porpoises, and toothed whales (Migaki *et al.*, 1971b; Dailey and Stroud, 1978; Dailey and Walker, 1978). These parasites are commonly found attached to the submucosal glands in the sinuses. They rarely cause problems in these animals. However, these flukes will occasionally migrate into the brain causing a severe parasitic encephalitis and death (Lewis and Berry, 1988; O'Shea *et al.*, 1991). The life cycle is unknown.

Halocercus and *Stenurus* spp. of metastrongyle parasites are found in the lungs of dolphins and porpoises. *Stenurus* spp. are most commonly found in the harbor porpoise and Dall's porpoise (*Phocoenoides dalli*) (Johnston and Ridgeway, 1969; Dailey and Stroud, 1978). *Halocercus* spp. inhabit the small bronchi and bronchioles of the lung. *Stenurus* spp. are located in bronchioles of the pulmonary parenchyma, causing subpleural nodules that are filled with parasites. Both can be incidental findings in animals with mild irritation of the bronchioles and an increase in mucus secretions. However, they can also be associated with a severe bacterial bronchopneumonia when the animal has large numbers of parasites in the bronchioles. The life cycle of *Halocercus* is unknown but believed to be direct. Infection of young dolphins also suggests transplacental transmigration of the parasite (Moser and Rhinehart, 1993).

Crassicauda species of nematodes are commonly found in the pterygoid air sinus of some toothed whales (Dailey and Stroud, 1978). Occasionally these parasites will migrate to the tympanic bulla or brain, leading to CNS lesions and stranding of the animals.

Braunina cordiformis and *Pholeter gastrophilus* are flukes commonly identified in the second chamber of the stomach of dolphins (Migaki *et al.*, 1971a). *Braunina* attach to the gastric mucosa and have a characteristic urn-shaped appearance in the lumen. *Pholeter gastrophilus* buries deep in the mucosa, forming small, black, cavitary nodules that can be identified on palpation.

Cyclorchis campula is a trematode that inhabits the bile and pancreatic ducts of numerous cetaceans (Migaki *et al.*, 1971a). This parasite may be an incidental finding or may cause extensive mucosal hyperplasia of the ductal mucosa and fibrosis of the periductal connective tissue. Parasites are usually found in the larger dilated pancreatic ducts.

Conclusions

Infectious disease, be it viral, bacterial, fungal, or parasitic, continues to be important in causing death in many marine mammals. Natural epizootics will continue to affect marine mammals. The reason for the catastrophic losses seen in such epizootics may be due to the accumulation of various xenobiotics in the animals, leading to immunosupression and increased susceptibility to infectious agents. It is not only important to evaluate dead and dying animals for infectious agents, but also to monitor these animals for possible accumulations of contaminants which may be causing immuno-suppression and increasing the risk of the animals to infection. These animals, like humans, are at the top of the food chain. Consequently, accumulation of these toxic compounds in marine mammals may reflect our own accumulations, which should be a concern to all about our environmental degradation.

References

Baker, J.R., McCann, T.S. 1989. Pathology and bacteriology of adult male Antarctic fur seals, *Arctocephalus gazella*, dying at Bird Island, South Georgia. *Br. Vet. J.* 145: 263–275.

Baker, J.R., Hall, A., Hiby, L., Munro, R., Robinson, I., Ross, H.M., Watkins, J.F. 1995. Isolation of *Salmonellae* from seals from U.K. waters. *Vet. Rec.* 136: 471–472.

Barr, B., Dunn, J.W., Daniel, M.D., Banford, A. 1989. Herpes-like viral dermatitis in a beluga whale (*Delphinapterus leucas*). *J. Wildlife Diseases* 25(4): 608–611.

Barrett, T., Visser, K.G., Mamaev, L., Goatley, L., Van Bressem, M.F., Osterhaus, A.D.M.E. 1993. Dolphin and porpoise morbilliviruses are genetically distinct from phocine distemper virus. *Virology* 193: 1010–1012.

Barrett, T., Blixenkrone-Moller, M., Di Guardo, G., Domingo, M., Duigan, P., Hall, A., Mamaev, L., Osterhaus, A.D.M.E. 1995. Morbillivirus in aquatic mammals: report on round table discussion. *Vet. Microbiol.* 44: 261–265.

Becker, P.R., Mackey, E.A., Demiralp, R., Schantz, M.M., Koster, B.J., Wise, S.A. 1997. Concentrations of chlorinated hydrocarbons and trace elements in marine mammal tissues archived in the U.S. National Biomonitoring Specimen Bank. *Chemosphere* 34(9/10): 2067–2098.

Bergman, A., Jarplid, B., Svensson, B.M. 1990. Pathological findings indicative of distemper in European seals. *Vet. Microbiol.* 23: 331–341.

Birkun, A., Kuiken, T., Krivokhizhin, S., Haines, D.M., Osterhaus, A.D.E.M., Van de Bildt, M.W.G., Joiris, C.R., Siebert, U., 1999. Epizootic of Morbilliviral disease in common dolphins (*Dephinus delphis ponticus*) from the Black Sea. *Vet. Rec.* 144: 85–92.

Bishop, L. 1979. Parasite related lesions in a bearded seal. *J. Wildlife Diseases* 15: 285–293.

Blanchard, T., Lipscomb, T., McFee, W., Gerber, R. 1998. Disseminated herpesvirus infection in a bottlenose dolphin. American College of Veterinary Pathologist 49th Annual Meeting. *Vet. Path.* 35: 435.

Boever, W.J., Thoen, C.O., Wallach, J.D. 1976. *Mycobacterium chelonei* infection in a Natterer manatee. *J. Am. Vet. Med. Ass.* 169: 927–929.

Borst, G.H.A., Walvoort, H.C., Reijnders, P.J.H., van der Kamp, J.S., Osterhaus, A.D.M.E. 1986. An outbreak of a herpesvirus infection in harbor seals (*Phoca vitulina*). *J. Wildlife Diseases* 22(1): 1–6.

Bossart, G.D., Brawner, T.A., Cobal, C., Kuhns, M., Eimstad, E.A., Caron, J., Trimm, M., Bradley, P. 1990. Hepatitis B-like infection in a Pacific white-sided dolphin (*Lagenorhynchus obliquidens*). *J. Am. Vet. Med. Ass.* 196(1): 127–130.

Bostock, C.J., Barrett, T., Crowther, J.R. 1990. Characterization of the European seal morbillivirus. *Vet. Microbiol.* 23: 351–360.

Britt, J.O., Howard, E.B. 1983. Virus diseases. In *Pathobiology of Marine Mammals*, Vol. I (ed. E.B. Howard). Boca Raton, Florida, CRC Press, pp. 47–67.

Britt, J.O., Nagy, A.Z., Howard, E.B. 1979. Acute viral hepatitis in California sea lions. *J. Am. Vet. Med. Ass.* 175(9): 921–923.

Brown, R.J., Smith, A.W., Keyes, M.C. 1974a. Sarcocystis in the northern fur seal. *J. Wildlife Diseases* 10: 53.

Brown, R.J., Smith, A.W., Keyes, U.C., Trevethan, W.P., Kupper, J.L. 1974b. Lesions associated with fatal hookworms infections in the northern fur seal. *J. Am. Vet. Med. Ass.* 165: 804–805.

Buck, J.D., Shepard, L.L., Spotte, S. 1987. *Clostridium perfringens* as the cause of death of a captive Atlantic bottle nosed dolphin (*Tursiops truncatus*). *J. Wildlife Diseases* 23(3): 488–491.

Buergelt, C.D., Bonde, R.K. 1983. Toxoplasma meningoencephalitis in a West Indian manatee. *J. Am. Vet. Med. Ass.* 1983: 1294–1296.

Burns, L.A., Meade, B.J., Manson, A.E. 1996. Toxic responses of the immune system. In *Casarett and Doull's Toxicology: The Basic Science of Poisons* (ed. C.D. Klaassen). New York, McGraw-Hill, pp. 373–374.

Caldwell, D.K., Caldwell, M.C., Woodard, J.C., Ajello, L., Kaplan, W., McClure, H.M. 1975. Lobomycosis as a disease of Atlantic bottle nosed dolphin. *Am. J. Trop. Med. Hyg.* 24(1): 105–114.

Callan, R.J., Early, G., Kida, H., Hinshaw, V.S. 1995. The appearance of H-3 influenza viruses in seals. *J. Gen. Virol.* 76: 199–203.

Carrol, J.M., Jasmin, A.M., Baucom, J.N. 1968. Pulmonary aspergillosis of the bottlenosed dolphin (*Tursiops truncatus*). *Am. J. Vet. Clin. Pathol.* 2: 139–140.

Conlogue, C.J., Ogden, J.A., Forryt, W.J. 1980. Pediculosis and severe heartworm infection in a harbor seal. *Vet. Med.* 75: 1184–1187.

Costa, D.P., Kooman, G.L. 1980. Effects of oil contamination in the sea otter, *Enhydra lutis*. Final report, research unit No. 71 Outer Continental Shelf Environmental Assessment Program NOAA contract No. 03-7-002-35130.

Costa, D.P., Kooman, G.L. 1982. Oxygen consumption, thermoregulation and the effect of fur oiling and washing on the sea otter, *Enhydra lutris*. *Can. J. Zool.* 60: 2761–2766.

Cousins, D.V., Francis, B.R., Gow, B.L. 1990. Tuberculosis in captive seals: bacteriological studies on an isolate belonging to the *Mycobacterium tuberculosis* complex. *Res. Vet. Sci.* 48: 196–200.

Cousins, D.V., Williams, S.N., Reuter, R., Forshaw, D., Chadwick, B., Coughran, D., Collins, P., Gales, N. 1993. Tuberculosis in wild seals and characterization of the seal bacillus. *Aust. Vet. J.* 70(3): 92–97.

Cowen, D. 1966. Pathology of the pilot whale, *Globicephala melaena*. *Arch. Path.* 82: 178–189.

Dailey, M.D., Brownell, R.L. 1972. A checklist of marine mammal parasites. In *Mammals of the Sea* (ed. S. Ridgway). Springfield, IL, Charles C. Thomas Publishers.

Dailey, M.D., Stroud, R. 1978. Parasites and associated pathology observed in cetaceans stranded along the Oregon Coast. *J. Wildlife Diseases* 14: 503–511.

Dailey, M.D., Walker, W.A. 1978. Parasitism as a factor in single strandings of Southern California cetaceans. *J. Parasitology* 64: 593–596.

Davis, J.L., Anderson, S.S. 1976. Effects of oil pollution on breeding grey seals. *Marine Pollution Bull.* 8: 115–118.

Diamond, S.S., Ewing, D.E., Cadwell, G.A. 1979. Fatal bronchopneumonia and dermatitis caused by *Pseudomonas aeruginosa* in an Atlantic bottlenosed dolphin. *J. Am. Vet. Med. Ass.* 175: 984–987.

Dierauf, L.A., Lowenstine, L.J., Jerome, C. 1981. Viral hepatitis (Adenovirus) in a California sea lion. *J. Am. Vet. Med. Ass.* 179(11): 1194–1197.

Dierauf, L.A., Vandenbroek, D.J., Roletto, J., Koski, M., Amaya, L., Gage, L.J. 1985. An epizootic of Leptospirosis in California sea lions. *J. Am. Vet. Med. Ass.* 187(11): 1145–1148.

Dilbone, R.P. 1965. Mycosis in a manatee. *J. Am. Vet. Med. Ass.* 147: 1095.

Domingo, M., Visa, J., Pumarola, M., Marco, A.J., Ferrer, L., Rabanal, R., Kennedy, S. 1992. Pathologic and immunocytochemistry studies of morbillivirus infection in striped dolphins (*Stenella coeruleoalba*). *Vet. Path.* 29: 1–10.

Duignan, P., House, C., Geraci, J.R., Early, G., Copland, A.G., Walsh, M.T., Bossart, G.D., Grey, C., Sadovc, S., St. Aubin, D.J., Moore, M. 1995a. Morbillivirus infections in two species of pilot whales (*Globicephala* sp.) from the Western Atlantic. *Marine Mammal Science* 11(2): 150–162.

Duignan, P., House, C., Walsh, M.T., Campbell, T., Bossart, G.D., Duffey, N., Fernandes, P.J., Rima, B.K., Wright, S., Geraci, J.R. 1995b. Morbillivirus infection in manatees. *Marine Mammal Science* 11: 441–451.

Dunn, J.L., Buck, J.D., Spotte, S. 1982. Candidiasis in captive cetaceans. *J. Am. Vet. Med. Ass.* 181: 1316–1321.

Dunn, J.L., Buck, J.D., Spotte, S. 1984. Candidiasis in captive pinnipeds. *J. Am. Vet. Med. Ass.* 185: 1328–1330.

Engelhardt, F.R. 1983. Petroleum effects on Marine Mammals. *Aquatic Toxicology* 4: 199–217.

Engelhardt, F.R., Geraci, J.R., Smith, T.G. 1977. Uptake and clearance of petroleum hydrocarbons in the ringed seal, *Phoca hispida. J. Fish Res. Board Canada* 34: 1143–1147.

Ewalt, D.R., Payeur, J.B., Martin, B.M., Cummins, D.R., Miller, W.G. 1994. Characteristics of a *Brucella* species from a bottlenosed dolphin (*Tursiops truncatus*). *J. Vet. Diag. Invest.* 6: 448–452.

Farnsworth, R.J., McKeever, P.J., Fletcher, J.A. 1975. Dermatomycosis in a harbor seal caused by *Microsporum canis. J. Zoo Animal Medicine* 6: 26–27.

Fauquier, D.A., Gulland, F.M.D., Trupkiewicz, J.G., Spraker, T.R., Lowenstine, L.J. 1996. Coccidioidomycosis in free-living California sea lions (*Zalophus californianus*) in Central California. *J. Wildlife Diseases* 32(4): 707–710.

Flom, J.O., Houk, E.J. 1979. Morphologic evidence of poxvirus in 'tattoo' lesions from captive bottlenose dolphins. *J. Wildlife Diseases* 15: 593–596.

Forshaw, D., Phelps, G.R. 1991. Tuberculosis in a captive colony of pinnipeds. *J. Wildlife Diseases* 23(2): 288–295.

Foster, G., Jahan, K.L., Reid, R.J., Ross, H.M. 1996. Isolation of *Brucella* species from cetaceans, seals, and an otter. *Vet. Rec.* 138: 583–586.

Frasca, S., Dunn, J.L., Cooke, J.C., Buck, J.D. 1996. Mycotic dermatitis in an Atlantic white-sided dolphin, a pygmy sperm whale and two harbor seals. *J. Am. Vet. Med. Ass.* 208(5): 727–729.

Frey, H.R., Liess, B., Haas, L., Lehmann, H., Marschall, H.J. 1989. Herpesvirus in harbor seals (*Phoca vitulina*): Isolation, partial characterization and distribution. *J. Vet. Med.* 36: 699–708.

Gage, L.J., Amaya-Sherman, L., Roletto, J., Bently, S. 1990. Clinical signs of San Miguel sea lion virus in debilitated California sea lions. *J. Zoo Wildlife Medicine* 21(1): 79–83.

Garner, M.M., Lambourn, D.M., Jeffries, S.J., Hall, P.B., Rhyan, J.C., Ewalt, D.R., Polzin, L.M., Cheville, N.F. 1997. Evidence of *Brucella* infections in *Parafilaroides* lungworms in a Pacific harbor seal (*Phoca vitalina richardsi*). *J. Vet. Diag. Invest.* 9: 298–303.

Geraci, J.R., St. Aubin, D.J. 1980. Offshore petroleum resource development and marine mammals: a review and research recommendation. *Marine Fisheries Review* 42: 1–12.

Geraci, J.R., St. Aubin, D.J. 1982. Study of the effects of oil on cetaceans, report for the U.S. Department of Interior, Bureau of Land Management. Contract AA-551-CT9-29 p. 274.

Geraci, J.R., Smith, T.G. 1976. Direct and indirect effects of oil on ringed seals (*Phoca hispida*) of the Beaufort Sea. *J. Fish Res. Board Canada* 33: 1976–1984.

Geraci, J.R., Sauer, R.M., Medway, W. 1966. Erysipelas in dolphins. *Am. J. Vet. Res.* 27: 597–606.

Geraci, J.R., Hicks, B.D., St. Aubin, D.J. 1979. Dolphin Pox: a skin disease of cetaceans. *Can. J. Comp. Med.* 43: 399–404.

Geraci, J.R., St. Aubin, D.J., Barker, I.K., Webster, R.G., Hinshaw, V.S., Bean, W.J., Ruhuke, J.H., Prescott, J.H., Early, G., Baker, A.S., Madoff, S., Schooley, R.T. 1982. Mass mortality of harbor seals: pneumonia associated with influenza A virus. *Science* 215: 1129.

Gilmartin, W.G., Vainik, P.M., Neill, V.A. 1979. *Salmonellae* in feral pinnipeds off the southern California coast. *J. Wildlife Diseases* 15: 511–514.

Grachev, M.A., Kumarev, V.P., Mamaev, L.V., Zorin, V.L., Baranova, L.V., Denikina, N.N., Belikov, S.I., Petrov, S.I., Petrov, E.A., Kolesnik, V.S., Dorfeev, R.S., Beim, V.M., Kudelin, V.M., Magieva, F.G., Sidorov, V.N. 1989. Distemper virus in Baikal seals. *Nature* 338: 209.

Greenwood, A.G., Taylor, D.C. 1978. Clostridial myositis in marine mammals. *Vet. Rec.* 103: 54–55.

Gulland, F.M.D., Koski, M., Lowenstine, L.J., Colagross, A., Morgan, L., Spraker, T. 1996. Leptospirosis in California sea lions (*Zalophus californianus*) stranded along the Central California Coast. 1981–1984. *J. Wildlife Diseases* 32(4): 572–580.

Gulland, F.M.D., Lowenstine, L.J., Lapointe, J.M., Spraker, T., King, D.P. 1997. Herpesvirus infection in stranded Pacific harbor seals of coastal California. *J. Wildlife Diseases* 33(3): 450–458.

Hall, A.J. 1995. Morbilliviruses in marine mammals. *Trends in Microbiology* 3: 4–9.

Harder, M., Vos, H.W., Kulonen, K., Kennedy-Stoskopf, S., Liess, B., Appel, M.J.G., Osterhaus, A.D.M.E. 1996. Characterization of phocid herpesvirus of North American and European pinnipeds. *J. Gen. Virol.* 77: 27–35.

Haebler, R., Moeller, R.B. 1993. Pathobiology of selected marine mammal diseases. In *Pathobiology of Marine and Estuarine Organism* (ed. J.A. Couch, J.W. Fournie). Boca Raton, Florida, CRC Press.

Hicks, B.D., Worthy, G.A.J. 1987. Seal pox in captive grey seals (*Halichoerus grypus*) and their handlers. *J. Wildlife Diseases* 23(1): 1–6.

Holshuh, H.J., Sherrod, A.E., Taylor, C.R., Andrews, B.F., Howard, E.B. 1985. Toxoplasmosis in a feral northern fur seal. *J. Vet. Med. Ass.* 187: 1229–1230.

Horvat, B., Willhaus, T., Frey, H.R., Liess, B. 1989. Herpesvirus in harbor seals (*Phoca vitulina*): Transmission in homologous host. *J. Vet. Med.* 36: 715–718.

Howard, E.B., Britt, J.O., Matsumoto, G.K., Itahara, R., Nagano, C.N. 1983a. Bacterial diseases. In *Pathobiology of Marine Mammal Disease*, Vol. I (ed. E.B. Howard). Boca Raton, Florida, CRC Press, pp. 70–117.

Howard, E.B., Britt, J.O., Matsumoto, G.K. 1983b. Parasitic diseases. In *Pathobiology of Marine Mammals*, Vol. II (ed. E.B. Howard). Boca Raton, Florida, CRC Press, pp. 92–162.

Inskeep, W., Gardiner, C.H., Harris, R.K., Dubey, J.P., Goldston, R.T. 1990. Toxoplasmosis in Atlantic bottle-nosed dolphins (*Tursiops truncatus*). *J. Wildlife Diseases* 26(3): 377–382.

Jahans, K.L., Foster, G., Broughton, E.S. 1997. The characterization of *Brucella* strains isolated from marine mammals. *Vet. Microbiol.* 57: 373–382.

Johnston, D.G., Ridgeway, S.H. 1969. Parasitism in some marine mammals. *J. Vet. Med. Ass.* 155: 1064–1072.

Joseph, B.E., Spraker, T.R., Migaki, G. 1990. Valvular endocarditis in a Northern sea otter (*Enhydra lutris*). *J. Zoo Wildlife Medicine* 21(1): 88–91.

Kannan, N., Tanabe, S., Ono, M., Tatsukana, R. 1989. Critical evaluation of polychlorinated biphenyl toxicity in terrestrial and marine mammals: Increasing impact on non-ortho and mono-ortho coplanar polychlorinated biphenyls from land to ocean. *Arch. Environ. Contam. Toxicol.* 18: 850–857.

Kennedy, S. 1990. A review of the 1988 European seal morbilliviral epizootic. *Vet. Rec.* 127: 563–567.

Kennedy, S. 1998. Morbillivirus infections in aquatic mammals. *J. Comp. Path.* 119: 201–225.

Kennedy, S., Smyth, J.A., McCullough, Allan, G.M., McNeilly, F., McQwaid, S. 1988. Confirmation of cause of recent seal deaths. *Nature* 336: 21.

Kennedy, S., Smyth, J.A., Cush, P.F., Duignan, P., Platten, M., McCullough, S.J., Allan, G.M. 1989. Histopathologic and immunocytochemical studies of distempers in seals. *Vet. Path.* 26: 97–103.

Kennedy, S., Smyth, J.A., Cush, P.F., McAliskey, M., McCullough, S.J., Rima, B.K. 1991. Histopathologic and immunocytochemical studies of distemper in harbor porpoises. *Vet. Path.* 28: 1–7.

Kennedy-Stoskopf, S., Stoskopf, M.K., Eckhaus, M., Strandberg, J.D. 1986. Isolation of a retrovirus and a herpesvirus from a captive California sea lion. *J. Wildlife Diseases* 22(2): 156–164.

Ketterer, P.J., Rosenfeld, L.E. 1974. Septic embolic nephritis in a dolphin caused by *Staphylococcus aureus*. *Aust. Vet. J.* 50: 123.

Keyes, M. 1965. Pathology of the northern fur seal. *J. Am. Vet. Med. Ass.* 147: 1090–1095.

Kim, K.C., Haas, V.L. 1980. Populations, microhabitat preference and effects of infestation of two species of *Orthohalarachne* (*Halarachnidae*: *Acarina*) in the northern fur seal. *J. Wildlife Diseases* 16(1): 45–51.

Kinsel, M.J., Boehm, J.R., Harris, B., Murnane, R.D. 1997. Fatal *Erysipelothrix rhusiopathiae* septicemia in a captive Pacific white-sided dolphin (*Lagenorhyncus obliquidens*). *J. Zoo Wildlife Medicine* 28(4): 494–497.

Lambertsen, R.H., Kohn, B.A., Sundberg, J.P., Buergelt, C.D. 1987. Genital papillomatosis in sperm whale bulls. *J. Wildlife Diseases* 23(3): 361–367.

Lewis, R.J., Berry, K. 1988. Brain lesions in a Pacific white-sided dolphin (*Lagenorhynchs obliquidens*). *J. Wildlife Diseases* 24: 577–581.

Lipscomb, T.P., Harris, R.K., Moeller, R.B., Plecher, J.M., Haebler, R.J., Ballachey, B.E. 1993. Histopathologic lesions in sea otters exposed to crude oil. *Vet. Path.* 30: 1–11.

Lipscomb, T.P., Kennedy, S., Moffett, D., Krafft, A., Klaunberg, B.A., Lichy, J.H., Regan, G.T., Worthy, C.A.J., Taubenberger, J.K. 1996. Morbillivirus epizootic in bottlenosed dolphins of the Gulf of Mexico. *J. Vet. Diagn. Invest.* 8: 282–290.

Lipscomb, T.P., Scott, D., Gulland, F., Lowenstine, L., Garber, R. 1998. Metastatic carcinoma of California sea lions: Evidence of genital origin and associated gamma herpesvirus infection. American College of Veterinary Pathologists 49th Annual Meeting. *Vet. Path.* 35: 421.

Lyons, E.T., DeLong, R.L., Melin, S.R., Tolliver, S.C. 1997. Uncinariasis in northern fur seals and California sea lion pups from California. *J. Wildlife Diseases* 33: 848–852.

Mamaev, L.V., Visser, I.K.C., Belikov, S.I., Denikina, N.N., Harder, T., Goatley, L., Rima, B., Edginton, B., Osterhaus, A.D.M.E., Barrett, T. 1996. Canine distemper virus in Lake Baikal seals. *Vet. Rec.* 138: 437–439.

Martineau, D., Lagace, A., Beland, P., Higgins, R., Armstrong, D., Shugart, L.R. 1988. Pathology of stranded beluga whales (*Delphinapterus leucas*) from the St. Lawrence estuary, Quebec, Canada. *J. Comp. Path.* 98: 287–311.

Martineau, D., DeGuise, S., Fournier, M., Shugart, L., Girard, C., Lagace, A., Belaud, P. 1994. Pathology and toxicology of beluga whales from the St. Lawrence estuary, Quebec, Canada. Past, present, and future. *Science of the Total Environment* 154: 201–215.

Medway, W. 1980. Some bacterial and mycotic diseases of marine mammals. *J. Am. Vet. Med. Ass.* 177: 831–834.

Mense, M.G., Dubey, J.P., Homer, B.L. 1992. Acute hepatic necrosis associated with a *Sarcocystis canis*-like protozoa in a sea lion (*Zalophus californianus*). *J. Vet. Diagn. Invest.* 4: 486–490.

Migaki, G. 1987. Selected dermatosis of marine mammals. *Clin. Dermatol.* 5(3): 155–164.

Migaki, G., Albert, T.F. 1980. Sarcosporidosis in the ringed seal. *J. Am. Vet. Med. Ass.* 177: 917–918.

Migaki, G., Jones, S.R. 1983. Mycotic diseases in marine mammals. In *Pathobiology of Marine Mammal Diseases*, Vol. II (ed. E.B. Howard). Boca Raton, Florida, CRC Press, pp. 1–127.

Migaki, G., Van Dyke, D., Hubbard, R. 1971b. Some histopathological lesions caused by helminthes in marine mammals. *J. Wildlife Diseases* 6: 152–154.

Migaki, G., Valerio, M.G., Irving, B., Garner, F.M. 1971a. Lobo's disease in an Atlantic bottlenosed dolphin. *J. Am. Vet. Med. Ass.* 159: 578–582.

Migaki, G., Allen, J.F., Casey, H.W. 1977. Toxoplasmosis in a California sea lion (*Zalophus californianus*). *Am. J. Vet. Res.* 38: 135–136.

Migaki, G., Gunnels, R.D., Casey, H.W. 1978a. Pulmonary Cryptococcosis in an Atlantic bottlenosed dolphin (*Tursiops truncatus*). *Laboratory Animal Science* 28: 603–606.

Migaki, G., Font, R.L., Kaplan, W., Asper, E.D. 1978b. Sporotrichosis in a Pacific white-sided dolphin (*Lagenorhynchus obliquiden*). *Am. J. Vet. Res.* 39: 1916–1919.

Miller, G.J. 1982. Ecotoxicity of petroleum hydrocarbons in the marine environment. *J. Appl. Toxicol.* 2(2): 88–97.

Minette, P.H. 1986. A review and commentary. Salmonellosis in marine mammals. *Int. J. Zoonoses* 13(2): 71–75.

Montali, R.J., Bush, M., Strandberg, J.D., Janssen, D.L., Boness, D.J., Whitla, J.C. 1981. Cyclic dermatitis associated with *Fusarium* sp. infection in pinnipeds. *J. Am. Vet. Med. Ass.* 179(11): 1198–1202.

Morales, P., Madin, S.H., Hunter, A. 1985. Systemic *Mycobacterium marinum* infection in an Amazon manatee. *J. Am. Vet. Med. Ass.* 187(11): 1230–1231.

Moser, M., Rhinehart, H. 1993. The lungworm, *Halocercus* spp. (Nematodea: Pseudaliidae) in cetaceans from California. *J. Wildlife Diseases* 29: 507–508.

Mossner, S., Ballschmiter, K. 1997. Marine mammals as global pollution indicators for organochlorines. *Chemosphere* 34(5–7): 1285–1296.

Nakeeb, S., Targowski, S.P., Spotte, S. 1977. Chronic cutaneous candidiasis in bottlenosed dolphins. *J. Am. Vet. Med. Ass.* 171: 961–965.

Neill, J.D., Seal, B.S. 1995. Development of PCR primers for specific amplification of two distinct regions of the genomes of San Miguel sea lion and vesicular exanthema of swine viruses. *Molec. Cell. Probes* 9: 33–38.

Nettleton, P.F., Munro, R., Pow, I., Gilray, J., Gray, E.W., Reid, H.W. 1995. Isolation of a parapoxvirus from a grey seal (*Halichoerus grypus*). *Vet. Rec.* 137: 562–564.

O'Hara, T.M., House, C., House, J.A., Suydam, R.S., George, J.C. 1998. Viral serologic survey of bowhead whales in Alaska. *J. Wildlife Diseases* 34(1): 39–46.

Olsen, R.W., Lyons, E.T. 1965. Life cycle of *Uncinaria lucasi stiles*, 1901 (Nematoda: *Ancylostomatidae*) of fur seals, *Callorhinus ursinus*, on the Pribilof islands, Alaska. *J. Parasitol.* 51: 689–700.

Olson, M.E., Roach, P.D., Stubler, M., Chan, W. 1997. Giardiasis in ringed seals from the western Artic. *J. Wildlife Diseases* 33: 646–648.

Oritsland, N.A., Engelhardt, F.R., Juck, R.J., Hurst, R.J., Watts, P.D. 1981. Effects of crude oil on polar bears. Department of Indian Affairs and Northern Development. Environmental Studies No. 24, p. 268.

O'Shea, T.J., Homer, B.L., Greiner, E.C., Layton, A.W. 1991. *Nasitrema* sp. associated encephalitis in a striped dolphin (*Stenella coeruleoalba*) stranded in the Gulf of Mexico. *J. Wildlife Diseases* 27: 706–709.

Osterhaus, A.D.M.E. 1988. Seal deaths. *Nature*, 334: 301–302.

Osterhaus, A.D.M.E., Vedder, E.J. 1988. Identification of virus causing recent seal death. *Nature* 335: 20.

Osterhaus, A.D.M.E., Yang, H., Spijkers, H.E.M., Groen, J., Teppema, J.S., van Steenis, G. 1985. The isolation and partial characterization of a highly pathogenic herpesvirus from the harbor seal (*Phoca vitulina*). *Arch. Virol.* 86: 239–251.

Osterhaus, A.D.M.E., Groen, J., Spijker, H.E.M., Broeders, H.W.J., Uytde Haag, F.G.C.M., de Vries, P., Teppema, J.S., Visser, I.K.G., van de Bildt, M.W.G., Vedder, E.J. 1990a. Mass mortality in seals caused by a newly discovered virus-like morbillivirus. *Vet. Microbiol.* 23: 343–350.

Osterhaus, A.D.M.E, Broeders, H.W.J., Visser, I.K.G., Teppema, J.S., Vedder, E.J. 1990b. Isolation of an orthopoxvirus from pox-like lesions in a grey seal (*Halichoerus grypus*). *Vet Rec.* 127: 91–92.

Osterhaus, A., Groen, J., Niester, H., Van de Bildt, M., Marina, B., Vedder, L., Vos, J., Van Egmond, H., Sidi, B.A., Barham, M.E., 1997. Morbillivirus in monk seal mortality. *Nature* 338: 838–839.

Osterhaus, A.D.M.E., Rimmelzwaan, G.F., Martina, B.E.E., Bestebroer, M. 2000. Influenza B virus in seals. *Science* 288: 1051–1053.

Owen, C.C., Kakulas, R.A. 1967. Sarcosporidiosis in the sperm whale. *Aust. J. Sci.* 31: 46–47.

Palmer, C.J., Schroeder, J.P., Fujioka, R.S., Douglas, J.T. 1991. *Staphylococcus aureus* infection in newly captured Pacific bottlenose dolphins (*Tursiops truncatus gilli*). *J. Zoo Wildlife Medicine* 22(3): 330–338.

Parson, E.C.M. 1998. Trace metal pollution in Hong Kong: Implications for the health of Hong Kong's Indo-Pacific hump-backed dolphins (*Sousa chinensis*). *The Science of the Total Environment* 214: 175–184.

Pier, A.C., Takayama, A.K., Miyahara, A.Y. 1970. Cetacean nocardiosis. *J. Wildlife Diseases* 6: 112–118.

Rahimtula, A.D., O'Brian, P.J., Payne, J.F. 1984. Induction of xenobiotic metabolism in rats on exposure to hydrocarbon based oils. In *Applied Toxicology* of Petroleum Hydrocarbons, *Advances in Modern Environmental Toxicology* (ed. M.A. Mehlman). Princeton, N.J, Princeton Scientific Publishing, Vol. VI, pp. 71–79.

Rand, C.S. 1975. Nodular suppurative cutaneous cellulitis in a Galapagos sea lion. *J. Wildlife Diseases* 11: 325–329.

Rausch, R. 1953. Studies on the helminth fauna of Alaska XIII. Diseases in the sea otter with special references to helminth parasites. *Ecology* 34: 584–604.

Rawson, A.J., Bradley, J.P., Teetsov, A., Rice, S.B., Haller, E.M., Patton, G.W. 1995. A role for airborne particles in high mercury levels of some cetaceans. *Ecotoxicity and Environmental Safety* 30: 309–314.

Reed, R.E., Migaki, G., Cummings, J.A. 1976. Coccidioidomycosis in a California sea lion (*Zalophus californianus*). *J. Wildlife Diseases* 12: 372–375.

Reidarson, T.H., Griner, L.A., Pappagianis, D., McBain, J. 1998. Coccidioidomycosis in a bottlenosed dolphin. *J. Wildlife Diseases* 34(3): 629–631.

Reijnders, P.J.H. 1986. Reproductive failure in common seals feeding on fish from polluted coastal waters. *Nature* 324: 456–458.

Ross, H.M., Jahans, K.L., MacMillian, A.P., Reid, R.J., Thompson, P.M., Foster, G. 1996. *Brucella* species infection in North Sea seal and cetacean populations. *Vet. Rec.* 138: 647–648.

Sayer, J.C. 1976. Vesicular Exanthema of swine and San Miguel Sea Lion Virus. *J. Am. Vet. Med. Ass.* 169(7): 707–709.

Schulman, F.Y., Lipscomb, T.P. 1999. Dermatitis with invasive ciliated protozoa in dolphins that died during the 1987–1988 Atlantic bottlenosed dolphin morbillivirus epizootic. *Vet. Path.* 36: 171–174.

Schulman, F.Y., Lipscomb, T.P., Moffett, D., Krafft, A.E., Lichy, J.H., Tsai, M.M., Taubenberger, J.K., Kennedy, S. 1997. Histologic, immunocytochemical, and polymerase chain reaction studies of bottlenose dolphins from the 1987–1988 United States Atlantic coast epizootic. *Vet. Path.* 34: 288–295.

Seal, B.S., House, J.A., Whetstone, C.A., Neill, J.D. 1995. Analysis of the serologic relationship among San Miguel sea lion virus and vesicular exanthema of swine isolates. Application of Western Blot assay for detection of antibodies in swine sera to these two virus types. *J. Vet. Diagn. Invest.* 7: 190–195.

Seibold, H.R., Neal, J.E. 1956. Erysipelothrix septicemia in the porpoise. *J. Am. Vet. Med. Ass.* 128: 537–539.

Simpson, C.F., Wood, F.G., Young, F. 1958. Cutaneous lesions on a porpoise with erysipelas. *J. Am. Vet. Med. Ass.* 133: 558–560.

Skaar, I., Gaustad, P., Tonjum, T., Holm, B., Stenwig, H. 1994. *Streptococcus phocae* sp., nov., a new species isolated from clinical specimens from seals. *Int. J. Syst. Bacteriol.* 44: 646–650.

Skilling, D.E., Barlough, J.E., Berry, E.S., Brown, R.F., Smith, A.W. 1987. First isolation of a Calicivirus from the Steller sea lion (*Eumetopias jubatus*). *J. Wildlife Diseases* 23(4): 534–538.

Smith, A.W., Boyt, P.M. 1990. Caliciviruses of ocean origin: A review. *J. Zoo Wildlife Medicine* 21(1): 3–23.

Smith, A.W., Skilling, D.E. 1979. Viruses and virus diseases of marine mammals. *J. Am. Vet. Med. Ass.* 175(9): 918–920.

Smith, A.W., Brown, R.J., Skilling, D.E., Delong, R.L. 1974. *Leptospira pomona* and reproductive failure in California sea lions. *J. Am. Vet. Med. Ass.* 165(11): 996–998.

Smith, A.W., Brown, R.J., Skilling, D.E., Bray, H.C., Keyes, M.C. 1977. Naturally occurring Leptospirosis in Northern fur seals (*Callorhinus ursinus*). *J. Wildlife Diseases* 13: 144–148.

Smith, A.W., Skilling, D.E., Dardiri, A.H., Latham, A.B. 1980. Calicivirus pathogenic for swine: a new serotype isolated from opaleye fish, *Girella nigricans*, an ocean fish. *Science* 209: 940–941.

Smith, A.W., Skilling, D.E., Latham, A.B. 1981. Isolation and identification of five new serotypes of Calicivirus from marine mammals. *Am. J. Vet. Res.* 42(4): 693–694.

Smith, A.W., Skilling, D.E., Ridgway, S. 1983a. Calicivirus-induced vesicular disease in cetaceans and probable interspecies transmission. *J. Am. Vet. Med. Ass.* 183(11): 1223–1225.

Smith, A.W., Skilling, D.E., Ridgway, S.H. 1983b. Regression of cetacean tattoo lesions concurrent with conversion of precipitin antibody against pox virus. *J. Am. Vet. Med. Ass.* 183(11): 1219–1222.

Smith, T.G., Geraci, J.R. 1975. The effects of contact and ingestion of crude oil on ringed seals of the Beaufort Sea Project. Technical Report #5, p. 67.

Stack, M.J., Simpson, V.R., Scott, A.C. 1993. Mixed poxvirus and calicivirus infections of grey seals (*Halichoerus grypus*) in Cornwall. *Vet Rec.* 132: 163–165.

Stroud, R.K., Roelke, M.E. 1980. Salmonella meningeoencephalitis in a northern fur seal (*Callorhinus ursinus*). *J. Wildlife Diseases* 16(1): 15–18.

Sweeney, J.C. 1974. Common diseases of pinnipeds. *J. Am. Vet. Med. Ass.* 165(9): 805–810.

Sweeney, J.C., Gilmartin, W.G. 1974. Survey of diseases in free living California sea lions. *J. Wildlife Diseases* 10: 370–376.

Sweeney, J.C., Migaki, G., Vainik, P.M., Conklin, R.H. 1976. Systemic mycosis in marine mammals. *J. Am. Vet. Med. Ass.* 169(9): 946–948.

Tanaka, E., Kimura, T., Wada, S., Hatai, K., Sonoda, S. 1994. Dermatophytosis in a Steller sea lion (*Eumetopias jubatus*). *J. Vet. Med. Sci.* 56(3): 551–553.

Thomas, R.E., Brodersen, C., Carls, M.G., Babcock, M., Rice, S.D. 1999. Lack of physiological response to hydrocarbon accumulation by *Mytilus trossulus* after 3–4 years chronic exposure to spilled Exxon Valdez crude oil in Prince William Sound. *Comp. Biochem. Physiol. C Pharmacol. Toxicol. Endocrinol.* 122: 153–163.

Tilbury, K.L., Stein, J.E., Meador, J.P., Krone, C.A., Chan, S.L. 1997. Chemical contaminants in harbor porpoise (*Phocoena phocoena*) from the North Atlantic coast: tissue concentrations and intra and inter organ distribution. *Chemosphere* 34(9/10): 2159–2181.

Van de Bildt, M., Vedder, E., Marina, B., Sidi, B.A., Jiddon, A.B., Barham, M., Andreoukaki, E., Komnenou, A., Niesters, H., Osterhaus, A. 1999. Morbillivirus in Mediterranean monk seals. *Vet. Microbiol.* 69: 19–21.

Van de Bildt, M., Marina, B., Vedder, E.J., Androukaki, E., Kotomatas, S., Komnenou, A., Sidi, B.A., Jiddon, A.B., Barham, M., Niesters, H., Osterhaus, A. 2000. Identification of Morbillivirus of probable cetacean origin in carcasses of Mediterranean monk seals (*Monachus monachus*). *Vet. Rec.* 146: 691–694.

Vandermeulen, J.H., Penrose, W.R. 1978. Absence of aryl hydrocarbon hydrolase (AHH) in three marine bivalves. *J. Fish Res. Board Canada* 35: 643–647.

Van Pelt, R.W., Dieterich, R.A. 1973. Staphylococcal infection and toxoplasmosis in a young harbor seal. *J. Wildlife Diseases* 9: 258–261.

Varaldo, P.E., Kilpper-Balz, R., Biavasco, F., Satta, G., Schleifer, K.H. 1988. *Staphylococcus delphini* sp. nov., a coagulase-positive species isolated from dolphins. *Int. J. System. Bacteriol.* 38(4): 436–439.

Vedros, N.A., Smith, A.W., Schonewald, J., Migaki, G. 1971. Leptospirosis epizootic among California sea lions. *Science* 172: 1250–1251.

Visser, I.K.G., Van Bressem, M.F., de Swart, R.L., Van de Bildt, M.W.G., Vos, H., Van der Heijden, R.W.J., Saliki, J.T., Orvell, C., Osterhaus, A.D.M.E. 1993a. Characterization of morbilliviruses isolated from dolphins and porpoises in Europe. *J. Gen. Virol.* 74: 631–641.

Visser, I.K.G., Van Bressem, M.F., Van de Bildt, M.W.G., Groen, J., Orvell, C., Raga, J.A., Osterhaus, A.D.M.E. 1993b. Prevalence of morbillivirus among pinniped and cetacean species. *Rev. Sci. Tech. Off. Int. Epiz.* 12(1): 197–202.

Wade, T.L., Chambers, L., Gardinali, P.R., Sericano, J.L., Jackson, T.J. 1997. Toxaphene, PCB, DDT, and Chlordane analysis of beluga whale blubber. *Chemosphere* 34(5–7): 1351–1357.

Warner, R.F. 1969. Environmental effects of oil pollution in Canada: an evaluation of problems and research needs. Can. Wildlife Service Ms Report Number 645, pp. 16–17.

Wells, S.K., Gutter, A., Van Meter, K. 1990. Cutaneous mycobacteriosis in an harbor seal: attempted treatment with hyperbaric oxygen. *J. Zoo Wildlife Medicine* 21(1): 73–78.

Williamson, W.M., Lombard, L.S., Getty, R.E. 1959. North American blastomycosis in a northern sea lion. *J. Am. Vet. Med. Ass.* 135: 513–515.

Wilson, T.M., Poglayen-Neuwall, I. 1971. Pox in South American sea lions (*Otaria byronia*). *Am. J. Comp. Med.* 35: 174–177.

Wilson, T.M., Cheville, N.F., Karstad, L. 1969. Seal pox. *Bull. Wildlife Dis. Ass.* 5: 412–418.

Wilson, T.M., Stockdale, P.H. 1970. The harp seal *Pogophilus groenlandicus* XI *Contracaecum* sp. infestation in a harp seal. *J. Wildlife Diseases* 6: 152–154.

Wilson, T.M., Dykes, R.W., Tsai, K.S. 1972. Pox in young, captive harbor seals. *J. Am. Vet. Med. Ass.* 161(6): 611–617.

Wilson, T.M., Kierstead, M., Long, J.R. 1974. Histoplasmosis in a harp seal. *J. Am. Vet. Med. Ass.* 165: 815–817.

Woods, R., Cousins, D.V., Kirkwood, R., Obendorf, D.L. 1995. Tuberculosis in a wild Australian fur seal (*Arctocephalus pusillus doriferus*) from Tasmania. *J. Wildlife Diseases* 31(1): 83–86.

Yantis, D., Dubey, J., Moeller, R., Braun, R., Aquirre, A., Gardiner, C. 1998. Hepatic sarcocystosis in a Hawaiian monk seal (*Monachus schauinslandi*). American College of Veterinary Patholgist 49th Annual Meeting. *Vet. Path.* 35: 453.

Zarnke, R.L., Harder, T.C., Vos, H.W., Ver Hoef, J.M., Osterhaus, A.D.M.E. 1997. Serologic survey for phocid herpesvirus 1 and 2 in marine mammals from Alaska and Russia. *J. Wildlife Diseases* 33(3): 459–465.

2 Contaminants and marine mammal immunotoxicology and pathology

Sylvain De Guise, Kimberlee B. Beckmen and Steven D. Holladay

Pollutants in marine mammals

Oceans have long been considered endless sinks where domestic and industrial wastes could be discarded. It is not surprising to now measure significant concentrations of such contaminants in the water, sediments and inhabitants of the marine environment. This is especially true for stable compounds, such as organochlorines, which often persist long after their production and use have been banned. Marine mammals, which are usually at the top of a complex food chain, often accumulate large amounts of those contaminants. Of special interest (because of their abundance and/or known toxicity) are organochlorines (PCBs, DDT, etc.), heavy metals, and polycyclic aromatic hydrocarbons (PAHs), which have been found in marine mammal tissues worldwide, including Europe (Holden and Marsden, 1967; Reijnders, 1980; Baumann and Martinsen, 1983), North America (Gaskin *et al.*, 1971; Addison *et al.*, 1973; Muir *et al.*, 1990), South America (Gaskin *et al.*, 1974), Asia (Taruski *et al.*, 1975), the Arctic (Addison and Smith, 1974; Born *et al.*, 1981) and Antarctica (Sladen *et al.*, 1966). Those studies demonstrated that pollution problems are now global, and that there are no more pristine environments. But the biological significance and potential health effects of these environmental contaminants still need to be addressed.

Potential effects of contaminants on marine mammals

Catastrophic viral epidemics have recently affected several populations of seals (Geraci *et al.*, 1982; Osterhaus *et al.*, 1988), porpoises (Kennedy *et al.*, 1991) and dolphins (Domingo *et al.*, 1990, 1992), all severely contaminated by industrial pollutants. No doubt, influenza virus or morbilliviruses were the direct causes for the deaths of thousands of these animals. However, it has long been known that experimental animals chronically exposed to PCBs are more susceptible to viral infections. The viruses used in these experiments represented a wide range, including duck hepatitis virus, a small RNA virus (picornavirus) (Friend and Trainer, 1970); a murine leukemia virus, an RNA virus (retrovirus) (Koller, 1977); and herpes simplex, a large DNA virus

(Imanishi *et al.*, 1980). Accordingly, a possible immunosuppressive role of organohalogens to explain the severity of the cetacean and pinniped epidemics has been suggested (Eis, 1989). Similarly, contaminant-induced immuno-suppression has been suggested to explain the high incidence, severity and diversity of lesions often caused by opportunistic and mildly pathogenic bacteria that were found on post-mortem examination of the endangered, small isolated population of beluga whales (*Delphinapterus leucas*) from the St. Lawrence estuary (Martineau *et al.*, 1988; De Guise *et al.*, 1995a). Although many contaminants of the marine environment (organochlorines, heavy metals, polycyclic aromatic hydrocarbons) are well characterized as immuno-toxicants in laboratory rodents, the demonstration of immunotoxic effects in marine mammals exposed to these agents represents a significant challenge. Such is true in these species because of the relatively limited immunologic database presently existing, relatively limited assay development and re-agent availability to evaluate immune function in marine mammals, genetic diversity naturally occurring in outbred populations such as marine mam-mals, and logistical and ethical considerations when working with marine mammals.

Immunotoxicology

Immunotoxicology is a relatively new scientific discipline, the aim of which is to detect, quantify and interpret direct or indirect alterations of the immune system that occur as a result of exposure to chemicals, pharmaceuticals, recombinant biologicals or environmental and occupational pollutants (Burleson and Dean, 1995). The immune system is a complex network of cells with diverse functions, communicating through a wide array of messenger molecules, which has evolved to protect the host from potentially pathogenic agents, including viruses, bacteria, parasites, fungi, neoplastic and non-self cells (Roitt *et al.*, 1993). Immunotoxicology studies deal with immune altera-tions, stimulatory or suppressive, their mechanisms, and their resulting effects on susceptibility or duration of infectious, allergic or autoimmune diseases (Burleson and Dean, 1995).

Considerable efforts have been made to standardize methods and assays. A comprehensive testing panel composed of two tiers has been developed and validated to characterize immune alterations following *in vivo* chemical exposure in mice (Luster *et al.*, 1988). Further studies demonstrated that immunophenotyping of splenic lymphocytes and the enumeration of plaque-forming cells (PFCs) after immunization with sheep red blood cells were the two individual assays most predictive of immunotoxicity (Luster *et al.*, 1992). In contrast, commonly employed measures such as leukocyte counts and lymphoid organ weights were fairly insensitive (Luster *et al.*, 1992). Several combinations of two assays could predict immunotoxicity with a percentage concordance (sum of specificity and sensitivity) of 90 per cent or more. These combinations included immunophenotyping (surface markers) of lymphocytes coupled to either PFC, natural killer (NK) cell activity, T-cell mitogen-induced proliferation, delayed hypersensitivity response, cytotoxic T-lymphocyte activity (CTL), thymus–body weight ratio, or lipopoly-saccharide (LPS)-induced B-lymphocyte proliferation, as well as PFC coupled to either NK cell activity, CTL or thymus–body weight ratio (Luster *et al.*, 1992). Further studies comparing the relationship between the immune function and host resistance showed a good correlation between changes in the immune tests and altered host resistance, with no instances where host resistance was altered without affecting an immune test (Luster *et al.*, 1993). These studies also showed that no single immune test could be identified which was fully predictive for altered host resistance, although several assays were relatively good individual indicators (Luster *et al.*, 1993).

Although significant progress has been made in chemical-induced immuno-toxicity in rodents, the predictive value of those studies for risk assessment in other species remains questionable. For this and other reasons, efforts have been made to develop models using *in vitro* exposure. Although such models do not allow evaluation of the pharmacokinetics of the compounds tested (including absorption, distribution, excretion and long-term accumulation), they have several advantages. They allow the evaluation of the direct effects of a compound on the immune system, with a relatively rapid detection of immunotoxic effects (compared to *in vivo* systems), at a much lower cost, and with fewer animals used (Munson and LeVier, 1995). These reasons have prompted an attempt to validate and use *in vitro* studies for regulatory purposes (Prochazkova, 1993; Jackson, 1998). But most importantly, *in vitro* exposures allow studies in species such as marine mammals, where obvious logistical and ethical considerations would make experimental *in vivo* studies very difficult, in addition to allowing comparisons among species. Nevertheless, several of the assays (such as CTL or PFC), for which predictive values were established *in vivo*, cannot be performed if the animals used have not been previously experimentally exposed or immunized.

In view of suspected contaminant-induced immune suppression underlying health problems and epizootics in marine mammals, we will briefly review the known immunotoxic effects of environmental pollutants demonstrated in laboratory species.

Immunotoxicity of organohalogens

Organohalogens are among the most abundant contaminants in tissues of marine mammals. Ample evidence that organohalogens have detrimental effects on the immune system of humans and animals has been collected over the past 25 years. These compounds alter both innate and acquired immune functions, and both cellular and humoral immunity. TCDD (2,3,7,8, -tetrachlorodibenzo-*p*-dioxin), the most immunotoxic of aromatic halogenated hydrocarbons (HAHs), induces thymic atrophy in all species evaluated

(Poland and Knutson, 1982; Davis and Safe, 1988; Kerkvliet *et al.*, 1990; Andersson *et al.*, 1991). PCBs, and most notably the coplanar congeners, have similar, albeit less severe, effects: they cause lymphoid depletion in chicks (Andersson *et al.*, 1991), reduce natural killer cell toxicity in rats (Exon *et al.*, 1985; Smialowicz *et al.*, 1989), decrease the number of T cells and the T-helper/T-suppresser cells ratio in non-human primates (Tryphonas *et al.*, 1989), and reduce T cell-mediated cytotoxic activity in mice (Kerkvliet *et al.*, 1990). PCBs decrease antibody production in response to injection of sheep red blood cells (SRBCs) in PCB-treated mice and non-human primates (Loose *et al.*, 1977; Thomas and Hinsdill, 1978; Tryphonas *et al.*, 1989). The reduction of serum IgA levels seems to be a consistent component of PCB immunotoxicity (Shigematsu *et al.*, 1971; Loose *et al.*, 1977; Loose *et al.*, 1978). B cells and particularly B-cell differentiation appear as important targets for halogenated hydrocarbons (Davis and Safe, 1988). Serum corticosteroid levels are also altered by PCBs (Wasserman *et al.*, 1973; Durham and Brouwer, 1990; Kerkvliet *et al.*, 1990). The immunotoxicity of various metabolites of PCBs has also been demonstrated: chlorinated diphenyl ethers, found in Great Lakes fish, significantly decrease circulating lymphocytes in male rats (Chu *et al.*, 1990). The complement system, a non-specific defense mechanism against infectious agents, is altered by PCBs (White *et al.*, 1986). Neutrophil functions are also modulated by *in vitro* exposure to PCBs (Ganey *et al.*, 1993).

Not surprisingly, PCB-induced immunosuppression results in a higher sensitivity of experimental animals to a wide variety of infectious agents: Gram-negative bacteria (endotoxin), protozoa and viruses. The sensitivity of PCB-treated mice to endotoxin, malaria (Loose *et al.*, 1978) and bacteria (Thomas and Hinsdill, 1978) is increased. Rabbits synthesize fewer antibodies after being challenged by pseudorabies virus (Koller and Thigpen, 1973) and mice are more sensitive to challenge by herpes simplex and ectromelia (mousepox) (Imanishi *et al.*, 1980). The resistance of PCB-treated ducks to duck hepatitis virus is also impaired (Friend and Trainer, 1970).

Immunotoxicity of heavy metals

The immunotoxicity of heavy metals in the aquatic environment has been reviewed relatively extensively (Bernier *et al.*, 1995). The general immunotoxic potential of different metals has been ranked as follows: mercury > copper > manganese > cobalt > cadmium > chromium (Lawrence, 1981a). The major immunotoxic effects of the most intensively studied heavy metals, including mercury, cadmium and lead, are discussed briefly below.

The immunotoxic potential of mercury has been very well documented. In addition to reducing the spleen and thymus weight in mice (Dieter *et al.*, 1983; Ilbäck, 1991), mercury exposure has been shown to decrease humoral immunity in several species, including the rabbit (Koller *et al.*, 1977), chicken (Thaxton and Parkhurst, 1973; Bridger and Thaxton, 1983) and mice (Dieter *et al.*, 1983). It was also demonstrated to reduce NK-cell activity in rats (Ilbäck *et al.*, 1991) and mouse (Ilbäck, 1991). In addition, repeated exposure to mercury induced autoimmune diseases in genetically susceptible laboratory animals (Aten *et al.*, 1988; Kubicka-Muranyi *et al.*, 1993).

NK-cell activity was reduced upon exposure to various doses of cadmium through different routes (Chowdhury and Chandra, 1989). For other parameters, such as PFCs, number of macrophages, phagocytosis and lymphocyte proliferation, results varied depending on the species, dose and route of administration (Bernier *et al.*, 1995). Host resistance to pathogens following exposure to cadmium appeared to be compromised with higher doses, but increased with lower doses (Bernier *et al.*, 1995).

Although the main target for lead toxicity appears to be the nervous system, immunotoxic effects have also been reported, although they appear to be quite variable. B lymphocytes appear to be a target for lead immunotoxicity. LPS-induced mouse B lymphocyte proliferation was reduced upon exposure to lead (Lawrence, 1981b; Burchiel *et al.*, 1987), whereas there were no dose-related changes in T-lymphocyte proliferation (Lawrence, 1981b). Reduced serum levels of IgG were also observed in rats exposed to lead (Luster *et al.*, 1978). The effects of lead on specific antibody production (PFCs) were not clear, with some studies showing decreased, enhanced or no effect on the response (Lawrence, 1981b). In contrast, the effects of lead on host resistance were characterized by a uniformly decreased resistance to infection (Cook *et al.*, 1975; Lawrence, 1981b; Laschi-Loquerie *et al.*, 1987).

Immunotoxicity of mixtures

The immunotoxicity of mixtures of toxic compounds has rarely been evaluated. Flipo *et al.* (1992) demonstrated a lack of suppression when mice were exposed to a mixture of dieldrin and carbofuran, two insecticides that separately induced a suppression of immune functions. Dolara *et al.* (1993) found no appreciable genotoxicity associated with a mixture of fifteen pesticides commonly found in the Italian diet. Consumption of the mixture of contaminants contained in Baltic Sea fishes was associated with a reduction of the number of NK cells in circulation in humans (Svensson *et al.*, 1994). Chu *et al.* (1980) did not observe any synergy between halogenated biphenyls and mirex-related compounds. Davis and Safe (1989) and Biegel *et al.* (1989) showed antagonist effects of commercial mixtures and individual PCB congeners on the immunotoxic effects of 2,3,7,8-TCDD as measured by splenic plaque-forming cell assay. In addition, Harper *et al.* (1995) demonstrated the non-additive antagonistic interactions of PCB congeners in commercial mixtures. They observed effects on mice splenic plaque-forming cell assay at a higher dose than that calculated by adding the effects of each component of the mixture using an immunotoxicity-derived toxic equivalent factor. On the other hand, in beluga whales, some PCB congeners appeared to have a synergistic effect in a mixture (see below; De Guise *et al.*, 1998a).

Marine mammal immunology

Marine mammal immunology is a relatively new, rapidly evolving discipline. Some of the reasons for its recently recognized importance include the realization that the immune system plays a central role in general health, and the fact that immune functions can be influenced by several factors such as pollution or stress, and the observation that the immune system is among the most sensitive of systems to certain HAHs, such as TCDD, and thus may make a good 'biomarker' system. Nevertheless, it is important to understand the complexity of the immune system and its functions. No single assay can pretend to evaluate the immune system as a whole, but different assays can evaluate different components, mechanisms or functions. Some laboratories have committed to the development and standardization of assays and reagents to evaluate different portions of the marine mammal immune system, such as humoral, cell-mediated and non-specific immunity. The following section will describe the different components of the mammalian immune system and briefly describe the assays and reagents available in marine mammals.

The immune system is composed of two relatively distinct portions, the innate or non-specific, and the acquired or specific immunity. The innate immunity, which comprises mechanisms mostly responsible for the first line of defense, such as phagocytic cells (neutrophils and macrophages) and NK-cell activity, is not enhanced upon re-exposure to an antigen. In contrast, the acquired immunity, which comprises lymphocytes and antibodies, is responsible for long-term protection which is enhanced upon re-exposure to an antigen, i.e. immunological memory. Furthermore, the acquired immune system has been classically divided into two arms, the cell-mediated immunity and the antibody-mediated humoral immunity, with intricately related functions.

Lymphocytes play a central role in all immune functions. The two major populations of lymphocytes are T cells (cell-mediated immunity) and B cells (humoral immunity). These subsets are morphologically indistinguishable although their functions differ profoundly. The identification of the different cell subsets proved to be a useful diagnostic and prognostic tool for disease processes such as human AIDS (Fahey *et al.*, 1984). Nevertheless, reagents for cetacean immunology are not readily available. In view of the current efforts to classify morphologically indistinguishable lymphocyte subpopulations in several species, and of the relative success obtained with some cross-reactive monoclonal antibodies (Jacobson *et al.*, 1993), attempts were first made to identify subclasses of lymphocytes in cetaceans using cross-reactive monoclonal antibodies. Romano *et al.* (1992) studied the cross-reactivity of an anti-human major histocompatibility (MHC) class II in bottlenose dolphin (*Tursiops truncatus*), and De Guise *et al.* (1997a) investigated the cross-reactivity of anti-bovine, anti-human, anti-ovine and anti-mouse monoclonal antibodies in beluga whale. Romano *et al.* (1992) also produced an antiserum to dolphin immunoglobulins for further use in identification of B lymphocytes. More recently, cetacean-specific monoclonal antibodies against bottlenose dolphin homologue to an epitope of CD2 (T cells), CD19 and CD21 (B cells), and CD45R (a marker of activation of T cells) and CD11/CD18 (a marker of activation of neutrophils, monocytes and B cells) have been characterized (De Guise *et al.*, 1997b, 1998b, 2002).

Beyond recognition of a cell, it is also important to determine its function. *In vitro* mitogen-induced lymphoblastic transformation, or blastogenesis, measures the ability of lymphocytes to proliferate in response to a polyclonal stimulation. This assay is interesting because it reflects the activation of the immune response after an antigenic stimulation *in vivo* (Kristensen *et al.*, 1982). Although this assay has been performed previously with peripheral blood from a variety of species (Mumford *et al.*, 1975; Colgrove, 1978; Shopp *et al.*, 1991; Romano *et al.*, 1992; Lahvis *et al.*, 1993; Ross *et al.*, 1993; de Swart *et al.*, 1993; Erickson *et al.*, 1995; DiMolfetto-Landon *et al.*, 1995; De Guise *et al.*, 1996a), the methods are getting more and more standardized. Also, methods were developed to evaluate the activation of T lymphocytes through the expression of the receptor for interleukin-2 (IL-2R) (DiMolfetto-Landon *et al.*, 1995; Erickson *et al.*, 1995).

Neutrophils represent the first line of defense of the innate immune system against invading agents, especially bacteria (van Oss, 1986). Ingestion of foreign material through the process of phagocytosis, and destruction of phagocytized particles through a series of biochemical events known as the respiratory burst, are the major function of neutrophils (Tizard, 1992). Quantitative assays have been developed to measure phagocytosis and respiratory burst in beluga whales using flow cytometry (De Guise *et al.*, 1995b).

NK cells represent a heterogeneous population of large granular lymphocytes, the activity of which is mainly directed against tumor cells and virus-infected cells (O'Shea and Ortaldo, 1992). NK-cell activity is not restricted to the major histocompatibility complex (MHC), does not need previous sensitization, and represents a first line of defense in the early phase of virus infection (O'Shea and Ortaldo, 1992). While NK cells remain difficult to define, owing to the heterogeneous nature of their subpopulations, they have been characterized on the basis of their activity. NK-cell activity is typically measured as the killing of target (usually tumor cell lines), and assays have been developed and standardized in harbor seals (*Phoca vitulina*) (Ross *et al.*, 1996) and beluga whales (De Guise *et al.*, 1997c).

Immunoglobulins, or antibodies, are soluble protein molecules that specifically recognize antigens in order to hasten their destruction or elimination. Immunoglobulins have been at least partially purified and characterized in several species of marine mammals (Boyden and Gemeroy, 1950; Travis and Sanders, 1972; Nash and Mack, 1977; Cavagnolo, 1979; Andresdottir *et al.*, 1987; Suer *et al.*, 1988). More recently, King *et al.* (1993) produced monoclonal antibodies to seal immunoglobulin subclasses, allowing their quantification (King *et al.*, 1994).

In addition to immunological assays, methods such as cryopreservation of marine mammal lymphocytes have been validated. This will allow the standardization of methods for studies using samples collected on different days and from different locations or populations, and for simultaneous analysis and therefore better comparison. These developments allowed for better quantification of the immune function of marine mammals, providing tools for immunotoxicology studies.

Marine mammal immunotoxicology

The direct determination of the effects of environmental contaminants on the immune system of wild marine mammals is difficult because of logistical and ethical considerations. Nevertheless, different approaches have been adopted, including *in vitro* exposure and animal models, in addition to semi-field and field studies.

In vitro exposures of Arctic beluga whale immune cells to heavy metals and organochlorines were performed. Proliferation of lymphocytes exposed to 10^{-5} M $HgCl_2$ and $CdCl_2$, but not to lower concentrations, was significantly reduced when compared to unexposed control cells (De Guise *et al.*, 1996b). These concentrations are within the range of mercury and cadmium found in tissues of wild belugas (De Guise *et al.*, 1996b). Similarly, proliferation of lymphocytes exposed to 20 ppm or more PCB 138, and 50 ppm or more *p,p'*-DDT, but not to PCB 153, 180, 169 or *p,p'*-DDE, was significantly reduced when compared to unexposed control cells (De Guise *et al.*, 1998a). Interestingly, some PCB congener concentrations which had no effect on splenocyte proliferation when tested individually (5 ppm), were found to significantly reduce proliferation when three of them were mixed together (De Guise *et al.*, 1998a). It is challenging to explain that the addition of PCB 169, a coplanar congener that is usually considered particularly immunotoxic, reduced the toxicity of the mixture on beluga cells. The above data suggest a synergistic effect of some organochlorine compounds on beluga whale splenocytes, and at concentrations that are well within the range of those measured in the wild in St. Lawrence beluga whale blubber (Muir *et al.*, 1990). Overall, the results of these *in vitro* assays demonstrated the susceptibility of beluga whale cells to toxic compounds at levels comparable to those found in tissues of wild belugas. The synergistic effects of *in vitro* exposure to mixtures of organochlorines at concentrations at which individual compounds had no effects on immune functions raises questions regarding the danger of environmentally relevant complex mixtures.

Animal models were developed to simulate exposure to complex, environmentally relevant mixtures of contaminants and to evaluate health effects of such exposure. In one of those models, rats were fed highly contaminated St. Lawrence beluga whale blubber versus much less contaminated Arctic beluga blubber (De Guise *et al.*, unpublished data). This resulted in a slight reduction of PFCs in response to sheep red blood cells in animals fed contaminated whale blubber (De Guise *et al.*, unpublished data). Similarly, rats were fed with oil extracted from highly contaminated Baltic Sea fish versus much less contaminated Atlantic Ocean fish (Ross *et al.*, 1997). This resulted in an impaired cellular immune response in rats exposed to contaminated fish oil, as evidenced by decreased mitogen-induced lymphocyte proliferation in spleen and thymus, lower CD4 : CD8 ratios in the thymus, and a lower viral infection-associated increase in NK-cell activity (Ross *et al.*, 1997). Those models proved to be helpful tools for assessing the immunotoxic potential associated with exposure to complex environmentally relevant mixtures of contaminants.

The *in vivo* effects of exposure to environmentally relevant concentrations of pollutants were demonstrated experimentally in semi-field conditions, where harbor seals were fed highly contaminated Baltic Sea fish versus much less contaminated Atlantic Ocean fish (de Swart *et al.*, 1994; Chapter 21 in this volume). This resulted in impaired NK-cell activity, T-lymphocyte function and delayed-type hypersensitivity in animals fed Baltic fish (de Swart *et al.*, 1994). This probably represents the best experimental study design to investigate the effects of exposure to environmental contaminants through food consumption. Nevertheless, the perinatal transfer of contaminants (through the placenta and milk) that occurs in nature and that affects the highly susceptible developing immune system is not accounted for in such a semi-field experiment.

Field studies in wild populations are logistically difficult. Nevertheless, they represent the best way to fully evaluate the effects of chronic exposure to environmental contaminants (see Ross *et al.*, Chapter 22 in this volume). In a study of free-ranging bottlenose dolphins in Florida, lymphocyte proliferation was correlated to blood concentrations of contaminants using regression analysis. Decreased lymphocyte proliferation was associated with increased blood concentrations of organochlorines in a group of randomly sampled animals (Lahvis *et al.*, 1995). Also, in Alaska, northern fur seal (*Callorhinus ursinus*) pups were captured and the mitogen-induced proliferation of blood lymphocytes, along with other immunoassays and health parameters, were correlated with whole-blood levels of organochlorine (OC) contaminants. Lymphocyte proliferation was decreased in correlation with increasing concentrations of nine PCB congeners (Beckmen, unpublished data). Additionally, it was found that pups of primiparous dams, having the highest blood levels of OC contaminants, had significantly lower antibody production in response to vaccination with tetanus toxoid compared to pups of old dams with low concentrations of OCs in their blood (Beckmen, unpublished data).

Taken together, these data strongly suggest that environmental contaminants with known immunotoxic effects in laboratory animals, and present in high concentrations in tissues of marine mammals, may have immunotoxic effects in marine mammals in nature.

Conclusions/future

The assessment of the effects of environmental contaminants on the immune response is essential to determine the impact of those contaminants on the health of wild populations. Deleterious effects of exposure to environmental contaminants on immune functions have been observed in marine mammals as well as in *in vitro* and animal models. Future studies in marine mammals should be directed toward contaminant exposure at critical life stages, such as neonatal exposure and subsequent effects on the developing immune system. Identifying a subpopulation or cohort at greatest risk (highest exposure and critical life stage) can allow for a focus of research and diagnostic efforts to determine if adverse health effects are ongoing or imminent. In addition, the study of the 'normal' and protective marine mammal immune response to common pathogens, as well as the mechanisms of action of those contaminants (individual and within mixtures) in marine mammals, should provide a better understanding of the relation between a potentially suppressed immune system and the development of diseases and epizootics.

References

Addison, R.F. and Smith, T.G. 1974. Organochlorine residue levels in Artic ringed seals: variation with age and sex. *OIKOS* 25: 335–337.

Addison, R.F., Kerr, S.R., Dale, J. and Sergeant, D.E. 1973. Variation of organochlorine residue levels with age in Gulf of St. Lawrence harp seals (*Pagophilus groenlandicus*). *Journal of the Fisheries Research Board of Canada* 30: 595–600.

Andersson, L., Nikolaidis, E., Brunstrom, B., Bergman, A. and Dencker, L. 1991. Effects of polychlorinated biphenyls with Ah receptor affinity on lymphoid development in the thymus and the bursa of Fabricius chick embryos in ovo. *Toxicology and Applied Pharmacology* 107: 183–188.

Andresdottir, V., Magnadottir, B., Andresson, O. and Petursson, G. 1987. Subclasses of IgG from whales. *Developmental and Comparative Immunology* 11: 801–806.

Aten, J., Bosman, C.B., Rozing, J., Stijnen, T., Hoedemaeker, P.J. and Weening, J.J. 1988. Mercuric chloride induced autoimmunity in the brown Norway rat: cellular kinetics and major histocompatibility complex antigen expression. *American Journal of Pathology* 133: 127–138.

Baumann, O.E. and Martinsen, K. 1983. Persistent organochlorine compounds in seals from Norwegian coastal waters. *Ambio* 12: 262–264.

Bernier, J., Brousseau, P., Krzystyniak, K., Tryphonas, H. and Fournier, M. 1995. Immunotoxicity of heavy metals in relation to Great Lakes. *Environmental Health Perspectives* 103 (Suppl. 9): 23–34.

Biegel, L., Harris, M., Davis, D., Rosengren, R., Safe, L. and Safe, S. 1989. 2,2',4,4',5,5'-Hexachlorobiphenyl as a 2,3,7,8-Tetrachlorodibenzo-P-dioxin antagonist in C57BL/6J mice. *Toxicology and Applied Pharmacology* 97: 561–571.

Born, E.W., Kraul, I. and Kristensen, T. 1981. Mercury, DDT and PCB in the Atlantic walrus (*Odobenus rosmarus rosmarus*) from the Thule district, North Greenland. *Arctic* 34: 255–260.

Boyden, A. and Gemeroy, D. 1950. The relative position of the cetacea among the order of mammalia as indicated by precipitin tests. *Zoologica* 35: 145.

Bridger, M.A. and Thaxton, J.P. 1983. Humoral immunity in chicken as affected by mercury. *Archives of Environmental Contamination and Toxicology* 12: 45–49.

Burchiel, S.W., Hadley, W.M., Cameron, C.L., Fincher, R.H., Lim, T.W., Elias, L. and Stewart, C.C. 1987. Analysis of heavy metal immunotoxicity by multiparameter flow cytometry: correlation of flow cytometry and immune function data in B6CF1 mice. *International Journal of Immunopharmacology* 9: 597–610.

Burleson, G.R. and Dean, J.H. 1995. Immunotoxicology: past, present, and future. In: Burleson G.R., Dean J.H. and Munson A.E. (eds), *Methods in immunotoxicology*. Wiley-Liss, New York, 3–10.

Cavagnolo, R.Z. 1979. The immunology of marine mammals. *Developmental and Comparative Immunology* 3: 245–257.

Chowdhury, B.A. and Chandra, R.K. 1989. Effect of zinc administration on cadmium-induced suppression of natural killer cell activity in mice. *Immunology Letters* 22: 287–291.

Chu, I., Villeneuve, D.C., Becking, G.C., Iverson, F. and Ritter, L. 1980. Short-term study of the combined effects of mirex, photomirex, and kepone with halogenated biphenyls in rats. *Journal of Toxicology and Environmental Health* 6: 421–432.

Chu, I., Villeneuve, D.C., Secours, V. and Valli, V.E. 1990. Toxicological assessment of chlorinated diphenyl ethers in the rat. *Journal of Environmental Science and Health* 25: 225–241.

Colgrove, G.S. 1978. Stimulation of lymphocytes from a dolphin (*Tursiops truncatus*) by phytomitogens. *Americal Journal of Veterinary Research* 39: 141–144.

Cook, J.A., Hoffmann, E.O. and DiLuzio, N.R. 1975. Influence of lead and cadmium on the susceptibility of rats to bacterial challenge. *Proceedings of the Society for Experimental Biology and Medicine* 150: 741–747.

Davis, D. and Safe, S. 1988. Immunosuppressive activities of polychlorinated dibenzofuran congeners: quantitative structure–activity relationship and interactive effects. *Toxicology and Applied Pharmacology* 94: 141–149.

Davis, D. and Safe, S. 1989. Dose–response immunotoxicities of commercial polychlorinated biphenyls (PCBs) and their interaction with 2,3,7,8-tetrachlorodibenzo-p-dioxin. *Toxicology Letters* 48: 35–43.

De Guise, S., Martineau, D., Béland, P. and Fournier, M. 1995a. Possible mechanisms of action of environmental contaminants on St. Lawrence beluga whales (*Delphinapterus leucas*). *Environmental Health Perspectives* 103 (Suppl. 4): 73–77.

De Guise, S., Flipo, D., Boehm, J., Martineau, D., Béland, P. and Fournier, M. 1995b. Immune functions in beluga whales (*Delphinapterus leucas*): Evaluation of phagocytosis and respiratory burst with peripheral blood using flow cytometry. *Veterinary Immunology and Immunopathology* 47: 351–362.

De Guise, S., Bernier, J., Dufresne, M.M., Martineau, D., Béland, P. and Fournier, M. 1996a. Immune functions in beluga whales (*Delphinapterus leucas*): Evaluation of mitogen-induced blastic transformation of lymphocytes from peripheral blood, spleen and thymus. *Veterinary Immunology and Immunopathology* 50: 117–126.

De Guise, S., Bernier, J., Martineau, D., Béland, P. and Fournier, M. 1996b. *In vitro* exposure of beluga whale lymphocytes to selected heavy metals. *Environmental Toxicology and Chemistry* 15: 1357–1364.

De Guise, S., Bernier, J., Dufresne, M.M., Martineau, D., Béland, P. and Fournier, M. 1997a. Phenotyping of beluga whale blood lymphocytes using monoclonal antibodies. *Developmental and Comparative Immunology* 21: 425–433.

De Guise, S., Erickson, K., Blanchard, M., Landon, L., Lepper, H., Wang, J., Stott, J.L. and Ferrick, D.A. 1997b. Characterization of monoclonal antibodies that recognize cell surface antigens for cetacean homologue to CD2, CD11/CD18, CD19, CD21, and CD45R. *Proceedings of the 28th annual IAAAM Conference.* Harderwijk, The Netherlands, p. 73.

De Guise, S., Ross, P.S., Osterhaus, A.D.M.E., Martineau, D., Béland, P. and Fournier, M. 1997c. Immune functions in beluga whales (*Delphinapterus leucas*): Evaluation of natural killer (NK) cell activity. *Veterinary Immunology and Immunopathology* 58: 345–354.

De Guise, S., Martineau, D., Béland, P., Fournier, M. 1998a. Effects of in vitro exposure of beluga whale leucocytes to selected organochlorines. *Journal of Toxicology and Environmental Health Part A* 55: 479–493.

De Guise, S., Erickson, K., Blanchard, M., DiMolfetto, L., Lepper, H., Wang, J., Stott, J.L. and Ferrick, D.A. 1998b. Characterization of a monoclonal antibody that recognizes lymphocyte surface antigens for cetacean homologue to CD45R. *Immunology* 94: 207–212.

De Guise, S., Erickson, K., Blanchard, M., DiMolfetto, L., Lepper, H., Wang, J., Stott, J.L. and Ferrick, D.A. 2002. Characterization of monoclonal antibodies that recognize lymphocyte surface antigens for cetacean homologue to CD2, CD19 and CD21. *Veterinary Immunology and Immunopathology* 84: 209–221.

de Swart, R.L., Kluten, R.M., Huizing, C.J., Vedder, L.J., Reijnders, P.J., Visser, I.K., UytdeHaag, F.G. and Osterhaus, A.D. 1993. Mitogen and antigen induced B and T cell responses of peripheral blood mononuclear cells from the harbour seal (*Phoca vitulina*). *Veterinary Immunology and Immunopathology* 37: 217–230.

de Swart, R.L., Ross, P.S., Vedder, L.J., Timmerman, H.H., Heisterkamp, S.H., van Loveren, H., Vos, J.G., Reijnders, P.J.H. and Osterhaus, A.D.M.E. 1994. Impairment of immune functions in harbour seals (*Phoca vitulina*) feeding on fish from polluted waters. *Ambio* 23: 155–159.

Dieter, M.P., Luster, M.I., Boorman, G.A., Jameson, C.W., Dean, J.H. and Cox, J.W. 1983. Immunological and biochemical responses in mice treated with mercuric chloride. *Toxicology and Applied Pharmacology* 68: 218–228.

DiMolfetto-Landon, L., Erickson, K.L., Blanchard-Channell, M., Jeffries, S.J., Harvey, J.T., Jessup, D.A., Ferrick, D.A., Stott, J.L. 1995. Blastogenesis and interleukin-2 receptor expression assays in the harbor seal (*Phoca vitulina*). *Journal of Wildlife Diseases* 31: 150–158.

Dolara, P., Vezzani, A., Caderni, G., Coppi, C. and Toricelli, F. 1993. Genotoxicity of a mixture of fifteen pesticides commonly found in the Italian diet. *Cell Biology and Toxicology* 9: 333–343.

Domingo, M., Ferrer, L., Pumarola, M., Marco, A., Plana, J., Kennedy, S. and McAliskey, M. 1990. Morbillivirus in dolphins. *Nature* 348: 21.

Domingo, M., Visa, J., Pumarola, M., Marco, A.J., Ferrer, L., Rabanal, R. and Kennedy, S. 1992. Pathologic and immunocytochemical studies of morbillivirus infection in striped dolphins (*Stenella coeruleoalba*). *Veterinary Pathology* 29: 1–10.

Durham, S.K. and Brouwer, A. 1990. 3,4,3',4'-tetrachlorobiphenyl distribution and induced effects in the rat adrenal gland. Localization in the zona fasciculata. *Laboratory Investigation* 62: 232–239.

Eis, D. 1989. Simplification in the etiology of recent seal deaths. *AMBIO* 18: 144.

Erickson, K.L., DiMolfetto-Landon, L., Wells, R.S., Reidarson, T., Stott, J.L. and Ferrick, D.A. 1995. Development of an Interleukin-2 receptor expression assay and its use in evaluation of cellular immune responses in bottlenose dolphin (*Tursiops truncatus*). *Journal of Wildlife Diseases* 31: 142–149.

Exon, J.H., Talcott, P.A. and Koller, L.D. 1985. Effect of lead, polychlorinated biphenyls and cyclophosphamide in rat natural killer cells, interleukin 2, and antibody synthesis. *Fundamental and Applied Toxicology* 5: 158–164.

Fahey, J.L., Prince, H., Weaver, M., Groopman, J., Visscher, B., Schwartz, K. and Detels, R. 1984. Quantitative changes in T helper or T-suppressor/cytotoxic lymphocyte subsets that distinguish acquired immune deficiency syndrome from other immune subset disorders. *American Journal of Medicine* 76: 95–100.

Flipo, D., Bernier, J., Girard, D., Krzystyniak, K. and Fournier, M. 1992. Combined effects of selected insecticides on humoral immune response in mice. *International Journal of Immunopharmacology* 14: 747–752.

Friend, M. and Trainer, D.O. 1970. Polychlorinated biphenyl: interaction with duck hepatitis virus. *Science* 170: 1314–1316.

Ganey, P.E., Sirois, J.E., Denison, M., Robinson, J.P. and Roth, R.A. 1993. Neutrophil function after exposure to polychlorinated biphenyls in vitro. *Environmental Health Perspectives* 101: 430–434.

Gaskin, D.E., Holdrinet, M. and Frank, R. 1971. Organochlorine pesticide residues in harbour porpoises from the Bay of Fundy region. *Nature* 233: 499–500.

Gaskin, D.E., Smith, G.J.D., Arnold, P.W. and Louisy, M.V. 1974. Mercury, DDT, dieldrin, and PCB in two species of odontoceti (Cetacea) from St. Lucia, Lesser Antilles. *Journal of the Fisheries Research Board of Canada* 31: 1235–1239.

Geraci, J.R., St. Aubin, D.J., Barker, I.K., Webster, R.G., Hinshaw, V.S., Bean, W.J., Ruhnke, H.R., Prescott, J.H., Early, G., Baker, A.S., Madoff, S. and Schooley, R.T. 1982. Mass mortality of harbor seals: pneumonia associated with influenza A virus. *Science* 215: 1129–1131.

Harper, N., Connor, K., Steinberg, M. and Safe, S. 1995. Immunosuppressive activity of polychlorinated biphenyl mixtures and congeners: nonadditive (antagonistic) interactions. Fundamental and Applied Toxicology 27: 131–139.

Holden, A.L. and Marsden, K. 1967. Organochlorine pesticides in seals and porpoises. *Nature* 216: 1274–1276.

Ilbäck, N.G. 1991. Effects of methyl mercury exposure on spleen and blood natural killer (NK) cell activity in the mouse. *Toxicology* 67: 117–124.

Ilbäck, N.G., Sundberg, J. and Oskarsson, A. 1991. Methylmercury exposure via placenta and milk impairs natural killer (NK) cell function in newborn rats. *Toxicology Letters* 58: 149–158.

Imanishi, J., Nomura, H., Matsubara, M., Kita, M., Won, S.-J., Mizutani, T. and Kishida, T. 1980. Effect of polychlorinated biphenyl on viral infections in mice. *Infection and Immunity* 29: 275–277.

Jackson, M.R. 1998. Priorities in the development of alternative methodologies in the pharmaceutical industry. *Archives of Toxicology* Suppl 20: 61–70.

Jacobson, C.N., Aasted, B., Broe, M.K. and Petersen, J.L. 1993. Reactivities of 20 anti-human monoclonal antibodies with leucocytes from ten different animal species. *Veterinary Immunology and Immunopathology* 39: 461–466.

Kennedy, S., Smyth, J.A., Cush, P.F., McAliskey, M., McCullough, S.J. and Rima, B.K. 1991. Histopathologic and immunocytochemical studies of distemper in harbor porpoises. *Veterinary Pathology* 28: 1–7.

Kerkvliet, N.I., Baecher-Steppan, L., Smith, B.B., Youngberg, J.A., Henderson, M.C. and Buhler, D.R. 1990. Role of the Ah locus in suppression of cytotoxic T lymphocyte activity by halogenated aromatic hydrocarbons (PCBs and TCDD): structure–activity relationships and effects in C57Bl/6 mice congenic at the Ah locus. *Fundamental and Applied Toxicology* 14: 532–541.

King, D.P., Hay, A.W., Robinson, I., Evans, S.W. 1993. The use of monoclonal antibodies specific for seal immunoglobulins in an enzyme-linked immunosorbent assay to detect canine distemper virus-specific immunoglobulin in seal plasma samples. *Journal of Immunological Methods* 160: 163–171.

King, D.P., Lowe, K.A., Hay, A.W. and Evans, S.W. 1994. Identification, characterisation, and measurement of immunoglobulin concentrations in grey (*Haliocherus grypus*) and common (*Phoca vitulina*) seals. *Developmental and Comparative Immunology* 18: 433–442.

Koller, L.D. 1977. Enhanced polychlorinated biphenyl lesions in Moloney leukemia virus-infected mice. *Clinical Toxicology* 11: 107–116.

Koller, L.D. and Thigpen, J.E. 1973. Biphenyl-exposed rabbits. *American Journal of Veterinary Research* 34: 1605–1606.

Koller, L.D., Exon, J.H. and Arbogast, B. 1977. Methylmercury: effect on serum enzymes and humoral antibody. *Journal of Toxicology and Environmental Health* 2: 1115–1123.

Kristensen, F., Kristensen, B. and Lazary, S. 1982. The lymphocyte stimulation test in veterinary immunology. *Veterinary Immunology Immunopathology* 3: 203–277.

Kubicka-Muranyi, M., Behmer, O., Uhrberg, M., Klonowski, H., Bister, J. and Gleichmann, E. 1993. Murine systemic autoimmune disease induced by mercuric chloride ($HgCl_2$): Hg-specific helper T-cells react to antigen stored in macrophages. *International Journal of Immunopharmacology* 15: 151–161.

Lahvis, G.P., Wells, R.S., Casper, D. and Via, C.S. 1993. In-vitro lymphocyte response of bottlenose dolphins (*Tursiops truncatus*): mitogen-induced proliferation. *Marine Environmental Research* 35: 115–119.

Lahvis, G.P., Wells, R.S., Kuehl, D.W., Stewart, J.L., Rhinehart, H.L. and Via, C.S. 1995. Decreased lymphocytes response in free-ranging bottlenose dolphins (*Tursiops truncatus*) are associated with increased concentrations of PCBs and DDT in peripheral blood. *Environmental Health Prespectives* 103 (Suppl. 6): 67–72.

Laschi-Loquerie, A., Eyraud, A., Morisset, D., Sanou, A., Tachon, P., Veysseyre, C. and Descotes, J. 1987. Influence of heavy metals on the resistance of mice toward infection. *Immunopharmacology and Immunotoxicology* 9: 235–241.

Lawrence, D.A. 1981a. Heavy metal modulation of lymphocyte activities. I: In vitro effects of heavy metals on primary humoral immune response. *Toxicology and Applied Pharmacology* 57: 439–451.

Lawrence, D.A. 1981b. In vivo and in vitro effects of lead on humoral and cell-mediated immunity. *Infection and Immunity* 31: 136–143.

Loose, L.D., Pittman, K.A., Benitz, K.-F. and Silkworth, J.B. 1977. Polychlorinated biphenyl and hexachlorobenzene induced humoral immunosuppression. *Journal of the Reticuloendothelial Society* 22: 253–271.

Loose, L.D., Pittman, K.A., Benitz, K.-F., Silkworth, J.B., Mueller, W. and Coulston, F. 1978. Environmental chemical-induced immune dysfunction. *Ecotoxicology and Environmental Safety* 2: 173–198.

Luster, M.I., Faith, R.E. and Kimmel, C.A. 1978. Depression of humoral immunity in rats following chronic developmental lead exposure. *Journal of Environmental Pathology and Toxicology* 1: 397–402.

Luster, M.I., Munson, A.E., Thomas, P.T., Holsapple, M.P., Fenters, J.D., White, K.L. Jr, Lauer, L.D., Germolec, D.R., Rosenthal, G.J. and Dean, J.H. 1988. Development of a testing battery to assess chemical-induced immunotoxicity: National Toxicology Program's guidelines for immunotoxicity evaluation in mice. *Fundamental and Applied Toxicology* 10: 2–19.

Luster, M.I., Portier, C., Pait, D.G., White, K.L. Jr, Gennings, C., Munson, A.E. and Rosenthal, G.J. 1992. Risk assessment in immunotoxicology. I. Sensitivity and predictability of immune tests. *Fundamental and Applied Toxicology* 18: 200–210.

Luster, M.I., Portier, C., Pait, D.G., Rosenthal, G.J., Germolec, D.R., Corsini, E., Blaylock, B.L., Pollock, P., Kouchi, Y., Craig, W., Munson, A.E. and White, K.L. 1993. Risk assessment in immunotoxicology. II. Relationships between immune and host resistance tests. *Fundamental and Applied Toxicology* 21: 71–82.

Martineau, D., Lagacé, A., Béland, P., Higgins, R., Armstrong, D. and Shugart, L.R. 1988. Pathology of stranded beluga whales (*Delphinapterus leucas*) from the St. Lawrence estuary, Québec, Canada. *Journal of Comparative Pathology* 98: 287–311.

Muir, C.G., Ford, C.A., Stewart, R.E.A., Smith, T.G., Addison, R.F., Zinck, M.E. and Béland, P. 1990. Organochlorine contaminants in belugas, *Delphinapterus leucas*, from Canadian waters. *Canadian Bulletin of Fisheries and Aquatic Sciences* 224: 165–190.

Mumford, D.M., Stockman, G.D., Barsales, P.B., Whitman, T. and Wilbur, J.R. 1975. Lymphocyte transformation studies of sea mammal blood. *Experientia* 31: 498–500.

Munson, A.E. and LeVier, D. 1995. Experimental design in immunotoxicology. In: Burleson, G.R., Dean, J.H. and Munson, A.E. (eds), *Methods in immunotoxicology*. Wiley-Liss, New York, 11–24.

Nash, D.R. and Mack, J.-P. 1977. Immunoglobulin classes in aquatic mammals characterized by serologic cross-reactivity, molecular size and binding of human free secretory component. *Journal of Immunology* 107: 1424–1430.

O'Shea, J. and Ortaldo, J.R. 1992. The biology of natural killer cells: insights into the molecular basis of function. In: Lewis, C.E. and McGee, J.O.D. (eds), *The natural killer cell*. IRL Press at Oxford University Press, New York, 2–40.

Osterhaus, A.D.M.E., Groen, J., De Vries, P., Uyt de Haag, F.G.C.M., Klingeborn, B. and Zarnke, R. 1988. Canine distemper virus in seals. *Nature* 335: 403–404.

Poland, A., Knutson, J.C. 1982. 2,3,7,8,-Tetrachlorodibenzo-p-dioxin and related halogenated hydro-carbons: examination of the mechanism of toxicity. *Annual Review of Pharmacology and Toxicology* 22: 517–554.

Prochazkova, J. 1993. Contribution of 'in vitro' assays to preclinical and premarketing testing in immunotoxicology. *Central Europe Journal of Public Health* 1: 101–105.

Reijnders, P.J.H. 1980. Organochlorine and heavy metal residues in harbour seals from the Wadden Sea and their possible effects on reproduction. *Netherlands Journal of Sea Research* 14: 30–65.

Roitt, I.M., Brostoff, J. and Male, D.K. (eds) 1993. *Immunology*, 3rd ed. Gower Medical, London.

Romano, T.A., Ridgway, S.H. and Quaranta, V. 1992. MHC class II molecules and immunoglobulins on peripheral blood lymphocytes of the bottlenosed dolphin, *Tursiops truncatus. Journal of Experimental Zoology* 263: 96–104.

Ross, P.S., Pohadjak, B., Bowen, W.D. and Addison, R.F. 1993. Immune function in free-ranging harbor seal, *Phoca vitulina,* mothers and their pups during lactation. *Journal of Wildlife Diseases* 29: 21–29.

Ross, P.S., de Swart, R.L., Timmerman, H.H., Vedder, L.J., van Loveren, H., Vos, J.G., Reijnders, P.J.H. and Osterhaus, A.D.M.E. 1996. Suppression of natural killer cell activity in harbour seals (*Phoca vitulina*) fed Baltic Sea herring. *Aquatic Toxicology* 34: 71–84.

Ross, P.S., de Swart, R.L., van der Vliet, H., Willemsen, L., de Klerk, A., van Amerongen, G., Groen, J., Brouwer, A., Schipholt, I., Morse, D.C., van Loveren, H., Osterhaus, A.D. and Vos, J.G. 1997. Impaired cellular immune response in rats exposed perinatally to Baltic Sea herring oil or 2,3,7,8-TCDD. *Archives of Toxicology* 71: 563–574.

Shigematsu, N., Norimatsu, Y., Ishibashi, T., Yoshida, M., Suetsugu, S., Kawatsu, T., Ikeda, T., Saito, R., Ishimaru, S., Shirakusa, T., Kido, M., Emori, K. and Toshimitsu, H. 1971. Clinical and experimental studies on impairment of respiratory organs in chlorobiphenyl poisoning. *Fukuoka Acta Medica* 62: 150.

Shopp, G.M., Galbreth, K., Gerber, J. and Shaw, S. 1991. A whole blood mitogen assay for the assessment of immune function in the elephant seal. *9th Biennial Conference on the Biology of Marine Mammals*, Chicago, IL, p. 63.

Sladen, W.J.L., Menzie, C.M. and Reichel, W.L. 1966. DDT residues in Adelie penguins and a crabeter seal from Antartica. *Nature* 218: 670–673.

Smialowicz, R.J., Andrews, J.E., Riddle, M.M., Rodgers, R.R., Loebke, R.W. and Copeland, C.B. 1989. Evaluation of immunotoxicity of low level PCB exposure in the rat. *Toxicology* 56: 197–211.

Suer, L.D., Vedros, N.A., Schroeder, J.P. and Dunn, J.L. 1988. *Erysipelothrix rhusiopathiae.* II. Enzyme immunoassay of sera from wild and captive marine mammals. *Diseases of Aquatic Organisms* 5: 7–13.

Svensson, B.G., Hallberg, T., Nilsson, A., Schultz, A. and Hagmar, L. 1994. Parameters of immunological competence in subjects with high consumption of fish contaminated with persistent organochlorine compounds. *International Archives of Occupational and Environmental Health* 65: 351–358.

Taruski, A.G., Olney, C.E. and Winn, H.E. 1975. Chlorinated hydrocarbons in cetaceans. *Journal of the Fisheries and Research Board of Canada* 32: 2205–2209.

Thaxton, P. and Parkhurst, C.R. 1973. Toxicity of mercury to young chickens. III: Changes in immunological responsiveness. *Poultry Science* 52: 761–764.

Thomas, P.T. and Hinsdill, R.D. 1978. Effect of polychloribnated biphenyls on the immune response of rhesus monkeys and mice. *Toxicology and Applied Pharmacology* 44: 41–51.

Tizard, I. (ed.) 1992. *Veterinary immunology, an introduction*, 4th ed. WB Saunders Company, Montreal.

Travis, J.C. and Sanders, B.G. 1972. Whale immunoglobulins II. Heavy chain structure. *Comparative Biochemistry and Physiology* 43B: 637.

Tryphonas, H., Hayward, S., O'Grady, L., Loo, J.C.K., Arnold, D.L., Bryce, F. and Zawidzka, Z.Z. 1989. Immuno-toxicity studies of PCB (Aroclor 1254) in the adult

Rhesus (*Macaca mulatta*) monkey – preliminary report. *International Journal of Immunopharmacology* 11: 199–206.

van Oss, C.J. 1986. Phagocytosis: an overview. *Methods in Enzymology* 132: 3–15.

Wasserman, D., Wasserman, M., Cucos, S. and Djavaherian, M. 1973. Function of adrenal gland zona fasciculata in rats receiving polychlorinated biphenyls. *Environmental Research* 6: 334–338.

White, K.L., Lysy, H., McCoy, J.A. and Anderson, A.C. 1986. Modulation of serum complement levels following exposure to polychlorinated dibenzo-p-dioxins. *Toxicology and Applied Pharmacology* 84: 209–219.

3 Reproductive and developmental effects of environmental organochlorines on marine mammals

Peter J.H. Reijnders

Introduction

A large number of xenobiotics with endocrine-disrupting properties have been detected in marine mammal tissue (Wagemann and Muir, 1984; Aguilar and Borrell, 1995; Colborn and Smolen, 1996; Reijnders, 1996). Although most of the species known to be contaminated in this way are coastal, considerable concentrations of such compounds have even been detected in at least one cetacean that forages in deep water, the sperm whale (*Physeter macrocephalus*) (de Boer *et al.*, 1998). Only in a few studies have observed reproductive disorders been found to be associated with certain chlorinated hydrocarbons and their metabolites. Among these studies are those involving ringed seals (*Pusa hispida*) and gray seals (*Halichoerus grypus*) in the Baltic Sea (Helle, 1980; Bergman and Olsson, 1985; see also Chapter 19 in this volume), beluga whales (*Delphinapterus leucas*) in the St. Lawrence River (Béland *et al.*, 1987), harbor seals (*Phoca vitulina*) in the Wadden Sea (Reijnders, 1980), and California sea lions (*Zalophus californianus*) in the eastern Pacific Ocean (DeLong *et al.*, 1973). The findings of these studies, although strongly suggestive, have not been conclusive. The etiology of the observed disorder has usually been uncertain, and proof of a causal relationship between exposure to a specific contaminant and an impact on the reproductive or endocrine system has remained elusive.

This chapter discusses the issue from an epidemiological point of view. My focus is on marine mammal species in which disorders in hormone concentrations, reproductive problems or pathological conditions associated with hormonal imbalance have been observed. An overview of associations between organochlorines and marine mammal reproduction and endocrinology is presented in Table 3.1. I conclude with some comments on the possibilities of monitoring and evaluating problems related to xenobiotic-induced reproductive impairment and endocrine disruption in marine mammals.

Table 3.1 Associations between organochlorines and marine mammal reproductive, hormonal and morphological disorders

Phenomenon	Where in process	Species	Location	Mode of action	Certainty associated with mode of action*	Contaminants	Certainty associated with contaminants*	Reference
Reproductive disorders								
Implantation failure	Implantation	Harbor seal	Wadden Sea	Enhanced hormone metabolism	2	PCBs/-metabolites	2	Reijnders (1986), This chapter
Failed implantation or fetal development	Implantation	Beluga whale	St. Lawrence River	Unknown	4	Organochlorines	4	Martineau et al. (1994)
Stenosis and occlusions	Postpartum	Gray and ringed seals	Baltic Sea	Organochlorine-induced uterine pathology	3	PCBs/DDE/ methylsulfones	3	Helle et al. (1976); Helle (1980)
Sterility	Unknown	Gray and ringed seals	Baltic Sea	Pathologic sterility	1	PCBs/DDT/ -metabolites	3	Bergman and Olsson (1985)
Premature pupping	Late gestation	California sea lion	Southern California Bight	Microsomal enzyme induction; steroid mimicking	3	PCBs/DDT	3	DeLong et al. (1973)
Hormonal disorders								
Low vit. A/thyroid hormones	All ages	Harbor seal	Wadden Sea	Binding competition	1	PCBs/-metabolites	2	Brouwer et al. (1989)
Reduced testosterone	Adults	Dall's porpoise	North Pacific Ocean	Unknown	4	PCBs/DDE	4	Subramanian et al. (1987)
Lowered estradiol level	Implantation	Harbor seal	Wadden Sea	Enhanced hormone metabolism/enzyme induction	3	PCBs/-metabolites	2	Reijnders, this chapter
Morphological disorders								
Skull lesions (osteoporosis, paradontitis)	Predominantly adults	Harbor and gray seals	Baltic Sea, Wadden Sea	Infection/ hyperadrenocortical	2	PCBs/DDT/ -metabolites	3	Stede and Stede (1990); Bergman et al. (1992)
Exostosis	All ages	Harbor seal	Baltic Sea, West coast of Sweden	Unknown	4	PCBs/DDT/ -metabolites	3	Mortensen et al. (1992)
Testis abnormalities	Immatures and adults	Minke whale	Southern Ocean	Unknown	4	Organochlorines	4	Fujise et al. (1998)
Adrenal hyperplasia	Unknown	Beluga whale	St. Lawrence River	Unknown	4	Organochlorines	4	Béland et al. (1991)
Hermaphroditism	Fetal	Beluga whale	St. Lawrence River	Genetic/environmental	4	PCBs/DDT	4	De Guise et al. (1994)

*1, Definite; 2, probable; 3, possible; 4, unknown.

Basic pharmacology and physiology of hormones in reproduction and early development

Hormones are messenger compounds. Their release leads to functional changes in an organism's cells, tissues and organs. They are produced by endocrine organs and delivered into the bloodstream. A small percentage of hormones circulate free, but the majority are bound to transport proteins. The free hormones diffuse into the tissues and cells. Target cells possess specific receptor molecules that bind to particular hormones, leading to activation of the receptor.

Steroid hormones, as well as thyroid hormones, play important roles in reproduction and early development. They are discussed separately here.

In vertebrates, sex hormones belong to a group of steroids that are synthesized from cholesterol. Steroids can be divided into four functional groups: the three sex hormone groups (androgens, estrogens and progesterone) and the glucocorticosteroids. The latter play a role in regulating metabolism and growth and in osmoregulation.

In mammals, steroid hormones are mainly synthesized in the adrenals, gonads and placenta. The production of each of the sex hormones is localized in a specific gland, such as testosterone in the testes, estrogens and progesterone in the ovaries, and progesterone, estrogens and testosterone in the adrenals. These organs can, in many species, also produce small quantities of the other sex hormones. The specific pathway of biosynthesis in mammals is in the following order: cholesterol → pregnenolone → progesterone → androstenedione → testosterone → estradiol (Figure 3.1).

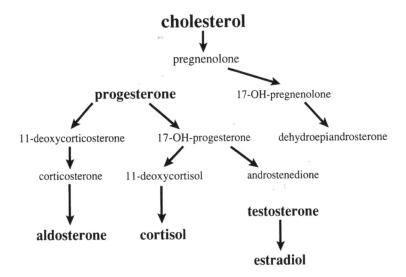

Figure 3.1 Major steroidogenic pathways in mammalian endocrine tissues. Black arrows indicate P450-mediated conversions.

The hypothalamus and pituitary are responsible in most vertebrates for the regulation of hormone concentrations and the timing of reproduction and sexual development. The hypothalamus produces a peptide, gonadotropin releasing hormone (GnRH), which stimulates the pituitary to synthesize follicle stimulating hormone (FSH) and luteinizing hormone (LH). These hormones regulate the synthesis of progesterone, estradiol and testosterone. Through the production of FSH and LH, the synthesis (positive and negative feedback) of sex hormones and gonadal development are regulated.

Thyroid hormones are important in the structural and functional development of sexual organs and the brain, both intra-uterine and postnatal. They are synthesized by the thyroid gland under stimulation through the thyroid stimulating hormone produced by the pituitary. The form that is most important as a biological parameter is thyroxine (T_4, tetraiodothyroxine).

Disorders of hormone concentrations

A negative correlation has been observed between testosterone levels and tissue concentrations of dichloroethylidene-chloro-benzene (DDE) in Dall's porpoises, *Phocoenoides dalli* (Subramanian *et al.*, 1987). In a semi-field experiment with harbor seals, Reijnders (1986, 1990) found that lower levels of 17β-estradiol occurred around the time of implantation.

A possible explanation for the observed lower hormone levels in both Dall's porpoises and harbor seals would be an increased breakdown of steroids as a consequence of polychlorinated biphenyl (PCB)- or PCB metabolite-induced enzyme activity. Enhanced steroidogenesis caused by some metals and a PCB mixture (Aroclor 1254) has been demonstrated *in vitro* for gray seals (Freeman and Sangalan, 1977). Furthermore, increased metabolism of PCBs as a result of cytochrome P450 enzyme induction has already been demonstrated in marine mammals (Tanabe *et al.* 1988; Boon *et al.*, 1992). A second explanation might be that PCB and DDE, or metabolites thereof, bind to hormone carrier proteins and/or hormone receptors. If such binding were to occur, either metabolism of steroids would be hindered or binding of steroids to receptor proteins in target tissue would be impeded. It is conceivable that both mechanisms – increased steroid breakdown and the binding of xenobiotic compounds to carrier proteins or hormone receptor – could operate in tandem.

With respect to the first mechanism, Troisi and Mason (2000) found experimentally that rates of progesterone and testosterone metabolism in harbor seals were negatively correlated with both PCB concentrations and the level of P450 enzyme induction. If this also holds for Dall's porpoise, it would imply that PCB/DDE-induced hydroxylation and subsequent elimination does not occur and hence could not explain the observed lower testosterone levels. However, the result of Troisi and Mason's experiment might be caused by a non-endocrine disrupting mechanism. They found particularly phenobarbital (PB)-type of induction (e.g. cytochrome P450 enzyme subfamilies CYP2B and CYP3A). This should lead to increased testosterone hydroxylation, but that was not observed. It is conceivable that in the presence of high PCB and DDE levels, lack of substrate occurs, and PCBs and DDE bind to active sites of P450 isoenzymes. This blocking of catalytic activity, in particular of the PB-type, would lead to decreased breakdown of testosterone in their experiment. The lowered levels of testosterone reported by Subramanian *et al.* (1987) could therefore still be explained by either PCB/DDE-induced breakdown of testosterone or its precursors, or by a hindrance in the transformation of precursors for testosterone by PCB, DDE or their metabolites. These postulated explanations obviously need further testing.

With respect to the lower levels of estradiol observed in the harbor seal experiment by Reijnders (1986), the finding by Troisi and Mason (2000) of decreasing metabolism of progesterone and testosterone in seal liver microsomes does not enable one to draw conclusions about a possible impediment of the transformation of both hormones, which could have, in turn, led to lower levels of estradiol as observed in harbor seals. This is because these transformation processes occur in reproductive organs and have to be tested in those tissues. With respect to a possible enhanced elimination of estradiol by PCB-induced metabolism, it has been found by Funae and Imaoka (1993) that in rats, sex-dependent cytochrome P450 isoenzyme patterns exist. Furthermore, it is known that strong induction of CYP1A2 occurs (Drenth *et al.*, 1994, table XV), which leads to increased excretion and hence lower levels of estradiol. Addisson *et al.* (1986), Troisi and Mason (2000) and Boon *et al.* (1992, 1997) have demonstrated that induction of CYP1A was also significant in harbor seals. Because CYP1A is female dominant (Drenth *et al.*, 1994), enhanced hydroxylation is to be expected. Feedback mechanisms will operate to counteract lowered estradiol levels and, in the course of time, lead to partial or complete recovery.

Concerning a second mechanism, it is known that PCB metabolites, in particular PCB-methylsulfones, bind to uteroglobin (Patnode and Curtis, 1994). Because we did not find any receptor interference in *in vitro* pilot experiments with harbor seal blood, this mechanism is not considered a likely explanation for the observed reproductive failure in this species.

Based on the foregoing, I postulate that enhanced breakdown of estradiol through enzyme-induced metabolism by PCBs, is a plausible mode of action to explain the temporary lower levels of estradiol as observed in harbor seals. Further studies are planned to investigate this possibility.

Estradiol has a priming effect on the proliferation of the endometrium, in effect preceding proliferation of the luminal and glandular epithelium under the influence of progesterone. The lower levels of estradiol could have impaired endometrial receptivity and prevented successful implantation of the blastocyst.

Decreased levels of thyroid hormones have been found in harbor seals (Brouwer *et al.*, 1989) as a consequence of competition between a hydroxylated metabolite of PCB-77 and thyroid hormones for binding to a transport protein, transthyretin (TTR). Such a hypothyroid condition can have significant effects on early development and later reproductive performance. Fetal accumulation of hydroxylated PCB metabolites has been found to occur in experimental animals and may also occur in seals (Brouwer *et al.*, 1998). Given that thyroids are involved in the development of Sertoli and Leydig cells (spermatogenesis), brain development and early development of the sex organs, further research in this area is warranted.

Green *et al.* (1996) and Green (1997) provide additional information on lactational transfer of PCB-methylsulfones (PCB-MSFs) from gray seals to their offspring. They report that the summed concentrations of PCB-MSFs (in lipid) are approximately 5 per cent of the total PCB concentration. A similar ratio was found in seal milk. The uptake of PCB-MSFs is therefore quantitatively important. Moreover, the pups excrete only *c.* 0.5 per cent from the amount they ingest. In contrast to the mobilization of PCB congeners from maternal blubber to milk, which is negatively correlated with congener lipophilicity, the metabolization of PCB-MSFs is independent from the degree of chlorination as well as the chlorination pattern. The ratio of PCB-MSFs:total PCBs is therefore higher in milk than in blubber. This metabolization process, offering a certain protection against the more lipophilic PCBs, obviously does not work in the case of the PCB-MSFs. The significance of this finding remains unclear.

Reproductive disorders

Clear cases of hermaphroditism were observed in 2 out of 94 examined beluga whales in the St. Lawrence River (De Guise *et al.*, 1994; Pierre Béland, 3 March 1999, personal communication). One animal was a true hermaphrodite. This condition has been attributed to hormonal disturbance in early pregnancy, whereby normal differentiation of male and female organs was disrupted. Research is ongoing to test that hypothesis and to acquire information on the underlying mechanism. Out of the 93 remaining animals, 48 females and 45 males, one beluga appeared to be a male pseudohermaphrodite. This phenomenon is less unique and is also observed in other cetaceans (see Philo *et al.*, 1993; Reijnders *et al.*, 1999).

In examined mature female Baltic seals, 30 per cent of the gray and 70 per cent of the ringed seals exhibited partial or complete sterility, caused by stenosis and occlusions. Recent studies suggest that PCB- and DDE-methylsulfones are the toxic compounds responsible for these abnormalities (Olsson *et al.*, 1994). A plausible hypothesis is that early pregnancy is interrupted, perhaps via decreased uteroglobin binding due to the methylsulfone occupation or low hormone levels, followed by development of pathological disorders. Toxic effects at several steps in the brain–hypothalamic–hypophyseal–adrenal–placental axis could be involved in the latter stage (Reijnders and Brasseur, 1992). Further investigations are needed to elucidate this phenomenon.

DeLong *et al.* (1973) found premature pupping in California sea lions to be associated with high PCB and DDE levels. The concurrent finding in the animals of pathogens with known potential to interfere with pregnancy rendered it impossible to attribute causation specifically to either of the organochlorines.

Abnormal testis, transformation of epididymal and testis tissue, has been observed in North Pacific minke whales (*Balaenoptera acutorostrata*) (Fujise *et al.*, 1998). A possible relationship with levels of organochlorines has been postulated. Further histological examinations and pathology are being conducted to investigate this phenomenon.

Other hormone-related reproductive or developmental disorders

Besides sterility, a suite of pathological disorders has been observed in seals of the Baltic Sea (see also Chapter 19 in this volume). These include a high incidence of uterine smooth muscle cell tumors (leiomyomas) in gray seals (Bergman, 1999), exostosis in harbor seal skulls (Mortensen *et al.*, 1992) and osteoporosis in gray seal skulls (Olsson *et al.*, 1994). Similar disorders have been found in harbor seals in the Wadden Sea (Stede and Stede, 1990). This disease complex is characterized as hyperadrenocorticism. It is unclear whether hyperadrenocorticism is manifest in early development.

The research field between reproductive biology and immunology is in an early stage of development. It is known that both the humoral (antibodies) and cellular (lymphocytes) immune systems are regulated by estrogens and androgens (Grossman, 1985). Disruption in steroid hormone balance might therefore lead to malfunctioning of the immune system. Of relevance in the present context is the role of progesterone and estradiol in preventing the maternal–fetal rejection response. This relationship could help explain the observed problems of harbor seals in the Wadden Sea, because these problems occurred at around the time of implantation. Also corticosteroids (Wilckens and de Rijk, 1997) and thyroid hormones (Brouwer *et al.*, 1989, 1998) are involved in immune functioning. The possible effects of xenobiotic-caused thyroidal and corticosteroidal hormone imbalances on early development and reproduction are insufficiently known.

Monitoring and evaluation of effects, and related research needs

There are serious impediments to the monitoring and evaluation of hormone-related xenobiotic effects. Actually, it is very well possible that many endocrine-like effects may not be observed at all. Some of the more prominent reasons for this are: gaps in knowledge on good population data, the mixture of compounds involved, the role of disease agents, the role of other environmental factors, the lack of readily available biomarkers for use in mammalian wildlife studies, lack of knowledge on endocrine and reproductive systems in mammalian wildlife species, and very few studies addressing problems occurring in early developmental stages. Particularly, monitoring of effects is hampered for several reasons. First, the majority of the present tests do not measure transgenerational influence, yet several disorders occur only in the adult stage or offspring. Second, gene *expression* is affected and not gene *constitution*. Therefore mutagenicity endpoint tests are not particularly relevant. Third, many tests are *in vitro*, and that complicates investigations of disruption in neurobehavioral function and reproductive morphological development. Finally, some xeno-estrogens only become biologically active after *in vivo* metabolism (e.g. methoxychlor).

Development of hormone-responsive cell cultures as biomarkers could provide a partial solution (Colborn *et al.*, 1993). Biomarkers are available to measure exposure to xenobiotics, including xeno-estrogens (Fossi, 1994). A series of recent studies describe techniques to investigate the metabolism of PCBs, PCDDs, PCDFs, and toxaphenes in marine mammals (Troisi and Mason, 1997; Boon *et al.*, 1997, 1998; Letcher *et al.*, 1998). These provide opportunities to measure exposure of marine mammals to, for example, xeno-estrogens and other endocrine-disrupting contaminants. However, preparatory research has to be carried out to adapt existing biomarker protocols for use in marine mammal studies. This includes sampling tissue, particularly of neonatal and juvenile animals, to analyze for:

- steroid hormones, thyroid hormones and vitamin A (important in cell differentiation) in blood;
- thyroid hormones and vitamin A in brain and liver;
- estrogen-receptor binding capacity of ovarian, brain and liver tissue;
- glial fibrillary acidic protein and synaptophysins in brain;
- P450 enzyme induction (i.e. CYP1A) in liver, brain and uterus; and
- levels of xeno-estrogens in blubber, blood, liver and brain.

In addition, a nondestructive biomarker of exposure to cytochrome P450-inducing organochlorines is the use of pattern analysis of metabolizable and nonmetabolizable PCB congeners in different tissues, including blood. The fraction of metabolizable congeners is negatively correlated with exposure to PCBs and can be modeled by a logistic curve. This relation can be used to define critical levels of exposure (van den Brink *et al.*, 2000).

To enhance the evaluation of the results obtained, in terms of biological significance, it is important to carry out investigations on animals from relatively unpolluted as well as highly polluted areas. Initially, a choice has to be made for model compounds as well as for model species, occurring over a gradient of pollution. Studies in relatively clean areas serve mainly to obtain reference values for exposure to contaminants and indicators for the status (functioning) of the studied population, and to follow trends in both. Studies in highly polluted areas allow pathological research to be carried out on neonatal and juvenile animals. In combination with analyses of contaminants and associated physiological and pathological responses, these studies will facilitate development of a *multiple response concept*, as described by Reijnders (1994) and Reijnders and de Ruiter-Dijkman (1995). The development of techniques to extrapolate observed individual responses and thereby evaluate population-level consequences, and possibly the ecosystem-level effects, is equally important.

Summary

Many endocrine-disrupting chemicals have been detected in marine mammal tissues. However, reproductive and developmental disorders have only been found to be associated with certain OCs and their metabolites in a few case studies. In this chapter an overview of associations between OCs and marine mammal reproductive, hormonal and morphological disorders is presented, and possible underlying mechanisms for observed disorders are discussed. This includes increased steroid breakdown through P450 enzyme induction and binding of xenobiotic compounds to carrier proteins or hormone receptors.

It is concluded that the etiology of most observed disorders remains uncertain. The development of hormone-responsive cell cultures as biomarkers could help to overcome the present impediments to the monitoring and evaluation of hormone-related xenobiotic effects. An additional approach is to use pattern analysis of metabolizable and nonmetabolizable PCB congeners as a nondestructive biomarker for exposure to cytochrome P450-inducing organochlorines. This technique enables us to define critical levels of exposure.

Enhancement of the biological significance of the findings obtained with the aforementioned techniques may be achieved by investigating model compounds in model species occurring over a gradient of pollution. The relation between tissue levels of contaminants and their physiological and pathological responses, obtained in this way, will facilitate the development of a more *predictive multiple response concept*. This can then be applied to species in other areas where data on either concentrations of contaminants or pathophysiological conditions are lacking.

References

Addisson, R.F., Brodie, P.F., Edwards, A. and Sadler, M.C. 1986. Mixed function oxidase activity in the harbour seal (*Phoca vitulina*) from Sable Is., N.S. *Comp. Biochem. Physiol. (C)* 85: 121–124.

Aguilar, A. and Borrell, A. 1995. Pollution and harbour porpoises in the eastern North Atlantic. In: A. Björge and G.P. Donovan (eds), *Biology of the Phocoenids*. Rep. Int. Whal. Commn Special Issue 16, 231–242.

Béland, P.R., Michaud, R. and Martineau, D. 1987. Recensements de la population de belugas du Saint-Laurent en 1985 par embarcations. *Rapp. Techn. Can. Sci. Halieut. Aquat.* No. 1545.

Béland, P., De Guise, S. and Plante, R. 1991. *Toxicology and pathology of St Lawrence Marine Mammals.* Report to the World Wildlife Fund, Inst. Nat. Ecotoxicol. Saint Laurent, Montreal.

Bergman, A. 1999. Prevalence of lesions associated with a disease complex in the Baltic grey seal (*Halichoerus grypus*) during 1977–1996. In: T.J. O'Shea, R.R. Reeves, and A. Kirk Long (eds), *Marine Mammals and Persistent Ocean Contaminants: Proceedings of the Marine Mammal Commission Workshop, Keystone, Colorado, 12–15 October 1998.* Marine Mammal Commission, Bethesda, USA, 139–143.

Bergman, A. and Olsson, M. 1985. Pathology of Baltic ringed seal and grey seal females with special reference to adrenocortical hyperplasia: is environmental pollution the cause of a widely distributed disease syndrome? *Finn. Game Res.* 44: 47–62.

Bergman, A., Olsson, M. and Reiland, S. 1992. Skull-bone lesions in the Baltic gray seal (*Halichoerus grypus*). *Ambio* 21: 517–519.

Boer, J. de, Wester, P.G., Klamer, H.J.C., Lewis, W.E. and Boon, J.P. 1998. Do flame retardants threaten ocean life? *Nature* 394: 28–29.

Boon, J.P., van Arnhem, E., Jansen, S., Kannan, N., Petrick, G., Duinker, J.C., Reijnders, P.J.H. and Goksöyr, A. 1992. The toxicokinetics of PCBs in marine mammals with special reference to possible interactions of individual congeners with the cytochrome P450-dependent monooxygenase system: an overview. In: C.H. Walker and D.R. Livingstone (eds), *Persistent Pollutants in Marine Ecosystems.* Pergamon, Oxford, 119–159.

Boon, J.P., van der Meer, J., Allchin, C.R., Law, R.J., Klungsöyr, J., Leonards, P.E.G., Splidd, H., Storr-Hansen, E., Mckenzie, C. and Wells, D.E. 1997. Concentration-dependent changes of PCB patterns in fish-eating mammals: structural evidence for induction of cytochrome P450. *Arch. Environ. Contam. Toxicol.* 33: 298–311.

Boon, J.P., Sleiderink, H.M., Helle, M.S., Dekker, M., van Schanke, A., Roex, E., Hillebrand, M.T.J., Klamer, H.J.C., Govers, B., Pastor, D., Morse, D.C., Wester, P.G. and de Boer, J. 1998. The use of a microsomal *in vitro* assay to study phase I biotransformation of chlorobornanes (toxaphene) in marine mammals and birds: possible consequences of biotransformation for bioaccumulation and genotoxicity. *Comp. Biochem. Physiol. (C)* 121: 385–403.

Brink, N.W. van den, de Ruiter-Dijkman, E.M., Broekhuizen, S., Reijnders, P.J.H. and Bosveld, A.T.C. 2000. Polychlorinated biphenyls pattern analysis: potential nondestructive biomarker in vertebrates for exposure to cytochrome P450-inducing organochlorines. *Environ. Toxicol. Chem.* 19: 575–581.

Brouwer, A., Reijnders, P.J.H. and Koeman, J.H. 1989. Polychlorinated biphenyl (PCB)-contaminated fish induces vitamin A and thyroid hormone deficiency in the common seal *Phoca vitulina. Aq. Toxicol.* 15: 99–106.

Brouwer, A., Morse, D.C., Lans, M.C., Schuur, G., Murk, A.J., Klason-Wehler, E., Bergman, A. and Visser, T.J. 1998. Interactions of persistent environmental organohalogens with the thyroid hormone system: mechanisms and possible consequences for animal and human health. *Toxicol. Industr. Health* 14 (1/2): 59–84.

Colborn, T. and Smolen, M.J. 1996. Epidemiological analysis of persistent organochlorine contaminants in cetaceans. *Rev. Environ. Contam. Toxicol.* 146: 91–171.

Colborn, T., vom Saal, F.S. and Soto, A.M. 1993. Developmental effects of endocrine-disrupting chemicals in wildlife and humans. *Environ. Health Perspect.* 101: 378–384.

De Guise, S., Lagace, A. and Béland, P. 1994. True hermaphroditism in a St. Lawrence beluga whale (*Delphinapterus leucas*). *J. Wildl. Dis.* 30: 287–290.

DeLong, R.L., Gilmartin, W.G. and Simpson, J.G. 1973. Premature births in Californian sealions: association with high organochlorine pollutant residue levels. *Science* 181: 1168–1170.

Drenth, H.J., van den Berg, M. and Bouwman, C. 1994. *Reproductie effecten van PCBs: de rol van cytochroom P450 inductie en steroid hormoon metabolisme.* Report, project 94230351, to Ministerie van VROM. RITOX, Utrecht, Netherlands.

Fossi, M.C. 1994. Nondestructive biomarkers in ecotoxicology. *Environ. Health Perspect.* 102: 49–54.

Freeman, H.C. and Sangalan, G.B. 1977. A study on the effects of methylmercury, cadmium, arsenic, selenium, and a PCB, (Arochlor 1254) on adrenal and testicular steroidogeneses *in vitro*, by the gray seal *Halichoerus grypus*. *Arch. Environ. Contam. Toxicol.* 5: 369–383.

Fujise, Y., Zenati, R. and Kato, H. 1998. An examination of the W-stock hypothesis for North Pacific minke whales, with special reference to some biological parameters using data collected from JARPN surveys from 1994 to 1997. Paper SC/50/RMP12 presented to the IWC Scientific Committee, Oman, 1998.

Funae, Y. and Imaoka, S. 1993. Cytochrome P450 in rodents. In: J.B. Schenkman and H. Greim (eds), *Cytochrome P450*. Springer Verlag, Berlin, Germany.

Green, N.J.L. 1997. A study of polychlorinated biphenyls in the lactating grey seal. PhD thesis, Lancaster University, Lancaster, UK.

Green, N., van Raat, P., Jones, K. and de Voogt, P. 1996. PCBs and their methyl sulfonyl metabolites in the maternal blubber, milk, pup blubber and faeces of grey seals. *Organohalogen Compounds* 29: 453–457.

Grossman, C.J. 1985. Interactions between the gonadal steroids and the immune system. *Science* 227: 257–261.

Helle, E. 1980. Lowered reproductive capacity in female ringed seals (*Pusa hispida*) in the Bothnian Bay, northern Baltic Sea, with special reference to uterine occlusions. *Ann. Zool. Fenn.* 17: 147–158.

Helle, E., Olsson, M. and Jensen, S. 1976. PCB levels correlated with pathological changes in seal uteri. *Ambio* 5: 261–263.

Letcher, R., Lewis, W., van den Berg, M., Seinen, W. and Boon, J. 1998. Biotransformation of coplanar PCBs, PCDDs, and PCDFs and specific cytochrome P450 isoenzyme activities in harbour seal (*Phoca vitulina*): selective inhibition of *in vitro* metabolism in hepatic microsomes. *Organohalogen Compounds* 37: 357–360.

Martineau, D., De Guise, S., Fournier, M., Shugart, L., Girard, C., Lagacé, A. and Béland, P. 1994. Pathology and toxicology of beluga whales from the St. Lawrence Estuary, Québec, Canada. Past, present and future. *Sci. tot. Environ.* 154: 201–215.

Mortensen, P., Bergman, A., Bignert, A., Hansen, H.-J., Härkönen, T. and Olsson, M. 1992. Prevalence of skull lesions in harbour seals (*Phoca vitulina*) in Swedish and Danish museum collections: 1835–1988. *Ambio* 21: 520–524.

Olsson, M., Karlsson, B. and Ahnland, E. 1994. Diseases and environmental contaminants in seals from the Baltic and the Swedish westcoast. *Sci. tot. Environ.* 154: 217–227.

Patnode, K.A. and Curtis, L.R. 1994. 2,2′,4,4′,5,5′- and 3,3′,4,4′,5,5′-hexachlorobiphenyl alteration of uterine progesterone and estrogen receptors coincides with embryotoxicity in mink (*Mustela vison*). *Toxicol. Appl. Pharmacol.* 127: 9–18.

Philo, M.L., Shotts, E.B. and George, J.C. 1993. Morbidity and mortality. In: J.J. Burns, J.J. Montague, and C.J. Cowles (eds), *The bowhead whale*. Allen Press, Lawrence, USA, 275–312.

Reijnders, P.J.H. 1980. Organochlorine and heavy metal residues in harbour seals from the Wadden Sea and their possible effects on reproduction. *Neth. J. Sea Res.* 14: 30–65.

Reijnders, P.J.H. 1986. Reproductive failure in common seals feeding on fish from polluted coastal waters. *Nature* 324: 456–457.

Reijnders, P.J.H. 1990. Progesterone and oestradiol-17β concentration profiles throughout the reproductive cycle in harbour seals (*Phoca vitulina*). *J. Reprod. Fert.* 90: 403–409.

Reijnders, P.J.H. 1994. Toxicokinetics of chlorobiphenyls and associated physiological responses in marine mammal, with particular reference to their potential for ecotoxicological risk assessment. *Sci. tot. Environ.* 154: 229–236.

Reijnders, P.J.H. 1996. Organohalogen and heavy metal contamination in cetaceans: observed effects, potential impact and future prospects. In: M.P. Simmonds and J.D. Hutchinson (eds), *The Conservation of Whales and Dolphins*. John Wiley & Sons Ltd, Chichester, UK, 205–217.

Reijnders, P.J.H. and Brasseur, S.M.J.M. 1992. Xenobiotic induced hormonal and associated developmental disorders in marine organisms and related effects in humans. In: T. Colborn and C. Clement (eds), *Chemically Induced Alterations in Sexual and Functional Development: The Wildlife/human Connection*. Princeton Sci. Publ., Princeton, NY, 131–146.

Reijnders, P.J.H. and de Ruiter-Dijkman, E.M. 1995. Toxicological and epidemiological significance of pollutants in marine mammals. In: A.S. Blix, L. Wallöe, and O. Ulltang (eds), *Whales, Seals, Fish and Man*. Elsevier Science BV, Amsterdam, 575–587.

Reijnders, P.J.H., Donovan, G.P., Aguilar, A. and Bjørge, A. 1999. Report of the Workshop on Chemical Pollutants and Cetaceans. In: P.J.H. Reijnders, A. Aguilar, and G.P. Donovan (eds), *Chemical Pollutants and Cetaceans*. J. Cetacean Res. & Mgmnt (Special Issue I). International Whaling Commission, Cambridge, UK, 1–42.

Stede, G. and Stede, M. 1990. Orientierende Untersuchungen von Seehundschädeln auf pathalogische Knochenveränderungen. In: *Zoologische und Ethologische Untersuchungen zum Robbensterben*. Inst. f. Haustierkunde, Kiel, Germany, 31–53.

Subramanian, A.N., Tanabe, S., Tatsukawa, R., Saito, S. and Myazaki, N. 1987. Reductions in the testosterone levels by PCBs and DDE in Dall's porpoises of Northwestern North Pacific. *Mar. Poll. Bull.* 18: 643–646.

Tanabe, S., Watanabe, S., Kan, H. and Tatsukawa, R. 1988. Capacity and mode of PCB metabolism in small cetaceans. *Mar. Mamm. Sci.* 4: 103–124.

Troisi, G.M. and Mason, C.F. 1997. Cytochromes P450, P420 and mixed-function oxidases as biomarkers of polychlorinated biphenyl (PCB) exposure in harbour seals (*Phoca vitulina*). *Chemosphere* 35: 1933–1946.

Troisi, G.M. and Mason, C.F. 2000. Steroid PCB-associated alteration of hepatic metabolism in harbour seals (*Phoca vitulina*). *J. Toxicol. Environm. Health* 61: 649–655.

Wagemann, R. and Muir, D.G.C. 1984. Concentrations of heavy metals and organochlorines in marine mammals of northern waters: overview and evaluation. *Can. Techn. Rep. Fish. Aquat. Sci.* 1297: 1–97.

Wilckens, T. and de Rijk, R. 1997. Glucocorticoids and immune function: unknown dimensions and new frontiers. *Immunol. Today* 18: 418–424.

4 Effects of environmental contaminants on the endocrine system of marine mammals

Mary Gregory and Daniel G. Cyr

Introduction

The effects of environmental contaminants on the endocrine system have received increased attention over the past several years. Much of this attention has stemmed from studies on fish and reptiles, which have demonstrated that organic compounds, and particularly organochlorinated compounds, can interact with the estrogen receptor to induce the production of yolk proteins in oviparous males and cause feminization of masculine traits. Although studies in mammals, particularly with rodents, have also shown that certain toxicants can interact with hormone receptors to mimic or inhibit hormone action, the concentrations of exogenous xenobiotics necessary to cause such an effect in adults are very high and this has cast doubt regarding the likelihood that these compounds actually represent a threat to mammalian species. Unlike other mammals, marine mammals represent a unique situation in that their large fat content can act as a pool for trapping contaminants (Colborn and Smolen, 1996). As such, these animals may be exposed to very high concentrations of environmental contaminants from the time of conception via exposure *in utero*, during breast feeding via the mother's milk, as well as through accumulation from the food chain during adolescence to adulthood.

To date few studies have demonstrated directly that environmental contaminants can alter the endocrine system of marine mammals. There are, however, several lines of evidence to suggest that contaminants may be having an effect on a variety of endocrine systems in different species of marine mammals. As indicated above, numerous studies have reported the presence of heavy metals, pesticide residues and metabolites, organochlorinated compounds, and other environmental contaminants in a variety of tissues and species of marine mammals. However, because there are several factors, such as temperature, habitat changes, etc., which can all influence endocrine systems, it is difficult to establish a causal relationship between an environmental contaminant and an effect on an endocrine system. None the less, there have been a number of studies which report some alarming endocrine disturbances, and potentially large and severe problems with respect to a variety of marine mammal populations may occur, or may have already occurred.

As a consequence of human activities, the environment is filled with a variety of chemicals, many of which are persistent and which may have unknown interactions. Some of the most widespread chemicals are organochlorines, particularly DDT and its metabolites, and polychlorinated biphenyls (PCBs), many of which are persistent and which may have unknown interactions. In addition, there are, in some areas, relatively high concentrations of certain heavy metals, such as mercury, cadmium and lead. However, much of the literature with respect to marine mammals regarding these substances relates to concentrations in tissues, primarily obtained as a result of strandings, beached carcasses, or massive die-offs resulting from epizootic infections.

An excellent comprehensive review of the literature regarding contaminant burdens and epidemiological analysis has been published by Colborn and Smolen (1996, Chapter 12 in this volume), and other reviews (Holden, 1978; Hutchinson and Simmonds, 1986; Danzo, 1998) have examined effects of environmental contaminants in marine mammals. This chapter will focus on those studies which have reported associations between environmental contaminants and observed endocrine and/or reproductive disruptions in marine mammals. First, however, we present a brief review of what is known regarding normal endocrine and reproductive systems in marine mammals.

The thyroid gland

The thyroid of marine mammals is a bilobate gland located on either side of the larynx. As with other vertebrates, the thyroid gland is composed of thyroid follicles which synthesize and store thyroid hormones (T_4 = L-thyroxine and T_3 = triiodo-L-thyronine). The parafollicular cells of the thyroid gland also produce calcitonin.

The synthesis and secretion of thyroid hormones from the thyroid gland is regulated by thyrotropin (TSH) which is produced by the adenohypophysis (St. Aubin, 1987; Figure 4.1). TSH binds to specific receptors located on the plasma membrane of thyroid follicles. The hormone–receptor complex activates the production of secondary messengers which stimulate the uptake of iodine from the circulation. The iodine is transported into the thyroid follicle and reacts with tyrosine residues located on thyroglobulin. This first reaction is catalyzed by a peroxidase and leads to the formation of mono- and diiodothyronine residues. A second reaction forms thyroxine (T_4) and triiodothyronine (T_3) residues (Figure 4.2). This complex is stored in the lumen of the follicle and upon stimulation by TSH is actively transported into the follicular cells, where the tyrosine residues are cleaved from thyroglobulin and T_4 and T_3 are secreted into circulation.

Thyroid hormones have low solubility in aqueous solutions and are therefore transported into circulation bound to transport proteins. In most

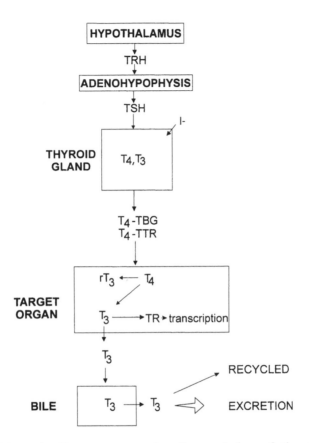

Figure 4.1 Schematic diagram representing the regulation of thyroid function in mammals. Thyroid releasing hormone (TRH) is released by the hypothalamus and acts on the adenohypophysis to stimulate the secretion of TSH. TSH acts on the thyroid gland to activate iodide uptake and the synthesis of both T_4 and T_3, although T_4 is produced and secreted in much greater quantity. Both hormones bind to the serum transport proteins TBG and transthyretin (TTR). Free circulating T_4 is taken up by the target cell and converted to T_3 by deiodinases. The T_3 binds to the thyroid hormone (TR) nuclear receptor and activates transcription. T_3 is then released into circulation. T_3 is conjugated with either sulfuric or glucuronic acid and excreted via the bile. The conjugates can be removed by the intestinal flora and the hormone absorbed by the intestine, to enter the circulation.

mammals the most important transporters of thyroid hormones are: thyroxine binding globulin (TBG) and transthyretin. Ridgway *et al.* (1970) have reported the presence of TBG in cetaceans. Once T_4 is transported within its target cell, it is deiodinated by microsomal deiodinases to its biologically active form, T_3. T_3 then crosses into the nucleus and binds to nuclear thyroid hormone receptors and activates specific gene transcription.

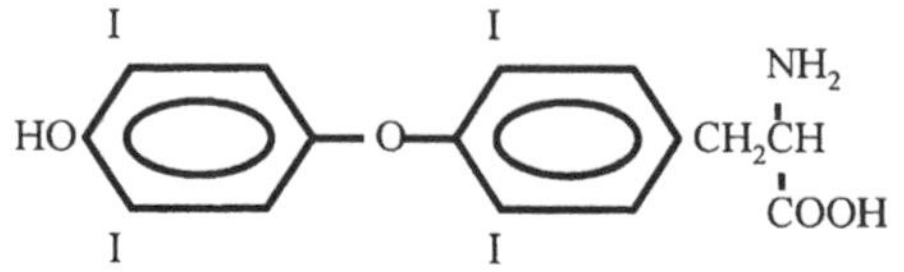

L-Thyroxine

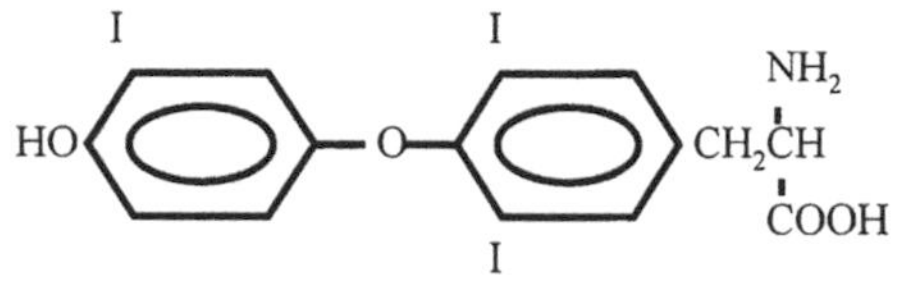

3,5,3'-triiodo-L-thyronine

Figure 4.2 Structure of L-thyroxine and triido-L-thyronine.

Thyroid hormones are important regulators of intermediary metabolism, growth, neural development and aspects of reproduction. Circulating levels of T_4 in cetaceans range from 1.5 to 19 µg/dl (5 to 190 µg/L) and 0.4 to 19.0 µg/dl (4 to 190 µg/L) in pinnipeds. Circulating levels of T_3 are much lower and range from 104 to 176 ng/dl (1.04 to 1.76 µg/L) in beluga whales (*Delphinapterus leucas*) and 10–330 ng/dl (0.1 to 3.3 µg/L) in pinnipeds (Leatherland and Ronald, 1979; John *et al.*, 1987; Kirby, 1990). In cetaceans they are also believed to play an important role in controlling heat loss (Crile and Quiring, 1940). As with other seasonal breeders, marine mammals undergo seasonal changes in circulating levels of thyroid hormones. Harrison (1960, 1969) has reported that in seals peak concentrations of serum thyroid hormone concentrations are observed during lactation as well as during fetal development.

Initiation of and early hair growth also appear to be stimulated by thyroxine (Chang, 1926; Baker, 1951), and studies of molting animals further suggests an association between plasma thyroid hormone levels and hair growth (Ebling and Johnson, 1964; Ridgway *et al.*, 1970; Engelhardt and Ferguson, 1980; Ashwell-Erickson *et al.*, 1986; John *et al.*, 1987).

Several studies have demonstrated that environmental contaminants can alter various aspects of thyroid function. Among the different classes of environmental contaminants that alter thyroid function are heavy metals (e.g. Pb, Cd and Hg) and organohalogens, in particular dioxins and PCBs.

Schumacher *et al.* (1993) investigated the thyroid glands of seals that had died during an epizootic of phocine distemper infection in the North Sea (1988–9). Mortality was high in areas with high levels of environmental pollution (Heidemannn, 1989) but was absent in areas around Iceland with little environmental pollution. Furthermore, far lower levels of chlorinated hydrocarbons were detected in tissues from Iceland seals. A positive correlation between high organochlorine levels and high mortality from phocine distemper virus has been reported (Hall *et al.*, 1992). In addition, some seals that died during the 1988–9 epizootic exhibited colloid depletion and fibrosis of the thyroid gland, a condition that has been associated with chronic PCB exposure (Schumacher *et al.*, 1990, 1991). Results from Schumacher's studies were similar to those of experiments with rats (Byrne *et al.*, 1987) and seals (Brouwer *et al.*, 1989) fed PCBs directly. In harbor seals (*Phoca vitulina*) fed PCB-contaminated fish, thyroid hormone levels were significantly decreased, further supporting the hypothesis that chronic PCB intoxication causes fibrosis of the thyroid gland and colloid depletion of thyroid follicles, resulting in impairment of thyroid function.

The adrenal gland

The structure of the adrenal gland in marine mammals is similar to that of other mammals, consisting of an outer cortex and inner medulla. In cetaceans and pinnipeds, however, the adrenal gland is pseudolobulated due to the presence of the adrenal medullary and connective tissue septae which extend into the cortex (Kirby, 1990). The adrenal cortex is responsible for steroid hormone synthesis, while catecholamines are produced by the medulla. Adrenal steroidogenesis is regulated by adrenocorticotropin (ACTH) which is produced by the adenohypophysis. As with other hypophyseal hormones, ACTH binds to a plasma membrane receptor and stimulates the production of secondary messengers such as cyclic adenosine monophosphate (cAMP). This leads to a sequestering of cholesterol from low-density lipoproteins in circulation and its mobilization to the cell mitochondria where steroidogenesis takes place. Steroidogenesis in terrestrial mammals is well characterized and is believed to be similar in marine mammals. The adrenal cortex produces three classes of steroid hormones: glucorticoids (cortisol, corticosterone), mineralocorticoids (aldosterone) and small amounts of sex steroids (androgens, estrogens and progestins) (Figure 4.3). In general, the function of these hormones in marine mammals is similar to that in terrestrial mammals.

In terms of effects of contaminants on other endocrine systems, there is a paucity of information. Haraguchi *et al.* (1992) reported fairly high concentrations of methylsulfate chlorobenzenes (MeSO4-CBs) and methylsulfate DDE residues in adrenal tissues of seals from the Baltic Sea. Bergman and Olsson (1986) observed adrenocortical hyperplasia among seals from the Baltic; it was suggested that this was possibly a consequence of environmental contaminants. DeGuise *et al.* (1995) also reported adrenal insufficiency in beluga whales and suggested an association between this condition and the present of environmental contaminants in tissues.

Lund (1994) studied the effects of DDT, PCB and their metabolites on adrenal function in the gray seal (*Halichoerus grypus*). Additionally, this study investigated the potential for *in vitro* bioactivation of DDT metabolites by the adrenal gland. Results indicated that *o,p′*-DDD, a metabolite of

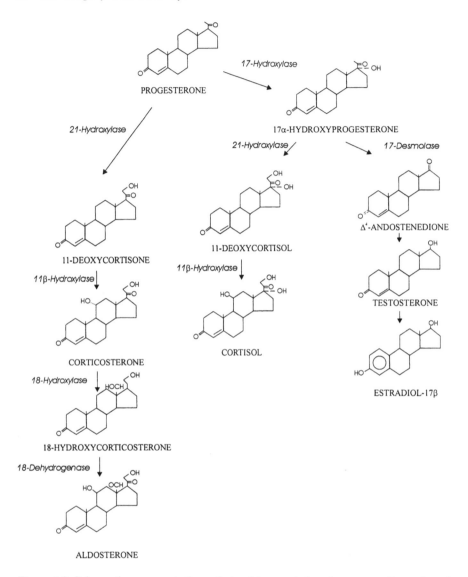

Figure 4.3 Schematic representation of steroidogenesis by the mammalian adrenal cortex, starting with progesterone and its metabolites. The enzymes involved in the steroidogenic pathway are indicated in italics.

DDT, was selectively bioactivated to protein-binding intermediates in the mitochondria of adrenal gland cells. This bioactivation was shown to be oxidative and dependent upon cytochrome P450, resulting in perturbation of adrenal function. These findings correspond to those found in mink (*Mustela vison*) (Jonsson *et al.*, 1993) and in dog (Hart *et al.*, 1973). The metabolite *o,p′*-DDD also exhibited inhibitory action on the enzyme steroid 11β-hydroxylase, suggesting an interference with corticosteroid metabolism (Lund, 1994).

Contradictory results from the few other studies on this subject demonstrate that the mode of action of organochlorines, such as DDT and PCBs, on the adrenal gland is unclear. In rats fed low levels of toxaphene, DDT, and technical PCB (Aroclor 1254) for 35–155 days, no hyperplasias were observed, but adrenal glucocorticoid production was inhibited (Young *et al.*, 1973; Mohammed *et al.*, 1985; Byrne *et al.*, 1988). Conversely, chronic exposure of dogs to DDT caused adrenal enlargement but no effects on steroid production (Copeland and Cranmer, 1974). Chronic exposure of mink to relatively high concentrations of planar PCBs caused enlarged adrenal glands (Aulerich *et al.*, 1985, 1987).

Hence, conflicting results indicate the need for further study, but suggest that observed adrenotoxicity may result from either the generation of cytotoxic reactive intermediates, and/or inhibition of glucocorticoid-synthesizing enzymes (Lund, 1994).

Reproductive systems

Perhaps the best-characterized endocrine systems of marine mammals are the reproductive endocrine cycles. The reproductive systems of marine mammals share the basic anatomical structures and organization found in terrestrial mammals, with minor adaptations to certain marine species (Anthony, 1922; Matthews, 1950; Brodie, 1971; Gaskin, 1988; DeGuise *et al.*, 1994). Gonadotropin releasing hormone (GnRH), produced and secreted by the hypothalamus, stimulates the synthesis and release of follicle stimulating hormone (FSH) by the adenohypophysis. This regulates the development of the ovarian follicle and the production of estradiol (Kirby, 1990). The ovarian follicle consists of thecal cells which synthesize androgens, and granulosa cells where testosterone and androstenedione are converted to estradiol by a cytochrome P450 enzyme referred to as aromatase. Estrogens released by the developing ovarian follicle are responsible for feminization traits. GnRH also stimulates the release of luteinizing hormone (LH), which acts upon the ovaries in females and triggers ovulation, as well as ovarian production of estradiol. In both cetacean and pinnipeds, follicular development normally leads to formation of a single Graafian follicle (Kirby, 1990). In the northern elephant seal (*Mirounga angustirostris*), ovulation occurs at the time of mating (Kirby, 1990). As with other mammals, a corpus luteum is formed in the ovary after ovulation. The corpus luteum is responsible for the production of progesterone, which acts on the endometrium to stimulate secretions and nutrients for the developing blastocyst (Kirby, 1990). In northern fur seals (*Callorhinus ursinus*) progesterone remains low during the 3.5 months of delayed implantation. An estradiol surge appears to act as the signal to initiate blastocyst reactivation (Daniel, 1974, 1975, 1981).

In the male, FSH acts on Sertoli cells of the seminiferous tubules to release androgen-binding protein, which stimulates spermatogenic cells to bind and concentrate testosterone, which then promotes spermatogenesis.

In males, LH binds to interstitial cells of the seminiferous tubule and stimulates secretion of testosterone. Seasonal changes in circulating testosterone concentrations suggest that for seals and cetaceans, the males impregnate the females as testosterone levels increase, although this may occur prior to reaching peak testosterone levels (Kirby, 1990). There is no information regarding the hormone requirement for maintaining spermatogenesis in marine mammals. In rodents, androgens can maintain spermatogenesis independent of hypophyseal hormones (Handelsman *et al.*, 1999).

Physiological control of reproduction in marine mammals is believed to be comparable, and current information is based primarily on those species that have been kept in captivity (Gaskin, 1988). The distinctive features of reproduction in marine mammals relate to temporal adaptations to lifestyle and climate. Most species are seasonal breeders, and exhibit highly synchronized pupping, mating and delayed blastocyst implantation (Fisher, 1954; Atkinson *et al.*, 1993). Generally, low concentrations of sex steroid hormones (i.e. testosterone, progesterone and estrogens) are present in immature animals, and higher but variable levels present in mature adults (Wells, 1984; Kjeld *et al.*, 1992; Desportes *et al.*, 1994).

Most species of whales attain sexual maturity at about 5 years of age for males, and 8 years for females; gestation lasts approximately 14.5 months, and lactation can last for up to 2 years (Laws, 1959; Brodie, 1971; Sargeant 1973). Similarly, the gestation period for the walrus (*Odobenus rosmarus*) is 15–16 months, and sexual maturation occurs between 9 and 10 years of age for both males and females; however, males usually do not breed until they are about 15 years of age, when they can successfully defend territories (Lentfer, 1988). Sea lions reach sexual maturity at about 5 years of age for females and 5–7 years for males but, like the walrus, males do not breed until they are 9–13 years of age.

Seals generally reach sexual maturity between 3 and 8 years for males and females, depending upon the species, and gestation periods are between 8.5 and 10.5 months (Lentfer, 1988). Kenyon (1969) reported that sea otters (*Enhydra lutris*) have a gestation period of about 4 months, and sexual maturity is attained at 5–6 years in males and 3–4 years in females. The polar bear (*Ursus maritimus*) usually first reproduces at between 4 and 8 years of age; normal litters consist of one or two cubs, and females usually reproduce only once every 2 years.

It is evident, then, that in most species of marine mammals, sexual maturity does not occur for at least 3–4 years after birth, and the reproductive cycles tend to be lengthy. Chronic exposure to even low levels of environmental contaminants during periods of growth, development and reproduction can, therefore, have significant consequences on a population.

Kihlstrom *et al.* (1992) performed various experiments in which female mink were fed daily doses of PCBs or fractions of PCBs for different lengths of time. Commercial PCBs are comprised of mixtures of chlorinated biphenyls and several contaminants, including polychlorinated dibenzofurans (PCDFs) and polychlorinated naphthalenes (PCNs), the latter having chemical properties similar to those of PCDDs (polychlorinated dibenzo-*p*-dioxins). Two commercial PCBs, Clophen A50 and Aroclor 1254, were separated into four fractions, as described by Athanasiadou *et al.* (1992):

1   2–4-ortho-CBs, with two or more chlorine atoms substituted in the ortho-position to the biphenyl bond;
2   1-ortho-CBs, with only one chlorine atom in the ortho-position to the bond;
3   0-ortho-CBs, with no chlorine atoms in an ortho-position of the biphenyl bond; these are also known as coplanar CBs; and
4   bi- and tri-cyclic contaminants, including PCN and PCDF.

Kihlstrom's group studied effects of these compounds on reproduction, and found that exposure to any single chlorobenzene fraction did not significantly influence zygote implantation rate; however, the fraction containing bi- and tri-cyclic contaminants (including PCDF and PCN) caused a significantly greater number of living pups (whelps) and increased survival rate up to day 5 after birth. In contrast, exposure to commercial PCB or mixtures of two or more fractions of PCBs caused an increase in the number of interrupted pregnancies, as well as a decrease in the number of living pups born and a decrease in litter size; implantation rate was not significantly affected. Exposure to PCBs or combinations of fractions also caused a significant increase in liver weight and a corresponding increase in hepatosomatic index. These effects are similar to those reported previously (Kihlstrom *et al.*, 1976 (cited in Kihlstrom *et al.*, 1992); Jensen *et al.*, 1977, 1979; Aulerich *et al.*, 1987) in which PCB caused reproductive failure in mink but did not affect the frequency of uterine implantation.

More recently, Sohoni and Sumpter (1998) reported evidence from yeast-based assays that several environmental estrogens, such as DDE and *o,p'*-DDT, also exhibit anti-androgenic activity. Thus, certain chemicals previously thought to be estrogenic may actually be blocking androgen action, resulting in changes normally associated with estrogen exposure.

Little is known regarding the effects of environmental pollutants on reproduction in marine mammals, and the link between pathological conditions and concentrations of various contaminants in tissues is difficult to ascertain. However, given the generally much longer reproductive cycles of these animals as compared to mink, the impact of reproductive failure on populations could be far more significant.

Reijnders (1980) analyzed polluted waters of the western coast of the Wadden Sea, The Netherlands and found that the major pollutants were PCBs. Subsequent studies (Reijnders, 1986; Boon *et al.*, 1987), in which harbor seals were fed PCB-contaminated fish for 2 years, reported reproductive failure of these animals. Examination of the hormone profiles of all treated and untreated females suggested that PCBs impaired reproduction at the postovulation stage, around the time of implantation. The author concluded that, from the evidence obtained, this reproductive impairment was responsible for the significant population decline in seals of that area. His findings substantiate those obtained from prior studies of PCBs in mink, in which ovulation, mating and implantation occurred but were followed by early abortion or resorption (Jensen *et al.*, 1977).

Several studies (DeLong *et al.*, 1973; Gilmartin *et al.*, 1976; Martin *et al.*, 1976; Addison, 1989) indicate that seals fed PCB-contaminated fish exhibit deficiencies in levels of thyroid hormone and vitamin A, resulting in increased susceptibility to microbial infections and reproductive disorders (Brouwer *et al.*, 1989). Indeed, Reijnders (1986) has suggested that PCBs are involved in poor reproduction among seals; however, in his studies, he does note that the effects of high DDT residues cannot be excluded. It is further suggested that there may be more than one mechanism involved in reproductive failure.

Gilmartin *et al.* (1976) found that California sea lions (*Zalophus californianus*) with relatively high levels of DDT and PCB gave birth prematurely, exhibited tissue imbalances of mercury, selenium, cadmium and bromine, and in some, carried disease-causing agents. In different studies, Helle *et al.* (1976) and Bergman *et al.* (1981) reported a low pregnancy rate, associated with pathological changes in various female reproductive structures, in ringed seals (*Pusa hispida*) in the Gulf of Bothnia in the Baltic Sea. These females had significantly higher concentrations of DDT and PCBs in tissues, compared with pregnant females and females with no reproductive pathologies. Reijnders (1984) suggested that PCBs may have interfered with steriod hormones, leading to an imbalance that may have triggered implantation failure or early resorption in Baltic seas (ringed seals, *Pusa hispida*, and gray seals, *Halichoerus grypus*).

High concentrations of organochlorines and PCBs have also been associated with a high incidence of prenatal and neonatal deaths, as well as birth defects, in harbor seals of southern Puget Sound (Arndt, 1973; Newby, 1973; Calambokidis *et al.*, 1978; Lentfer, 1988).

Freeman and Sangalang (1977) demonstrated that some metals and Aroclor 1254 (a PCB mixture) caused an increase in steroidogenesis *in vitro* in gray seals (*Halichoerus grypus*). Increased PCB metabolism as a result of their induction of P450 enzymes has also been demonstrated in marine mammals (Tanabe *et al.*, 1988; Boon *et al.*, 1992). Furthermore, decreases in steroid hormone concentrations in male Dall's porpoises (*Phocoenoides dalli*) (Subramanian *et al.*, 1987) and female harbor seals (Reijnders, 1986, 1999) have been associated with concentrations of DDE and PCBs in tissues. Similar findings regarding negative correlations between PCB concentrations and progesterone and testosterone metabolism have been reported for harbor seals (Reijnders, 1999).

DeGuise *et al.* (1994) observed hermaphroditism in two of 120 beluga whales examined from the St. Lawrence River. Abnormal testes and changes in epididymal tissue in North Pacific minke whales (*Balaenoptera acutorostrata*) have been reported and may be associated with organochlorines (Reijnders, 1999).

Conclusions

A significant problem in determining the effects of environmental contaminants on the endocrine system of marine mammals is the relative lack of information on normal endocrine physiology. Not only is it necessary to understand endocrine cycles but also hormone action at the level of the cell. This latter information is particularly important because it would allow for the development of highly sensitive endpoints that could be used as indicators of specific endocrine dysfunction. Given the important differences between species of marine mammals and their reproductive strategies, significant effort will be required to establish normal physiological parameters in a sufficiently large population in order to demonstrate deviants from the norm conclusively. Such information is needed particularly quickly because some marine mammals have shown signs of endocrine and reproductive impairment, and these populations should require immediate attention by the scientific community.

References

Anthony, R. 1922. Recherches anatomiques sur l'appareil génito-urinaire male du *Mesoplodon*. *Mem. Inst. Espanol. Oceanog.* 3: 35–115.

Arndt, D.P. 1973. DDT and PCB levels in three Washington State harbor seal (*Phoca vitulina richardii*) populations. M.S. Thesis, University of Washington, Seattle. In: J.W. Lentfer (ed.), *Selected Marine Mammals of Alaska*, Marine Mammal Commission, Washington, DC, 1988.

Athanasiadou, M., Jensen, S. and Klasson-Wehler, E. 1992. Preparative fractionation of a commercial PCB product. *Chemosphere* 23: 957–970.

Atkinson, S., Gilmartin, W.G. and Lasley, B.L. 1993. Testosterone response to a gonadotrophin-releasing hormone agonist in Hawaiian monk seals (*Monachus schauinslandi*). *J. Reprod. Fertil.* 97: 35–38.

Aulerich, R.J., Bursian, S.J., Breslin, W.J., Olsson, B.A. and Ringer, R.K. 1985. Toxicological manifestations of 2,4,5,2′,4′,5′-, 2′3,6,2′3′,6′-, and 3,4,5,3′,4′,5′-hexachlorobiphenyl and Aroclor 1254 in mink. *J. Toxicol. Environ. Health* 15: 63–79.

Aulerich, R.J., Bursian, S.J., Evans, M.G., Hochstein, J.R., Koudele, K.A., Olson, B.A. and Napolitano, A.C. 1987. Toxicity of 3,4,5,3′,4′,5′-hexachlorobiphenyl to mink. *J. Toxicol. Environ. Health* 16: 53–60.

Baker, B.L. 1951. The relationship of the adrenal, thyroid, and pituitary glands to the growth of hair. *Ann. N.Y. Acad. Sci.* 53: 690.

Bergman, A. and Olsson, M. 1986. Pathology of Baltic grey seal and ringed seal females with special reference to adrenocortical hyperplasia: Is environmental pollution the cause of a widely distributed disease syndrome? Proc. From the Symposium on the Seals in the Baltic and Eurasian Lakes, Savonlinna. 1984-06-05 – 08. *Finn. Game Res.* 44: 47–62.

Boon, J.P., Reijnders, P.J.H., Dols, J., Wensvoort, P. and Hillebrand, M.T.J. 1987. The kinetics of individual polychlorinated biphenyl congeners in female harbour seals (*Phoca vitulina*), with evidence for structure-related metabolism. Aquat. Toxicol. 10: 307–324.

Boon, J.P., van Arnhem, E., Jansen, S., Kannan, N., Petrick, G., Schulz, D., Duinker, J.C., Reijnders, P.J.H. and Goksoyr, A. 1992. The toxicokinetics of PCBs in marine mammals with special reference to possible interactions of individual congeners with the cytochrome P-450-dependent monoxygenase system: an overview. In: C.H. Walker and D.R. Livingstone (eds), *Persistent Pollutants in Marine Ecosystems*. Pergamon Press, Oxford, pp. 119–159.

Brodie, P.F. 1971. A reconsideration of aspects of growth, reproduction, and behavior of the white whale (*Delphinapterus leucas*) with reference to the Cumberland Sound, Baffin Island population. *Can. J. Fish. Aquat. Sci.* 28: 1309–1318.

Brouwer, A., Reijnders, P.J.H. and Koeman, J.H. 1989. Polychlorinated biphenyl (PCB)-contaminated fish induces vitamin A and thyroid hormone deficiency in the common seal *Phoca vitulina. Aquat. Toxicol.* 15: 99–106.

Byrne, J.E., Carbone, J.P. and Hanson, E. 1987. Hypothyroidism and abnormalities in the kinetics of thyroid hormone metabolism in rats treated chronically with polychlorinated biphenyl and polybrominated biphenyl. *Endocrinology* 121: 520–527.

Byrne, J.J., Carbone J.P. and Pepe M.G. 1988. Suppression of serum adrenal cortex hormones by low-dose polychlorobiphenyl or polybromobiphenyl treatments. *Arch. Environ. Contam. Toxicol.* 17: 47–53.

Calambokidis, J., Bowman, K., Carter, S., Cubbage, J., Dawson, P., Fleischner, T., Skidmore, J. and Taylor, B. 1978. Chlorinated hydrocarbon concentrations and the ecology and behavior of harbor seals in Washington State waters. Unpubl. manuscr., Evergreen State College, Olympia, Washington. In: J.W. Lentfer (ed.), *Selected Marine Mammals of Alaska*, Marine Mammal Commission, Washington, DC, 1988.

Chang, H.C. 1926. Specific influence of the thyroid gland on hair growth. *Am. J. Physiol.* 77: 562.

Colborn, T. and Smolen, M.J. 1996. Epidemiological analysis of persistent organochlorine contaminants in cetaceans. *Rev. Environ. Contam. Toxicol.* 146: 91–171.

Copeland, M.F. and Cranmer, M.F. 1974. Effects of o,p'-DDT on the adrenal gland and hepatic microsomal enzyme system in the beagle dog. *Toxicol. Appl. Pharmacol.* 27: 1–10.

Crile, G.C. and Quiring, D.P. 1940. A comparison of the energy-releasing organs of the white whale and the thorough-bred horse. *Growth* 4: 291–298.

Daniel, J.C. Jr 1974. Circulating levels of oestradiol-17beta during early pregnancy in the Alaskan fur seal showing an oestrogen surge preceding implantation. *J. Reprod. Fertil.* 37: 425–428.

Daniel, J.C. Jr 1975. Concentrations of circulating progesterone during early pregnancy in the northern fur seal, *Callorhinus ursinus. Can. J. Fish. Aquat. Sci.* 321: 65–70.

Daniel, J.C. Jr 1981. Delayed implantation in the northern fur seal (*Callorhinus ursinus*) and other pinnipeds. *J. Reprod. Fertil.* 29 (Suppl): 35–50.

Danzo, B.J. 1998. The effects of environmental hormones on reproduction. *Cell. Mol. Life Sci.* 54: 1249–1264.

DeGuise, S., Lagace, A. and Beland, P. 1994. True hermaphroditism in a St. Lawrence beluga whale (*Delphinapterus leucas*). *J. Wildl. Dis.* 30: 287–290.

DeGuise, S., Martineau, D., Beland, P. and Fournier, M. 1995. Possible mechanisms of action of environmental contaminants on St. Lawrence beluga whales (*Delphinapterus leucas*). *Environ. Hlth. Perspect.* 103 (Suppl. 4): 73–77.

Desportes, G., Saboureau, M. and Lacroix, A. 1994. Growth-related changes in testicular mass and plasma testosterone concentrations in long-finned pilot whales, *Globicephala melas*. *J. Reprod. Fertil.* 102: 237–244.

Ebling, F.J. and Johnson, E. 1964. The control of hair growth. *Symp. Zool. Soc. Lond.* 12: 97.

Engelhardt, F.R. and Ferguson, J.M. 1980. Adaptive hormone changes in harp seals, *Phoca groenlandica*, and gray seals, *Halichoerus grypus*, during the postnatal period. *Gen. Comp. Endocrinol.* 40: 434–445.

Fisher, H.D. 1954. Delayed implantation in the harbour seal, *Phoca vitulina* L. *Nature* (Lond.) 173: 879–880.

Freeman, H.C. and Sangalang, G.B. 1977. A study on the effects of methylmercury, cadmium, arsenic, selenium, and a PCB, (Arochlor 1254) on adrenal and testicular steroidogenesis in vitro, by the gray seal *Halichoerus grypus*. *Arch. Environ. Contam. Toxicol.* 5: 369–383.

Gaskin, D.E. 1988. *Whales, Dolphins and Seals*. Heinemann Education Books Ltd., Toronto, Canada.

Gilmartin, W.G., DeLong, R.L., Smith, A.W., Sweeney, J.C., DeLappe, R.W., Risebrough, R.W., Griner, L.A., Dailey, M.D. and Peakall, D.B. 1976. Premature parturition in the California sea lion. *J. Wildl. Dis.* 12: 104–115.

Hall, A.J., Law, R.J., Wells, D.E., Harwood, J., Ross, H., Kennedy, S., Allchin, C.R., Campbell, L.A. and Pomeroy, P.P. 1992. Organochlorine levels in common seals (*Phoca vitulina*) which were victims and survivors of the 1988 phocine distemper epizootic. In: E. Hamilton (ed.), *Sci. Total Envir.* Special Issue on phocid distemper, Elsevier Science Publ., Amsterdam, Netherlands.

Handelsman, D.J., Spaliviero, J.A., Simpson, J.M., Allan, C.M., Singh, J. 1999. Spermatogenesis without gonadotropins: maintenance has a lower testosterone threshold than initiation. *Endocrinology* 140: 3938–3946.

Haraguchi, K., Athanasiadou, M., Bergman, A., Hovander, L. and Jensen, S. 1992. PCB and PCB methyl sulphones in selected groups of seals from Swedish waters. *Ambio* 21(8): 546–549.

Harrison, R.J. 1960. Reproduction and reproductive organs in common seals (*Phoca vitulina*) in the Wash, East Anglia. *Mammalia* 24: 372–385.

Harrison, R.J. 1969. Endocrine organs: Hypophysis, thyroid and adrenal. In: H.T. Andersen (ed.), *The Biology of Marine Mammals*, Academic Press, New York, pp. 349–90.

Hart, M.M., Reagan, R.L. and Adamson, R.H. 1973. The effect of isomers of the DDD on the ACTH-induced steroid output, histology and ultrastructure of the dog adrenal cortex. *Toxicol. Appl. Pharmacol.* 24: 101–113.

Heidemann, G. 1989. Okologische Probleme des Seehundbestandes. *Arbeiten des deutschen Fischerei-Verbandes* 48: 76–87.

Helle, E. 1980. Lowered reproductive capacity in female ringed seals (*Phoca hispida*) in the Bothnian Bay, northern Baltic Sea, with special reference to uterine occlusions. *Ann. Zool. Fenn.* 17: 147–158.

Helle, E., Olsson, M. and Jensen, S. 1976. PCB levels correlated with pathological changes in seal uteri. *Ambio* 5: 261–263.

Holden, A.V. 1978. Pollutants and seals – a review. *Mamm. Rev.* 8: 53–66.

Hutchinson, J.D. and Simmonds, M.P. 1994. Organochlorine contamination in pinnipeds. *Rev. Environ. Contam. Toxicol.* 136: 123–167.

Jensen, S., Kihlstrom, J.E., Olsson, M., Kundberg, C. and Orberg, J. 1977. Effects of PCB and DDT on mink (*Mustela vison*) during the reproductive season. *Ambio* 6: 239.

Jensen, S., Jansson, B. and Olsson, M. 1979. Number and identity of anthropogenic substances known to be present in Baltic seals and their possible effects on reproduction. *Ann. N.Y. Acad. Sci.* 320: 436–448.

John, T.M., Ronald, K. and George, J.C. 1987. Blood levels of thyroid hormones and certain metabolites in relation to moult in the harp seal (*Phoca groenlandica*). *Comp. Biochem. Physiol.* A88: 655–657.

Jonsson, C.-J., Lund, B.-O. and Brandt, I. 1993. Adrenocorticolytic DDT-metabolites: Studies in mink, *Mustela vison*, and otter, *Lutra lutra*. *Ecotoxicology* 2: 41–53.

Kenyon, K.W. 1969. The sea otter in the eastern Pacific Ocean. *N. Am. Fauna* 68.

Kihlstrom, J.E., Olsson, M., Jensen, S., Johansson, A., Ahlbom, J. and Bergman, A. 1992. Effects of PCB and different fractions of PCB on the reproduction of the mink (*Mustela vison*). *Ambio* 21: 563–569.

Kirby, V.L. 1990. Endocrinology of marine mammals. In: L.A. Dierauf (ed.), *CRC Handbook of Marine Mammal Medicine: Health, Disease, and Rehabilitation*, CRC Press, Boca Raton, Florida, pp. 303–351.

Kjeld, J.M., Sigurjonsson, J. and Arnason, A. 1992. Sex hormone concentrations in blood serum from the north Atlantic fin whale (*Balaenoptera physalus*). *J. Endocrinol.* 134: 405–413.

Laws, R.M. 1959. The foetal growth rates of whales with special reference to the fin whale *Balenoptera physalus* Linn. *Discovery Rep.* 29: 281–308.

Leatherland, J.F. and Ronald, K. 1979. Thyroid activity in adult and neonate harp seals, *Pagophilus groenlandicus*. *J. Zool. (London)* 189: 399–409.

Lentfer, J.W. (ed.) 1988. *Selected Marine Mammals of Alaska.* Marine Mammal Commission, Washington, DC.

Lund, B.O. 1994. In vitro adrenal bioactivation and effects on steroid metabolism of DDT, PCBs and their metabolites in the gray seal (*Halichoerus grypus*). *Environ. Toxicol. Chem.* 13: 911–917.

Matthews, L.H. 1950. The male urogenital tract in *Stenella frontalis* (G. Cuvier). *Atlantide Repts.* 1: 223–247.

Mohammed, A., Hallberg, E., Rydstrom, J. and Slanina, P. 1985. Toxaphene: Accumulation in the adrenal cortex and effect on ACTH-stimulated corticosteroid synthesis in the rat. *Toxicol. Lett.* 24: 137–143.

Newby, T.C. 1973. Observations on the breeding behavior of the harbor seal in the State of Washington. *J. Mammal.* 54: 540–543.

Reijnders, P.J.H. 1980. Organochlorine and heavy metal residues in harbour seals from the Wadden Sea and their possible effects on reproduction. *Neth. J. Sea Res.* 14: 30–65.

Reijnders, P.J.H. 1986. Reproductive failure in common seals feeding on fish from polluted coastal waters. *Nature* (London) 324: 456–457.

Reijnders, P.J.H. 1999. Reproductive and developmental effects of endocrine-disrupting chemicals on marine mammals. In: *Marine Mammals and Persistent*

Ocean Contaminants: Proceedings of the Marine Mammal Commission Workshop, Keystone, Colorado, 12–15 October 1998. Marine Mammal Commission, pp. 93–100.

Ridgway, S.H., Simpson, J.G., Patton, G.S. and Gilmartin, J. 1970. Hematologic findings in certain small cetaceans. *J. Am. Vet. Med. Assoc.* 157: 566–575.

Sargeant, D.E. 1973. Biology of white whales (*Delphinapterus leucas*) in Western Hudson Bay. *J. Fish. Res. Bd. Can.* 30(8): 1065–1090.

Schumacher, U., Horny, H.-P., Heidemann, G., Schultz, W. and Welsch, U. 1990. Histopathological findings in harbour seals (*Phoca vitulina*) found dead on the German North Sea Coast. *J. Comp. Pathol.* 102: 299–309.

Schumacher, U., Horny, H.-P., Heidemann, G., Skirnisson, K. and Welsch, U. 1991. Histological and biochemical investigations into the cause of the recent seal death epidemic. *Prog. Histochem. Cytochem.* 23: 390–394.

Schumacher, U., Zahler, S., Horny, H.P., Heidemann, G., Skirnisson, K. and Welsch, U. 1993. Histological investigations on the thyroid glands of marine mammals (*Phoca vitulina*, *Phocoena phocoena*) and the possible implications of marine pollution. *J. Wildl. Dis.* 29: 103–108.

Sohoni, P. and Sumpter, J.P. 1998. Several environmental oestrogens are also anti-androgens. *J. Endocrinol.* 158: 327–339.

St. Aubin, D. 1987. Stimulation of thyroid hormone secretion by thyrotroppin in Beluga whales. *Can. J. Vet. Res.* 51: 409–412.

Subramanian, A., Tanabe, S., Tatsukawa, R., Saito, S. and Miyazaki, N. 1987. Reduction in the testosterone levels by PCBs and DDE in Dall's porpoises of Northwestern North Pacific. *Mar. Pollut. Bull.* 18: 643–646.

Tanabe, S., Tanabe, S., Khan, H. and Tatsukawa, R. 1988. Capacity and mode of PCB metabolism in small cetaceans. *Mar. Mammal Sci.* 4: 103–124.

Wells, R.S. 1984. Reproductive behavior and hormonal correlates in Hawaiian spinner dolphins, *Stenella longirostris*. In: W.F. Perrin, R.L. Brownell, Jr, and D.P. DeMasters (eds), *Reproduction in Whales, Dolphins and Porpoises*. Report of the International Whaling Commission, Special Issue No. 6. Cambridge, England, pp. 465–472.

Young, R.B., Bryson, M.J., Sweat, M.L. and Street, J.C. 1973. Complexing of DDT and o,p′-DDD with adrenal cytochrome P-450 hydroxylating systems. *J. Steroid Biochem.* 4: 585–591.

5 Opportunities for environmental contaminant research: What we can learn from marine mammals in human care

Michelle L. Reddy and Sam H. Ridgway

> Even more than before, marine mammals in captivity should be used to obtain a set of reference data to interpret values obtained from animals expected to be affected by contaminants.
>
> Reijnders (1988)

Most reported levels of organochlorine (OC) contamination in marine mammals are derived from deceased or moribund animals for which there is little or no relevant biological information. In the absence of data for reproductive and medical histories, parentage, and feeding habits, it is difficult to understand the dynamics and consequences of contaminant exposure (Bignert *et al.*, 1993; Skaare, 1996). In addition, the time between death and sampling of tissues has been shown to affect analysis (Borrell and Aguilar, 1990). Stranded specimens are generally recovered after some, often extensive, degradation has taken place. Even when samples are fresh, it is difficult to ascribe the cause of the event to a disease process or toxic exposure without knowing anything about the animal's history, the progression of clinical illness, or the mechanism leading to the mortality event. Thus, the scope of OC investigations on wild animals is limited. Logistical problems can place significant restrictions on the experimental design of longitudinal studies to monitor contaminants in wild populations over time. What is needed is a population of animals that can be readily accessed for sampling and observation and for which long-term health and reproductive histories are available. Marine mammals in the care of humans may fulfill the need.

Holden and Marsden first reported OCs in marine mammals in 1967 (Holden and Marsden, 1967). Over the next three decades, many studies addressed the contaminant burdens of OCs in wild marine mammals (see Aguilar and Borrell, 1997 for an annotated bibliography). However, most of these studies lacked biologic data, indicators of exposure, and health data. Their scope has generally been limited to assessing contaminant levels in tissues, which does not equate to contaminant toxicity (Bayne, 1984). Carefully done studies that incorporate biological, physiological, and clinical endpoints will be required to evaluate properly the effects of potentially toxic contaminants on marine mammals. Studies on marine mammals in collections should comprise an integral component of this research. We must employ every means at our disposal to meet the challenge head-on and beat the prediction by Klamer *et al.* (1991), 'If the increase in ocean PCB concentration continues, it may ultimately result in the extinction of fish-eating marine mammals.'

Benchmark population

The lack of comprehensive studies and the use of varying methodologies in the study of OCs make it difficult to assimilate the results of all investigations. Such integration would be helpful in mapping global and species distribution patterns, and linking findings of different studies from the suborganismal level (subcellular and/or *in vitro* systems) to the organismal (individual) level, up through the population level (Keith, 1996). The development of standard reference materials with certified concentrations of contaminants has allowed for interlaboratory calibration (Schantz *et al.*, 1993, 1995; Wise *et al.*, 1993; de Boer *et al.*, 1994) to improve comparability of sample analysis procedures. Standardized sample collection and handling protocols also enhance comparability between studies (Wise *et al.*, 1993).

Even with the most stringent collection and analysis protocols, there are often broad differences between OC studies, some of which are known, such as species differences, and some of which are not known, such as reproductive histories. Because of these differences, it would be beneficial to establish a reference population of living marine mammals to serve as a link between various studies. In laboratory studies, this link is the control population. However, because of the ubiquitous distribution of environmental contaminants, it is unlikely that there are any marine mammal populations that are unexposed, which could be used as a control and from which 'normal' OC levels can be derived. Because any measure of a synthetic compound in these animals would not be 'normal', it is highly unlikely that any pristine 'baseline' population exists anywhere. However, it is possible to establish a living benchmark population – one that is readily available for sample collection, observation, and health assessment in a long-term monitoring program of OC effects. This population could then be used as a point of reference for other studies that are more limited in scope and duration.

Marine mammal collections are ideal benchmark populations for contaminant studies. There are established research programs at many research and display facilities, with the needed expertise and animal accessibility to conduct serial, long-term investigations. Collaborations between these facilities, toxicologists, epidemiologists, chemists, biologists, veterinarians, pathologists, and other experts, markedly improve our ability to evaluate the effects of OCs at various concentrations. Some collection animals are maintained in traditional marine mammal enclosures, while others are housed in open ocean enclosures and are exposed to many of the same conditions as wild animals. Collection animals are often conditioned for behaviors that

Figure 5.1 Voluntary husbandary behaviors facilitate sample collection without possible artifacts from capture. Reproduced with kind permission of the US Navy.

facilitate medical procedures and serial sampling, including blood (Figure 5.1) and milk collection (Kamolnick *et al.*, 1994; Ridgway *et al.*, 1995); ultrasound; fecal, urine, and semen collection (Ridgway, 1968; Schroeder and Keller, 1989, 1990; Sweeney *et al.*, 1999). Such behaviors support a regular sample collection regimen while reducing the risk of stress-induced alterations in measured health parameters as a result of capture. Clinical health of collection animals is assessed on a regular basis, and in many cases, extensive databases of clinical data, reproductive history, parentage, and other such data are available for creating relational databases. Findings from studies of collection animals can be used to assess the threat of contaminant loads found in live animals or to evaluate the role of OC contaminants in mortality events. Additionally, breeding programs that have become increasingly successful in these populations (Asper *et al.*, 1990; Duffield and Wells, 1991; also see NMFS inventory) are a crucial component for monitoring effects on reproduction, growth and development, and transgenerational effects of maternal exposure (Figure 5.2).

The Marine Mammal Protection Act of 1972 recognizes the usefulness of exhibit marine mammals for conducting research and raising environmental awareness by specifically allowing for the collection of marine mammals for such purposes. In its declaration of policy the MMPA states, '. . . (3) there is inadequate knowledge of the ecology and population dynamics of such marine mammals and of the factors which bear upon their ability to reproduce themselves successfully; (4) negotiations should be undertaken immediately to encourage the development of international arrangements for research on, and conservation of, all marine mammals' (MMPA section 2, p. 2).

Information obtained from collection animals has already yielded extensive data on dolphin behavior, physiology (brain function, sleep, sound process-

Figure 5.2 Breeding programs, which have become increasingly successful with collection populations, are a crucial component for monitoring the effects of environmental contaminants on reproduction, growth and development, and transgenerational effects of maternal exposure. Reproduced with kind permission of the US Navy.

ing, hearing), psychology, cognition, husbandry, communication, nutrition and medicine (cf. Anonymous, 1998). Facilities caring for these animals and the veterinary expertise acquired as a result of their care have contributed to the success of rescue and rehabilitation programs for marine mammals. Many studies begun with collection animals have been extended to studies of wild animals. Martin and Smith (1999) attest to this, stating, 'Experiments on captive animals have answered some of the questions related to physiological adaptations and strategies, and trained animals have extended the experimental laboratory into the open sea.' In reference to studies on wild and collection animals, Norris (1984) stated, 'both kinds of study are crucial, really'. Unfortunately, and despite repeated recommendations to utilize collection animals in OC studies (Reijnders, 1988; Erickson *et al.*, 1995; Marine Mammal Commission, 1999), populations of collection animals have been largely ignored by most funding agencies and scientific investigators.

The use of marine mammals maintained in display and research facilities is controversial from the standpoint of bioethics. Decisions to use marine mammals should be based on a risk/benefit approach, considering the likelihood of untoward events and reactions as well as the potential benefit to the species if the research is conducted.

Case study

The importance of biological data in assessment of contaminant assessment effects is demonstrated by unpublished data collected by the authors from pregnant bottlenose dolphins (*Tursiops truncatus*). Two female bottlenose dolphins, both approximately 7 years old, were collected from the wild within

Figure 5.3 Long-term multigenerational studies are necessary to understand the transfer of OCs from one generation to the next. Such studies can best be conducted on marine mammals in collections, like the calf, mother, grandmother trio seen here. Reproduced with kind permission of the US Navy.

several days of each other, at approximately the same location. They were maintained in the same enclosure at the same facility for the same length of time and had similar diets. Both became pregnant for the first time within days or weeks of each other, and gave birth to calves within the same week. However, one calf died 6 days after birth, while the other calf was still alive more than 6 years later. Blubber biopsies collected from the two dams prior to parturition showed a marked difference in levels of ΣPCB (IUPAC # 52, 101, 118, 128, 153, 105, 138 + 158, 170, 180, 187) and tDDT (*p,p'*-DDD, *o,p'*-DDD, *p,p'*-DDE, *o,p'*-DDE, *p,p'*-DDT, *o,p'*-DDT). The animal whose calf died had about four times greater concentrations of ΣPCB (22.8 versus 5.1 ppm, lipid weight) and tDDT (38.8 versus 9.8 ppm, lipid weight) than the female whose calf survived. Because these animals were collected from the wild, nothing is known about their orders of birth or the body burdens of their mothers, both of which would impact the tissue levels measured. First-born calves inherit a greater proportion of their mother's residue load of OCs than subsequent siblings (Fukushima and Kawai, 1981; Tanabe *et al.*, 1981a; Cockcroft *et al.*, 1989). In this case, data from the previous generation may have helped explain the difference between the two reproductive outcomes. Long-term multigenerational studies are necessary to understand the transfer of OCs from one generation to the next (Figure 5.3). Such studies can best be conducted on marine mammals in collections.

Model species

Toxicologists often use animal models, such as rats and mice, to conduct controlled investigations to assess the effects of contaminant exposure. While such studies can produce valuable information regarding mechanisms of action, risk assessment, and dose–response relationships, extrapolation of these data to other species or populations must be made very judiciously (Marine Mammal Commission, 1999). Studies with laboratory animals such as rodents cannot be used as a substitute for assessing the effects of OCs on marine mammals because of the species specificity that is characteristic of some contaminants (Jacobson *et al.*, 1983; Marine Mammal Commission, 1999). Therefore, it is necessary to conduct studies on representative marine mammal species.

At an international workshop on marine mammals and persistent ocean contaminants (Marine Mammal Commission, 1999), it was agreed that there is a need for multidisciplinary studies on the significance of ocean contaminants in relation to the health and well being of marine mammals. The model species suggested for such investigations were those that have been studied in the wild to a considerable extent and that are also well represented in marine mammal facilities. None of these species: the California sea lion (*Zalophus californianus*), the harbor seal (*Phoca vitulina*), the bottlenose dolphin, or the beluga or white whale (*Delphinapterus leucas*) are endangered. In a recent census of zoos, aquaria, and marine zoological parks in the United States and Canada, three of the model species, bottlenose dolphins, California sea lions and harbor seals represented 78 per cent of the animals at these facilities (Andrews *et al.*, 1997). Within this inventory, 70 per cent of California sea lions, 56 per cent of harbor seals and 43 per cent of bottlenose dolphins were captive-born, all species having second- and third-generation births (Andrews *et al.*, 1997).

Bottlenose dolphins

Of all the models, more is known about the medicine, pathology, and biology of bottlenose dolphins than perhaps any other species (Ridgway, 1968; reviewed by Leatherwood and Reeves, 1990; Wells and Scott, 1999). They have a global distribution in warm and temperate waters that permits for comparison of different tissue levels from various marine populations. Additionally, bottlenose dolphins have both onshore and offshore populations that can be distinguished hematologically and morphologically (Duffield *et al.*, 1983). The existence of two distinct populations within the same species allows comparison of levels of OCs in coastal and pelagic ecosystems. Among cetaceans, this species has the greatest representation in public display and research facilities (Andrews *et al.*, 1997). Although this species has not been collected from the wild for display in US facilities since 1989, as a result of successful breeding programs (Duffield and Wells, 1991; see also NMFS inventory), the collection population is growing. Additionally, the Sarasota Bay bottlenose dolphin population off the Gulf Coast of Florida is the subject of the longest and most intensive long-term catch and release study of wild marine mammals in the world (Scott *et al.*, 1990). Since 1970, regular surveys and health assessments have been conducted to monitor this population.

Worldwide, the largest collection of bottlenose dolphins, and the one that has been monitored most extensively for OC contaminants, resides in open water enclosures in San Diego Bay as part of the US Navy's marine mammal program. Because these animals are housed in netted enclosures, they are physically exposed to fluctuations in temperature and other environmental variables common to the natural environment. They are provided high-quality fish that has been caught, fresh frozen and thawed under controlled conditions to assure a high-grade diet. Fish eaten from the bay, if any, represent a minor component of their diet. Their medical and reproductive histories can be tracked with a newly developed relational database. Biological data and tissue samples can be collected regularly; through gestation, lactation, and transgenerationally. Immunological studies can be conducted on-site at a newly developed molecular laboratory dedicated to marine mammal studies. Neurological assessment can be incorporated using a protocol developed to measure reaction time to sound and/or light stimulus (Ridgway *et al.*, 1991; Ridgway and Carder, 1997) in animals conditioned for such studies.

Measuring the effects of exposure to OC contaminants

Several studies have suggested a relationship between OCs and a wide variety of reproductive, developmental, neurologic, and immune dysfunctions (Barsotti *et al.*, 1976; Colborn *et al.*, 1993; Ankley and Giesy, 1998). Cause and effect relationships are difficult to establish for environmental contaminants, and evidence of a high body burden of contaminants in a beached carcass is not proof that they are the causative agent in the demise of the animal. Addison (1989) points out that much of the current cause and effect evidence is weakened due to lack of control data, especially relevant biological data. To strengthen the link between cause and effect, Colborn and Smolen (1996) suggest that it is necessary to demonstrate that the incidence of effects in more highly contaminated populations is greater than in less-contaminated populations. They also suggest that this be done on live populations and not in response to a mass mortality. Collections of living animals are available for sampling and can be monitored on a long-term basis for any physiological effects of measured levels of OCs.

Reproduction and growth/development

Supervised breeding programs are crucial for measuring the effects of OCs on reproduction and parental exposure in offspring. Many contaminants have developmental and transgenerational effects (Colborn *et al.*, 1993) that may not manifest themselves until later in life (Arai *et al.*, 1983; Gray, 1992; Guillette *et al.*, 1995; Ankley *et al.*, 1997). Collection populations permit

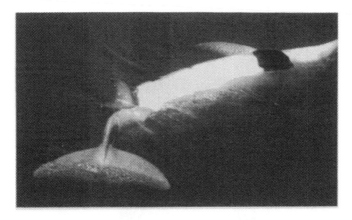

Figure 5.4 Collection populations permit subjects to be monitored from conception through development, birth, growth and throughout adulthood. Reproduced with kind permission of the US Navy.

subjects to be monitored from conception through adulthood (Figure 5.4). Investigation of the effects of maternal blubber OC concentrations on pregnancy is difficult on wild populations, where an unsuccessful pregnancy is not likely to be detected. In marine mammal collections, ultrasound and serial hormone levels are commonly used to diagnose early pregnancy. Ultrasound can be used to confirm pregnancy within the first month, therefore an early miscarriage or resorbed fetus could be detected in a carefully monitored pregnancy. In addition, serial blood and/or urine collections would allow for circulating OCs and hormonal and ovarian cycles (e.g. progesterone and estradiol) to be monitored. Because many OCs are believed to be endocrine disruptors, it is imperative to establish normal hormone levels. These values can be correlated with the ovarian cycle, which can be monitored using ultrasound examinations. Deviations from the normal endocrine and/or ovarian cycle can be correlated with circulating levels of OC contaminants to identify alterations that may result from OC exposure. Pregnancies can be monitored via ultrasound and blood collection. Products of failed pregnancies can be used to develop histologic, cellular, and morphometric 'normals'. Currently, there are few such data for assessing evidence of contaminant exposure in marine mammals. Offspring can be followed through development into adulthood, and endocrine-sensitive systems can be monitored for normal development and function (Ankley *et al.*, 1997). Such comprehensive investigations would be impossible to conduct on wild animals.

The US Navy's bottlenose dolphin population has already been the subject of some preliminary OC investigations. Although it had been suggested that as lactation progresses, contaminant levels in milk decrease (Aguilar and Borrell, 1994a), the Navy animals were used to document the phenomenon

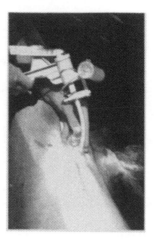

Figure 5.5 To monitor levels of organochlorines in milk, dolphins were conditioned
to allow milk samples to be collected with a modified human breast pump.
Reproduced with kind permission of the US Navy.

for the first time. In a study conducted with lactating bottlenose dolphins
trained for serial milk collection, Ridgway and Reddy (1995) measured the
levels of OCs over the course of lactation. From animals conditioned for
volunteer collection, milk was sampled with a modified breast pump (Figure
5.5). From day 94 to day 615 of lactation, levels of PCB and DDE (lipid
weight) decreased by 69 per cent and 82 per cent, respectively.

Lactation is the primary route of excretion for OC contaminants (Kurzel
and Cetrulo, 1985) and it can significantly reduce body burden. In recent
years, spontaneous lactation in bottlenose dolphins has been observed at
marine mammal facilities (Ridgway *et al.*, 1995). It may be possible to induce
lactation in an animal showing high levels of OCs, thereby reducing body
burden through repeated collection. Such investigations could employ animals
at a marine mammal facility.

Immunocompetence

While many laboratory studies may suggest cause and effect relationships
between OCs and immunosuppression, 'there have been no systematic efforts
to evaluate the impact of pollutants on the immune functions of these animals even though the available data suggest that such an impact might
be substantial' (Martineau *et al.*, 1994). An understanding of the cetacean
immune system is critical for evaluation of the effects and impact of environmental contaminants on natural resistance to disease. Erickson *et al.* (1995)
stated, 'Establishing correlations between environmental conditions and
immuno-competence in marine mammals will be possible only when baseline values for healthy animals have been determined.' Because of ongoing

observation and regular health assessments, the best populations for such studies are maintained in collections.

Prior to 1990, few intensive investigations on the cetacean immune system were conducted, due, in part, to limited access to live animals and tissue samples, the lack of cetacean-specific immunological reagents, and the challenge of designing and conducting non-invasive experiments for functional studies on the cetacean immune system. Immune parameters such as lymphoid function and subpopulations should be included in toxicity assessments for these chemicals. Collaborative studies at the US Navy marine mammal laboratory have facilitated significant progress in the area of cetacean immunology (Romano *et al.*, 1999). They have resulted in the development of cetacean-specific reagents and assays to study and evaluate the immune system, a protocol for generating monoclonal antibodies against cetacean lymphoid cell surface markers, and an enzyme-linked immunosorbent assay (ELISA) to use on cetacean lymphocytes to screen these cells for monoclonal antibodies that recognize cetacean lymphocytes (Romano *et al.*, 1999).

Live animals with known life, dietary, and medical histories provide a source of blood samples for *in vitro* experiments investigating effects of OCs on immune function. These studies allow for more direct evaluations of the marine mammal immune system and assessment of the impact of marine contaminants on immune function. Such studies will also help establish critical time points and methodologies for optimal preservation of field samples. It may be possible to extrapolate associations between individual contaminant levels and individual immune function to population effects.

Dietary studies

Because the food chain is the most significant route of exposure of cetaceans to OCs, studies on dietary exposure are crucial. While most of the dose–response studies on OCs have been conducted on laboratory animals, Reijnders (1986) and Ross *et al.* (1996a, 1996b) report on various dietary intake responses involving studies of harbor seals in a semi-field situation. To date, no such studies have been conducted on cetaceans, although such studies are necessary for understanding exposure effects. Carnivorous marine mammals eat food items such as fish and squid that contain marine contaminants. Therefore, feeding studies where animals are exposed to single contaminants are difficult to conduct due to lack of availability of a contaminant-free diet. Use of single chemical exposures also fails to reflect the natural exposure levels of these animals. However, because the diets of collection animals are consistent and monitored for nutritional content, it is feasible to monitor contaminant levels and conduct long-term 'incidental dosing' studies with these animals. Additionally, natural diets involve mixtures of contaminants as opposed to single contaminants, and therefore more accurately resemble dietary exposure in the wild. Serial samples of blood can be collected and health and reproductive parameters can be monitored to assess the impact of these contaminants on the endocrine and immune systems.

Blood/blubber

Because blubber is the major storage site for lipids (Aguilar and Borrell, 1990), it is the predominant repository of lipophillic OCs (Tanabe *et al.*, 1981b; Cockcroft *et al.*, 1990; Jenssen *et al.*, 1996). Contaminants are transported throughout the body by the circulatory system (Aguilar and Borrell, 1994b). Blood samples are not generally available from dead animals, and it may not always be possible to collect blubber samples from live wild animals. Both samples can be collected from live collection animals. To aid in correlating blubber and blood levels, Reddy *et al.* (1998) collected blood (RBC) and blubber samples from clinically healthy, live animals. They assessed OC levels in all the samples and used the data to develop regression tables for use in estimating blubber levels of OCs from levels found in blood samples. Because a recent meal can affect the blood concentrations of the compounds as well as lipid profiles, samples were collected before the first feeding of the day. These data will be helpful for correlating results of studies of wild animals, those from which blood is collected and those from which blubber is sampled. It will also help in studies of compartmentalization of contaminants in various body tissues.

Conclusion

Marine mammals in collections are an underutilized yet valuable tool in the study of marine contaminant effects. They are a natural benchmark population for which biological data are easily monitored and from which samples can be readily collected to investigate many suspected endpoints of contaminant exposure. Biomarker effectiveness, comparability of various tissue levels, immune response, reproductive effects, transgenerational effects, dietary exposure, and neurological effects may be more readily assessed with marine mammals in collections because of the availability of animals for long-term monitoring. Used in conjunction with contaminant levels found in wild populations, these data can provide an important tool for environmental risk assessment.

Acknowledgments

The authors would like to thank the veterinarians and training staffs of the Navy Marine Mammal Program and at Six Flags Marine World for their cooperation and support with organochlorine investigations. We would like to extend a special note of thanks to Dr John Reif for his continuing collaborative efforts with this work and for his helpful suggestions on the manuscript.

References

Addison, R.F. 1989. Organochlorines and marine mammal reproduction. *Can. J. Fish. Aquat. Sci.* 46: 360–368.

Aguilar, A. and Borrell, A. 1990. Patterns of lipid content and stratification in the blubber of fin whales (*Balaenoptera physalus*). *Journal of Mammalogy* 71: 544–554.

Aguilar, A. and Borrell, A. 1994a. Abnormally high polychlorinated biphenyl levels in striped dolphins (*Stenella coeruleoalba*) affected by the 1990–1992 Mediterranean epizootic. *Science of the Total Environment* 154: 237–247.

Aguilar, A. and Borrell, A. 1994b. Assessment of organochlorine pollutants in cetaceans by means of skin and hypodermic biopsies. In: *Nondestructive Biomarkers in Vertebrates*, M.C. Fossi and C. Leonzio (eds). Lewis Publishers, Boca Raton, pp. 245–267.

Aguilar, A. and Borrell, A. 1997. *Marine Mammal and Pollutants: An Annotated Bibliography*. Fundació pel Desenvolupament Sostenible, Barcelona, Spain.

Andrews, B.F., Duffield, D.A. and McBain, J.F. 1997. Marine mammal management: Aiming at year 2000. *International Marine Biological Research Institute* 7: 125–130.

Ankley, G.T. and Giesy, J.P. 1998. Endocrine disruptors in wildlife: A weight of evidence perspective. In: *Principles and Processes for Assessing Endocrine Disruption in Wildlife*, R. Kendall, R. Dickerson, W. Suk and J. Giesy (eds). SEATAC Press, Pensacola, Florida, pp. 349–368.

Ankley, G.T., Johnson, R.D., Detenbeck, N.E., Bradbury, S.P., Toth, G. and Folmar, L.C. 1997. Development of a research strategy for assessing the ecological risk of endocrine disruptors. *Reviews in Toxicology* 1: 71–106.

Anonymous, 1998. *Annotated Bibliography of Publications from the U.S. Navy's Marine Mammal Program*. SSC San Diego TD 627, Revision D.

Arai, Y., Mori, T., Suzuki, Y. and Bern, H.A. 1983. Long-term effects of perinatal exposure to sex steroids and diethylstilbestrol on the reproductive system of male mammals. *International Review of Cytology* 84: 235–268.

Asper, E.D., Duffield, D.A., Dimeo-Ediger, N. and Shell, D. 1990. Marine mammals in zoos, aquaria and marine zoological parks in North America: 1990 census report. *International Zoo Year Book* 29: 179–187.

Barsotti, D.A., Marlar, R.J. and Allen, J.R. 1976. Reproductive dysfunction in rhesus monkeys exposed to low levels of polychlorinated biphenyls (Aroclor 1248). *Food and Cosmetics Toxicology* 14: 99–103.

Bayne, B.L. 1984. Strategies and advanced techniques for marine pollution studies. In: *Strategies and Advanced Techniques for Marine Pollution Studies: Mediterranean Sea*, C.S. Giam and H.J.-M. Dou (eds). NATO ASI Series. Springer-Verlag, Berlin, Preface.

Bignert, A., Gothberg, A., Jensen, F., Litzen, K., Odsjo, T., Olsson, M. and Reutergard, L. 1993. The need for adequate biological sampling in ecotoxicological investigations: a retrospective study of twenty years pollution monitoring. *Science of the Total Environment* 128: 121–139.

Borrell, A. and Aguilar, A. 1990. Loss of organochlorine compounds in the tissues of a decomposing stranded dolphin. *Bulletin of Environmental Contamination and Toxicology* 45: 46–53.

Cockcroft, V.G., DeKock, A.C., Lord, D.A. and Ross, G.J.B. 1989. Organochlorines in bottlenose dolphins *Tursiops truncatus* from the east coast of South Africa. *South African Journal of Marine Science* 8: 207–217.

Cockcroft, V.G., DeKock, A.C., Ross, G.J.B. and Lord, D.A. 1990. Organochlorines in common dolphins caught in shark nets during the Natal 'sardine run'. *South African Journal of Zoology* 25: 144–148.

Colborn, T. and Smolen, M.J. 1996. Epidemiological analysis of persistent organochlorine contaminants in cetaceans. In: *Reviews of Environmental Contamination and Toxicology*, G.W. Ware (ed.), Vol. 146. Springer, New York, pp. 91–172.

Colborn, T., von Saal, F.S. and Soto, A.M. 1993. Developmental effects of endocrine-disrupting chemicals in wildlife and humans. *Environmental Health Perspectives* 101: 378–384.

de Boer, J., van der Meer, J., Reutergårdh, L. and Calder J. A. 1994. Determination of chloribiphenyls in cleaned-up seal blubber and marine sediment extracts: Interlaboratory study. *Journal of AOAC International* 77: 1411–1422.

Duffield, D.A. and Wells, R.S. 1991. Bottlenose dolphins: Comparison of census data from dolphins in captivity with a wild population. *Soundings* Spring, 1991: 11–15.

Duffield, D.A., Ridgway, S.H. and Cornell, L.H. 1983. Hematology distinguishes coastal and offshore forms of dolphins (*Tursiops*). *Canadian Journal of Zoology* 61: 930–933.

Erickson, K.L., DiMolfetto-Landon, L., Wells, R.S., Reiderson, T., Stott, J.L. and Ferrick, D.A. 1995. Development of an interleukin-2 receptor expression assay and its use in evaluation of cellular immune responses in bottlenose dolphin (*Tursiops truncatus*). *Journal of Wildlife Diseases* 31: 142–149.

Fukushima, M. and Kawai, S. 1981. Variation of organochlorine residue concentration and burden in striped dolphin (*Stenella coeruleoalba*) with growth. In: *Studies on the levels of organochlorine compounds and heavy metals in the marine organisms. Report for the fiscal year of 1980*, T. Fujiyama (ed.). Grant-in-aid for Scientific Research. University of Ryukyus, Okinawa, Japan, pp. 97–114.

Gray, L.E. Jr 1992. Chemical-induced alterations of sexual differentiation: a review of effects in humans and rodents. In: *Chemically Induced Alterations in Sexual and Functional Development: The Wildlife/Human Connection*, T. Colborn and C. Clement (eds). Princeton Scientific Publishing, Princeton, N.J., pp. 203–230.

Guillette, L.J., Crain, D.A., Rooney, A.A. and Pickford, D.B. 1995. Organization versus activation: the role of endocrine-disrupting contaminants (EDCs). *Environmental Health Perspectives* 103: 157–164.

Holden, A.V. and Marsden, K. 1967. Organochlorine pesticides in seals and porpoises. *Nature* 216: 1274–1276.

Jacobson, S.W., Jacobson, J.L., Swartz, P.M. and Fein, G.G. 1983. Intrauterine exposure of human newborns to PCBs: Measures of exposure. In: *Human and Environmental Hazards*, F.M. D'Itri and M. Kamrin (eds). Butterworth, Boston, pp. 311–343.

Jenssen, B.M., Skaare, J.U., Ekker, M., Vongraven, D. and Lorentsen, S.H. 1996. Organochlorine compounds in blubber, liver and brain in neonatal grey seal pups. *Chemosphere* 32: 2115–2125.

Kamolnick, T., Reddy, M.L., Miller, D., Curry, C. and Ridgway, S. 1994. Conditioning a bottlenose dolphin (*Tursiops truncatus*) for milk collection. *Marine Mammals: Public Display and Research* 1: 22–25.

Keith, J.O. 1996. Residue analyses: How they were used to assess the hazards of contaminants to wildlife. In: *Environmental Contaminants in Wildlife: Interpreting Tissue Concentrations*, W.N. Beyer, G.H. Heinz, and A.W. Redmon-Norwood (eds). Lewis Publishers, Boca Raton.

Klamer, J.C., Laane, R.W.P.M. and Marquenie, J.M. 1991. Sources and fate of PCBs in the North Sea: A review of available data. *Water Science and Technology* 24: 77–85.

Kurzel, R.B. and Cetrulo, C.L. 1985. Chemical teratogenesis and reproductive failure. *Obstetrical and Gynecological Survey* 40: 397–424.

Leatherwood, S. and Reeves, R.R. 1990. *The Bottlenose Dolphin.* Academic Press, San Diego, CA.

Marine Mammal Commission 1999. *Marine Mammals and Persistent Ocean Contaminants: Proceedings of the Marine Mammal Commission Workshop, Keystone, Colorado, 12–15 October, 1998.*

Martin, A.R. and Smith, T.G. 1999. Strategy and capability of wild belugas, *Delphinapterus leucas*, during deep, benthic diving. *Canadian Journal of Zoology* 77: 1783–1793.

Martineau, D., De Guise, S., Fournier, M., Shugart, L., Girard, C., Lagacé, A. and Béland, P. 1994. Pathology and toxicology of beluga whales from the St. Lawrence Estuary, Quebec, Canada. Past, present and future. *Science of the Total Environment* 154: 201–215.

Norris, K. 1984. The dolphin science sabbatical. *Whalewatcher* 17–18.

Reddy, M., Echols, S., Finklea, B., Busbee, D., Reif, J. and Ridgway, S. 1998. PCBs and chlorinated pesticides in clinically healthy *Tursiops truncatus*: relationships between levels in blubber and blood. *Marine Pollution Bulletin* 36: 892–903.

Reijnders, P.J.H. 1986. Reproductive failure in common seals feeding on fish from polluted coastal waters. *Nature* 324: 456–457.

Reijnders, P.J.H. 1988. Ecotoxicological perspectives in marine mammalogy: research principles and goals for a conservation policy. *Marine Mammal Science* 4: 91–102.

Ridgway, S.H. 1968. The bottlenosed dolphin in biomedical reseach. In: *Methods of Animal Experimentation*, Volume 3, W.I. Gay (ed.). Academic Press, New York, pp. 387–440.

Ridgway, S.H. and Carder, D.A. 1997. Hearing deficits measured in some *Tursiops truncatus*, and discovery of a deaf/mute dolphin. *Journal of the Acoustical Society of America* 101: 590–593.

Ridgway, S.H. and Reddy, M.L. 1995. Residue levels of several organochlorines in *Tursiops truncatus* milk collected at varied stages of lactation. *Marine Pollution Bulletin* 30: 609–614.

Ridgway, S.H., Carder, D.A., Kamolnick, P.L., Skaar, D.J. and Root, W.A. 1991. Acoustic response times (RTs) for *Tursiops truncatus*. *Journal of the Acoustical Society of America* 89: 1967–1968.

Ridgway, S., Kamolnick, T., Reddy, M., Curry, C. and Tarpley, R. 1995. Orphan-induced lactation in *Tursiops* and analysis of collected milk. *Marine Mammal Science* 11: 172–182.

Romano, T.A., Ridgway, S.H., Felten, D.L. and Quaranta, V. 1999. Molecular cloning and characterization of CD4 in an aquatic mammal, the white whale, *Delphinapterus leucas*. *Immunogenetics* 49: 376–383.

Ross, P.S., De Swart, R.L., Addison, R.F., Van Loveren, H., Vos, J.G. and Osterhaus, A.D. 1996a. Contaminant-induced immunotoxicity in harbour seals: wildlife at risk? *Toxicology* 112: 157–169.

Ross, P.S., De Swart, R.L., Timmerman, H.H., Reijnders, P.J.H., Vos, J.G., Van Loveren, H. and Osterhaus, A.D.M.E. 1996b. Suppression of natural killer cell activity in harbour seals (*Phoca vitulina*) fed Baltic Sea herring. *Aquatic Toxicology* 34: 71–84.

Schantz, M.M., Parris, R.M. and Wise, S.A. 1993. NIST standard reference materials (SRMs) for polychlorinated biphenyl (PCB) determinations and their applicability to toxaphene measurements. *Chemosphere* 27: 1915–1922.

Schantz, M.M., Koster, B.J., Oakley, L.M., Schiller, S.B. and Wise, S.A. 1995. Certification of polychlorinated biphenyl congeners and chlorinated pesticides in a whale blubber standard reference material. *Analytical Chemistry* 67: 901–910.

Schroeder, J.P. and Keller, K.V. 1989. Seasonality of serum testosterone levels and sperm density in *Tursiops truncatus*. *Journal of Experimental Zoology* 249: 316–321.

Schroeder, J.P. and Keller, K.V. 1990. Artificial Insemination of Bottlenosed Dolphins. In: *The Bottlenose Dolphin*, J.S. Leatherwood and R.R. Reeves (eds). Academic Press, San Diego, CA, pp. 447–460.

Scott, M.D., Wells, R.S. and Irvine, A.B. 1990. A long-term study of bottlenose dolphins on the west coast of Florida. In: *The Bottlenose Dolphin*, J.S. Leatherwood and R.R. Reeves (eds). Academic Press, San Diego, CA, pp. 235–244.

Skaare, J.U. 1996. Environmental pollutants in marine mammals from the Norwegian coast and Arctic. *Science of the Total Environment* 186: 25–27.

Sweeney, J.C., Reddy, M.L., Lipscomb, T.P., Bjorneby, J.M. and Ridgway, S.H. 1999. *Handbook of Cetacean Cytology*. Dolphin Quest, San Diego, CA.

Tanabe, S., Tanaka, H. and Tatsukawa, R. 1981a. Ecology and bioaccumulation of *Stenella coeruleoalba*: considerations on the several factors related to the variation of tDDT (p,p'-DDE + p,p'-DDD + p,p'-DDT) residue levels with age of striped dolphins (*Stenella coeruleoalba*). In: *Studies on the Levels of Organochlorine Compounds in Marine Organisms*, T. Fujiyama (ed.). Report for the fiscal Year of 1980. Grant-in-Aid for Scientific Research. University of Ryukyus, Okinawa, Japan, pp. 123–132.

Tanabe, S., Tatsukawa, R., Tanaka, H., Maruyama, K., Miyazaki, N. and Fujiuama, T. 1981b. Distribution and total burdens of chlorinated hydrocarbons in bodies of striped dolphins (*Stenella coeruleoalba*). *Agricultural and Biological Chemistry* 45: 2569–2578.

Wells, R.S. and Scott, M.D. 1999. Bottlenose Dolphin – *Tursiops truncatus* (Montagu, 1821). In: *Handbook of Marine Mammals, Vol. 6, The Second Book of Dolphins and the Porpoises*, S.H. Ridgway and R.J. Harrison (eds). Academic Press, London, pp. 137–182.

Wise, S.A., Schantz, M.M., Koster, B.J., Demiralp, R., Mackey, E.A., Greenverg, T.T., Burow, M., Ostapczuk, P. and Lillestolen, T.I. 1993. Development of frozen whale blubber and liver reference materials for the measurement of organic and inorganic contaminants. *Fresenius Journal of Analytical Chemistry* 345: 270–277.

Part II

An overview of contamination of marine mammals and their environment

6 Persistent ocean contaminants and marine mammals: A retrospective overview

Thomas J. O'Shea and Shinsuke Tanabe

Introduction

The study of marine mammal toxicology is a relatively recent area of research. This chapter provides a chronological summary of advances in knowledge about marine mammals and persistent ocean contaminants, beginning with the 1940s. We summarize major findings and advances for each decade in the context of related events in environmental policy, toxicology and technology. We provide indices of the cumulative amount of information on contaminants and marine mammals reached by the end of each decade, based on our own compilations from the literature and those in other recent reviews (e.g. Fossi and Marsili, 1997; O'Shea, 1999). We also illustrate progress made in the application of the scientific method to problems in the field. This overview does not consider impacts of oil on marine mammals, which has been a topic treated in depth in other sources (Geraci and St. Aubin, 1990; Loughlin, 1994).

Methods

This chapter is a literature review emphasizing the history of studies of persistent ocean contaminants and marine mammals. Our objectives are to provide an overview of major developments in the field during the past half century. As part of this overview we provide arithmetic summaries of information on numbers and groups of marine mammals sampled for contaminant residues and biomarkers each decade, and also illustrate growth in the numbers and the kinds of contaminants reported in tissues of marine mammals. These summaries are based on data gleaned from: 289 references listed in appendices to the recent summary by O'Shea (1999), generally extending through 1996; over 40 references on biomarkers summarized by Fossi and Marsili (1997); and an additional 90 recent references from primary sources extending through most of 1998 (listed below). (The large number of published references in just the past 2 years attests to the growing interest in the field of marine mammal toxicology.) We do not claim that numbers we tabulate are absolute totals, because we undoubtedly have failed to review all published reports in existence. However, we consider these totals to be good relative indices of the amount of research carried out for a particular time period. Full citations for the 289 papers summarized through 1996 can be found in O'Shea (1999) and references on biomarkers in Fossi and Marsili (1997). The more recent papers not summarized in those sources are: Aono *et al.* (1997), Beck *et al.* (1997), Becker *et al.* (1997), Beckmen *et al.* (1997), Bernhoft *et al.* (1997), Borrell *et al.* (1997), Calambokidis and Barlow (1991), Cardellichio (1995), de Moreno *et al.* 1997), Dietz *et al.* (1996, 1998), Estes *et al.* (1997), Fossi *et al.* (1997a, b), Gauthier *et al.* (1997a, b, 1998a, b), Gerpe *et al.* (1990), Green *et al.* (1996), Hall *et al.* (1998), Hayteas and Duffield (1997), Holsbeek *et al.* (1998), Hyvärinen *et al.* (1998), Iwata *et al.* (1997, 1998), Jarman *et al.* (1996), Kannan and Falandysz (1997), Kannan *et al.* (1996, 1997, 1998), Kiceniuk *et al.* (1997), Kim *et al.* (1996a, b), Kleivane and Skaare (1998), Koistinen *et al.* (1997), Klobes *et al.* (1998), Krahn *et al.* (1997), Law *et al.* (1997, 1998), Letcher *et al.* (1996, 1998), Loewen *et al.* (1998), Madhusree *et al.* (1997), Marsili *et al.* (1997a, b, 1998), Mathieu *et al.* (1997), McKenzie *et al.* (1997), Medvedev *et al.* (1997), Monaci *et al.* (1998), Moreno (1994), Moreno *et al.* (1984), Mössner and Ballschmiter (1997), Nakata *et al.* (1997, 1998a, b), Norstrom *et al.* (1998), Oehme *et al.* (1996), Outridge *et al.* (1997), Parsons (1998), Peña *et al.* (1988), Prudente *et al.* (1997), Rogan and Berrow (1996), Roots and Talvari (1997), Schantz *et al.* (1996), Sepulveda *et al.* (1997), Storelli *et al.* (1998), Strandberg *et al.* (1998), Sydeman and Jarman (1998), Szefer *et al.* (1998), Tanabe *et al.* (1997a, b, 1998), Tarasova *et al.* (1997), Tilbury *et al.* (1997), Tirpenou *et al.* (1998), Troisi and Mason (1997), Troisi *et al.* (1998), Vetter *et al.* (1996), Wade *et al.* (1997), Wagemann *et al.* (1996, 1998), Watanabe *et al.* (1996, 1998), Weis and Muir (1997), Westgate *et al.* (1997), Wiberg *et al.* (1998), Wolkers *et al.* (1998), Young *et al.* (1998), and Zitko *et al.* (1998).

The 1940s and 1950s

During this period much of the world was involved with and recovering from massive military conflicts. Experience in mobilization of technology to meet these conflicts was followed by great faith in the applications of research in chemistry and physics to advance global social and economic development. Tremendous growth in chemical industries occurred in the decades immediately following the Second World War (Arora and Rosenberg, 1998; Hikino *et al.*, 1998). As examples pertaining to impacts on marine mammals, production of industrial polychlorinated biphenyls (PCBs) escalated markedly (de Voogt and Brinkman, 1989), and the era of large-scale use of organochlorine pesticides was also ushered in during this period; DDT was released from restricted military applications to general public use in the US in 1945 (Linduska, 1952). Worldwide expansion in the production and use of other organochlorines also began by the late 1940s and increased through the 1950s and 1960s (Stickel, 1968; Keith, 1991). The importance of such developments at the time is underscored by the awarding of a Nobel Prize in 1948 for discovery of the insecticidal properties of DDT, a compound which saved millions of people from death due to insect-borne pathogens (Metcalf, 1973). However, DDT and metabolites were to be found in blubber of marine mammals in remote reaches of the world, beginning nearly 20 years later, and this chemical was subsequently to become one of our most intensively studied persistent contaminants.

The widespread use of such compounds had effects that were not predicted at the time of their introduction. None the less, population impacts of DDT and dieldrin in some groups of terrestrial and freshwater vertebrates were apparent from field surveys of mortality beginning in the 1940s (e.g. Linduska and Surber, 1948; Scott *et al.*, 1959). In subsequent decades knowledge of direct mortality and reproductive effects, especially in birds, grew from data gathered in several parts of the world. Concurrently, studies of common laboratory mammals in the late 1940s and early 1950s gave the first evidence of the accumulation of organochlorines in fatty tissues, their transfer through milk during lactation, and their estrogenic activity (Keith, 1991). With the exception of radionuclide fallout from use and testing of nuclear weapons, however, there was little widespread public recognition or concern about environmental contaminants during this period, particularly with regard to marine mammals. The scientific literature of the time is absent of information on contaminants in marine mammals, although retrospective studies of commercial oils show that seals and whales contained metabolites of DDT in the 1950s (Addison *et al.*, 1972), and that mercury was present at low levels in Southern Ocean fin whales (*Balaenoptera physalis*) in the 1940s (Nagakura *et al.*, 1974).

The 1960s

This decade witnessed a broad awakening of concern about the impacts of persistent environmental contaminants on human health as well as on wildlife and ecosystems, eventually leading to the first studies of the topic in marine mammals. In Japan, local effects on human health from contamination of sea foods with mercury at Minamata and Niigata Bays (Hammond, 1971) attracted worldwide attention, as did human illnesses resulting from contamination of rice oil with PCBs and polychlorinated dibenzofurans (PCDFs) in Japan (Yusho disease) at around the time PCBs were first recognized in environmental samples in 1966 (de Voogt and Brinkman, 1989). Direct effects of DDT in avian mortality were confirmed by establishment of diagnostic lethal concentrations in brains through controlled experiments in captivity (Stickel *et al.*, 1966). Evidence of indirect effects of DDT and metabolites on reproduction in birds was supported by strong correlations with eggshell thinning and reproductive failure in field studies (e.g. Ratcliffe, 1967), confirmed by direct experimental evidence from captive feeding studies (e.g. Wiemeyer and Porter, 1970).

In 1966 the first publications on contaminants in marine mammals reported DDT and metabolites in seals from the Antarctic (George and Frear, 1966; Sladen *et al.*, 1966), and The Netherlands (Koeman and van Genderen, 1966), providing the initial documentation of this long-lasting issue. By the end of the decade, the presence of mercury, DDT, DDE [2,2-bis-(*p*-chlorophenyl)-1,1-dichloroethylene, the principal metabolite of DDT], DDD [2,2-bis-(*p*-chlorophenyl)-1,1-dichloroethane, an additional metabolite of DDT], dieldrin, and PCBs (Holden and Marsden, 1967; Jensen *et al.*, 1969) had been documented in tissues of marine mammals, with published organochlorine data based on about 52 individual pinnipeds in six species, four harbor porpoises (*Phocoena phocoena*) and one common dolphin (*Delphinus delphis*). Mercury was the only toxic element reported in marine mammals in the 1960s, based primarily on examination of 32 ringed seals (*Pusa hispida*) from Lake Saimma, Finland (Helminen *et al.*, 1968; Henriksson *et al.*, 1969). In addition to residue determinations for organochlorines and mercury, a few marine mammals were examined for radionuclides (Osterberg *et al.*, 1964; Holtzman, 1969). Although the findings of contaminants in marine mammals raised hypotheses about possible effects, no detailed follow-up investigations or experiments had taken place during this decade.

The 1970s

The 1970s were times of much formal worldwide recognition of the dangers of chemical contamination of the environment, centered primarily on human health concerns, and were marked by many advances in related legislation and policy. Examples in the US included formation of the Environmental Protection Agency in 1970, banning of the use of DDT in 1972, and passage of the National Environmental Policy Act in 1975 and the Toxic Substances Control Act in 1976. Recognition of the widespread contamination of the environment by PCBs resulted in dramatic drops in production during this decade. Total PCB production in France, Germany, Italy, Spain, UK, Japan and the US dropped from 235 000 metric tons the first half of the decade to 121 000 in the second half (de Voogt and Brinkman, 1989). In Japan, where PCB production began in 1954, manufacture and import of these chemicals was terminated in 1972; in the US manufacture of PCBs was voluntarily ended in 1977 (de Voogt and Brinkman, 1989).

Research on contamination of marine mammals expanded. The number of potentially toxic elements reported for marine mammal tissues increased to 18 (Table 6.1), and data on elements were reported for at least 1410 individual marine mammals, including 10 species of pinnipeds and 17 species of cetaceans (Table 6.2). Reports of the occurrence of lead and of very high concentrations of cadmium and mercury were made for marine mammals in various parts of the world (Braham, 1973; Anas, 1974a; Buhler *et al.*, 1975; Smith and Armstrong, 1975; Martin *et al.*, 1976; Roberts *et al.*, 1976; Duinker *et al.*, 1979). However, associated pathology was not observed, and it was

Table 6.1 Potentially toxic elements reported in tissues of marine mammals by decade

1960s	1970s					1980s					1990s				
											MBT	DBT	TBT	DPT	TPT
Hg	As	Ag	Br	Cd	Cu	As	Ag	Ba	Be	Br	Ag	Al	As	Au	B
	Cr	Ca	Fe	Hg	K	Ca	Cd	Co	Cr	Cu	Ba	Be	Br	Ca	Cd
	Mg	Mn	Na	Pb	Sb	Fe	Hg	K	Mg	Mn	Ce	Cl	Co	Cr	Cs
	Se	Ti	Zn			Mo	Na	Ni	P	Pb	Cu	Eu	Fe	Ha	Hf
						Sb	Se	Sr	Ti	V	Hg	I	K	La	Li
						Zn	Zr				Mg	Mn	Mo	Na	Ni
											P	Pb	Rb	Sa	Sb
											Sc	Se	Sm	Sn	Sr
											Ta	Tb	Th	Ti	Tl
											U	V	W	Zn	Zr

Abbreviations for organotin compounds: MBT, monobutyltin; DBT, dibutyltin; TBT, tributyltin; DPT, diphenyltin; TPT, triphenyltin.

Table 6.2 Approximate numbers of marine mammals analyzed for toxic elements by decade, based on the literature reviewed for this chapter

	1960s		1970s		1980s		1990s		Total	
	Species	Individuals	Species	Individuals	Species	Individuals	Species	Individuals	Species	Individuals
Pinnipeds	1	32	10	1144	12	718	23	1945	25	3839
Cetaceans			17	264	15	622	38	4551	41	5437
Others			1	2	3	431	3	537	4	970
Total	1	32	28	1410	30	1771	64	7033	70	10 246

'Others' represent sirenians, sea otters, and the polar bear.

discovered that marine mammals have mechanisms that detoxify some of these naturally occurring substances. These include metallothionein proteins which sequester cadmium (Olafson and Thompson, 1974; Lee *et al.*, 1977), and a protective action of selenium against mercury in the liver (Arima and Nagakura, 1979; Koeman *et al.*, 1973, 1975; van de Ven *et al.*, 1979). The only toxicological experiments ever conducted on captive marine mammals with toxic elements were carried out by exposing a very small number of seals to mercury to determine excretion rates (Tillander *et al.*, 1972), possible hearing organ pathology (Ramprashad and Ronald, 1977), tissue distribution (Freeman *et al.*, 1975; van de Ven *et al.*, 1979) and pathology and concentrations in tissues associated with lethal exposure (Ronald *et al.*, 1975). *In vitro* studies of steroid synthesis, enzymatic demethylation and P450 enzymes were also carried out on seals (Freeman *et al.*, 1975; Freeman and Sangalang, 1977; van de Ven *et al.*, 1979).

Organochlorines were reported for 1230 marine mammals during the 1970s, including 11 species of pinnipeds and 26 species of cetaceans (Table 6.3). The list of organochlorines reported in marine mammals expanded to about 30 compounds or groups of compounds (Tables 6.4 and 6.5). Organochlorine residue studies began to show patterns of accumulation with sex and age, geographic trends and 'hot spots', and characteristic statistical distributions wherein most values were low or moderate (except in a few highly contaminated locales), with a few more heavily contaminated individuals. Organochlorine concentrations in blubber of males typically increased throughout life, whereas females reached a plateau or actually showed decreasing concentrations with age, beginning at the age of first reproduction (Figure 6.1).

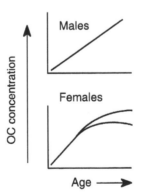

Figure 6.1 Graphical model of the differences in the dynamics of accumulation with age documented for some organochlorines in blubber of male and female marine mammals. Concentrations of some organochlorines in blubber tend to plateau or decrease at reproductive maturity in females, due to transfer to offspring, primarily via milk. Females that fail to reproduce and lactate, regardless of cause, can have patterns of accumulation with age that are similar to those in males.

Table 6.3 Approximate numbers of marine mammals analyzed for organochlorine contaminant residues by decade, based on the literature reviewed for this chapter

	1960s		1970s		1980s		1990s		Total	
	Species	Individuals	Species	Individuals	Species	Individuals	Species	Individuals	Species	Individuals
Pinnipeds	6	52	11	921	12	1143	20	1870	26	3986
Cetaceans	2	5	26	255	33	1364	41	2741	53	4365
Others			4	54	2	187	3	1086	4	1327
Total	8	57	41	1230	47	2694	64	5697	83	9678

'Others' represent sirenians, sea otters and the polar bear.

Table 6.4 Organochlorine pesticides and metabolites reported in tissues of marine mammals by decade, based on the literature reviewed for this chapter

1960s	1970s	1980s	1980s (cont.)	1990s	1990s (cont.)
Dieldrin	Aldrin	HCBs	Aldrin	p,p′-DDE	Tox 50
p,p′-DDE	Chlordanes	HCHs	Dieldrin	p,p′-DDD	Aldrin
p,p′-DDD	Dieldrin	Heptachlor	Endrin	p,p′-DDT	Dieldrin
p,p′-DDT	p,p′-DDE	Heptachlor epoxide	Mirex	o,p′-DDE	Endrin
∑ DDT	p,p′-DDD	Lindane	HCHs	o,p′-DDD	Lindane (γ-BHC)
	p,p′-DDT	Mirex	α-HCH	o,p′-DDT	Methoxychlor
	o,p′-DDE	Toxaphene	β-HCH	∑ DDT	Mirex
	o,p′-DDD	Methylsulfones of DDE	γ-HCH	3-MeSO2-p, p′- DDE	CBz
	o,p′-DDT	DDMU	CBz	HCHs	1,2,3,4-tetrachlorobenzene
	∑ DDT	DdP	Pentachlorobenzene	α-HCH	1,2,4,5-tetrachlorobenzene
	DDE phenolics	DDA	HCB	β-HCH	Pentachlorobenzene
	α-HCH	trans-Nonaclor	C	γ-HCH	HCB
	β-HCH	Oxychlordane	C1	δ-HCH	C
	Nonachlor	trans-Chlordane	C2	Chlordane	C1A
		cis-Chlordane	C3	Oxychlordane	C2/U5
		trans-Nonachlor	C4	trans-Chlordane	C3
		cis-Nonachlor		cis-Chlordane	C5
		2-Chlorochlordene		Photo-cis-Chlordane	U1
		Heptachlor		trans-Nonachlor	U3
		Heptachlor epoxide		cis-Nonachlor	U81
		Photoheptachlor		Heptachlor	U82
		Nonachlor – III (MC6)		Heptachlor epoxide	U83
		Toxaphene		Photoheptachlor	B7-1453
				Nonachlor – III (MC6)	B8-2229
				MC4	B8-1412
				MC5	B8-1413
				MC6	B8-1414
				Toxaphene	B8-1945
				Tox8/T2 (Parlar 26)	B9-1679
				Tox9/T2 (Parlar 50)	CHB32
					CHB62
					TS2 (Parlar 39) B8-531

Table 6.5 Industrial organohalogens and metabolites reported in tissues of marine mammals by decade, based on the literature reviewed for this chapter

1960s	1970s	1980s	1990s
PCBs	Methylsulfones of PCBs	TCP	EOCls
	PCBs	PCP	EOBrs
	PCB phenolics	PCBs	EOIs
	PCTs	PCB16/32	TCP
		PCB17/18	tris(4-Chlorophenyl)-methanol
		PCB28	TCP-Methane
		PCB31/28	Octachlorostyrene
		PCB41/71	PBBs
		PCB44	PBDPEs
		PCB47	PCNs
		PCB49	PCBs
		PCB52	PCB8
		PCB56/60	PCB18
		PCB66/95	PCB26
		PCB70/76	PCB28
		PCB74	PCB31
		PCB77	PCB37
		PCB79	PCB40
		PCB80	PCB41/71/64
		PCB84/89	PCB42
		PCB85	PCB44
		PCB87	PCB46
		PCB89	PCB47
		PCB91	PCB49
		PCB95/66	PCB52
		PCB97	PCB56/60
		PCB99	PCB58/74
		PCB101	PCB60
		PCB105	PCB61/74
		PCB106	PCB66
		PCB108	PCB70
		PCB110	PCB74
		PCB114	PCB77
		PCB118	PCB81
		PCB126	PCB82/151
		PCB128	PCB83
		PCB129	PCB84/89
		PCB132	PCB85
		PCB135	PCB87
		PCB136	PCB90/101
		PCB137	PCB91
		PCB138	PCB95
		PCB141/179	PCB97
		PCB146	PCB99
		PCB147	PCB101
		PCB148	PCB105
		PCB149	PCB107
		PCB151	PCB110
		PCB153	PCB114
		PCB156	PCB118
			PCB119
			PCB120
			PCB126
			PCB128
			PCB129

PCB157	2,3,7,8-TCDF	PCB130/176	PCB179
PCB158	1,3,7,8-TCDF	PCB131	PCB180
PCB161	1,3,4,7-TCDF	PCB132	PCB182
PCB163	1,2,4,7-TCDF	PCB134	PCB183
PCB169	1,2,3,7-TCDF	PCB135/144	PCB185
PCB170	1,6,7,8-TCDF	PCB136	PCB187
PCB172	1,2,3,8-TCDF	PCB137	PCB187/182
PCB173	1,3,4,9-TCDF	PCB138	PCB188
PCB174	1,4,6,9-TCDF	PCB138/163/164	PCB189
PCB177	2,4,6,7-TCDF	PCB141	PCB191
PCB178	1,2,3,9-TCDF	PCB146	PCB193
PCB179	2,3,6,7-TCDF	PCB148	PCB194
PCB180	1,2,4,7,8-PeCDF	PCB149	PCB195
PCB183	1,2,4,8,9-PeCDF	PCB151	PCB196/203
PCB185	1,2,3,7,8-PeCDF	PCB153	PCB198
PCB187	2,3,4,7,8-PeCDF	PCB154	PCB199
PCB194	1,2,4,6,8-PeCDF	PCB156	PCB200
PCB196/203	1,3,4,7,9-PeCDF	PCB157	PCB201
PCB199	2,3,4,6,8-PeCDF	PCB158	PCB202
PCB201	1,2,6,7,8-PeCDF	PCB159	PCB203
PCB206	1,2,3,4,9-PeCDF	PCB169	PCB205
PCB209	1,2,3,4,7,8-HxCDF	PCB170	PCB206
PCDDs	1,2,3,6,7,8-HxCDF	PCB170/190	PCB207
OCDD	2,3,4,6,7,8-HxCDF	PCB171	PCB208
PCDFs	1,2,3,7,8,9-HxCDF	PCB172	PCB209
		PCB174	ΣMeSO$_2$-PCBs
		PCB175	3-MeSO$_2$-PCB
		PCB177	4-MeSO$_2$-PCB
		PCB178	4-MeSO$_2$-2,2',4',5,5'-pentachlorobiphenyl

Table 6.5 (cont'd)

1960s	1970s	1980s	1990s
		1,2,4,6,8,9-HxCDF	3-MeSO$_2$-2,2',4',5,5'-pentachlorobiphenyl
		1,2,4,6,7,9-HxCDF	3-MeSO$_2$-2,2',4',5-tetrachlorobiphenyl
		1,2,3,4,7,9-HxCDF	4-MeSO$_2$-2,2',4',5-tetrachlorobiphenyl
		1,2,3,4,6,8-HxCDF	3-MeSO$_2$-2,2',3',4',5-pentachlorobiphenyl
		1,2,3,4,6,9-HxCDF	4-MeSO$_2$-PCB174
		1,2,3,6,8,9-HxCDF	3-MeSO$_2$-PCB174
		1,2,3,4,6,8,9-HpCDF	+ 16 Methyl sulfones
		1,2,3,4,6,7,8-HpCDF	PCDEs
		1,2,3,4,7,8,9-HpCDF	PCDE47
		OCDF	PCDE85
			PCDE99
			PCDE100
			PCDE118
			PCDE137
			PCDE138
			PCDE140/167
			PCDE147/153
			PCDE154
			PCDE170
			PCDE180/181
			PCDE182
			PCDE184
			PCDE187
			PCDE190
			PCDE194
			PCDE195
			PCDE196
			PCDE197
			PCDE201
			PCDE203
			PCDE204
			PCDE206
			PCDDs
			2,3,7,8-TCDD
			1,2,3,7,8-PeCDD
			1,2,3,4,7,8-HxCDD
			1,2,3,6,7,8-HxCDD
			1,2,3,7,8,9-HxCDD
			1,2,3,4,7,8,9-HpCDD
			1,2,3,4,6,7,8-HpCDD
			1,2,3,4,6,7,9-HpCDD
			OCDD
			PCDFs
			2,3,7,8-TCDF
			1,2,4,7,8-PeCDF
			1,2,4,8,9-PeCDF
			1,2,3,7,8-PeCDF
			2,3,4,7,8-PeCDF
			1,2,3,4,7,8-HxCDF
			1,2,3,6,7,8-HxCDF
			2,3,4,6,7,8-HxCDF
			1,2,3,7,8,9-HxCDF
			1,2,4,6,8,9-HxCDF
			1,2,3,6,8,9-HxCDF
			1,2,3,4,6,8,9-HpCDF
			1,2,3,4,6,7,8-HpCDF
			1,2,3,4,7,8,9-HpCDF
			1,2,3,4,6,7,9-HpCDF
			OCDF

Once reproductively active, females had the opportunity to excrete these compounds to their young through lactation, whereas males did not. These patterns were demonstrated in seals (e.g. Frank *et al.*, 1973; Addison and Smith, 1974; Helle *et al.*, 1976a, b; Addison and Brodie, 1977) and the harbor porpoise (Gaskin *et al.*, 1971), and were later verified with many other species in the 1980s and 1990s. Extraordinarily high contamination of marine mammals with organochlorines was noted in some coastal environments, most notably the Baltic Sea and southern California (e.g. LeBouef and Bonnell, 1971; Helle *et al.*, 1976a).

Two important sets of observations were made during the 1970s that suggested possible reproductive effects of organochlorines on females in wild populations of California sea lions (*Zalophus californianus*) (DeLong *et al.*, 1973) and ringed and gray seals (*Halichoerus grypus*) in the Baltic Sea (e.g. Helle, 1976a, b). California sea lions had a high incidence of premature pups, and those which successfully raised young had lower DDE concentrations in blubber. The sea lion observations had complications to their interpretation, including the presence of abortion-causing disease agents, as noted in various subsequent findings (reviewed by Addison, 1989; O'Shea and Brownell, 1998). Interpretation of this situation, and the observations of higher organochlorines in Baltic Sea female ringed seals with uterine occlusions and stenosis, also suffered from the problem that non-reproductive females will have higher organochlorines regardless of cause because they cannot excrete these chemicals through lactation (Figure 6.1). Thus interpretations of cause and effect were frustrating.

The scientific method that underlies most fields of modern research usually matures from early stages in which observations are made and questions perceived, to phases wherein hypotheses are generated and tested to form a body of theory and principles. Hypothesis-testing in wildlife toxicology typically proceeds through gathering data in controlled experiments (as was done regarding DDT/DDE and avian reproduction and mortality in the 1960s), or by carefully designed field sampling and epidemiological approaches. By the close of the 1970s, the science of marine mammal toxicology had generated much observational data, leading to many questions, but formal hypothesis-testing had been very limited. Other areas of research in toxicology of marine organisms which had developed to experimental stages by this time showed that impacts of contaminants followed typical dose–response relationships: with increasing exposure organisms progressed through stages of normal adjustment with no effects, physiological compensation with detectable but subtle effects (e.g. biomarkers), to stages with major disturbance in function and ultimately death at high exposure (Sindermann, 1995). Such relationships had not been established for any substance and marine mammals, but observations of exceptionally high concentrations of contaminants in tissues engendered much speculation that stages of major disturbances to normal physiological and perhaps population-level function had been breached.

The 1980s

Continued changes in policy towards the use of some persistent contaminants progressed in the 1980s. China, for example, was the world's largest producer and user of hexachlorocyclohexane (HCH), but banned its production in 1983 (Li *et al.*, 1998). In the study of marine mammal toxicology, this decade was marked by:

1   landmark experimental approaches with captive harbor seals (*Phoca vitulina*);
2   more sophisticated analytical technologies allowing quantification of individual PCB congeners and corresponding hypotheses about metabolic capacities of marine mammals, and the possible vulnerability of their endocrine physiology;
3   application of biomarkers to the study of contaminants in marine mammals;
4   formation of hypotheses on adrenocortical function impairment due to organochlorines; and
5   an ever-expanding set of information on chemical residues in the bodies of marine mammals (primarily organochlorines in blubber and elements in livers and kidneys).

In the first toxicological experiment with marine mammals and organochlorines, Reijnders (1986) showed that reproduction in a group of 12 captive female harbor seals was impaired when the diet consisted of fish with higher concentrations of mixed organochlorines, in contrast to a group of 12 controls fed fish low in organochlorines. Reproductive failure was thought to occur at the implantation stage, and retinol and thyroid hormone concentrations were lowered (Reijnders, 1986; Brouwer *et al.*, 1989).

Tanabe (1988), Tanabe *et al.* (e.g. 1987, 1988), and Boon *et al.* (e.g. 1987, 1992) pioneered congener-specific quantification of PCBs in marine mammals during the 1980s, and showed that, due to structural configurations, certain PCB congeners (e.g. PCB 153) were especially recalcitrant to metabolism in marine mammals (but with differences among species) based on comparison of residue profiles in food and terrestrial mammals, and this was linked with possible enzyme-induction deficiencies in mixed-function oxidase (MFO) systems and possible endocrinological effects. Polar bears (*Ursus maritimus*), in contrast, had congener profiles suggestive of a remarkable capacity to metabolize PCBs (Norstrom *et al.*, 1988; Norstrom and Muir, 1994). Biomarker studies began to develop in the latter half of this decade, with some form of biomarker information (including P450 and MFO activities) gathered on at least 83 individuals in eight species of cetaceans and 90 individuals in five species of pinnipeds (Fossi and Marsili, 1997). Most of these studies, however, did not link enzyme activities with organochlorine concentrations in blubber. A strong hypothesis on the role of organochlorines as interfering with adrenocortical metabolism, supported by pathological observations, was also established during this decade, based on observations of Baltic seals that included reproductive tract lesions, adrenocortical hyperplasia, hyperkeratosis, and claw deformations, some of which paralleled captive PCB-feeding experiments on other mammals (Bergman and Olsson, 1985). The first reports of elevated PCB residues in belugas (*Delphinapterus leucas*) in the St. Lawrence River and estuary in Canada were also associated with reproductive pathology (Martineau *et al.*, 1987, 1988). A modest correlation ($r^2 = 0.44$) was observed between testosterone levels in the blood and DDE concentrations in blubber of 12 male Dall's porpoises (*Phocoenoides dalli*), but this correlation was not significant for PCBs (Subramanian *et al.*, 1987).

Much data on residues in tissues accrued in the 1980s, and many more extensive studies verified strong patterns of accumulation of organochlorines by sex and age across species for some compounds. Notable examples include studies of ringed seals (Muir *et al.*, 1988b), harbor porpoises (Gaskin *et al.*, 1982, 1983), Antarctic minke whales (*Balaenoptera bonaerensis*) (Tanabe *et al.*, 1986) and fin whales (Aguilar and Borrell, 1988). Observational data on organochlorine contaminant residues continued to overshadow experimental investigations, with our tabulations for the 1980s embracing another 2700 individuals in 47 species of marine mammals (Table 6.3). With advances in analytical chemistry allowing capability for congener determination, the number of organochlorines found in marine mammals (usually in blubber and often at very low concentrations) expanded greatly to at least 148, and included the polychlorinated benzodioxins (PCDDs) and PCDFs (Tables 6.4 and 6.5). Toxic element determinations were carried out on tissues of about 1770 individuals in 30 species (Table 6.2), and expanded to include at least 27 elements (Table 6.1).

The 1990s

In the decades between the Second World War and the present, the chemical industry grew dramatically. It has become the largest manufacturing industry in the United States, with 40 per cent of the global market shared by the US and Japan (Arora *et al.*, 1998). The list of chemical substances registered by the American Chemical Society had also grown phenomenally and numbered nearly 20 million by the end of this decade (Figure 6.2). Attention in studies of marine mammals, however, continued to focus on the organochlorines. General environmental concerns in the 1990s widened to embrace the concept of endocrine disruptors (Colborn and Clement, 1992; Reijnders, Chapter 3 in this volume), which included several organochlorines that had been repeatedly found in marine mammal tissues over the previous two decades. Policy and legislation continued to develop internationally, including an announcement by Mexico in 1997 of a plan to phase out the use of DDT and chlordane. Concern and debate about the influence of

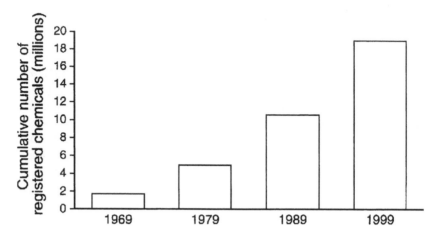

Figure 6.2 Cumulative numbers of chemical substances registered by the Chemical Abstracts Service (1999) at or near the end of each decade, 1969–1999 (1999 data are through 4 April).

organochlorines on disease processes in marine mammals came to the fore-front early in this decade, on the heels of highly publicized die-offs of seals and small cetaceans, and this concern was coupled with additional focus on the possible importance of endocrine disruptors on marine mammal population dynamics. Although the amount of information on contaminants in tissues of marine mammals grew at an accelerating rate in the 1990s, the decade is highlighted by increased efforts at hypothesis-testing.

New reports on beluga whales in the highly polluted St. Lawrence River (e.g. Béland *et al.*, 1993; Martineau *et al.*, 1994) and pinnipeds in the Baltic (e.g. Bergman *et al.*, 1992) showed various lesions (including skull bone asymmetry in seals) strongly suggestive of contaminant impacts. Studies designed to further elucidate relationships between organochlorines and the adrenocortical function hypothesis in harbor porpoises, however, were not conclusive but suggested that in this case adrenal hyperplasia was generally related to chronic causes of death such as disease and starvation rather than organochlorine exposure (Kuiken *et al.*, 1993). Examination of harbor porpoises found dead due to various diseases, in comparison with those killed by trauma in fisheries, showed no significant differences in PCB concentrations (Kuiken *et al.*, 1994). Young northern elephant seals (*Mirounga angustirostris*) on the Pacific coast of the USA, however, were reported with a skin disease syndrome that may be related to organochlorine exposure, with ΣDDT in serum and blubber and PCBs in serum of diseased seals significantly higher than in normal seals (Beckmen *et al.*, 1997).

Die-offs of bottlenose dolphins (*Tursiops truncatus*) on the Atlantic Coast of the US in the late 1980s (Geraci, 1989; Lipscomb *et al.*, 1994; Duignan *et al.*, 1996) were followed by additional dramatic events involving pinnipeds and small cetaceans in Europe and the Gulf of Mexico during the late 1980s and 1990s (Domingo *et al.*, 1990; Lipscomb *et al.*, 1996; Taubenberger *et al.*, 1996). Studies of these events included sampling designs that statistically partitioned cause of death and organochlorine residue data among variables such as age and sex in order to test hypotheses about linkages with disease and contaminants. Results were sometimes conflicting. Hall *et al.* (1992) designed an analysis comparing organochlorine concentrations in blubber of harbor seals which succumbed to a morbillivirus epidemic in 1988 with survivors sampled by biopsy in 1989, but determined that 'data are not sufficient to conclude that there was a direct link between mortality from PDV infection and OC contamination'. Aguilar and Borrell (1994), however, reported much higher PCBs in striped dolphins (*Stenella coeruleoalba*) that succumbed to a morbillivirus outbreak in the Mediterranean in comparison with individuals sampled before and after the outbreak. Other studies were unable to confirm a relationship between organochlorine concentrations and morbillivirus outbreaks in seals (Blomkvist *et al.*, 1992; Kendall *et al.*, 1992), or in experimental susceptibility of seals with high or low PCB contamination to dosing with cell culture phocine distemper virus (Harder *et al.*, 1992). Die-offs associated with morbillivirus infections were very dramatic and attracted much attention and speculation about links with contaminants, but such infections are typically associated with a high mortality in naïve mammalian populations even when contaminants are not present.

Other milestones in hypothesis-testing were achieved during this decade in an experiment on immune function and organochlorines in the diet of captive harbor seals (Ross *et al.*, 1995, 1996a, b; de Swart *et al.*, 1993, 1994, 1996), and a creative field experiment on mother–pup pairs of gray seals. In the captive study, vitamin A levels were lower in the group of seals fed a herring diet about five times higher in PCBs and DDTs than in the reference group. Counts of total white blood cells and granulocytes (but not lymphocytes or monocytes) were higher, and natural killer cell activity and lymphocyte function assays after stimulation with mitogens were significantly lower in the seals with higher dietary organochlorine exposure (see Vos *et al.*, Chapter 21, and Ross *et al.*, Chapter 20, in this volume for greater details on results and conclusions from these experiments). Hall *et al.* (1997) measured PCBs in the milk of wild gray seal mothers, and challenged their pups with morbillivirus vaccines and stimulated them with mitogens to determine immunocompetence. No relationships were found between the prevalence of infection in pups (as an indicator of possible immunosuppression) and cumulative exposure to PCBs in mothers' milk. Pups exposed to higher PCBs in milk did not show any biochemical, hematological or immunological abnormalities. *In vitro* study of mitogen-induced proliferation responses of lymphocyte cultures and organochlorines in the blood of five free-ranging bottlenose dolphins showed an inverse correlation between these immune responses and organochlorine concentrations (Lahvis *et al.*, 1995).

Biomarker studies continued to increase in the 1990s, with some form of biomarker sampling published for at least 184 cetaceans in nine species and 408 pinnipeds in ten species (Fossi and Marsili, 1997 and references noted in the methods therein). Biomarker studies began to include organochlorine determinations from biopsies or tissue samples from the same individuals, and results generally showed a positive correlation with MFO activity (e.g. Letcher *et al.*, 1996; Fossi *et al.*, 1997a, b; Troisi and Mason, 1997; Boon *et al.*, 1998; Marsili *et al.*, 1998; Wolkers *et al.*, 1998). Other observation-level data continued to mount during this decade, with analytical chemistry techniques continuing to expand the numbers of individual compounds (especially industrial chemical congeners) and elements reported in marine mammal tissues at even finer levels of sensitivity. During this decade at least 5695 marine mammals in 64 species were examined for organochlorine residues (Table 6.3), including determinations of well over 200 compounds (Tables 6.4 and 6.5). Some 50 toxic elements were reported in marine mammals in the 1990s, based on surveys of over 7000 individuals in 64 species (Tables 6.1 and 6.2). The latter represents a doubling of the cumulative historical number of samples in just 9 years. Additionally, towards the close of the 1990s hypothesis-testing had broadened to include more sophisticated *in vitro* laboratory studies on specific contaminants and various physiological and immunological response functions in marine mammal tissues and cell cultures (e.g. De Guise *et al.*, 1996, 1998; Gauthier *et al.*, 1998b).

A few reports were published on polycyclic aromatic hydrocarbons (PAHs) in marine mammals during the 1990s (Hellou *et al.*, 1990, 1991; Law and Whinnet, 1992; Zitko *et al.*, 1998), including determination of up to 14 individual compounds (Marsili *et al.*, 1997b, Table 6.6). No evidence of toxic effects was noted. Butyltins in tissues of marine mammals were first reported in the 1990s (Iwata *et al.*, 1994) and are becoming commonly determined (e.g. Kannan *et al.*, 1998; Tanabe *et al.*, 1998; see Tanabe 1999 for a review), but are of unknown toxicological significance to these species. One *in vitro* study showed that MFO activity in hepatic microsomes of a Steller's sea lion (*Eumetopias jubatus*) and a Dall's porpoise was inhibited by tributyltins (Kim *et al.*, 1998). A few additional studies of radionuclides also appeared

Table 6.6 Polyaromatic hydrocarbon compounds reported in tissues of marine mammals, 1990s

Acenapthene	Chryseine
Anthracene	Fluoranthene
Benzo(a)anthracene	Fluorene
Benzo(a)pyrene	Naphthalene
Benzo(g,h,i)perylene	C1–C4 naphthalenes
Benzo(b)fluoranthene	Phenanthrene
Benzo(k)fluoranthene	C1–C2 phenanthrenes
Dibenzo(a,h)anthracene	Pyrene
C1–C2-dibenzothiophenes	

in this decade (Anderson *et al.*, 1990; Calmet *et al.*, 1992; Berrow *et al.*, 1998) but concentrations reported have not been considered detrimental.

Discussion

Despite what is becoming a very large database on contaminant residues in marine mammals (organochlorine and toxic element data have been documented for nearly 20 000 individuals in reports cited in our methods section, with numbers sampled for organochlorines doubling each decade), there are major limitations to our knowledge. Even the basic observational residue data are biased by geography and taxon. About 90 per cent of the samples in references we examined are from the northern hemisphere (primarily Europe and Canada; only 10 per cent are from US waters). More than three-quarters of the pinnipeds sampled for organochlorines are from four species (*Halichoerus grypus*, 629; harp seal (*Pagophilus groenlandicus*), 497; *Phoca vitulina*, 956; and *Pusa hispida*, 909) with nearly two-thirds of the cetaceans sampled in just five species (*Balaenoptera physalis*, 555; long-finned pilot whale (*Globicephala melas*), 495; *Phocoena phocoena*, 951; *Stenella coeruleoalba*, 430; and *Tursiops truncatus*, 381). Similar disparities exist in toxic element analyses. Although 10 246 marine mammals in 70 species have been examined for elements (there are about 78 cetaceans and 33 pinniped species worldwide), 70 per cent of the pinnipeds are in three species (*Pagophilus groenlandicus*, 397; *Phoca vitulina*, 586; and *Pusa hispida*, 1692); about half (49 per cent) of the cetaceans are represented by just five species (*Delphinapterus leucas*, 852; *Globicephala melas*, 355; *Phocoena phocoena*, 525; sperm whale (*Physeter macrocephalus*), 435; and *Stenella coeruleoalba*, 488). Despite this extensive sampling for organochlorine residues and trace element concentrations, there have never been any deaths in marine mammals that have been attributed directly to organochlorine poisoning, and to our knowledge lethal or other serious toxic effects of elements have not been observed and linked with concentrations in tissues of wild marine mammals. Diagnostic concentrations of organochlorines in blubber that can be linked to detrimental health effects in marine mammals have also not been established. Indeed, the rate at which organochlorines are concentrated in the blubber as lipid reserves are utilized is not well known, despite being critical to interpretations of residues in ill animals in comparison with healthy individuals. Concentrations of DDE reported in blubber or milk of marine mammals in many parts of the world are no higher than in contemporary human adipose tissue or milk in some countries (Ejobi *et al.*, 1996; Waliszewski *et al.*, 1996; Chikuni *et al.*, 1997; Çok *et al.*, 1997, 1998; Kinyamu *et al.*, 1998; Nasir *et al.*, 1998; Pardio *et al.*, 1998), or to those in humans in many parts of the world in past decades (Jensen, 1983; Matsumara, 1985). Results of different sampling designs in field studies executed to date sometimes show conflicting results about the actual impacts of organochlorines on various aspects of immune function, physiology or reproduction in marine mammals.

How can the science of marine mammal toxicology progress in the future to correct these deficiencies? And what are the practical applications that should be sought from this research? Clearly, towards the end of the twentieth century there was a much-needed advance towards the stage of testing hypotheses about contaminant effects on marine mammals. The emphasis should continue in this direction as much as possible, and steps should be taken to increase the number of carefully designed non-lethal controlled experiments in captivity, while field sampling designs should encompass many more factors of relevance to interpreting impacts than simply residue determinations. These two points were at the core of recommendations by a recent international workshop on marine mammals and persistent ocean contaminants (O'Shea *et al.*, 1999; see Chapter 23 in this volume for a synopsis). Although there are now two sets of feeding experiments on captive seals exposed to mixtures of organochlorines in fish that show reproductive and immunological effects, it has not been determined experimentally what the toxic effects of a single organochlorine (such as the ubiquitous DDE) are to any marine mammal species. It can certainly be argued that mixtures of contaminants are what these animals are naturally exposed to in the 'real world', lending relevance to these experimental treatments. However, many developing countries allow large-scale use of several organochlorine compounds because they are cheaper than alternatives that may be preferable for marine mammal health. It is much more sensible to expect regulators to restrict compounds based on experimental results rather than to disallow the entire class of compounds based on suspicions about mixtures, particularly when it is known from studies of reproduction and mortality in other wildlife that different organochlorines can vary widely in their toxic effects. Presumably one goal of this field beyond basic questions about toxicology in a fascinating set of animal models is to apply results to help alleviate unnecessary ocean pollution. Well-controlled experiments would be of especial value, considering that human impacts on the marine environment in our world today are very complex, and that marine mammal populations in the wild are subject to many different stressors and their interactions (Figure 6.3). These include harvesting and bycatch impacts and histories, fisheries depletion and other food-chain shifts, eutrophication in nearshore environments, global temperature shifts, exotic species introductions, fluxes in greenhouse gases, changes in incoming radiation, ship traffic, noise pollution and toxic algal blooms. Given such complexities, scientists should also caution policy makers that persistent contaminants may not always be the root cause for every die-off, reproductive anomaly, or failure for a marine mammal to thrive.

Despite these problems and inconsistencies, and the scientific desire for more clear-cut cause-and-effect experiments, there is already ample justification for concern. Research accomplished to date shows that organochlorines can be expected to be found in every marine mammal on Earth. Laboratory studies on other species of mammals show that organochlorines and toxic elements can cause the kinds of effects seen or suspected in marine mammals

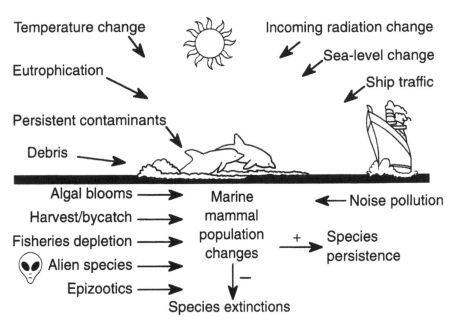

Figure 6.3 Marine mammal populations in the wild are subject to numerous stressors and complex interactions, making interpretation of impacts of contaminants alone very difficult.

at certain locations under certain circumstances. (However, they also show that such effects are dose-related, and can vary with species across the mammals.) In terms of policy implications, it is clear that these substances are unlikely to be considered benign to marine mammals. In an enlightened world, it might be hoped that restrictions on pollution of the marine environment could be based on a 'precautionary principle' that considers a weight of evidence approach (Gray and Bewers, 1996), and that there is ample justification to stem contaminant inputs for the benefit of marine mammals without waiting for results of difficult but definitive experiments on cause–effect relationships (e.g. Sinderman, 1997). In contrast, there is grave concern that no such actions will be taken in time to avoid ecological catastrophe. Room for optimism seems narrow in a world where the number of people have doubled to some 6 billion over the time span considered by this review, and the number of registered chemical substances has increased exponentially (from less than 2 million in 1970 to nearly 20 million by 1999 in the American Chemical Society Chemical Abstract Services Registry; Figure 6.2), as have human impacts on the marine environment. Indeed, as noted by Reeves and Leatherwood (1994): 'Concern about impacts on human health should be adequate incentive [for taking actions to stem marine pollution] . . . If it is not, then there is little cause for optimism with regard to limiting the exposure of cetaceans to environmental contaminants.'

References

Addison, R.F. 1989. Organochlorines and marine mammal reproduction. *Canadian Journal of Fisheries and Aquatic Sciences* 46: 360–368.

Addison, R.F. and Brodie, P.F. 1977. Organochlorine residues in maternal blubber, milk, and pup blubber from grey seals (*Halichoerus grypus*) from Sable Island, Nova Scotia. *Journal of the Fisheries Research Board of Canada* 34: 937–941.

Addison, R.F. and Smith, T.G. 1974. Organochlorine residue levels in Arctic ringed seals: variation with age and sex. *Oikos* 25: 335–337.

Addison, R.F., Zinck, M.E. and Ackman, R.G. 1972. Residues of organochlorine pesticides and polychlorinated biphenyls in some commercially produced Canadian marine oils. *Journal of the Fisheries Research Board of Canada* 29: 349–355.

Aguilar, A. and Borrell, A. 1988. Age- and sex-related changes in organochlorine compound levels in fin whales (*Balaenoptera physalus*) from the eastern North Atlantic. *Marine Environmental Research* 25: 195–211.

Aguilar, A. and Borrell, A. 1994. Abnormally high polychlorinated biphenyl levels in striped dolphins (*Stenella coeruleoalba*) affected by the 1990–1992 Mediterranean epizootic. *The Science of the Total Environment* 154: 237–247.

Anas, R.E. 1974a. Heavy metals in the northern fur seal, *Callorhinus ursinus*, and harbor seals, *Phoca vitulina richardi*. *Fishery Bulletin* 72: 133–137.

Anderson, S.S., Livens, F.R. and Singleton, D.L. 1990. Radionuclides in grey seals. *Marine Pollution Bulletin* 21: 343–345.

Aono, S., Tanabe, S., Fujise, Y., Kato, H. and Tatsukawa, R. 1997. Persistent organochlorines in minke whale (*Balaenoptera acutorostrata*) and their prey species from the Antarctic and the North Pacific. *Environmental Pollution* 98: 81–89.

Arima, S. and Nagakura, K. 1979. Mercury and selenium content of Odontoceti. *Bulletin of the Japanese Society of Scientific Fisheries* 45: 623–626.

Arora, A. and Rosenberg, N. 1998. Chemicals: a U.S. success story. In: A. Arora, R. Landau, and N. Rosenberg, eds. *Chemicals and long-term economic growth: insights from the chemical industry*. John Wiley and Sons, Inc., New York, pp. 71–102.

Arora, A., Landau, R. and Rosenberg, R. 1998. Introduction. In: A. Arora, R. Landau, and N. Rosenberg, eds. *Chemicals and long-term economic growth: insights from the chemical industry*. John Wiley and Sons, Inc., New York, pp. 3–24.

Beck, K.M., Fair, P., McFee, W. and Wolf, D. 1997. Heavy metals in livers of bottlenose dolphins stranded along the South Carolina coast. *Marine Pollution Bulletin* 34: 734–739.

Becker, P.R., Mackey, E.A., Demiralp, R., Schantz, M.M., Koster, B.J. and Wise, S.A. 1997. Concentrations of chlorinated hydrocarbons and trace elements in marine mammal tissues archived in the U.S. National Biomonitoring Specimen Bank. *Chemosphere* 34: 2067–2098.

Beckmen, K.B., Lowenstine, L.J., Newman, J., Hill, J., Hanni, K. and Gerber, J. 1997. Clinical and pathological characterization of northern elephant seal skin disease. *Journal of Wildlife Diseases* 33: 438–439.

Béland, P., DeGuise, S., Girard, C., Lagacé, A., Martineau, D., Michaud, R., Muir, D.C.G., Norstrom, R.J., Pelletier, É., Ray, S. and Shugart, L.R. 1993. Toxic compounds and health and reproductive effects in St. Lawrence beluga whales. *Journal of Great Lakes Research* 19: 766–775.

Bergman, A. and Olsson, M. 1985. Pathology of Baltic grey seal and ringed seal females with special reference to adrenocortical hyperplasia: Is environmental pollution the cause of a widely distributed disease syndrome? *Finnish Game Research* 44: 47–62.

Bergman, A., Olsson, M. and Reiland, S. 1992. Skull-bone lesions in the Baltic Grey seal (*Halichoerus grypus*). *Ambio* 21: 517–519.

Bernhoft, A., Wiig, O. and Skaare, J.U. 1997. Organochlorines in polar bears (*Ursus maritimus*) at Svalbard. *Environmental Pollution* 95: 159–175.

Berrow, S.D., Long, S.C., McGarry, A.T., Pollard, D., Rogan, E. and Lockyer, C. 1998. Radionuclides (^{137}Cs and ^{40}K) in harbour porpoises *Phocoena phocoena* from British and Irish coastal waters. *Marine Pollution Bulletin* 36: 569–576.

Blomkvist, G., Roos, A., Jensen, S., Bignert, A. and Olsson, M. 1992. Concentrations of sDDT and PCB in seals from Swedish and Scottish waters. *Ambio* 21: 539–545.

Boon, J.P., Reijnders, P.J.H., Dols, J., Wensvoort, P. and Hillebrand, M.T.J. 1987. The kinetics of individual polychlorinated biphenyl congeners in female harbour seals (*Phoca vitulina*), with evidence for structure-related metabolism. *Aquatic Toxicology* 1: 307–324.

Boon, J.P., Van Arnhem, E., Jansen, S., Kannan, N., Petrick, G., Schulz, D., Duinker, J.C., Reijnders, P.J.H. and Goksøyr, A. 1992. The toxicokinetics of PCBs in marine mammals with special reference to possible interactions of individual congeners with the cytochrome P450-dependent monooxygenase system: an overview. In: C.H. Walker and D.R. Livingstone, eds. *Persistent pollutants in marine ecosystems*. Pergamon Press, Oxford, England, pp. 119–159.

Boon, J.P., Sleiderink, H.M., Helle, M.S., Dekker, M., van Schanke, A., Roex, E., Hillebrand, M.T.J., Klamer, H.J.C., Govers, B., Pastor, D., Morse, D., Wester, P.G. and de Boer, J. 1998. The use of a microsomal in vitro assay to study phase I biotransformation for bioaccumulation of Chlorobornanes (Toxaphene®) in marine mammals and birds. Possible consequences of biotransformation and genotoxicity. *Comparative Biochemistry and Physiology Part C* 121: 385–403.

Borrell, A., Aguilar, A. and Pastor, T. 1997. Organochlorine pollutant levels in Mediterranean monk seals from the Western Mediterranean and the Sahara coast. *Marine Pollution Bulletin* 34: 505–510.

Braham, H.W. 1973. Lead in the California sea lion (*Zalophus californianus*). *Environmental Pollution* 5: 253–258.

Brouwer, A., Reijnders, P.J.H. and Koeman, J.H. 1989. Polychlorinated biphenyl (PCB)-contaminated fish induces vitamin A and thyroid hormone deficiency in the common seal (*Phoca vitulina*). *Aquatic Toxicology* 15: 99–106.

Buhler, D.R., Claeys, R.R. and Mate, B.R. 1975. Heavy metal and chlorinated hydrocarbon residues in California sea lions (*Zalophus californianus californianus*). *Journal of the Fisheries Research Board of Canada* 32: 2391–2397.

Calambokidis, J. and Barlow, J. 1991. Chlorinated hydrocarbon concentrations and their use for describing populations discreteness in harbor porpoises from Washington, Oregon, and California. In: J.E. Reynolds, III and D.K. Odell, eds. *Marine Mammal Strandings in the United States: Proceedings of the Second Marine Mammal Stranding Workshop*. National Oceanic and Atmospheric Administration Technical Report NMFS 98, pp. 101–110.

Calmet, D., Woodhead, D. and André, J.M. 1992. ^{210}Pb, ^{137}Cs and ^{40}K in three species of porpoises caught in the Eastern Tropical Pacific Ocean. *Journal of Environmental Radioactivity* 15: 153–169.

Cardellichio, N. 1995. Persistent contaminants in dolphins: an indication of chemical pollution in the Mediterranean Sea. *Water Science and Technology* 32: 331–340.

Chemical Abstracts Service 1999. CAS databases. http://info.cas.org/casdb.html. Accessed April 4, 1999.

Chikuni, O., Polder, A., Skaare, J.U. and Nhachi, C.F.B. 1997. An evaluation of DDT and DDT residues in human breast milk in the Kariba Valley of Zimbabwe. *Bulletin of Environmental Contamination and Toxicology* 58: 776–778.

Çok, I., Bilgili, A., Özdemir, M., Özbek, H., Bilgili, N. and Burgaz, S. 1997. Organochlorine pesticide residues in human breast milk from agricultural regions of Turkey, 1995–1996. *Bulletin of Environmental Contamination and Toxicology* 59: 577–582.

Çok, I., Bilgili, A., Yarsan, E., Bağci, C. and Burgaz, S. 1998. Organochlorine pesticide residue levels in human adipose tissue of residents of Manisa (Turkey), 1995–1996. *Bulletin of Environmental Contamination and Toxicology* 61: 311–316.

Colborn, T. and Clement, C. (eds) 1992. *Chemically-induced alterations in sexual and functional development: the wildlife/human connection.* Princeton Scientific Publishing, Princeton, New Jersey.

De Guise, S., Bernier, J., Martineau, D., Béland, P. and Fournier, M. 1996. Effects of in vitro exposure of beluga whale splenocytes and thymocytes to heavy metals. *Environmental Toxicology and Chemistry* 15: 1357–1364.

De Guise, S., Martineau, D., Béland, P. and Fournier, M. 1998. Effects of in vitro exposure of beluga whale leukocytes to selected organochlorines. *Journal of Toxicology and Environmental Health, Part A* 55: 479–493.

DeLong, R.L., Gilmartin, W.G. and Simpson, J.G. 1973. Premature births in California sea lions: association with high organochlorine pollutant residue levels. *Science* 181: 1168–1170.

de Moreno, J.E.A., Gerpe, M.S., Moreno, V.J. and Vodopivez, C. 1997. Heavy metals in Antarctic organisms. *Polar Biology* 17: 131–140.

De Swart, R.L., Kluten, R.M.G., Huizing, C.J., Vedder, E.J., Reijnders, P.J.H., Visser, I.K.G., UytdeHaag, F.G.C.M. and Osterhaus, A.D.M.E. 1993. Mitogen and antigen induced B cell and T cell responses of peripheral blood mononuclear cells from the harbour seal (*Phoca vitulina*). *Veterinary Immunology and Immunopathology* 37: 217–230.

De Swart, R.L., Ross, P.S., Vedder, E.J., Timmerman, H.H., Heisterkamp, S.H., van Loveren, H., Vos, J.G., Reijnders, P.J.H. and Osterhaus, A.D.M.E. 1994. Impairment of immunological functions in harbour seals (*Phoca vitulina*) feeding on fish from polluted coastal waters. *Ambio* 23: 155–159.

De Swart, R.L., Ross, P.S., Vos, J.G. and Osterhaus, A.D.M.E. 1996. Impaired immunity in harbour seals (*Phoca vitulina*) exposed to bioaccumulated environmental contaminants: review of a long-term feeding study. *Environmental Health Perspectives* 104 (Suppl. 4): 823–828.

de Voogt, P. and Brinkman, U.A.T. 1989. Production, properties and usage of polychlorinated biphenyls. In: R. Kimbrough and S. Jensen, eds. *Halogenated biphenyls, terphenyls, napthalenes, dibenzodioxins and related products.* Elsevier Science Publishers, B.V. (Biomedical Division), Amsterdam, The Netherlands, pp. 3–45.

Dietz, R., Riget, F. and Johansen, P. 1996. Lead, cadmium, mercury and selenium in Greenland marine animals. *The Science of the Total Environment* 186: 67–94.

Dietz, R., Norgaard, J. and Hansen, J.C. 1998. Have arctic marine mammals adapted to high cadmium levels? *Marine Pollution Bulletin* 36: 490–492.

Domingo, M., Ferrer, L., Pumarola, M., Marco, A., Plana, J., Kennedy, S. and McAliskey, M. 1990. Morbillivirus in dolphins. *Nature* 348: 21.

Duignan, P.J., House, C., Odell, D.K., Wells, R.S., Hansen, L.J., Walsh, M.T., St. Aubin, D.J., Rima, B.K. and Geraci, J.R. 1996. Morbillivirus infection in bottlenose dolphins: evidence for recurrent epizootics in the Western Atlantic and Gulf of Mexico. *Marine Mammal Science* 12: 499–515.

Duinker, J.C., Hillebrand, M.T.J. and Nolting, R.F. 1979. Organochlorines and metals in harbour seals (Dutch Wadden Sea). *Marine Pollution Bulletin* 10: 360–364.

Ejobi, F., Kanja, L.W., Kyule, M.N., Müller, P., Krüger, J. and Latigo, A.A.R. 1996. Organochlorine pesticide residues in mothers' milk in Uganda. *Bulletin of Environmental Contamination and Toxicology* 56: 873–880.

Estes, J.A., Bacon, C.E., Jarman, W.M., Norstrom, R.J., Anthony, R.G. and Miles, A.K. 1997. Organochlorines in sea otters and bald eagles from the Aleutian Archipelago. *Marine Pollution Bulletin* 34: 486–490.

Fossi, M.C. and Marsili, L. 1997. The use of non-destructive biomarkers in the study of marine mammals. *Biomarkers* 2: 205–216.

Fossi, M.C., Marsili, L., Junin, M., Castello, H., Lorenzani, J.A., Casini, S., Savelli, C. and Leonzio, C. 1997a. Use of nondestructive biomarkers and residue analysis to assess the health status of endangered species of pinnipeds in the South-West Atlantic. *Marine Pollution Bulletin* 34: 157–162.

Fossi, M.C., Savelli, C., Marsili, L., Casini, S., Jimenez, B., Junin, M., Castello, H. and Lorenzani, J.A. 1997b. Skin biopsy as a nondestructive tool for the toxicological assessment of endangered populations of pinnipeds: preliminary results on mixed function oxidase in *Otaria flavescens*. *Chemosphere* 35: 1623–1635.

Frank, R., Ronald, K. and Braun, H.E. 1973. Organochlorine residues in harp seals (*Pagophilus groenlandicus*) caught in eastern Canadian waters. *Journal of the Fisheries Research Board of Canada* 30: 1053–1063.

Freeman, H.C. and Sangalang, G.B. 1977. A study of the effects of methyl mercury, cadmium, arsenic, selenium, and a PCB, (Aroclor 1254) on adrenal and testicular steroidogeneses *in vitro*, by the gray seal *Halichoerus grypus*. *Archives of Environmental Contamination and Toxicology* 5: 369–383.

Freeman, H.C., Sanglang, G., Uthe, J.F. and Ronald, K. 1975. Steroidogenesis *in vitro* in the harp seal (*Pagophilus groenlandicus*) without and with methyl mercury treatment *in vivo*. *Environmental Physiology and Biochemistry* 5: 428–439.

Gaskin, D.E., Holdrinet, M. and Frank, R. 1971. Organochlorine pesticide residues in harbour porpoises from the Bay of Fundy region. *Nature* 233: 499–500.

Gaskin, D.E., Holdrinet, M. and Frank, R. 1982. DDT residues in blubber of harbour porpoise, *Phocoena* (L.), from eastern Canadian waters during the five-year period 1969–1973. In: *Mammals in the Seas*. FAO Fisheries Series No. 5, Vol. 4, pp. 135–143.

Gaskin, D.E., Frank, R. and Holdrinet, M. 1983. Polychlorinated biphenyls in harbor porpoises *Phocoena phocoena* (L.) from the Bay of Fundy, Canada and adjacent waters, with some information on chlordane and hexachlorobenzene levels. *Archives of Environmental Contamination and Toxicology* 12: 211–219.

Gauthier, J.M., Metcalfe, C.D. and Sears, R. 1997a. Chlorinated organic contaminants in blubber biopsies from Northwestern Atlantic balaenopterid whales summering in the Gulf of St. Lawrence. *Marine Environmental Research* 44: 201–223.

Gauthier, J.M., Metcalfe, C.D. and Sears, R. 1997b. Validation of the blubber biopsy technique for monitoring of organochlorine contaminants in balaenopterid whales. *Marine Environmental Research* 43: 157–179.

Gauthier, J.M., Pelletier, E., Brochu, C., Moore, S., Metcalfe, C.D. and Béland, P. 1998a. Environmental contaminants in tissues of a neonate St. Lawrence beluga whale (*Delphinapterus leucas*). *Marine Pollution Bulletin* 36: 102–108.

Gauthier, J.M., Dubeau, H. and Rassart, É. 1998b. Mercury-induced micronuclei in skin fibroblasts of beluga whales. *Environmental Toxicology and Chemistry* 17: 2487–2493.

George, J.L. and Frear, D.E.H. 1966. Pesticides in the Antarctic. In: N.W. Moore, ed. Pesticides in the environment and their effects on wildlife. *Journal of Applied Ecology* 3 (suppl): 155–167.

Geraci, J.R. 1989. Clinical investigation of the 1987–88 mass mortality of bottlenose dolphins along the U.S. central and south Atlantic coast. Final report to National Marine Fisheries Service, U.S. Navy Office of Naval Research, and U.S. Marine Mammal Commission.

Geraci, J.R. and St. Aubin, D.J. (eds) 1990. *Sea mammals and oil: confronting the risks*. Academic Press, San Diego, CA.

Gerpe, M., Moreno, J., Pérez, A., Bastida, R., Rodríguez, D. and Marchovecchio, J. 1990. Trace metals in the South American fur seal, *Arctocephalus australis* (Zimmermann, 1783). In: J. Barceló, ed. *Environmental contamination*. CEP consultants Ltd, UK, pp. 591–593.

Gray, J.S. and Bewers, J.M. 1996. Towards a scientific definition of the precautionary principle. *Marine Pollution Bulletin* 32: 768–771.

Green, N.J.L., Jones, K.C. and Harwood, J. 1996. Contribution of coplanar and non-coplanar polychlorinated biphenyls to the toxic equivalence of grey seal (*Halichoerus grypus*) milk. *Chemosphere* 33: 1273–1281.

Hall, A.J., Law, R.J., Wells, D.E., Harwood, J., Ross, H.M., Kennedy, S., Allchin, C.R., Campbell, L.A. and Pomeroy, P.P. 1992. Organochlorine levels in common seals (*Phoca vitulina*) which were victims and survivors of the 1988 phocine distemper epizootic. *The Science of the Total Environment* 115: 145–162.

Hall, A., Pomeroy, P., Green, N., Jones, K. and Harwood, J. 1997. Infection, haematology and biochemistry in grey seal pups exposed to chlorinated biphenyls. *Marine Environmental Research* 43: 81–98.

Hall, A.J., Green, N.J.L., Jones, K.C., Pomeroy, P.P. and Harwood, J. 1998. Thyroid hormones as biomarkers in grey seals. *Marine Pollution Bulletin* 36: 424–428.

Hammond, A.L. 1971. Mercury in the environment: natural and human factors. *Science* 171: 788–789.

Harder, T.C., Willhaus, T., Leibold, W. and Liess, B. 1992. Investigations on course and outcome of phocine distemper virus infection in harbor seals (*Phoca vitulina*) exposed to polychlorinated biphenyls. *Journal of Veterinary Medicine B* 39: 19–31.

Hayteas, D.L. and Duffield, D.A. 1997. The determination by HPLC of PCB and *p,p*′-DDE residues in marine mammals stranded on the Oregon coast, 1991–1995. *Marine Pollution Bulletin* 34: 844–848.

Helle, E., Olsson, M. and Jensen, S. 1976a. DDT and PCB levels and reproduction in ringed seal from the Bothnian Bay. *Ambio* 5: 188–189.

Helle, E., Olsson, M. and Jensen, S. 1976b. PCB levels correlated with pathological changes in seal uteri. *Ambio* 5: 261–263.

Hellou, J., Stenson, G., Ni, I.H. and Payne, J.F. 1990. Polycyclic aromatic hydrocarbons in muscle tissue of marine mammals from the Northwest Atlantic. *Marine Pollution Bulletin* 21: 469–473.

Hellou, J., Upshall, C., Ni, I.H., Payne, J.F. and Huang, Y.S. 1991. Polycyclic aromatic hydrocarbons in harp seals (*Phoca groenlandica*) from the Northwest Atlantic. *Archives of Environmental Contamination and Toxicology* 21: 135–140.

Helminen, M., Karppanen, E. and Koivisto, I. 1968. Saimaan norpan elohopeapitoisuudesta 1967. *Suomen Eläinlääkärilehti Finsk Veterinärtidskrift* 74: 87–89.

Henriksson, K., Karppanen, E. and Helminen, M. 1969. Kvicksilverhalter hos insjö-ocn haussälar. *Nordisk Hygienisk Tidskrift* 50: 54–59.

Hikino, T., Harada, T., Tokuhisa, Y. and Yoshida, J.A. 1998. The Japanese puzzle: rapid catch-up and long struggle. In: A. Arora, R. Landau, and N. Rosenberg, eds. *Chemicals and long-term growth: insights from the chemical industry*. John Wiley and Sons, Inc., New York, pp. 103–135.

Holden, A.V. and Marsden, K. 1967. Organochlorine pesticides in seals and porpoises. *Nature* 216: 1274–1276.

Holsbeek, L., Siebert, U. and Joiris, C.R. 1998. Heavy metals in dolphins stranded on the French Atlantic coast. *The Science of the Total Environment* 217: 241–249.

Holtzman, R.B. 1969. Concentrations of the naturally occurring radionuclides ^{226}Ra, ^{210}Pb and ^{210}Po in aquatic fauna. In: D.J. Nelson and F.C. Evans, eds. *Proceedings of the second national symposium, Symposium on radioecology*. The Clearinghouse for Federal Scientific and Technical Information, National Bureau of Standards, Springfield, VA, pp. 535–546.

Hyvärinen, H., Sipilä, T., Kunnasranta, M. and Koskela, J.T. 1998. Mercury pollution and the Saimaa ringed seal (*Phoca hispida saimensis*). *Marine Pollution Bulletin* 36: 76–81.

Iwata, H., Tanabe, S., Miyazaki, N. and Tatsukawa, R. 1994. Detection of butyltin compound residues in the blubber of marine mammals. *Marine Pollution Bulletin* 28: 607–612.

Iwata, H., Tanabe, S., Mizuno, T. and Tatsukawa, R. 1997. Bioaccumulation of butyltin compounds in marine mammals: the specific tissue distribution and composition. *Applied Organometallic Chemistry* 11: 257–264.

Iwata, H., Tanabe, S., Iida, T., Baba, N., Ludwig, J.P. and Tatsukawa, R. 1998. Enantioselective accumulation of α-hexachlorocyclohexane in northern fur seals and double-crested cormorants: effects of biological and ecological factors in the higher trophic levels. *Environmental Science and Technology* 32: 2244–2249.

Jarman, W.M., Bacon, C.E., Estes, J.A., Simon, M. and Norstrom, R.J. 1996. Organochlorine contaminants in sea otters: the sea otter as a bioindicator. *Endangered Species Update* 13: 20–22.

Jensen, A.A. 1983. Chemical contamination in human milk. *Residue Reviews* 89: 1–128.

Jensen, S., Johnels, A.G., Olsson, M. and Otterlind, G. 1969. DDT and PCB in marine animals from Swedish waters. *Nature* 224: 247–250.

Kannan, K. and Falandysz, J. 1997. Butyltin residues in sediment, fish, fish-eating birds, harbour porpoise and human tissues from the Polish coast of the Baltic Sea. *Marine Pollution Bulletin* 34: 203–207.

Kannan, K., Corsolini, S., Focardi, S., Tanabe, S. and Tatuskawa, R. 1996. Accumulation pattern of butyltin compounds in dolphin, tuna, and shark collected from Italian coastal waters. *Archives of Environmental Contamination and Toxicology* 31: 19–23.

Kannan, K., Senthilkumar, K., Loganathan, B.G., Takahashi, S., Odell, D.K. and Tanabe, S. 1997. Elevated accumulation of tributyltin and its breakdown products in bottlenose dolphins (*Tursiops truncatus*) found stranded along the U.S. Atlantic and Gulf coasts. *Environmental Science and Technology* 31: 296–301.

Kannan, K., Guruge, K.S., Thomas, N.J., Tanabe, S. and Giesy, J.P. 1998. Butyltin residues in southern sea otters (*Enhydra lutris nereis*) found dead along California coastal waters. *Environmental Science and Technology* 32: 1169–1175.

Keith, J.O. 1991. Historical perspectives. In: T.J. Peterle, ed. *Wildlife toxicology.* Van Nostrand Reinhold, New York, pp. 1–21.

Kendall, M.D., Safieh, B., Harwood, J. and Pomeroy, P.P. 1992. Plasma thymulin concentrations, the thymus and organochlorine contaminant levels in seals infected with phocine distemper virus. *The Science of the Total Environment* 115: 133–144.

Kiceniuk, J.W., Holzecher, J. and Chatt, A. 1997. Extractable organohalogens in tissues of beluga whales from the Canadian arctic and the St. Lawrence estuary. *Environmental Pollution* 97: 205–211.

Kim, G.B., Tanabe, S., Iwakiri, R., Tatsukawa, R., Amano, M., Miyazaki, N. and Tanaka, H. 1996a. Accumulation of butyltin compounds in Risso's dolphin (*Grampus griseus*) from the Pacific Coast of Japan: comparison with organochlorine residue pattern. *Environmental Science and Technology* 30: 2620–2625.

Kim, G.B., Tanabe, S., Tatsukawa, R., Loughlin, T.R. and Shimazaki, K. 1996b. Characteristics of butyltin accumulation and its biomagnification in Steller sea lion (*Eumetopias jubatus*). *Environmental Toxicology and Chemistry* 15: 2043–2048.

Kim, G.B., Nakata, H. and Tanabe, S. 1998. In vitro inhibition of hepatic cytochrome P450 and enzyme activity by butyltin compounds in marine mammals. *Environmental Pollution* 99: 255–261.

Kinyamu, J.K., Kanja, L.W., Skaare, J.U. and Maitho, T.E. 1998. Levels of organochlorine pesticides residues in milk of urban mothers in Kenya. *Bulletin of Environmental Contamination and Toxicology* 60: 732–738.

Kleivane, L. and Skaare, J.U. 1998. Organochlorine contaminants in northeast Atlantic minke whales (*Balaenoptera acutorostrata*). *Environmental Pollution* 101: 231–239.

Klobes, U., Vetter, W., Luckas, B., Skírnisson, K. and Plötz, J. 1998. Levels and enantiomeric ratios of α-HCH, oxychlordane, and PCB 149 in blubber of harbour seals (*Phoca vitulina*) and grey seals (*Halichoerus grypus*) from Iceland and further species. *Chemosphere* 37: 2501–2512.

Koeman, J.H. and van Genderen, H. 1966. Some preliminary notes on residues of chlorinated hydrocarbon insecticides in birds and mammals in the Netherlands. In: N.W. Moore, ed. Pesticides in the environment and their effects on wildlife. *Journal of Applied Ecology* 3 (suppl.), 99–106.

Koeman, J.H., Peeters, W.H.M. and Koudstaal-Hol, C.H.M. 1973. Mercury-selenium correlations in marine mammals. *Nature* 245: 385–386.

Koeman, J.H., van de Ven, W.S.M., de Goeij, J.J.M., Tjioe, P.S. and van Haaften, J.L. 1975. Mercury and selenium in marine mammals and birds. *The Science of the Total Environment* 3: 279–287.

Koistinen, J., Stenman, O., Haahti, H., Suonperä, M. and Paasivirta, J. 1997. Polychlorinated diphenyl ethers, dibenzo-p-dioxins, dibenzofurans and biphenyls in seals and sediment from the Gulf of Finland. *Chemosphere* 35: 1249–1269.

Krahn, M.M., Becker, P.R., Tilbury, K.L. and Stein, J.E. 1997. Organochlorine contaminants in blubber of four seal species: integrating biomonitoring and specimen banking. *Chemosphere* 34: 2109–2121.

Kuiken, T., Höfle, U., Bennett, P.M., Allchin, C.R., Kirkwood, J.K., Baker, J.R., Appleby, E.C., Lockyer, C.H., Walton, M.J. and Sheldrick, M.C. 1993. Adrenocortical hyperplasia, disease and chlorinated hydrocarbons in the harbour porpoise (*Phocoena phocoena*). *Marine Pollution Bulletin* 26: 440–446.

Kuiken, T., Bennett, P.M., Allchin, C.R., Kirkwood, J.K., Baker, J.R., Lockyer, C.H., Walton, M.J. and Sheldrick, M.C. 1994. PCB's, cause of death and body condition in harbour porpoises (*Phocoena phocoena*) from British waters. *Aquatic Toxicology* 28: 13–28.

Lahvis, G.P., Wells, R.S., Kuehl, D.W., Stewart, J.L., Rhinehart, H.L. and Via, C.S. 1995. Decreased lymphocyte responses in free-ranging bottlenose dolphins (*Tursiops truncatus*) are associated with increased concentrations of PCBs and DDT in peripheral blood. *Environmental Health Perspectives* 103 (suppl. 4): 62–67.

Law, R.J. and Whinnett, J.A. 1992. Polycyclic aromatic hydrocarbons in muscle tissue of harbour porpoises (*Phocoena phocoena*) from UK waters. *Marine Pollution Bulletin* 24: 550–553.

Law, R.J., Allchin, C.R., Jones, B.R., Jepson, P.D., Baker, J.R. and Spurrier, C.J.H. 1997. Metals and organochlorines in tissues of a Blainville's beaked whale (*Mesoplodon densirostris*) and a killer whale (*Orcinus orca*) stranded in the United Kingdom. *Marine Pollution Bulletin* 34: 208–212.

Law, R.J., Blake, S.J., Jones, B.R. and Rogan, E. 1998. Organotin compounds in liver tissue of harbour porpoises (*Phocoena phocoena*) and grey seals (*Halichoerus grypus*) from the coastal waters of England and Wales. *Marine Pollution Bulletin* 36: 241–247.

LeBouef, B.J. and Bonnell, M.L. 1971. DDT in California sea lions. *Nature* 234: 108–109.

Lee, S.S., Mate, B.R., von der Trenck, K.T., Rimerman, R.A. and Buhler, D.R. 1977. Metallothionein and the subcellular localization of mercury and cadmium in the California sea lion. *Comparative Biochemistry and Physiology* 57C: 45–53.

Letcher, R.J., Norstrom, R.J., Lin, S., Ramsay, M.A. and Bandiera, S.M. 1996. Immunoquantitation and microsomal monoxygenase activities of hepatic cytochromes P4501A and P4502B and chlorinated hydrocarbon contaminant levels in polar bear (*Ursus maritimus*). *Toxicology and Applied Pharmacology* 137: 127–140.

Letcher, R.J., Norstrom, R.J. and Muir, D.C.G. 1998. Biotransformation versus bioaccumulation: sources of methyl sulfone PCB and 4,4'-DDE metabolites in the polar bear food chain. *Environmental Science and Technology* 32: 1656–1661.

Li, Y.F., Cai, D.J. and Singh, A. 1998. Technical hexachlorocyclohexane use trends in China and their impact on the environment. *Archives of Environmental Contamination and Toxicology* 35: 688–697.

Linduska, J.P. 1952. Wildlife in a chemical world. *Audubon Magazine* 54: 144–149, 190, 248–252.

Linduska, J.P. and Surber, E.W. 1948. *Effects of DDT and other insecticides on fish and wildlife: summary of investigations during 1947.* U.S. Fish and Wildlife Service, Circular 15. Washington DC.

Lipscomb, T.P., Schulman, F.Y., Moffett, D. and Kennedy, S. 1994. Morbilliviral disease in Atlantic bottlenose dolphins (*Tursiops truncatus*) from the 1987–1988 epizootic. *Journal of Wildlife Diseases* 30: 567–571.

Lipscomb, T.P., Kennedy, S., Moffett, D., Krafft, A., Klaunberg, B.A., Lichy, J.H., Regan, G.T., Worthy, G.A.J. and Taubenberger, J.K. 1996. Morbilliviral epizootic in bottlenose dolphins of the Gulf of Mexico. *Journal of Veterinary Diagnostic Investigations* 8: 283–290.

Loewen, M.D., Stern, G.A., Westmore, J.B., Muir, D.C.G. and Parlar, H. 1998. Characterization of three major toxaphene congeners in arctic ringed seal by electron ionization and electron capture negative ion mass spectrometry. *Chemosphere* 36: 3119–3135.

Loughlin, T.R. (ed.) 1994. *Marine mammals and the Exxon Valdez.* Academic Press, San Diego, CA.

Madhusree, B., Tanabe, S., Öztürk, A.A., Tatsukawa, R., Miyazaki, N., Özdamar, E., Aral, O., Samsun, O. and Özturk, B. 1997. Contamination by butyltin compounds in harbour porpoise (*Phocoena phocoena*) from the Black Sea. *Fresenius Journal of Analytical Chemistry* 359: 244–248.

Marsili, L., Casini, C., Marini, L., Regoli, A. and Focardi, S. 1997a. Age, growth and organochlorines (HCB, DDTs and PCBs) in Mediterranean striped dolphins *Stenella coeruleoalba* stranded in 1988–1994 on the coasts of Italy. *Marine Ecology Progress Series* 151: 273–282.

Marsili, L., Fossi, M.C., Casini, S., Savelli, C., Jimenez, B., Junin, M. and Castello, H. 1997b. Fingerprint of polycyclic aromatic hydrocarbons in two populations of southern sea lions (*Otaria flavescens*). *Chemosphere* 34: 759–770.

Marsili, L., Fossi, M.C., Notarbartolo di Sciara, G., Zanardelli, M., Nani, B., Panigada, S. and Focardi, S. 1998. Relationship between organochlorine contaminants and mixed function oxidase activity in skin biopsy specimens of Mediterranean fin whales (*Balaenoptera physalus*). *Chemosphere* 37: 1501–1510.

Martin, J.H., Elliott, P.D., Anderlini, V.C., Girvin, D., Jacobs, S.A., Risebrough, R.W., DeLong, R.L. and Gilmartin, W.G. 1976. Mercury–selenium–bromine imbalance in premature parturient California sea lions. *Marine Biology* 35: 91–104.

Martineau, D., Béland, P., Desjardins, C. and Lagacé, A. 1987. Levels of organochlorine chemicals in tissues of beluga whales (*Delphinapterus leucas*) from the St. Lawrence estuary, Quebec, Canada. *Archives of Environmental Contamination and Toxicology* 16: 137–147.

Martineau, D., Lagacé, A., Béland, P., Higgins, R., Armstrong, D. and Shugart, L.R. 1988. Pathology of stranded beluga whales (*Delphinapterus leucas*) from the St. Lawrence estuary, Quebec, Canada. *Journal of Comparative Pathology* 98: 287–311.

Martineau, D., De Guise, S., Fournier, M., Shugart, L., Girard, C., Lagacé, A. and Béland, P. 1994. Pathology and toxicology of beluga whales from the St. Lawrence estuary, Quebec, Canada. Past, present and future. *The Science of the Total Environment* 154: 201–215.

Mathieu, A., Payne, J.F., Fancey, L.L., Santella, R.M. and Young, T.L. 1997. Polycyclic aromatic hydrocarbon–DNA adducts in beluga whales from the arctic. *Journal of Toxicology and Environmental Health* 51: 1–4.

Matsumara, F. 1985. *Toxicology of insecticides*, 2nd edn. Plenum Press, New York.

McKenzie, C., Rogan, E., Reid, R.J. and Wells, D.E. 1997. Concentrations and patterns of organic contaminants in Atlantic white-sided dolphins (*Lagenorhynchus acutus*) from Irish and Scottish coastal waters. *Environmental Pollution* 98: 15–27.

Medvedev, N., Panichev, N. and Hyvärinen, H. 1997. Levels of heavy metals in seals of Lake Ladoga and the White Sea. *The Science of the Total Environment* 206: 95–105.

Metcalf, R.L. 1973. A century of DDT. *Journal of Agricultural and Food Chemistry* 21: 511–519.

Monaci, F., Borrell, A., Leonzio, C., Marsili, L. and Calzada, N. 1998. Trace elements in striped dolphins (*Stenella coeruleoalba*) from the western Mediterranean. *Environmental Pollution* 99: 61–68.

Moreno, V.J. 1994. Baseline studies on total mercury content in marine biota of Argentina: a review toward a legal limit fixation. In: T.C. Hutchinson, C.A. Gordon, and K.M. Meema, eds. *Global perspectives on lead, mercury, and cadmium cycling in the environment*. Wiley Eastern, Ltd, New Delhi, pp. 247–257.

Moreno, V.J., Pérez, A., Bastida, R.O., de Moreno, J.E.A. and Malaspina, A.M. 1984. Distribución de mercurio total en los tejidos de un delfín nariz de botella (*Tursiops gephyreus* Lahille, 1908) de la provincia de Buenos Aires (Argentina). *Revista del Instituto Nacional de Investigación y Desarollo Pesquero Mar del Plata* 4: 93–102.

Mössner, S. and Ballschmiter, K. 1997. Marine mammals as global pollution indicators for organochlorines. *Chemosphere* 34: 1285–1296.

Muir, D.C.G., Norstrom, R.J. and Simon, M. 1988b. Organochlorine contaminants in Arctic marine food chains: accumulation of specific polychlorinated biphenyls and chlordane-related compounds. *Environmental Science and Technology* 22: 1071–1079.

Nagakura, K., Arima, S., Kurihara, M., Koga, T. and Fujita, T. 1974. Mercury content of whales. *Bulletin of Tokai Registry of Fisheries Research Laboratory* 78: 41–46.

Nakata, H., Tanabe, S., Tatsukawa, R., Amano, M., Miyazaki, N. and Petrov, E.A. 1997. Bioaccumulation profiles of polychlorinated biphenyls including coplanar congeners and possible toxicological implications in Baikal seal (*Phoca sibirica*). *Environmental Pollution* 95: 57–65.

Nakata, H., Kannan, K., Jing, L., Thomas, N., Tanabe, S. and Giesy, J.P. 1998a. Accumulation pattern of organochlorine pesticides and polychlorinated biphenyls in southern sea otters (*Enhydra lutris nereis*) found stranded along coastal California, USA. *Environmental Pollution* 103: 45–53.

Nakata, H., Tanabe, S., Tatsukawa, R., Koyama, Y., Miyazaki, N., Belikov, S. and Boltunov, A. 1998b. Persistent organochlorine contaminants in ringed seals (*Phoca hispida*) from the Kara Sea, Russian Arctic. *Environmental Toxicology and Chemistry* 17: 1745–1755.

Nasir, K., Bilto, Y. and Al-Shuraiki, Y. 1998. Residues of chlorinated hydrocarbon insecticides in human milk of Jordanian women. *Environmental Pollution* 99: 141–148.

Norstrom, R.J. and Muir, D.C.G. 1994. Chlorinated hydrocarbon contaminants in arctic marine mammals. *The Science of the Total Environment* 154: 107–128.

Norstrom, R.J., Simon, M., Muir, D.C.G. and Schweinsburg, R.E. 1988. Organochlorine contaminants in arctic marine food chains: identification, geographical distribution and temporal trends in polar bears. *Environmental Science and Technology* 22: 1063–1071.

Norstrom, R.J., Belikov, S.E., Born, E.W., Garner, G.W., Malone, B., Olpinski, S., Ramsay, M.A., Schliebe, S., Stirling, I., Stishov, M.S., Taylor, M.K. and Wiig, Ø. 1998. Chlorinated hydrocarbon contaminants in polar bears from Eastern Russia, North America, Greenland, and Svalbard: biomonitoring of arctic pollution. *Archives of Environmental Contamination and Toxicology* 35: 354–367.

Oehme, M., Schlabach, M., Kallenborn, R. and Haugen, J.E. 1996. Sources and pathways of persistent polychlorinated pollutants to remote areas of the North Atlantic and levels in the marine food chain: a research update. *The Science of the Total Environment* 186: 13–24.

Olafson, R.W. and Thompson, J.A.J. 1974. Isolation of heavy metal binding proteins from marine vertebrates. *Marine Biology* 28: 83–86.

O'Shea, T.J. 1999. Environmental contaminants and marine mammals. In: J.E. Reynolds, III and S.A. Rommel, eds. *Biology of marine mammals*. Smithsonian Institution Press, Washington, DC, pp. 485–564.

O'Shea, T.J. and Brownell, R.L. Jr. 1998. California sea lion (*Zalophus californianus*) populations and ΣDDT contamination. *Marine Pollution Bulletin* 36: 159–164.

O'Shea, T.J., Reeves, R.R. and Long, A.K. (eds) 1999. Marine mammals and persistent ocean contaminants. *Proceedings of the Marine Mammal Commission workshop, Keystone, Colorado, 12–15 October 1998.*

Osterberg, C., Pearcy, W. and Kujala, N. 1964. Gamma emitters in the fin whale. *Nature* 204: 1006–1007.

Outridge, P.M., Evans, R.D., Wagemann, R. and Stewart, R.E.A. 1997. Historical trends of heavy metals and stable lead isotopes in beluga (*Delphinapterus leucas*) and walrus (*Odobenus rosmarus rosmarus*) in the Canadian Arctic. *The Science of the Total Environment* 203: 209–219.

Pardío, V.T., Waliszewski, S.M., Aguirre, A.A., Coronel, H., Burelo, G.V., Infanzon R. R.M. and Rivera, J. 1998. DDT and its metabolites in human milk collected in Veracruz City and suburban areas (Mexico). *Bulletin of Environmental Contamination and Toxicology* 60: 852–857.

Parsons, E.C.M. 1998. Trace metal pollution in Hong Kong: implications for the health of Hong Kong's Indo-Pacific hump-backed dolphins (*Sousa chinensis*). *The Science of the Total Environment* 214: 175–184.

Peña, N.I., Moreno, V.J., Marcovecchio, J.E. and Pérez, A. 1988. Total mercury, cadmium and lead distribution in tissues of the southern sea lion (*Otaria flavescens*) in the ecosystem of Mar del Plata, Argentina. In: U. Seeliger, L.D. Lacerda, and S.R. Patchineelam, eds. *Metals in coastal environments of Latin America*. Springer-Verlag, Berlin, pp. 140–146.

Prudente, M., Tanabe, S., Watanabe, M., Subramanian, A., Miyazaki, N., Suarez, P. and Tatsukawa, R. 1997. Organochlorine contamination in some Odontoceti species from the North Pacific and Indian Ocean. *Marine Environmental Research* 44: 415–427.

Ramprashad, F. and Ronald, K. 1977. A surface preparation study on the effect of methyl mercury on the sensory hair cell population in the cochlea of the harp seal (*Pagophilus groenlandicus* Erxleben, 1777). *Canadian Journal of Zoology* 55: 223–230.

Ratcliffe, D.A. 1967. Decrease in eggshell weight in certain birds of prey. *Nature* 215: 208–210.

Reeves, R.R. and Leatherwood, S. 1994. *Dolphins, porpoises, and whales: 1994–1998 action plan for the conservation of cetaceans.* IUCN, Gland, Switzerland.

Reijnders, P.J.H. 1986. Reproductive failure in common seals feeding on fish from polluted coastal waters. *Nature* 324: 456–457.

Roberts, T.M., Heppleston, P.B. and Roberts, R.D. 1976. Distribution of heavy metals in tissues of the common seal. *Marine Pollution Bulletin* 7: 194–196.

Rogan, E. and Berrow, S.D. 1996. A review of harbour porpoises, *Phocoena phocoena*, in Irish waters. *Reports of the International Whaling Commission* 46: 595–605.

Ronald, K., Uthe, J.F. and Freeman, H. 1975. Effects of methyl mercury on the harp seal. International Council for the Exploration of the Sea. C.M. 1975/N: 9. Mimeo.

Roots, O. and Talvari, A. 1997. Bioaccumulation of toxic chlororganic compounds and their isomers into the organism of Baltic grey seal. *Chemosphere* 35: 979–985.

Ross, P.S., De Swart, R.L., Reijnders, P.J.H., Loveren, H.V., Vos, J.G. and Osterhaus, A.D.M.E. 1995. Contaminant-related suppression of delayed-type hypersensitivity and antibody responses in harbor seals fed herring from the Baltic Sea. *Environmental Health Perspectives* 103: 162–167.

Ross, P.S., De Swart, R.L., Timmerman, H.H., Reijnders, P.J.H., Vos, J.G., Van Loveren, H. and Osterhaus, A.D.M.E. 1996a. Suppression of natural killer cell activity in harbour seals (*Phoca vitulina*) fed Baltic Sea herring. *Aquatic Toxicology* 34: 71–84.

Ross, P., De Swart, R., Addison, R., Van Loveren, H., Vos, J. and Osterhaus, A. 1996b. Contaminant-induced immunotoxicity in harbour seals: wildlife at risk? *Toxicology* 112: 157–169.

Schantz, M.M., Porter, B.J., Wise, S.A., Segstro, M., Muir, D.C.G., Mössner, S., Ballschmiter, K. and Becker, P.R. 1996. Interlaboratory comparison study for PCB congeners and chlorinated pesticides in beluga whale blubber. *Chemosphere* 33: 1369–1390.

Scott, T.G., Willis, Y.L. and Ellis, J.A. 1959. Some effects of a field application of dieldrin on wildlife. *Journal of Wildlife Management* 23: 409–427.

Sepulveda, M.S., Ochoa-Acuna, H. and Sundlof, S.F. 1997. Heavy metal concentrations in Juan Fernandez fur seals (*Arctocephalus philippii*). *Marine Pollution Bulletin* 34: 663–665.

Sindermann, C.J. 1995. *Ocean pollution: effects on living resources and humans.* CRC Press, Boca Raton, Florida.

Sindermann, C.J. 1997. The search for cause and effect relationships in marine pollution studies. *Marine Pollution Bulletin* 34: 218–221.

Sladen, W.J.L., Menzie, C.M. and Reichel, W.L. 1966. DDT residues in Adelie penguins and a crabeater seal from Antarctica. *Nature* 210: 670–673.

Smith, T.G. and Armstrong, F.A.J. 1975. Mercury in seals, terrestrial carnivores, and principal food items of the Inuit, from Holman, N.W.T. *Journal of the Fisheries Research Board of Canada* 32: 795–801.

Stickel, L.F. 1968. *Organochlorine pesticides in the environment.* U.S. Fish and Wildlife Service, Special Scientific Report – Wildlife, No. 119.

Stickel, L.F., Stickel, W.H. and Christensen, R. 1966. Residues of DDT in brains and bodies of birds that died on dosage and survivors. *Science* 151: 1549–1551.

Storelli, M.M., Ceci, E. and Marcotrigiano, G.O. 1998. Comparison of total mercury, methylmercury, and selenium in muscle tissues and in the liver of *Stenella coueruleoalba* (Meyen) and *Caretta caretta* (Linnaeus). *Bulletin of Environmental Contamination and Toxicology* 61: 541–547.

Strandberg, B., Strandberg, L., Bergqvist, P.-A., Falandysz, J. and Rappe, C. 1998. Concentrations and biomagnification of 17 chlordane compounds and other organochlorines in harbour porpoise (*Phocoena phocoena*) and herring from the southern Baltic Sea. *Chemosphere* 37: 2513–2523.

Subramanian, A., Tanabe, S., Tatsukawa, R., Saito, S. and Miyazaki, N. 1987. Reduction in the testosterone levels by PCBs and DDE in Dall's porpoises of northwestern North Pacific. *Marine Pollution Bulletin* 18: 643–646.

Sydeman, W.J. and Jarman, W.M. 1998. Trace metals in seabirds, Steller sea lion, and forage fish and zooplankton from central California. *Marine Pollution Bulletin* 36: 828–832.

Szefer, P., Rokicki, J., Frelek, K., Skóra, K. and Malinga, M. 1998. Bioaccumulation of selected trace elements in lung nematodes, *Pseudalius inflexus*, of harbor porpoise (*Phocoena phocoena*) in a Polish zone of the Baltic Sea. *The Science of the Total Environment* 220: 19–24.

Tanabe, S. 1988. PCB problems in the future: foresight from current knowledge. *Environmental Pollution* 50: 5–28.

Tanabe, S. 1999. Butyltin contamination in marine mammals—a review. *Marine Pollution Bulletin* 39: 62–72.

Tanabe, S., Miura, S. and Tatsukawa, R. 1986. Variations of organochlorine residues with age and sex in Antarctic minke whale. *Memoirs of the National Institute for Polar Research, Special Issue* 44: 174–181.

Tanabe, S., Kannan, N., Subramanian, A., Watanabe, S. and Tatsukawa, R. 1987. Highly toxic coplanar PCBs: occurrence, source, persistency and toxic implications to wildlife and humans. *Environmental Pollution* 47: 147–163.

Tanabe, S., Watanabe, S., Kan, H. and Tatsukawa, R. 1988. Capacity and mode of PCB metabolism in small cetaceans. *Marine Mammal Science* 4: 103–124.

Tanabe, S., Madhusree, B., Öztürk, A.A., Tatsukawa, R., Miyazaki, N., Özdamar, E., Aral, O., Samsun, O. and Öztürk, B. 1997a. Persistent organochlorine residues in harbour porpoise (*Phocoena phocoena*) from the Black Sea. *Marine Pollution Bulletin* 34: 338–347.

Tanabe, S., Madhusree, B., Öztürk, A.A., Tatsukawa, R., Miyazaki, N., Özdamar, E., Aral, O., Samsun, O. and Öztürk, B. 1997b. Isomer-specific analysis of polychlorinated biphenyls in harbour porpoise (*Phocoena phocoena*) from the Black Sea. *Marine Pollution Bulletin* 34: 712–720.

Tanabe, S., Prudente, M., Mizuno, T., Hasegawa, J., Iwata, H. and Miyazaki, N. 1998. Butyltin contamination in marine mammals from North Pacific and Asian coastal waters. *Environmental Science and Technology* 32: 193–198.

Tarasova, E.N., Mamontov, A.A., Mamontova, E.A., Klasmeier, J. and McLachlan, M.S. 1997. Polychlorinated dibenzo-p-dioxins (PCDDs) and dibenzofurans (PCDFs) in Baikal seal. *Chemosphere* 34: 2419–2427.

Taubenberger, J.K., Tsai, M., Krafft, A.E., Lichy, J.H., Reid, A.H., Schulman, F.Y. and Lipscomb, T.P. 1996. Two morbilliviruses implicated in bottlenose dolphin epizootics. *Emerging Infectious Diseases* 2: 213–216.

Tilbury, K.L., Stein, J.E., Meador, J.P., Krone, C.A. and Chan, S. 1997. Chemical contaminants in harbor porpoise (*Phocoena phocoena*) from the North Atlantic Coast: tissue concentrations and intra- and inter-organ distribution. *Chemosphere* 34: 2159–2181.

Tillander, M., Miettinen, J.K. and Koivisto, I. 1972. Excretion rate of methyl mercury in the seal (*Pusa hispida*). In: M. Ruivo, ed. *Marine pollution and sea life.* Fishing News (Books) Ltd., Surrey, England, pp. 303–305

Tirpenou, A.E., Tsigouri, A.D. and Gouta, E.H. 1998. Residues of organohalogen compounds in various dolphin tissues. *Bulletin of Environmental Contamination and Toxicology* 60: 216–224.

Troisi, G.M. and Mason, C.F. 1997. Cytochromes P450, P420 and mixed-function oxidases as biomarkers of polychlorinated biphenyl (PCB) exposure in harbour seals (*Phoca vitulina*). *Chemosphere* 35: 1933–1946.

Troisi, G.M., Haraguchi, K., Simmonds, M.P. and Mason, C.F. 1998. Methyl sulphone metabolites of polychlorinated biphenyls (PCBs) in cetaceans from the Irish and Aegean Seas. *Archives of Environmental Contamination and Toxicology* 35: 121–128.

van de Ven, W.S.M., Koeman, J.H. and Svenson, A. 1979. Mercury and selenium in wild and experimental seals. *Chemosphere* 8: 539–555.

Vetter, W., Luckas, B., Heidemann, G. and Skírnisson, K. 1996. Organochlorine residues in marine mammals from the northern hemisphere – a consideration of the composition of organochlorine residues in the blubber of marine mammals. *The Science of the Total Environment* 186: 29–39.

Wade, T.L., Chambers, L., Gardinali, P.R., Sericano, J.L., Jackson, T.J., Tarpley, R.J. and Suydam, R. 1997. Toxaphene, PCB, DDT, and chlordane analyses of beluga whale blubber. *Chemosphere* 34: 1351–1357.

Wagemann, R., Innes, S. and Richard, P.R. 1996. Overview and regional and temporal differences of heavy metals in Arctic whales and ringed seals in the Canadian Arctic. *The Science of the Total Environment* 186: 41–66.

Wagemann, R., Trebacz, E., Boila, G. and Lockhart, W.L. 1998. Methylmercury and total mercury in tissues of arctic marine mammals. *The Science of the Total Environment* 218: 19–31.

Waliszewski, S.M., Pardío, V.T.S., Chantiri, N.P., Infanzón, R.M.R. and Rivera, J. 1996. Organochlorine pesticide residues in adipose tissue of Mexicans. *The Science of the Total Environment* 181: 125–131.

Watanabe, I., Ichihashi, H., Tanabe, S., Amano, M., Miyazaki, N., Petrov, E.A. and Tatsukawa, R. 1996. Trace element accumulation in Baikal seal (*Phoca sibirica*) from the Lake Baikal. *Environmental Pollution* 94: 169–179.

Watanabe, I., Tanabe, S., Amano, M., Miyazaki, N., Petrov, E.A. and Tatsukawa, R. 1998. Age-dependent accumulation of heavy metals in Baikal seal (Phoca sibirica) from the Lake Baikal. *Archives of Environmental Contamination and Toxicology* 35: 518–526.

Weis, I.M. and Muir, D.C.G. 1997. Geographical variation of persistent organochlorine concentrations in blubber of ringed seal (*Phoca hispida*) from the Canadian arctic: univariate and multivariate approaches. *Environmental Pollution* 96: 321–333.

Westgate, A.J., Muir, D.C.G., Gaskin, D.E. and Kingsley, M.C.S. 1997. Concentrations and accumulation patterns of organochlorine contaminants in the blubber of harbour porpoises, *Phocoena phocoena*, from the coast of Newfoundland, the Gulf of St. Lawrence, and the Bay of Fundy/Gulf of Maine. *Environmental Pollution* 95: 105–119.

Wiberg, K., Oehme, M., Haglund, P., Karlsson, H., Olsson, M. and Rappe, C. 1998. Enantioselective analysis of organochlorine pesticides in herring and seal from the Swedish marine environment. *Marine Pollution Bulletin* 36: 345–353.

Wiemeyer, S.N. and Porter, R.D. 1970. DDE thins eggshells of captive American kestrels. *Nature* 227: 737–738.

Wolkers, J., Burkow, I.C., Lydersen, C., Dahle, S., Monshouwer, M. and Witkamp, R.F. 1998. Congener specific PCB and polychlorinated camphene (toxaphene) levels in Svalbard ringed seals (*Phoca hispida*) in relation to sex, age, condition and cytochrome P450 enzyme activity. *The Science of the Total Environment* 216: 1–11.

Young, D., Becerra, M., Kopec, D. and Echols, S. 1998. GC/MS analysis of PCB congeners in blood of the harbor seal *Phoca vitulina* from San Francisco Bay. *Chemosphere* 37: 711–733.

Zitko, V., Stenson, G. and Hellou, J. 1998. Levels of organochlorine and polycyclic aromatic compounds in harp seal beaters (*Phoca groenlandica*). *The Science of the Total Environment* 221: 11–29.

7 Heavy metals in marine mammals

Krishna Das, Virginie Debacker, Stéphane Pillet and Jean-Marie Bouquegneau

Introduction

During the past few decades, increasing concern about environmental pollution has led to many investigations on heavy metals and their distribution in the sea, air or biological materials. The distribution of xenobiotics in the marine environment is not homogeneous and a considerable variation of the concentrations may occur regionally and temporally. The use of bioindicators offers a useful alternative for pollution monitoring studies. Marine mammals appear to be potentially valuable indicators of the level of heavy metals accumulated in the marine environment: according to their top position in the trophic network, their long life span and their long biological half-time of elimination of pollutants, these animals accumulate high levels of chemicals such as organochlorines (Kamrin and Ringer, 1994; Tanabe *et al.*, 1994), or heavy metals (Bouquegneau and Joiris, 1988, 1992; André *et al.*, 1991a; Dietz *et al.*, 1998). The increased interest in studying contaminants in marine mammals is due to large-scale die-off (Sarokin and Schulkin, 1992; Forcada *et al.*, 1994) or impaired reproduction (De Guise *et al.*, 1995), which could lead to population declines of some pinniped and small cetacean species in Europe and North America, and the finding of relatively large contaminant burdens in these animals. In many cases, morbillivirus infections were the primary cause of the disease outbreaks (Heide-Jørgensen *et al.*, 1992; Thompson and Hall, 1993; de Swart *et al.*, 1995a). These mass mortalities among seals and dolphins inhabiting contaminated marine areas have led to speculation about the possible involvement of immunosuppression associated with environmental pollution.

Controlled experiments are unavailable to establish any definite causal relation between these pollutant concentrations and any physiological problem. Moreover, the data are always difficult to interpret because of the presence of other chemical contaminants and other stressors. The physiological status of the organisms (e.g. pregnancy, moulting, fasting, etc.) also modulates the toxicity of heavy metals. In addition, the available measurements have almost been all performed on animals found dead, which leaves doubts about the general applicability of collected values at which an effect at individual or population level might be expected.

Several investigations have been carried out in an attempt to evaluate contaminant effects at ambient environmental levels (Reijnders, 1986; Aguilar and Borrell, 1994; De Guise *et al.*, 1995; de Swart *et al.*, 1995b, 1996). For example, it has been demonstrated that seals fed polluted fish from the Dutch Wadden Sea showed reduced pup production when compared to those fed much less polluted fish from the Northeast Atlantic (Reijnders *et al.*, 1986). This study was the first sign of a causal relationship between naturally occurring levels of pollutants and a physiological response in marine mammals.

A more recent study over a 2-year period by de Swart *et al.* (1994, 1995b, 1996) has demonstrated an impairment of several immune parameters in harbour seals (*Phoca vitulina*) fed herring from the polluted Baltic Sea when compared to those fed with fish from the Atlantic Ocean. Among impaired parameters, natural killer cell activity plays an important role in the first line of defence against viral infections (de Swart *et al.*, 1996). Moreover, those seals consuming contaminated herring accumulated higher body burdens of potentially immunotoxic organochlorines than seals fed relatively uncontaminated herring. In the latter study, heavy metal levels have unfortunately not been determined either in fish or in seals.

Possible immunosuppressive actions of other groups of environmental contaminants, such as heavy metals, cannot be ruled out. Indeed, many laboratories and epidemiological studies have demonstrated the immunotoxic effects of heavy metals in a variety of species (Zelikoff and Thomas, 1998). Direct cause and effect links between a single kind of contaminant and possible population declines has not been established so far, and many researchers have proposed the possibility of a synergistic role of different substances in increasing the susceptibility of affected animals to diseases or biotoxins.

In this chapter we will focus on heavy metals and their possible effects on marine mammals. Heavy metals are usually divided into essential (Zn, Cu, Cr, Se, Ni, Al) and non-essential metals (Hg, Cd, Pb), the latter being potentially toxic even at low concentrations. Nickel and copper hazards to wildlife have been reviewed extensively by Eisler (1997, 1998). When considering marine mammals, there are limited data about heavy metals, except mercury. Chromium, nickel and lead concentrations are generally low, rarely exceeding a few µg/g dry weight (dw) in marine mammal tissues. No meaningful assessment of their toxicity in marine mammals can be made as yet (Law, 1996). However, investigations carried out on a ringed seal population (*Pusa hispida saimensis*) from Finland showed a clear connection between stillbirth of the pups and nickel concentrations in the air (Hÿvarinen and Sipilä, 1984). These authors have underlined the considerable nickel input in the environment from industrial activity in that particular area. On the other hand, zinc, copper, cadmium and mercury concentrations often exceed several tens of µg/g dw and so will be discussed in particular, as well as their levels in the different marine mammal groups, detoxification mechanisms, potential hazards and ecological implications.

Factors affecting heavy metal concentrations in marine mammals

Some reviews about heavy metal contamination of marine mammals have been published (Wagemann and Muir, 1984; Thompson, 1990; Kemper *et al.*, 1994; Law, 1994, 1996). Tables 7.1 to 7.5 present some selected references. It appears that metal concentrations vary greatly within marine mammals, especially those of non-essential metals such as cadmium and mercury (Table 7.1).

The large variation in these data (Table 7.1) illustrates the numerous physiological and ecological factors that might affect heavy metal contamination: geographic location, diet, age, sex, the tissues considered and metabolic rates. Information concerning metabolic rates is largely unavailable for the different marine mammals species so will not be discussed here. However it must be kept in mind that ingestion and assimilation rates differ between all the marine mammals species in relation with their weight and their migration or physiological status (fasting).

Geographic location

Heavy metal contamination sources can be both anthropogenic and natural, and distinguishing between the two can be very difficult. The natural

Table 7.1 Examples of maximum and minimum metal concentrations for marine mammals

Metal	Minimum	Maximum
Hg	0.2	13 156
	Muscle	Liver
	Pusa hispida	*Tursiops truncatus*
	Western Arctic	Mediterranean Sea
	Wagemann *et al.* (1996)	Leonzio *et al.* (1992)
Cd	0.007	2324
	Muscle	Kidney
	Pusa hispida (<1 year)	*Pusa hispida* (8 years)
	Northwest Greenland	Northwest Greenland
	Dietz *et al.* (1998)	Dietz *et al.* (1998)
Zn	2	4183
	Blubber	Liver
	Leptonychotes weddellii (13 years)	*Dugong dugon* (>30 years)
	Antarctic	Australia
	Yamamoto *et al.* (1987)	Denton *et al.* (1980)
Cu	0.4	600
	Blood	Liver
	Leptonychotes weddellii (13 years)	*Tursiops truncatus*
	Antarctic	Argentina
	Yamamoto *et al.* (1987)	Marcovecchio *et al.* (1990)

When available, the age of the animals is given (all the data are expressed in µg/g dry weight, assuming a mean water content of 75 per cent of the tissue).

Table 7.2 Heavy metal mean and range concentrations in livers (and kidney for Cd) from odontocetes: selected references

Family	Species	Location	n (age estimation)	Hg-total	CH_3-Hg
Pontoporiidae	*Pontoporia blainvillei*	Argentina	2 (2 and 3 years)	15 ± 6	nd
Platanistidae	*Platanista gangetica*	India	4	nd	nd
		Argentina	1 (10 years)	344 ± 29	nd
	Tursiops truncatus	South Carolina, USA	34	71 <2–586	nd
		Gulf of Mexico	10 Adult males	180 20–351	nd
		Alaska	11	180 4–448	nd
Delphinidae	*Globicephala melas*	Faroë Islands	Hg: n = 8 Cd: n = 28	852 ± 776	17 ± 15%
		Newfoundland, Canada	26	63 0.3–298	nd
	Stenella coeruleoalba	Northeast Atlantic	Hg: n = 8 Others: n = 22	206 5–348	nd
		Pacific Ocean, Japan	Mature dolphins n = 15	820 ± 408	3.4%
	Delphinus delphis	Northeast Atlantic	28 Stranding	128 3–631	7%
	Lagenorhynchus albirostris	Newfoundland	26	3 0.5–6	nd
Monodontidae	*Monodon monoceros*	Baffin Island	38	24 ± 12	/
		West Greenland	n > 48	21[a] <0.02–171	/
	Delphinapterus leucas	West Greenland	40	7 0.3–123	/
		St. Lawrence, Canada	30	134 1.5–808	/
		Canadian western Arctic	77	108 1–464	/
		Canadian eastern Arctic	73	41 5–154	/
Kogiidae	*Kogia breviceps*	Argentina	1	47	/
Physeteridae	*Physeter macrocephalus*	North Sea	6	41 9–61	5% 2–8%
Phocoenidae	*Phocoena phocoena*	North Sea	5	170 1–504	37%
		Baltic sea	4	/	/
		West Greenland	44	16 2–80	/
	Phocoenoides dalli	Northwestern Pacific	3	6 (n = 1; fetus)	/

[a] Median.
When the range was not available, standard deviation was used. All the data are expressed in dry weight assuming a mean water content of 75 per cent of the tissues.

Cd liver	Cd kidney	Se	Cu	Zn	Source
13 ± 6	40 ± 16	nd	64 ± 9	330 ± 160	Marcovecchio *et al.* (1990)
0.1	2	nd	207	126	Kannan *et al.* (1993)
<0.04–0.15	<0.04–6		9–400	64–210	
3 ± 1	114 ± 17	nd	310 ± 1	785 ± 136	Marcovecchio *et al.* (1990)
0.2	nd	38	43	227	Beck *et al.* (1997)
0.02–1		0.7–189	5–316	46–1084	
2	nd	74	nd	nd	Kuehl and Haebler (1995)
0.4–5		41–133			
nd	nd	52	nd	nd	
		6–114			
308	344	nd	nd	nd	Caurant *et al.* (1996)
6–668	6–976	0–480	8–60	100–900	Caurant and Amierd-Triquet (1995)
78	43	31	17	264	Muir *et al.* (1988)
0–190	0–102	3–113	9–35	68–716	
17	91	nd	43	167	Das *et al.* (2000)
0.2–51	0.1–199		7–272	33–385	André *et al.* (1991b)
nd	nd	194 ± 115	nd	nd	Itano *et al.* (1984)
6	13	nd	12	143	Holsbeek *et al.* (1998)
0–96	0.4–81		3–32	65–293	
2	14	8	20	100	Muir *et al.* (1988)
0.2–8	2–44	4–12	3–32	43–136	
133 ± 130	298 ± 192	16 ± 7	21 ± 13	151 ± 40	Wagemann *et al.* (1983)
43[a]	156[a]	13[a]	/	144[a]	Hansen *et al.* (1990)
<0.06–295	<0.06–500	<0.8–144		53–271	
9	41	15	/	114	Hansen *et al.* (1990)
<0.06–34		2–111		87–181	
0.6	6	/	/	:	Wagemann *et al.* (1990)
0.004–2	0.004–15				
9	39	75	45	112	Wagemann *et al.* (1996)
0.5–27	15–88	3–235	3–140	43–185	
26	90	21	77	115	Wagemann *et al.* (1996)
0.03–103	0.3–314	1.5–91	3–1324	41–361	
30	1650	/	40	652	Marcovecchio *et al.* (1990)
82	258	18	8	104	Bouquegneau *et al.* (1997b)
52–175	133–426	6–43	5–12	90–125	Holsbeek *et al.* (1999)
/	/	/	/	/	Joiris *et al.* (1991)
0.3	1.5	/	24	120	Szefer *et al.* (1994)
0.3–0.4	0.2–3		18–6	96–144	
13	53	11	48	200	Paludan-Müller *et al.* (1993)
0.2–45	0.4–290	2–36	20–200	145–370	
/	/	/	/	/	Fujise *et al.* (1988)
0–84	0–136		20–329	110–186	

Table 7.3 Heavy metal mean and range concentrations in livers from mysticetes: selected references

Family	Species	n	Location	Hg-total	CH₃-Hg	Cd liver	Cd kidney	Se	Cu	Zn	Source
Eschrichtiidae	Eschrichtius robustus	10	Western North America	0.06 0.01–0.1	nd	4 0.06–6	4 0.1–6	2 0.3–3	9 0.6–25	99 2–160	Varanasi et al. (1994)
Balaenopteridae	Balaenoptera acutorostrata	135	Antarctic	0.2 0.1–0.5	nd	38 9–133	nd	nd	17 9–34	146 99–232	Honda et al. (1986, 1987)
		17	West Greenland	2[a] 1–11	nd	4[a] 2–6	15[a] 7–22	6[a] 4–10	nd	138[a] 106–192	Hansen et al. (1990)
	Balaenoptera physalis	11	Northeast Atlantic	2 0.6–5	40%	nd	nd	nd	nd	nd	Sanpera et al. (1993)
Balaenidae	Balaena mysticetus	20	Alaska	0.2 0.08–0.4	nd	31 2–88	nd	4 1–9	20 12–40	137 88–261	Krone et al. (1999)

[a] Median.
Cadmium concentrations are also given for the kidney.
All the data are expressed in dry weight, assuming a mean water content of 75 per cent of the tissues.
nd: not determined.

Table 7.4 Heavy metal mean and range concentrations in livers from pinnipeds: selected references

Family	Species	Location	n	Hg-total	CH₃-Hg	Cd liver	Cd kidney	Se	Cu	Zn	Source
Phocidae	*Pusa hispida*	Northwest Greenland	5	/	/	232 108–436	1596 1036–2324	/	/	/	Dietz et al. (1998)
		Swedish coasts	4	176 19–348	/	2.6 0.6–3.4	8 5–22	76 11–112	18 15–19	128 72–180	Frank et al. (1992)
	Phoca vitulina	Swedish coasts	8	104 5–264	/	0.4 0.2–0.7	2 1–3	44 16–104	34 6–52	216 76–248	Frank et al. (1992)
		Arctic	13	24 0.9–173	/	12 1–39	77 2–222	14 2–66	26 12–84	171 121–287	Wagemann (1989)
	Halichoerus grypus	Swedish coasts	9	104 92–368	/	0.7 0.3–0.8	5 2–11	56 39–208	56 28–72	184 140–304	Frank et al. (1992)
		Northwest England (Liverpool Bay)	12	590 6–1720	/	2 <0.06–4	/	/	53 9–112	218 88–356	Law et al. (1992)
	Leptonychotes weddellii	Antarctica	3	16 0.2–34	/	3 <0.02–5	17 <0.02–40	/	80 76–103	191 166–220	Yamamoto et al. (1987)
Otariidae	*Arctocephalus gazella*	Georgia	11	215 52–334	/	350 55–684	/	/	263 132–438	384 259–643	Malcolm et al. (1994)
	Arctocephalus philippii	Chile	Pups: n = 27 Adult: n = 1	9 ± 6 75	/	0.5 ± 0.2 2	/	/	/	/	Sepulveda et al. (1997)
Odobenidae	*Odobenus rosmarus*	Arctic	114	4 0.03–19	/	38 0.1–137	244 0.1–564	10 2–20	32 6–137	151 50–300	Wagemann and Stewart (1994)

Cadmium concentrations are also given for the kidney.
When the range was not available, standard deviation was used.
All the data are expressed in dry weight, assuming a mean water content of 75 per cent of the tissues.

Table 7.5 Heavy metals in livers from other marine mammals: selected references

Order	Family	Species	n	Location	Hg-total	CH₃-Hg	Cd liver	Cd kidney	Se	Cu	Zn	Source
Sirenia	Dugongidae	*Dugong dugon*	6	Australia	/	/	<0.1–59	0.2–309	/	9–608	219–4183	Denton *et al.* (1980)
Carnivora	Ursidae	*Ursus maritimus* (<7 years)	8	Northwest Greenland	86 ± 5	/	7 ± 6	79 ± 6	36 ± 5	/	/	Dietz *et al.* (1996)

All the data are expressed in dry weight.
Because mean values were not available for dugong, only the range are displayed.
The concentrations found in polar bears are expressed as mean ± standard deviation.

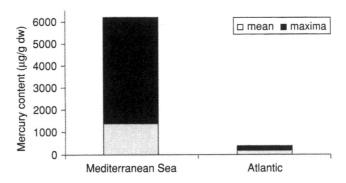

Figure 7.1 Mercury levels in livers from striped dolphins (*Stenella coeruleoalba*) from the north-east Atlantic and the Mediterranean Sea. Concentrations are expressed per gram dry weight, assuming a mean water content of 75 per cent of the tissues. (After André *et al.*, 1991a, b.)

background component of the input of heavy metals in marine ecosystems may be as important as the anthropogenic one, and, in some areas, it appears to be the major source. This is important because it emphasises that marine mammals have been exposed to heavy metals long before the development of human activities. For example, this is the case for the Mediterranean Sea and the Arctic, which are known for their high natural metal levels: mercury in the Mediterranean Sea and cadmium in the Arctic.

The mercury levels measured in dolphins from the Mediterranean Sea are higher than those encountered in dolphins from Pacific coasts of Japan or the north-east Atlantic (Figure 7.1) (Honda *et al.*, 1983; André *et al.*, 1991a; Leonzio *et al.*, 1992). According to André *et al.* (1991b), the origin of high mercury levels observed in Mediterranean dolphins is natural because of the large natural sources present in the Mediterranean basin.

The current state of knowledge of concentrations, spatial and temporal trends of contaminants, including heavy metals, have been described extensively in the Arctic (Muir *et al.*, 1992; Dietz *et al.*, 1996; AMAP, 1998). It seems that cadmium concentrations have always been high in the Greenland Arctic regions, as indicated by the lack of obvious temporal trends in sediment cores, as well as historic hair samples from the fifteenth century from both seals and Inuits (Dietz *et al.*, 1998). Johansen *et al.* (1980) first reported that cadmium levels in tissues of ringed seals from Greenland were higher than previously reported for seals from the North Sea. Cadmium seems to accumulate to higher levels in seals living in unpolluted Arctic waters than in those taking their preys in the North Sea area, which receives large inputs of pollutants. A similar observation can be made for porpoises (Figure 7.2).

Wagemann *et al.* (1996) have investigated the heavy metal distribution in belugas (*Delphinapterus leucas*), narwhals (*Monodon monoceros*) and ringed seals within the Arctic region. Mean mercury concentrations in the livers of

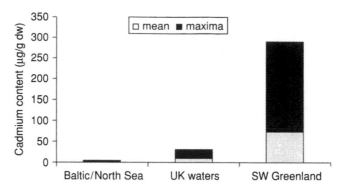

Figure 7.2 Cadmium levels in kidneys from porpoises (*Phocoena phocoena*) from the Baltic and North seas, and waters around the UK and south-western Greenland. Concentrations are expressed per gram dry weight, assuming a mean water content of 75 per cent of the tissues. (After Harms *et al.*, 1978; Falconer *et al.*, 1983; Paludan-Müller *et al.*, 1993.)

belugas and ringed seals were significantly higher in the western Arctic than in the eastern Arctic. This comparison was not possible for narwhals, as they are not found in the western Canadian Arctic. On the contrary, cadmium as well as zinc and copper concentrations in tissues (liver and kidney) of belugas and ringed seals were higher in the eastern than in the western Arctic. These differences in metal levels between marine mammals of the eastern and western Arctic corresponded to the different geological settings and sediments of these two regions (Wagemann *et al.*, 1995, 1996).

Routes of entry

There exist several different routes of entry of heavy metals in marine mammals: uptake from the atmosphere through the lungs, absorption through the skin, across the placenta before birth, via milk through lactating, ingestion of sea water and ingestion of food. Nevertheless, the major route of heavy metal contamination for marine mammals seems to be via feeding (André *et al.*, 1990a, b; Augier, 1993b; Law, 1996). Considering this, the following discussion will mainly refer to the diet (including suckling) and its influence on metal uptake.

Obviously, mysticetes are less contaminated by heavy metals than odontocetes and pinnipeds (which are located at higher trophic levels in the marine food web). Cadmium seems to be transferred to the highest trophic levels of the marine food chain mainly by molluscs, particularly cephalopods, which concentrate cadmium in their viscera (Honda and Tatsukawa, 1983; Bouquegneau and Joiris, 1988; Miles and Hills, 1994). Long-finned pilot whales (*Globicephala melas*), for example, are known to concentrate cadmium in relation to a preferential cephalopod diet. Elevated levels of cadmium in Pacific walrus (*Odobenus rosmarus divergens*) and northern fur seals (*Callorhinus ursinus*) have been reported in a population of the Bering Sea, a remote area away from industrial activity (Miles and Hills, 1994). In an attempt to find out which prey may transfer cadmium to the walrus, the most common preys (mainly bivalves) found in the stomach contents were analysed for their metal concentrations. Amongst the bivalves analysed, *Mya* sp. showed the highest cadmium content, suggesting that this prey could be a cadmium transmitter for walruses (Miles and Hills, 1994). Wagemann and Stewart (1994) have studied heavy metal concentrations of walruses (*Odobenus rosmarus rosmarus*) from the eastern Canadian Arctic in relation to their food, mostly bivalves (*Mya* sp. and *Serripes* sp.). These authors showed that bivalves could also be a source of lead for walruses, judging from the correspondence between the high levels of lead in both bivalves and walruses (Wagemann and Stewart, 1994).

However, some high cadmium levels have been reported in some marine mammals that do not eat cadmium-contaminated prey (Denton *et al.*, 1980). Trace metals have been determined in the dugong (*Dugong dugon*), which mostly feeds on seagrasses and algae. High cadmium and zinc concentrations have been detected in their livers and kidneys. Denton *et al.* (1980) have reported renal cadmium and hepatic zinc concentrations reaching, respectively, 309 µg/g and 4183 µg/g dry weight, while low levels of these metals are found in the seagrasses analysed. The authors suggested that the low levels of copper in seagrasses may influence intestinal absorption of zinc and cadmium in the dugong. Higher than normal zinc and cadmium absorption through the intestinal tract occurs when dietary copper intake is deficient, due to competition between the metals for available binding sites on carrier proteins within the intestine (Denton *et al.*, 1980). Unlike dugongs of coastal Australia, which display a copper deficiency due to their seagrass diet, Florida manatees (*Trichechus manatus*) are considered to be facing the opposite problem: copper concentrations in the livers of Florida manatees were significantly elevated in areas of high herbicidal copper usage (O'Shea *et al.*, 1984). Manatees might be inefficient at maintaining copper homeostasis in the face of dietary excess. The death of a captive dugong was attributed to exposure to copper sulphate added to its tank as an algicide (Oke, 1967).

The position of top predators in the food web influences mercury levels in marine mammals as this highly toxic metal is biomagnified, when available as methylmercury, through the food web (Bouquegneau and Joiris, 1988). Diets, and especially those of marine mammals relying on fish, are responsible for mercury contamination (Svensson *et al.*, 1992; Nakagawa *et al.*, 1997). Much higher mercury concentrations have been reported in minke whales (*Balaenoptera acutorostra*) from Greenland (Hansen *et al.*, 1990) compared to Antarctic minke whales (*Balaenoptera bonaerensis*), which are several times less contaminated (Honda *et al.*, 1986, 1987). Hansen *et al.* (1990) attributed the lower mercury concentrations found in tissues of Antarctic minke whales to differences in trophic levels and subsequently in the mercury content of the food items: northern minke whales feed mainly on fish while southern minke whales feed on krill. Indeed, the food web in the Antarctic ecosystem is rather simple, as the major food component is the Antarctic krill. The low trophic position of the Antarctic minke whale is reflected in the very low bioconcentration factor of mercury in this animal.

Age

Cadmium and mercury accumulated strongly with age in most marine mammal tissues analysed (e.g. Honda and Tatsukawa, 1981, 1983; Hamanaka *et al.*, 1982; Honda *et al.*, 1983; Augier *et al.*, 1993b). This increase is enhanced in the liver in which the excretion rate might be extremely low due to the fossilisation of mercury under a detoxified form (see below). However, some exceptions have been reported in the literature. Honda *et al.* (1986, 1987) have studied the heavy metal distribution in Antarctic minke whales, and compared it to their food habit and age. The age of these whales varies between 1 and 45 years. They found maximum concentrations of cadmium and mercury in the livers of 20-year-old minke whales. Both concentrations increase with age until about 20 years and thereafter decrease. Because there is no evidence that half-lives of cadmium and mercury change with age, Honda *et al.* (1986, 1987) suggested a higher food intake for the younger animals compared to older. They also suggested that these changes may be due to the significant decrease in stocks of blue whales (*Balaenoptera musculus*) and fin whales (*Balaenoptera physalis*). Both these species used to occupy ecological niches similar to those of the minke whales in the Antarctic marine ecosystems. The young minke whales would be less exposed to feeding competition from blue and fin whales, than the older ones. Accumulations of toxic metals such as cadmium or mercury may also have been influenced by this ecosystem disruption (Honda *et al.*, 1986, 1987).

High copper and zinc concentrations have also been observed in very young animals and neonates (e.g. Julshamn *et al.*, 1987; Wagemann *et al.*, 1988; Caurant *et al.*, 1994). These essential elements are known to increase in tissues undergoing rapid development and differentiation (Baer and Thomas, 1991). It has been suggested that these higher levels encountered in young might reflect a specific requirement in newborns or a very low excretion rate of these metals by the fetus (Wagemann *et al.*, 1998).

Sex

Reproductive activities such as pregnancy, parturition and lactation can modify the metal levels. Several studies have demonstrated that metal transfer from females to pups occurs through the placenta or lactation: Honda *et al.* (1987) reported a hepatic iron, cobalt, lead and nickel transfer from mother to pup. As a consequence, these metal concentrations decrease in the mature female with the progress of gestation.

Canella and Kitchener (1992) found significantly lower levels of mercury in pregnant and lactating sperm whales (*Physeter macrocephalus*) when compared with non-breeding females. They suggested that this may be due to hormonal changes or stress causing the redistribution of mercury in body tissues.

Distribution within organs

The pattern of distribution of metals within the organism is tissue and metal specific. For example, mercury is mostly concentrated in the liver, with kidney and muscle having successively lower levels. On the contrary, the highest cadmium concentrations are usually encountered in the kidney, due to the presence of metal-binding proteins. This pattern prevails in most marine mammals (Wagemann and Muir, 1984).

Yamamoto *et al.* (1987) have studied the distribution of heavy metals in the whole organism of three Weddell seals (*Leptonychotes weddellii*) from the Antarctic. These authors estimated the whole metal body burden, which was calculated from the weight of the different tissues and their respective concentrations. If whole-body burdens of metal are estimated for an adult Weddell seal, zinc is mostly located in muscles and in bones, copper in muscles and liver, mercury in liver and muscles, and cadmium in liver and kidney. The apparent contrast between the low concentration of mercury generally measured in marine mammal muscles and the high burden encountered here is due to the high muscle mass of these animals. The redistribution of mercury from highly contaminated organs such as liver or kidney through muscles seem to be a protection mechanism against mercury toxicity (Cuvin-Aralar and Furness, 1991).

Skin contains generally low mercury concentrations (Yamamoto *et al.*, 1987). However, the skin of marine mammals is not a homogeneous tissue. It consists of four distinct layers, in which mercury increases progressively outwards, with a concentration of 6 µg/g (estimation in dry weight) in the outermost layer of Arctic belugas and narwhals. During the process of moulting, the outermost and the underlying layers are shed, thus approximately 20 per cent of the total mercury in the skin is lost annually in belugas and 14 per cent in narwhals (Wagemann *et al.*, 1996).

Toxicity of mercury and cadmium

As quoted above, the accumulation through the food chain is a major risk for top predators. The accumulation of chemical substances may result in toxic concentrations in organisms (secondary poisoning) even if the concentration in the environment remains below the threshold level for direct toxicity (Nendza *et al.*, 1997). The finding of high concentrations of metals such as cadmium or mercury have raised the question about their toxicity.

Reliable toxicity data for predating marine mammals are scarce. Instead, threshold levels are often extrapolated from terrestrial species, i.e. interspecies correlations are assumed to hold for rats or captive seals. The validity of these extrapolations is highly questionable and can only be justified by the current lack of better data and by ethical considerations. Indeed, potential effects of toxic metals cannot be tested in free-living animals because experimental manipulations are undesirable. *In vitro* experimentations or systematic post-mortem investigations to establish the disease status of contaminated animals in a relatively large number of individuals from the same species are complementary and valuable alternative ways of understanding the numerous processes involved.

Mercury toxicity

Mercury exists in several interchangeable forms in the biosphere (Kaiser and Tölg, 1980), but the mercury accumulation through the food web mainly occurs in an organic form, methylmercury (MeHg), as a result of its lipid solubility and preferential assimilation during zooplankton grazing (Mason *et al.*, 1995). Above zooplankton, organic mercury is biomagnified along the food chain up to marine mammals.

Total mercury concentration is a poor indicator of toxic effects, as organic mercury compounds seem to be considerably more toxic to animals than inorganic mercury. The biological and toxicological activity of mercury depends on the form that is taken up, the route of entry in the body (skin, inhalation or ingestion), and on the extent to which mercury is absorbed (Kaiser and Tölg, 1980). Wolfe *et al.* (1998) have recently reviewed the toxicity on mercury on wildlife.

In mammals, methylmercury toxicity is manifested primarily as central nervous system damage, including sensory and motor deficits and behavioural impairment. Animals become anorexic and lethargic. Methylmercury is easily transferred across the placenta (Wagemann *et al.*, 1988) and thus concentrates in the fetal brain (Wolfe *et al.*, 1998). This reproductive effect ranges from development alterations in the fetus to fetal death.

Methylmercury is also absorbed by the gastrointestinal tract, while inorganic salts of mercury are less readily absorbed. Mercury is absorbed from fish mainly as the methylated form since almost all the mercury present in fish is methylated (Svensson *et al.*, 1992).

Experimental seal intoxication has led the animals to lethargy, weight loss and finally death (Ronald and Tessaro, 1977). Four harp seals (*Pagophilus groenlandicus*) were intoxicated with methylmercury by a daily oral intake. Two seals were fed with mercury doses of 0.25 mg/kg of body weight per day for 60 and 90 days. These two seals did not show abnormal blood concentrations but exhibited a reduction in appetite and consequent weight loss. Two others seals fed 25 mg/kg of body weight per day died on day 20 and day 26 of exposure. The measurements of blood parameters indicated toxic hepatitis, uraemia and renal failure. These pathologies have been related to high accumulation of mercury in these tissues. The liver concentrations reached more than 500 µg/g dry weight after the death. Almost 90 per cent of the mercury analysed in the liver was methylmercury. No detoxification mechanisms were described in this case.

However, this experimental study did not reproduce the real daily food intake of marine mammals in the wild. Nigro and Leonzio (1993) have calculated the mean daily food intake of small cetaceans to be approximately 3 kg of fish and cephalopods, with an average mercury concentration of 0.3 mg/kg fresh weight, from which the mean dietary mercury intake for an adult specimen can be estimated at 0.9 mg mercury/day for the whole animal. This is quite far from the 25 mg/kg of body weight/day administrated in gel caps to seals in the study of Ronald and Tessaro (1977). Another feature that might explain the absence of detoxification is the absence of selenium added to the food. In the wild, if fish are a source of exposure to mercury, they are also a source of selenium (Svensson *et al.*, 1992), and in marine mammals demethylation mechanisms occur in the presence of selenium (see below). In this experimental study, the lack of additional selenium in the diet was probably the limiting factor to detoxification.

Very few studies have tried to link metal concentrations measured in free-ranging marine mammals and health status (Hÿvarinen and Sipilä, 1984; Rawson *et al.*, 1993, 1995; Siebert *et al.*, 1999; Bennet *et al.*, 2001). Only one case of mercury toxicity has been reported by Helminen *et al.* (1968): the ringed seal suspected of mercury intoxication was from an area of heavy industrial mercury dumping. Chronic mercury accumulation was associated with liver abnormalities observed in stranded bottlenose dolphins from the Atlantic. Large deposits of a brown pigment, identified as lipofuscin, in the portal areas of the liver were observed in the livers of nine animals with high hepatic mercury levels (>60 µg/g fresh weight). Analytical electron microscopy carried out on these pigments demonstrated that mercuric selenide (HgSe) was the predominant material (Rawson *et al.*, 1995). Lipofuscin is believed to be derived from damaged subcellular membranes. This pigment accumulation strongly correlated with mercury concentrations. Mercury would have inhibited the activity of lysosomal digestive enzymes and, therefore, reduced degradation of proteins, leading to excessive accumulation of lipofuscin within cells and finally cell death (Rawson *et al.*, 1993).

More recently, Siebert *et al.* (1999) examined the possible relationship between mercury tissue concentrations and disease in harbour porpoises from the German waters of the North and Baltic seas. A higher mercury content has been measured in organs of harbour porpoises from the North Sea compared to those of the Baltic Sea, indicating that mercury is a more important threat for animals of the North Sea than for those of the Baltic Sea. High mercury concentrations were associated with prevalence of parasitic infection and pneumonia.

Bennet *et al.* (2001) have also used this indirect approach to investigate the prediction that increased exposure to toxic metals results in lowered resistance to infectious disease in harbour porpoises from the coasts of England and Wales. Mean liver concentrations of mercury, selenium, mercury : selenium ratio and zinc were significantly higher in the porpoises that died of infectious diseases (parasitic, bacterial, fungal and viral pathogens such as pneumonia), compared to porpoises that died from physical trauma (most frequently entrapment in fishing gear). Liver concentrations of lead, cadmium, copper, and chromium did not differ between the two groups.

In some cases, balances between elements seem to be more important than the absolute concentration when the possibility of toxic effects is considered. High premature birth rates have been observed and studied between 1968 and 1972 in the Californian sea lion (*Zalophus californianus*) from the southern California Channel Island rookeries (Martin *et al.*, 1976). These premature pups were ataxic, had difficulties in breathing and died shortly after birth. Heavy metals were analysed and compared between normal and premature pups and between their respective mothers. The results revealed that severe imbalance in the mercury : selenium : bromine ratio occurred in the livers of the abnormal mothers. The absolute concentrations seemed not to be involved in this case as mercury, selenium and bromine were in higher concentrations in the livers of normal mothers compared to abnormal mothers. This suggests that the selenium : mercury balance is a very complex phenomenon and might be more important for general health status than absolute concentrations.

Some *in vitro* studies have also been carried out to evaluate the potential hazard of mercury in marine mammals. Freeman and Sangaland (1977) demonstrated that methylmercury alters the *in vitro* synthesis of steroid hormones which play an important role in reproduction. Genetic effects of methylmercury on lymphocytes of one bottlenose dolphin (*Tursiops truncatus*) have also been evaluated *in vitro* by Betti and Nigro (1996). Lymphocytes were isolated from blood obtained from a 15-year-old dolphin (Adriatic Sea). Methylmercury induces DNA single-strand breaks and cytotoxicity in a dose-dependent manner. The doses of MeHg used in this study are likely to be in the range of concentrations (between 1 and 10 µg/ml) naturally occurring in the blood of wild dolphins found in the Mediterranean Sea (Betti and Nigro, 1996). It appears that dolphin lymphocytes have a greater resistance both to the genotoxic and cytotoxic effects of MeHg when compared to human or rat cells. This feature can be interpreted as an adaptation acquired by dolphins to counter the methylmercury exposure.

Cadmium toxicity

Cadmium is regarded as one of the most toxic metals. High dietary concentrations of cadmium in humans can lead to well-known heavy skeletal deformities ('itai-itai' disease), kidney lesions (mainly on the proximal tubules) usually preceding lung damage, dysfunction of cardiovascular and haematopoietic systems, as well as carcinogenic, mutagenic and teratogenic effects (Förstner, 1980; Lamphère *et al.*, 1984; Jonnalagadda and Prasada Rao, 1993). The effect of cadmium on marine ecosystems has been reviewed recently (AMAP, 1998). The renal concentrations can reach levels as high as 2000 µg/g dry weight in some Arctic ringed seals (Dietz *et al.*, 1998). This is much higher than the critical concentrations of approximately 800 µg/g dry weight (200 µg/g fresh weight) associated with kidney damage in mammals including humans (WHO, 1992). Moreover, following Elinder and Järup (1996), this critical concentration has been largely overestimated, as cadmium-induced renal dysfunctions have been displayed with kidney cortex concentrations of the order of 200 µg/g dry weight (50 µg/g fresh weight). For comparison, in human adults, the renal cadmium concentrations amongst non-smokers is less than 5 µg/g fresh weight (Pesch *et al.*, 1989). No obvious cadmium toxic effect has yet been registered in marine mammals, despite the high levels encountered in several species, suggesting highly efficient detoxification mechanisms (Dietz *et al.*, 1998).

Detoxification mechanisms

The exposure of marine mammals to heavy metals has occurred throughout their evolutionary history, during which they have developed mechanisms either to control the internal concentration of certain elements or to mitigate their toxic effects. The most obvious case is that of mercury in dolphins.

Compared to other terminal consumers, such as tunas or seabirds, some marine mammals accumulate much higher levels of mercury, with biomagnification factors in respect to prey of 500 in dolphins compared to 30 in predatory fish, for example (Leonzio, 1996). This can be explained by physiological differences not only in the involved uptake and release, but also – and sometimes mainly – in detoxification processes.

Mercury detoxification

Wagemann and Muir (1984) found mercury and selenium concentrations reaching up to 510 µg/g fresh weight (approximately 2000 µg/g dry weight). Despite such extremely high values, the animals did not show any overt signs of mercury or selenium poisoning because the presence of the two elements together provided protection to the animal. Many studies have demonstrated the mutual antagonism between mercury and selenium (Pelletier, 1985; Cuvin-Aralar and Furness, 1991). The mutual antagonism between these two elements has become one of the strongest and most general examples of interactions between heavy metals. This phenomenon occurs throughout the animal kingdom from oysters and shrimps to marine mammals and human beings. Koeman *et al.* (1973, 1975) first reported the strong correlation between mercury and selenium in livers of marine mammal species. A molar

Hg : se ratio of approximately 1 has been observed, suggesting mercury detoxification mechanisms in the presence of selenium.

Different forms of mercury coexist in the environment. Methylmercury is known to be one of the most toxic. Mercury is transferred up to marine mammals in a methylated form. However, very small amounts of methylmercury are generally found in the livers of marine mammals: less than 10 per cent of mercury is present in a methylated form in the livers of adult marine mammals. Mercury also occurs in an inorganic form (Wagemann *et al.*, 1998), which implies that a demethylation process occurs (Joiris *et al.*, 1991). The fate of this inorganic mercury has been mainly elucidated by histological studies carried out in livers from specimens of Cuvier's beaked whale (*Ziphius cavirostris*) and bottlenose dolphins (Martoja and Viale, 1977; Martoja and Berry, 1980). These authors first observed granules composed of mercuric selenide (HgSe). Successively, similar granules were also described in the striped dolphin (*Stenella coeruleoalba*). Mercury and selenium occurred as dense intracellular granules, located mainly in the liver macrophages, the Kupffer cells, and in the proximal tubules of the kidney. These granules appear as spherical or polygonal particles ranging from 15 to 80 Å (Augier *et al.*, 1993a; Nigro and Leonzio, 1993, 1996).

More recently, Rawson *et al.* (1995) found HgSe crystals in both the liver and respiratory system of the bottlenose dolphin and short-finned pilot whale and reported HgSe in the lung and hilar lymph nodes associated with soot particles. In both the liver and hepatic lymph nodes, these crystals were small, averaging 50 Å. In the lung and hilar lymph nodes, the crystals were much larger, measuring 250–500 Å. Abundant carbon was present in the hilar nodes, while only very small amounts were found in the hepatic nodes and in the liver. These findings suggest an alimentary and a respiratory entry for mercury in cetaceans: mercury in the liver is likely to be trophically acquired, passing through the gastrointestinal tract and carried to the liver by the way of portal veins. In the liver, it may be converted into HgSe which accumulates as an end product (Martoja and Berry, 1980). In animals producing large amount of HgSe, some of this may be carried to the hepatic lymph nodes and even to the spleen (Rawson *et al.*, 1995). HgSe in the lungs and the hilar nodes appears to be closely associated with carbon, suggesting an atmospheric association between these elements. Indeed, mercury and selenium pollution is largely attributed to the burning of fossil fuel or waste incinerations and these elements tend to aggregate as particles (Rawson *et al.*, 1995). However, *in vivo* precipitation of HgSe into the surfaces of inhaled soot particles cannot be ruled out, and further investigations are needed to understand this lung accumulation process. These mercuric selenide granules seem to be the last step of a very efficient detoxification mechanism leading to high but non-toxic concentrations in the organs. The molar ratio of 1 has been generally observed in marine mammals. However, the Hg : Se molar ratio found in the liver can vary within the range of 0.2 (Hansen *et al.*, 1990) to 2.49 (Caurant *et al.*, 1996).

Palmisano *et al.* (1995) have explained this variation: a Hg : Se molar ratio of approximately 1 has been observed in the livers of striped dolphins only after a certain threshold in the total mercury concentration (approximately 100 µg Hg/g fresh weight) has been exceeded. Palmisano *et al.* (1995) have proposed a two-stage mechanism for the demethylation and accumulation process. At low mercury levels (first stage) the metal is retained mainly in its methylated form. At higher mercury levels (second stage) demethylation, with a concurrent accumulation of selenium, seems to be the prevailing mechanism. Moreover, above the threshold only a small fraction of mercury is present in a labile form as Hg^{2+} and $MeHg^+$ bound to the cysteine residue. These authors have determined that 63 per cent of the total mercury analysed in the liver of one dolphin is involved in the formation of a very insoluble selenocompound, certainly present as HgSe (tiemannite), but in addition, mercury can be involved in the formation of other selenocompounds as Hg-selenoproteins. This hypothesis of a threshold has also been suggested in other studies dealing with mercury speciation (Sanpera *et al.*, 1993; Caurant *et al.*, 1996). Sanpera *et al.* (1993) have found no decrease in the organic mercury fraction with increasing total mercury concentration in the livers of fin whales (*Balaenoptera physalis*). On the contrary, Caurant *et al.* (1996) found a decreasing correlation between organic mercury and total mercury in the livers of long-finned pilot whales. Organic mercury was lower than 2 per cent when the total mercury concentration was higher than 100 µg/g fresh weight (400 µg/g dry weight). Only a small fraction (<1 per cent) of total mercury is bound to heat-stable compounds (which include the metallothioneins). These heat-stable proteins are able to bind bivalent metals and are believed to participate in the heavy metal detoxification processes (see above). More than 90 per cent of the mercury was in the insoluble fraction, except in two animals. In these two individuals, insoluble mercury in the liver was less than 90 per cent and 14 and 35 per cent of the mercury was bound to metallothioneins. These individuals exhibited total mercury concentrations below 50 µg/g fresh weight (200 µg/g dry weight) in their livers. When total mercury was higher than this value, the percentage bound to metallothioneins was low, always less than 1 per cent, whatever the total mercury concentration (Caurant *et al.*, 1996).

In some species, mercury concentrations can be low during their entire life span. Fin whales feed at the bottom of the trophic web (Sanpera *et al.*, 1993) and their mercury levels stay relatively low. In their livers, the mean ratio of organic/total mercury is about 40 per cent. This result seems very high compared to other data (see Table 7.6). The authors suggest that demethylation is carried out at a constant rate, probably because concentrations remain low throughout their life span, between 50 and 500 µg/g dry weight (Sanpera *et al.*, 1993). It seems in this case that the specific threshold has not been reached and so high percentages of organic mercury are observed in the livers.

Table 7.6 Comparison of mean percentage (%) methylmercury to total mercury in suckling marine mammals compared to adults

Species	Pups	Adults	Sources
Pagophilus groenlandicus	70	14	Wagemann *et al.* (1988)
Stenella coeruleoalba	45	2.5	Itano *et al.* (1984)
Globicephala melas	55	17	Caurant *et al.* (1996)

Young marine mammals can display a high percentage of methylmercury (Table 7.6), suggesting that they do not reach a specific threshold (Palmisano *et al.*, 1995). A second hypothesis is that young individuals are still unable to demethylate mercury efficiently (Caurant *et al.*, 1996).

Cadmium detoxification

As quoted above, the concentrations of cadmium in kidneys of marine mammals can reach levels more than two times the critical concentrations of approximately 800 µg/g dry weight (200 µg/g fresh weight) associated with kidney damage in terrestrial mammals, including humans (WHO, 1992). This raises questions about the health status of such heavily cadmium-contaminated marine mammals. Dietz *et al.* (1998) have compared kidneys with low and high cadmium contamination from ringed seals from northwest Greenland, by macroscopic and light microscopic examination. No differences in renal morphology could be observed between experimental groups. These investigations indicate that marine mammals appear to be able to maintain considerable concentrations of cadmium without showing renal damage. Dietz *et al.* (1998) have therefore postulated that ringed seals are adapted to the naturally high cadmium levels of the Greenland Arctic regions.

Role of metallothioneins

Marine mammals might mitigate the toxic effects of cadmium through its binding to metallothioneins (reviewed by Das *et al.*, 2000). The role of metallothioneins in cadmium detoxification has often been raised since the first isolation of this protein in horse kidney cortex by Margoshes and Vallee (1957). The presence of these low molecular weight proteins has been further demonstrated in many organisms, ranging from blue-green algae to primates (Kägi, 1991). These soluble and heat-stable proteins are found in the cytosolic fraction of the tissues. They are quite unique as they are characterised by a high cysteine content, divalent ion inducibility, such as Cu^{2+} and Zn^{2+}, and high affinity in the binding of these cations (Kägi, 1991; Roesijadi, 1992, 1996). The primary function of metallothioneins is the homeostasis of essential heavy metals, specifically zinc and copper. Zinc-thionein levels increase in tissues undergoing rapid development and differentiation, such as the neonatal rat liver. Zinc-thioneins are also able to transfer zinc to metal-dependent enzymes in the case of high metabolic demand (Baer and Thomas, 1991). As a result of this capacity to bind cations, metallothioneins are able to bind non-essential metals such as Cd^{2+}, Ag^+, Hg^{2+} and Pb^{2+}. In this way, these proteins reduce the bioavailability of these ions and, consequently, their toxicity. Metallothioneins are produced by several tissues, but liver and kidney cells are the more potent producers and consequently preferential accumulation of heavy metals occurs in these tissues (Kägi, 1991). According to Cosson *et al.* (1991), the participation of metallothioneins in detoxification mechanisms would be due to nothing more than fortuitous interactions of foreign cations with the normal homeostasis mechanisms for zinc and, perhaps, copper. However, these proteins are able to bind important amounts of cadmium and so prevent the cellular damage in the organism (Viarengo, 1989; Roesijadi, 1992).

Metallothioneins have already been characterised from livers and kidneys of different marine mammals such as grey seals (*Halichoerus grypus*) and northern fur seals (Olafson and Thompson, 1974), California sea lions (Ridlington *et al.*, 1981), striped dolphins (Kwohn *et al.*, 1986), harbour seals (Mochizuki *et al.*, 1985; Tohyama *et al.*, 1986), narwhals (Wagemann and Hobden, 1986) and sperm whales (Bouquegneau *et al.*, 1997b; Holsbeek *et al.*, 1998). The role of metallothioneins has been reviewed recently by Das *et al.* (2000). A study performed on harbour seals caught on Japanese coasts showed that metallothionein concentrations are significantly correlated with the level of cadmium and copper in the livers and cadmium, zinc, copper and inorganic mercury in the kidneys (Tohyama *et al.*, 1986). Correlations were also found with age.

The percentage of the cytosolic cadmium bound to metallothioneins varies from 9 to almost 100 per cent (Table 7.7). It is interesting to note that this 9 per cent value has been measured in highly debilitated sperm whales found stranded on the Belgian coast (Bouquegneau *et al.*, 1997b; Holsbeek *et al.*, 1998). This implies that cadmium did not occur in a detoxified form. These animals were seriously debilitated, as indicated by their reduced blubber thickness and body weight (Jauniaux *et al.*, 1998). Cadmium, which is known to induce debilitation in mammals, can be considered as one of the factors responsible for the debilitation of these animals, a condition which could have favoured their stranding, in addition to stress and starvation (Bouquegneau *et al.*, 1997b).

Other results described in Table 7.7 have been obtained from captured and not stranded animals. In the narwhal, more than 80 per cent of the hepatic cadmium and 90 per cent of the renal cadmium was located in the cytosolic fraction of the tissues in which metallothioneins are located. Almost all this cytosolic cadmium was bound to these proteins, indicating that the metallothionein mechanism was not saturated (Wagemann *et al.*, 1984, 1988). This could be an adaptation of this Arctic species to the high cadmium

Table 7.7 Cadmium speciation in the whole tissue, the cytosolic fraction and metallothioneins

Species	Tissues	Cd (µg/g dw)	(%) of Cd in cytosolic fraction	% Cd bound to cytosolic metallothioneins	Sources
Stenella coeruleoalba	Kidney	87	58	98	Kwohn *et al.* (1986)
Zalophus californianus	Kidney	37	68	nd	Ridlington *et al.* (1981)
	Liver	<dl	<dl	<dl	
Globicephala melas	Kidney	548	nd	54	Amiard-Triquet and Caurant (1997)
	Liver	312	nd	51	
Physeter macrocephalus	Liver	50	90	nd	Ridlington *et al.* (1981)
		95	49	9	
	Kidney	258	83	40	Bouquegneau *et al.* (1997b), Holsbeek *et al.* (1999)
Monodon monoceros	Kidney	332	92	72	Wagemann *et al.* (1984, 1986)
	Liver	176	88	77	

Cadmium concentrations are estimated in µg/g dry weight.
nd, not determined; dl, detection limit.

concentrations present in its environment. More data on cadmium speciation in marine mammals should be collected to better understand the precise role of metallothioneins in detoxification processes. It has been suggested that cadmium toxicity occurs when available metallothioneins are insufficient to bind all the cadmium. Recent experiments with mice genetically deprived of metallothioneins due to the loss of functional *MTI* and *MTII* genes (coding for the two main isoforms of metallothionein involved in the detoxification process) confirm the protective role of these proteins against cellular damage from metals such as cadmium or inorganic mercury (Satoh *et al.*, 1997; Klaassen and Liu, 1998).

Cadmium spherocrystals

Cadmium-containing granules have been observed in the kidney of two white-sided dolphins (Gallien *et al.*, 2001). These two individuals, with high cadmium concentrations, exhibited electron-dense mineral concretions with diameters up to 300 nm in the basal membranes of the proximal tubule. These spherocrystals are made up of numerous strata mineral deposits of calcium, phosphorus and cadmium. Cadmium has been detected with a molar ratio of Ca : Cd of 10 : 1 in the middle of these concretions. The occurrence of metal-containing granules is well documented in invertebrates (Simkiss, 1976) but this is the first report of granules containing cadmium in wild vertebrates. In these marine mammals exhibiting high levels of cadmium, these granules could constitute a way of immobilisation and detoxication.

Limits to detoxification

Remarkable tolerance of marine mammals to heavy metals has been suggested through several detoxification processes such as tiemmanite storage and binding to metallothioneins, but is there a limit to the detoxification process and, if so, what is the actual hazard of heavy metals?

The ratios between different metals appear more important than their absolute concentrations (Martin *et al.*, 1976; Becker *et al.*, 1995). Pups are more affected by these metals as they exhibit a higher methylmercury ratio compared to total mercury, due to their poorly efficient detoxification mechanism (Wagemann *et al.*, 1988). Moreover, a depressed molar ratio of bromine: mercury: selenium in premature pups of California sea lions was suggested to be a main death factor (Martin *et al.*, 1976).

Caurant *et al.* (1996) have proposed that detoxification of mercury could be limited in lactating long-finned pilot whales. Indeed, compared to other females, mercury concentrations were much higher in lactating females, while selenium concentrations were lower. Squid is the major food item of pilot whales, but a greater quantity and variety of fish species have been observed in the diet of lactating females. The authors suggested that the energy value of fish is higher than squid and a higher consumption of fish would cover the increasing need for energy to produce milk. The percentage of selenites (the inorganic form of selenium) and the inorganic forms of mercury seem to be higher in squids than in fish muscles, where methylated mercury is dominant. Most of selenium found in fish occurs in an organic form, which is less efficient than selenites in preventing the toxicity of mercury. The different diet of lactating females could induce both higher levels of mercury and a lower efficiency of the detoxification process (Caurant *et al.*, 1996).

Binding of cadmium to metallothioneins could also be a limiting factor to detoxification. In debilitated sperm whales stranded on the Belgian coast, only a small amount of cadmium was bound to metallothioneins (Bouquegneau *et al.*, 1997b; Holsbeek *et al.*, 1999). As quoted above, this could explain the observed debilitation which could have favoured their stranding. Precipitation of cadmium in a granule form could also lead to some toxic effect (Gallien *et al.*, 2001). The authors underlined the fact that lesions could be associated with these granules, especially in older animals exhibiting high cadmium concentrations, and so detoxification processes could lead to some toxic effects.

We have to keep in mind that any detoxification process has a cost for the cell or the organism involved and might have a threshold. This threshold can not be fully defined in terms of tissue concentration because of the number of parameters that can interact to limit physiological pathways that lead to detoxification. For example, gender and hormonal activity can modulate the synthesis of metallothioneins (Blazka and Shaikh, 1991). Moreover, detoxification processes can lead to the formation of compounds which could have toxic effects. In mice, the accumulation and degradation of cadmium–metallothionein complex (CdMT) in the renal tubular epithelial cells can induce nephrotoxicity that is counteracted by zinc, which has a protective effect against this CdMT-induced nephrotoxicity (Liu *et al.*, 1996; Tang *et al.*, 1998). As a result of their physiological function in the homeostasis of essential metals, metallothioneins could be involved in many cellular pathways. Thus, they could modulate physiological process as an indirect effect of heavy metal exposure. For example, metallothioneins have been demonstrated as potential modulators of some parameters of the immune response (Leibbrandt *et al.*, 1994; Borghesi *et al.*, 1996). Detoxification pathways could therefore lead to more subtle toxic effects underlying the complexity of the toxic effects of heavy metals.

Conclusions

The actual toxic effects of heavy metals on marine mammals remain unclear. Are they responsible, even in part, for the decline of some marine mammal species? This decline is obviously multifactorial: past overfishing, present increasing human activities and accumulation of pollutants, among which heavy metals can not be neglected. The role of marine mammals on the whole marine ecosystem is still poorly understood: their contribution in the recycling of nutrients is not very important, but their part in structuring marine communities and in modifying benthic habitats is more and more evident (Bowen, 1997). Marine mammals presently consume at least three times more prey than do human fisheries, but that could be low compared to their ecological role, which is still poorly understood. Some species compete with fisheries, while others obviously do not, and yet others partly compete but could be useful to fisheries by regulating the development of non-commercial species, thereby limiting excessive competition with commercial ones (Bouquegneau *et al.*, 1997a). It is our belief that marine mammals deserve their place in the oceans and are worth protecting.

Acknowledgements

This work was supported by a grant from the Belgian Office for Scientific, Technical and Cultural Affairs (Contract MN/DD/50). The authors wish to thank A. Distèche, K. Finnson and C. Beans for reading the manuscript.

References

Aguilar A. and Borrell A. (1994). Abnormally high polychlorinated biphenyl levels in striped dolphins (*Stenella coeruleoalba*) affected by the 1990–1992 Mediterranean epizootic. Sci. Total. Environ. 154: 237–247.

AMAP (1998). AMAP Assessment Report: Arctic Pollution Issues. Artic Monitoring and Assessment Programme (AMAP), Oslo, Norway.

Amiard-Triquet C. and Caurant, F. (1997). Adaptation des organismes marins à la contrainte métallique: l'exemple du cadmium chez *Globicephala melas*, mammifère Odontocète. Bull. Soc. zool. Fr. 122(2): 127–136.

André J., Amiard J.C., Amiard-Triquet C., Boudou A., Ribeyre F. (1990a). Cadmium contamination of tissues and organs of Delphinid species (*Stenella attenuata*) – Influence of biological and ecological factors. Ecotoxicological and Environmental Safety 20: 290–306.

André J.M., Ribeyre F., Boudou A. (1990b). Mercury contamination levels and distribution in tissues and organs of Delphinids (*Stenella attenuata*) from the Eastern Tropical Pacific, in relation to biological and ecological factors. Mar. Environ. Res. 30: 43–72.

André J.M., Boudou A., Ribeyre F. (1991a). Mercury accumulation in Delphinidae. Water, Air and Soil Pollut. 56: 187–201.

André J.M., Boudou A., Ribeyre F., Bernhard M. (1991b). Comparative study of mercury accumulation in dolphins (*Stenella coeruleoalba*) from French Atlantic and Mediterranean coasts. Sci. Total Environ. 104: 191–209.

Augier H., Benkoël L., Brisse J., Chamlian A., Park W.K. (1993a). Microscopic localization of mercury–selenium interaction products in liver, kidney, lung, and brain of Mediterranean striped dolphins (*Stenella coeruleoalba*) by silver enhancement kit. Cell. Mol. Biol. 39(7): 765–772.

Augier H., Benkoël L., Chamlian A., Park W.K., Ronneau C. (1993b). Mercury, zinc and selenium bioaccumulation in tissues and organs of Mediterranean striped dolphins *Stenella coerulealba* Meyen. Toxicological result of their interaction. Cell. Mol. Biol. 39(6): 621–634.

Baer K.N. and Thomas P. (1991). Isolation of novel metal-binding proteins distinct from metallothionein from spotted seatrout (*Cynoscion nebulosus*) and Atlantic croaker (*Micropogonias undulatus*) ovaries. Mar. Biol. 108: 31–37.

Beck K.M., Fair P., McFee W., Wolf D. (1997). Heavy metals in livers of bottlenose dolphins stranded along the south Carolina coast. Mar. Pollut. Bull. 34(9): 734–739.

Becker P.R., Mackey E.A., Demiralp R., Suydam R., Early G., Koster B.J., Wise S.A. (1995). Relationship of silver with selenium and mercury in the liver of two species of toothed whales (Odontocetes). Mar. Pollut. Bull. 30(4): 262–271.

Bennet P.M., Jepson P.D., Law R.J., Jones B.R., Kuiken Y., Baker J.R., Rogan E., Kirkwood J.K. (2001). Exposure to heavy metals and infectious disease mortality in harbour porpoises from England and Wales. Environ. Pollut. 112: 33–40.

Betti C. and Nigro M. (1996). The Comet assay for the evaluation of the genetic hazard of pollutants in cetaceans: preliminary results on the genotoxic effects of methyl-mercury on the bottle-nosed dolphin (*Tursiops truncatus*) lymphocytes *in vitro*. Mar. Pollut. Bull. 32(7): 545–548.

Blazka M.E. and Shaikh Z.A. (1991). Sex difference in hepatic and renal cadmium accumulation and metallothionein induction. Biochem. Pharmacol. 41(5): 775–780.

Borghesi L.A., Youn J., Olson E.A., Lynes M.A. (1996). Interactions of metallothionein with murine lymphocytes: plasma membrane binding and proliferation. Toxicol. 108: 129–140.

Bouquegneau J.M. and Joiris C. (1988). The fate of stable pollutants – heavy metals and organochlorines in marine organisms. Advances in Comparative and Environmental Physiology 2: 219–247.

Bouquegneau J.M. and Joiris C. (1992). Ecotoxicology of stable pollutants in cetaceans: organochlorines and heavy metals. In: Symoens J.J., ed. Whales: Biology – Threats – Conservation. Symposium Proceedings, Brussels, 5–7 June 1991. Royal Academy of Overseas Sciences (Brussels), pp. 247–250.

Bouquegneau J.M., Debacker V., Gobert S. (1997a). Biological oceanography and marine food webs: role of marine mammals and seabirds. In: Jauniaux T., Bouquegneau J.M., Coignoul F., eds. Marine mammals, seabirds and pollution of marine systems. Presse de la Faculté de Médecine Vétérinaire de l'Université de Liège, pp. 15–20.

Bouquegneau J.M., Debacker V., Gobert S., Nellissen J.P. (1997b). Toxicological investigations on four sperm whales stranded on the Belgian coast: inorganic contaminants. In: Jacques T.G. and Lambertsen R.H., eds. Sperm whale deaths in the North Sea: Science and Management. Bull. Inst. Roy. Sci. Nat. Belg. 67 (suppl): 75–78.

Bowen W.D. (1997). Role of marine mammals in aquatic ecosystems. Mar. Ecol. Prog. Ser. 158: 267–274.

Canella E.J. and Kitchener D.J. (1992). Differences in mercury levels in female sperm whale, *Physeter macrocephalus* (Cetacea: Odontoceti). Aust. Mamm. 15: 121–123.

Caurant F. and Amiard-Triquet C. (1995). Cadmium contamination in pilot whales *Globicephala melas*: source and potential hazard to species. Mar. Pollut. Bull. 30(3): 207–210.

Caurant F., Amiard J.C., Amiard-Triquet C., Sauriau P.G. (1994). Ecological and biological factors controlling the concentrations of trace elements (As, Cd, Cu, Hg, Se, Zn) in delphinids *Globicephala melas* from the North Atlantic Ocean. Mar. Ecol. Prog. Ser. 103: 207–219.

Caurant F., Navarro M., Amiard J.C. (1996). Mercury in pilot whales: possible limits to the detoxification process. Sci. Total Environ. 186: 95–104.

Cosson R.P., Amiard-Triquet C., Amiard J.C. (1991). Metallothioneins and detoxication. Is the use of detoxication proteins for MTs a language abuse? Water Air Soil Pollut. 57–58: 555–567.

Cuvin-Aralar A. and Furness R.W. (1991). Mercury and selenium interaction: A Review. Ecotoxicol. Environ. Safety 21: 348–364.

Das K., Debacker V., Bouquegneau J.M. (2000). Metallothioneins in marine mammals. Cell. Mol. Biol. 46: 283–294.

Das K., Lepoint G., Loizeau V., Debacker V., Dauby P., Bouquegneau J.M. (2000). Tuna and dolphin associations in the North-East Atlantic: Evidence of different ecological niches from stable isotope and heavy metal measurements. Mar. Pollut. Bull. 40: 102–109.

De Guise S., Martineau D., Béland P., Fournier M. (1995). Possible mechanisms of action of environmental contaminants on St. Lawrence beluga whales (*Delphinapterus leucas*). Environmental Health Perspectives 103(4): 73–77.

Denton G.R.W., Marsh H., Heinsohn G.E., Burdon-Jones C. (1980). The unusual metal status of the Dugong *Dugong dugong*. Mar. Biol. 57: 201–219.

de Swart R.L., Ross P.S., Timmerman H.H., Vedder L.J., Heisterkamp S., Van Loveren H., Vos J.G., Reijnders P.J.H. and Osterhaus A.D.M.E. (1994). Impairment of immune function in harbor seals (*Phoca vitulina*) feeding on fish from polluted waters. Ambio 23(2): 155–159.

de Swart R.L, Harder T.C., Ross P.S., Vos H.W., Osterhaus A.D.M.E. (1995a). Morbilliviruses and morbilliviruses diseases of marine mammals. Infect. Agents Dis. 4: 125–130.

de Swart R.L., Ross P.S., Timmerman H.H., Hijman W.C., de Ruiter E.M., Liem A.K.D., Brouwer A., van Loveren H., Reijnders P.J.H., Vos J.G., Osterhaus A.D.M.E. (1995b). Short term fasting does not aggravate immunosuppression in harbour seals (*Phoca vitulina*) with high body burdens of organochlorines. Chemosphere 31(10): 4289–4306.

de Swart R.L., Ross P.S., Vos J.G., Osterhaus A.D.M.E. (1996). Impaired immunity in harbour seals (*Phoca vitulina*) exposed to bioaccumulated environmental contaminants: review of long-term feeding study. Environmental Health Perspectives 104(4): 823–828.

Dietz R., Riget F., Johansen P. (1996). Lead, cadmium, mercury and selenium in Greenland marine animals. Sci. Total Environ. 186: 67–93.

Dietz R., Nørgaard J., Hansen J.C. (1998). Have arctic mammals adapted to high cadmium levels? Mar. Pollut. Bull. 36(6): 490–492.

Eisler R. (1997). Copper hazards to fish, wildlife and invertebrates: A synoptic review. U.S. Geological survey, Biological Resources Division, Biological Science Report USGS/BRD/BSR-1997-0002.

Eisler R. (1998). Nickel hazards to fish, wildlife and invertebrates: A synoptic review. U.S. Geological survey, Biological Resources Division, Biological Science Report USGS/BRD/BSR-1998-0001.

Elinder C.J. and Järup L. (1996). Cadmium exposure and health risks: recent finding. Ambio 25: 370–373.

Falconer C.R., Davies I.M., Topping G. (1983). Trace metals in common porpoises (*Phocoena phocoena*). Mar. Environ. Res. 8: 119–127.

Forcada J., Aguilar A., Hammond P.S., Pastor X., Aguilar R. (1994). Distribution and numbers of striped dolphins in the Western Mediterranean sea after the 1990 epizootic outbreak. Mar. Mam. Sc. 10(2): 137–150.

Förstner U. (1980). Cadmium. In: The Handbook of Environmental Chemistry. 3(A): 59–107.

Frank A., Galgan V., Roos A., Olsson M., Petersson L.R., Bignert A. (1992). Metal concentrations in seals from Swedish waters. Ambio 21(8): 529–538.

Freeman H.C. and Sangalang G.B. (1977). A study of the effects of methylmercury, cadmium, arsenic, selenium, and a PCB (Aroclor 1254) on adrenal and testicular steroidogeneses *in vitro*, by the gray seal *Halichoerus grypus*. Arch. Environ. Contam. Toxicol. 5: 369–383.

Fujise Y., Honda K., Tatsukawa R., Mishima S. (1988). Tissue distribution of heavy metals in Dall's porpoise in the northwestern Pacific. Mar. Pollut. Bull. 19(5): 226–230.

Gallien I., Caurant F., Bordes M., Bustamante P., Fernandez B., Quellard N., Babin P. (2001). Cadmium-containing granules in kidney tissue of the Atlantic white-sided dolphin (*Lagenorhynchus acutus*) off the Faroe Islands. Comp. Biochem. Phys. C Toxicol. Pharmacol. 130(3): 389–395.

Hamanaka T., Itoo T., Mishima S. (1982). Age-related change and distribution of cadmium and zinc concentrations in the steller sea lion (*Eumetopias jubata*) from the coast of Hokkaido, Japan. Mar. Pollut. Bull. 3(2): 57–61.

Hansen C.T., Nielsen C.O., Dietz R., Hansen M.M. (1990). Zinc, cadmium, mercury and selenium in minke whales, belugas and narwhals from West Greenland. Polar Biol. 10: 529–539.

Harms U., Drescher H.E., Huschenbeth E. (1978). Further data on heavy metals and organochlorines in marine mammals from German coastal waters. Meeresforshung 26(3–4): 153–161.

Heide-Jørgensen M.P., Härkönen T., Dietz R., Thompson P.M. (1992). Retrospective of the 1988 European seal epizootic. Dis. Aquat. Org. 13: 37–62.

Helminen M., Karppanen E., Koivisto J.I. (1968). Mercury content of the ringed seal of Lake Saimaa. Suom. Laaklehti 74: 87–9 (in Finnish).

Holsbeek L., Siebert U., Joiris C.R. (1998). Heavy metals in dolphins stranded on the French Atlantic coast. Sci. Total Environ. 217: 241–249.

Holsbeek L., Joiris C., Debacker V., Ali I., Roose P., Nellissen J.P., Gobert S., Bouquegneau J.M., Bossicart M. (1999). Heavy metals, organochlorines and polycyclic aromatic hydrocarbons in sperm whales stranded in the Southern North Sea during the 1994/1995 winter. Mar. Pollut. Bull. 38: 304–313.

Honda K. and Tatsukawa R. (1981). Ecology and bioaccumulation of *Stenella coeruleoalba* – heavy metal concentration in the muscle and liver tissue of *Stenella coeruleoalba*. In: Fujiyama ed. Studies of the level of organochlorine compounds and heavy metals in the marine organisms. Univ. Ryukyus, pp. 13–23.

Honda K. and Tatsukawa R. (1983). Distribution of cadmium and zinc in tissues and organs, and their age-related changes in striped dolphins, *Stenella coeruleoalba*. Arch. Environ. Contam. Toxicol. 12: 543–550.

Honda K., Tatsukawa R., Itano K., Miyazaki N., Fujiyama T. (1983). Heavy metal concentrations in muscle, liver, and kidney tissue of striped dolphin, *Stenella coeruleoalba* and their variations with body length, weight, age and sex. Agric. Biol. Chem. 47(6): 1219–1228.

Honda K., Yamamoto Y., Tatsukawa R. (1986). Heavy metal accumulation in the liver of Antarctic minke whale (*Balaenoptera acutorostrata*). Mem. Natl. Inst. Polar. Res., Spec. Issue, 44: 182–184.

Honda K., Yamamoto Y., Kato H., Tatsukawa R. (1987). Heavy metal accumulations and their recent changes in Southern minke whales *Balaenoptera acutorostrata*. Arch. Environ. Contam. Toxicol. 16: 209–216.

Hyvärinen H., Sipilä T. (1984). Heavy metals and high pup mortality in the Saimaa ringed seal population in Eastern Finland. Mar. Pollut. Bull. 15(9): 335–337.

Itano K., Kawai S., Miyazaki N., Tatsukawa R., Fujiyama T. (1984). Mercury and selenium levels in striped dolphins caught off the Pacific coasts of Japan. Agric. Biol. Chem. 48: 1109–1116.

Jauniaux T., Brosens L., Jacquinet E., Lambrigts D., Addink M., Smeenk M., Coignoul F. (1998). Postmortem investigations on winter stranded sperm whales from the coasts of Belgium and the Netherlands. J. Wild. Dis. 34: 99–109.

Johansen P., Kapel F.O., Kraul I. (1980). Heavy metals and organochlorines in marine mammals from Greenland. Int. Council Explor. Sea, ICES CM 1980/E: 32.

Jonnalagadda S.B. and Prasada Rao P.V.V. (1993). Toxicity, bioavailability and metal speciation. Comp. Biochem. Physiol. 106C(3): 585–595.

Joiris C.R., Holsbeek L., Bouquegneau J.M., Bossicart M. (1991). Mercury contamination of the harbour porpoise *Phocoena phocoena* and other cetaceans from the North Sea and the Kattegat. Water, Air, and Soil Pollut. 56: 283–293.

Julshamn K., Andersen A., Ringdal O., Mørkøre J. (1987). Trace elements intake in the Faroe Islands. I. Element levels in edible parts of pilot whales (*Globicephalus meleanus*). Sci. Total Environ. 65: 53–62.

Kägi J.H.R. (1991). Overview of metallothionein. In: Riordan J.F. and Vallee B.L., eds. Methods in Enzymology 205: 613–626.

Kaiser G. and Tölg G. (1980). Mercury. In: The Hand book of Environmental Chemistry. 3(A): 1–58.

Kamrin M.A. and Ringer R.K. (1994). PCB residues in mammals: a review. Toxicology and Environmental Chemistry 41: 63–84.

Kannan K., Sinha R.K., Tanabe S., Ichihashi H., Tatsukawa R. (1993). Heavy metals and organochlorine residues in Ganges river dolphins from India. Mar. Pollut. Bull. 26(3): 159–162.

Kemper C., Gibbs P., Obendorf D., Marvanek S., Lenghaus C. (1994). A review of heavy metal and organochlorine levels in marine mammals in Australia. Sci. Total Environ. 154: 129–139.

Klaassen C.D. and Liu J. (1998). Metallothionein transgenic and knock-out mouse models in the study of cadmium toxicity. J. Toxicol. Sci. 2: 97–102.

Koeman J.H., Peeters W.H.M., Koudstaal-Hol C.H.M., Tjioe P.S., De Goeij J.J.M. (1973). Mercury–selenium correlations in marine mammals. Nature 245: 385–386.

Koeman J.H., van de Ven W.S.M., Goeij J.J.M., Tjioe P.S., van Haaften J.L. (1975). Mercury and selenium in marine mammals and birds. Sci. Total Environ. 3: 279–287.

Krone C.A., Robisch P.A., Tilbury K.L., Stein J.E., Mackey A., Becker P., O'Hara T.M., Philo L.M. (1999). Elements in tissues of bowhead whales (*Balaena mysticetus*). Mar. Mammal Sci. 15: 123–142.

Kuehl D. and Haebler R. (1995). Organochlorine, organobromine, metal, and selenium residues in bottlenose dolphins (*Tursiops truncatus*) collected during an unusual mortality event in the Gulf of Mexico, 1990. Arch. Environ. Contam. Toxicol. 28: 494–499.

Kwohn Y.T., Yamazaki S., Okubo A., Yoshimura E., Tatsukawa R., Toda S. (1986). Isolation and characterization of metallothionein from kidney of striped dolphin, *Stenella coeruleoalba*. Agric. Biol. Chem. 50(11): 2881–2885.

Lamphère D.N., Dorn C.R., Reddy C.S., Meyer A.W. (1984). Reduced cadmium body burden in cadmium-exposed calves fed supplemental zinc. Environ. Res. 33: 119–129.

Law R.J. (1994). Collaborative UK marine mammal project: summary of data produced 1988–1992. Fisheries Research Technical Report 97.

Law R.J. (1996). Metals in marine mammals. In: Beyer W.N., Heinz G.H., Redmond-Norwood A.W. eds. Environmental contaminants in wildlife. Interpreting Tissues Concentrations. SETAC Special Publication Series. CRC Press Inc., Lewis Publishers INC., Boca Raton, FL, pp. 357–376.

Law R.J., Jones B.R., Baker J.R., Kennedy S., Milnes R., Morris R.J. (1992). Trace metals in the livers of marine mammals from the Welsh coast and the Irish sea. Mar. Pollut. Bull. 24(6): 296–304.

Leibbrandt M.E.I., Khokha R., Koropatnick, J. (1994). Antisense down-regulation of metallothionein in a human monocytic cell line alters adherence, invasion, and the respiratory burst. Cell Growth and Differentiation 5: 17–25.

Leonzio C. (1996). Terminal consumers and mercury in the marine ecosystem. In: The development of science for the improvement of human life. Third Kyôto-Siena Symposium, section: Natural Sciences. Ed. Rolando Barbucci, Edizioni Cadmo, pp. 27–33.

Leonzio C., Focardi S., Fossi C. (1992). Heavy metals and selenium in stranded dolphins of Northern Tyrrhenian (NW Mediterranean). Sci. Total. Environ. 119: 77–84.

Liu J., Liu Y., Michalska A.E., Choo K.H., Klaassen C.D. (1996). Metallothionein plays less of a protective role in cadmium–metallothionein–induced nephrotoxicity than in cadmium chloride-induced hepatotoxicity. J. Pharmacol. Exp. Ther. 276 (3): 1216–1223.

Malcolm H.M., Boyd I.L., Osborn D., French M.C., Freestone P. (1994). Trace metals in Antarctic fur seal (*Arctocephalus gazella*) livers from Bird Island, South Georgia. Mar. Pollut. Bull. 28(6): 375–380.

Marcovecchio J.E., Moreno V.J., Bastida R.O., Gerpe M.S., Rodriguez D.H. (1990). Tissue distribution of heavy metals in small cetaceans from the Southwestern Atlantic ocean. Mar. Pollut. Bull. 21(6): 299–304.

Margoshes M. and Vallee B.L. (1957). A cadmium protein from equine kidney cortex. J. Am. Chem. Soc. 79: 4813.

Martin J.H., Elliott P.D., Anderlini V.C., Girvin D., Jacobs S.A., Risebrough R.W., Delong R.L., Gilmartin W.G. (1976). Mercury–Selenium–Bromine imbalance in premature parturient California sea lions. Mar. Biol. 35: 91–104.

Martoja R. and Berry J-P. (1980). Identification of tiemannite as a probable product of demethylation of mercury by selenium in cetaceans. A complement to the scheme of the biological cycle of mercury. Vie Millieu, 30(1): 7–10.

Martoja R. and Viale D. (1977). Accumulation de granules de séléniure mercurique dans le foie d'Odontocètes (Mammifères, Cétacés): un mécanisme possible de détoxication du méthylmercure par le sélénium. C.R. Acad. Sc. Paris, t. 285: 109–112.

Mason R.P., Reinfelder J.R., Morel F.M.M. (1995). Bioaccumulation of mercury and methylmercury. Water, Air, and Soil Pollution 80: 915–921.

Miles A.K. and Hills S. (1994). Metal in diet of Bering Sea walrus: *Mya* sp. as a possible transmitter of elevated cadmium and other metals. Mar. Pollut. Bull. 28 (7): 456–458.

Mochizuki Y., Suzuki K.T., Sunaga H., Kobayashi T., Doi R. (1985). Separation and characterization of metallothionein in two species of seals by high performance liquid chromatography–atomic absorption spectrophotometry. Comp. Biochem. Physiol. 82C(2): 249–254.

Muir D.C.G., Wagemann R., Grift N.P., Norstrom R.J., Simon M., Lien J. (1988). Organochorine chemical and heavy metal contaminants in white-beaked dolphins (*Lagenorhynchus albirostris*) and pilot whales (*Globicephala melaena*) from the coast of Newfoundland, Canada. Arch. Environ. Contam. Toxicol. 17: 613–629.

Muir D.C.G., Wagemann R., Hargrave B.T., Thomas D.J., Peakall D.B., Norstrom R.J. (1992). Arctic marine ecosystem contamination. Sci. Total Environ. 122: 75–134.

Nakagawa R., Yumita Y., Hiromoto M. (1997). Total mercury intake from fish and shellfish by Japanese people. Chemosphere 35(12): 2909–2913.

Nendza M., Herbst T., Kussatz C., Gies A. (1997). Potential for secondary poisoning and biomagnification in marine organisms. Chemosphere 35(9): 1875–1885.

Nigro M. and Leonzio C. (1993). Mercury selenide accumulation in dolphins. European Research on Cetaceans 17: 212–215.

Nigro M. and Leonzio C. (1996). Intracellular storage of mercury and selenium in different marine vertebrates. Mar. Ecol. Prog. Ser. 135: 137–143.

Oke V.R. (1967). A brief note on the dugong Dugong dugon at Cairns Oceanarium. Int. Zoo Yearb. 7: 220–221.

Olafson R.W. and Thompson J.A.J. (1974). Isolation of heavy metal binding proteins from marine vertebrates. Mar. Biol. 28: 83–86.

O'Shea T.J., Moore J.F., Kochman H.I. (1984). Contaminant concentrations in manatees in Florida. J. Wildl. Manage. 48(3): 741–748.

Palmisano F., Cardellichio N., Zambonin P.G. (1995). Speciation of mercury in dolphin liver: a two-stage mechanism for the demethylation accumulation process and role of selenium. Mar. Environ. Res. 40(2): 109–121.

Paludan-Müller P., Agger C.T., Dietz R., Kinze C.C. (1993). Mercury, cadmium, zinc, copper and selenium in harbour porpoise (*Phocoena phocoena*) from west Greenland. Polar Biol. 13: 311–320.

Pelletier E. (1985). Mercury–selenium interactions in aquatic organisms: a review. Mar. Environ. Res. 18: 111–132.

Pesch H.J., Palesch T., Seibold H. (1989). The increase of the Cd-burden in man. Post-mortem examinations in Franconia by absorption spectroscopy. In: Vernet, J.P. ed. Heavy metals in the environment. 2: 111–114.

Rawson A.J., Patton G.W., Hofmann S., Pietra G.G., Johns L. (1993). Liver abnormalities associated with chronic mercury accumulation in stranded bottlenose dolphins. Ecotoxicol. Environ. Saf. 25: 41–47.

Rawson A.J., Bradley J.P., Teetsov A., Rice S.B., Haller E.M., Patton G.W. (1995). A role for airborne particulates in high mercury levels of some cetaceans. Ecotoxicol. Environ. Saf. 30: 309–314.

Reijnders P.J.H. (1986). Reproductive failure in common seals feeding on fish from polluted coastal waters. Nature 324: 456–457.

Ridlington J.W., Chapman D.C., Goeger D.E., Whanger P.D. (1981). Metallothionein and Cu-chelatin: characterization of metal-binding proteins from tissues of four marine animals. Comp. Biochem. Physiol. 70B: 93–104.

Roesijadi G. (1992). Metallothioneins in metal regulation and toxicity in aquatic animals. Aquatic Toxicology 22: 81–114.

Roesijadi G. (1996). Metallothionein and its role in toxic metal regulation. Comp. Biochem. Physiol. 113C(2): 117–123.

Ronald K. and Tessaro S.V. (1977). Methylmercury poisoning in the harp seal (*Pagophilus groenlandicus*). Sci. Total Environ. 8: 1–11.

Sanpera C., Capelli R., Minganti V., Jover L. (1993). Total and organic mercury in North Atlantic fin whales. Distribution pattern and biological related changes. Mar. Pollut. Bull. 26(3): 135–139.

Sarokin D. and Schulkin J. (1992). The role of pollution in large-scale population disturbances. Part. 1: Aquatic populations. Environ. Sci. Technol. 26(8): 1478–1484.

Satoh M., Nishimura N., Kanayama Y., Naganuma A., Suzuki T., Tohyama C. (1997). Enhanced renal toxicity by inorganic mercury in metallothionein-null mice. J. Pharmacol. Exp. Ther. 283(3): 1529–1533.

Sepulveda M.S., Ochoa-Acuña H., Sundlof S.F. (1997). Heavy metal concentrations in Juan Fernández Fur Seals (*Arctocephalus philippii*). Mar. Pollut. Bull. 34(8): 663–665.

Siebert U., Joiris C., Holsbeek L., Benke H., Failing K., Failing K., Frese K., Petzinger E. (1999). Potential Relation between mercury concentrations and necropsy findings in Cetaceans from German waters of the North and Baltic seas. Mar. Pollut. Bull. 38: 285–295.

Simkiss K. (1976). Intracellular and extracellular routes in biomineralization. Symposia of the Society for Experimental Biology, Vol. 30. University Press, Cambridge, UK, pp. 423–444.

Svensson B.G., Schütz A., Nilsson A., Åkesson I., Åkesson B., Skerfving S. (1992). Fish as a source of exposure to mercury and selenium. Sci. Total Environ. 126: 61–74.

Szefer P., Malinga M., Skora K., Pempkowiak J. (1994). Heavy metals in harbour porpoises from Puck Bay in the Baltic Sea. Mar. Pollut. Bull. 28(9): 570–571.

Tanabe S., Iwata H., Tatsukawa R. (1994). Global contamination by persistent organochlorines and their ecotoxicological impact on marine mammals. Sci. Total Environ. 154: 163–177.

Tang W., Sadovic S., Shaikh Z.A. (1998). Nephrotoxicity of cadmium–metallothionein: protection by zinc and role of glutathione. Toxicol. Appl. Pharmacol. 151(2): 276–282.

Thompson D.R. (1990). Metal levels in marine vertebrates. In: Furness R.W. and Rainbow P.S., eds. Heavy metals in the marine environment. CRC Press: Boca Raton, pp. 143–182.

Thompson P.M. and Hall A.J. (1993). Seals and epizootics – what factors might affect the severity of mass mortalities? Mammal Rev. 23(3/4): 147–152.

Tohyama C., Himeno S.I., Watanabe C., Suzuki T., Morita M. (1986). The relationship of the increased level of metallothionein with heavy metal levels in the tissue of the harbor seal (*Phoca vitulina*). Ecotoxicol. Environ. Safety 12(1): 85–94.

Varanasi U., Stein J.E., Tilbury K.L., Meador J.P., Sloan C.A., Clarck R.C., Chan S.L. (1994). Chemical contaminants in gray whales (*Eschrichtius robustus*) stranded along the west coast of North America. Sci. Total Environ. 145: 29–53.

Viarengo A. (1989). Heavy metals in marine invertebrates: mechanisms of regulation and toxicity at the cellular level. Rev. Aquat. Sci. 1: 295–317.

Wagemann R. (1989). Comparison of heavy metals in two groups of ringed seals (*Pusa hispida*) from the Canadian Artic. Can. J. Fish. Aquat. Sci. 46: 1558–1563.

Wagemann R. and Hobden B. (1986). Low molecular weight metalloproteins in tissues of the narwhal (*Monodon monoceros*). Comp. Biochem. Physiol. 84C(2): 325–344.

Wagemann R. and Muir D.C.G. (1984). Concentrations of heavy metals and organochlorines in marine mammals of Northern waters: overview and evaluation. Can. Tech. Rep. Fish. Aquat. Sc. N°1279.

Wagemann R., Snow N.B., Lutz A., Scott D.P. (1983). Heavy metals in tissues and organs of the narwhal (*Monodon monoceros*). Can. J. Fish. Aquat. Sci. 40 (suppl. 2): 206–214.

Wagemann R., Hunt R., Klaverkamp J.F. (1984). Subcellular distribution of heavy metals in liver and kidney of a narwhal whale (*Monodon monoceros*): an evaluation for the presence of metallothionein. Comp. Biochem. Physiol. 78C(2): 301–307.

Wagemann R., Stewart R.E.A., Lockhart W.L., Stewart B.E., Poveldo M. (1988). Trace metals and methylmercury: associations and transfer in harp seals (*Phoca groenlandica*) mothers and their pups. Mar. Mam. Sci. 4(4): 339–355.

Wagemann R., Stewart R.E.A., Beland P., Desjardins C. (1990). Heavy metals in tissues of belugas whales from various locations in the Canadian Arctic and the St. Lawrence River. In: J.Geraci and T.G. Smith (eds). Can. Bull. Fish. Aquat. Sci. 224: 191–206.

Wagemann R. and Stewart R.E.A. (1994). Concentrations of heavy metals and selenium in tissues and some foods of walrus (*Odobenus rosmarus rosmarus*) from the eastern Canadian Arctic and sub-Arctic, and associations between metals, age and gender. Can. J. Fish. Aquat. Sci. 51: 426–436.

Wagemann R., Lockhart W.L., Welch H., Innes S. (1995). Arctic marine mammals as integrators and indicators of mercury in the Arctic. Water, Air, and Soil Pollution 80: 683–693.

Wagemann R., Innes S., Richard P.R. (1996). Overview and regional and temporal differences of heavy metals in Arctic whales and ringed seals in the Canadian Arctic. Sci. Total Environ. 186: 41–66.

Wagemann R., Trebacz E., Boila G., Lockhart W.L. (1998). Methylmercury and total mercury in tissues of Arctic marine mammals. Sci. Total Environ. 218: 19–31.

WHO (1992) Cadmium. Environmental Health Criteria, 134. World Health Organization, Geneva.

Wolfe M.F., Schwarzbach S., Sulaiman R.A. (1998). Effects of mercury on wildlife: a comprehensive review. Environ. Toxicol. Chem. 17(2): 146–160.

Yamamoto Y., Honda K., Hidaka H., Tatsukawa R. (1987) Tissue distribution of heavy metals in Weddels seals (*Leptonychotes weddellii*). Mar. Pollut. Bull. 18(4): 164–169.

Zelikoff, J.T. and Thomas, P.T. (1998). Immunotoxicology of environmental and occupational metals. Taylor and Francis, London.

8 Persistent organic contaminants in Arctic marine mammals

Todd M. O'Hara and Paul R. Becker

Introduction

This chapter will address the concentrations and possible effects of selected persistent organic contaminants (POCs) in marine mammals of the Arctic region. Other recent sources of information concerning POCs in Arctic marine mammals include Letcher *et al.* (1996, 1998), Becker *et al.* (1997), Bernhoft *et al.* (1997), Muir *et al.* (1997), Weis and Muir (1997), Addison and Smith (1998), De March *et al.* (1998), Norstrom *et al.* (1998), Wiberg *et al.* (1998 and 2000), Krahn *et al.* (1999), O'Hara *et al.* (1999), Becker (2000), Kucklick *et al.* (2001) and Seagars and Garlich-Miller (2001). Arctic marine mammals are long-lived, develop large lipid or fat depots, and many occupy high trophic levels in the lipid-rich, Arctic marine food webs. These factors are important in the uptake and magnification of persistent, lipophilic organic contaminants. Bowhead whales (*Balaena mysticetus*), which feed almost exclusively on copepods, euphausiids and amphipods (Lowry, 1993), account for some of the lowest tissue concentrations of 'bioaccumulating' organochlorines (OCs), and correspondingly occupy the lowest trophic level of marine mammals addressed in this chapter (Hoekstra *et al.*, 2000a, b). Pacific walrus (*Odobenus rosmarus divergens*) and Atlantic walrus (*Odobenus rosmarus rosmarus*) typically feed on mollusks, but also may occasionally eat ringed seals (*Pusa hispida*) (Fay, 1960; Lowry and Fay, 1984). Other Arctic marine mammals feed on a combination of invertebrates, fish and/or mammals. One class of POCs, the polycyclic aromatic hydrocarbons (PAHs), are readily metabolized by mammals and do not accumulate in their tissues. Assessing the exposure of marine mammals to these compounds is not as straightforward as is the case for lower vertebrates (e.g. fish) and invertebrates. The latter have little capacity for metabolizing this class of aromatic compounds; therefore PAHs will accumulate in their tissues (Livingstone *et al.*, 1992). Although specific cases of marine mammal response to benzo[*a*]pyrene exposure will be discussed, we will not address PAHs or petroleum effects in this review. Concentrations of contaminants provided in this chapter are on a wet weight basis, except when noted otherwise.

Persistent organic contaminants of concern

The following chlorinated industrial compounds and pesticides are of concern for marine mammals: polychlorinated biphenyls (PCBs), 2,3,7,8-tetrachlorodibenzo-*p*-dioxin (dioxin) as a member of the polychlorinated dibenzodioxins (PCDD), polychlorinated dibenzofurans (PCDFs), 1,1,1-trichloro-2,2-*bis*(*p*-chlorophenyl) ethane and metabolites (DDTs), chlorinated cyclodienes (chlordane, dieldrin, mirex/kepone), chlorinated bornanes (toxaphenes), chlorinated benzenes (including hexachlorobenzene or HCB), chlorinated cyclohexanes (HCHs) and, in very specific cases, benzo[*a*]pyrene (a PAH). POC exposure and concentrations in tissues are affected by biological factors (age, sex, body condition, reproductive status and season), spatial distribution (habitat or home range, global region) and temporal relationships (i.e. based on historic data, are levels decreasing, increasing or not changing?).

Potential sources and transport mechanisms

Generally, anthropogenic contamination of the Arctic with POCs is thought to be the result of long-range atmospheric transport, with minor contributions from local sources (Wania and Mackay, 1993, 1996). Atmospheric contamination is visible in the form of Arctic haze (Shaw, 1995) and may threaten the health of Arctic residents (Barrie, 1986; Hoff and Chan, 1986; Heintzenberg, 1989; Hoff *et al.*, 1992; Kinloch *et al.*, 1992; Dewailly *et al.*, 1993; Ayotte *et al.*, 1995; Barrie *et al.*, 1997). Many contaminants are transported in the gaseous phase, on particles and in aerosols, and are deposited during precipitation events. These compounds may be trapped by the colder temperatures associated with the Arctic, due to the relatively large surface areas of both the sources and the receiver (the Arctic). Seasonal changes in temperature can also affect atmospheric levels of contaminants (Manchester-Neesvig and Andren, 1989; Hoff *et al.*, 1992; Barrie *et al.*, 1997; Stern *et al.*, 1997) that are delivered to the Arctic in snow (Gregor and Gummer, 1989).

Physical-chemical partitioning of the contaminant is key to the introduction of such compounds into the Arctic food chain. All of the POCs have ringed molecular structures. Positions of the chlorine (or other halogen) atoms on these ring structures, and whether the chlorine is bound to aromatic or aliphatic carbons, are major factors affecting bioaccumulation and toxicity (Boon *et al.*, 1992, 1994). For example, the planar molecules (coplanar congeners) of PCBs and dioxins are the most toxic of these compounds. This characteristic generally holds true for other chlorinated hydrocarbons and for PAHs as well. It should also be stressed that 'super lipophilic' ($K_{ow} > 10^6$) agents have a decreased bioaccumulative potential because they have an increased time to effective equilibrium, i.e. dissolution time (Livingstone *et al.*, 1992).

The POCs are well known for their lipophilicity and ability to bioconcentrate, and the concentrations of these compounds are many times higher in biota than in the aqueous and atmospheric media. Persistent OCs bind to organic particles (including detritus, plankton and algae) and enter the base of the food chain. This bioaccumulation (higher concentrations in biota than in water) sets the stage for biomagnification (increased concentrations in the predator versus the prey). The biomagnification factor, or BMF, is commonly expressed as [X]predator/[X]prey (when > 1.0 it is 'magnified'), where [X] = contaminant concentration in lipid weight. Fish, pinnipeds or baleen whales consume invertebrates, many of which consume organic particles. Arctic cod (*Boreogadus saida*) appear to play a major role in biomagnification of POCs in the Arctic. This small fish feeds extensively on invertebrates and is a major prey of marine mammals in the Arctic.

PCB Aroclor mixtures (Monsanto Chemical Company) 1221, 1232, 1242, 1248, 1254, 1260 and 1268 contain 21, 32, 42, 48, 54, 60 and 68 per cent chlorine, by weight, respectively, and theoretically contain 209 congeners (O'Hara and Rice, 1996). As heat-stable oils, PCBs were used around the world: in electrical transformers and capacitors, in hydraulic fluids, as flame retardants, as plasticizers in waxes, in paper manufacturing and other uses. PCBs are known to be transported in the atmosphere (Hoff *et al.*, 1992) and to interact with surface waters (Bidleman *et al.*, 1989). PCBs are adsorbed and carried on sediments and particles in water, as well as by migratory organisms in both freshwater and marine systems (O'Hara and Rice, 1996; Ewald *et al.*, 1998). Fish represent a significant dietary source of PCBs to many marine mammals. Adverse effects on fish as prey (decreased reproduction) could have effects on food availability.

Additional classes of POCs that originate from industrial activities are the halogenated dibenzo-*p*-dioxins and dibenzofurans. Chlorinated or brominated forms of dioxins and furans can result from combustion, the production of other chlorinated organic compounds (e.g. chlorinated pesticides and PCBs) or from industrial processes that mix halogens with organic material (i.e. pulp-mill effluents). These compounds are easily distributed to the aquatic and atmospheric environments. Brominated organic compounds are widely used (i.e. flame retardants) as well, and have been reported in marine mammal tissues (de Boer *et al.*, 1998). The occurrence of brominated fire retardents in biota of the Arctic will require further evaluation.

More than 30 kinds of POCs were developed intentionally for release into the environment as pesticides. These chlorinated pesticides, which have been used worldwide for the past 50 years, include DDT (*o,p'*- and *p,p'*-forms) and associated metabolites (*o,p'*- and *p,p'*-forms of DDD and DDE), chlorinated cyclodienes (chlordane, endrin, dieldrin, mirex/kepone), chlorinated camphenes and bornanes (including toxaphene), chlorinated benzenes and chlorinated cyclohexanes (primarily α-HCH, β-HCH, and γ-HCH, also known as lindane). Industrial- or technical-grade versions of chemicals such as chlordane and toxaphene contain many isomers or chemicals closely related to the parent or intended synthesized chemical. A mixture's toxicity is often related to these 'by-products'. Toxaphene is produced by the chlorination of camphene to 67–69 per cent, producing hundreds of congeners. Many studies have shown its wide global distribution and presence in biota (Voldner and Schroeder, 1989; Saleh, 1991; Muir and deBoer, 1995; Wade *et al.*, 1995). Chlorinated pesticides are well known to be of low water solubility and high solubility in oils and organic solvents. They tend to accumulate and to be stored in lipids, are resistant to metabolic breakdown and have relatively long half-lives in tissues of biota (Blus *et al.*, 1996).

Movement of POCs from one medium to another is determined by their vapor pressure, octanol/air partition coefficient, temperature of condensation and octanol/water (o/w) partition coefficient (high lipophilicity = high octanol/water partitioning coefficient) (Bidleman *et al.*, 1989). Compounds such as HCB, dieldrin, HCHs, three-ringed PAHs, and PCB congeners having six chlorine atoms or fewer per molecule have relatively high vapor pressures and relatively low octanol/air partition coefficients. They are therefore easily transported via the atmosphere from lower latitudes to the Arctic where, due to their relatively low temperatures of condensation, they can be preferentially deposited (Wania and Mackay, 1993, 1996). Other compounds, such as mirex, and octa- and nonachlorinated biphenyls (PCB congeners containing 8–9 chlorine atoms per molecule) have relatively high octanol/ air partition coefficients, high temperature of condensation and relatively low vapor pressures, and are not easily transported via the atmosphere. Compounds such as chlordane, DDTs and toxaphene exhibit characteristics mid-way between these two extremes; therefore, preferential deposition and accumulation might be expected in the mid-latitudes rather than in the high Arctic.

The relatively high octanol/water partition coefficients of these POCs explain their accumulation in lipid-rich tissues (adipose, blubber, brain) and partitioning from water to the surfaces of particles and sediments. The kind of lipid also affects the degree of accumulation. For example, accumulation in brain is lower than in blubber or adipose tissue, due to the high phospholipid content of brain. Determination of concentrations of POCs in brains is critical for diagnosis, because the brain is the target organ of acute toxicosis. The trophic level of a species determines its respective exposure to these lipophilic agents. Species such as seals, beluga whales (*Delphinapterus leucas*) and narwhals (*Monodon monoceros*), that feed at the middle level (i.e. consuming fish) have moderate concentrations in their tissues; and the polar bears (*Ursus maritimus*), which consume mammals dependent upon fish (i.e. seals), tend to have the highest exposures. Tissue burdens at these higher trophic levels will also depend on the species-specific ability to absorb, metabolize and/or eliminate these compounds. In addition to trophic level, body size and metabolic rate also play a role in the accumulation of POCs. Smaller species can have higher concentration in tissues but sometimes lower total body burdens (Aguilar *et al.*, 1999).

Factors affecting POC concentrations and interpretation of effects

Many biological factors must be considered when interpreting POC concentrations and their possible effects (Aguilar *et al.*, 1999). Parturition and lactation are important excretion mechanisms for sexually mature and active females. However, this also results in exposure of offspring *in utero* and during lactation at critical phases in the development of neonatal organs [especially the central nervous system (CNS), reproductive and immune systems]. In general, POC concentrations are lower in mature females than in mature males for polar bear, ringed seals and bowhead, beluga and narwhal whales (Muir *et al.*, 1992a, b; O'Hara *et al.*, 1999). This reproduction-dependent route of excretion for the adult and exposure for the neonate represents a critical pathway. Although this neonatal exposure may be the highest and most significant for the entire life of the animal, indications of adverse effects may be delayed until later life stages. Dysfunction in the neonate may be expressed acutely, at maturation or in the next generation (transgenerational). Transgenerational effects have only been noted in laboratory animals. Body burdens or concentrations in tissues usually increase with age for both sexes until sexual maturity, when a difference based on gender is usually detected (Boon *et al.*, 1992). However, exceptions have been reported. Stern *et al.* (1992) found no correlation of DDTs and PCBs with increased age in male belugas from Greenland. Muir *et al.* (1996a) suggest that this lack of correlation could result from a shift in the diet of older males.

In most cases (the exceptions being marine mammals from highly contaminated areas), males continue to accumulate POCs as they increase in age, whereas concentrations in females reach a plateau and stabilize or decrease (Boon *et al.*, 1992). This sex difference is due to the production of offspring and lactation by the female, as described above. The largest POC mobilization and subsequent excretion is likely to occur during lactation. This varies by compound, with the higher lipophilic compounds probably removed to a lesser degree by lactation (Boon *et al.*, 1992). This should be considered when comparing populations and assessing potential impacts to critical life stages.

Blubber or fat is the main repository for POCs; hence changes in an animal's nutritional status may affect the concentrations of these compounds in these lipid-rich tissues. However, measurement of nutritional status is difficult, and reliable condition indices have not been developed as part of contaminant studies for most cetaceans (IWC, 1995), pinnipeds or polar bears. Blubber thickness is unreliable as a sole indicator of condition and, instead, the use of blubber weight and lipid content of the blubber in standardized body locations is recommended. Lipid content is important, particularly in emaciated individuals, because lipid mobilization may be coupled with increased water content of blubber, as is known for some pinnipeds (Beck *et al.*, 1993; Gales and Renouf, 1994; Fadely, 1997), and variations in lipid types could affect partitioning of the POCs. Collecting these biological data is important for the toxicological interpretation of the results (Law, 1994), especially with respect to compounds that are lipophilic and susceptible to changes with change in lipid status of the animal. Seasonal differences are mostly reflective of feeding and reproductive behavior, and can influence the concentrations of POCs (Boon *et al.*, 1992). These can be caused by migratory behavior (differing concentrations in regional prey) or by opportunistic feeding strategies (Boon *et al.*, 1992).

Metabolism or biotransformation of POCs within an animal can affect POC concentration patterns (Boon *et al.*, 1992, 1994). Some PCB congeners may be lower in concentration than expected simply based upon types and concentrations in prey. This indicates possible metabolic elimination (Tanabe *et al.*, 1988, 1994). Research has emphasized the inducible Phase I mixed function oxidase (MFO) system of the liver. The MFO system utilizes cytochrome P450 enzymes (CYP) that occur as many different types (CYP1A, CYP2, etc.). The P450 enzymes are well known for oxidizing many natural (i.e. steroid hormones, PAHs) and some synthetic compounds (some POCs) (Safe, 1990). Cytochrome P450 enzymes have not been studied to any great extent in Arctic cetaceans. Of those whales that have been studied (i.e. odontocetes) mostly CYP1A activity was detected and there was evidence for the CYP3 subfamily, but no evidence of CYP2 (Boon *et al.*, 1992). Hepatic CYP1A1 isozymes have been characterized for use in monitoring POC exposure in Arctic belugas (White *et al.*, 1994). There have been many differences seen in P450 activity with respect to age and sex for each species that has been studied (Boon *et al.*, 1992), thus complicating our understanding of the role of MFOs in eliminating and detoxifying POCs.

Marine mammals can apparently metabolize some PCBs, especially the CYP1A substrates (e.g. PCBs with vicinal H atoms in the *o,m* positions and a maximum of one *ortho*-Cl); however, odontocetes may lack, or have very low activity of, CYP2B (Boon *et al.*, 1992; Norstrom *et al.*, 1992). Metabolic indices suggest lower levels of PB (phenobarbital or CYP2B)-inducible or MC (3-methylchalanthrene or CYPIA)-inducible enzyme activities in piscivorous cetaceans, seals and marine birds than those of other animals which are not piscivores (Tanabe and Tasukawa, 1992). This phenomenon was noted in both marine and terrestrial organisms and may reflect the evolutionary loss of a specific P450-based oxidative capability (i.e. isozyme) in piscivores. It has been speculated that perhaps this is because herbivorous or non-fish prey may excrete specific types of lipophilic agents (thus need the oxidative capacity), whereas the piscivores (carnivores) had no need to maintain this specific P450 system if the lipophilic agents were not present in the prey (Tanabe and Tasukawa, 1992). Carvan *et al.* (Chapter 16 in this volume) discuss the cytochrome P450 system and interactions with aromatic hydrocarbons in greater detail.

Physiologic and morphologic measures of exposure or response to contaminants in free-ranging animals are occasionally referred to as 'biomarkers'.

Biomarkers are needed for Arctic marine mammals to better assess impacts of 'high' contaminant concentrations. Proposed toxicant biomarkers for cetaceans include an assessment of DNA adducts, induction of the MFO system, plasma hormone levels, immune responses (IWC, 1995) and lesions (i.e. tumors). These biomarkers can show a documentable biological response to measured contaminants. However, sample collection logistics and the multiple biological variables that can affect these systems make interpretation very difficult. This is especially true for non-lethal sampling, where sample size and type are extremely limited (i.e. skin/blubber biopsies, blood samples). Neoplastic lesions are useful gross indicators, considering that PCBs and the related PCDDs and PCDFs are known to cause preneoplastic lesions, neoplastic nodules and hepatocellular carcinomas. Female mammals tend to be more susceptible to these tumorigenic potentials through free-radical generation, Ah-receptor interactions, lipid peroxidation, degeneration of cell defense mechanisms and altered vitamin A metabolism (Livingstone *et al.*, 1992). *In vitro* immunoassays are being developed for Arctic belugas and St. Lawrence estuary beluga whales (De Guise *et al.*, 1997; Brousseau *et al.*, Chapter 14 in this volume) and other assays (Gauthier *et al.*, 1999, Chapter 15 in this volume) also show promise as biomarkers. Martineau *et al.* (1999, and Chapter 13 in this volume) and De Guise *et al.* (1994a) emphasized the use of neoplasia and non-neoplastic lesions (De Guise *et al.*, 1995; Lair *et al.*, 1997) as biomarkers, used for the St. Lawrence estuary beluga whales as examples (Martineau *et al.*, 1988, 1994).

Some POCs have been implicated as endocrine disrupters by affecting sexual or gonadal development in laboratory animals, as early as *in utero*. Deficiency of the CYP2B activity and low activity of CYP1A in cetaceans may limit their metabolic capabilities and the elimination of POCs, making these animals more susceptible to reproductive derangement (Livingstone *et al.*, 1992). True hermaphroditism has been reported in the beluga whale (De Guise *et al.*, 1994b) and pseudohermaphroditism has been found in the bowhead whale (Tarpley *et al.*, 1995). However, cause–effect relationships for these conditions have not been determined but are discussed further by Reijnders (Chapter 3 in this volume).

Assessment of effects of contaminants has lagged behind analytical chemistry (biotic and abiotic) and to date there has been little progress in determining effects of contaminants in marine mammals. Biomarkers have an important role in health assessments; however, there is still a large gap between determining a 'biomarker' response and having a clear understanding of how this really affects the health of an animal or a population. Recognizing the known *in utero* and neonatal exposures to POCs in marine mammals, several recent reports (IWC, 1995; O'Shea *et al.*, 1999; Reijnders *et al.*, 1999) include recommended measures for assessing their effects (Table 8.1).

No laboratory test can replace a thorough necropsy and/or clinical examination, where one conducts gross (must include life history) and histologic assessments (as complete as possible) to address general health and possible

Table 8.1 Some endpoints for assessing possible effects of known *in utero* and neonatal exposures to POCs in marine mammals (after IWC, 1995; O'Shea *et al.*, 1999; Reijnders *et al.*, 1999)

Endpoints	Matrix
Thyroid hormone	p, b, l
Vitamin A	p
Steroid hormones	p
Estrogen receptors	o, b, l
Clinical signs, histopathology + biochemistry	CNS, e, i
Gross and histopathology	Routine*

p, Plasma; o, ovary; b, brain; l, liver; CNS, central nervous system; e, endocrine glands; i, immune; * for 'ruling in' and 'ruling out' other disease factors.

confounding variables related to disease ('ruling in' and 'ruling out'). True examination must provide for long-term interdisciplinary studies to encourage collaboration of multiple perspectives, and report findings in a timely manner (this helps future experimental designs and allows managers to respond). Standardizing field and laboratory protocols, and statistically 'powerful' experimental designs will enhance interpretative potential and allow for comparisons. There is great need to link exposure to effects as expressed at the whole animal and population levels (beyond classical biochemical biomarkers) in a 'dose–response' manner.

Persistent organic contaminants in Arctic cetaceans

Arctic cetaceans are long lived, especially the bowhead whale, which is estimated to live for more than 100 years (George *et al.*, 1999). Bowhead whales feed almost exclusively on copepods, euphausiids and amphipods (Lowry, 1993) and occupy the lowest trophic level of the cetaceans addressed here. The beluga and narwhal feed on invertebrates and fish. The summer diet of the narwhal was investigated on northern Baffin Island, Canada, where Arctic cod and Greenland halibut (*Reinhardtius hippoglossoides*) made up 51 per cent and 37 per cent (by weight) of the diet, respectively. Squid (*Gonatus fabricii*) beaks were noted but squid mass could not be quantified. Deep-water fish [halibut, redfish (*Sebastes marinus*), and polar cod (*Arctogadus glacialis*)] were found in the stomachs of males and may be from deeper dives than in females (Finley and Gibb, 1982). This high trophic level feeding (piscivory) sets the stage for POC biomagnification.

Bowhead whales

Bowhead whales are baleen whales (suborder Mysticeti) and feed mainly on invertebrates in Arctic waters for most of the year. Occasionally, bowhead whales have been known to consume small fish, crabs, mollusks and even sediments, but these are likely infrequent and represent insignificant pathways for contaminant exposure (Bratton *et al.*, 1993). The few POC studies in the bowhead whale do not address possible effects of these pollutants, but only pollutant concentrations in tissues at the time of death (Bratton *et al.*, 1993). This hampers assessment of effects, which will require extrapolation from other species. Although the bowhead whale is a mysticete and is assumed to be of low risk for bioaccumulation (O'Shea and Brownell, 1994) as a 'filter feeder' (as compared to a 'fish eater'), we should not be completely unconcerned about biomagnification and potential effects of POCs on this species. Bratton *et al.* (1993) noted that the bowhead whale is endangered, and the unknown but potential effects of contaminants on their reproduction and on their young (lactational exposure) could be critical. The bowhead whale is also a major subsistence resource for Arctic people. Considering the lack of knowledge about effects of POCs, we point out the documentation of male pseudohermaphroditism in two bowhead whales (Tarpley *et al.*, 1995) for which the cause is unknown. In addition to POCs, exposure of bowhead whales to petrogenic hydrocarbons (PAHs, benzene, etc.) is of great concern considering the exploration and development of oil reserves in the Arctic, especially in northern Alaska. Exposure could be by direct external contact, ingestion or inhalation of volatile components during a spill (Bratton *et al.*, 1993). Petroleum avoidance has been a controversial issue, but oil exposure can affect a variety of systems including the dermis, baleen, eyes, respiratory, gastrointestinal, hematopoietic (blood formation) and hepatic systems. Surprisingly, studies are very limited with respect to petroleum contamination and assessment of potential target organs.

A few reports have described POC concentrations in bowhead whale blubber or liver sampled in Alaska (McFall *et al.*, 1986; O'Hara *et al.*, 1999). Mean concentration of the sum of PCB congeners (ΣPCBs) was 0.212 µg/g (McFall *et al.*, 1986), indicating low concentrations as compared to other marine mammals. Mean concentrations of chlorinated pesticide in blubber were: sum of DDT, DDD and DDE (ΣDDTs) 0.032 µg/g; chlordane, 0.007 µg/g; dieldrin, 0.017 µg/g; heptachlor epoxide, 0.008 µg/g; and lindane, 0.009 µg/g (McFall *et al.*, 1986). Both unsubstituted (parent) and alkyl-substituted PAHs were detected in the blubber at concentrations of 0.028 and 0.025 µg/g, respectively, and would be considered background exposure (McFall *et al.*, 1986). As part of the Alaska Marine Mammal Tissue Archival Project (AMMTAP) liver and blubber have been archived from more than 70 bowhead whales (Becker *et al.*, 1997 and recent sampling efforts) and the POC analyses have been completed on some of these animals (O'Hara *et al.*, 1999; Hoekstra *et al.*, 2000a, b). More recent data indicate that concentrations are low compared to some odontocetes, and are similar to concentrations reported by McFall *et al.* (1986). Female bowhead whales do not accumulate POCs with age, but 7 of 11 POCs that were measured accumulate in males (O'Hara *et al.*, 1999).

The enantiomeric fractions (EFs) of α-HCH and chlordane-related compounds were near racemic (1 : 1) in water and plankton samples off the coast of northern Alaska. The (+)-α-HCH and (+)-*cis*-heptachlor epoxide were selectively enriched in bowhead blubber and liver relative to plankton and water samples, and independent of sex, season and age. The EFs of other chlordane-related compounds were near racemic (Hoekstra *et al.*, 2000a). Although previous studies have determined PCB concentrations in tissues of bowhead whales, Hoekstra *et al.* (2000b) analyzed chiral PCB congeners. Blubber and liver samples were analyzed for several chiral PCB congeners (PCBs 91, 95, 136, 149, 174 and 176) and for many achiral PCB congeners. PCB concentrations in bowhead whales were dominated by penta- and hexachlorobiphenyl congeners and were relatively low (mean: 671 ng/g lipid). Female bowhead whales (regardless of age) and shorter (<13 m) male bowhead whales were characterized by near-racemic enantiomeric ratios (ER = 1.0) for many chiral PCB congeners. Nonracemic enantiomeric ratios (ER > or < 1.0) for PCBs 91, 95 and 149 were found in larger (>13 m) male whales. Hoekstra *et al.* (2000a) suggest that PCB biotransformation or uptake from the marine environment is enantioselective and may be age- and/or sex-dependent for the bowhead whale.

For consumers of bowhead whale blubber there is little, if any, significant exposure to POCs that warrants any dietary restriction, based on our current understanding (O'Hara *et al.*, 1999). However, concentrations of contaminants in tissues are not a sole determining factor for evaluating animal toxicosis. Other histologic, biochemical, immune and physiologic assessments (i.e. health assessment) would better evaluate whether or not there is an effect. Some of these 'effects endpoints' are currently being evaluated as part of an ongoing study in northern Alaska, but only in the short term (1998–2001).

Beluga and narwhal whales

Belugas are generally circumpolar in distribution, but some populations occur outside of the Arctic. The latter include populations in the St. Lawrence estuary (between Canada and USA in the Great Lakes region; see Chapters 13–15 in this volume) and in Cook Inlet (near Anchorage, Alaska, USA). The narwhal has a more limited range, encompassing the eastern Arctic of Canada and Greenland. Both species are important animals for people of the north, both culturally and nutritionally.

The PCB and DDT concentrations in blubber of Arctic beluga whales from Alaska are similar to those reported from the Canadian Arctic and Greenland belugas, narwhals and polar bears, but at least an order of magnitude lower than in beluga whales of the St. Lawrence estuary (Becker *et al.*, 1997; Krahn *et al.*, 1999). The mean concentrations of ΣDDTs and ΣPCBs in blubber of narwhal (NWT, Canada) were 3.51 and 10.09 ug/g, respectively (Wagemann and Muir, 1984). However, total chlordane

(Σchlordanes) and toxaphene concentrations in blubber of beluga whales from Alaska are similar to those in belugas, narwhal and polar bears from Arctic Canada, and belugas of western Greenland, and are within the same order of magnitude as in blubber of beluga whales of the St. Lawrence estuary (Becker *et al.*, 1997). There appears to be enrichment or biological accumulation of specific PCCs as compared to a toxaphene standard. Based on an evaluation of narwhal blubber and burbot (*Lota lota*) liver, octachloro- and nonachlorobornanes are increased (enriched) in the narwhal (Bidleman *et al.*, 1993). The concentrations of total chlordanes, HCHs and chlorobenzenes in Arctic beluga are not significantly different from those in beluga whales of the St. Lawrence estuary, suggesting that the similar concentrations are a result of the relatively higher volatility of these compounds and atmospheric transport to regions more distant from the sources (Muir *et al.*, 1990, 1992).

For the less volatile (and more chlorinated) PCBs and DDT, the concentrations were 25- and 30-fold lower, respectively, for Arctic belugas as compared to beluga whales of the St. Lawrence estuary, which may be related to their higher chlorination (Muir *et al.*, 1990; Letcher, 1996). The beluga whales of the St. Lawrence estuary had about 100 times higher concentrations of mirex than beluga of the Canadian Arctic; however, the BMFs for mirex and PCB are similar in both locations (Muir *et al.*, 1990, 1996a). The BMF for mirex and ΣPCBs from fish to beluga ranged from 11 to 16 (Muir *et al.*, 1996a). The TEQs (2,3,7,8-TCDD toxic equivalents) averaged 330 ng/kg in females and 1400 ng/kg in males, and were dominated by PCB 126, 105 and 118 for the St. Lawrence estuary beluga whales. It would appear that local sources of DDT, PCBs and mirex from the Great Lakes area outweigh atmospheric inputs, and the St. Lawrence estuary beluga whales have higher loading values for DDT, PCBs and mirex when analyzed using principal components analysis (Muir *et al.*, 1996a). Total concentrations of non-ortho PCBs 77, 126 and 169 in blubber of St. Lawrence estuary beluga whales (8000 ng/g in males) were 10–20 times higher than in Arctic belugas (Muir *et al.*, 1996a). In the case of the Greenland beluga whales, Muir *et al.* (1996a), found no correlation with age for POC concentrations in blubber, which may be explained by a shift in diets of older males.

Becker *et al.* (1995, 1997) reported concentrations of POCs in beluga whales from Alaska as part of the AMMTAP. The ratio of 4,4'-DDE to ΣDDTs concentrations in blubber from the Alaska whales was similar to that of Canadian and Greenlandic beluga whales. For blubber from beluga whales of Alaska, the descending order of PCB congener concentrations was 153, 138, 149, 118, 101, 180, 187, 52 and 66/95. Becker *et al.* (1995 and 1997) concluded that toxaphene and total chlordanes represent a more significant proportion of the total chlorinated hydrocarbons in these Arctic animals. Becker *et al.* (1997) reported lower concentrations of POCs (most notably the chlordanes) in beluga whales of the Cook Inlet stock (Alaska), which may be exposed to different anthropogenic sources of POCs than beluga stocks in Chukchi and Beaufort Seas, Alaska. However, toxaphene concentrations were in the same range for all Alaska stocks studied (Becker *et al.*, 1997), and ranged from 500 to 6620 ng/g. Wade *et al.* (1995) reported toxaphene concentrations ranging from 1.35 to 8.21 µg/g (1350–8210 ng/g) and 3.20 to 15.9 µg/g lipid weight (3200–15 940 ng/g) in blubber of beluga whales from Point Lay, Alaska. Krahn *et al.* (1999) indicated that concentrations and patterns of POC vary by stock in Alaska.

It is evident that POC contamination differences by region vary by specific class (PCBs, DDT, toxaphene, etc.). There was a surprising lack of PCDDs in beluga blubber (detection limit 2 ng/kg). In narwhal blubber the chlorobornanes were the predominant OCs, ranging from 2990 to 13 200 ng/g in males and from 1910 to 8390 ng/g in females. These were mostly octachlorobornane and nonachlorobornane, which are two- to four-fold lower than in St. Lawrence estuary beluga whales (Muir *et al.*, 1992b). Toxaphene in Alaska beluga whales ranged from 1.49 to 8.94 µg/g dw (Wade *et al.*, 1995), which indicates significant transport to the Arctic as well. ΣPCBs ranged from 2250 to 7290 ng/g in males and 894–5710 ng/g in female Canadian Arctic belugas; concentrations were about 15-fold lower than in beluga whales of the St. Lawrence estuary (Muir *et al.*, 1992a). However, ΣHCHs, dieldrin and Σchlorobenzenes differed by less than twofold between these two groups (Muir *et al.*, 1992a).

Comparative studies of PCB metabolic capacity in mammals indicate that the capacity is greatest in terrestrial species, moderate in some pinnipeds, and lowest in whales (Tanabe *et al.*, 1988, 1994; Boon *et al.*, 1992). There is evidence for a lack of bioaccumulation of 2,3,7,8-TCDD in Arctic beluga, even though ringed seals from the same area with lower concentrations of ΣPCBs have detectable 2,3,7,8-TCDD. These findings may indicate that belugas have the ability to metabolize PCDDs but not PCDFs (Muir *et al.*, 1996a, b). However, they may also be explained by varying dietary sources. Eels represented a likely source of the contaminants detected in the St. Lawrence estuary beluga whales (Muir *et al.*, 1996a, b). The presence of PCDF congeners not fully chlorinated at positions 2, 3, 7 and 8 may imply a diet dominated by invertebrates, as opposed to fish (Muir *et al.*, 1996a). The relative abundance of certain PCDD/F and non-*ortho*-substituted PCB congeners can be easily explained by metabolic elimination, because the POC profiles (TCDF greater than TCDD) and non-*ortho* PCBs (126>77>169) are similar in St. Lawrence estuary beluga whales and the Arctic belugas (Muir *et al.*, 1996a). The non-*ortho* and mono-*ortho* PCBs (PCBs 105, 114, 118 and 156) contributed 98 per cent to total TCDD equivalents in narwhal and beluga (Ford *et al.*, 1993). PCB 126/153 ratios indicated that belugas were depleted in PCB 126, as compared to fish (Norstrom *et al.*, 1992). This is additional evidence that CYP1A isozyme activity is present and likely high in beluga whales (Muir *et al.*, 1996a). Muir *et al.* (1996a) concluded that the differences of PCB52/153 and PCB126/153 ratios between beluga whales of the Arctic and the St. Lawrence estuary might be due to local sources versus long-range transport, as well as due to a greater (i.e. induced) metabolic capability in the animals with the higher exposure. This was shown by Muir *et al.* (1996a), using the congener classification developed by Boon *et al.* (1994), where the relative ratios (R_{rel}) for St. Lawrence estuary beluga whales had values that averaged 0.14 and 0.11 for groups III and IV, respectively (indicating metabolism); and values of 1.16, 3.61 and 1.90 for groups I, II and V (indicating a lack of metabolism) of *meta, para*-substituted congeners or congeners with more than three *ortho*-chlorines. The results indicate that the St. Lawrence estuary beluga whales have a much greater ability to metabolize congeners with vicinal unsubstituted *meta, para*-positions, as well as *ortho, meta*-positions, than the Arctic whales. This implies greater CYP2B and CYP1A activity, which may be a physiologic response to elevated PCB burdens (Muir *et al.*, 1996a). Correlations of CYP1A content and ethoxy-resorufin-*O*-deethylase (EROD) activity in the liver, and non-*ortho/mono-ortho* PCB content in the blubber of beluga whale supports this hypothesis (White *et al.*, 1994). However, there is the possibility of effects due to multiple confounding variables when interpreting metabolic potential using residue data alone.

This allows us to speculate as to why St. Lawrence estuary beluga whales have elevated concentrations of PCBs and DDT, and to a lesser extent chlorinated benzenes/hexanes and Σchlordanes, as compared to other beluga whale populations (Muir *et al.*, 1996a). Local sources are likely playing a significant role. The BMFs for mirex and ΣPCBs from fish to beluga were similar to those for Arctic beluga whales, and the lack of age correlations with ΣPCBs and ΣDDTs is also similar (Muir *et al.*, 1996a). This apparent enhanced metabolic capability of the St. Lawrence estuary beluga whales may lead to detoxification. However, it should be recognized that in some cases more potent compounds might be produced, resulting in injury to different target organs than by the parent POCs. The toxicity of these POC metabolites has been poorly characterized.

Temporal changes in POC concentrations have been noted in beluga whales of the St. Lawrence estuary. Based on lipid-normalized and age-adjusted contaminant data, a decline over time was noted for ΣDDTs, Aroclor PCBs, hexachlorohexane and Σchlorobenzenes in males from 1982–5 to 1993–4 (Muir *et al.*, 1996b). Mean concentrations rose for toxaphene, whereas concentrations of dieldrin, Σchlordanes and mirex showed no trend (Muir *et al.*, 1996b). Declines in POCs were not detected in females. This type of temporal assessment is required for beluga and narwhal populations of the Arctic to determine whether increases or decreases are occurring for these persistent contaminants. If atmospheric and other sources have been reduced, we may see this reflected as lower concentrations in biota over time. However, if we have not yet reached a steady state and/or loading of the Arctic continues, concentrations in tissues will likely continue to increase.

Beluga whales span the range of exposure and possible effects and should be recognized as a critical species for study within and outside the Arctic.

Persistent organic contaminants in Arctic pinnipeds

The suborder Pinnipedia consists of a diverse group of marine mammals classified into three families: Phocidae, the 'true' or 'hair' seals, Otariidae, the 'eared' pinnipeds (sea lions and fur seals), and Odobenidae, the walruses. One view is that all pinnipeds are descended from a common ancestor, while others suggest that the otariids and odobenids are descended from the dog–bear stock and the phocids from otter-like carnivores (Mitchell, 1975; Repenning, 1976). The following species of phocids are considered to be Arctic species and will be discussed in this section: ringed seal, spotted or largha seal (*Phoca largha*), ribbon seal (*Histriophoca fasciata*), harp seal (*Pagophilus groenlandicus*), bearded seal (*Erignathus barbatus*), and hooded seal (*Cystophora cristata*). Two species of otariids, the northern fur seal (*Callorhinus ursinus*) and Steller's sea lion (*Eumetopias jubatus*), occur seasonally (spring and summer) in the southern portion of the Arctic (central and southern Bering Sea), but are considered to be North Pacific rather than Arctic species and are not discussed in this chapter. The single species of odobenid, the walrus (*Odobenus rosmarus*), is considered to be a true Arctic species.

Almost all species of arctic pinnipeds are considered to be important subsistence food resources for coastal indigenous human populations. The concentrations of POCs in the tissues of these animals have important implications regarding human health and risk to humans from consumption of meat (muscle) and organs (liver and kidney), and oil derived from the rendering of blubber. As a group, pinnipeds feed at various trophic levels and overlap both baleen whales (including gray whales, *Eschrichtius robustus*) and odontocetes in their principal food resources. Generally speaking, for those pinnipeds that feed near the top of the food web, the potential for accumulation of POCs is probably not as great as that for the Arctic odontocetes (beluga whale and narwhal) and polar bear. The Arctic pinnipeds generally are not as long-lived as cetaceans; therefore, the potential for accumulation may be somewhat less. Most pinnipeds are characteristic of near-shore waters, where they have the potential for being exposed to coastally derived contaminants. However, this is probably not as important in the Arctic as it is in temperate regions because the principal source of POCs in the Arctic is apparently derived from atmospheric transport from lower latitudes (Wania and Mackay, 1993). Evidence also suggests that metabolism and excretion of DDT and PCBs may be less efficient in odontocetes than in pinnipeds, leading to greater magnification in the former (Tanabe *et al.*, 1988, 1994).

Ringed seals

Ringed seals are the most abundant and most widely distributed of all of the Arctic phocids, and most of the POC data for Arctic pinnipeds has been generated on this species. Through use of their long front claws, ringed seals are able to maintain breathing holes in the Arctic land-fast ice throughout the winter. They are the only pinnipeds to remain in the high Arctic during the entire winter, feeding on Arctic cod and invertebrates. They are the principal winter food for polar bears. As such, they are an important component of the Arctic marine food web leading to the top predators, the polar bear and humans. Although the data are not geographically uniform throughout the range of this species, the ringed seal has the largest POC database of any of the Arctic pinnipeds. In addition, reference data on physiological effects of POCs, such as PCBs and DDT, has been generated on this species in some areas of its range, particularly the Baltic Sea (Helle *et al.*, 1976a, b; Helle *et al.*, 1983; Olsson *et al.*, 1992, 1994).

The geographical data on POCs in ringed seals are most abundant for the North American Arctic, particularly for the Canadian Arctic, although some data are also available from Spitzbergen (Svalbard) and the South Kara Sea in the Russian Arctic (Muir *et al.*, 2000). Surveys in the mid-1980s (Muir *et al.*, 1988; Weis and Muir, 1997) showed similar concentrations in ringed seals throughout the Canadian Arctic, although differences were found in patterns of the individual POCs. For example, there were higher proportions of penta- and hexachlorobiphenyls in seals from Hudson Bay compared to seals from other more northerly and westerly regions. Additional data from the late 1980s and early 1990s showed higher concentrations of POCs in ringed seals from Hudson Bay. Muir *et al.* (1997) suggested this might be due to these populations being nearer contaminant sources in southern Canada and the eastern United States. Concentrations of ΣPCBs and ΣDDTs were higher in the blubber of ringed seals from Spitzbergen than in those animals from the North American Arctic (Muir *et al.*, 1992a). Additional data on ringed seals from Alaska (Krahn *et al.*, 1997; Kucklick *et al.*, 2001) suggest similar concentrations of PCBs and DDTs in ringed seals across the western North American Arctic. However, Kucklick *et al.* (2001) found the concentrations of PCBs to be higher in ringed seals from the Chukchi Sea near Barrow, Alaska, than concentrations reported by Krahn *et al.* (1997) in ringed seals from Norton Sound in the Bering Sea (e.g. 710 ± 182 versus 249 ± 75 ng/g wet weight in blubber of males when summing the same 17 dominant PCB congeners). ΣDDTs and Σchlordane were also higher in the Chukchi Sea animals (e.g. 685 ± 205 versus 188 ± 63 ng/g and 885 ± 612 versus 157 ± 67 ng/g wet weight in blubber of males, respectively). In the Canadian Arctic the non-*ortho* and mono-*ortho* PCBs (PCBs 105, 114, 118 and 156) were important contributors (about 50 per cent) to total TCDD equivalents in ringed seals (Ford *et al.*, 1993). These PCB congeners have also been found in the tissues of ringed seals from Alaska (Becker *et al.*, 1997; Krahn *et al.*, 1997; Kucklick *et al.*, 2001).

At least for one geographical location, the ringed seal has the longest temporal database on persistent organic contaminants for any Arctic marine mammal (Addison *et al.*, 1986). Recently, Addison and Smith (1998) published the results from a 20-year study of organochlorine residues in ringed seals from Holman, Northwest Territories, Canada. This study included the re-analysis of blubber samples collected in the early 1970s from the same study area to assess methodological errors in POC analyses, which has improved over time. They found that the concentrations of ΣDDT in blubber were about the same between 1972 and 1981, but after that period these compounds began to decrease. In 1991 4,4'-DDE was about one-third and 4,4'-DDT was about 20 per cent of the 1972 values (mean 1972 concentrations for 4,4'-DDE were 0.3 mg/g for females and 0.8 mg/g for males, whereas 4,4'-DDT concentrations were 0.3 mg/g for females and 0.5 mg/g for males). Concentrations of PCBs appeared to decrease early (between 1972 and 1981) and remained relatively constant between 1981 and 1991 (one-third of the 1972 concentrations, which were 1.8 mg/g for females and 3.8 mg/g for males). Although during the 1981–1991 period the concentrations of total PCBs remained relatively constant, the relative proportions of individual congeners changed. This suggests possible metabolic effects on these compounds or other mechanisms of redistribution of congeners. Between 1981 and 1991, the concentrations of HCHs did not change, but HCB decreased by 50 per cent in both males and females.

Historically, the most success in linking contaminants to health effects and population declines in pinnipeds occurred in studies of ringed seals, gray seals (*Halichoerus grypus*) and harbor seals (*Phoca vitulina*) in the Baltic Sea during the 1980s (Olsson *et al.*, 1992, 1994; see Bergman *et al.*, Chapter 19 in this volume). A decline in pinniped populations in the Baltic was first observed in the 1950s and the suspicion that it was linked to organochlorine pollution was formulated during the 1960s (Olsson *et al.*, 1992). Key to these studies was the indication of reproductive impairment and symptoms suggesting immune dysfunction in these animals, bone deterioration (particularly in the area around the teeth), gastrointestinal lesions and proliferation of gastrointestinal parasites. Reproductive impairment was first noticed by the decreases in fecundity, followed by the documentation of abnormalities in the reproductive organs of the females (i.e. uterine stenosis or occlusions) (Olsson, 1972, 1978; Helle *et al.*, 1976a, 1983; Olsson *et al.*, 1994). A 60 per cent decrease in the number of females becoming pregnant occurred in ringed seals from PCB-contaminated Bothnian Bay (Helle *et al.*, 1976a).

Spotted seals

Spotted seals range in distribution in the western Arctic from the eastern Beaufort Sea to the western Chukchi Sea, south to the Aleutian Islands and along the western Pacific coast down through the Sea of Japan. In the winter and spring this species occurs along the ice front, and during the summer months it is very common in small groups in the Chukchi Sea. For areas where they occur in abundance, these animals are an important subsistence food resource. The only data on concentrations of POCs in this species is from Tanabe *et al.* (1988), who reported ΣPCBs concentrations in blubber of spotted seals from waters off Hokkaido, Japan, as ranging between 0.67 and 1.5 µg/g (mean = 0.98 µg/g; n = 4). This is within the range of concentrations reported for harbor seals from the Gulf of Alaska and ringed seals in the Arctic (Krahn *et al.*, 1997). Because of its similarity to ringed seals and harbor seals in trophic position, one would expect concentrations of these compounds in spotted seal tissues to be similar to the other two species, particularly where the ranges of these three species overlap (Alaska).

Ribbon seals

Ribbon seals are somewhat solitary. They range in distribution in the western Arctic from the eastern Beaufort Sea to central Chukchi Sea and south to the Aleutian Islands, and along the western Pacific coast down through the Sea of Okhotsk. Not much is known about this species. It is associated with the sea ice and maintains itself away from coastal areas. Although occasionally taken by indigenous subsistence hunters in the Arctic, it is a very minor component of the marine subsistence resources, as compared to ringed seals, bearded seals, spotted seals and walrus. Tanabe *et al.* (1988) reported ΣPCBs in the blubber of ribbon seals from waters off Hokkaido, Japan, ranging from 1.0 to 1.5 µg/g (mean = 1.3 µg/g; n = 5).

Harp seals

Harp seals occur in the open sea of the eastern Arctic from the Kara Sea to Greenland, and from the high Arctic of eastern Canada south to Newfoundland and the Gulf of St. Lawrence. In some regions, harp seals are used for food. This species occurs in gregarious migratory groups associated with sea ice. The most recent data on POCs in harp seals come from papers by Addison *et al.* (1984), Ronald *et al.* (1984), Beck *et al.* (1994), Oehme *et al.* (1995, 1988) and Zitko *et al.* (1998). These are based on harp seals from the Gulf of St. Lawrence, Hudson Strait and Greenland Sea.

Concentrations of PCBs in the blubber of harp seals from the Gulf of St. Lawrence declined by about 50 per cent between 1972 and 1981, as did PCBs in the blubber of Arctic ringed seals (Addison *et al.*, 1984). Zitko *et al.* (1998) also reported declines in PCBs and DDTs for harp seals over a 20-year period. PCDD and PCDF concentrations in blubber of harp seals from the Greenland Sea were slightly lower than those of ringed seals from the Arctic, but concentrations of PCBs were comparable (Oehme *et al.*, 1995). Concentrations of PCDD, PCDF, non-*ortho* substituted (or coplanar) PCBs (77, 126 and 169), and di-*ortho* substituted PCBs in brain and blubber of harp seals were also compared, and concentrations in brains were one to two orders of magnitude lower. A highly significant positive correlation existed between age and the concentrations of 4,4'-DDE and di-*ortho* PCBs (suggesting metabolism) and between single PCDD and coplanar PCBs in blubber (Oehme *et al.*, 1995). Zitko *et al.* (1998) reported concentrations of dibenzo-*p*-dioxins, dibenzofurans, PCBs and organochlorine pesticides to be lowest in the muscle of harp seals, intermediate in kidney and liver, and highest in the blubber. Concentrations of PCBs and chlorinated pesticides in blubber were equivalent in males and females; however, concentrations of these compounds in muscle, liver and kidney were 2–3 times higher in males.

Bearded seals

Bearded seals are circumpolar in distribution, but do not remain in the high Arctic during winter. They tend to follow the ice edge, moving to lower latitudes during the fall and winter, and into the high Arctic during the spring and summer. Due to its large size, use of hide for constructing skin boats (*umiaq*), and palatability, it is one of the most popular subsistence pinnipeds in the Arctic, especially Alaska. The bearded seal feeds on bottom invertebrates (e.g. crabs, shrimp, clams, octopus) and near-bottom fish in relatively shallow water. It is also an important component of the polar bear diet during periods when it is available.

The bearded seal occupies a very different position in the food web as compared to the ringed seal and, therefore, one would expect concentrations in tissues and patterns of POCs to be different. Concentrations of PCBs, DDTs, ΣHCHs and chlordane in blubber have been reported for bearded seals sampled at Baffin Island in the mid-1980s (Thomas and Hamilton, 1988). Concentrations of each of these groups of compounds were similar to what was reported for the same period for Arctic ringed seals. A comparison between individuals of these two species sampled in the same region, season and year would be more informative. Krahn *et al.* (1997) compared concentrations of PCBs, DDTs, chlordane, HCB and dieldrin in the blubber of bearded seals and ringed seals sampled in Norton Sound (Alaska) in 1993–5. For both species, concentrations in adult males ranged from 0.05 to 0.36 mg/kg for PCBs, 0.01–0.36 mg/kg for DDTs, and 0.01–0.45 mg/kg for chlordane. PCBs and DDTs were slightly lower in the bearded seals, but this difference was not significant; differences in concentrations of chlordane were not apparent, but concentrations of HCB and dieldrin in the bearded seals were an order of magnitude lower than in the ringed seals. Considering the subsistence value of bearded seals for indigenous people of the Arctic, additional information on the POC concentrations in this animal is warranted. The proposed difference in trophic level as compared to ringed seals must be questioned, based on the similarity of contaminant concentrations in bearded to ringed seals. Studies currently under way in northern Alaska will test this comparison using concentrations of POC, carbon and nitrogen stable isotopes, and heavy metals.

Hooded seals

Hooded seals occur in the open sea of the eastern Arctic from Spitzbergen to Greenland and from the high Arctic of eastern Canada south to Newfoundland and the Gulf of St. Lawrence. This species prefers deep water and thick, drifting ice flows. Somewhat solitary, it forms small family groups during the spring breeding season. Although Clausen *et al.* (1974) and Johansen *et al.* (1980) reported PCB and DDT concentration data for this species in Greenland, the hooded seal is an Arctic pinniped for which we have no recent POC information.

Walruses

Walruses are circumpolar, shallow-water pinnipeds that feed on benthic organisms such as clams and gastropods, but some specialize in feeding on seals. Two subspecies are recognized, *Odobenus rosmarus rosmarus* occurs in the eastern Arctic and *O. rosmarus divergens* occurs in the western Arctic (King, 1983). This species prefers areas of moving pack ice. This allows individuals to feed and rest over large geographical areas, and provides a means of transport during the spring and fall, while minimizing energy expenditure. The walrus is an extremely important subsistence food resource to the indigenous population of the Arctic coast.

Earlier studies reported very low concentrations of POCs in the blubber of walrus over all geographical areas (Born *et al.*, 1981; Thomas and Hamilton, 1988; Taylor *et al.*, 1989). However, Muir *et al.* (1995) reported concentrations of PCBs and chlorinated pesticides orders of magnitude higher in walrus blubber from some Hudson Bay groups, and attributed this to the possible consumption of seals, a habit reported by Lowry and Fay (1984) for some of the older male Alaska walrus. The Hudson Bay males had median concentrations of ΣPCBs and ΣDDTs (11.5 and 2.2 mg/kg, respectively) that were two orders of magnitude higher, and chlordane concentrations (6.3 mg/kg) one order of magnitude higher, than reported for other Arctic areas (Muir *et al.*, 1995; de March *et al.*, 1998). Elevated concentrations similar to those in Hudson Bay walrus were found by Skaare *et al.* (1994) in skin/blubber biopsies of walrus from Svalbard. The hypothesis of seal-eating to explain these relatively high POC concentrations was supported by carbon and nitrogen isotope analysis of the walrus muscle tissue in comparison with that of ringed seals and other walrus prey (Muir *et al.*, 1995).

These elevated concentrations have not been found in any of the Alaska walrus so far. Seagars and Garlich-Miller (2001) report levels of POCs measured in 27 walrus sampled in the Bering Sea in 1991 to be similar to the low levels previously reported by Taylor *et al.* (1989). Efforts are also under way at the National Institute of Standards and Technology (NIST) Charleston to selectively analyze blubber tissues from older male walruses that have been archived at the National Biomonitoring Specimen Bank, to see if a pattern of elevated POCs can be identified in any of these samples from the western Arctic walrus population. Preliminary results indicate very low POC concentrations in these samples, similar to those reported by Seagars and Garlich-Miller (2001) and J. Kucklick (NIST-Charleston, personal communication).

Polar bear

POC concentrations in tissues and associated biological factors

Polar bears are considered a critical species for biomonitoring programs because they occur throughout the Arctic and Subarctic and are top predators that accumulate and integrate many contaminants (Norstrom *et al.*, 1998). However, many factors can affect POC concentrations in polar bears, including diet, fasting (i.e. pregnant sows), region or stock, season, sex, age, presence or absence of cubs, available prey, forage preferences of prey (food-web structure) and others (Polischuk *et al.*, 1995; Norstrom *et al.*, 1998). The difficulty in collecting adequate samples is also restrictive. The primary prey species, ringed seal, is also Holarctic and will integrate contaminants over a limited range (Norstrom *et al.*, 1998). Differences in POC accumulation and possible physiologic and metabolic roles as compared to other marine mammals are emphasized here.

Kucklick *et al.* (2001) determined the 'apparent bioaccumulative factor (ABAF)' for polar bears and ringed seals from Barrow, Alaska, based on the ratio of the lipid-based organochlorine concentrations in fat of polar bear divided by that in ringed seal, and compared these with ABAFs for polar bears and ringed seals from several Canadian Arctic locations. The lower ABAFs for Barrow, Alaska, for compounds that substantially biomagnify in polar bear, suggest that the polar bears from the western North American Arctic (and Alaska in particular) may prey on a much wider variety of prey than polar bears from the eastern North American Arctic, which may be more restricted to preying on ringed seals.

Chlordanes

The largest contributor to total chlordane in polar bear fat and liver is the metabolite oxychlordane (56 per cent and 61 per cent, respectively) (Letcher, 1996). Total chlordane was dominated by oxychlordane (62 per cent) in polar bear fat, with approximately equal concentrations of nonachlor-III (13 per cent), *trans*-nonachlor (11 per cent), and heptachlor epoxide (11 per cent) (Zhu *et al.*, 1995). In studies by Zhu *et al.* (1995) of POCs in ringed seals, polar bears, and the blood plasma from humans from northern Quebec, it was found that polar bear fat contained 5–6 times more total chlordane (4287 ng/g lipid) than did human plasma (840 ng/g lipid) or seal blubber (706 ng/g lipid). Photoheptachlor in polar bear was 3.4 per cent of total chlordanes and had a similar BMF as oxychlordane. The oxychlordane contribution was approximately 40 per cent for ringed seal blubber and 10 per cent for whole Arctic cod, indicating the metabolic capacity (formation of oxychlordane) may be cod < seal < bear (Letcher, 1996), or variability in absorption and biomagnification. Nonachlor-III is an original component of chlordane that is not easily biotransformed, and is biomagnified nearly ten-fold from seal blubber (59 ng/g lipid) to polar bear fat (545 ng/g lipid) (Zhu *et al.*, 1995). The median concentration of total chlordane was 2.3 µg/g lipid, 72 per cent of which was represented by the metabolite oxychlordane.

Young and subadult bears had significantly higher chlordane concentrations than adult and old male bears (Bernhoft *et al.*, 1997), and concentrations in matched sows and cubs were twice as high in cubs (Polischuk *et al.*, 1995; Norstrom *et al.*, 1998). The sum chlordanes (CHL) in fat of polar bears from Hudson Bay was higher for cubs and subadults (<5 years old) than adults, and decreased from young of the year (YOY) to 5 years of age (Norstrom *et al.*, 1998). Norstrom *et al.* (1998) indicated that sum chlordanes (geometric mean) was 30 per cent lower in males than females (with and without cubs). The decrease in CHL over approximately 4 years for subadults would fit with a half-life of 1 year (nearly achieving a steady state in 4 years), suggesting a very slow clearance rate (Norstrom *et al.*, 1998) if one makes many toxicokinetic assumptions. For chlordane, the decrease with age in male bears and the lower concentration in male than in female adults may be due to sex differences in metabolism. However, Norstrom *et al.* (1998) also considered the higher concentrations of ΣPCBs as inducers of hepatic CYP2B, which may increase clearance of CHL. Whatever the mechanism, polar bears do appear to metabolize chlordane when concentrations in prey are compared to concentrations in the bear (Letcher *et al.*, 1996; Wiberg *et al.*, 2000; Kucklick *et al.*, 2001). These results are in contrast to the findings of higher chlordane concentrations in males than females of other marine mammals, which do not have this metabolic capacity (Bernhoft *et al.*, 1997).

PCBs and DDTs

In general, the ratio of higher to lower chlorinated PCBs decreased from west to east in polar bears from the Bering Sea to Greenland, indicating there may be a higher input of higher chlorinated PCBs to the eastern Arctic (Norstrom *et al.*, 1998). Declines in pinnipeds, critical prey species for polar bears, have been causally linked with PCBs elsewhere, based upon known reproductive and immunologic effects of this class of POCs (Borrell, 1993; Kuiken *et al.*, 1994), and ringed seal blubber makes up the majority of the polar bear diet (Best, 1985). Therefore any decreases in seal productivity could affect the polar bear. ΣPCBs concentrations in bear fat (6819 ng/g lipid) were more than ten times that of seal blubber and human plasma, and congener patterns were similar (Zhu *et al.*, 1995). In previous studies of polar bears in arctic Canada, Norstrom *et al.* (1988) found that the ΣPCBs and total chlordane accounted for more than 80 per cent of the total organochlorines in polar bear adipose tissue.

The PCB concentrations present in the polar bears of Svalbard are extremely high. The POC patterns in yearlings reflect the low transfer of the highest chlorinated PCB congeners into maternal milk, and the other PCB congeners were at higher concentrations in the yearlings compared to their mothers (Bernhoft *et al.*, 1997). The median concentration of ΣPCBs in subcutaneous fat for 85 polar bears was found by Bernhoft *et al.* (1997) to be 15.5 µg/g lipid, 62 per cent of which was contributed by PCB 153 and 180. The highest concentration of the PCBs was found in adult male bears, and was significantly higher than in young and adult female bears. Nine of 14 congeners showed significantly higher concentrations in adult males than in one or more of the other age or sex groups studied. Concentration of the PCBs seems to increase with age until about 7 years in females and 14 years in males, and thereafter declines (Bernhoft *et al.*, 1997). However, the ΣPCB was higher for cubs and subadults (<5 years) than adults; and concentrations were 46 per cent higher in male than female adults (Norstrom *et al.*, 1998). In lactating bears, similar concentrations of chlordane, HCB and ΣHCHs were found in lipids of milk, subcutaneous tissue and plasma. The concentrations of DDE, and penta- and hexa-chlorinated PCBs in milk lipid were similar to the corresponding concentrations in plasma lipids. The POC concentrations in subcutaneous fat of yearlings were generally higher than in their mothers. However, the concentrations of the hepta- to decachlorinated biphenyls (PCB 187, 194, 206, 209) in yearlings compared to their mothers decreased gradually with increased chlorination and were below maternal concentrations. PCB 105 and 118, and DDE were found in similar concentrations in mothers and yearlings. Mothers that had been lactating for 1.25 years contained significantly less chlordane, HCB, ΣHCHs, and ΣPCBs than females that had been lactating for 2.25 years or had no young.

Denning status is regarded as a measure of polar bear reproduction success. The POC concentrations in denning bears were not significantly different from those not denning (Bernhoft *et al.*, 1997). Concentrations of ΣPCBs in adult males from Svalbard were about six and three times higher than the averages in polar bears in Alaska and Canada (Bernhoft *et al.*, 1997). ΣPCBs were significantly higher (11–16 mg/kg) in eastern Greenland, Svalbard and M'Clure Strait, above 14 other study areas which averaged 3.7 ± 1.6 mg/kg. Bernhoft *et al.* (1997) associated this with geographical differences in POC distribution, but we should consider differences in feeding behavior of bears and seals in these regions as well. In ringed seals, higher concentrations of PCBs are found in the Svalbard area as compared to the Canadian Arctic (Muir *et al.*, 1992a). Biomagnification of oxychlordane, HCB, HCHs and PCBs from ringed seals to polar bears is evident, with a BMF of about 10 for PCBs. Metabolism of lower chlorinated PCBs and biomagnification of the more highly chlorinated forms can explain the species differences observed (Bernhoft *et al.*, 1997).

Tri- and tetrachlorinated biphenyls, in addition to penta- and hexachlorinated biphenyls with *meta-para* adjacent hydrogen atoms, are not found in polar bears (Bernhoft *et al.*, 1997). These are typically present in other marine mammals. The age- and sex-based differences for most of the PCBs and the HCHs (disproportional accumulation with age in males) is opposite that of chlordane. The deviation in females from the PCB concentrations seen in males at 7–11 years of age coincides with sexual maturation and breeding, with the first offspring occurring at 6–7 years of age. This represents the initiation of reduction in female body burdens of POCs through transfer to the offspring during pregnancy and lactation (Bernhoft *et al.*, 1997). Milk has a high fat content and lactation may occur for as long as 2.5 years. The incomplete transfer of lipid from the circulatory system may reduce the efficiency of the POC transfer, particularly for the most lipophilic compounds. This is supported by the correlations between plasma and milk lipids for most POCs, particularly for the higher chlorinated PCBs and DDE (Bernhoft *et al.*, 1997).

Concentrations of POCs in subcutaneous fat of yearling polar bears reflect a similar pattern to that of the mother's milk. Most of the POCs are present at higher concentrations in the yearling offspring than in the mother (Bernhoft *et al.*, 1997). Seal blubber gradually becomes a more available food source for young polar bears and is less contaminated than maternal milk (Bernhoft *et al.*, 1997). Higher concentrations have been found in cubs in other studies (Polishuck *et al.*, 1995). A difference in POC content between pregnant versus non-pregnant females was not detected, but the sample size was limited. There is concern, because high cub mortality has been observed in polar bears at Svalbard (Bernhoft *et al.*, 1997). PCBs have been suggested as potential causative agents in seven cases of vestigial male reproductive organs (both male and female genital structures) in 'genetically' female polar bears of Svalbard, some of which have cubs (Wiig *et al.*, 1998). The capability to metabolize several toxic POCs may protect the polar bear. However, the generation of toxic metabolites, as well as age-based accumulation and rather high exposure of offspring via lactation, are of concern. Mean PCB concentrations in fat are difficult to interpret because they are 4–10 times lower than concentrations associated with reproductive effects based on laboratory studies (Norstrom *et al.*, 1998), but if the polar bear is sensitive to these compounds, some individuals may be approaching levels of concern. The unique metabolism of polar bears may enhance excretion and elimination of these compounds, but may also produce active metabolites with affinity for critical tissues (e.g. liver, adrenal gland). Occupying the highest trophic level and the potential for high metabolic activity related to OCs makes the polar bear a critical animal for improving understanding of the impact(s) of PCBs in arctic marine mammals.

Polychlorinated dibenzo-p-dioxins and dibenzofurans

Polychlorinated dibenzo-*p*-dioxins and dibenzofurans were determined in ringed seals and polar bears. No apparent biomagnification (BMF < 1.0) of TCDD, OCDD or TCDF occurred from seal to bear fat; in contrast, the BMFs for PCBs and HCB were 5–17 and 14–21, respectively (Norstrom *et al.*, 1990). All polar bears studied had detectable concentrations of TCDD and OCDD, but no TCDF or HxCDD. The polar bear must have a mechanism for efficient elimination of TCDD and TCDF. OCDD may accumulate in bears, but the limitation of the detection level hampered this assessment (Norstrom *et al.*, 1990). PCDD and PCDF have been detected in blubber of ringed seals from the eastern northern Atlantic (Oehme *et al.*, 1988; Bignert *et al.*, 1989). Norstrom *et al.* (1990) indicated that the highest concentrations were in bears from Barrow Strait and Larsen Sound, and lowest from Hudson Bay, similar to patterns in ringed seals. This distribution is the opposite of that reported for ΣPCBs, Σchlordanes, ΣDDTs, ΣHCHs, and dieldrin (Norstrom *et al.*, 1988, 1990; Norstrom and Muir, 1994).

Toxaphene and HCH

Fifteen congeners of toxaphene were analyzed in polar bear fat (total 1.0 mg/kg) and ringed seal blubber (total 0.25 mg/kg) (Zhu and Norstrom, 1993). Proportions of toxaphene compounds were quite different for polar bears and ringed seals compared to other marine organisms. Only 8–11 per cent of total toxaphene was octachloroborane (T2) and nonachlorobornane (T12) in ringed seal and polar bear (Zhu and Norstrom, 1993), whereas these typically dominate in Arctic amphipods, burbot, narwhal and beluga (Stern *et al.*, 1992; Bidleman *et al.*, 1993). A better understanding of toxaphene pathways in Arctic biota is needed.

The median concentrations of DDE, HCB and ΣHCH in blubber were 272, 146 and 240 ng/g lipid, respectively. Males had higher concentrations of ΣHCH and ΣPCBs than females. The lower HCH concentration in females corresponds with the finding of relatively high concentrations of HCHs in milk (Bernhoft *et al.*, 1997). In males, ΣHCH increased with age until about 12 years, whereas in females no change with age was evident. The HCH accumulation was not expected, because ΣHCH has been noted to decrease up the food chain. However, β-HCH, which is the most persistent isomer of HCH, constituted 81 per cent of the ΣHCH (Bernhoft *et al.*, 1997). Tanabe *et al.* (1996) proposes that higher β-HCH/ΣHCH indicates higher metabolic capacity of a species to metabolize α-HCH and γ-HCH. Furthermore, both HCH and PCB patterns differ between polar bear and ringed seal: β-HCH dominating in the polar bear, whereas α-HCH was the dominant form in ringed seal, and γ-HCH (lindane) was found in the seal but not the polar bear.

POC metabolism

POC metabolism and mixed function oxidases (MFO)

Polar bears were shown to metabolize congeners of PCBs and 4,4'-DDE that are normally recalcitrant or not biotransformed (Letcher, 1996; Letcher *et al.*, 1996) in other mammals addressed here. This is presumably based upon differences in hepatic CYP metabolic capacity and differential transfer of some PCB congeners from one trophic level to the next (cod → ringed seal → polar bear) (Figure 8.1). PCB 153 represented 44 per cent, 14 per cent and 4 per cent of the ΣPCBs of the polar bear, ringed seal and cod, respectively, indicating a proportional magnification of PCB 153 and/or loss of non-PCB 153 congeners. This PCB congener is known to be resistant to metabolism and can be used to determine metabolic indices (Letcher, 1996). Many congeners present in cod are not present in polar bears (Figure 8.1) and this indicates likely metabolic degradation and/or excretion by the polar bear.

The presence in ringed seals of PCBs with *meta-para* vicinal hydrogen atoms indicate low CYP2B-type metabolism for seals as compared to polar bears. CYP1A-type metabolism of PCBs occurs in ringed seals, but at a lower activity in ringed seals than in polar bears (Letcher, 1996; Letcher *et al.*, 1996). This high CYP1A activity seems to be unique in the polar bear as compared to other marine mammals in the Arctic regions of the western hemisphere. CYP2B-type activity was also high, as evidenced by the lack of PCBs with *meta-para* vicinal hydrogen atoms and high concentrations of methylsulfone polychlorinated biphenyls (MeSO$_2$-PCBs) (Letcher, 1996; Letcher *et al.*, 1996). There is clear evidence of mobilization of MeSO$_2$-PCBs from adipose tissue, and lactational transfer from mother to cub, which may be more rapid during fasting (Letcher, 1996). The accumulation of methylsulfone metabolites of PCBs has been shown to be enantioselective for ringed seals and polar bears (Wiberg *et al.*, 1998). This can involve selective formation, uptake, transport, storage and/or clearance. The presence of MeSO$_2$-CBs in polar bears results from both formation by the bear (e.g. 3- and 4-MeSO$_2$-PCB 91 and 4-MeSO$_2$-PCB 149) and accumulation from

Figure 8.1 (opposite) Ratios of various PCB congeners to PCB 153 in blubber or fat of Arctic cod (*Boreogadus saida*), ringed seals (*Pusa hispida*) and polar bears (*Ursus maritimus*). Polar bears metabolize congeners of PCBs that are normally recalcitrant or not biotransformed in other mammals. This is presumably based upon differences in hepatic CYP metabolic capacity and differential transfer of some PCB congeners from one trophic level to the next (cod → ringed seal → polar bear). PCB 153 represented 44, 14 and 4 per cent of the ΣPCBs of the polar bear, ringed seal and cod, respectively, indicating a proportional magnification of PCB 153 and/or loss of non-PCB 153 congeners. Many congeners present in cod are not present in polar bears and this indicates likely metabolic degradation and/or excretion in the polar bear. (After Letcher, 1996; Letcher *et al.*, 1996.)

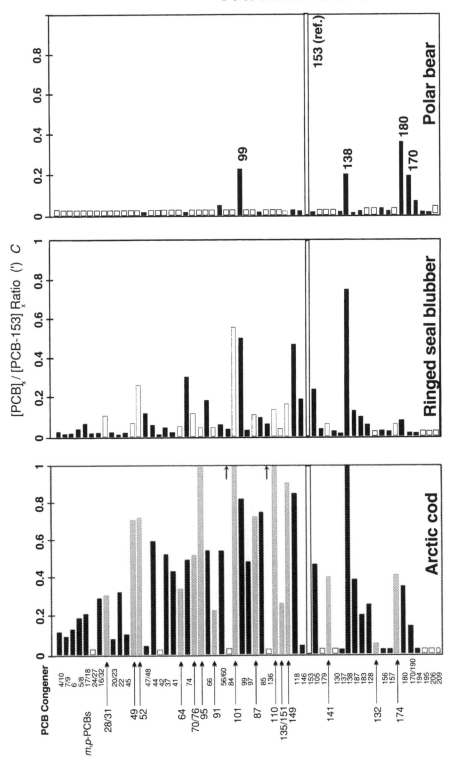

the seal (3- and 4-MeSO$_2$-PCB 132). However, the Arctic cod does not produce these metabolites (Wiberg *et al.*, 1998).

Bandiera *et al.* (1995) immunochemically evaluated rat subfamily cytochrome P450s and detected homologues for 1A, 2B, 2C and 3A, and rat epoxide hydroxylase in polar bear liver. The bears had 'high' concentrations of 1A and 2B compared to other marine mammals, and this could be a consequence of induction by environmental contaminants (Bandiera *et al.*, 1995). Letcher *et al.* (1996) showed CYP1A (P450 expression) was correlated with concentrations of PCBs, PCDDs and PCDFs, and that CYP2B (P450 expression) was correlated with *ortho*-Cl substituted PCBs and chlordanes. A better understanding of polar bear metabolism is needed. This will be key to recognizing their unique detoxification/excretion mechanisms, and to better recognize the potential target organs (e.g. adrenal glands) for these metabolites.

DDT and PCB methylsulfones

The considerably lower DDE concentrations in polar bears than in ringed seals possibly reflects the capacity of the polar bear to metabolize the most persistent of the DDT compounds. Similar DDE concentrations were found in polar bears from Svalbard, Canadian bears (Bernhoft *et al.*, 1997) and Alaskan bears (Kucklick *et al.*, 2001). The tendency of decreasing DDE concentrations from young to subadults may suggest a lower capacity for metabolism of DDE in younger bears (Bernhoft *et al.*, 1997). The findings in polar bears are in contrast to the age-related accumulation of DDTs seen in other male marine mammals. The lack of significant age-class or sex differences in DDE concentrations in polar bears may be due to a higher metabolic capacity of DDT-based compounds in polar bears than in other mammals (Kucklick *et al.*, 2001). Females with cubs have 32 per cent higher sum DDTs than males, and Norstrom *et al.* (1998) reports 22 per cent higher DDE concentrations in females with cubs as compared to solitary females and males. Polar bear fat contained more than 98 per cent of the total body burden of POCs, except for 38.5 per cent of the methylsulfone-DDE (S-MeSO$_2$-4,4'-DDE) which was in the liver (liver mass makes up only 2.6 per cent of the total body mass of polar bears) (Letcher, 1996).

Polar bears are exposed to high concentrations and different MeSO$_2$-PCB and -4,4'-DDE patterns, indicating that the potential toxicological risks from chronic exposure may be different in polar bears relative to ringed seals. Methylsulfone-DDE formation in bears was high; however, another unknown metabolic path must be substantial for 4,4'-DDE. The retention of MeSO$_2$-PCBs and -4,4'-DDE may be a consequence of stellate cells specialized in the retention of polar retinols and retinol esters or binding to specific proteins. This clearly favors pentachloro-3- and MeSO$_2$-PCB 87 (Norstrom *et al.*, 1998). Letcher *et al.* (1998) showed that Arctic cod had no detectable MeSO$_2$-PCBs or 4,4'-DDE (detection level of <0.01 ng/g lipid). In contrast, ringed seals contained 0.4 ng/g 3-MeSO$_2$-4,4′-DDE and 13 ng/g 3- and 4-MeSO$_2$-PCB isomers, and polar bear fat contained 432 ng/g of MeSO$_2$-PCB isomers and 2.0 ng/g 3-MeSO$_2$-4,4′-DDE. Fifteen MeSO$_2$-PCB congeners are likely bioaccumulated, whereas seven are completely or partially formed in the bear (Letcher *et al.*, 1998). However, this may also result in the production of toxic metabolites, and may be related to the presence and amount of CYP.

High activity of CYP1A (loss of PCB 105 and 118, small change in PCB 99), and high CYP2B activity (loss of PCB congeners with *meta-para* vicinal hydrogen atoms in bear versus seal) were evident in polar bears (Letcher, 1996). Induction may enhance this enzymatic process, and the presence of POCs may be critical for the maintenance of the CYP system at this 'high' activity level. All PCBs with significant bioaccumulation potential in polar bears have 2,2′,4,4′ Cl substitutions (Letcher, 1996; Letcher *et al.*, 1996). According to Letcher (1996), the residual anthropogenic compound patterns demonstrate that, relative to ringed seals and cod, the polar bear has a high capacity to depurate, with or without metabolism, all persistent OCs measured.

Methylsulfones of six PCB congeners (99, 153, 138, 180, 170, 194) made up about 93 per cent of the PCB metabolites in polar bears and beluga whales (Bergman *et al.*, 1994). Based on PCB and MeSO$_2$-PCB patterns, ringed seals have low cytochrome P450 (CYP) 1A and CYP2B-type enzyme activities, whereas polar bears apparently have higher levels of both types (Letcher, 1996; Letcher *et al.*, 1996). In polar bears these enzyme levels correlated with the presence of some chlorinated hydrocarbons. All precursor PCBs had 2,5- or 2,5,6-chlorination on the MeSO$_2$-substituted ring. The polar bear is capable of metabolizing some POCs because the BMF is less than 1.0 for 4,4′-DDT and some PCB congeners are altered (Muir *et al.*, 1988; Norstrom *et al.*, 1988). Muir *et al.* (1988) showed that ringed seals did not accumulate (concentrations were less in seals than in cod) PCB 177, 183, 187, 194 and 201, and these would not reach the polar bear. Polar bears can biotransform additional congeners, reducing further the number of congeners (those in seals plus PCB 99, 128, 138 and 177) that it accumulates (Muir *et al.*, 1988). The congeners PCB 153, 180 and 194 are persistent (Muir *et al.*, 1988).

Reproduction implications and associations

Risk from exposure to POCs in polar bears may be greatest at two phases in the reproduction cycle: lactation and delayed implantation. Polar bear mating occurs in spring and the fertilized eggs do not implant until September–October. This implantation is usually concurrent with denning. Only pregnant females tend to den. Species with delayed implantation may be more susceptible to POCs (Sandell, 1990). The cubs are born in late December or early January and the mother and cubs emerge from the den in March or April. At this point the mother has fasted for approximately 6 months, but continues to lactate. Polar bears can enter a hyperphagic phase in the late spring and early summer, and as a result more than 50 per cent of their body weight can be lipid. During phases of fasting and lactation these lipid stores are mobilized (Polischuck *et al.*, 1995). As fasting takes place, a higher concentration of some persistent OCs (e.g. PCBs) is found in milk and fat of sows, whereas others (such as 4,4'-DDT) do not change and may be affected by metabolism (Polischuk *et al.*, 1995). The cubs of the year had the highest OCs of any life stage, and this indicates a significant transfer of contaminants to young. This could be a significant phase to examine for adverse health effects or exposure leading to delayed effects of parent compounds or metabolites of POCs. *In utero* and/or neonatal exposure to persistent OCs has been implicated as a possible cause of female pseudohermaphrodites at Svalbard (north of Scandinavia) at a rate of 1.5 per cent (4 of 269 examined animals) (Wiig *et al.*, 1998). However, investigations of other polar bear stocks (i.e. less contaminated) are needed to determine the 'baseline' occurrence of pseudohermaphroditism in relation to POC exposure.

Special significance of POCs in polar bears

In summary, the polar bear may be quite unique in its exposure, metabolism and expression of possible adverse effects, compared to other marine mammals in the Arctic. This species merits special consideration for effects assessment, not simply residue analysis, because it represents the top of the food chain and integrates these accumulative compounds of the Arctic. Evidence indicates that the most contaminated of the polar bears (i.e. Svalbard animals) are suffering suspicious effects that may be related to POC exposure.

Conclusions

This chapter reviewed the occurrence and possible effects of POCs in Arctic marine mammals. Recent work has clearly indicated that many differences in ecologic, regional and physiologic factors are related to exposure and potential adverse effects of POCs for these species. Unfortunately, we do not understand how these differences relate to individual and population health and responses. Current efforts will only detect gross, dramatic changes, and at that point we are unlikely to be able to reverse unwanted trends. A better understanding of POC effects and toxicoses is needed for these marine mammals from the Arctic, especially if we are concerned about cetacean (i.e. beluga whale), pinniped (i.e. ringed seal) and polar bear populations and health.

Acknowledgements

The authors thank Derek Muir and John Kucklick for reviewing a draft of this chapter.

Bibliography

Addison, R.F. and Smith, T.G. 1998. Trends in organochlorine residue concentrations in ringed seal (*Phoca hispida*) from Holman, Northwest Territories, 1972–91. *Arctic* 51: 253–261.

Addison, R.F., Brodie, P.F., Zinck, M.E. and Sergeant, D.E. 1984. DDT has declined more than PCBs in eastern Canadian seals during the 1970s. *Environmental Science and Technology* 18: 935–937.

Addison, R.F., Zinck, M.E. and Smith, T.G. 1986. PCBs have declined more than DDT-group residues in Arctic ringed seals (*Phoca hispida*) between 1972 and 1981. *Environmental Science and Technology* 20: 253–256.

Aguilar, A., Borrell, A. and Pastor, T. 1999. Biological factors affecting variability of persistent pollutant levels in cetaceans. *Journal of Cetacean Research and Management* (Special Issue) 1: 83–116.

Ayotte, P., Dewailly, E., Bruneau, S., Careau, H. and Vezina, A. 1995. Arctic air pollution and human health: What effects should be expected? *Science of the Total Environment* 160/161: 529–537.

Bandiera, S.M., Torok, S.M., Lin, S., Ramsay, M.A. and Norstrom, R.J. 1995. Catalytic and immunologic characterization of hepatic and lung cytochrome P450 in the polar bear. *Biochemical Pharmacology* 49: 1135–1146.

Barrie, L.A. 1986. Arctic air pollution: an overview of current knowledge. *Atmospheric Environment* 20: 643–663.

Barrie, L.A., Macdonald, R., Bidleman, T., Diamond, M., Gregor, D., Semkin, R., Strachan, W., Alaee, M., Backus, S., Bewers, M., Gobeil, C., Halsall, C., Hoff, J., Li, A., Lockhart, L., Mackay, D., Muir, D., Pudykiewicz, J., Reimer, K., Smith, J., Stern, G., Schroeder, W., Wagemann, R., Wania, F. and Yunker, M. 1997. Sources, occurrence, and pathways. In: J. Jensen, K. Adare, and R. Shearer (eds). *Canadian Arctic Contaminants Assessment Report*. Indian and Northern Affairs Canada, Ottawa, pp. 35–166.

Beck, G.G., Smith, T.G. and Hammill, M.O. 1993. Evaluation of body condition in the northwest Atlantic harp seal. *Canadian Journal of Fisheries and Aquatic Science* 50: 13772–1381.

Beck, G., Smith, T.G. and Addison, R.F. 1994. Organochlorine residues in harp seals, *Phoca groenlandica*, from the Gulf of St. Lawrence and Hudson Strait: an evaluation of contaminant concentrations and burdens. *Canadian Journal of Zoology* 72: 174–182.

Becker, P.R. 2000. Concentrations of chlorinated hydrocarbons and heavy metals in Alaska Arctic marine mammals. *Marine Pollution Bulletin* 40/10: 819–829.

Becker, P.R., Mackey, E.A., Schantz, M.M., Greenberg, R.R., Koster, B.J., Wise, S.A. and Muir, D.C.G. 1995. Concentrations of chlorinated hydrocarbons, heavy metals and other elements in tissues banked by the Alaska marine mammal tissue archival project, NISTIR 5620. National Institute of Standards and Technology, Gaithersburg, Maryland, p. 115.

Becker, P.R., Mackey, E.A., Schantz, M., Koster, B.J. and Wise, S.A. 1997. Concentrations of chlorinated hydrocarbons and trace elements in marine mammal tissues archived in the US National Biomonitoring Specimen Bank. *Chemosphere* 34 (9/10): 2067–2098.

Bergman, A., Norstrom, R.J., Haraguchi, K., Kuroki, H. and Béland, P. 1994. PCB and DDE methyl sulfones from Canada and Sweden. *Environmental Toxicology and Chemistry* 13: 121–128.

Bernhoft, A., Wiig, O. and Skaare, J.U. 1997. Organochlorines in polar bears at Svalbard. *Environmental Pollution* 96: 230–243.

Best, R.C. 1985. Digestibility of ringed seals by the polar bear. *Canadian Journal of Zoology* 63: 1033–1036.

Bidleman, T.F., Patton, G.W., Walla, M.D., Hargrave, B.T., Vass, W.P., Erickson, P., Fowler, B., Scott, V. and Gregor, D.J. 1989. Toxaphene and other organochlorines in Arctic Ocean fauna: evidence of atmospheric delivery. *Arctic* 42: 307–313.

Bidleman, T.F., Walla, M.D., Muir, D.C.G. and Stern, G.A. 1993. Selective accumulation of polychlorocamphenes in aquatic biota from the Canadian arctic. *Environmental Toxicology and Chemistry* 12(4): 701–709.

Bignert, A., Olsson, M., Bergqvist, P.-A., Bergek, S., Rappe, C., De Wit, C. and Jansson, B. 1989. Polychlorinated dibenzo-*p*-dioxins (PCDD) and dibenzofurans (PCDF) in seal blubber. *Chemosphere* 19: 551–556.

Blus, L.J., Weimeyer, S.N. and Henny, C.J. 1996. Organochlorine pesticides. In: A. Fairbrother, L. Locke, and G. Hoff (eds). *Noninfectious Diseases of Wildlife*, 2nd edition. Iowa State University Press, Ames Iowa, USA, pp. 61–70.

Boon J.P., van Arnhem, E., Jansen, S., Kannan, N., Petrick, G., Duinker, J.C., Reijnders, P.J.H. and Goksoyr, A. 1992. The toxicokinetics of PCBs in marine mammals with special reference to possible interactions of individual congeners with the cytochrome P450-dependent monoxygenase system: An overview. In: C.H. Walker and D.R. Livingstone (eds). SETAC Special Publications Series. *Persistent Pollutants in Marine Ecosystems*, 1st edition. Pergamon Press, Oxford, United Kingdom, pp. 119–160.

Boon, J.P., Oostingh, I., Van der Meer, J. and Hillebrand, M.T.J. 1994. A model for the bioaccumulation of chlorobiphenyl congeners in marine mammals. *European Journal of Pharmacology* 270: 237–251.

Born, E.W., Kraul, I. and Kristensen, T. 1981. Mercury, DDT and PCB in the Atlantic Walrus (*Odobenus rosmarus rosmarus*) from the Thule District, North Greenland. *Arctic* 34: 255–260.

Borrell, A. 1993. PCB and DDTs in blubber of cetaceans from the northeastern Atlantic. *Marine Pollution Bulletin* 26(3): 146–151.

Bratton, G.R., Spainhour, C.B., Flory, W., Reed, M. and Jayko, K. 1993. Presence and potential effects of contaminants. In: Burns, Monague, and Crower (eds). *The Bowhead Whale*. Special Publication No. 2. The Society of Marine Mammalogy, pp. 701–744.

Clausen, J., Braestrup, L. and Berg, O. 1974. The content of polychlorinated hydrocarbons in Arctic mammals. *Bulletin of Environmental Contamination and Toxicology* 12: 529–534.

de Boer, J., Wester, P.G., Klamer, H.J.C., Lewis, W.E. and Boon, J.P. 1998. Do flame retardants threaten ocean life? *Nature* 394: 28–29.

De Guise, S., Lagacé, A. and Béland, P. 1994a. Tumors in St. Lawrence beluga whales. *Veterinary Pathology* 31: 444–449.

De Guise, S., Lagacé, A. and Béland, P. 1994b. True hermaphroditism in a St. Lawrence beluga whale. *Journal of Wildlife Diseases* 30: 287–290.

De Guise, S., Lagacé, A., Béland, P., Girard, C. and Higgins, R. 1995. Non-neoplastic lesions in beluga whales and other marine mammals from the St. Lawrence Estuary. *Journal of Comparative Pathology* 112: 257–271.

De Guise, S., Ross, P.S., Osterhaus, A.D.M.E., Martineau, D., Béland, P. and Fournier, M. 1997. Immune functions in beluga whales: evaluation of natural killer cell activity. *Veterinary Immunology and Immunopathology* 58: 345–354.

De March, B.G.E., de Wit, C.A., Muir, D.C.G., Braune, B.M., Gregor, D.J., Norstrom, R.J., Olsson, M., Skaare, J.U. and Stange, K. 1998. Persistent organic pollutants. In: *AMAP Assessment Report: Arctic Pollution Issues.* Arctic Monitoring and Assessment Programme (AMAP), Oslo, Norway, pp. 183–371.

Dewailly, E., Ayotte, P., Bruneau, S., Laliberte, C., Muir, D.C.G. and Norstrom, R.J. 1993. Inuit exposure to organochlorines through the aquatic food chain in arctic Quebec. *Environmental Health Perspectives* 101: 618–620.

Ewald, G., Larsson, P., Linge, H., Okla, L. and Szarzi, N. 1998. Biotransport of organic pollutants to an inland Alaska lake by migrating sockeye salmon (*Oncorhyncus nerka*). *Arctic* 51: 40–47.

Fadely, B.S. 1997. Investigations of harbor seal health status and body condition in the Gulf of Alaska. PhD Thesis, University of Alaska, Fairbanks, Alaska.

Fay, F. 1960. Carnivorous walrus and some Arctic zoonoses. *Arctic* 13: 11–122.

Finley, K.J. and Gibb, E.J. 1982. Summer diet of the narwhal in Pond Inlet, northern Baffin Island (Canada). *Canadian Journal of Zoology* 60(12): 3353–3363.

Ford, C.A., Muir, D.C.G., Norstrom, R.J., Simon, M. and Mulvihill, M.J. 1993. Development of a semi-automated method for non-ortho PCBs: application to Canadian arctic marine mammal tissues. *Chemosphere* 26: 1981–1991.

Gales, R. and Renouf, D. 1994. Assessment of body condition of harp seals. *Polar Biology* 14: 381–387.

Gauthier, J.M., Dubeau, H. and Rassart, E. 1999. Induction of micronuclei in vitro by organochlorine compounds in beluga whale skin fibroblasts. *Mutation Research – Genetic Toxicology and Environmental Mutagenesis* 439: 87–95.

George, J.C., Bada, J., Zeh, J., Scott, L., Brown, S. E., O'Hara, T. and Suydam, R. 1999. Age and growth estimates of bowhead whales (*Balaena mysticetus*) using aspartic acid racemization. *Canadian Journal of Zoology* 571–580.

Gregor, D.J. and Gummer, W.D. 1989. Evidence of atmospheric transport and deposition of organochlorine pesticides and polychlorinated biphenyls in Canadian Arctic Snow. *Environmental Science Technology* 23: 561–565.

Heintzenberg, J. 1989. Arctic haze: air pollution in polar regions. *Ambio* 18: 50–55.

Helle, E., Olsson, M. and Jensen, S. 1976a. DDT and PCB levels and reproduction in ringed seal from the Bothnian Bay. *Ambio* 5: 188–189.

Helle, E., Olsson, M. and Jensen, S. 1976b. PCB levels correlated with pathological changes in seal uteri. *Ambio* 5: 261–263.

Helle, E., Hyvarinen, P.H. and Wickstrom, K. 1983. Levels of organochlorine compounds in inland seal population in Eastern Finland. *Marine Pollution Bulletin* 14: 256–260.

Hoekstra, P.F., Karlsson, H.M., Muir, D.C.G., O'Hara, T.M. and Solomon, K.R. 2000a. Biomagnification of chiral and achiral organochlorine pesticides in the bowhead whale (*Balaena mysticetus*) food chain. *21st Annual Meeting of the Society of Environmental Toxicology and Chemistry, Nashville, Tennessee, November 12–17.*

Hoekstra, P.F., Wong, C.S., Muir, D.C.G., O'Hara, T.M., Mabury, S.A. and Solomon, K.R. 2000b. Enantiomeric composition of PCB atropisomers in bowhead whales (*Balaena mysticetus*). *21st Annual Meeting of the Society of Environmental Toxicology and Chemistry, Nashville, Tennessee, November 12–17.*

Hoff, R.M. and Chan, K.W. 1986. Atmospheric concentrations of chlordane at Mould Bay, N.W.T., Canada. *Chemosphere* 15: 449–452.

Hoff, R.M., Muir, D.C.G. and Grift, N.P. 1992. Annual cycle of PCBs and organohalogen pesticides in air in southern Ontario. 1. Air concentration data. *Environmental Science and Technology* 26: 266–275.

IWC (International Whaling Commission) 1995. Report of the Workshop on Chemical Pollution and Cetaceans, SC/47/Rep2, Chairman Peter Reijnders, Bergen, Norway: 34.

Johansen, P., Kapel, F.P. and Kraul, I. 1980. Heavy metals and organochlorines in marine mammals from Greenland. *Int. Coun. Explor. Sea. Mar. Environ. Qual. Comm*: 15.

King, J.E. 1983. *Seals of the World*, 2nd edition. Natural History Museum Publications, Cornell University Press.

Kingsley, M.C.S. 1996. Population index estimate for the belugas of the St. Lawrence in 1995. *Canadian Technical Report of Fisheries and Aquatic Sciences* No. 2117.

Kingsley, M.C.S. 1998. Population index estimates for the St. Lawrence belugas, 1973–1995. *Marine Mammal Science* 14: 508–530.

Kinloch, D., Kuhnlein, H. and Muir, D.C.G. 1992. Inuit foods and diet: a preliminary assessment of benefits and risks. *Science of the Total Environment* 122: 247–278.

Krahn, M.M., Becker, P.R., Tilbury, K.L. and Stein, J.E. 1997. Organochlorine contaminants in blubber of four seal species: integrating biomonitoring and specimen banking. *Chemosphere* 34(9/10): 2109–2121.

Krahn, M.M., Burrows, D.G., Stein, J.E., Becker, P.R., Schantz, M.M., Muir, D.C.G., O'Hara, T.M. and Rowles, T. 1999. White whales from three Alaskan stocks – concentrations and patterns of persistent organochlorine contaminants in blubber. *Journal of Cetacean Research and Management* 1: 239–250.

Kucklick, J.R., Struntz, W.D.J., Becker, P.R., York, G.W., O'Hara, T.M. and Bohonowych, J.E. 2002. Persistent organochlorine pollutants in ringed seals and polar bears collected from Northern Alaska. *The Science of the Total Environment* 287: 45–59.

Kuiken, T., Bennet, P.M., Allchin, C.R., Kirkwood, J.K., Baker, J.R., Lockyer, C., Walton, M.J. and Sheldrick, M.C. 1994. PCBs, cause of death and body condition in harbor porpoises from British waters. *Aquatic Toxicology* 28: 13–28.

Lair, S., Béland, P., De Guise, S. and Martineau, D. 1997. Adrenal hyperplastic and degenerative changes in beluga whales. *Journal of Wildlife Diseases* 33(3): 430–437.

Law, R.J. 1994. Collaborative UK marine mammal project. Summary of data produced 1988–1992. Fisheries Research Technical Report No. 97. Directorate of Fisheries Research, Lowestoft.

Letcher, R.J. 1996. The ecological and analytical chemistry of chlorinated hydrocarbon contaminants and methyl sulfonyl-containing metabolites in the polar bear (*Ursus maritimus*) food chain. PhD Thesis, Carleton University, Ottawa, Ontario.

Letcher, R.J., Norstrom, R.J., Lin, S., Ramsay, M.A. and Bandiera, S.M. 1996. Immunoquantitation and microsomal monooxygenase activities of hepatic cytochromes P450 1A and P450 2B and chlorinated hydrocarbon contaminant levels in polar bear. *Toxicology and Applied Pharmacology* 137: 127–140.

Letcher, R.J., Norstrom, R.J. and Muir, D.C.G. 1998. Biotransformation versus bioaccumulation: Sources of methyl sulfone PCB and 4,4'-DDE metabolites in the polar bear food chain. *Environmental Science and Technology* 32: 1656–1661.

Livingstone, D.R., Donkin, P. and Walker, C.H. 1992. Pollutants in marine ecosystems: A review. In: C.H. Walker and D.R. Livingstone (eds). *Persistent Pollutants in Marine Ecosystems*. Pergamon Press, New York, pp. 235–264.

Lowry, L.F. 1993. Foods and feeding ecology. In: J.J. Burns, J.J. Montague, and C.J. Cowles (eds). *The Bowhead Whale*, Special Publication Number 2, The Society for Marine Mammalogy. Lawrence, KS Allen Press, pp. 201–234.

Lowry, L.F. and Fay, F.H. 1984. Seal eating by walruses in the Bering and Chukchi seas. *Polar Biology* 3: 11–18.

Manchester-Neesvig, J.B. and Andren, A.W. 1989. Seasonal variation in the atmospheric concentration of polychlorinated biphenyl congeners. *Environmental Science and Technology* 23: 1138–1148.

Martineau, D., Lagace, A., Béland, P., Higgins, R., Armstrong, D. and Shugart, L.R. 1988. Pathology of stranded beluga whales from the St. Lawrence Estuary, Quebec. *Canadian Journal of Comparative Pathology* 98: 287–311.

Martineau, D., De Guise, S., Fournier, M., Shugart, L., Girard, C., Lagace, A. and Béland, P. 1994. Pathology and toxicology of beluga whales from the St. Lawrence Estuary, Quebec, Canada – past, present and future. *Science of the Total Environment* 154: 201–215.

Martineau, D., Lair, S., De Guise, S., Lipscomb, T.P. and Béland, P. 1999. Cancer in beluga whales from the St Lawrence Estuary, Quebec, Canada: A potential biomarker of environmental contamination. *Journal of Cetacean Research and Management* (Special Issue) 1: 247–265.

McFall, J.A., Antoine, S.R. and Overton, E.B. 1986. Organochlorine compounds and polynuclear aromatic hydrocarbons in tissues of subsistence harvested bowhead whales. Final Report to Department of Wildlife of Management, North Slope Borough, PO Box 69, Barrow, AK.

Mitchell, E.D. 1975. Parallelism and convergence in the evolution of Otariidae and Phocidae. *Rapports et Procés-Verbauz des Reunions Conseil International pourl Exploration de la Mer* 169: 12–26.

Muir, D.C.G. and de Boer, J. 1995. Recent developments in the analysis and environmental chemistry of toxaphene with emphasis on the marine environment. *Trends in Analytical Chemistry* 14: 56–66.

Muir, D.C.G., Norstrom, R.J. and Simon, M. 1988. Organochlorine contaminants in arctic marine foods chains: accumulation of specific PCB congeners and chlordane-related compounds. *Environmental Science and Technology* 22: 1071–1079.

Muir, D.C.G., Ford, C.A., Stewart, R.E.A., Smith, T.G., Addison, R.F., Zinck, M.E. and Béland, P. 1990. Organochlorine contaminants in belugas from Canadian waters. In: T.G. Smith, D.J. St. Aubin, and J.R. Geraci (eds). Advances in Research on the Beluga Whale. *Canadian Bulletin of Fisheries and Aquatic Science* 224: 165–190.

Muir, D.C.G., Wagemann, R., Hargrave, B.T., Thomas, D.J., Peakall, D.B. and Norstrom, R.J. 1992a. Arctic marine ecosystem contamination. *Science of the Total Environment* 122: 75–134.

Muir, D.C.G., Ford, C.A., Grift, N.P. and Stewart, R.E.A. 1992b. Organochlorine contaminants in narwhal from the Canadian Arctic. *Environmental Pollution* 75: 397–416.

Muir, D.C.G., Segstro, M.D., Hobson, K.A., Ford, C.A., Stewart, R.E.A. and Olpinski, S. 1995. Can seal eating explain elevated levels of PCBs and organochlorine pesticides in walrus blubber from eastern Hudson Bay (Canada)? *Environmental Pollution* 90: 335–348.

Muir, D.C.G., Ford, C.A., Rosenberg, B., Norstrom, R.J., Simon, M. and Béland, P. 1996a. Persistent organochlorines in beluga whales from the St. Lawrence River Estuary – I. Concentrations and patterns of specific PCBs, chlorinated pesticides, and polychlorinated dibenzo-*p*-dioxins and dibenzofurans. *Environmental Pollution* 93: 219–234.

Muir, D.C.G., Koczanski, K., Rosenberg, B. and Béland, P. 1996b. Persistent organochlorines in beluga whales from the St. Lawrence River Estuary – II. Temporal trends, 1982–1994. *Environmental Pollution* 93: 235–245.

Muir, D., Becker, P., Koczanski, K., Stewart, R. and Innes, S. 1997. Spatial and temporal trends of persistent organochlorines in marine mammals from the North American Arctic. *AMAP International Symposium on Environmental Pollution of the Arctic, Tromso, Norway, June 1–5.* Extended Abstracts 1: 121–123.

Muir, D.C., Riget, F., Cleemann, M., Skaare, J., Kleivane, L., Nakata, H., Dietz, R., Severinsen, T. and Tanabe, S. 2000. Circumpolar trends of PCBs and organochlorine pesticides in the Arctic marine environment inferred from levels in ringed seals. *Environmental Science and Technology* 34: 2431–2438.

Norstrom, R.J. and Muir, D.C.G. 1994. Chlorinated hydrocarbon contaminants in arctic marine mammals. *Science of the Total Environment* 154(2–3): 107–128.

Norstrom, R.J., Simon, M., Muir, D.C.G. and Schweinsberg, R.E. 1988. Organochlorine contaminants in arctic marine foods chains: identification, geographical distribution, and temporal trends in polar bears. *Environmental Science and Technology* 22: 1063–1071.

Norstrom, R.J., Simon, M. and Muir, D.C.G. 1990. Polychlorinated dibenzo-*p*-dioxins and dibenzofurans in marine mammals in the Canadian North. *Environmental Pollution* 66: 1–19.

Norstrom, R.J., Muir, D.C.G., Ford, C.A., Simon, M., MacDonald, C.R. and Béland, P. 1992. Indications of P450 monoxygenase activities in beluga and narwhal from patterns of PCB, PCDD, and PCDF accumulation. *Marine Environmental Research* 34: 267–272.

Norstrom, R.J., Belikov, S.E., Born, E.W., Garner, G.W., Malone, B., Olpinski, S., Ramsay, M.A., Schliebe, S., Stirling, I., Stishov, M.S., Taylor, M.K. and Wiig, D. 1998. Chlorinated hydrocarbon contaminants in polar bears from eastern Russia, North America, Greenland and Svalbard: biomonitoring of arctic pollution. *Archives of Environmental Contamination and Toxicology* 35: 354–367.

Oehme, M., Furst, P., Kruger, C., Meemken, A. and Groebel, W. 1988. Presence of polychlorinated dibenzo-*p*-dioxins, dibenzofurans and pesticides in Arctic seal from Spitzbergen. *Chemosphere* 18: 1291–1300.

Oehme, M., Schlabach, M., Hummert, K., Luckas, B. and Nordoy, E.S. 1995a. Determination of levels of polychlorinated dibenzo-*p*-dioxins, dibenzofurans, biphenyls, and pesticides in harp seals from the Greenland Sea. *Science of the Total Environment* 162: 75–91.

Oehme, M., Biseth, A., Schlabach, M. and Wiig, O. 1995b. Concentrations of polychlorinated dibenzo-p-dioxins, dibenzofurans, and non-ortho substitutes biphenyls in polar bear milk from Svalbard (Norway). *Environmental Pollution* 90: 401–407.

O'Hara, T.M. and Rice, C. 1996. Polychlorinated biphenyls. In: A. Fairbrother, L. Locke, and G. Hoff (eds). *Noninfectious Diseases of Wildlife*, 2nd edition. Iowa State University Press, Ames Iowa, USA, pp. 71–86.

O'Hara, T.M., Krahn, M. M., Boyd, D., Becker, P. R. and M. Philo, L. 1999. Organochlorine contaminant levels in Eskimo harvested bowhead whales of arctic Alaska. *Journal of Wildlife Diseases* 35(4): 741–752.

Olsson, M. 1972. PCBs in the Baltic environment. In: *PCBs and the Environment*, Volume III. Department of Agriculture, Commerce and Interior, Washington, DC, Chapter 7, pp. 181–208.

Olsson, M. 1978. PCB and reproduction among Baltic seals. *Finnish Game Research* 37: 40–45.

Olsson, M., Andersson, O., Bergman, A., Blomkvist, G., Frank, A. and Rappe, C. 1992. Contaminants and diseases in seals from Swedish waters. *Ambio* 21: 561–562.

Olsson, M., Karlsson, B. and Ahnland, E. 1994. Diseases and environmental contaminants in seals from the Baltic and Swedish west coast. *Science of the Total Environment* 154: 217–227.

O'Shea, T.J. and Brownell, R.L., Jr 1994. Organochlorine and metal contaminants in baleen whales: a review and evaluation of conservation implications. *Science of the Total Environment* 154: 179–200.

O'Shea, T.J., Reeves, R.R. and Long, A.K. (eds) 1999. *Marine mammals and persistent ocean contaminants*. Proceedings of a workshop held in Keystone, Colorado, October 12–15 1998. US Marine Mammal Commission, Washington, DC, p. 150.

Polischuk, S.C., Letcher, R.J., Nortstrom, R.J. and Ramsay, M.A. 1995. Preliminary results on the kinetics of organochlorines in western Hudson polar bear. *Science of the Total Environment* 160/161: 465–472.

Reijnders, P.J.H., Aguilar, A. and Donovan, G.P. 1999. Chemical pollutants and cetaceans. *The Journal of Cetacean Research and Management* Special Issue 1: 273.

Repenning, C.A. 1976. Adaptive evolution of sea lions and walruses. *Systematic Zoology* 25: 375–390.

Ronald, K., Frank, R.J., Dougan, J.L., Frank, R. and Baun, H.E. 1984a. Pollutants in harp seals (*Phoca groenlandica*). I. Organochlorines. *Science of the Total Environment* 38: 133–152.

Ronald, K., Frank, R.J., Dougan, J., Frank, R. and Braun, H.E. 1984b. Pollutants in harp seals. II. Heavy metals and selenium. *Science of the Total Environment* 38: 153–166.

Safe, S.H. 1990. Polychlorinated biphenyls (PCBs), dibenzo-*p*-dioxins (PCDDs), dibenzofurans (PCDFs), and related compounds: environmental and mechanistic considerations which support the development of toxic equivalency factors (TEFs). *Critical Reviews in Toxicology* 21: 51–88.

Saleh, M.A. 1991. Toxaphene: chemistry, biochemistry, toxicity and environmental fate. *Reviews in Environmental Contamination and Toxicology* 118: 1–85.

Sandell, M. 1990. The evolution of seasonal delayed implantation. *Quarterly Review of Biology* 65: 23–42.

Seagars, D.J. and Garlich-Miller, J. 2001. Organochlorine compounds and aliphatic hydrocarbons in Pacific walrus blubber. *Marine Pollution Bulletin* 43: 122–131.

Shaw, G.E. 1995. The Arctic haze phenomenon. *Bulletin of the American Meteorological Society* 76: 2403–2413.

Skaare, J.U., Espeland, O., Ugland, K.I., Berhoft, A., Wiig, O. and Kleivane, L. 1994. Organochlorine contaminants in marine mammals from the Norwegian Arctic. Report to the International Council for Exploration of the Sea. ICES C.M. 1994/E+N:3.

Stern, G.A., Muir, D.C.G., Grift, N.P., Dewailly, E., Bidleman, T.F. and Wallace, M.D. 1992. Isolation and identification of two major recalcitrant toxaphene congeners. *Environmental Science and Technology* 26(9): 1838–1840.

Stern, G.A., Halstall, C.J., Barrie, L.A., Muir, D.C.G., Fellin, P., Rosenberg, B., Rovinsky, F.Y., Kononov, E. Y. and Pastuhov, B. 1997. Polychlorinated biphenyls in Arctic air. 1. Temporal and spatial trends 1992–1994. *Environmental Science and Technology* 31: 3619–3628.

Tanabe, S. and Tasukawa, R. 1992. Chemical modernization and vulnerability of cetaceans: increasing toxic threat of organochlorine contaminants. In: C.H. Walker and D.R. Livingstone (eds). *Persistent Pollutants in Marine Ecosystems*, 7. Pergamon Press Ltd., Oxford, pp. 161–177.

Tanabe, S., Watanabe, S., Kan, H. and Tatsukawa, R. 1988. Capacity and mode of PCB metabolism in small cetaceans. *Marine Mammal Science* 4: 103–124.

Tanabe, S., Sung, J., Choi, D., Baba, N., Kiyota, M., Yoshida, K. and Tatsukawa, R. 1994a. Persistent organochlorine residues in northern fur seal from the Pacific coast of Japan since 1971. *Environmental Pollution* 85: 305–314.

Tanabe S., Iwata, H. and Tatsukawa, R. 1994b. Global contamination by persistent organochlorines and their ecotoxicological impact on marine mammals. *Science of the Total Environment* 154: 163–177.

Tanabe, S., Kumaran, P., Iwata, H., Tatsukawa, R. and Miyazaki, N. 1996. Enantiomeric ratios of alpha-hexachlorocyclohexane in blubber of small cetaceans. *Marine Pollution Bulletin* 32: 27–31.

Tarpley, R., Jarrell, G., George, J., Cubbage, J. and Stott, G. 1995. Male pseudohermaphroditism in the bowhead whale. *Journal of Mammalogy* 76(4): 1267–1275.

Taylor, D.L., Schliebe, S. and Metsker, H. 1989. Contaminants in blubber, liver, and kidney tissue of Pacific walruses. *Marine Pollution Bulletin* 20: 465–468.

Thomas, D.J. and Hamilton, M.C. 1988. Organochlorine residues in biota of the Baffin Island region. Report prepared for Indian and Northern Affairs Canada, Ottawa. SeaKem Oceanography Ltd., Sidney, B.C.

Voldner, E.C. and Schroeder, W.H. 1989. Modeling of atmospheric transport and deposition of toxaphene in the Great Lakes ecosystem. *Atmospheric Environment* 23: 1949–1961.

Wade, T.L., Chambers, L., Gardinall, P.R., Sericano, J.L., Jackson, T.J., MacDonald, T.J., Tarpley, R.J. and Suydam, R.S. 1995. High resolution mass spectrometry analyses of toxaphene in the blubber of beluga whales. *Organohalogen Compounds* 26: 339–344.

Wagemann, R. and Muir, D.G.C. 1984. Concentrations of heavy metals and organochlorines in marine mammals of northern waters: overview and evaluation. *Canadian Technical Report on Fisheries and Aquatic Science 1279*, Ottawa, Canada, 97p.

Wania, F. and Mackay, D. 1993. Global fractionation and cold condensation of low volatility organochlorine compounds in polar regions. *Ambio* 22: 10–18.

Wania, F. and Mackay, D. 1996. Tracking the distribution of persistent organic pollutants. *Environmental Science and Technology* 30: 390A–396A.

Weis, I.M. and Muir, D.C.G. 1997. Geographical variation of persistent organochlorine concentrations in blubber of ringed seal (*Phoca hispida*) from the Canadian Arctic: Univariate and multivariate approaches. *Environmental Pollution* 96(3): 321–333.

White, R.D., Hahn, M.E., Lockhart, W.L. and Stegeman, J.J. 1994. Catalytic and immunochemical characterization of hepatic microsomal P450 in beluga whale. *Toxicology and Applied Pharmacology* 126: 45–57.

Wiberg, K., Letcher, R., Sandau, C., Duffe, J., Norstrom, R., Haglund, P. and Bidleman, T. 1998. Enantioselective gas chromatography/mass spectrometry of methylsulfonyl PCBs with application to arctic marine mammals. *Analytical Chemistry* 70(18): 3845–3852.

Wiberg, K., Letcher, R., Sandau, C.D., Norstrom, R.J., Tysklind, M. and Bidleman, T.F. 2000. The enantioselective bioaccumulation of chiral chlordane and α-HCH contaminants in the polar bear food chain. *Environmental Science and Technology* 34: 2668–2674.

Wiig, O., Derocher, A.E., Cronin, M.M. and Skaare, J.U. 1998. Female pseudohermaphrodite polar bears at Svalbard. *Journal of Wildlife Diseases* 34: 792–796.

Zhu, J. and Norstrom, R.J. 1993. Identification of polychlorocamphenes (PCCs) in the polar bear food chain. *Chemosphere* 27(10): 1923–1936.

Zhu, J., Norstrom, R.J., Muir, D.C.G., Ferron, L.A., Weber, J.P. and DeWailley, E. 1995. Persistent chlorinated cyclodiene compounds in ringed seal blubber, polar bear fat, and human plasma from northern Quebec, Canada: Identification and concentrations of Photoheptachlor. *Environmental Science and Technology* 29: 267–271.

Zitko, V., Stenson, G. and Hellou, J. 1998. Levels of organochlorine and polycyclic aromatic compounds in harp seal beaters (*Phoca groenlandica*). *Science of the Total Environment* 221: 11–29.

9 Inorganic pollutants in Arctic marine mammals

Todd M. O'Hara, Victoria Woshner and Gerald Bratton

Introduction

Species and elements

This chapter addresses two specific areas: Arctic marine mammals and selected inorganic pollutants. The species covered include cetaceans [bowhead (*Balaena mysticetus*), beluga (*Delphinapterus leucas*), and narwhal whales (*Monodon monoceros*)], pinnipeds [ringed (*Pusa hispida*), bearded (*Erignathus barbatus*), spotted or largha (*Phoca largha*), ribbon (*Histriophoca fasciata*), harp (*Pagophilus groenlandicus*), and hooded (*Cystophora cristata*) seals; Pacific (*Odobenus rosmarus divergens*) and Atlantic walrus (*O. rosmarus rosmarus*)] and polar bears (*Ursus maritimus*). Arctic marine mammals are long-lived, develop large fat depots and frequently occupy high trophic levels. Bowhead whales are known to live more than 100 years (George *et al.*, 1999). These biological factors are key to the entry and magnification of the persistent and lipophilic forms of metals in this lipid-dependent food web, and for elements with non-lipid-dependent accumulation mechanisms. High trophic-level feeding (piscivory) sets the stage for biomagnification and bioaccumulation (increased levels in the predator versus the prey), where some chemicals can biomagnify (BMF = [X]predator/[X]prey > 1.0; where BMF = biomagnification factor and [X] = contaminant concentration). Polar bears feed on seals and are apex predators.

Minerals discussed in this chapter include inorganic and organic forms of mercury (Hg), cadmium (Cd), selenium (Se), silver (Ag), arsenic (As), lead (Pb), copper (Cu), zinc (Zn) and others. Radionuclides addressed include cesium-137 (^{137}Cs), strontium-90 (^{90}Sr), plutonium (^{238}Pu, ^{239}Pu, ^{240}Pu), polonium (^{210}Po) and potassium-40 (^{40}K). Concentrations of contaminants are given on a wet weight basis, except where noted otherwise. There are many influences on exposure to, and concentrations of, inorganic contaminants in tissues, including biological factors (age, sex, body condition, reproductive status, season, trophic level), spatial distribution (habitat or home range, global region) and temporal relationship (trends with time). In this report we consider available data with respect to potential adverse health effects for each species. However, the lack of basic biologic and health assessment information pertaining to these Arctic species hampers interpretation of contaminant data. As a subsistence resource, marine mammals have significant cultural and nutritional value and are sources of contaminant exposure to humans who consume them.

It is necessary to determine the chemical form of potentially toxic trace elements and the various mechanisms of detoxification [International Whaling Commission (IWC), 1995] in each species. Investigation into elements other than Cd and Hg is warranted, as is consideration of within-population variation in the ability to detoxify metals (Caurant *et al.*, 1994). Tissues other than kidney and liver should be examined, because metals could alter the effects of many other substances (other metals, trace minerals) (IWC, 1995), and other tissues (i.e. muscle and blubber) are important sources of food for many northern peoples. The Inuit in the coastal communities of arctic Canada and Greenland have always hunted marine mammals for food, and metals in the diet are of concern to these people (e.g. Dietz *et al.*, 1996; Wagemann *et al.*, 1996). They commonly eat the flesh (meat) of marine mammals, and preferentially the skin (*muktuk*) of whales, which can be raw, cooked or fermented. *Muktuk*, for example, is rich in essential elements such as Se and Zn (Wagemann *et al.*, 1996), and the associated fat is rich in omega-3 fatty acids. Concerns about metals in marine mammal tissue arose when studies showed that Hg concentrations were well above the Canadian export guideline for Hg in fish muscle (0.5 µg/g; Health and Welfare Canada, 1979; Wagemann *et al.*, 1996), and exceeded various limits developed by other organizations and agencies. Clark (1989) showed that regulatory levels for Hg range from 0.7 to 1.5 µg/g in edible tissue, depending on the country, and more recently, the United States Environmental Protection Agency (USEPA) has established a reference dose for methylmercury of 0.1 µg/g (USEPA, 2001).

Compared to the variety and number of organic compounds, the metals are fewer and generally of a simpler chemical structure. Although not considered biodegradable, metals can occur in several valence states and can be major components of organic molecules (i.e. organometallic compounds), and once absorbed may undergo very complicated metabolism and distribution. Valence state and organic versus inorganic form also have a direct effect on bioavailability and toxicity. Metals frequently induce complicated toxicoses because they tend to affect a number of cellular processes in more than one organ system. Many elements undergo significant interactions. A deficiency of one element may predispose an organism to toxic effects of another element at lower levels of exposure than normally expected. For example, laboratory studies have established that Hg and Se are mutually protective, in that each offsets the toxicity of the other. A significant correlation between these two elements is common in the liver of some marine mammals, where they increase in parallel. Absolute or functional mineral deficiencies may increase susceptibility to other agents, as well as produce outright signs of deficiency, but we found no documentation of this in Arctic marine mammals. The bioavailability (absorption) of metals may range from essentially no uptake (no concern) to nearly complete absorption and/or prolonged retention in some organs (Cd in kidney, Hg in liver or muscle). We refer readers to Table 7.5 of the Arctic Monitoring and Assessment Program (AMAP, 1998) with respect to these bioavailability comparisons.

Acute or chronic heavy metal toxicosis, diagnosed on the basis of mortality or morbidity through clinical signs or lesions, has not occurred in any wild cetacean and most marine mammal species. We have no examples of target organs, lesions, effects or levels of concern for metals in these Arctic species. Few studies (for example, Ronald *et al.*, 1977) of adverse effects due to heavy metals in marine mammals have been conducted, making the health impact of exposure to metal difficult to assess, especially during strandings or mortality events. Consequently, we have completely depended upon extrapolations from very unrelated laboratory animal models or domestic species. Only recently have attempts been undertaken to correlate pathology with levels of inorganic contaminants in cetaceans (Wagemann *et al.*, 1990; Béland *et al.*, 1993; Rawson *et al.*, 1993; Siebert *et al.*, 1999; Woshner, 2000; Woshner *et al.*, 2002) and seals (Dietz *et al.*, 1998).

Potential sources and transport mechanisms

Metals

In general, metal concentrations are low in sea water. However, some Arctic Ocean surface waters contain higher concentrations of some metals (Cd, Zn and Cu), probably because of the dearth of organic material that can bind with them and cause them to settle at greater depths. Therefore, metals can be more bioavailable to the invertebrates at the base of the Arctic Ocean food web in comparison with warmer areas. It is very likely these metals in Arctic surface waters are from natural sources with minimal anthropogenic input. Phytoplankton and zooplankton are well known to bioaccumulate metals and are the principal prey for baleen whales and fish, which in turn are prey for toothed whales and most pinnipeds. Cadmium accumulates more effectively in the bowhead whale (Bratton *et al.*, 1997; Woshner *et al.*, 2001a) and other baleen whales (via invertebrates; Honda *et al.*, 1990), whereas Hg and Se tend to increase in beluga (Woshner *et al.*, 2001a) and narwhal via fish and some invertebrates (Friberg *et al.*, 1986). Concentrations in water have very little direct effect on exposure and subsequent body burdens in the Arctic marine mammals, because exposure occurs mostly by ingestion.

Anthropogenic contamination can be through long-range transport (Ayotte *et al.*, 1995; Shaw, 1995) or from local sources (Jaffe *et al.*, 1995), with some studies indicating that the latter mode contributes more significantly to body burdens of trace elements in marine animals than the former (Muir *et al.*,

1992; Ford *et al.*, 1995; Wagemann *et al.*, 1996). Natural sources of trace elements include erosion, fires, sediment disruption and volcanic eruption. Cadmium is commonly associated with Cu and Zn deposits and, while releases in the Arctic probably occur naturally, there is increased potential for anthropogenic releases with growth of human populations and associated industrial activities (smelting, drilling, mining, oil production, etc.) (Barrie, 1986). Effects of temperature and prevailing air and ocean currents may favor deposition of some contaminants in the Arctic (Bidleman *et al.*, 1989; Gregor and Gummer, 1989; Manchester-Neesvig and Andren, 1989).

Radionuclides

Specific releases of radionuclides in the Arctic, such as atmospheric emissions associated with the Chernobyl accident and marine discharges of nuclear waste by the former Soviet Union, have increased concerns. Nevertheless, anthropogenic radionuclide content of sediments collected in 1993 in the Beaufort Sea was predominantly due to global fallout (Efurd *et al.*, 1996). Very little anthropogenic radioactivity was detected in fish, birds and marine mammals sampled from north-western Alaska (Efurd *et al.*, 1996). Although there is some evidence of radiocesium accumulation by Bering and Chukchi Sea benthic organisms (Cooper *et al.*, 1995) and marine mammals (Efurd *et al.*, 1996; Cooper *et al.*, 2000), marine sediments from the Bering and Chukchi Seas, including the Yukon River delta, showed low levels of ^{137}Cs compared to tundra or lake sediments (Cooper *et al.*, 1995).

Global fallout is the major source of Pu in biota of the former Soviet Union and US, with virtually no detectable Pu from European close-in fallout or from dumped nuclear reactors in the Kara Sea (Baskaran *et al.*, 1996; Efurd *et al.*, 1996). However, a significant portion of radiocesium and a small portion of Pu released from the Sellafield reprocessing plant (UK) are transported to the North Sea, Norwegian Sea, Barents Sea, Arctic Ocean and Greenland (Baskaran *et al.*, 1996). Marine sources affecting the Barents and Greenland seas are estimated to be 26–46 per cent of the total Pu from European reprocessing plants (Holm *et al.*, 1986). Baskaran and Naidu (1995) showed that the East Chukchi Sea (Alaskan Arctic) sediment ^{137}Cs concentrations were as expected based on fallout patterns, and that former Soviet Union nuclear dump sites have not contaminated this region as of the time of sampling. Although Pu in benthic marine invertebrates is 100 times greater than that in free-swimming organisms (Pillai and Mathew, 1976, cited in Baskaran *et al.*, 1996), in general the concentrations appear to decrease significantly up the food chain (Marshal *et al.*, 1974).

At present, marine food chains do not appear to be affected by radionuclides. The US Food and Drug Administration's screening limit for ^{137}Cs is 370 Bq/kg (Efurd *et al.*, 1996) and is well above any reasonable consumption rate and subsequent exposure (Cooper *et al.*, 2000) via Arctic marine mammals.

Basic toxicity and interactions of elements

General considerations

Metals bind to proteins and alter enzymatic activity. This results in the cellular damage that underlies clinical signs at the gross level. Metals can also interfere with the 'normal' function of essential elements, some of which are required cofactors crucial to enzymes for proper function. When rendered deficient, either by inadequate intake or by displacement or inhibition by a toxic element (toxicoses), deleterious effects may be seen. However, this has not been documented in Arctic cetaceans or pinnipeds. It is very likely that marine mammals have been exposed to high concentrations of metals for millennia and have evolved adaptive mechanisms to reduce their toxicity. However, many of these mechanisms are as yet unknown (Dietz *et al.*, 1998; Sonne-Hansen *et al.*, 2000). Unanswered questions are: Does a mechanism exist for metal detoxification that is unique to marine mammals? Has a well-conserved system (i.e. metallothionein) been enhanced? Have both occurred to reduce toxicity of the 'high' metal levels that confront some marine mammals?

Metal and radionuclide toxicoses are well described for domestic and laboratory mammals, but not well understood for marine mammals. Many studies have determined concentrations of metals in tissues of marine mammals, but very few have attempted any assessment of health effects. Monitoring studies attempt to explain 'elevated' concentrations by comparisons to terrestrial species or to other marine mammals, but this is inadequate. Marine mammals differ markedly from terrestrial mammals in many aspects of their physiology, as well as with respect to routes and types of exposure to contaminants. The differences in metal concentrations among marine mammals are most likely based on feeding ecology and region (i.e. eastern versus western Arctic for some elements). This hampers our understanding of adaptive mechanisms in marine animals, which may differ among species that occupy diverse ecological and geographical niches. Without adequate study of the species of concern, we will continue to extrapolate from species that may have very little in common with Arctic marine mammals. Concern about elevated metal exposure is warranted, however, in light of the plethora of studies that have documented numerous adverse effects of metals on laboratory and domestic animals, as well as humans. These include impacts on viability, reproductive performance, immune function, genetic components (mutagenicity), growth and carcinogenicity, and are mostly due to severe functional alterations resulting from the interaction of metals with enzyme systems and proteins (Friberg *et al.*, 1986; Furness and Rainbow, 1990; Bratton *et al.*, 1997). Studies of metals in Arctic marine mammals have been published for bowhead whale (Byrne *et al.*, 1985; Bratton *et al.*, 1993, 1997; Krone *et al.*, 1998; Woshner *et al.*, 2001a), polar bear (Eaton and Faurant, 1982; Norstrom *et al.*, 1986; Lentfer and Galster, 1987;

Braune *et al.*, 1991; Norheim *et al.*, 1992; Dietz *et al.*, 1995; Woshner *et al.*, 2001b), ringed seal (Smith and Armstrong, 1978; Mackey *et al.*, 1996; Goessler *et al.*, 1998; Woshner *et al.*, 2001b), harp seal (Jones *et al.*, 1976; Ronald *et al.*, 1984), bearded seal (Smith and Armstrong, 1978), walrus (Born *et al.*, 1981; Taylor *et al.*, 1989), narwhal (Wagemann *et al.*, 1983; Hansen *et al.*, 1990) and beluga (Hansen *et al.*, 1990; Wagemann *et al.*, 1990; Becker *et al.*, 1992; Zeisler *et al.*, 1993; Mackey *et al.*, 1996; Woshner *et al.*, 2001a). Very limited literature exists for radionuclides in these species, but initial reports indicate that the levels are very low for marine mammals of northern Alaska (Cooper *et al.*, 2000) and other regions of the Arctic (AMAP, 1998) compared to terrestrial mammals in the same region.

Variations in concentrations of metals in cetaceans are dependent on the species, tissue sampled, portion of the tissue sampled (i.e. kidney medulla or cortex), age, location (geographic), predominant forage or prey and other factors. Thus multiple factors must be considered when making comparisons of concentrations of metals in marine mammals. O'Shea and Brownell (1994) and Bratton *et al.* (1997) reported that a comparison of the available data showed that, in general, baleen whales have lower concentrations of metal residues in tissues than odontocetes, except for Cd. Cadmium is well known to accumulate mostly in kidney, followed by liver, with much lower concentrations in muscle and blubber. This accumulation is age dependent, but appears to be independent of sex (Bratton *et al.*, 1997; Woshner *et al.*, 2001a, b). No lesions (i.e. histopathology) or effects have been documented with respect to Cd in marine mammals (Dietz *et al.*, 1998; Sonne-Hansen *et al.*, 2000; Woshner, 2000; Woshner *et al.*, 2002). Only small amounts of Cd are likely to be transferred to the fetus (Wagemann *et al.*, 1988).

Copper and iron

Copper is an essential element required by mammals for numerous processes. Copper in liver and kidney of free-ranging Alaska beluga whales was at concentrations that would be in the marginal range compared to domestic species (Woshner *et al.*, 2001a). Such species comparisons are difficult to interpret in apparently healthy animals because concentrations are likely to be 'normal'. Iron is an essential element needed for production of the oxygen-binding molecules hemoglobin and myoglobin, as well as other substances. Blood contains variable and large quantities of Fe. This can make interpretation of analytic data difficult when organs (i.e. liver) are congested with blood, resulting in highly variable concentrations.

Mercury and selenium

The origin of mercury that is transferred through the food web to marine mammals can be natural, anthropogenic or some combination thereof (Vandal *et al.*, 1993). Mercury (Hg) can occur in inorganic and organic forms. Toxic effects of Hg are well known but vary according to form. In general, ingested inorganic Hg tends to be poorly absorbed and distributed throughout the body, and as a consequence is less toxic than organic Hg (i.e. methylmercury). Bacteria and marine invertebrates are able to convert inorganic Hg to organic Hg (Furness and Rainbow, 1990; Sadiq, 1992). Organic and inorganic forms of Hg coexist in varying ratios in the diet, complicating the interpretation of Hg concentrations in tissues associated with exposure and toxicity. In addition, the disparate chemical properties of the two forms result in differences in absorption, metabolism, tissue-tropism, target organs and susceptibility of different phases of the life cycle (fetal, neonatal, adult, reproductively active). In general, the majority of Hg in muscle of marine mammals is in the organic form (50–100 per cent), whereas the inorganic form is a higher proportion of the total Hg (70–90 per cent) in liver and kidney (Gaskin *et al.*, 1979; Itano *et al.*, 1984a, b, c; Julshamn *et al.*, 1987; Dietz *et al.*, 1990; Law *et al.*, 1991; Zeisler *et al.*, 1993; Woshner *et al.*, 2001a, b).

The most common form of Hg in marine fish and invertebrates is methylated (Falandysz, 1990; Bloom, 1992). Methylmercury is easily absorbed from the gut (90 per cent) and is well known to bioaccumulate in marine organisms and biomagnify through the food chain (Francesconi and Lenanton, 1992). Mercury concentrations greater than 0.15 µg/g in muscle of marine fish are not unusual and are, as a rule, much higher in visceral organs (Clark, 1989). Some pelagic fish accumulate high concentrations of Hg (exceeding 5.0 µg/g), whereas the levels in fish from polluted areas in the North Sea can exceed 1.3 µg/g. Levels in Polar cod (*Boreogadus saida*) ranged from 0.01 to 0.41 µg/g, depending on geographical region (AMAP, 1998).

The toothed whales are likely to have enhanced exposure to mercury, the majority of which is probably in the more bioavailable form (methylmercury), through the consumption of fish and certain invertebrates. Because methylmercury (MeHg) can be excreted in milk (Piotrowski and Coleman, 1980, cited in Bratton *et al.*, 1997) and passes to the fetus via the placenta (Andre *et al.*, 1990; Muir *et al.*, 1992; Meador *et al.*, 1993) these physiologic processes could decrease Hg concentrations in the mature female. Whole body half-life of Hg in cetaceans has been variously estimated to be roughly 10 years (Wagemann *et al.*, 1990) and 1000 days (Law *et al.*, 1992). In the liver of young whales a higher proportion of Hg tends to be organic, as compared to the proportion in the liver of older whales (Zeisler *et al.*, 1993), which may result from the accumulation of inorganic Hg following demethylation of MeHg. In addition, it has been suggested that the ability to demethylate organic Hg is less efficient in younger walruses as compared to mature animals (Born *et al.*, 1981).

The positive correlation between Hg and Se in marine mammals has been observed for many years, but the reason for it remains unclear (Krone *et al.*, 1998; Woshner *et al.*, 2001a, b). Based on molar ratios, Hg and Se appear to accumulate in a 1 : 1 fashion in liver and kidney, and it has been proposed that each element offsets the toxicity of the other. This may involve demethylation of organic Hg, a process that may require Se (Se can increase demethylation) and result in the formation of inert complexes (inorganic and/or proteinaceous) that can be stored in hepatic lysosomes (Friberg *et al.*, 1986; Cuvin-Aralar and Furness, 1991; Woshner *et al.*, 2002). The role of other elements requires investigation. For example, Cd may accumulate in parallel with Hg and Se in liver of beluga whale (Becker *et al.*, 1995a; Woshner *et al.*, 2001a). Caurant *et al.* (1996) indicated that pilot whales (*Globicephala melas*) have a remarkable tolerance to heavy metal contamination (mean hepatic Hg = 64 mg/kg) that may reflect effective detoxification (i.e. demethylation) processes. Although the presence of a small molecular weight metallothionein-like peptide was shown, 95 per cent of the Hg was in the high molecular weight, insoluble, cellular fraction. They suggested the formation of Hg–Se complexes that would undergo subsequent demethylation. This process probably occurs predominantly in the liver and to only a very limited extent in muscle (Caurant *et al.*, 1996).

There is also evidence that Hg and Se form a complex, known as tiemannite, that is stored as inert concretions in marine mammal hepatocytes, thus detoxifying these elements by insolubilization (Martoja and Berry, 1980). This is clearly important, because most Arctic marine mammals are principally exposed to methylmercury through their prey, but the major form of Hg in marine mammal liver is inorganic. This suggests that the liver is the major site of demethylation and accumulation of Hg and Se (Caurant *et al.*, 1996). However, in the polar bear a 1 : 1 ratio is also found in the kidney, which in this species has even higher Hg and Se concentrations than liver (Dietz, personal communication). This detoxification is more evident with increased age, by comparing percent methylmercury for immature versus mature animals as follows: 55 and 3–18 per cent in pilot whales (Caurant *et al.*, 1996); 45 and 2.5 per cent in striped dolphins (*Stenella coeruleoalba*); 70 and 14 per cent in harp seals (Wagemann *et al.*, 1988); and 45 and 19 per cent in beluga whales and narwhal (Dietz *et al.*, 1990). As the concentration of total Hg increased, the methylated fraction decreased, so that only 2 per cent of the Hg was methylated in pilot whales with total Hg levels above 100 mg/kg.

Selenium has been well documented to influence the distribution, kinetics and toxic effects of mercury (Cuvin-Aralar and Furness, 1991). It has been theorized that the antioxidative property of Se plays a role, as part of the glutathione peroxidase system. This system quenches free radicals and reactive intermediates (reactive oxygen species) and, in turn, affects demethylation of Hg. It has been proposed that the hydroxy free-radical (OH·) (Suda *et al.*, 1991) and myeloperoxides (Suda and Takahashi, 1992) may remove the methyl group. This role of free radicals in demethylation not only implicates Se, but also vitamin E (rarely analyzed) in marine mammals, and vitamin E has been shown to be protective in rats (Welsh, 1979) and hamsters (Chang *et al.*, 1978) exposed to MeHg. Once demethylated, the more water-soluble

Hg is trapped in the cell, where it can bind other ligands, including sulfur or selenium. Demethylation is known to occur in liver, kidney and gastrointestinal cells, and to be slow in the brain (Hansen and Danscher, 1995).

Metallothionein (MTH) is important for detoxification of Hg in terrestrial mammals, but appears to be much less so for marine mammals (Caurant *et al.*, 1996). In narwhals, 5 and 10 per cent of Hg was bound to MTH in liver and kidney, respectively (Wagemann and Hobden, 1986).

Clearly, Hg is a very important contaminant with respect to consumers of marine fish and mammals across the Arctic. Suggested exposure levels (provisional tolerable weekly intake, or PTWI) and the 0.5 µg/g limit in fish tissue (Canada) are exceeded in some cases. Consumers of pilot whale meat from the Faroe Islands have been estimated to consume 1200–2900 µg Hg/week (Andersen *et al.*, 1987) and Hg concentrations in this population were elevated compared to those in Norwegians (Julshamn *et al.*, 1989; Grandjean *et al.*, 1992). This led to a consumption advisory that whale meat and blubber should only be eaten once a week (Simmonds *et al.*, 1994), and that internal organs should be avoided (Gibson-Londale, 1990). Ohi *et al.* (1980) fed rats a diet high in Se and MeHg originating from sperm whale (*Physeter macrocephalus*) and sebastes (*Sebastes iracundus*), a deep sea fish. They showed that protection from neurotoxicity (delay in signs of 7 weeks) was not solely related to conjugation of Se with MeHg, but probably also involved other protective agents, such as vitamin E, As, and Zn. The complicated interaction of Se and the two forms of mercury is very important to interpreting the impact of these elements on Arctic marine mammals (which display high levels) and on those dependent upon them for food. If Se truly allows for protection in the marine mammals, as well as in humans who consume them, then we will be much less concerned about adverse health effects, and will better understand mechanisms of detoxification that may be used in the future to treat or prevent Hg, or other metal-induced, toxicoses.

Methylmercury determinations in harp seal pups caught shortly after birth and fasted postnatally showed that placental transfer of MeHg occurred. Moreover, analysis of non-fasted pups showed only slight increases in MeHg burdens, suggesting that transfer via milk is low; however, the effects of 'growth dilution' need to be considered. Total Hg distribution was similar among pups and dams, being highest in liver (0.33 µg/g and 7.65 µg/g in pups and dams, respectively) and lowest in brain (0.063 µg/g and 0.135 µg/g). Among 20 mother–pup pairs from the Gulf of St. Lawrence, Canada, the distribution of Hg mirrored that observed in other studies, with tissues in order of highest to lowest concentration being liver, kidney and muscle (Wagemann *et al.*, 1988). Although mean hepatic Hg was 30 times lower in pups than dams, a far greater percentage of the Hg was in the methylated form (70 per cent in pups versus 14 per cent in dams). Total Hg burdens in pups were greater than would have been expected based upon Hg concentrations in milk and estimated intake, leading Wagemann *et al.* (1988) to concur with Jones *et al.*'s (1976) assertion that the majority of Hg in pup tissues was acquired transplacentally as MeHg rather than through milk, and to conclude that pups may not have been able to demethylate this form of Hg effectively. This supposition was supported by Se concentrations, which showed a significantly different tissue distribution between mothers and pups (Wagemann *et al.*, 1988).

Ronald *et al.* (1977) examined the effects of methylmercuric chloride administered orally at low (0.25 mg/kg/day) and high (25 mg/kg/day) dosages to harp seals. The two seals in the high-dose group died on day 20 and 26 of the experiment, after exhibiting signs of renal failure and acute uremia, including vomiting and convulsions. The only clinical signs exhibited by the two seals in the low-dose group during the experiment were decreased activity, appetite and body weight in comparison to controls. No neurologic lesions were found in any of the study animals, but lesions typical of inorganic Hg were observed in the renal convoluted tubular epithelium. Cochlear surface preparations from this same group of seals exhibited a dose-dependent toxicity of MeHg to sensory hair cells (Ramprashad and Ronald, 1977), the significance of which is difficult to interpret. In the two low-dose seals, given 0.25 mg/kg of MeHg for 61 days, significant demethylation occurred, with over 70 per cent of total mercury in most tissues (including liver and kidney) being inorganic, whereas muscle was much higher in MeHg (Freeman *et al.*, 1975).

Lesions assumed to be associated with Hg have been reported in liver and kidneys of bottlenose dolphins (*Tursiops truncatus*) from the Atlantic (Rawson *et al.*, 1993;) at concentrations generally above 50 µg/g. Among the changes noted was increased hepatic lipofuscin deposition (Rawson *et al.*, 1993). However, Woshner *et al.* (2002) found no such correlation between lipofuscin deposition and liver Hg concentrations in cetaceans. Moreover, discrepancies between tissue localization of Hg versus lipofuscin were observed using histochemical methods (Woshner *et al.*, 2002). *In vitro* exposure of beluga whale skin fibroblasts to Hg induced micronuclei formation (DNA damage) (Gauthier *et al.*, 1998). Reports of disease occurrence that implicate con-taminants (organochlorines, metals, etc.) in the St. Lawrence Estuary beluga whale include Béland *et al.* (1991), De Guise *et al.* (1994a, b) and Lair *et al.* (1997). However, Leonzio *et al.* (1992) and Itano *et al.* (1984a, b, c) did not detect any significant lesions at similar levels in dolphins. Most studies of Arctic species have not included a rigorous assessment of lesions.

Hansen and Danscher (1995) described Hg concentrations and histology/ histochemistry in sledgedogs that were on a marine animal diet (Thule, Greenland) and showed that, in most organs, Hg concentrations increased with age of the dog. Mesenteric lymph nodes had the highest concentrations of Hg (24.8 µg/g in 10–15-year-old dogs) followed by liver (14.2 µg/g). This led to the suggestion that the lymphatic system may circulate Hg to target organs (Hansen and Danscher, 1995). This assertion complements the findings of Suda and Takahashi (1990), that the reticuloendothelial system is the main site for MeHg biotransformation. Hansen and Danscher (1995) also showed a decreased proportion of MeHg in comparison to inorganic Hg with increasing age, except in cardiac and skeletal muscle, where all Hg was methylated. Histochemical techniques (autometallography and electron microscopy) revealed that Hg was primarily in lysosomes and occasionally in secretory vesicles in most organs, as well as in macrophages in the lymph nodes and livers of these dogs. The autometallographic technique detects inorganic Hg present in sulfide or selenide crystal lattices, which was evident in all tissues except skeletal muscle and brains of pups, where all Hg was in the methylated form. Autometallography was used to localize inorganic Hg in liver and kidney of Alaska beluga whales (Woshner *et al.*, 2002). Inorganic Hg was evident in cytoplasm of the entire uriniferous tubular epithelium. In liver, inorganic Hg was concentrated in stellate macrophages and hepatocytes in the bile cannalicular domain of the cytoplasm. In the liver, inorganic Hg was concentrated primarily in periportal regions, followed by pericentral and, finally, by midzonal areas of the hepatic lobule (Woshner *et al.*, 2002).

Although it is an important nutrient, Se is capable of inducing toxicoses in a variety of species at excessive concentrations. Human consumption of Se at 5 mg/day has resulted in chronic toxicoses (WHO, 1987) and diets containing 5–10 µg/g can be toxic. Selenium has been established to increase with age or length for some cetaceans, including beluga whales (Becker *et al.*, 1995a; Woshner *et al.*, 2001a), seals (Becker *et al.*, 1995b, 1997; Smith and Armstrong, 1978; Woshner *et al.*, 2001b) and polar bears (Norstrom *et al.*, 1986; Braune *et al.*, 1991; Woshner *et al.*, 2001b). High concentrations of Se in marine mammals may result from its accumulation in conjunction with Hg during detoxification of the latter element, or from high dietary concentrations (i.e. fish).

Silver

Silver (Ag) has not been measured often in marine mammals and is not a metal of major concern for toxicity. Silver is currently not considered an essential element for mammals. However, Ag has been shown to play a role in mimicking Se deficiency in vitamin E-limited animals, which was accomplished by Se and Ag binding to form insoluble complexes (Hammond and Beliles, 1980). Little is known regarding the metabolic pathways of Ag in mammalian systems and, of the limited studies of Arctic marine mammals thus far, only the beluga whale appears to amass appreciable concentrations of Ag (Mackey *et al.*, 1996). Silver in the livers of beluga whales ranged from 6–40 µg/g, with an extreme value of 107 µg/g (Mackey *et al.*, 1996). There are conflicting data pertaining to Ag, with one study (Becker *et al.*, 1995a) implying that it accumulates with age in the livers of belugas, whereas another study (Woshner *et al.*, 2001a) found that it does not.

Lead

Lead is not essential for any known biological process. Deposition of Pb in the Arctic from increased anthropogenic activity has been documented (Ng and Patterson, 1981) but current concentrations in tissues of Arctic marine mammals do not appear to be of concern.

Vanadium

Vanadium (V) has been reported to accumulate to relatively high concentrations in some marine mammals in comparison to terrestrial mammals (maximum of 1.2 µg/g in bowhead liver), and it appears to be higher in Alaskan marine mammals compared to those from the eastern USA. Vanadium was found to accumulate in the livers of Alaskan beluga whales, ringed seals, bowhead whales and bearded seals (Mackey *et al.*, 1996). Mackey *et al.* (1996) concluded that the relatively high concentrations of V in these marine mammals might reflect a unique dietary source, geochemical source or anthropogenic input to the marine system in Alaska. Hepatic V was shown to increase (range 0.1–1.0 µg/g) with size or age in these four species of Arctic marine mammals, and appears to be higher than in other marine mammals outside of the Arctic (Mackey *et al.*, 1996). Hepatic age-related accumulation of V has been documented in three species of pinnipeds from the Swedish coast (Frank *et al.*, 1992) and four species of pinnipeds from the northern Pacific (Saeki *et al.*, 1999). Saeki *et al.* (1999) reported that approximately 90 per cent of the body burden of V in pinnipeds was distributed in liver, hair and bone. However, most studies do not address V.

Vanadium is a rather abundant element, with sources including crude oil, geochemical disturbance, terrestrial run-off, atmospheric transport (minimal input) and local or regional petroleum seepage, most of which are recognized in Alaska (Mackey *et al.*, 1996). Accumulation has been documented in organisms at lower trophic levels (Robertson and Abel, 1979), and has been shown to be 10^6 times for some ascidians (orders Aplousobranchia and Phlebobranchia) (Michibata and Sakuri, 1990). Ingestion of these invertebrates may lead to higher exposure in Arctic marine mammals (Mackey *et al.*, 1996). Ascidians and other benthic invertebrates are eaten by walrus (Fay *et al.*, 1977), and walrus had the highest V concentrations of Arctic marine mammals (Warburton and Seagers, 1993). Piscivorous marine mammals, such as the beluga whale, may be exposed the least; beluga whales had the lowest concentration among ringed seals, bearded seals, bowhead and beluga whales, as determined by Mackey *et al.* (1996). Vanadium toxicity does not appear to be an issue for marine mammals or subsistence consumers of them (Mackey *et al.*, 1996), but further research is required to determine pathways of exposure and the role of vanadium in health.

Zinc

Zinc is an essential element with no apparent accumulation in any particular soft tissue. Zinc toxicoses or deficiency would be considered very unlikely in marine mammals. However, we emphasize that Zn is likely to interact with many elements and requires further study. A Zn–Cu relationship was evident in polar bear liver (Norstrom *et al.*, 1986), and in various tissues of cetaceans (Koeman *et al.*, 1973; Honda *et al.*, 1983; Wagemann *et al.*, 1983, 1988; Julshamn *et al.*, 1987; Muir *et al.*, 1992; Woshner *et al.*, 2001a), including bowhead whale blubber (Bratton *et al.*, 1997; Woshner *et al.*, 2001a). A Zn–Cd correlation was evident in cetacean liver and kidney (Honda *et al.*, 1983) and in kidney of beluga whale, bowhead whale, ringed seal, polar bear and narwhal (Wagemann *et al.*, 1983; Hansen *et al.*, 1990; Woshner *et al.*, 2001a, b); whereas Zn, Cd and Hg were correlated in livers of northern minke whales (*Balaenoptera acutorostrata*) (Honda *et al.*, 1990). A Zn–Hg correlation was noted in bowhead whale muscle (Bratton *et al.*, 1997) and in liver of the Antarctic minke whale (*Balaenoptera bonaerensis*) (Honda *et al.*, 1987).

Butyltin

The butyltin compounds are of increasing interest, based on recent findings that these compounds accumulate in marine mammals (Yang *et al.*, 1998). Butyltins do not appear to accumulate with respect to lipid content, but more so with protein binding capabilities, as determined in Steller's sea lions (*Eumetopias jubatus*) (Kim *et al.*, 1996). This class of contaminants will certainly receive more attention in Arctic species and should also be considered for toxicologic assessment.

Interpreting metal concentrations in tissues

Metals (or any other contaminant or stressor) might affect Arctic marine mammals in many ways. Directly, one might see metal-induced lesions or pathology. Indirect effects could include immunosuppression, decreased and adversely affected metabolic or physiologic function, altered organ function, neurologic (behavioral) effects, decreased prey/forage, or decreased foraging and reproductive success. From an ecological perspective, metal contamination might change community structure and alter health or availability of suitable prey. These are rarely considered in contaminants research, in which metals typically are evaluated in harvested or animals found dead, with little basic biological data (e.g. age, gender, stock, body condition) or pathologic effects assessments included. Varying concentrations among and within tissues should also be addressed; for instance, the renal cortex has higher concentrations of Cd than the medulla. Sampling procedures and handling can be critical, especially considering some of the logistical constraints associated with collecting specimens from Arctic marine mammals. Strict protocols, as discussed in Becker *et al.* (1991) and Bratton *et al.* (1997), must be carefully followed. If there is suspicion of iatrogenic contamination, the analyses and interpretation of data will *always* be suspect. Sources of metal contamination include: blood, feces, state of decomposition, blood congestion (especially hepatic), stomach contents, knives, glass or plastic containers, table tops, fumes (exhaust), soil/sand, method of killing (lead bullet, gun powder, shrapnel) and others. This quality control is essential and also extends to the laboratory (Bratton *et al.*, 1997).

Human use and exposure

Wheatley (1996) clearly points out that researchers must consider the social and cultural importance of marine mammal consumption and not just the direct clinical effects of chemical contamination. Aboriginal peoples are influenced by a holistic concept of health and environment. Perceptions of change and the disruption (i.e. contamination) of this environmental concept have a considerable impact on social, cultural, spiritual and economic well being. Scientists and policy-makers who establish formal guidelines must be aware of their impacts on the lifestyle, and therefore on the health, of aboriginal peoples (Wheatley, 1996). Certain cultures (e.g. Inuit, Norwegians, Greenlanders, Russians) still depend heavily on Arctic marine mammals for nutritional, cultural and spiritual reasons. This cultural reliance on marine mammals must be balanced with concern for marine mammal populations and the effects of contaminants on them. Consequently, we advocate the holistic health approach that Wheatley (1996) applies to aboriginal peoples when considering Arctic marine mammals and indigenous peoples.

Access of animals for study

There continue to be numerous obstacles to collection and transport of marine mammal tissues to proper facilities for evaluation (IWC, 1995). This is sometimes compounded by difficulty in involving appropriately trained experts. Obstacles occur at many levels, including international (treaties), national (federal authorities and regulations), local (animal rights activists) and sometimes individuals (scientists). It is a difficult topic to address, but some recent events have shown clearly that scientific investigation and diagnostic efforts can be hindered severely by lack of cooperation and political or bureaucratic obstacles. Scientists need to recognize that, in the majority of cases, they are funded by the public and are working for the welfare of the natural resource. In most cases, they are not 'independent' and do not 'own' the data, nor have sole jurisdiction over a species or population. There are many professional and user conflicts that affect the management, study and care of marine mammals, and these conflicts hamper our attempts to understand the role of heavy metals and radionuclides in Arctic marine mammal health. We encourage open discussion and a multidisciplinary approach to unraveling the 'Gordian knot' of contaminant–marine mammal interactions. All groups that value marine mammals – scientists, conservationists, subsistence users, wildlife professionals and countless others – need to cooperate in their management. It is the only chance for these marine mammals.

Inorganics in Arctic cetaceans

Bowhead whales

Bowhead whales are baleen whales (suborder Mysticeti) that feed in Arctic waters throughout the year. The bowhead whale is a large cetacean of considerable cultural and nutritional importance to the indigenous cultures of arctic Alaska, Russia and Canada, as well as a highly endangered species in some regions of the world (i.e. Sea of Okhotsk, Davis Strait). Among the marine mammals addressed in this chapter, bowhead whales exhibit some of the lowest concentrations of metals (i.e. Hg) and radionuclides. This is because they occupy the lowest trophic level, feeding almost exclusively on copepods, euphausiids and amphipods (Lowry, 1993). Much of the data concerning contaminants in bowhead whales has been acquired from animals taken in the subsistence hunt in northern Alaska by the Inuit. Five reports have addressed heavy metals in bowhead whales: Bratton *et al.* (1993, 1997), Byrne *et al.* (1985), Krone *et al.* (1998) and Woshner *et al.* (2001a). Bratton *et al.* (1997) and Woshner *et al.* (2001a) have presented the most comprehensive work related to metal concentrations in the bowhead whale.

Copepods and euphausiids have low Hg concentrations (Honda *et al.*, 1983) that may account for the low Hg concentrations found in bowhead whale tissues (mean in liver = 0.059 µg/g and kidney = 0.052 µg/g) as compared to other Arctic marine mammals (Bratton *et al.*, 1997). Consequently, Hg is not considered a significant risk for this species, or for consumers of bowhead whale tissues. Muscle was the only tissue in which Hg concentrations correlated with body length (used as a surrogate indicator of age; Woshner *et al.*, 2001a). Among samples collected from 1983 to 1990 (Bratton *et al.*, 1997) and 1995–7 (Woshner *et al.*, 2001a) there was considerable variation in concentrations of elements among individuals and tissues. Furthermore, in general: As was highest in blubber, followed by liver ≈ kidney ≈ muscle; Cd was highest in kidney > liver > muscle > blubber > epidermis (in which concentrations were below the detection limit); Cu was highest in liver > kidney ≈ muscle ≈ blubber ≈ epidermis; Pb concentrations in liver ≈ kidney ≈ muscle > blubber; Hg in liver ≈ kidney > blubber ≈ muscle ≈ epidermis; Se in liver ≈ kidney > epidermis > muscle ≈ blubber; and Zn in liver ≈ muscle > kidney > blubber ≈ epidermis (Woshner *et al.*, 2001a). Iron was highest in liver > muscle > kidney > blubber (Bratton *et al.*, 1997). Total body length (age-related) was positively correlated with: Cd concentrations in liver, kidney and muscle; Se in liver and epidermis; Zn in blubber and Hg in muscle. Body length was negatively correlated with Cu and Ag in liver. Hepatic Se showed a strong positive correlation with Hg, Cd and Zn. Mercury was correlated with Cd in liver and kidney, probably due to mutual accretion with age (Woshner *et al.*, 2001a). Arsenic, Pb and Hg were low in bowhead whales in comparison to other cetaceans (Bratton *et al.*, 1997; Woshner *et al.*, 2001a). Concentrations of Mo, Co, Mg and Mn in tissues were within normal ranges for domesticated species (Woshner, 2000).

Of the metals analyzed, only Cd in renal tissue was noted to be elevated (occasionally 50 µg/g or higher) in bowhead whales. However, Cd concentrations appear to be elevated in a variety of Arctic marine mammals. For bowhead whales, the average concentration of Cd was 20.0 and 9.6 µg/g in kidney and in liver (Woshner *et al.*, 2001a), respectively, which is similar to most baleen whales and narwhal (Wagemann *et al.*, 1983; Hansen *et al.*, 1990). Toxicologically, these concentrations would be considered at the high end of a normal range in some domestic livestock. Although the Cd concentrations in kidneys noted here are not unusual for some marine mammals, the threshold for Cd-induced chronic renal toxicosis is not known for these species. According to Eisler (1985), Cd concentrations of 10 µg/g indicate contamination, and concentrations above 15 µg/g represent a potential hazard; others have claimed the hazard range is 100–400 µg/g (described in Dietz *et al.*, 1998). Among terrestrial mammals, including humans, a renal cortical concentration of 200 µg/g has been acceded as the critical threshold for chronic Cd toxicosis, seen clinically as a low molecular weight proteinuria (WHO, 1992).

It is important to recognize that the studies of bowhead whales cited herein, as well as most other studies in marine mammals, have documented Cd concentrations in entire renicules, including medulla. Cd concentrates in proximal tubular cells of the cortex. Because the cortex averages 22 per cent of the thickness of the entire bowhead whale renicule (Haldiman and Tarpley, 1993), it might be assumed that cortical Cd concentrations are proportionately greater. Dietz *et al.* (1998) conjectured that analysis of intact marine mammal renicules would underestimate cortical Cd concentrations by about 25 per cent. Nevertheless, the concentrations reported here are very similar to those in other marine mammals in which no adverse health effects were shown (although in most studies of Cd in marine mammals the investigation of effects is typically not extensive).

Fujise *et al.* (1988) indicated that the kidney of the Dall's porpoise (*Phocoenoides dalli*) might develop lesions at 20 µg/g. Histologic examination of renal tissue from five bowhead whales revealed a generalized, noninflammatory, periglomerular and interstitial fibrosis that was mild in three immature animals, but moderate in two adults (Woshner, 2000). Although Cd was at elevated concentrations (range 12.2–44.1 µg/g), the lesion observed was not typical of a Cd-induced nephropathy. Moreover, lesion severity did not track Cd concentrations but was believed to be a normal aging change for this species. Continuing research addressing bowhead whale histopathology should help to clarify the significance of kidney changes in relation to renal Cd concentrations in this species. Due to the complicated interactions of Cd with other elements (Zn, Cu, Fe, Se, etc.) and the possibility of increased exposure due to human activities, we should enhance our understanding of the subcellular Cd distribution and of the systems (metallothionein) that protect the cell and, ultimately, the organ. For example, Se correlated positively with Cd (Bratton *et al.*, 1997; Woshner *et al.*, 2001a) in bowhead whale liver. In some laboratory experiments Se has been shown to decrease Cd toxicity (Ridlington and Whanger, 1981; Caurant *et al.*, 1994). Extrapolation from the laboratory animal model to cetaceans may not be appropriate, because the observed Cd : Se correlation may be spurious if both elements simply accumulate with age.

Cadmium concentrations in bowhead whale kidney are of potential concern for subsistence users. Kidneys of 57 per cent of the whales analyzed were above the weekly allowable intake established by the World Health Organization (WHO), based on consumption of small portions (Bratton *et al.*, 1997). However, we must be very cautious in applying this safety level to kidneys of bowhead whales because it includes assumptions regarding weekly intake, life-long exposure, Cd from other sources (i.e. not mammalian renal tissue), safety factors and other considerations. The nutritional (Cu, Zn, Se and Fe) and cultural importance of this subsistence resource must be weighed against the 'apparent' or 'potential' risks of Cd exposure.

Bratton *et al.* (1997) measured eight metals (As, Cd, Cu, Fe, Hg, Pb, Se and Zn) in the tissues (liver, kidney, blubber, muscle and visceral fat) of 41 bowhead whales. Additional data were added from 20 bowhead whales in which As, Cd, Co, Cu, Pb, Mg, Mn, Hg, Mo, Se, Ag and Zn were analyzed in liver, kidney, muscle, blubber and epidermis (Woshner *et al.*, 2001a). There were no significant differences between gender and year of sampling for the metals studied. The concentrations of Zn in bowhead whale were similar to those reported for other cetaceans (Bratton *et al.*, 1997). Concentrations of Se in bowhead whales were low (Bratton *et al.*, 1997) compared to other Arctic marine mammals. For example, mean Se was 1.6 µg/g in both bowhead liver ($n = 54$) and kidney ($n = 46$), but roughly two and four times higher in livers and kidneys of ringed seals, respectively (Woshner *et al.*, 2001b), and 20 and four times higher in belugas of Alaska (Woshner *et al.*, 2001a). Woshner (2000) observed acute myodegeneration (in cardiac and/or skeletal muscle) in 3 of 6 whales, which was consistent with an etiology of exertional myopathy that may have occurred when the animals were obtained (subsistence hunt). Whales with this lesion had muscle Se concentrations below the mean for the study, and one of these whales had the lowest muscle Se concentration (0.08 µg/g) observed among 42 whales sampled. Although no signs of chronic Se deficiency were observed, Woshner (2000) suggested that exertion and associated oxidative stress imposed by the hunt, diving, migration, fasting, etc. might have overwhelmed the protection afforded by Se concentrations, and culminated in the observed myodegeneration.

Arsenic concentrations were relatively low in the bowhead whale and present mostly in the organic form (98 per cent) as arsenobentaine (Bratton *et al.*, 1997), and highest in the blubber (range 0.005–1.92 µg/g) (Woshner *et al.*, 2001a). Studies of other cetaceans have also found highest As concentrations in blubber (Wagemann *et al.*, 1983; Bloch *et al.*, 1985; Julshamn *et al.*, 1987). In bowhead whale epidermis, a preferred food item of the Inuit, As concentrations are quite low (mean = 0.08 µg/g). Arsenic is of little concern as a contaminant for the bowhead whale or to human consumers of bowhead whale tissue.

Very few reports in cetaceans have included Cu. Copper concentrations in the bowhead are below what would be considered the normal range for terrestrial domesticated species (Puls, 1994), but are consistent with reports in cetaceans (Kannan *et al.*, 1993; Wagemann *et al.*, 1996). Copper was negatively correlated with: body length (age) in liver, Se and Zn in kidney, and Zn in muscle, while exhibiting a strong positive correlation with Ag in liver and Se in muscle (Woshner *et al.*, 2001a).

Analysis for radionuclides in tissues (liver, kidney, muscle, blubber and epidermis) of 14 Alaskan bowhead whales revealed lower burdens of the γ-emitter ^{137}Cs (mean activity < 0.3 Bq/kg) than those observed for marine mammals in the North Atlantic (Cooper *et al.*, 2000). The α- and β-emitters, ^{90}Sr and ^{239,240}Pu, were at or below levels of detection; thus radionuclides do not currently merit concern with respect to bowhead whales or subsistence consumers of them (Cooper *et al.*, 2000).

The authors of this chapter, among other researchers, are continuing studies on bowhead whale metal toxicology, including analysis of metals in tissues, histologic assessment (light microscopy, special staining and electron microscopy), characterization of metallothionein RNA and kidney and liver cell culture (Goodwin *et al.*, 1999). Research on metals in the bowhead whale is evolving and requires a multidisciplinary approach to answer some basic questions with respect to effects of metals on ecological, animal and human health.

Beluga whales

Beluga whales in eastern and western Arctic Canada spend summers in nine common areas: James Bay, South Hudson Bay, West Hudson Bay, East Hudson Bay, Northern Hudson Bay, Ungava Bay, Southeast Baffin, Baffin Bay and the Beaufort Sea (Wagemann *et al.*, 1996). Mitochondrial and nuclear DNA studies have confirmed some of these stock identifications (Brown *et al.*, 1994, cited in Wagemann *et al.*, 1996). The St. Lawrence estuary beluga whale population is isolated and not considered a part of the Arctic beluga whales. In Alaska, there are five concentrations of beluga whales in summer: Cook Inlet, Bristol Bay, Norton Sound, the eastern

Chukchi Sea and the eastern Beaufort Sea. These concentrations represent geographically separate stocks, based on differentiation in mitochondrial DNA and, to a lesser degree, on differences in body size, contaminant profiles and population trends (O'Corry-Crowe *et al.*, 1997). Beluga whale feeding can be quite varied and likely is affected by the availability of, or preference for, various prey in the above summering areas (Becker *et al.*, 1995a). Known prey include numerous species of fish, cephalopods, shrimp, euphausiids and many other invertebrates (Seaman *et al.*, 1986).

Beluga whales in Alaska have much higher concentrations of hepatic Hg (Zeisler *et al.*, 1993; Mackey *et al.*, 1996; Woshner *et al.*, 2001a), than bowhead whales harvested from the same geographical locations (Bratton *et al.*, 1997; Woshner *et al.*, 2001a). The source of Hg is likely fish, but may include some invertebrates that the bowhead whale may not commonly consume. However, very little is known about the feeding ecology of the beluga whale, and possible sources of contaminants require careful interpretation. Livers of belugas contain Hg concentrations of concern to consumers, but our experience in Alaska indicates that liver is not commonly consumed by people, and that the majority of Hg is in the inorganic form. Nevertheless, consumption patterns need to be further investigated, because local food preferences exist for certain tissue types.

Methylmercury reported for livers of Alaska beluga whales ranges from 3 to 29 per cent of total (inorganic and organic) mercury (Becker *et al.*, 1995a), and organic mercury ranges from 18 to 39 per cent of total mercury in beluga whales from Greenland (Dietz *et al.*, 1990), with the proportion of MeHg apparently decreasing as total mercury increases (Becker *et al.*, 1995a). Organic forms other than MeHg occur in liver and may constitute more than 50 per cent of the organic fraction (Wagemann *et al.*, 1997). Becker *et al.* (1995a) stated that the relatively high concentrations of Se and total Hg in livers of beluga whales from Point Lay (north-western Alaska) as compared to whales from the Canadian Arctic (Wagemann *et al.*, 1990) could not be explained by age differences alone. A spatial relationship was noted for Hg concentrations in beluga and ringed seals, with concentrations of Hg in livers higher in the western versus the eastern Canadian Arctic (Wagemann *et al.*, 1996). This difference was attributed to differences in the geological formations of the two regions. A similarly different (i.e. higher) environmental Hg exposure may be encountered by beluga whales known to migrate from southern Alaska to the Mackenzie River delta, and is consistent with their designation as a separate stock (O'Corry-Crowe *et al.*, 1997). Hansen *et al.* (1990) and Wagemann *et al.* (1996) examined temporal trends from the early 1980s to early 1990s and noted that Hg increased in livers of beluga whales from both the eastern and western Canadian Arctic during this period. Beluga whales from the St. Lawrence River had significantly higher mean Hg concentrations as compared to whales from Arctic Canada (Wagemann *et al.*, 1996) or Alaska. Wagemann *et al.* (1996) noted a significant positive correlation for Hg and Se in livers of beluga whales.

Wagemann *et al.* (1996) reported that total Hg concentrations in beluga whale muscle ranged from 0.70 to 1.34 µg/g and increased with age, such that animals above 4 years of age had mean concentrations greater than the Canadian guidelines for human consumption, of 0.5 µg/g for Hg in fish (Wagemann *et al.*, 1996). In comparison, beluga whales from Greenland had total mercury concentrations (mean 1.31 µg/g for animals > 14 years; Hansen *et al.*, 1990) similar to those seen in the eastern Canadian Arctic, and also accumulated Hg in conjunction with Se (Hansen *et al.*, 1990). Beluga whales sampled during 1993–4 had mean total Hg concentrations of 0.78 µg/g (western Arctic) and 0.59 µg/g (eastern Arctic) in *muktuk* (skin), with 50–60 per cent having concentrations greater than 0.5 µg/g (Wagemann *et al.*, 1996). In addition, Hg concentrations were highest closer to the surface of the skin, so that an estimated 20 per cent of the skin Hg burden is lost during the annual molt (Wagemann *et al.*, 1996). Woshner *et al.* (2001a) noted an increase with age for total Hg (including both divalent Hg and MeHg) in beluga liver, although the percentage of total Hg present as MeHg declined. In beluga epidermis and muscle (where virtually all Hg is MeHg), total Hg, divalent Hg and MeHg all accumulated with age. No significant correlation was observed between Hg and Se in epidermis; whereas in muscle, Se was positively correlated with divalent Hg, but not with any other form of Hg. Methylmercury was negatively correlated with Cu in both epidermis and muscle, and with Zn in epidermis (Woshner *et al.*, 2001a).

Arsenic concentrations in liver, kidney and muscle (Zeisler *et al.*, 1993) were generally below 1 µg/g. Arsenic was positively correlated with Se in beluga liver and epidermis (Woshner *et al.*, 2001a), as has also been reported in pilot whale kidney (Caurant *et al.*, 1994) and liver (Julshamn *et al.*, 1987). Arsenic in beluga whales occurs mostly in the organic form, as arsenobetaine (Goessler *et al.*, 1998). Among Alaskan beluga whales, As was highest in blubber (mean = 1.41) followed by epidermis (mean = 0.30), whereas kidney, liver and muscle all had similarly low concentrations (Woshner *et al.*, 2001a). This type of tissue distribution of As is similar to that seen in other cetaceans (including bowhead whales) and supports evidence indicating that As is in the organic (non-toxic) form (Vahter *et al.*, 1983; Sabbioni *et al.*, 1991).

Most marine animals have Ag concentrations < 1 µg/g, whereas Alaskan beluga whales had mean concentrations in liver of > 10 µg/g (Becker *et al.*, 1995a; Woshner *et al.*, 2001a). Silver can be accumulated by shellfish and other invertebrates (Rouleau *et al.*, 2000), with reported concentrations of 0.65–2.22 µg/g in crab muscle and 3.4–85 µg/g dry weight (dw) in a gastropod (Robertson and Abel, 1979), whereas fish show lower concentrations. Among Arctic marine mammals, only the beluga whale appears to accumulate Ag (Mackey *et al.*, 1996). This includes a rarely detected form, silver-108, which is a radionuclide solely of anthropogenic origin that was detected in bomb fallout following atmospheric weapons tests (Cooper *et al.*, 2000). Silver concentrations were one to three orders of magnitude higher in livers of beluga whales as compared to pilot whales as well as to previous reports in other marine mammals (Becker *et al.*, 1995a). Although Hg, Se and Ag were all positively correlated, Ag correlated more strongly with Se than did Hg in these beluga whales (Becker *et al.*, 1995a). In light of these relationships, Becker *et al.* (1995a) suggested a possible role for Se in Ag accumulation and the potential for Ag to interfere with free-radical-scavenging enzyme systems, as well as with the normal detoxification of Hg in belugas. It was indicated that Hg, Se and Ag may increase in parallel with age, but differences in accumulation of these elements based upon sex, stock (geographical and/or genetic separation), feeding behavior or other factors were not ruled out (Becker *et al.*, 1995a). More recent research in Alaskan belugas found no accumulation of Ag in liver, but on the contrary, a slight decrease (not statistically significant) of Ag in liver with age, although Ag did amass in the kidney (Woshner *et al.*, 2001a). Moreover, a correlation between Ag and Se was not evident in livers. In kidney of beluga, Ag was positively correlated with Se, Cd and Hg. A positive correlation between hepatic Ag and Cu was observed in beluga (and bowhead) whales, which Woshner *et al.* (2001a) speculated might connote binding by a common ligand.

Mean Cd concentration in liver and kidney was higher in eastern (mean kidney and liver = 22.4 and 6.51 µg/g, respectively) as compared to western (mean kidney and liver = 9.68 and 2.27 µg/g, respectively) Canadian Arctic beluga whales, whereas concentrations in *muktuk* and muscle were low and did not differ between regions (Wagemann *et al.*, 1996). Although positive linear correlations for hepatic Cd and length or age are known for beluga whales (Mackey *et al.*, 1996), this geographic difference was attributed to differences in geological formations. In addition, Zn was higher in *muktuk* and kidney, whereas Cu was higher in muscle, liver and kidney of eastern as compared to western Canadian Arctic belugas. There was no temporal trend associated with Cd in the tissues of belugas, ringed seals and narwhal over a 10–12-year period in the Canadian Arctic (Wagemann *et al.*, 1996). Lead concentrations in tissues for both eastern and western Canadian Arctic beluga whales were low as compared to the St. Lawrence River population, which probably is reflective of anthropogenic inputs in the latter population (Wagemann *et al.*, 1990, 1996).

Mean Se concentrations in liver of belugas were higher in the western Canadian Arctic, which was expected because Hg was also higher (Wagemann *et al.*, 1996). The Se concentrations in *muktuk* and muscle were similar for eastern and western Canadian Arctic belugas. In Alaskan beluga whales, Se concentrations were similar to those in western Canadian Arctic belugas and were approximately 16–20 times that in bowhead whale liver, and approximately four times higher than in bowhead kidney (Tarpley *et al.*, 1995; Woshner *et al.*, 2001a). In beluga and narwhal whales, Zn concentrations were highest in the epidermis (*muktuk*), where concentrations were 2–3 times greater than in other tissues (Wagemann *et al.*, 1996). In eastern as compared to western Canadian belugas, Zn concentrations were higher in kidney and *muktuk*, whereas Cu was higher in muscle, liver and kidney (Wagemann *et al.*, 1996).

Butyltin compounds (mono-, di and tri-butyltin) were measured in liver and blubber of beluga whales from the St. Lawrence River population (Yang *et al.*, 1998). Butyltin concentrations were highest in liver. Concentrations in blubber were affected by lipid content. The authors concluded that butyltins are accumulating in these belugas, which is cause for concern due to the continued use of butyltins as ship antifouling agents and the known potential for their bioaccumulation and toxicity.

Woshner (2000) examined subsistence-harvested Alaskan beluga whales, both grossly and histologically, and found no lesions supportive of a diagnosis of chronic metal toxicosis. Similarly, Siebert *et al.* (1999) noted no specific lesions attributable to Hg toxicosis among harbor porpoises (*Phocoena phocoena*) and whitebeaked dolphins (*Lagenorhynchus albirostris*) obtained as bycatch or stranded in the North and Baltic Seas. However, Siebert *et al.* (1999) associated Hg burdens (Hg concentrations in liver ranged from 0.2 to 130 µg/g) with increased prevalence of parasitism and severity of pulmonary disease (pneumonia), presumably through immunosuppressive mechanisms. Among the 25 beluga whale lungs examined by Woshner (2000), all displayed inflammatory changes, which ranged from mild to severe and focal to multifocal. However, there was no apparent link between Hg concentrations in tissues (total Hg in liver ranged from 0.33 to 83.5 µg/g) and pulmonary lesions, with a few belugas with the highest concentrations of total Hg in livers among those with the mildest lung disease (Woshner, 2000).

Autometallography (AMG) was used to localize inorganic Hg in kidney and liver of beluga whales, and to compare deposition patterns to bowhead whales (Woshner *et al.*, 2002). AMG granules reflecting Hg deposition were observed in the cytoplasm of epithelial cells throughout the entire length of the uriniferous tubule (kidney) of belugas. In beluga livers, AMG granules were concentrated in stellate macrophages and in the bile cannalicular domain of hepatocytes. In belugas with lower hepatic Hg concentrations, AMG granules aggregated in periportal regions, whereas AMG granule deposition extended to pericentral and midzonal regions of liver lobules among whales with higher hepatic Hg burdens. Mean areas (measured surface areas, µm^2) occupied by AMG granules correlated well with hepatic Hg concentrations. Locations of Hg-induced AMG granules were compared with those of lipofuscin (as an index of oxidative damage) in beluga liver. The AMG granules and lipofuscin were occasionally co-localized, but more often were not, implying that Hg was not a component of the lipofuscin deposition in belugas (Woshner *et al.*, 2002). These findings contrasted with those of Rawson *et al.* (1993), who implicated liver Hg accumulation in the pathogenesis of lipofuscin deposition and hepatic disease in stranded bottlenose dolphins in Florida.

In experiments using stimulated (mitogenic response) splenocytes and thymocytes of beluga whales that were exposed to various concentrations

$(10^{-4}-10^{-7}$ M) of the chloride salts of Pb, Hg and Cd *in vitro*, the highest experimental concentrations of Hg and Cd decreased the proliferative response (De Guise *et al.*, 1996). The authors claimed that concentrations employed were relevant, based on known concentrations in liver of wild belugas. However, we caution that this interpretation may be unwarranted because the liver is known to accumulate these elements and detoxify them, probably by complexation with Se over a long period (years). These *in vitro* assays were conducted over a 66-hour period (De Guise *et al.*, 1996). Although we do not discourage the use of cetacean *in vitro* models, careful interpretation and extrapolation to *in vivo* is imperative.

Narwhal

Two narwhal stocks have been identified, based on areas utilized in summer in eastern Canada: the Baffin Bay and the North Hudson Bay stocks (Wagemann *et al.*, 1996). Narwhal are not commonly found in the western Canadian Arctic. The summer diet of narwhal off northern Baffin Island, Canada, was predominantly polar cod (*Arctogadus glacialis*) and Greenland halibut (*Reinhardtius hippoglossoides*), which made up 51 and 37 per cent, respectively, of dietary intake by weight. Squid (*Gonatus fabricii*) beaks were observed, but squid mass could not be quantified. In stomachs of male narwhal, deep-ocean fish [halibut, redfish (*Sebastes marinus*) and arctic cod (*Boroegadus saida*)] were found and may be evidence of diving to greater depths than is usual in females (Finley and Gibb, 1982).

Wagemann *et al.* (1996) reported the following mean total Hg concentrations in tissues of narwhal sampled in 1992–4: *muktuk* 0.59 µg/g, muscle 1.03 µg/g, liver 10.8 µg/g and kidney 1.93 µg/g. As seen in belugas, Hg was not evenly distributed in the skin (*muktuk*) and it was predicted that 14 per cent of the total skin Hg burden would be lost during the annual molt (Wagemann *et al.*, 1996). Wagemann *et al.* (1996) noted a significant positive correlation for Hg with Se concentrations in liver of narwhal. Most of the total Hg in kidney and liver was in the pelleted fraction of cell lysates, indicating that only a small fraction was associated with the cytosol. This is in contrast to Cu, Zn and Cd, which were mostly in the cytosolic fraction (Wagemann *et al.*, 1983). In general, Cd was associated with low molecular weight proteins, whereas Hg was apparently bound to high molecular weight proteins (Wagemann and Hobden, 1986).

Cadmium concentrations in narwhal kidney were among the highest known in Canadian Arctic marine mammals, with means of 38–81 µg/g in different groups, and an overall mean of 54.1 µg/g (Wagemann *et al.*, 1996). Wagemann *et al.* (1983) described a bimodal distribution of Cd with age in the narwhal, such that Cd was lower in the youngest and oldest animals compared to mature animals of middle age. Cadmium has also been shown to increase with age in polar bears (Norstrom *et al.*, 1986), seals (Ronald *et al.*, 1984) and many other cetaceans (as reviewed by Bratton *et al.*, 1997). Although the concentrations noted may be of concern, the kidneys were not examined histologically and nearly all the cytoplasmic Cd was associated with metallothionein (Wagemann and Hobden, 1986; Wagemann *et al.*, 1996). Cadmium concentrations in narwhal liver and muscle were 29.7 and 0.21 µg/g, respectively (Wagemann *et al.*, 1996).

Concentrations of Pb in tissues of Arctic Canadian narwhal are low (Wagemann *et al.*, 1996). The mean concentration of Se in narwhal *muktuk* was more than 10 times higher than in muscle, whereas that of Zn was 2.6 times higher. Hepatic Zn was higher and Cu was lower in narwhal as compared to belugas (Wagemann *et al.*, 1996). Wagemann *et al.* (1983, 1990) indicated that Cu concentrations decreased in narwhal liver with increasing body length. Arsenic was highest in the blubber of narwhal (Wagemann *et al.*, 1983), a finding that was mirrored in studies of pilot, beluga and bowhead whales (Bloch *et al.*, 1985; Julshamn *et al.*, 1987; Woshner *et al.*, 2001a).

Inorganics in Arctic pinnipeds

The Arctic pinnipeds represent a variety of species, niches and regions that affect the interpretation of metal residue data. The ringed seal may be best suited for circumpolar comparison because adequate data are available on this species for some parts of the Arctic. Mercury has been shown to accumulate with age in a variety of species of seals (Smith and Armstrong, 1975, 1978; Ronald *et al.*, 1984). Total Hg in kidneys ranged from 0.08 to 0.45 µg/g in Arctic seals and was about five times higher in gray (*Halichoerus grypus*) and harbor seals (*Phoca vitulina*) than in other species (Skaare *et al.*, 1994). Selenium concentrations in livers of Arctic seals ranged from 0.8 to 3.7 µg/g, and were 1.0–23.3 µg/g for harbor and gray seals. For seals with concentrations above 15 µg/g in liver, the Hg : Se molar ratio was close to unity (Skaare *et al.*, 1994; Dietz *et al.*, 1998). Cadmium concentrations in livers were also shown to increase with age in seals, whereas Cu did not differ over length or between sexes (Ronald *et al.*, 1984). A positive correlation for Cu and Pb was reported in liver of harp seals (Ronald *et al.*, 1984).

Ringed seal

With population estimates in the millions (Wagemann *et al.*, 1996), ringed seals are very common, and represent a significant food source to the polar bear, as well as to indigenous people of the North. A circumpolar species associated with land-fast ice, the ringed seal feeds on polar cod (*Boreogadus saida*) and other fish during ice-cover months, and mostly on crustaceans (*Parathemisto* sp., *Thysanoessa* sp. and *Mysis* sp.) during open-water periods (Smith and Armstrong, 1978). This species has been identified as essential for contaminant monitoring (AMAP, 1998).

Ringed seals have Hg concentrations intermediate to those in beluga and bowhead whales. Total Hg concentrations in ringed seals from Norway were 30 per cent of that reported by Wagemann *et al.* (1989) for Arctic Canadian animals and less than 1 per cent of those reported for Baltic Sea seals. Mercury concentrations in muscle of ringed seals from Arctic Canada were determined for both eastern (mean 0.39 µg/g) and western (mean 0.41 µg/g) groups (Wagemann *et al.*, 1996). These concentrations of Hg in ringed seal flesh are close to the Canadian human consumption guideline level of 0.5 µg/g. Mean Hg concentrations in liver were much higher, at 32.9 and 8.34 µg/g for western and eastern Canadian Arctic groups, respectively. Skaare *et al.* (1994) detected low Hg concentrations in Arctic ringed seal, ranging from 0.2–0.7 µg/g, as compared to more coastal (Norwegian) species. Ringed seals showed a similar geographic pattern for Hg accumulation as that described for Arctic Canadian belugas (Wagemann *et al.*, 1996). The rates of accumulation for the seals were 2.54 and 0.750 µg/g/year for the western and eastern regions, respectively (Wagemann *et al.*, 1996). Narwhal and ringed seals showed similar temporal increases, such that, in ringed seals, the annual rate of hepatic Hg accretion 15–20 years ago was 0.866 µg/g/year, whereas more recently it was shown to be 2.54 µg/g/year (Wagemann *et al.*, 1996). Riget and Dietz (2000) found no consistent temporal trends in Cd or Hg concentrations among ringed seals sampled in Greenland from the late 1970s to the mid-1990s. However, they noted that Cd concentrations in ringed seal liver increased to a maximum among animals approximately 6–8 years of age and declined in older animals.

Most of the total Hg in liver was in the inorganic form and likely associated with Se, with higher concentrations in the west probably reflecting differing geology (Wagemann *et al.*, 1996) or possibly food habits, because shifts in prey (invertebrates versus fish) could alter exposure to Hg and Cd (Dietz, personal communication). As would be expected, concentrations of Se paralleled those of Hg in the liver and were higher in seals from the western versus the eastern Canadian Arctic (Wagemann *et al.*, 1996). The mean Se concentration in livers of the western Arctic seals was comparable to that in livers of seals from the Baltic Sea (median = 19 µg/g) (Frank *et al.*, 1992; Wagemann *et al.*, 1996). In ringed seals, the reported Se concentrations in kidney and in liver are roughly two and four times higher, respectively, than in the same organs of bowhead whales. As in the beluga whale, the source of Hg to ringed seals is probably fish, but could include some invertebrates. Ringed seal liver (Hg content 8.34–27.5 µg/g), which is commonly consumed by Canadian Native people in the Arctic, significantly exceeds a Hg guideline for human food consumption (Wagemann *et al.*, 1996). However, the major proportion of this hepatic total Hg is in inorganic form (probably complexed with Se), which is believed to render it non-bioavailable to consumers (Martoja and Berry, 1980).

A regional trend similar to that observed for Canadian belugas also was seen in ringed seals, where Zn in the kidney and Cu in the liver and kidney were greater in animals taken from eastern areas as compared to those from western areas (Wagemann *et al.*, 1996). There appeared to be no difference in Cd concentrations in liver and kidney among ringed seals from the eastern and western Canadian Arctic, in contrast to the east–west gradient noted for Cd in beluga whales (Wagemann *et al.*, 1996). Positive linear correlations for hepatic Cd and length or age have been documented in ringed seals (Mackey *et al.*, 1996). As with Hg, concentrations of Pb in ringed seals were higher in western than eastern Canada (Wagemann *et al.*, 1996). However, Pb concentrations in harvested species must be interpreted with caution, because bullets can be a major source of contamination. The higher concentrations of Pb in livers of Canadian seals were comparable to those in seals sampled from the Baltic Sea (0.11–0.27 µg/g) (Frank *et al.*, 1992; Wagemann *et al.*, 1996).

Oral dosing of a ringed seal with radioactive methylmercury proteinate followed by whole-body counting resulted in maximum activity detected at 27 days: this was attributed to an initial phase of redistribution to fat near the body surface (Tillander *et al.*, 1972). This was succeeded by bimodal excretion, in which the rapid phase accounted for 55 per cent of the dose with a half-life of 20 days whereas the half-life of the slow component was 500 days.

Bearded seal

Mean concentrations (µg/g) of total Hg in livers were determined for 64 bearded seals from three locations across Canada: 143.0 ($n = 6$) at W. Victoria Island, 26.2 ($n = 56$) at E. Hudson Bay and 44.3 mg/kg at Barrow Strait ($n = 2$). These concentrations were positively correlated with Se (Smith and Armstrong, 1978). Lower concentrations were reported in livers of three Alaskan bearded seals by Mackey *et al.* (1996), ranging from 1.4 to 9.4 µg/g for Hg and 0.43–5.3 µg/g for Se. They also reported concentrations of hepatic Cd from 0.99 to 2.0, and of Ag from 0.075 to 0.17 µg/g. Among the variety of elements analyzed in these three bearded seals, Mackey *et al.* (1996) also noted an apparent increase in hepatic V with age (0.15–1.04 µg/g).

Dietz *et al.* (1990) evaluated Hg in three bearded seal females from Greenland, two of which were juveniles. Hepatic total Hg ranged from 5.0 (7.2 per cent of which was organic) to 27.3 (1.6 per cent of which was organic) µg/g. In kidney and muscle total Hg ranged from 0.94 (12.8 per cent organic) to 2.74 (6.7 per cent organic) µg/g, and from 0.21 (78.5 per cent organic) to 0.23 (71.7 per cent organic) µg/g, respectively (Dietz *et al.*, 1990).

Spotted seal

Numerous trace elements (including Cd, As, Ag, Se and Cs) were analyzed in liver samples from a single spotted seal archived in the National Biomonitoring Specimen Bank. Concentrations were typical of those observed in other phocids (Becker *et al.*, 1997).

Ribbon seal

Very little data were available from the published literature.

Harp seal

Harp seals are an Arctic pelagic seal occurring in Norwegian waters and the White Sea. The mean concentrations of Hg in livers of harp seals caught in Jarfjord (east Barents Sea stock) was 3 per cent of that for harp seals from St. Lawrence Bay, and 7 per cent of that of a seal from NW Greenland (Newfoundland stock) (Ronald *et al.*, 1984). Skaare *et al.* (1994) detected low concentrations of Hg in livers of harp seals as compared to more coastal (Norwegian) seals, ranging from 0.04–1.0 µg/g. Among tissues of ten female harp seals and their pups taken within 1 week postpartum in eastern Canada, highest total Hg concentrations were found in the liver, but were ten times lower in livers of pups (mean of 0.46 µg/g in pups versus 4.6 µg/g in mothers) (Freeman and Horne, 1973). The discrepancy was much narrower in muscle, where the mean total Hg concentration was 0.46 and 0.22 µg/g for mothers and pups, respectively (Freeman and Horne, 1973). Three harp seals from Greenland had total Hg concentrations in livers ranging from 0.31 µg/g (40.1 per cent of which was organic) to 1.01 µg/g (65.5 per cent of which was organic). In kidney and muscle, total Hg ranged from 0.23 (28 per cent organic) to 1.07 (19.5 per cent organic) µg/g, and from 0.09 (63.7 per cent organic) to 0.25 (67.1 per cent organic) µg/g, respectively (Dietz *et al.*, 1990).

Mean total Hg in muscle and liver of 30 harp seals taken in Newfoundland was 1.7 and 2.33 µg/g, respectively (Botta *et al.*, 1983). Total Hg concentrations in muscle were invariably lower than the Canadian consumption guideline of 0.5 µg/g. Total and MeHg residue concentrations were analyzed in blubber, liver, kidney and brain of four adult lactating females and 16 neonatal (ten male, six female) harp seals from the Gulf of St. Lawrence (Jones *et al.*, 1976). Mercury was not detectable in blubber. Concentrations in the liver, kidney and brain did not differ between sexes in 10-day-old pups, although MeHg in blood was slightly higher in females than in males (0.06 µg/g versus 0.03 µg/g). Although mean Se concentrations in kidney and muscle did not differ between dams and pups, mean Se concentrations in livers were an order of magnitude higher in mothers. Liver displayed the highest Se concentrations among adults, but kidney had the highest concentrations of this element among tissues of pups. Se correlated positively with Hg in livers and kidneys of dams and pups. Additionally, Se concentrations correlated strongly between dams and pups for all tissues, signifying that Se crosses the placenta in harp seals.

In contrast, Cd was not detectable in any neonatal tissues, despite being relatively elevated in maternal tissues, in which concentrations correlated positively with age (Wagemann *et al.*, 1988). Tissue distribution of Cd in these harp seal dams was similar to that observed in other marine mammals, with highest concentrations observed in kidney (27.7, 6.6 and 0.04 µg/g in kidney, liver and muscle, respectively). Among tissues, liver of both adults and pups had the most Cu and Zn; also, these two elements were significantly greater in tissues of pups as compared to dams, in which they were negatively correlated with age. In maternal kidney tissue, positive correlations existed for Cd and Hg with Zn, and for Cd and Zn with Cu. These associations were believed to be due to the concurrent induction and binding of Cu–Zn metallothionein by Cd and coincidental renal accumulation of Cd and inorganic Hg with age (Wagemann *et al.*, 1988).

Strontium-90 and cesium-137 were measured in muscle and mammary glands of seven Canadian harp seal females and in the muscle of their pups (Samuels *et al.*, 1970). Strontium-90 was generally below detection limits; the mean concentration of cesium-137 was higher in muscle of pups (3.6 pCi/g ash) than in their dams (2.3 pCi/g ash), with milk being the apparent source to pups (Samuels *et al.*, 1970).

Hooded seal

Among four hooded seals from Greenland, the mean total Hg concentrations in liver ranged from 5.32 (4.7 per cent of which was organic) to 91.8 µg/g (1.5 per cent of which was organic). In kidney and muscle, total Hg ranged from 1.63 (8.7 per cent organic) to 3.90 (14.9 per cent organic), and from 0.10 (103 per cent organic) to 0.75 (88 per cent organic) µg/g, respectively (Dietz *et al.*, 1990; and reviewed in AMAP, 1998).

Walrus

Walrus are associated with the pack ice circumpolarly, and are known to consume large amounts of mollusks (Riedman, 1990) as well as some seals (Muir *et al.*, 1995). Measurement of $^{15}N/^{14}N$ ($\delta^{15}N$) and $^{13}C/^{12}C$ ($\delta^{13}C$) has been used to determine trophic level, by assessing $\delta^{15}N$, which shows enrichment in predators, and food source by $\delta^{13}C$ (Muir *et al.*, 1995).

Between 1981 and 1989 the US Fish and Wildlife Service (USFWS) evaluated metals in adult (11–32-year-old) Pacific walrus harvested in the spring by Alaska natives (Warburton and Seagers, 1993). Cadmium increased with age in both liver and kidney. Zinc and As in kidney, and As in liver, also increased with age. Mercury concentrations were not considered elevated and correlated with Se concentrations (Warburton and Seagers, 1993). Concentrations of Se in liver were higher in females (mean 7.0 mg/kg dw) versus males (mean 4.7 mg/kg dw), whereas mean As concentrations in liver and kidney of males (liver = 0.49, kidney = 1.31 mg/kg dw) were higher than in females (liver = 0.31, kidney 0.70 mg/kg dw) (Warburton and Seagers, 1993). Arsenic was considered to be very low in liver (Norstrom *et al.*, 1986; Taylor *et al.*, 1989). Mercury concentrations did not appear to increase with age in liver or kidney; however, only adults were sampled (Warburton and Seagers, 1993). Data from walrus are contradictory, with Hg concentrations increasing with age in one study (Born *et al.*, 1981), but not in others (Taylor *et al.*, 1989; Warburton and Seagers, 1993). Born *et al.* (1981) observed that MeHg concentrations in walrus pups were similar to that of their mothers, indicating maternal transfer of MeHg. Warburton and Seagers (1993) claimed a weak correlation for age and Cd concentrations in liver and kidney, but only adults were sampled. Some mollusks have been implicated as significant sources of Cd for walrus (Miles and Hills, 1994). Warburton and Seagers (1993) reported that mean Ag concentration was 1.49 µg/g (dw) and ranged from 0.70 to 5.11 µg/g (dw), two orders of magnitude less than that reported for belugas from the same region (Becker *et al.*, 1995a).

Inorganics in polar bears

Polar bears are considered critical for Arctic biomonitoring programs because they are well distributed throughout the Arctic and Subarctic and are top predators that accumulate many contaminants (Norstrom *et al.*, 1998). The primary prey species, ringed seal, is also Holarctic and will integrate contaminants over a limited range (Best, 1985; Norstrom *et al.*, 1998). This food chain is crucial for understanding how elements move to apex predators in the Arctic, including humans.

Polar bears have concentrations of Hg in their livers several times higher than that of their primary prey, ringed seals (Woshner *et al.*, 2001b). Mercury has been shown to accumulate with age in polar bears (Eaton and Faurant, 1982; Norstrom *et al.*, 1986; Lentfer and Galster, 1987). The same geographical distribution of Hg described in belugas and ringed seals is also reflected in the tissues of polar bears (Norstrom *et al.*, 1986; Braune *et al.*, 1991). Geometric mean concentrations of Hg and Se in livers by location ranged from 36.7 to 99.8 µg/g for Hg, and 14.7 and 32.8 µg/g for Se, from east to west (Norstrom *et al.*, 1986; Muir *et al.*, 1992). Mercury concentrations in livers of polar bears from Greenland were lower than in those from more western areas, and ranged from 2.83 to 23.9 µg/g (Dietz *et al.*, 1990, 1995, 1998; Muir *et al.*, 1992). Mercury correlated with an accumulation of Se in polar bear liver (Norstrom *et al.*, 1986; Braune *et al.*, 1991; Norheim *et al.*, 1992) and kidney, which is the primary tissue for Hg accumulation in this species (Dietz *et al.*, 1995). Mean concentrations of Hg adjusted for age in polar bear liver ranged from 18.5 to 200.3 µg/g dw (6.0–71.1 µg/g ww) (Braune *et al.*, 1991). However, mean Se concentrations adjusted for age only ranged from 3.9 to 63.4 µg/g dw (1.27–14.5 µg/g ww) in liver while Ag was 0.21 to 0.54 µg/g (0.07–0.20 µg/g ww) (Braune *et al.*, 1991).

Cadmium increased with age in polar bears, with mean concentrations of 0.61 µg/g (Norstrom *et al.*, 1986) and 0.16–1.2 µg/g (Braune *et al.*, 1991) in liver. Regional differences in Cd concentrations in livers of polar bears were detected across the Arctic (Norstrom *et al.*, 1986). The spatial trend observed for belugas for Cd was also noted for polar bears by Norstrom *et al.* (1986) and Braune *et al.* (1991). Vanadium concentrations in liver were low, at 0.07 µg/g (Norstrom *et al.*, 1986). In general, tissue disposition of metals in polar bears is more like that of terrestrial mammals than cetaceans or pinnipeds, with Pb and Hg reaching greatest concentrations in the kidney, whereas As tends to accumulate in liver (Dietz *et al.*, 1995; Woshner *et al.*, 2001b).

Radionuclide levels are low in polar bears of Alaska, and are similar to those in other marine mammals sampled in Alaska (Cooper *et al.*, 2000). Polar bear milk sampled from two animals in 1968 was evaluated for selected radionuclides. Sr-90 was not detected and Cs-137 was low (0–71 pCi/l) as compared to human samples collected from northern Alaska (0–220 pCi/l), indicating low exposure and low transfer to cubs (Baker *et al.*, 1970).

Conclusions

Marine mammals, including Arctic species, accumulate tissue burdens of some elements, particularly Cd, Hg and Se, that are elevated in comparison to levels in terrestrial mammals. Interactions between elements can affect their bioavailability and effects in the body, with such interactions potentially being antagonistic or synergistic. One well-known example of an antagonistic interaction in marine mammals is the protective effect of Se on the toxicity of Hg. Trophic transfer of metals through diet appears to be the primary mode of bioaccumulation in these species. To date, there is contradictory evidence as to whether anthropogenic increases in global metal mobilization are reflected in Arctic marine mammals, with geographic and trophic variation proving more influential factors in metal exposure. Although marine mammals appear well adapted to tolerate high concentrations of some toxic metals in tissues, few studies have attempted to relate metal concentrations in tissues to effects (i.e. histopathology, enzymology, disease prevalence) and associated threshold levels. The potential for increased anthropogenic input of toxic metals (especially Hg) to the Arctic marine environment, the possibility of synergistic effects with other pollutants, and the dearth of information on both normal and abnormal responses to these compounds by marine mammals underscore the imperative for more research into the mechanisms and effects of persistent toxicants upon these irreplaceable species in the vulnerable Arctic environment.

Acknowledgments

Paul Becker and Rune Dietz reviewed this chapter and provided very helpful comments.

Bibliography

AMAP (Arctic Monitoring and Assessment Program) 1998. *AMAP Assessment Report: Arctic Pollution Issues*. Arctic Monitoring and Assessment Program, Oslo, Norway.

Andersen, A.K., Julshamn, K., Ringdal, O. and Mørkøre, J. 1987. Trace elements intake in the Faroe Islands, II. Intake of mercury and other elements by consumption of pilot whales. *The Science of the Total Environment* 65: 63–68.

Andre, J.M., Ribeyre, F. and Boudou, A. 1990. Mercury contamination levels and distribution in tissues and organs of delphinids from eastern tropical Pacific, in relation to biological and ecological factors. *Marine Environmental Research* 30: 43–72.

Ayotte, P., Dewailly, E., Bruneau, S., Careau, H. and Vezina, A. 1995. Arctic air pollution and human health: What effects should be expected. *The Science of the Total Environment* 160/161: 529–537.

Baker, B.E., Neilson, C.H. and Samuels, E.R. 1970. Strontium-90 and cesium-137 in human and other milks collected in Alaska. *Journal of Dairy Science* 53: 241–244.

Barrie, L.A. 1986. Arctic air pollution: An overview of current knowledge. *Atmospheric Environment* 20: 643–663.

Baskaran, M. and Naidu, A.S. 1995. Pb-210-derived chronology and the fluxes of Pb-210 and Cs-137 isotopes into continental shelf sediments, East Chukchi Sea, Alaskan Arctic. *Geochimica et Cosmochimica Acta* 59: 4435–4448.

Baskaran, M., Asbill, S., Santschi, P., Davis, T., Brooks, J., Champ, M., Makeyev, V. and Khlebovich, V. 1995. Distribution Pu-239/240 and Pu-238 concentrations in sediments from the Ob and Yenisey Rivers and the Kara Sea. *Applied Radiation and Isotopes* 46: 1109–1119.

Baskaran, M., Asbill, S., Santschi, P., Brooks, J., Champ, M., Adkinson, D. Colmer, M.R. and Makeyev, V. 1996. Pu, Cs-137, and excess Pb-210 in Russian Arctic sediments. *Earth and Planetary Science Letters* 140: 243–257.

Becker, P.R., Wise, S.A., Koster, B.J. and Zeisler, R. 1991. *Alaska marine mammal tissue archival project: Revised collection protocol*, NISTIR 4529. National Institute of Standards and Technology, Gaithersburg, Maryland.

Becker, P.R., Wise, S.A., Schantz, M.M., Koster, B.J. and Zeisler, R. 1992. *Alaska Marine Mammal Tissue Archival Project: sample inventory and results of selected samples for organic compounds and trace elements*. NISTIR.

Becker, P.R., Mackey, E.A., Demiralp, R., Suydam, R., Early, G., Koster, B.J. and Wise, S.A. 1995a. Relationship of silver and selenium and mercury in the liver of two species of toothed whales (Odontocetes). *Marine Pollution Bulletin* 30: 262–271.

Becker, P.R., Mackey, E.A., Schantz, M.M., Greenberg, R.R., Koster, B.J., Wise, S.A. and Muir, D.C.G. 1995b. *Concentrations of chlorinated hydrocarbons, heavy metals and other elements in tissues banked by the Alaska Marine Mammal Tissue Archival Project*, NISTIR 5620. National Institute of Standards and Technology, Gaithersburg, Maryland.

Becker, P.R., Mackey, E.A., Schantz, M., Koster, B.J. and Wise, S.A. 1997. Concentrations of chlorinated hydrocarbons and trace elements in marine mammal tissues archived in the U.S. National Biomonitoring Specimen Bank. *Chemosphere* 34: 2067–2098.

Béland, P., DeGuise, S. and Plante, R. 1991. *Toxicology and pathology of St. Lawrence marine mammals.* Final report, Wildlife Toxicology Fund, St. Lawrence National Institute of Ecotoxicology. World Wildlife Fund, Toronto.

Béland, P., De Guise, S., Girard, C., Lagace, A., Martineau, D., Michaud, R., Muir, D.C.G., Norstrom, R.J., Pelletier, E., Ray, S. and Shugart, L.R. 1993. Toxic compounds and health and reproductive effects in St. Lawrence beluga whales. *Journal of Great Lakes Research* 19: 766–775.

Best, R.C. 1985. Digestibility of ringed seals by the polar bear. *Canadian Journal of Zoology* 63: 1033–1036.

Bidleman, T.F., Patton, G.W., Walla, M.D., Hargrave, B.T., Vass, W.P., Erickson, P., Fowler, B., Scott, V. and Gregor, D.J. 1989. Toxaphene and other organochlorines in Arctic ocean fauna: evidence of atmospheric delivery. *Arctic* 42: 307–313.

Bloch, D., Hanusardottir, M. and Davidsen, A. 1985. The contamination of mercury and persistent organochlorines in the marine environment of the Faroe Islands. Presented to International Whaling Commission Scientific Committee, pp. 1–11.

Bloom, N.S. 1992. On the chemical form of mercury in edible fish and marine invertebrate tissue. *Canadian Journal of Fisheries and Aquatic Science* 49: 1010–1017.

Born, E.W., Kraul, I. and Kristensen, T. 1981. Mercury, DDT and PCB in the Atlantic walrus (*Odobenus rosmarus rosmarus*) from the Thule District, North Greenland. *Arctic* 34: 255–260.

Botta, J.R., Arsenault, E. and Ryan, H.A. 1983. Total mercury content of meat and liver from inshore Newfoundland-caught harp seal. *Bulletin of Environmental Contamination and Toxicology* 30: 28–32.

Bratton, G.R., Spainhour, C.B., Flory, W., Reed, M. and Jayko, K. 1993. Presence and potential effects of contaminants. In: J.J. Burns, J.J. Montague, and C.J. Cowles (eds). *The bowhead whale.* Special Publication Number 2, The Society for Marine Mammalogy. KS Allen Press, Inc. Lawrence, Kansas, pp. 701–744.

Bratton, G.R., Flory, W., Spainhour, C.B. and Haubold, E.M. 1997. Assessment of selected heavy metals in liver, kidney, muscle, blubber, and visceral fat of Eskimo harvested bowhead whales from Alaska's north coast, Final Report, North Slope Borough, Department of Wildlife Management, p. 233.

Braune, B.M., Norstrom, R.J., Wong, M.P., Collins, B.T. and Lee, J. 1991. Geographical distribution of metals in livers of polar bears from the Northwest Territories, Canada. *The Science of the Total Environment* 100: 283–299.

Byrne, C.R., Balasubramanian, R., Overton, E.B. and Albert, T.F. 1985. Concentrations of trace metals in bowhead whale. *Marine Pollution Bulletin* 16: 497–498.

Caurant, F., Amiard, J.C., Amiard-Triquet, C. and Sauriau, P.G. 1994. Ecological and biological factors controlling the concentrations of trace elements (As, Cd, Cu, Hg, Se, Zn) in delphinids from the North Atlantic Ocean. *Marine Ecology Progress Series* 103: 207–219.

Caurant, F., Navarro, M. and Amiard, J.C. 1996. Mercury in pilot whales: possible limits to the detoxification process. *The Science of the Total Environment* 186: 95–104.

Chang, L.W., Gilbert, M. and Sprecher, J. 1978. Modification of methyl mercury neurotoxicity by Vitamin E. *Journal of Environmental Research* 17: 356–366.

Clark, R.B. 1989. *Marine pollution*, 2nd edition. Clarendon Press, Oxford.

Cooper, L.W., Grebmeier, J.M., Larsen, I.L., Solis, C. and Olsen, C.R. 1995. Evidence for re-distribution of Cs-137 in Alaskan tundra, lake and marine sediments. *The Science of the Total Environment* 160/161: 295–306.

Cooper, L.W., Larsen, I.L., O'Hara, T.M., Dolvin, S., Woshner, V. and Cota, G. 2000. Radionuclide contaminant burdens in arctic marine mammals harvested during subsistence hunting. *Arctic* 53(2): 174–182.

Cuvin-Aralar, M.L.A. and Furness, R.W. 1991. Mercury–selenium interaction: A review. *Ecotoxicology Environmental Safety* 21: 348–364.

De Guise, S., Lagace, A. and Béland, P. 1994a. True hermaphroditism in a St. Lawrence beluga whale. *Journal of Wildlife Disease* 30: 287–290.

De Guise, S., Lagace, A. and Béland, P. 1994b. Tumors in St. Lawrence beluga whales. *Veterinary Pathology* 31: 444–449.

De Guise, S., Bernier, J., Martineau, D., Béland, P. and Fournier, M. 1996. Effects of in vitro exposure of beluga whale splenocytes and thymocytes to heavy metals. *Environmental Toxicology and Chemistry* 15: 1357–1364.

Dietz, R., Nielsen, C.O., Hansen, M.M. and Hansen, C.T. 1990. Organic mercury in Greenland birds and mammals. *The Science of the Total Environment* 95: 41–51.

Dietz, R., Born, E.W. and Agger, C.T. 1995. Zinc, cadmium, mercury, selenium in polar bears from central east Greenland. *Polar Biology* 15: 175–185.

Dietz, R., Riget, F. and Johansen, P. 1996. Lead, cadmium, mercury and selenium in Greenland marine animals. *The Science of the Total Environment* 186: 67–93.

Dietz, R., Norgaard, J. and Hansen, J.C. 1998. Have arctic marine mammals adapted to high cadmium levels? *Marine Pollution Bulletin* 36: 490–492.

Eaton, R.D.P. and Faurant, J.P. 1982. The polar bear as a biological indicator of the environmental mercury burden. *Arctic* 35: 422–425.

Efurd, D., Miller, G., Rokop, D., Roensch, F., Attrep, M., Thompson, J., Inkert, W., Poths, H., Banar, J., Musgrave, J., Rios, E., Fowler, M., Gritzo, R., Headstream, J., Dry, D., Hameedi, M., Robertson, A., Valette-Silver, N., Dolvin, S., Thorsteinson, L., O'Hara, T. and Olsen, R. 1996. Evaluation of the anthropogenic radionuclide concentrations in sediments and fauna collected in the Beaufort Sea and northern Alaska. Los Alamos National Laboratory, Los Alamos, New Mexico USA, LAMS-13115-MS UC-721.

Eisler, R. 1985. Cadmium hazards to fish, wildlife, and invertebrates: a synoptic review. *US Fish and Wildlife Service Biol. Report* 85(15): 46.

Falandysz, J. 1990. Mercury content of squid *Loligo opalescens*. *Food Chemistry* 38: 171–177.

Fay, F.H., Feder, H.M. and Stoker, S.W. 1977. *An estimation of the impact of the Pacific walrus population on its food resources in the Bering Sea.* Marine Mammal Commission Report MMC-75/06, 75/03.

Finley, K.J. and Gibb, E.J. 1982. Summer diet of the narwhal in Pond Inlet, northern Baffin Island (Canada). *Canadian Journal of Zoology* 60(12): 3353–3363.

Ford, J., Landers, D., Kugler, D., Lasorsa, B., Allen-Gill, S., Crecelius, E. and Martinson, J. 1995. Inorganic contaminants in Arctic Alaskan ecosystems: long-range atmospheric transport or local point sources? *The Science of the Total Environment* 160/161: 323–335.

Francesconi, K. and Lenanton, R.C.J. 1992. Mercury contamination in semi-enclosed marine embayment: Organic and inorganic mercury content of biota, and factors influencing mercury levels in fish. *Marine Environmental Research* 33: 189–212.

Frank, A., Galgan, V., Roos, A., Olsson, M., Peterson, L.R. and Bignert, A. 1992. Metal concentrations in seals from Swedish waters. *Ambio* 21: 529–538.

Freeman, H.C. and Horne, D.A. 1973. Mercury in Canadian seals. *Bulletin of Environmental Contamination and Toxicology* 10(3): 172–180.

Freeman, H.C., Sanglang, G., Uthe, J.F. and Ronald, K. 1975. Steroidogensis in vitro in the harp seal without and with methyl mercury treatment in vivo. *Environmental Physiology and Biochemistry* 5: 428–439.

Friberg, L., Nordberg, G.F. and Vouk, V.K. (eds) 1986. *Handbook on the toxicology of metals.* Volume II, 2nd edition. *Specific metals.* Elsevier Applied Science Publishers, New York.

Fujise, Y., Honda, K., Tatsukawa, R. and Mishima, S. 1988. Tissue distribution of heavy metals in Dall's porpoise in the northwestern Pacific. *Marine Pollution Bulletin* 19: 226–230.

Furness, R.W. and Rainbow, P.S. (eds) 1990. *Heavy metals in the marine environment.* CRC Press, Boca Raton, FL.

Gaskin, D.E., Stonefield, K.I., Suda, P. and Frank, R. 1979. Changes in mercury levels in harbor porpoises from the Bay of Fundy, Canada, and adjacent waters during 1967–1977. *Archives of Environmental Contamination Toxicology* 8: 733–762.

Gauthier, J.M., Dubeau, H. and Rassart, E. 1998. Mercury-induced micronuclei in skin fibroblasts of beluga whales. *Environmental Toxicology and Chemistry* 17: 2487–2493.

George, J.C., Bada, J., Zeh, J., Scott, L., Brown, S.E., O'Hara, T. and Suydam, R. 1999. Age and growth estimates of bowhead whales (*Balaena mysticetus*) using aspartic acid Racemization. *Canadian Journal of Zoology* 77: 571–580.

Gibson-Londale, J.J. 1990. Pilot whaling in the Faroe Islands – its history and present significance. *Mammal Review* 20: 44–52.

Goessler, W., Rudorfer, A., Mackey, E.A., Becker, P.R. and Irgolic, K.J. 1998. Determination of arsenic compounds in marine mammals with high performance liquid chromatography and inductively coupled plasma spectrometry as element detector. *Applied Organometal Chemistry* 12: 491–501.

Goodwin, T., Hammond, T., Li, L., Linnehan, R., Kaysen, J., Albert, T.F. and O'Hara, TM. 1999. Preliminary Report on the Molecular Analysis and Culture of Tissues from Subsistence Harvested Bowhead Whales. IWC Meeting (51st) Grenada May 1–15. SC/51/AS13.

Grandjean, P., Weihe, P., Jorgensen, P.J., Clarkson, T., Cernichiari, E. and Videro, T. 1992. Impact of maternal seafood diet on fetal exposure to mercury, selenium, and lead. *Archives of Environmental Health* 47: 185–195.

Gregor, D.J. and Gummer, W.D. 1989. Evidence of atmospheric transport and deposition of organochlorine pesticides and polychlorinated biphenyls in Canadian Arctic Snow. *Environmental Science and Technology* 23: 561–565.

Haldiman, J.T. and Tarpley, R.J. 1993. Anatomy and Physiology. In: J.J. Burns, J.J. Montague, and C.J. Cowles (eds). *The bowhead whale.* Special Publication Number 2, The Society for Marine Mammalogy. Allen Press, Inc., Lawrence, Kansas, pp. 71–156.

Hammond, P.B. and Beliles, R.P. 1980. Metals. In: J. Doull, C.D. Klassen, and M.O. Amdur (eds). *Casarret and Doull's Toxicology: The basic science of poisons,* 2nd edition. MacMillan, New York, pp. 409–467.

Hansen, J.C. and Danscher, G. 1995. Quantitative and qualitative distribution of mercury in organs from arctic sledgedogs: an atomic absorption spectrophotometric and histochemical study of tissue samples from natural long-termed high dietary organic mercury-exposed dogs from Thule, Greenland. *Pharmacology and Toxicology* 77: 189–195.

Hansen, J.C., Tarp, U. and Bohm, J. 1990. Prenatal exposure to methyl mercury among Greenlandic polar Inuits. *Archives of Environmental Health* 45: 355–358.

Health and Welfare Canada 1979. *Methylmercury in Canada: Exposure of Indian and Inuit residents to methylmercury in the Canadian environment.* Health and Welfare Canada, Medical Services Branch, Ottawa.

Holm, E., Aakrog, A., Ballestra, S. and Dahlgaard, H. 1986. Origin and isotopic ratios of plutonium in the Barents and Greenland Seas. *Earth Planet Science Letters* 79: 27–32.

Honda, K., Tatsukawa, R., Itano, K., Miyazaki, N. and Fujiyama, T. 1983. Heavy metal concentration in muscle, liver, and kidney tissue of striped dolphin, *Stenella coeruleoalba*, and their variations with body length, weight, age and sex. *Agricultural Biology and Chemistry* 47: 1219–1228.

Honda, K., Yamamoto, Y., Kato, H. and Tatsukawa, R. 1987. Heavy metal accumulations and their recent changes in southern minke whales. *Archives of Environmental Contamination and Toxicology* 16: 209–216.

Honda, K., Marcovecchio, J.E., Kan, S., Tatsukawa, R. and Ogi, H. 1990. Metal concentrations in pelagic seabirds from the North Pacific Ocean. *Archives of Environmental Contamination and Toxicology* 19: 704–711.

International Whaling Commission (IWC) 1995. Report of the Workshop on Chemical Pollution and Cetaceans, SC/47/Rep2, Chairman Peter Reijnders, Bergen, Norway.

Itano, K., Kawai, S., Miyazaki, N., Tatsukawa, R. and Fujiyama, T. 1984a. Body burdens and distribution of mercury and selenium in striped dolphins. *Agricultural Biology and Chemistry* 48: 1117–1121.

Itano, K., Kawai, S., Miyazaki, N., Tatsukawa, R. and Fujiyama, T. 1984b. Mercury and selenium levels in striped dolphins caught off the Pacific Coast of Japan. *Agricultural Biology and Chemistry* 48: 1109–1116.

Itano, K., Kawai, S., Miyazaki, N., Tatsukawa, R. and Fujiyama, T. 1984c. Mercury and selenium levels in the fetal and suckling stages of striped dolphins. *Agricultural Biology and Chemistry* 48: 1691–1698.

Jaffe, D., Cerundolo, B., Rickers, J., Stolzberg, R. and Baklanov, A. 1995. Deposition of sulfate and heavy metals on the Kola Peninsula. *The Science of the Total Environment* 160/161: 127–134.

Jones, D., Ronald, K., Lavigne, D.M., Frank, R., Holdrinet, M. and Uthe, J.F. 1976. Organochlorine and mercury residues in the harp seal (*Pagophilus groenlandicus*). *Science of the Total Environment* 5: 181–195.

Julshamn, K., Andersen, A., Ringdal, O. and Morkore, J. 1987. Trace elements intake in the Faroe Islands. Part I. Element levels in edible parts of pilot whales. *The Science of the Total Environment* 65: 53–62.

Julshamn, K., Andersen, K.J., Svendsen, E., Ringdal, O. and Ogholm, M. 1989. Trace elements intake in the Faroe Islands III. Element concentrations in human organs in populations from Bergen Norway and the Faroe Islands. *The Science of the Total Environment* 84: 25–33.

Kannan, K., Sinha, R.K., Tanabe, S., Ichihashi, H. and Tatsukawa, R. 1993. Heavy metals and organochlorine residues in Ganges River dolphins from India. *Marine Pollution Bulletin* 26: 159–162.

Kim, G.B., Lee, J.S., Tanabe, S., Iwata, H., Tatsukawa, R. and Shimazaki, K. 1996. Specific accumulation and distribution of butyltin compounds in various organs and tissues of the Steller sea lion: comparison with organochlorine accumulation pattern. *Marine Pollution Bulletin* 32: 558–563.

Koeman, J.H., Peeters, W.H.M. and Koudstaal-Hol, C.H.M. 1973. Mercury-selenium correlations in marine mammals. *Nature* 245: 271–274.

Krone, C.A., Robisch, P.A., Tilbury, K.L., Stein, J.E., Mackey, E., Becker, P., O'Hara, T.M. and Philo, L.M. 1998. Elements in liver tissues of bowhead whales (*Balaena mysticetus*). *Marine Mammal Science* 15: 123–142.

Lair, S., Béland, P., De Guise, S. and Martineau, D. 1997. Adrenal hyperplastic and degenerative changes in beluga whales. *Journal of Wildlife Diseases* 33: 430–437.

Lasorsa, B. and Allen-Gill, S. 1995. The methylmercury to total mercury ratio in selected marine, freshwater, and terrestrial organisms. *Water, Air and Soil Pollution* 80: 905–913.

Law, R.J., Fileman, C.F., Hopkins, A.D., Baker, J.R., Harwood, J., Jackson, D.B., Kennedy, S., Martin, A.R. and Morris, R.J. 1991. Concentrations of trace metals in the livers of marine mammals (seals, porpoises, and dolphins) from waters around the British Isles. *Marine Pollution Bulletin* 22: 183–191.

Law, R.J., Jones, B.R., Baker, J.R., Kennedy, S., Milne, R. and Morris, R.J. 1992. Trace metals in the livers of marine mammals from the Welsh Coast and the Irish Sea. *Marine Pollution Bulletin* 24: 296–304.

Lentfer, J.W. and Galster, W.A. 1987. Mercury in polar bears from Alaska. *Journal of Wildlife Diseases* 23: 338–341.

Leonzio, C., Focardi, S. and Fossi, C. 1992. Heavy metals and selenium in stranded dolphins of the northern Tyrrhenian (NW Mediterranean). *The Science of the Total Environment* 119: 77–84.

Lowry, L.F. 1993. Foods and feeding ecology. In: J.J. Burns, J.J. Montague, and C.J. Cowles (eds). *The bowhead whale*. Special Publication Number 2, The Society for Marine Mammalogy. KS Allen Press, Inc., Lawrence, Kansas, pp. 201–234.

Mackey, E.A., Becker, P.R., Demiralp, R., Greenberg, R.R., Koster, B.J. and Wise, S.A. 1996. Bioaccumulation of vanadium and other trace metals in livers of Alaskan cetaceans and pinnipeds. *Archives of Environmental Contamination and Toxicology* 30: 503–512.

Manchester-Neesvig, J.B. and Andren, A.W. 1989. Seasonal variation in the atmospheric concentration of polychlorinated biphenyl congeners. *Environmental Science and Technology* 23: 1138–1148.

Marshal, J.S., Waller, B.J. and Yaguchi, E.M. 1974. Plutonium in the Laurentian Great lakes: Food chain relationships. In: *Proceedings of the XIX Congress of the Association of Limnology, Winnipeg, Manitoba, Canada, Aug. 23–28*.

Martoja, R. and Berry, J.P. 1980. Identification of tiemannite as a probable product of demethylation of mercury by selenium in cetaceans. *Vie Milieu* 30: 7–10.

Meador, J.P., Varanasi, U., Robisch, P.A. and Chan, S.L. 1993. Toxic metals in pilot whales from strandings in 1986 and 1990 on Cape Cod, Massachusetts. *Canadian Journal of Fisheries and Aquatic Science* 50: 2698–2706.

Michibata, H. and Sakuri, H. 1990. Vanadium in ascidians. In: N.D. Chasteen (ed.). *Vanadium in biological systems*. Kluwer Academic Press, Dordecht, The Netherlands, pp. 152–171.

Miles, A.K. and Hills, S. 1994. Metals in diet of Bering Sea walrus: *Mya* sp. As a possible transmitter of elevated cadmium and other metals. *Marine Pollution Bulletin* 28: 456–458.

Muir, D.C.G., Wagemann, R., Hargrave, B.T., Thomas, D.J., Peakall, D.B. and Norstrom, R.J. 1992. Arctic marine ecosystem contamination. *The Science of the Total Environment* 122: 75–134.

Muir, D.C.G., Segstro, M.D., Hobson, K.A., Ford, C.A., Stewart, R.E.A. and Olpinski, S. 1995. Can seal eating explain elevated levels of PCBs and organochlorine pesticides in walrus blubber from eastern Hudson Bay (Canada)? *Environmental Pollution* 90: 335–348.

Ng, A. and Patterson, C. 1981. Natural concentrations of lead in ancient Arctic and Antarctic ice. *Geochimica et Cosmochemica Acta* 45: 2109–2121.

Norheim, G., Skaare, J.U. and Wiig, O. 1992. Some heavy metals, essential elements, and chlorinated hydrocarbons in polar bear at Svalbard. *Environmental Pollution* 77: 51–57.

Norstrom, R.J., Sweinsberg, R.E. and Collins, B.T. 1986. Heavy metals and essential elements in liver of polar bears in the Canadian Arctic. *The Science of the Total Environment* 48: 195–212.

Norstrom, R.J., Belikov, S.E., Born, E.W., Garner, G.W., Malone, B., Olpinski, S., Ramsay, M.A., Schliebe, S., Stirling, I., Stishov, M.S., Taylor, M.K. and Wiig, D. 1998. Chlorinated hydrocarbon contaminants in polar bears from eastern Russia, North America, Greenland and Svalbard: biomonitoring of arctic pollution. *Archives of Environmental Contamination and Toxicology* 35: 354–367.

O'Corry-Crowe, G.M., Suydam, R.S., Rosenberg, A., Frost, K.J. and Dizon, A.E. 1997. Phylogeography, population structure, and dispersal patterns of the beluga whale in the western Nearctic by mitochondrial DNA. *Molecular Ecology* 6: 955–970.

Ohi, G., Nishigaki, S., Seki, H., Tamura, Y., Maki, T., Minowa, K., Shimamura, Y., Mizoguchi, I., Inaba, Y., Takizawa, Y. and Kawanishi, Y. 1980. The protective potency of marine animal meat against the neurotoxicity of methylmercury: Its relationship with the organ distribution of mercury and selenium in the rat. *Food and Cosmetic Toxicology* 18: 139–145.

O'Shea, T.J. and Brownell, R.L. Jr 1994. Organochlorine and metal contaminants in baleen whales: a review and evaluation of conservation implications. *The Science of the Total Environment* 154: 179–200.

Piotrowski, J.K. and Coleman, D.O. 1980. *Environmental hazards of heavy metals: Summary evaluation of lead, cadmium, and mercury.* A General Report. MARC (The Monitoring and Assessment Research Center) Publications, Chelsea College, University of London.

Puls, R. 1994. *Mineral levels in animal health: diagnostic data*, 2nd edition. Sherpa International, Clearbrook, BC, Canada.

Ramprashad, F. and Ronald, K.A. 1977. Surface preparation study on the effect of methyl mercury on the sensory hair cell population in the cochlea of the harp seal (*Pagophilus groenlandicus* Erxleben, 1777). *Canadian Journal of Zoology* 55: 223–230.

Rawson, A.J., Patton, G.W., Hofmann, S., Pietra, G.G. and Johns, L. 1993. Liver abnormalities associated with chronic mercury accumulation in stranded Atlantic bottlenose dolphins. *Ecotoxicology and the Environment* 25: 41–47.

Ridlington, J.W. and Whanger, P.D. 1981. Interactions of selenium and antioxidants with mercury, cadmium, and silver. *Fundamental and Applied Toxicology* 1: 368–375.

Riedman, M. 1990. *The pinnipeds: seals, sea lions, and walruses.* University of California Press, Berkeley, California, USA.

Riget, F. and Dietz, R. 2000. Temporal trends of cadmium and mercury in Greenland marine biota. *The Science of the Total Environment* 245: 49–60.

Robertson, D.E. and Abel, K.H. 1979. Natural distribution and environmental background of trace metals in Alaskan shelf and estuarine areas. US Dept of Commerce, NOAA, Environmental Assessment of the Alaskan Continental Shelf, Annual Reports 5, pp. 660–698.

Ronald, K., Tessaro, S.V., Uthe, J.F., Freeman, H.C. and Frank, R. 1977. Methylmercury poisoning in the harp seal (*Pagophilus groenlandicus*). *The Science of the Total Environment* 8: 1–11.

Ronald, K., Frank, R.J., Dougan, J., Frank, R. and Braun, H.E. 1984. Pollutants in harp seals. II. Heavy metals and selenium. *The Science of the Total Environment* 38: 153–166.

Rouleau, C., Gobeil, C. and Tjälve, H. 2000. Accumulation of silver from the diet in two marine benthic predators: the snow crab (*Chionoecetes opilio*) and American plaice (*Hippoglossoides platessoides*). *Environmental Toxicology and Chemistry* 19(3): 631–637.

Sabbioni, E., Fischbach, M., Pozzi, G., Pietra, R., Gallorini, M. and Piette, J.L. 1991. Cellular retention, toxicity and carcinogenic potential of seafood arsenic. I. Lack of cytotoxicity and transforming activity of arsenobetaine in the BALB/3T3 cell line. Carcinogenesis 12: 1287–1291.

Sadiq, M. 1992. *Toxic metal chemistry in marine environments.* Marcel Dekker, New York.

Saeki, K., Nakajima, M., Noda, K., Loughlin, T.R., Baba, N., Kiyota, M., Tatsukawa R. and Calkins, D.G. 1999. Vanadium accumulation in pinnipeds. *Archives of Environmental Contamination and Toxicology* 36: 81–86.

Samuels, E.R., Cawthorn, M., Lauer, B.H. and Baker, B.E. 1970. Strontium-90 and cesium-137 in tissues of fin whales and harp seals. *Canadian Journal of Zoology* 48: 267–269.

Sanpera, C., Capelli, R., Minganti, V. and Jover, L. 1993. Total and organic mercury in North Atlantic fin whales: distribution pattern and biological related changes. *Marine Pollution Bulletin* 26: 135–139.

Seaman, G.A., Frost, K., Lowry, L. and Burns, J.J. 1986. Investigations of belukha whales in coastal waters of western and northern Alaska. Final report 56 US Dept of Commerce NOAA, OCSEAP, pp. 153–391.

Shaw, G.E. 1995. The Arctic haze phenomenon. *Bulletin of the American Meteorological Society* 76: 2403–2413.

Siebert, U., Joiris, C., Holsbeek, L., Benke, H., Failing, K., Frese, K. and Petzinger, E. 1999. Potential relation between mercury concentrations and necropsy findings in cetaceans from German waters of the North and Baltic Seas. *Marine Pollution Bulletin* 38(4): 285–295.

Simmonds, M.P., Johnston, P.A., French, M.C., Reeve, R. and Hutchinson, J.D. 1994. Organochlorines and mercury in pilot whale blubber consumed by Faroe islanders. *The Science of the Total Environment* 149: 97–111.

Skaare, J.U., Degre, E., Aspholm, P.E. and Ugland, K.I. 1994. Mercury and selenium in Arctic and coastal seals off the coast of Norway. *Environmental Pollution* 85: 153–160.

Smith, T.G. and Armstrong, F.A.J. 1975. Mercury in seals, terrestrial carnivores, and principal food items of the Inuit, from Holman, NWT. *Journal of Fisheries Research Board of Canada* 32: 795–801.

Smith, T.G. and Armstrong, F.A.J. 1978. Mercury and selenium in ringed and bearded seal tissues from Arctic Canada. *Arctic* 31(2): 75–84.

Sonne-Hansen, C., Dietz, R., Leifsson, P.S., Hyldstrup, L. and Riget, F.F. 2000. Cadmium toxicity to ringed seals (*Phoca hispida*). An epidemiologic study of possible cadmium induced nephropathy and osteodysatrophy in ringed seals from Qaanaaq in Northwest Greenland. National Environmental Research Institute, Denmark. NERI Technical Report No. 307.

Suda, I. and Takahashi, H. 1990. Effect of the reticuloendothelial system blockade on the biotransformation of methyl mercury in the rat. *Bulletin of Environmental Contamination and Toxicology* 44: 609–615.

Suda, I. and Takahashi, H. 1992. Degradation of methyl and ethyl mercury into inorganic mercury by other reactive species besides hydroxyl radical. *Archives of Toxicology* 66: 34–39.

Suda, I., Totoki, S. and Takahashi, H. 1991. Degradation of methyl and ethyl mercury into inorganic mercury by oxygen free radical producing systems: involvement of hydroxyl radical. *Archives of Toxicology* 65: 129–134.

Tarpley, R.J., Wade, T.L. and Haubold, E.M. 1995. Toxicological studies in tissues of the beluga whale *Delphinapterus leucas* along northern Alaska with an emphasis on public health implications of subsistence utilization. Final Report to the Alaska Beluga Whale Committee, North Slope Borough Department of Wildlife Management, Barrow, AK.

Taylor, D.L., Schliebe, S. and Metsker, H. 1989. Contaminants in blubber, liver and kidney tissue of Pacific walruses. *Marine Pollution Bulletin* 20: 465–468.

Tiffany-Castiglioni, E., Zmudzki, J., Wu, J. and Bratton, G. 1987. Effects of lead treatment on intracellular iron and copper concentrations in cultures astroglia. *Metabolic Brain Disease* 2: 61–79.

Tillander, M., Miettinen, J.K. and Koivisto, I. 1972. Excretion rate of methyl mercury in the seal (*Pusa hispida*). In: M. Ruivo (ed.). *Marine pollution and sea life.* Fishing News (Books) Ltd, Surrey, England, pp. 303–305.

United States Environmental Protection Agency. 2001. *Water quality criteria: notice of availability of water quality criterion for the protection of human health: methylmercury.* Federal Register 66: 1344.

Vahter, M., Marafante, E. and Dencker, L. 1983. Metabolism of arsenobetaine in mice, rats, and rabbits. *The Science of the Total Environment* 30: 197–211.

Vandal, G.M., Fitzgerald, W.F., Boutron, C.F. and Candelone, J.P. 1993. Variations in mercury deposition to Antarctic over the past 34,000 years. *Nature* 362: 621–623.

Wagemann, R. 1989. Comparison of heavy metals in two groups of ringed seals (*Phoca hispida*) from the Canadian Arctic. *Canadian Journal of Fisheries and Aquatic Science* 46: 1558–1563.

Wagemann, R. and Hobden, B. 1986. Low-molecular weight metalloproteins in tissues of the narwhal. *Comparative Biochemistry and Physiology* 84C: 325–344.

Wagemann, R., Snow, N.B., Lutz, A. and Scott, P. 1983. Heavy metals in tissue and organs of the narwhal. *Canadian Journal of Fisheries and Aquatic Science* 40: 206–214.

Wagemann, R., Stewart, R.E.A., Lockhart, W.L. and Stewart, B.E. 1988. Trace metals and methyl mercury: associations and transfer in harp seal (*Phoca groenlandica*) mothers and their pups. *Marine Mammal Science* 4: 339–355.

Wagemann, R., Stewart, R.E.A., Béland, P. and Desjardins, C. 1990. Heavy metals and selenium in tissues of beluga whales from the Canadian Arctic and St. Lawrence Estuary. In: T.G. Smith, D.J. St. Aubin, and J.R. Geraci (eds). Advances in research on the beluga whale. *Canadian Bulletin of Fisheries and Aquatic Science* 224: 191–206.

Wagemann, R., Innes, S. and Richard, P.R. 1996. Overview and regional and temporal differences of heavy metals in Arctic whales and ringed seals in the Canadian Arctic. *Science of the Total Environment* 186: 41–66.

Wagemann, R., Trebacz, E., Hunt, R. and Bolia, G. 1997. Percent methylmercury and organic mercury in tissues of marine mammals and fish using different experimental calculation methods. *Environmental Toxicology and Chemistry* 16: 1859–1866.

Warburton, J. and Seagers, D.J. 1993. Metal concentrations in liver and kidney tissues of Pacific walrus: continuation of a baseline study. USFWS Technical Report R7/MMM 93-1. Marine Mammals Management USWFS, Anchorage Alaska.

Welsh, S.D. 1979. The protective effect of Vitamin E and N,N'-diphenyl-*p*-phenylenediamine (DDPD) against methyl mercury toxicity in the rat. *Journal of Nutrition* 109: 1673–1681.

Wheatley, M.A. 1996. The importance of social and cultural effects of mercury on aboriginal peoples. *Neurotoxicology* 17: 251–256.

World Health Organization (WHO) 1987. *Environmental health criteria 58, Selenium.* World Health Organization, Geneva, Switzerland.

World Health Organization (WHO) 1992. *Cadmium. Environmental Health Criteria,* Vol. 134. World Health Organization, Geneva, Switzerland.

Woshner, V.M. 2000. Concentrations and interactions of selected elements in tissues of four marine mammal species harvested by Inuit hunters in arctic Alaska, with an intensive histologic assessment, emphasizing the beluga whale. PhD Dissertation, University of Illinois at Urbana-Champaign, Urbana, Illinois.

Woshner, V.M., O'Hara, T.M., Bratton, G.R., Suydam, R.S. and Beasley, V.R. 2001a. Concentrations and interactions of selected essential and non-essential elements in bowhead and beluga whales of arctic Alaska. *Journal of Wildlife Diseases* 37(4): 693–710.

Woshner, V.M., O'Hara, T.M., Bratton, G.R. and Beasley, V.R. 2001b. Concentrations and interactions of selected essential and non-essential elements in ringed seals and polar bears of arctic Alaska. *Journal of Wildlife Diseases* 37(4): 711–721.

Woshner, V.M., O'Hara, T.M., Eurell, J.A., Wallig, M.A., Bratton, G.R., Suydam, R.S. and Beasley, V.R. 2002. Distribution of inorganic mercury in liver and kidney of beluga and bowhead whales through autometallographic development of light microscopic tissue sections. *Toxicologic Pathology* 30(2): 209–215.

Yang, F., Chau, Y.K. and Maguire, R.J. 1998. Occurrence of butyltin compounds in beluga whales (*Delphinapterus leucas*). *Applied Organometallic Chemistry* 12(8–9): 651–656.

Zeisler, R., Demiralp, R., Koster, B.J., Becker, P.R., Burow, M., Ostapczuk, P. and Wise, S.A. 1993. Determination of inorganic constituents in marine mammal tissues. *Science of the Total Environment* 139/140: 365–386.

10 Impacts of algal toxins on marine mammals

Frances M. Van Dolah, Gregory J. Doucette, Frances M.D. Gulland, Teri L. Rowles and Gregory D. Bossart

Introduction

Diatoms and dinoflagellates are microalgae that make up the base of the marine food webs that support marine mammals. However, certain species produce potent toxins that are responsible for extensive fish kills and human illnesses. Acute morbidity or mortality events occur almost annually among marine mammals that also appear to correlate with the presence of toxic algal blooms. Four classes of algal toxin have been associated to date with marine mammal morbidity and mortality events (Figure 10.1, Table 10.1). All are neurotoxins, but are structurally diverse and act on a variety of different cellular receptors, leading to expected differences in symptomology, pathology and epidemiology in exposed animals. Exposure to algal toxins may be direct, through respiratory exposure, or indirect via food-web transfer. Thus, the potential for exposure of marine mammals to these toxin classes is dependent upon the occurrence of toxin-producing microalgae within the habitat of the mammals in question and, in the case of food-web transfer, the co-occurrence of appropriate prey species with the toxin producer. This chapter reviews our current understanding of the impacts of algal toxins on marine mammals.

Historically, the identification of an algal toxin as the causative agent in marine mammal events has been difficult due to inadequate detection methods for these toxins, the lack of hallmark symptoms in marine mammals associated with intoxication by a particular algal toxin class, and difficulty in obtaining fresh tissues from stranded animals for pathological and toxicological analyses. However, with the improvement in toxin detection methods over the past decade, evidence is mounting to support the hypothesis that marine algal toxins may play a significant role in previously unexplained episodic mass mortalities of marine mammals. The global occurrence of harmful algal blooms has expanded over the past half century, in terms of both frequency and geographic distribution (Smayda, 1990; Hallegraeff, 1993; Van Dolah, 2000), due, at least in part, to anthropogenic impacts on the coastal environment, including the effects of global warming. This suggests that the impacts of algal toxins on marine mammals may likewise increase.

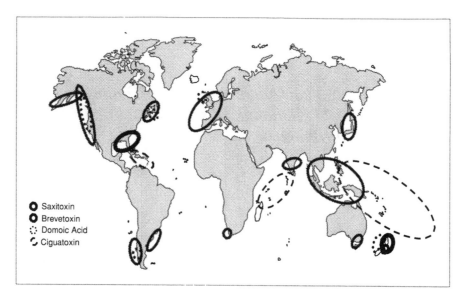

Figure 10.1 Global distributions of algal toxins associated with marine mammal morbidity or mortality events.

Table 10.1 Marine algal toxins implicated in marine mammal morbidity and mortality

Toxin	Causative organism	Known or (potential) vector for marine mammals	Pharmacologic target
Saxitoxin	*Alexandrium* spp. *Gymnodinium catenatum* *Pyrodinium bahamense* var. *compressum*	Mackerel, (zooplankton, krill??)	Voltage-dependent sodium channel, site 1
Brevetoxin	*Karenia brevis* (formerly *Gymnodinium breve*) *Chattonella* spp.	Aerosol, water, fish Fish, invertebrates	Voltage-dependent sodium channel, site 5
Ciguatoxin	*Gambierdiscus toxicus*	Reef fish	Voltage-dependent sodium channel, site 5
Domoic acid	*Pseudo-nitzschia* spp.	Anchovy, mole crab (krill, squid)	Glutamate receptors (NMDA and kainate subtypes)

NMDA, *N*-methyl-D-aspartate.

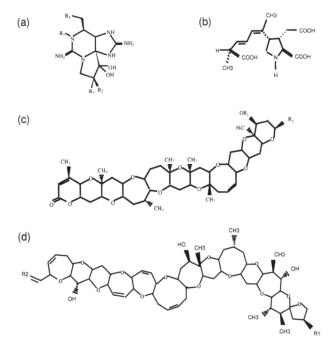

Figure 10.2 Structures of algal toxins associated with marine mammal morbidity or mortality events: (a) saxitoxin; (b) domoic acid; (c) brevetoxin; and (d) ciguatoxin.

The currently known global distribution of toxic algal blooms implicated in marine mammal mortalities is summarized in Figure 10.2, and serves as a guide to which toxin class(es) may be anticipated to be involved in marine mammal events in different regions of the world.

Saxitoxins

Saxitoxins (STXs) are a suite of water-soluble, heterocyclic guanidinium toxins (Figure 10.1a), of which there are currently more than 21 recognized congeners. These toxins are produced in varying combinations by several gonyaulacoid and gymnodinioid dinoflagellate species in three genera: *Alexandrium*, *Gymnodinium* and *Pyrodinium*. In humans, STXs are responsible for paralytic shellfish poisoning (PSP), with symptoms that include: tingling and numbness of the perioral area and extremities, loss of motor control, drowsiness, incoherence and, in extreme cases, death. Saxitoxins bind with high affinity (K_d ~2 nM) to site 1 on the voltage-dependent sodium channel, inhibiting channel opening, and thereby causing a neuronal blockade. The polarity of the STX molecule largely excludes it from traversing the blood–brain barrier in mammals; therefore, the primary site of STX action appears to be the peripheral nervous system, with respiratory paralysis being the primary cause of death. Clearance of STXs from the blood is rapid (<24 h in humans), with the primary route being via the kidney in humans (Gessner et al., 1997), rats (Hines et al., 1993) and cats (Andrinolo et al., 1999).

STXs have been implicated in the mortality of humpback whales (*Megaptera novaeangliae*), in Cape Cod Bay, Massachusetts, between November 1987 and January 1988 (Geraci et al., 1989). During late fall, humpback whales usually prey on sand lance (*Ammodytes* spp.) on Stellwagen Bank, off the Massachusetts coast (Anderson and White, 1989). However, in the fall of 1987, sand lance were largely absent from this area, and humpbacks were apparently feeding on Atlantic mackerel (*Scomber scombrus*). Fourteen whales died within 5 weeks in Cape Cod Bay and northern Nantucket Sound. Baleen whales have not previously been found to mass strand, and therefore the mortality of 14 humpback whales was highly unusual. The whales affected in this event were in robust form, with abundant blubber. Furthermore, death appeared to occur quickly, based on the observation of one whale that exhibited normal behavior 90 minutes before it was found dead. The stomachs of six of nine carcasses examined contained incompletely digested fish, indicating that the whales had been feeding not long before death. Therefore, an acutely toxic substance was suspected. In this region the STX-producing dinoflagellate, *Alexandrium tamarense*, blooms annually and therefore STXs were investigated as a potential causative agent. STX-like activity was detected in whale stomach contents, liver and kidney, as determined by mouse bioassay; however, the presence of toxin could not be corroborated by high performance liquid chromatography (HPLC) analysis. Thus the identification of STX as the causative agent remains circumstantial. None the less, STX was found by both mouse bioassay and HPLC in the viscera and livers of mackerel caught in local waters during the same time frame, with an average body burden of approximately 80 µg/100 g. Curiously, no congeners other than STX were found in the mackerel livers, although subsequent studies have reported multiple STX congeners in the livers of the same species during the 1988 *Alexandrium* bloom season (Haya et al., 1989). At the toxin concentration found in the mackerel, Geraci et al. (1989) estimated that a whale consuming 4 per cent of its body weight daily would have ingested 3.2 µg STX/kg body weight. By comparison, the lethal dose of STX in humans is estimated at 1–4 mg (c. 6–24 µg/kg) (Levin, 1992). Large mammals are generally found to be more sensitive to the effects of bioactive compounds, and thus linear extrapolation from human to humpback whale cannot be assumed for STX (Stoskopf et al., 2001). Two additional physiological adaptations may make the humpback whale highly susceptible to the toxic effects of STXs: (1) approximately 30 per cent of their body weight is blubber, into which the water-soluble STXs would not partition, thus being more highly concentrated in metabolically sensitive tissues; and (2) the diving physiology of the humpback whale concentrates blood in the heart and brain and away from those organs required for detoxification, further concentrating a neurotoxin in sensitive tissues (Geraci *et al.*, 1989). This proposed enhancement of toxicity during a dive is consistent with the extreme sensitivity of the cetacean respiratory system to anesthetics (Ridgway and McCormick, 1971; Haulena and Heath, 2001).

STXs are also suspected in the mortality of highly endangered Mediterranean monk seals (*Monachus monachus*) on the coast of Mauritania, west Africa, during May–June 1997. In this event, over 100 monk seals died, making up over 70 per cent of the local population and one-third of the world population of Mediterranean monk seals. A previously undescribed morbillivirus (WAMV-WA) was isolated from three monk seal carcasses from this event (Osterhaus *et al.*, 1997), indicating that this virus was active in the population at the time. Morbilliviruses have been identified as causative agents of mass mortality events of other marine mammal species, and therefore were identified as a likely causative agent in this event. However, unlike previous morbillivirus-associated events, these animals appeared to die quickly, with no signs of long-term illness. Dying seals exhibited lethargy, motor incoordination and paralysis in the water, symptoms similar to those associated with STX exposure. Histopathological analysis indicated that the monk seals suffered severe respiratory distress, with congestion of the lungs a consistent finding. Hernandez *et al.* (1998) identified three toxic dinoflagellate species present at moderate concentrations in the waters near the seal colony: *Alexandrium minutum*, *Gymnodinium catenatum* and *Dinophysis acuta*. Fish collected from seal feeding grounds were positive for STX-like activity as assessed by the Association of Official Analytical Chemists (AOAC) mouse bioassay and by HPLC using fluorometric detection (Hernandez *et al.*, 1998), with up to 90 µg decarbamoyl STX (dcSTX)/100 g viscera. Seal liver and a composite mussel sample from the same area were below the detection limit (35 µg/100 g tissue) in the mouse bioassay (Osterhaus *et al.*, 1998). However, seal tissue tested positive at concentrations below the detection limit of the mouse bioassay, when analyzed by HPLC (Hernandez *et al.*, 1998). STX congeners, decarbamoyl STX, neoSTX and GTX1, were identified in liver, kidney, skeletal muscle and brain. Decarbamoyl STX was present in liver at approximately 12 µg/100 g and in brain at approximately 3 µg/100 g. The presence of dcSTX and neoSTX toxins in prey and in seal livers was further confirmed by high performance liquid chromatography – mass spectrometry (HPLC–MS) (Reyero *et al.*, 1999). As in the case of the humpback whales, whether the levels of STXs observed were sufficient to have caused mortality is difficult to determine without further insight into effect levels in these animals.

Episodic mass mortalities from PSP toxins could be extremely important to the population biology of long-lived mammals such as the Mediteranean monk seal (Forcada *et al.*, 1999). The 1997 mortality event was age specific, with adults being mostly affected, and resulted in a significant change in the age structure of the population (12 per cent juveniles prior to the event versus 29 per cent after). The event reduced the breeding population to less than 77 individuals, possibly altering genetic heterogeneity of the population, and ultimately the survival of this highly endangered species.

In addition to their role in epizootic events, PSP toxins may also have a long-term influence on the population distribution of predator species, as hypothesized by Kvitek and Beitler (1991) for the distribution of sea otters (*Enhydra lutris*) in Alaskan coastal waters. Sea otters are voracious predators of bivalves, consuming 20–30 per cent of their body weight per day. In Alaskan waters, the butter clam (*Saxidomus giganteus*) accounts for the majority of prey eaten by otters (Kvitek *et al.*, 1991). The butter clam is capable of sequestering STXs at high concentrations in its siphon, where they are retained for as long as 1 year, presumably as a chemical defense against predation (Kvitek and Beitler, 1991). In the Kodiak region of Alaska, butter clams are abundant and sea otters and butter clams are co-distributed in both inside passage waters and on the outer coast. Butter clams are also abundant throughout south-eastern Alaska, but become toxic only in inside passage waters, where they remain toxic year round due to seasonal *Alexandrium* blooms. However, in south-eastern Alaska, butter clams on the outer coast are never toxic. Correspondingly, sea otter distribution in south-eastern Alaska is limited to the outer coast. To test the hypothesis that the distribution of sea otters in south-eastern Alaska is influenced by butter clam toxicity, Kvitek *et al.* (1991) carried out feeding studies on caged sea otters. Otters fed live butter clams both decreased their rate of consumption when offered toxic clams and selectively discarded the most toxic tissues, the siphons and kidneys, of clams containing even low levels (<40 µg/100 g) of toxin. These findings indicate that sea otters have the ability to detect and avoid toxic clams, and such behavioral avoidance of toxic prey is consistent with the hypothesis that sequestration of STXs by the butter clam is a chemical defense against predation that may influence sea otter distribution.

Domoic acid

Domoic acid (Figure 10.1b) is a water-soluble tricarboxylic amino acid that acts as an analog of the neurotransmitter glutamate and is a potent glutamate receptor agonist. Domoic acid (DA) is produced by several species of diatom in the genus *Pseudo-nitzschia* (Subba-Rao *et al.*, 1988; Bates *et al.*, 1989; Bates, 1998). Seven congeners to domoic acid have been identified, of which three, isodomic acids D, E and F, and the C5′ diasteriomer are found in small amounts, in addition to domoic acid, in both the diatom and in shellfish tissue (Wright *et al.*, 1990; Walter *et al.*, 1994). The first insight into domoic acid as an algal toxin resulted from a human intoxication event in Prince Edward Island, Canada in 1987, when approximately 100 people became ill after consuming contaminated mussels. A unique hallmark of this intoxication event was permanent loss of short-term memory in certain victims; domoic acid intoxication in humans has therefore been termed amnesic shellfish poisoning (ASP). The symptoms of ASP in humans include gastrointestinal effects (e.g. nausea, vomiting, diarrhea) and neurological effects, including dizziness, disorientation, lethargy, seizures and permanent loss of short-term memory. Domoic acid binds with high affinity to both kainate ($K_d \sim$ 5 nM) and alpha-amino-3-hydroxy-5-methyl-4-isoxazolepropionic acid (AMPA) ($K_d \sim$ 9 nM) subtypes of glutamate receptor (Hampson *et al.*, 1992; Hampson and Manalo, 1998). Persistent activation of the kainate glutamate receptor results in greatly elevated intracellular Ca^{2+}, through co-operative interactions with *N*-methyl-D-aspartate (NMDA) and non-NMDA glutamate receptor subtypes, followed by activation of voltage-dependent calcium channels (Xi and Ramsdell, 1996). Neurotoxicity due to domoic acid thus results from toxic levels of intracellular calcium, leading to neuronal cell death and lesions in areas of the brain where glutaminergic pathways are heavily concentrated. The CA1 and CA3 regions of the hippocampus, areas responsible for learning and memory processing, are particularly susceptible to domoic acid toxicity, and display extensive lesions in experimental animals (Peng *et al.*, 1994). In rats, the LD_{50} (intraperitoneal) for domoic acid is 4 mg/kg, whereas the oral potency is substantially lower (35–70 mg/kg) (Clayton *et al.*, 1999). However, in the 1987 outbreak, human toxicity occurred at 1–5 mg/kg, indicating greater sensitivity to oral exposure.

In 1991, the first evidence of domoic acid on the west coast of North America was obtained as a consequence of a mass mortality of pelicans (*Pelecanus occidentalis*) and cormorants (*Phalacrocorax penicillatus*) in Monterey Bay, California. Affected birds exhibited neurological symptoms similar to those reported in experimentally exposed mice, including scratching and head weaving (Work *et al.*, 1993). Domoic acid was present in the stomach contents of the affected birds, of which northern anchovies (*Engraulis mordax*) were the main component. Moreover, frustules (siliceous cell walls) of *Pseudo-nitzschia* spp. were found in the stomach contents of the fish, confirming the trophic transfer of the toxin through this vector species.

The first confirmed domoic acid poisoning of marine mammals occurred on the California coast in 1998. In May and June, 1998, 70 California sea lions (*Zalophus californianus*) were stranded along the central California coast, from San Luis Obispo to Santa Cruz (Scholin *et al.*, 2000). All animals were in good nutritional condition and displayed clinical symptoms that were predominantly neurological, including head weaving, scratching, tremors and convulsions (Gulland, 2000). Of the 70 animals that stranded, 57 died. The majority of clinically affected animals were adult females, of which 50 per cent were pregnant. Juveniles were of both sexes. No adult males were affected, which may reflect normal differences in sex distribution along the California coast. Many of the fetuses of pregnant females were found to be dead and two pups born during the episode did not survive. Domoic acid was identified in serum, urine and feces of many (but not all) of the sea lions that exhibited clinical symptoms, with the highest concentrations found in urine and feces. The highest concentration observed in serum (0.17 µg DA equiv./ml) was of similar magnitude to those in mice treated with 4 mg/kg DA, the LD_{50} dose of toxin (Peng and Ramsdell, 1996). The highest concentration measured in urine was 3.7 μg DA equiv./ml. No measurable toxin was found in other tissues. This is consistent with a rapid, renal route of toxin clearance, as observed in rodents (Iverson *et al.*, 1989). Clearance of domoic acid from rodents and primates following intravenous (i.v.) injection occurs in under 4 hours (Truelove and Iverson, 1994).

All California sea lions that died within 24 h of stranding (*n* = 27) had significant histologic lesions in the brain, consistent with domoic acid poisoning. The predominant lesion observed was neuronal necrosis that was most severe in the CA3 and CA4, followed by CA1 and CA2, regions of the hippocampus and in the dentate gyrus (Gulland, 2000). The heart was the second most frequently affected organ, with myofiber necrosis and edema present in 21 of the animals that died early in the event.

The origin of the domoic acid responsible for this mortality event appears to be a bloom of *Pseudo-nitzschia australis* that developed in Monterey Bay during May 1998, reaching its peak at about May 22 (Scholin *et al.*, 2000). The greatest concentration of cells recorded was ~200 000 cells/l. Anchovies collected from the bay on May 22 had levels of domoic acid of approximately 70 μg/g in their gut (Scholin *et al.*, 2000), and *P. australis* frustules were visible in their stomach contents. Anchovy vertebrae, and otoliths and frustules from *P. australis*, were also found in California sea lion fecal samples, which contained toxin levels exceeding 182 μg DA equiv./g (Lefebvre *et al.*, 1999).

Overall, this mortality event is likely to have affected larger numbers of sea lions than recorded, due to the extensive areas of coastline in central California that are inaccessible for monitoring and the lack of post-mortem examinations on animals that stranded dead and were decomposed. None the less, as the overall population of California sea lions is estimated at 167 000 individuals, an event causing mortality of less than 0.0004 per cent does not constitute a threat to the overall population. Moreover, the 1998 sea lion mortality event does not appear to be unique. Retrospective analysis reveals that clusters of stranded animals showing symptoms similar to those animals in the 1998 event have been reported at other times along the California coast, although no causes of death were determined. In 1978, 40 California sea lions stranded in Ventura County, displaying similar neurologic effects (Gilmartin *et al.*, 1980); in 1986, 1988 and 1992 sea lions with similar symptoms were admitted for rehabilitation to the Marine Mammal Center in Saulsalito, CA; and in 1992 18 sea lions stranded in San Luis Obispo County, also displaying the same symptoms (Beckmen *et al.*, 1995).

A similar event occurred in 2000 involving 187 California sea lions that stranded on the central coast of California, primarily in San Luis Obispo County, beginning in June and continuing through December. Analysis of serum and urine samples confirmed the involvement of domoic acid (Ch'ne and VanDolah, unpubl.), and blooms of *P. australis* were found in Monterey Bay at different times during the extended mortality event.

Approximately 1 month following the peak of strandings of California sea lions in the 1998 event, an increased number of southern sea otter (*Enhydra lutris nereis*) deaths was observed in the same region. The southern sea otter is a threatened species with a distribution limited to a 250 mile range along the California coast, from Santa Cruz to Purisima Point, that supports a current population of approximately 1900 individuals. Despite its protected status, the population of southern sea otters remains well below its optimum sustainable population (Laidre *et al.*, 2001). Analysis of serum, urine, kidneys and stomach contents collected from southern sea otters that died from unidentified causes between 1995 and 2000 ($n = 50$) revealed 28 positive for domoic acid-like activity in one or more fluids/tissues (Powell *et al.*, unpublished). Most of the domoic acid-positive animals died during the 1998 (15) or 2000 (13) *Pseudo-nitzschia* bloom events. Unlike the California sea lion, the southern sea otter feeds primarily upon benthic invertebrates, including crabs, clams and mussels. Many of the animals positive for domoic acid were found to have been feeding upon the spiny mole crab (*Blepharipoda occidentalis*) and, less frequently, another mole crab species (*Emerita analoga*). Although there are no data on domoic acid levels in *B. occidentalis*, *E. analoga* is an intertidal suspension feeder that can consume *Pseudo-nitzschia* spp., and was shown to have accumulated domoic acid during *Pseudo-nitzschia* blooms in Monterey Bay (Ferdin *et al.*, 2002). This study identifies *E. analoga* as a potential benthic vector of domoic acid transfer to marine mammals that, given its position in the swash zone, can become toxic either by exposure to an ongoing *Pseudo-nitzschia* bloom event or following the sinking out of a decaying bloom.

The gray whale (*Eschrichtius robustus*) is another benthic-feeding marine mammal that appears to have been impacted by the 2000 *P. australis* bloom in Monterey Bay. Abnormally high numbers of gray whale strandings occurred in 2000, primarily during their northward migration from their Baja Mexico calving grounds to their northern feeding grounds. A total of 350 gray whale deaths were documented along the migration route in 2000, compared with 273 in 1999 and less than 50 in previous years. Many of the gray whales stranded in emaciated condition. Among the whales that stranded in California, there was a cluster of 25 dead whales in the San Francisco Bay area (San Mateo to Marin counties) in April–May. Their strandings followed a bloom of *P. australis* in Monterey Bay, through which the whales would have migrated. Although gray whales generally feed in the pelagic zone, they can consume benthic invertebrates by disturbing the sediment and then sieving the disrupted benthos through their baleen. Of the 25 whales that stranded in this area, samples for domoic acid analysis were obtained from only 12, due to decomposition. Of these, domoic acid was confirmed in the serum (0.8 μg/ml), urine (0.5 μg/ml) and feces (0.4 μg/g) of only one young animal that stranded live on the beach near Santa Cruz, at concentrations sufficient to implicate domoic acid toxicity (Ch'ne and VanDolah, unpublished). Among the other gray whales tested, all of which stranded within

San Francisco Bay, we cannot be certain that either toxin clearance from these fluids, or degradation of DA in the decomposing tissues, did not occur prior to sample collection. None the less, the individual positive-testing animal confirms the exposure and potential impact of domoic acid on a baleen whale. It is likely that malnutrition was a predisposing factor in this case. In addition to enhancing susceptibility to toxic effects, malnutrition may alter feeding behavior, so that gray whales feed on prey not normally consumed. Although it was previously believed that gray whales do not forage significantly during their migrations from the nursery grounds to their summer feeding grounds, records as early as the 1970s in California and Washington indicated that gray whales, including juveniles, feed along the migration route. Observations of their feeding behavior indicate that domoic acid could have been acquired by feeding on kelp-associated invertebrates, krill swarms or from the benthos (R. Kvitek, personal communication). Recent studies on krill, *Euphasia pacifica* and *Thysanoessa spinifera*, suggest that they readily consumed toxic *P. australis* during the 2000 blooms in Monterey Bay, accumulating domoic acid at levels up to 44 µg/g (Bargu *et al.*, in press). Krill serve as a major prey item for baleen whales, in particular the blue whale (*Balaenoptera musculus*), which can consume up to 2 tons of krill/day. Although domoic acid has not been implicated in blue whale deaths, Bargu *et al.* (in press) estimate that, based on the domoic acid levels accumulated in the 2000 Monterey Bay *P. australis* bloom, krill could convey up to 85 g domoic acid to a blue whale per day (0.85 µg/kg. Because *P. australis* frustules have been identified in fecal samples of both gray whales and blue whales during the 2000 *P. australis* bloom season (Bargu, unpublished), it is apparent that the latter was also exposed to domoic acid, perhaps at sub-acute or even sub-effect levels.

Brevetoxins

Brevetoxins (PbTx) are a suite of ladder-like polycyclic ether toxins, of which there are nine known congeners (Figure 10.1c). Brevetoxins are best known as the Florida red tide toxin, produced by the dinoflagellate *Karenia brevis* (formerly *Gymnodinium breve*). However, PbTxs have also been shown to be produced by newly identified *K. brevis*-like species in New Zealand (Haywood *et al.*, 1996) and by raphidophytes, *Chattonella marina* and *C. cf. verruculosa* in Japan (Khan *et al.*, 1995a, b) and *C. cf. verruculosa* in the mid-Atlantic coast of the US (Bordelais *et al.*, 2002).

Brevetoxins bind with high affinity (K_d = 1–50 nM) to site 5 on the voltage-dependent sodium channel (Poli *et al.*, 1986), such that they interact with both the voltage sensor, on the extracellular side of the channel, and the inactivation gate, on the intracellular side (Trainer *et al.*, 1994; Gawley *et al.*, 1995; Baden and Adams, 2000). Binding to this site therefore alters both the voltage sensitivity of the channel, resulting in inappropriate opening of the channel under conditions in which it is normally closed, and inhibits channel inactivation, resulting in prolonged channel opening. One route of brevetoxin exposure in humans is through the consumption of toxic shellfish, and it is thus termed neurotoxic shellfish poisoning (NSP), with symptoms that include nausea, tingling and numbness of the perioral area, severe muscular aches, loss of motor control and, in particularly severe cases, seizure (Poli *et al.*, 2000). NSP has not been documented as a fatal intoxication in humans. The LD_{50} (i.p.) of PbTx3 in the mouse is 200 µg/kg, whereas the oral potency is approximately fivefold lower, around 1 mg/kg.

A second route of exposure to brevetoxins is through aerosolization of the toxin, due to lysis of fragile *K. brevis* cells by wind or wave action. In humans, exposure to aerosolized toxin results in coughing, gagging and burning of the upper respiratory tract (Baden, 1988). Fish are particularly susceptible to brevetoxins, probably because the fragile *K. brevis* cells are lysed as water passes by their gills, permitting passage of these lipophilic toxins across the gill epithelium directly into the bloodstream (Baden, 1988). Consequently, Florida red tides are generally associated with extensive fish kills.

Karenia brevis red tides initiate off the west coast of Florida essentially annually in late summer/early fall, and may persist for several months. Offshore blooms may be carried into coastal waters, where they have their most significant impacts. They may also become entrained in the Loop Current which carries them around the base of Florida into the Atlantic Ocean, where they are carried northward by the Gulf Stream (Steidinger *et al.*, 1998). Based on circumstantial evidence, PbTxs have long been suspected to be responsible for mortalities of Florida manatees (*Trichechus manatus latirostris*) and bottlenose dolphins (*Tursiops truncatus*) in the Gulf of Mexico and on the south Atlantic coast of the US. With improvements in detection methods for PbTx over the past decade, the presence of PbTx in both manatees and dolphins during red tide-associated mortality events has now been confirmed.

Layne (1965) first reported a potential link between a Florida red tide and the mortalities of seven manatees during March–April 1963 in the Fort Myers area of south-western Florida. Further circumstantial evidence of the involvement of brevetoxins in manatee deaths was documented for an epizootic that occurred between February and April 1982, in which 39 manatees died in the lower Caloosahatchee River, its estuary, and nearby bays, near Fort Meyers (O'Shea *et al.*, 1991). Timing of the mortality event coincided with the presence of a persistent *K. brevis* bloom and associated fish kills and cormorant (*Phalacrocorax auritus*) mortalities. Both male (18) and female (21) manatees, adults and calves, were affected. Behavior of the affected individuals included disorientation and inability to submerge or maintain a horizontal position, listlessness, flexing of the back, lip flaring and labored breathing. Histological lesions were found in three of five brains examined, which consisted of congestion of blood vessels in the cerebellum or choroid plexus, or hemorrhage associated with the meninges. No consistent lesions were found in other tissues. Neither microbial nor chemical contaminants were identified in tissues from these animals that could explain the mortalities. In most cases, the stomachs were full, indicating recent feeding, the contents consisting largely of seagrasses and filter feeding tunicates (*Molgula* spp.) that are found in association with seagrasses. However, tunicates tested for PbTx, using the standard mouse bioassay, did not contain measurable amounts of PbTx. Thus, although PbTx was certainly linked to this event by circumstantial evidence, conclusive identification of PbTx as the causative agent was not made. Interestingly, a resident population of bottlenose dolphins appeared to be unaffected during the 1982 epizootic of manatees (O'Shea *et al.*, 1991).

In a remarkably parallel episode, 149 manatees died during March–April 1996, along approximately an 80-mile stretch of south-western Florida, centering around the mouth of the Caloosahatchee River. At the time a significant, persistent *K. brevis* red tide was present in the same geographic area. As in the 1982 epizootic, stomach contents of the manatees consisted largely of seagrass. However, tunicates were not a consistent component. PbTx activity was found to be present in stomach contents, liver, lung and kidney from ten manatees tested, using a receptor binding assay (Baden, 1996) which has a detection limit approximately two orders of magnitude lower than the mouse bioassay used in the previous epizootic. All tissues varied widely in concentration, from background to a maximum of over 450 ppb in stomach, and maximum values of 158 ppb in lung, 138 ppb in kidney and 51 ppb in liver. The receptor assay reports the combined activity of any PbTx-like compounds present. The chemical identification of PbTx3 in liver and lung from several manatees affected in the 1996 epizootic was independently confirmed by high performance liquid chromatography – tandem mass spectrometry (HPLC-MS/MS) (Moeller *et al.*, unpublished). Based on extrapolation from the human symptomatic dose of 1.6 mg/68 kg person (*c*. 24 µg/kg), Baden (1996) estimated that a dose of less than 16 mg could be sufficient to cause symptoms in a 700-kg manatee (*c*. 22 µg/kg). Based on stomach contents, it was estimated that, by oral exposure alone, a manatee eating 7 per cent of its body weight/day could obtain this level of PbTx. However, the high concentrations found in lung tissue suggest that these animals acquired toxin by both oral and respiratory routes.

Histopathological analysis of tissues from the affected manatees showed consistent, severe congestion in nasopharyngeal tissues, bronchi, lungs, kidney and brain (Bossart *et al.*, 1998). Lungs, liver, kidney and brain displayed edema and hemorrhage, whereas the gastrointestinal (GI) tract showed no gross lesions. Immunohistochemical staining of liver, lung and lymphoid tissues with an anti-PbTx antibody showed intense positive staining of lymphocytes and macrophages in the lung, liver and lymphoid tissues (Bossart *et al.*, 1998). These results further support the involvement of PbTx in the 1996 epizootic, and furthermore, that its toxic effect in the manatees may not have been due to its acute neurotoxic effects alone, but rather may have resulted from chronic inhalation. In addition to its activity at the voltage-dependent sodium channel, indirect evidence supports the hypothesis that PbTx may also inhibit the activity of cysteine cathepsins, a class of lysosomal proteases that function in antigen presentation in B cells (Baden, 1996; Bossart *et al.*, 1998). Cathepsin inhibition would likely result in immune suppression. It is also of note that immunohistochemical staining for the interleukin-1b-converting enzyme (ICE) in these tissues showed positive staining in the same cells. This confirms that the PbTx-immunopositive cells were activated macrophages. ICE is a key mediator of release of inflammatory cytokines from activated macrophages, which can result in fatal toxic shock (Bossart *et al.*, 1998). Retrospective immunohistochemical analysis of tissues from animals affected in the 1982 epizootic also showed positive staining for PbTx in tissue macrophages. Further insight into the implications of the extensive PbTx immunostaining in tissue macrophages will require exposure studies in an appropriate experimental model.

The 1982 and the 1996 manatee epizootics occurred under similar environmental conditions. In both cases, the outflow of the Caloosahatchee River was low, due to drought and water management practices, with resulting high salinity in the bays and lower reaches of the river. Salinity in these embayments, usually less than 26 parts per thousand (ppt), was well within the 31–37 ppt salinity optimal for the maintenance of coastal blooms of *K. brevis* during both the 1982 and 1996 events. Because river outflow is regulated for water management purposes, it is possible that modification of existing water management strategies during drought years, to maintain lower salinities in the estuaries and lower river reaches, may provide some protection to manatees in the future (O'Shea *et al.*, 1991; Bossart *et al.*, 1998).

Like the manatees, mortalities of bottlenose dolphins have long been linked circumstantially to red tides. The earliest report documenting the co-occurrence of a red tide with mortalities of bottlenose dolphins was in south-western Florida in 1946–7 (Gunter *et al.*, 1948), where an extensive Florida red tide persisted for 8 months, November 1946–August 1947, from Florida Bay to St. Petersburg. At the time, the identity of *K. brevis* (then known as *Gymnodimium brevis*) as the causative organism was tenuous and the toxin was unidentified.

Brevetoxin was also proposed as a causative agent in an unprecedented mortality of over 740 bottlenose dolphins, which occurred from June 1987 to February 1988. The strandings began in New Jersey and continued southward during fall and winter. During the fall of 1987, a rare bloom of *K. brevis* was carried via the Loop Current from the Gulf of Mexico into the Atlantic, where it became entrained in the Gulf Stream and moved northward as far as North Carolina. A shoreward intrusion of warm Gulf Stream water onto the narrow continental shelf in North Carolina carried the bloom shoreward in late October and permitted its maintenance in coastal waters until March 1988, where it resulted in toxic shellfish and human NSP intoxications (Tester *et al.*, 1991). The unusual presence of a PbTx-producing bloom in mid-Atlantic coastal waters prompted the investigation of PbTx in the dolphin mortalities.

The evidence that PbTx was involved in this event remains equivocal. PbTx was identified in fish from the stomach contents of one dolphin that stranded in Florida in January 1988, as assessed by fish bioassay of a crude chloroform extract and of thin-layer chromatography (TLC)-separated fractions, and by HPLC analysis using ultraviolet (UV) detection at 215 nm. Livers of 17 beached dolphins were similarly examined by fish bioassay of extracts and by HPLC; eight livers tested positive for PbTx by these criteria at levels of 0.083–15.82 µg/g (Baden, 1989), whereas livers from 17 control bottlenose dolphins did not contain measurable PbTx. Five of the eight positive livers were from bottlenose dolphins that stranded in Virginia in August–September, prior to the *K. brevis* bloom that was documented in North Carolina beginning in late October; whereas the other three positive dolphins were from Florida and stranded in January–February 1988 (Geraci, 1989). In order for the animals from Virginia to have died from PbTx exposure, Geraci (1989) propose that a bloom of *K. brevis* may have existed in the Gulf Stream unnoticed prior to when the North Carolina bloom occurred. However, based on the calculated transport rates of the Gulf Stream system Tester *et al.* (1991) suggest that the origin of the October North Carolina bloom was more likely a bloom that occurred in Charlotte Harbor in September 1987. Transport to North Carolina would have taken 22–54 days, placing it in North Carolina by October 3–23. In the absence of more definitive analytical methods at that time, HPLC-UV analysis suggested that the toxin activity present in livers was PbTx-2. However, more recent data suggest that PbTx-2 is rapidly metabolized to PbTx-3 and to polar metabolites by both shellfish and mammals (Dickey *et al.*, 1999; Poli *et al.*, 2000). Thus it is unlikely that PbTx-2 alone would be present. This further calls into question the identity of PbTx-2 in livers of dolphins impacted in the 1987 mortality event.

Most of the stranded dolphins in the 1987 epizootic exhibited a wide range of pathologies associated with chronic physiological stress, including fibrosis of the liver and lung, adhesions of abdominal and thoracic viscera, and secondary fungal and microbial infections associated with immune suppression, which was evidenced by pathological changes in lymph nodes (Geraci, 1989). Levels of organic contaminants in the blubber were among the highest recorded for a cetacean (Geraci, 1989). However, extensive analyses for organic contaminants and infectious agents did not identify a causative agent at the time of the investigation. In a retrospective study, however, histologic examinations of lung and lymph nodes using an immunoperoxidase staining technique identified morbillivirus antigen in 42 of 79 dolphins examined (Lipscomb *et al.*, 1994). Morbillivirus infection in other mammals causes immunosuppression that commonly promotes opportunistic infections; the frequent occurrence of fungal and bacterial infections in the dolphins was consistent with an immunocompromised condition. Although this was the first report of morbillivirus in bottlenose dolphins, morbilliviruses have been suggested to be a major causative agent of pinniped and cetacean mortality events worldwide (Lipscomb *et al.*, 1994; see also Chapter 1 in this volume).

Analysis of the involvement of PbTx in the most recent dolphin epizootic associated with red tide benefited from improvements in analytical methods for PbTx that have occurred over the past decade. During August 1999–February 2000, over 120 bottlenose dolphins stranded along the Florida panhandle, a fourfold increase over historical records for this area. Two peaks of stranding coincided with a persistent *K. brevis* bloom in the same region, which also came ashore as two pulses, the first around St. Josephs Bay in September 1999 and the second around Choctawhatchee Bay in November 1999–January 2000. The strandings were evenly distributed among sexes and age classes. Most animals were in good physical condition. However, autolysis of tissues following stranding made only 24 animals suitable for analysis. Histopathological examination showed significant upper respiratory tract lesions, which consisted of lymphoplasmacytic oropharyngitis and tracheitis (Mase *et al.*, 2000). Other changes included lymphoplasmacytic intersititial pneumonia and lymphoid tissue depletion. Polymerase chain reaction (PCR) analysis for morbillivirus was negative in the animals tested. Brevetoxin-specific staining was found in the lung and spleen of two fresh-dead animals suitable for analysis, using the immunoperoxidase method of Bossart *et al.* (1998). In addition, brevetoxin was identified by receptor binding assay and confirmed by LC-MS/MS in 29 per cent of the stranded animals, with the highest concentrations found in stomach contents (undetected to 474 ng/g), followed by liver (undetected to 163 ng/g) and kidney (undetected to 4.8 ng/g) (Leighfield and Van Dolah, unpublished). Brevetoxin was not found in either the spleen or lung by these methods. This differs from the tissue distribution of PbTx observed in manatees in the 1996 epizootic, in which brevetoxin was present in lung (3.7–158 ng/g) and kidney (1.75–138 ng/g) at higher concentrations than liver (5.7–51.4 ng/g) (Baden, 1996). Because little was known about the metabolism of brevetoxins, and nothing was known about their metabolism in marine mammals, liver and stomach extracts were fractionated by HPLC, and column fractions were tested for cross-reactivity using a radioimmunoassay (RIA) (method of Poli *et al.*, 2000). PbTx-3 was predominant in stomach contents, made up primarily of fish. Since PbTx-2 is the predominant toxin produced by *K. brevis*, this implies that PbTx-2 had been metabolized to PbTx-3 in the fish, or that PbTx-3 had been selectively retained. Of PbTx-positive livers ($n = 12$) examined by RIA, ten contained only PbTx-3, while two samples had both PbTx-3 and a minor peak co-eluting with PbTx-2 (Leighfield and VanDolah, unpublished). None of the polar metabolites of PbTx that have been identified in shellfish (Dickey *et al.*, 1999; Poli *et al.*, 2000) were present.

These data collectively provide evidence of PbTx involvement in the 1999–2000 dolphin mortality event. However, neither acute nor chronic adverse effect levels have been defined for brevetoxin exposure in bottlenose dolphins, and the lethal dose is yet to be determined. In this mortality event, cell counts of *K. brevis* were as high as 4×10^6 cells/l in adjacent waters, with brevetoxin concentrations of up to 26 µg/l (Leighfield, unpublished). This cell concentration is no higher than concentrations frequently encountered farther south on the west coast of Florida, where bottlenose dolphins are essentially exposed to *K. brevis* red tides annually. The affected animals were found to be coastal animals, as opposed to animals from an offshore population, based on DNA analysis (P. Rosel, personal communication). As in both the 1982 and 1996 manatee events, the *K. brevis* blooms associated with the bottlenose dolphins mortalities on the Florida panhandle were caught in semi-enclosed embayments. Possible explanations for the unusually severe impact of this event on the bottlenose dolphins population may include the involvement of another underlying agent that resulted in increased susceptibility to PbTx, or the possibility that these animals were physiologically or behaviorally 'naïve' to *K. brevis* blooms, which are historically rare in this part of the Florida coast.

Ciguatera

Ciguatera fish poisoning (CFP) is another intoxication caused by a suite of ladder-like polyether toxins known as ciguatoxins (CTX; Figure 10.1d), of which there are currently more than 18 known congeners. CTXs are produced by the dinoflagellate, *Gambierdiscus toxicus*, which grows as an epiphyte on filamentous macroalgae associated with coral reefs and reef lagoons worldwide (for a review, see Lewis and Holmes, 1993). CTX enters the food web when these algae are grazed upon by herbivorous fishes and invertebrates, and the toxins are biotransformed and bioaccumulated in the highest trophic levels of reef fishes. Although ciguatera occurs persistently at certain locations, outbreaks are sporadic and unpredictable at others. Reef disturbance due to storm damage or human activities frequently precedes ciguateric conditions (Ruff, 1989; Kaly and Jones, 1994). The overgrowth of corals by macroalgae, due to coral bleaching, overfishing or nutrient enrichment, may also promote ciguateric conditions, by providing increased substrate for the epiphytic *G. toxicus* (Bagnis, 1987; Kohler and Kohler, 1992).

The CTXs are structurally related to the brevetoxins and compete with brevetoxin for binding to site 5 on the voltage-dependent sodium channel with a high affinity ($K_d \sim 0.04$–4 nM) (Vernoux and Lewis, 1997). However, the potency of CTX is much greater than that of PbTx: the minimum toxicity level to humans is estimated at 0.5 ng/g fish flesh (Legrand, 1998). The symptoms of ciguatera in humans include nausea, vomiting and diarrhea, which may be followed by a variety of neurological symptoms, including numbness of the perioral area and extremities, temperature dysthesia, muscle and joint aches, headache, itching, tachycardia, hypertension, blurred vision, paralysis and, in extreme cases, death. Following acute intoxication, chronic symptoms often persist for weeks, months or even years.

Evidence for the involvement of ciguatera in the morbidity or mortality of marine mammals currently remains speculative. Ciguatera has been proposed as one potential factor in the decline in populations of the highly endangered Hawaiian monk seal (*Monachus schauinslandi*). A number of other factors have also been attributed to its decline: habitat disturbance as a result of human activity, starvation due to fishery competition, mobbing behavior, predation by sharks, and a decadal-scale decrease in the productivity of the central North Pacific, resulting in decreased food sources (Craig and Ragen, 1999; Ragen and Lavigne, 1999). For the past several decades, the largest population of monk seals has been located at French Frigate Shoals, an atoll approximately 950 km north-west of Oahu, which reached a peak in numbers in the 1980s, when it is believed the population was at its carrying capacity. The population decline that has occurred over the past decade has been attributed primarily to the poor survival rates among juveniles, associated with reduced sizes of weaned pups, emaciation and slower growth rates of juveniles (Craig and Ragen, 1999). Populations elsewhere in the atoll chain are also diminished, including those at Kure Atoll and Midway Island. The relocation of young female monk seals from French Frigate Shoals to Kure Atoll, as part of the Hawaiian monk seal recovery plan, has been successful in increasing that population. Midway Island, the former location of a naval air station, is the site of the most depleted monk seal population (Gilmartin and Antonelis, 1998). This most likely reflects habitat disturbance due to human inhabitation. With the closure of the air station, a recovery plan for the Midway monk seal populations was put into place, in which young females were relocated from French Frigate Shoals to Midway. However, only 2 of 18 introduced animals survived beyond 1 year (Gilmartin and Antonelis, 1998). The reasons for this mortality rate are not clear, because prey populations were believed to be adequate, but one hypothesis is that the reefs at Midway support a high incidence of ciguatera. This suspicion was based in part on the possible involvement of ciguatera in the deaths of 50 seals on Laysan Island in 1978 (Gilmartin *et al.*, 1980) and a historical record of human outbreaks of ciguatera on Midway (Banner and Helfrich, 1964; Wilson and Jokiel, 1986). Preliminary surveys of known prey fish species from the Midway lagoon were carried out in 1986 (Wilson and Jokiel, 1986) and in 1992 (Vanderlip and Sakumoto, 1993), using the 'stick test' immunoassay for CTX (Hokama, 1985). Both surveys yielded similar results, in which more than half of all fish tested were positive or borderline-positive for the presence of CTX. Unfortunately, this assay and its revisions are known to yield false positives, making the results ambiguous (Vanderlip and Sakumoto, 1993; Dickey *et al.*, 1994). Therefore, further evaluation of the occurrence of ciguateric fish in Midway atoll, as well as its potential impact on monk seals, is recommended as a research agenda for the recovery plan for the Hawaiian monk seal population at Midway Island (Gilmartin and Antonelis, 1998).

Conclusions

Investigations into marine mammal mortality events over the past two decades have yielded considerable insight into the ecological conditions associated with exposure of marine mammals to harmful algal blooms, the food web vectors that carry toxins to marine mammals, the symptoms associated with algal toxin exposure, pathology of toxin exposure and, to some degree, toxicokinetics of specific toxin classes in marine mammals. Compelling evidence points to the involvement of three of the major algal toxin classes in marine mammal morbidity and mortality: saxitoxins in mass mortalities of humpback whales and Mediterranean monk seals; domoic acid in mortalities of California sea lions, sea otters and gray whales; and brevetoxins in mortalities of bottlenose dolphins and Florida manatees. Circumstantial evidence suggests that ciguatoxins may be involved in the poor survival of Hawaiian monk seals. However, confirmation of any of these toxins as the sole causative agent remains difficult because acute, subacute and chronic adverse effects levels have not been defined for any toxin or toxin metabolite in any species of marine mammal. Coordinated, multidisciplinary responses to unusual mortality events, in combination with establishing the background body burdens, particularly of the lipophilic toxins, will be critical to resolving these questions.

References

Anderson, D.M. and White, A.W. 1989. *Toxic dinoflagellates and marine mammal mortalities.* Proceedings of an Expert Consultation held at the Woods Hole Oceanographic Institution. Woods Hole Oceanographic Institute Tech. Rept. WHOI-89-36.

Andrinolo, D., Michea, L.F. and Lagos, N. 1999. Toxic effects, pharmacokinetics and clearance of saxitoxin, a component of paralytic shellfish poison (PSP), in cats. *Toxicon* 37: 447–464.

Baden, D.G. 1988. Public health problems of red tides. In: A.T. Tu (ed.). *Marine toxins and venoms. Handbook of Natural Toxins*, Vol. 3. Marcel Dekker, New York, pp. 259–278.

Baden, D.G. 1989. Brevetoxin analysis. In: D.M. Anderson and A.W. White (eds). *Toxic dinoflagellates and marine mammal mortalities.* Proceedings of an Expert Consultation held at the Woods Hole Oceanographic Institution. Woods Hole Oceanographic Institute Tech. Rept. WHOI-89-36, Appendix 3c, pp. 47–52.

Baden, D.G. 1996. *Analysis of brevetoxins (red tide) in manatee tissues.* Report No. MR148. Marine and Freshwater Biomedical Sciences Center. National Institute of Environmental Health Sciences, Rosenstiel School of Marine and Atmospheric Sciences, University of Miami.

Baden, D.G. and Adams, D.J. 2000. Brevetoxins: chemistry, mechanism of action and methods of detection. In: L. Botana (ed.). *Seafood and freshwater toxins.* Marcel Dekker, New York, pp. 505–532.

Bagnis, R.S. 1987. Ciguatera fish poisoning: an objective witness of coral reef stress. In: B. Salvat (ed.). *Human impacts on coral reefs: facts and recommendations.* Antenne Museum, French Polynesia, pp. 241–253.

Banner, A.H. and Helfrich, P. 1964. *The distribution of ciguatera in the tropical Pacific*. Hawaii Marine Laboratory Tech Rep. No. 3.

Bargu, S., Powell, C.L., Coale, S.L., Busman, M., Doucette, G.J. and Silver, M.W. (in press) Krill: a potential vector for domoic acid in marine food webs. *Marine Ecol. Prog. Ser.*

Bates, S.S. 1998. Bloom dynamics and physiology of domoic acid producing *Pseudo-nitzschia* species. In: D.M. Anderson, A.D. Cembella, and G.M. Hallegraeff (eds). *Physiological ecology of harmful algal blooms*. Springer-Verlag, New York, pp. 267–292.

Bates, S.S., Bird, C.J., Defrietas, A.S.W., Foxall, R., Gilgan, M., Hanic, L.A., Johnson, G.R., McCulloch, A.W., Odense, P., Pocklington, R., Quilliam, M.A., Sim, P.G., Smith, J.C., Subba Rao, D.V., Todd, E.C.D., Walter, J.A. and Wright, J.L.C. 1989. Pennate diatom *Nitzschia pungens* as the primary source of domoic acid, a toxin in shellfish from eastern Prince Edward Island, Canada. *J. Fish Aquat. Sci.* 46: 1203–1215.

Beckmen, K., Lowenstein, L.J. and Galey, F. 1995. Epizootic seizures of California sea lions. *Proceedings 11th Biennial Conference of the Marine Mammal Society, Orlando, FL.*

Bordelais, A.J., Tomas, C.R., Naar, J., Kubanek, J. and Baden, D.G. 2002. New fish-killing alga in coastal Delaware produces neurotoxins. *Environ. Health Persp.* 110: 465–470.

Bossart, G.D., Baden, D.G., Ewing, R.Y., Roberts, B. and Wright, S.D. 1998. Brevetoxicosis in manatees (*Trichechus manatus latriostris*) from the 1996 epizootic: gross, histologic, and immunohistochemical features. *Environ. Toxicol. Pathol.* 26: 276–282.

Ch'ng, M.M., Leighfield, T.A., Busman, M.A., Gulland, F., Matassa, M., Chechowitz, M., Rowles, T. and Van Dolah, F.M. submitted. *Analysis of domoic acid involvement in marine mammal mortalities on the west coast of the U.S.: February – August 2000*. Report to the National Marine Fisheries Service Working Group on Unusual Marine Mammal Mortality Events. US Dept. Commer. NOAA Tech. Memo.

Clayton, E.C., Peng, Y-G., Means, L.W. and Ramsdell, J.S. 1999. Working memory deficit induced by single, but not repeated exposures to domoic acid. *Toxicon* 37: 1025–1039.

Craig, M.P. and Ragen, T.J. 1999. Body size, survival, and decline of juvenile Hawaiian monk seals *Monachus schauinslandi*. *Marine Mammal Sci.* 15: 786–809.

Dickey, R.L., Granade, H.R. and McClure, F.D. 1994. Evaluation of a solid-phase immunobead assay for detection of ciguatera-related biotoxins in Caribbean finfish. *Mem. Queensland Museum* 34: 481–488.

Dickey, R., Jester, E., Granade, R., Mowdy, D., Montcreiff, C., Rebarchik, D., Robl, M. and Poli, M. 1999. Monitoring brevetoxins during a *Gymnodinium breve* red tide: comparison of sodium channel specific toxicity and the mouse bioassay for determination of shellfish toxins in shellfish extracts. *Natural Toxins* 7: 157–165.

Ferdin, M.E., Krivtek, R.G., Bretz, C.K., Powell, C.L., Doucette, G.J., LeFebvre, K.A., Coale, S., Silver, M.W. 2002. *Emerita analoga* (Stimpson) – possible new indicator species for the phycotoxin domoic acid in California coastal waters. *Toxicon*: in press.

Forcada, J., Hammond, P.S. and Aguilar, A. 1999. Status of the Mediterranean monk seal *Monachus monachus* in the western Sahara and implications of a mass mortality event. *Marine Ecol. Prog. Ser.* 188: 249–261.

Gawley, R.E., Rein, K.S., Jeglitsch, G., Adams, D.J., Theodorakis, E.A., Tiebes, J., Nicolau, K.C. and Baden, D.G. 1995. The relationship of brevetoxin 'length' and a-ring functionality to binding and activity in neuronal sodium channels. *Chem. Biol.* 2: 533–541.

Geraci, J.R. 1989. *Clinical investigations of the 1987–1988 mass mortality of bottlenose dolphins along the U.S. central and south Atlantic coast.* Final report. US Marine Mammal Commission, Washington, DC.

Geraci, J.R., Anderson, D.M., Timperi, R.J., St. Aubin, D.J., Early, G.A., Prescott, J.H. and Mayo, C.A. 1989. Humpback whales (*Megaptera novaeangeliae*) fatally poisoned by dinoflagellate toxin. *Can. J. Fish. Aquat. Sci.* 46: 1895–1898.

Gessner, B.D., Bell, P., Doucette, G.J., Moczydlowski, E., Poli, M.A., Van Dolah, F.M. and Hall, S. 1997. Hypertension and identification of toxin in human urine and serum following a cluster of mussel-associated paralytic shellfish poisoning outbreaks. *Toxicon* 35: 711–722.

Gilmartin, W.G. and Antonelis, G.A. 1998. *Recommended recovery actions for the Hawaiian monk seal population at Midway Island.* NOAA Tech. Mem. NMFS NOAA-NMFS-SWFSC-253.

Gilmartin, W.G., DeLong, R.L., Smith, L.A., Griner, L.A. and Dailey, M.D. 1980. *An investigation into an unusual mortality event in the Hawaiian Monk seal, Monachus schauinslandi.* Proc. Symp. Status of Resource Investigations in the Northwestern Hawaiian Islands. UNIHI-SEAGRANT Report No. MR-80-04, pp. 32–41.

Gulland, F. 2000. *Domoic acid toxicity in California sea lions* (Zalophus californianus) *stranded along the central California coast, May – October, 1998.* Report to the National Marine Fisheries Service Working Group on Unusual Marine Mammal Mortality Events. US Dept. Commer. NOAA Tech. Memo. NMFS-OPR-17.

Gunter, G., Williams, R.H., Davis, C.C. and Walton Smith, F.G. 1948. Catastrophic mass mortality of marine mammals and coincident phytoplankton bloom on the west coast of Florida, November 1946 to August 1947. *Ecological Monographs* 18: 309–324.

Hallegraeff, G.M. 1993. A review of harmful algal blooms and their apparent global increase. *Phycologia* 32: 79–99.

Hampson, D.R. and Manalo, J.L. 1998. The activation of glutamate receptors by kainic and domoic acid. *Natural Toxins* 6: 153–158.

Hampson, D.R., Huang, X., Wells, J.W., Walter, J.A. and Wright, J.L.C. 1992. Interaction of domoic acid and several derivatives of kainic acid and AMPA binding sites in rat brain. *Eur. J. Pharmacol.* 218: 1–8.

Haulena, M. and Heath, R.B. 2001. Marine mammal anaesthesia. In: L. Dierauf and F. Gulland (eds). *CRC handbook of marine mammal medicine.* CRC Press, Ohio, pp. 655–684.

Haya, K., Martin, J.L., Waiwood, B.A., Buridge, L.E., Hungerford, J. and Zitko, V. 1989. Identification of paralytic shellfish toxins in mackerel from Southwest Bay of Fundy, Canada. In: *4th Int. Conf. on Toxic Marine Phytoplankton, Lund, Sweden.*

Haywood, A., MacKenzie, L., Garthwaite, I., Towers, N. 1996. *Gymnodinium breve* 'look-alikes': three *Gymnodinium* isolates from New Zealand. In: T. Yasumoto, Y. Oshima and Y. Fukuyo (eds). *Harmful and toxic algal blooms.* International Oceanographic Committee of UNESCO, Paris, pp. 227–230.

Hernandez, M., Robinson, I., Aguilar, A., Gonzalez, L.M., Lopez-Jurado, L.F., Reyero, M.I., Cacho, E., Franco, J., Lopez-Rodas, V. and Costas, E. 1998. Did algal toxins cause monk seal mortality? *Nature* 393: 28–29.

Hines, H.B., Naseem, S.M. and Wannamacher, R.W. Jr. 1993. ³H saxitoxinol metabolism and elimination in the rat. *Toxicon* 31: 905–908.

Hokama, Y. 1985. A simplified enzyme immunoassay stick test for the detection of ciguatoxin and related polyethers from fish tissue. *Toxicon* 23: 939–946.

Iverson, F., Truelove, J., Nera, E., Tryphonas, L., Campbell, J. and Lok, E. 1989. Domoic acid poisoning and mussel-associated intoxication: preliminary investigations into the response of mice and rats to toxic mussel extract. *Food Chem. Toxicol.* 27: 377–384.

Kaly, U.L. and Jones, G.P. 1994. Test of the effect of disturbance on ciguatera in Tuvalu. *Mem. Qd. Mus.* 34: 523–532.

Khan, S., Ahmed, M.S., Arakawa, O. and Onoue, Y. 1995a. Properties of neurotoxins separated from harmful red tide organism *Chattonella marina. Israeli J. Aquaculture* 47: 137–140.

Khan, S., Haque, M., Arakawa, O. and Onoue, Y. 1995b. Toxin profiles and ichthyotoxicity of three phytoflagellates. *Bangladesh J. Fish* 15: 73–81.

Kohler, S.T. and Kohler, C.C. 1992. Dead bleached coral provides new surfaces for dinoflagellates implicated in ciguatera fish poisonings. *Environ. Biol. Fishes* 35: 413–416.

Kvitek, R.R. and Beitler, M.K. 1991. Relative insensitivity of butter clam neurons to saxitoxin: a pre-adaptation for sequestering paralytic shellfish poisoning toxins as a chemical defense. Mar. Ecol. Prog. Ser. 69: 47–54.

Kvitek, R.R., DeGange, A.R. and Beitler, M.K. 1991. Paralytic shellfish poisoning toxins mediate feeding behavior in sea otters. *Limnol. Oceanogr.* 36: 393–404.

Laidre, K., Jamieson, R.J. and DeMaster, D.P. 2001. An estimation of the carrying capacity for sea otters along the California coast. *Marine Mammal Science* 17: 294–309.

Layne, J.N. 1965. Observations on marine mammals in Florida waters. *Bull. Fl. State Mus.* 9: 131–181.

Lefebvre, K.A., Powell, C.L., Busman, M., Doucette, G.J., Moeller, P.D.R., Silver, J.B., Miller, P.E., Hughes, M.P., Singaram, S., Silver, M.W. and Tjeerdema, R.S. 1999. Detection of domoic acid in Northern anchovies and California sealions associated with an unusual mortality event. *Nat. Toxins* 7: 85–92.

Legrand, A.M. 1998. Ciguatera toxins: origin, transfer through food chain and toxicity to humans. In: B. Reguera, J. Blanco, M.L. Fernandez, and T. Wyatt (eds). *Harmful algae.* Xunta de Galacia, Santiago del Compostella, pp. 39–43.

Levin, R.E. 1992. Paralytic shellfish toxins: their origins, characteristics, and methods of detection: a review. *J. Food Biochem.* 15: 405–417.

Lewis, R.J. and Holmes, M.J. 1993. Origin and transfer of toxins involved in ciguatera. *Comp. Biochem. Physiol.* 106C: 615–628.

Lipscomb, T.P., Schulman, F.Y., Moffett, D. and Kennedy, S. 1994. Morbilliviral disease in Atlantic bottlenose dolphins (*Tursiops truncatus*) from the 1987–1988 epizootic. *J. Wildlife Diseases* 30: 567–571.

Mase, B., Jones, W., Ewing, R., Bossart, G., Van Dolah, F., Leighfield, T., Busman, M., Litz, J., Roberts, B. and Rowles, T. 2000. Epizootic of bottlenose dolphins in the Florida panhandle: 1999–2000. *Proc. Amer. Assoc. Zoo Vet. and Intern. Assoc. Aquat. Anim. Med. Joint Conference, New Orleans, LA.*

O'Shea, T.J., Rathbun, G.B., Bonde, R.K., Buergelt, C.D. and Odell, D.K. 1991. An epizootic of Florida manatees associated with a dinoflagellate bloom. *Marine Mammal Science* 7: 165–179.

268   F.M. Van Dolah et al.

Osterhaus, A., Groen, J., Niesters, H., van de Bildt, M., Martina, B., Vedder, L., Vos, J., van Egmond, H., Sidi, B.A. and Barham, M.E.O. 1997. Morbillivirus in monk seal mortality. *Nature* 388: 838–839.

Osterhaus, A., van de Bildt, M., Vedder, L., Martina, B., Niesters, H., Vos, J., van Egmond, H., Liem, D., Baumann, R., Androukaki, E., Kotomatas, S., Komnenou, A., Sidi, B.A., Jiddou, A.B. and Barham, M.E.O. 1998. Monk seal mortality: virus or toxin? *Vaccine* 16: 979–981.

Peng, Y.-G. and Ramsdell, J.S. 1996. Brain fos is a sensitive biomarker for the lowest observed neuroexcitatory effects of domoic acid in mice. *Fundam. Appl. Phrmacol.* 31: 162–168.

Peng, Y.-G., Taylor, T.B., Finch, R.E., Switzer, R.C. and Ramsdell, J.S. 1994. Neuroexcitatory and neurotoxic actions of the amnesic shellfish poison, domoic acid. *Neuroreport* 5: 981–985.

Poli, M.A., Mende, T.J. and Baden, D.G. 1986. Brevetoxins, unique activators of voltage-sensitive sodium channels, bind to specific sites in rat brain synaptosomes. *Mol. Pharmacol.* 30: 129–135.

Poli, M.A., Musser, S.M., Dickey, R.W., Eilers, P.P. and Hall, S. 2000. Neurotoxic shellfish poisoning and brevetoxin metabolites: a case study. *Toxicon* 38: 981–993.

Ragen, T.J. and Lavigne, D.M. 1999. The Hawaiian monk seal. In: J.T. Twiss and R.R. Reeves (eds). *Conservation and management of marine mammals.* Smithsonian Institute Press, Washington, DC, pp. 224–245.

Reyero, M., Cacho, E., Martinez, A., Vasquez, J., Marina, A., Fraga, S. and Franco, J.M. 1999. Evidence of saxitoxin derivatives as causative agents in the 1997 mass mortality of monk seals in the Cape Blanc Penninsula. *Natural Toxins* 7: 311–315.

Ridgway, S.H. and McCormick, J.G. 1971. Anesthesia of the porpoise. In: L.R. Soma (ed.). *Textbook of veterinary anesthesia.* Williams and Wilkinson, Baltimore, pp. 394–403.

Ruff, T.A. 1989. Ciguatera in the Pacific: a link with military activities. *Lancet* 8631: 201–205.

Scholin, C.A., Gulland, F., Doucette, G.J., Benson, S., Busman, M., Chavez, F.P., Cordaro, J., De Long, R., de Vogelaere, A., Harvey, J., Haulena, M., Lefebvre, K., Limscomb, T., Luscatoff, S., Lowenstine, LJ., Marin, R., Miller, P., McLellan, W.A., Moeller, P.D.R., Powell, C., Rowles, T., Silvagni, P., Silver, M., Spraker, T., Trainer, V. and Van Dolah, F.M. 2000. Mortality of sea lions along the central California coast linked to a toxic diatom bloom. *Nature* 403: 80–84.

Smayda, T.J. 1990. Novel and nuisance phytoplankton blooms in the sea: Evidence for a global epidemic. In: E. Graneli, B. Sundstrom, L. Edler and D.M. Anderson (eds). *Toxic marine phytoplankton.* Elsevier, New York, pp. 29–40.

Steidinger, K.A., Vargo, G.A., Tester, P.A. and Tomas, C.R. 1998. Bloom dynamics and physiology of *Gymnodinium breve* with emphasis on the Gulf of Mexico. In: D.M. Anderson, A.D. Cembella, and G.M. Hallegraeff (eds). *Physiological ecology of harmful algal blooms.* Springer-Verlag, Berlin, pp. 133–154.

Stoskopf, M.K., Willens, S. and McBain, J.F. 2001. Pharmaceuticals and formularies. In: L. Dierauf and F. Gulland (eds). *CRC handbook of marine mammal medicine.* CRC Press, Ohio, pp. 703–727.

Subba-Rao, D.V., Quilliam, M.A. and Pocklington, R. 1988. Domoic acid – a neurotoxic amino acid produced from the marine daitom *Nitzschia pungens* in culture. *Can. J. Fish. Aquat. Sci.* 45: 2076–2079.

Tester, P.A., Stumpf, R.P., Vukovich, F.M., Fowler, P.K. and Turner, J.T. 1991. An expatriate red tide bloom: transport, distribution, and persistance. *Limnol. Oceanogr.* 36: 1053–1061.

Trainer, V.L., Baden, D.G. and Caterall, W.A. 1994. Identification of peptide segments of the brevetoxin receptor site of rat brain sodium channels. *J. Biol. Chem.* 269: 19904–19909.

Truelove, J. and Iverson, F. 1994. Serum domoic acid clearance and clinical observations in the cynomolgus monkey and Sprague–Dawley rat following i.v. dose. *Bull. Environ. Contam. Toxicol.* 52: 479–486.

Vanderlip, C. and Sakumoto, D. 1993. Ciguatera assessment at Midway atoll. Student skill project report to the University of Hawaii Marine Option Program, under guidance of Gilmartin, W., NOAA/NMFS/SWFSC/Honolulu Laboratory.

Van Dolah, F.M. 2000. Marine algal toxins: origins, health effects, and their increased occurrence. *Environmental Health Perspectives* 108S1: 133–141.

Vernoux, J.-P. and Lewis, R.L. 1997. Isolation and characterization of Caribbean ciguatoxins from the horse-eye jack (*Caranx latus*). *Toxicon* 35: 889–900.

Walter, J.A., Falk, M. and Wright, J.L.C. 1994. Chemistry of the shellfish toxin domoic acid: characterization of related compounds. *Can. J. Chem.* 72: 430–436.

Wilson, M.T. and Jokiel, P.J. 1986. *Ciguatera at Midway: an assessment using the Hokama 'stick test' for ciguatoxin.* NOAA Tech. Rept. NOAA-SWFSC-86-1.

Work, T.M., Barr, B., Beale, A.M., Fritz, L., Quilliam, M.A. and Wright, J.L.C. 1993. Epidemiology of domoic acid poisoning in brown pelicans (*Pelicanus occidentalis*) and Brandt's cormorants (*Phalacrocorax penicillatus*) in California. *J. Zoo Wild. Med.* 24: 54–62.

Wright, J.L.C., Falk, M., McInnes, A.G. and Walter, J.A. 1990. Identification of isodomoic acid D and two new geometrical isomers of domoic acid in toxic mussels. *Can. J. Chem.* 68: 22–25.

Xi, D. and Ramsdell, J.S. 1996. Glutamate receptors and calcium entry mechanisms for domoic acid in hippocampal neurons. *Neuroreport* 7: 1115–1120.

11 Toxicology of sirenians

Thomas J. O'Shea

Introduction

Marine mammals of the order Sirenia differ greatly from cetaceans, pinnipeds, sea otters and polar bears. Sirenians have an evolutionary lineage distinct from these groups, and among living mammals are more closely related to the terrestrial elephants and hyraxes than to other aquatic species. Sirenians are herbivores that feed primarily on marine and freshwater plants, and are thus exposed to toxic substances through ingestion solely at lower levels of food webs. Unlike other groups of marine mammals, they occur only in warm, tropical or subtropical coastal waters and associated freshwater systems, primarily in the developing world (Marsh and Lefebvre, 1994). Their distribution is limited to areas where macroscopic aquatic plants occur, and by an intolerance to cool water dictated by low metabolic rates and high thermal conductance (Whitehead, 1977; Gallivan and Best, 1980; Irvine, 1983). The four species of extant sirenians are all protected worldwide by various laws and treaties because of concern about their population status. In many areas populations are considered declining or have been extirpated (Marsh and Lefebvre, 1994). Most of the negative human impacts on sirenian populations have been a result of hunting, incidental entanglement in artisanal fisheries, habitat change and accidental collisions with boats.

There is limited evidence for direct impacts of toxic substances on sirenian populations, with the exception of biotoxins (Bossart *et al.*, 1998, see Van Dolah *et al.*, Chapter 10 in this volume for review). Nevertheless, because sirenian populations are naturally limited in growth potential by slow reproductive rates and require high survival to maintain stable or increasing populations (O'Shea *et al.*, 1995), any added impacts of toxic substances on reproduction or mortality of these animals could have serious consequences for population recovery and would be cause for concern. For this reason, a number of investigators have examined tissues of sirenians from various parts of the world for the presence of contaminants, or have otherwise evaluated them from a toxicological standpoint. This chapter provides an overview of this previous work and its implications. Modern sirenians are classified in two families (Rathbun, 1984). The Dugongidae includes the dugong (*Dugong dugon*) and the recently extinct Steller's sea cow (*Hydrodamalis gigas*). The Trichechidae comprises three species of manatees.

Dugongs

Although the dugong is distributed across the warm marine waters of the coastal Indo-Pacific from Malagasy to eastern Australia, toxicologic information is limited to individuals from Australia and Sulawesi Island, Indonesia (Tables 11.1 and 11.2). Polychlorinated biphenyls (PCBs) have been reported in one study of dugongs from Australia at concentrations ranging from not detected to 3 ng/g in fat (Table 11.1; Haynes *et al.*, 1999a), and from unquantifiable concentrations to 209 µg/kg (lipid weight) in a second study (Vetter *et al.*, 2001). These concentrations are very low in comparison with those in marine mammals other than sirenians, and are several orders of magnitude less than maximum values found in pinnipeds or odontocete cetaceans globally (which can range to 1000 µg/g or more in exceptional cases, as summarized in O'Shea, 1999). Maximum concentrations of ΣPCBs in blubber of dugongs from north-eastern Queensland were two orders of magnitude less than in blubber of bottlenose dolphins (*Tursiops truncatus*)

Table 11.1 Summary of toxic organic contaminant residue surveys in tissues of dugongs (wet weight basis, unless otherwise noted)

Contaminant	Tissue	n	Concentration mean (range)	Reference and comments
PCBs				
ΣPCBs	Muscle	2	ND	A
ΣPCBs (17 congeners)	Liver	3	ND	B
ΣPCBs (10 congeners)	Blubber	7	NQ-209 µg/kg lipid	D
PCB 28	Fat	3	2.14 (0.89–3.00) ng/g	B
PCB 170	Fat	3	0.06 (0.03–0.07) ng/g	B
	Blubber	7	NQ–29 µg/kg lipid	D
PCBs 52, 77, 99, 101, 105, 118, 126, 138, 153, 156, 169, 180, 199, 202, 209	Fat	3	ND	B
PCB 138	Blubber	7	13–127 µg/kg lipid	D
PCB 153	Blubber	7	34–68 µg/kg lipid	D
PCB 180	Blubber	7	10–31 µg/kg lipid	D
PAHs	Fat	3	ND	B
PCDDs/PCDFs				
2,3,7,8-TCDD	Blubber	3	3.7 (2.3–6.0) ng/kg	C
1,2,3,7,8-PeCDD	Blubber	3	8.4 (5.8–10) ng/kg	C
1,2,3,4,7,8-HxCDD	Blubber	3	9.9 (7.6–11) ng/kg	C
1,2,3,6,7,8-HxCDD	Blubber	3	12.2 (8.7–14) ng/kg	C

Table 11.1 (cont'd)

Contaminant	Tissue	n	Concentration mean (range)	Reference and comments
1,2,3,7,8,9-HxCDD	Blubber	3	11.3 (7.9–13) ng/kg	C
1,2,3,4,6,7,8-HpCDD	Blubber	3	59.3 (44–72) ng/kg	C
OCDD	Blubber	3	206 (170–250) ng/kg	C
2,3,7,8-TCDF	Blubber	3	ND	C
1,2,3,7,8-PeCDF	Blubber	3	ND	C
2,3,4,7,8-PeCDF	Blubber	3	ND	C
1,2,3,4,7,8-HxCDF	Blubber	3	ND	C
1,2,3,6,7,8-HxCDF	Blubber	3	ND	C
1,2,3,7,8,9-HxCDF	Blubber	3	ND	C
2,3,4,6,7,8-HxCDF	Blubber	3	ND	C
1,2,3,4,6,7,8-HpCDF	Blubber	3	5.2 (2.5–6.7) ng/kg	C
1,2,3,4,7,8,9-HpCDF	Blubber	3	ND	C
OCDF	Blubber	3	20.3 (3.5–9.5) ng/kg	C
OC Pesticides				
p,p′-DDE	Liver	3	0.5 (ND–1.2) µg/g	B
	Fat	3	4.0 (0.9–8.0) µg/g	B
	Muscle	2	ND	A
	Blubber	7	NQ–161 µg/kg lipid	D
p,p′-DDD	Liver, Fat	3	ND	B
	Muscle	2	ND	A
	Blubber	7	NQ–5.4 µg/kg lipid	D
p,p′-DDT	Liver, Fat	3	ND	B
	Muscle	2	ND	A
	Blubber	7	NQ–6.5 µg/kg lipid	D
Dieldrin	Liver	3	0.3 (ND–0.5) µg/g	B
	Fat	3	1.6 (0.9–2.1) µg/g	B
	Muscle	2	ND	A
	Blubber	7	NQ–14	D
Aldrin	Liver, Fat	3	ND	B
HCB	Liver, Fat	3	ND	B
	Muscle	2	ND	A
	Blubber	7	NQ	D
Υ-BHC	Liver, Fat	3	ND	B
	Muscle	2	ND	A
Heptachlor	Liver, Fat	3	ND	B
Heptachlor epoxide	Liver, Fat	3	ND	B
Oxychlordane, *trans-*, *cis*-chlordane, *trans-*, *cis*-nonachlor	Blubber	7	NQ	D
Toxaphene compound	Blubber	7	NQ	D

ND, not detected, entered as 0 in computations; NQ = not quantified due to low level or interference.

References and comments: A, Sulawesi Island, Indonesia, 1975 (Miyazaki *et al.*, 1979); B, Great Barrier Reef region, Australia, 1996 (Haynes *et al.*, 1999a); C, Great Barrier Reef region, Australia, 1996 (Haynes *et al.*, 1999b); D, coastal Queensland, Australia, 1996–1999 (Vetter *et al.*, 2001).

Table 11.2 Summary of toxic element residue surveys in tissues of dugongs (μg/g dry weight basis except as noted)

Element	Tissue	n	Concentration mean (range)	Reference and comments
Ag	Muscle	25	(<0.1–<0.2)	F
	Liver	42	(0.2–38.8)	F
	Kidney	28	(<0.1–8.0)	F
	Brain	3	(0.8–1.4)	F
Al	Muscle	3	(6.4–54.9)	B
	Liver	3	(10.8–27.4)	B
As	Muscle	2	0.03 (0.015–0.05)	A
	Muscle	3	(0.44–0.605)	B
	Liver	3	(2.2–3.1)	B
	Muscle	1	<0.12	C
	Muscle	4	(0.12–0.29)	D
Cd	Muscle	2	0.08 (0.03–0.12)	A
	Muscle	3	(0.046–0.081)	B
	Muscle	1	<0.6	C
	Muscle	4	0.04–0.12	D
	Muscle	1	<0.16	E
	Muscle	25	(<0.1–<0.2)	F
	Liver	3	(3.6–5.9)	B
	Liver	2	14 (13–15)	C
	Liver	4	(18–36)	D
	Liver	2	26.5 (16.4–36.7)	E
	Liver	42	(0.1–58.8)	F
	Kidney	1	57	E
	Kidney	28	(0.2–309)	F
	Brain	3	(0.1–0.2)	F
Co	Muscle	2	0.41 (0.40–0.43)	A
	Muscle	25	(<0.2–<0.5)	F
	Muscle	3	(0.098–0.36)	B
	Liver	42	(0.5–72.0)	F
	Liver	3	(22–40)	B
	Kidney	28	(0.1–6.5)	F
	Brain	3	<0.4	F
Cr	Muscle	25	(<0.3–<0.5)	F
	Muscle	3	(1.1–1.95)	B
	Liver	42	(<0.2–<0.5)	F
	Liver	3	(0.62–0.93)	B
	Kidney	28	(<0.2–<0.3)	F
	Brain	3	<0.3	F
Cu	Muscle	2	0.86 (0.74–0.97)	A
	Muscle	3	(1.2–1.7)	B
	Muscle	25	(0.4–2.9)	F
	Muscle	1	0.95	E
	Muscle	1	<1.1	C
	Muscle	4	(0.53–1.15)	D
	Liver	42	(9.1–608)	F
	Liver	2	64 (21.9–107)	E
	Liver	2	37 (8.6–28)	C
	Liver	4	(83–1406)	D

Table 11.2 (*cont'd*)

Element	Tissue	n	Concentration mean (range)	Reference and comments
	Liver	3	(77–168)	B
	Kidney	28	(2.7–16.6)	F
	Brain	3	(7.8–11.7)	F
Fe	Muscle	2	25 (18–32)	A
	Muscle	25	(28–337)	F
	Muscle	1	77	E
	Muscle	3	(90–100)	B
	Liver	42	(778–82 363)	F
	Liver	2	(25 879–69 377)	E
	Liver	3	(2360–19 800)	B
	Kidney	28	(222–3059)	F
	Brain	3	(115–138)	F
Hg	Muscle	2	(0.002–0.005)	A
	Muscle	1	0.01	G
	Muscle	1	<0.02	C
	Muscle	4	<0.04	D
	Muscle	3	<0.005	B
	Liver	2	(0.05–0.05)	G
	Liver	2	(<0.02–0.03)	C
	Liver	4	(0.08–0.15)	D
	Liver	3	(0.069–0.1)	B
	Kidney	2	(ND–0.05)	G
MeHg	Muscle	2	(ND–0.004)	A
Mn	Muscle	2	0.063 (0.031–0.094)	A
	Muscle	25	(0.1–3.6)	F
	Muscle	1	<0.2	C
	Muscle	3	(0.6–7)	B
	Liver	3	(8.7–34.9)	B
	Liver	42	(1.3–9.2)	F
	Liver	2	(4.3–9.9)	C
	Kidney	28	(1.3–8.9)	F
	Brain	3	(1.1–2.2)	F
Ni	Muscle	2	0.9 (0.55–1.25)	A
	Muscle	25	(<0.2–<0.5)	F
	Muscle	3	(0.16–0.31)	B
	Liver	42	(<0.1–<0.3)	F
	Liver	3	(0.28–2.1)	B
	Kidney	28	(<0.1–<0.3)	F
	Muscle	1	ND	E
	Liver	2	ND	E
	Kidney	1	ND	E
	Brain	3	<0.4	F
Pb	Muscle	25	(<0.3–<0.5)	F
	Muscle	2	0.22 (0.20–0.25)	A
	Muscle	1	0.13	C
	Muscle	4	(0.08–0.16)	D
	Muscle	3	(0.076–0.34)	B
	Liver	42	(<0.1–<0.30)	F
	Liver	2	(0.17–0.38)	C

Table 11.2 (cont'd)

Element	Tissue	n	Concentration mean (range)	Reference and comments
	Liver	4	(0.19–0.38)	D
	Liver	3	(0.15–0.49)	B
	Kidney	28	(<0.1–<0.3)	F
	Muscle	1	ND	E
	Liver	2	ND	E
	Kidney	1	ND	E
	Brain	3	<0.5	F
Se	Muscle	2	0.08 (0.074–0.087)	A
	Muscle	4	(0.25–0.90)	D
	Muscle	3	(0.34–0.72)	B
	Liver	4	(4.2–6.1)	D
	Liver	3	(1.3–2.4)	B
Zn	Muscle	2	22.2 (14.8–29.5)	A
	Muscle	25	(32–113)	F
	Muscle	1	0.39	E
	Muscle	1	82	C
	Muscle	4	(5–98)	D
	Muscle	3	(31–73)	B
	Liver	2	(1378–1928)	E
	Liver	42	(219–4183)	F
	Liver	2	(1205–1380)	C
	Liver	4	(380–2736)	D
	Liver	3	(2010–3470)	B
	Kidney	28	(74.4–278)	F
	Kidney	1	164	E
	Brain	3	(35.7–36.3)	F

ND, not detected, entered in computations as 0.

References and comments: A, Sulawesi Island, Indonesia, 1975, wet weight basis (Miyazaki *et al.*, 1979); B, Great Barrier Reef region, Australia, 1996 (Haynes *et al.*, 1999a); C, McArthur River, Northern Territory, Australia, 1992 (Parry and Munksgaard 1992, 1993, cited in Haynes *et al.*, 1999a); D, Boigu Island, Torres Strait 1991–1993 (Dight and Gladstone, 1993; Gladstone, 1996, cited in Haynes *et al.*, 1999a); E, McArthur River, Northern Territory, Australia 1984 (Marsh, 1989); F, northern Queensland, Australia, 1974–1978 (Denton *et al.*, 1980); G, northern Queensland, Australia, 1977–1980, wet weight basis (Denton and Breck, 1981).

sampled in the same study (Vetter *et al.*, 2001). Interestingly, however, the congener-specific profile reported for a small number of dugongs sampled in one study (Haynes *et al.*, 1999a) was very different from that found in cetaceans or pinnipeds. In these other taxa, more highly chlorinated congeners predominate, such as the di-ortho PCBs 138, 153 and 180 (e.g. Marsili and Focardi, 1996; Tanabe *et al.*, 1997; Wolkers *et al.*, 1998). These congeners were not detected in dugongs from the northern Queensland coast, whereas PCBs 28 and 170 were the only two congeners found of the 17 analyzed; these are usually only minor contributors to the total PCB profiles in other marine mammals. These preliminary findings may imply that exposure to individual congeners may be very different at the dugong trophic level, at least in north-eastern Australia, or that dugongs have a very different metabolic capacity for PCBs (PCBs that are very recalcitrant to metabolism in cetaceans and pinnipeds were not found in dugongs). In a second study of PCBs in blubber of dugongs from north-eastern Queensland, however, Vetter *et al.* (2001) detected minor amounts of PCBs 138, 153 and 180, as well as PCB 170. In the latter study, highest concentrations were reached in females rather than males, unlike the typical situation in other marine mammals. As in other marine mammals, low concentrations of PCBs (2.6–19 µg/kg lipid) and DDT and metabolites (<8 µg/kg lipid) found in blubber of a fetus verify that these compounds can cross the placenta in dugongs (Vetter *et al.*, 2001).

In contrast to low concentrations of PCBs and organochlorine pesticides and metabolites in dugongs, a recent report (Haynes *et al.*, 1999b) reveals dioxins and dibenzofurans in blubber of dugongs from north-eastern Australia at concentrations ranging up to 390 ng/kg (Table 11.1), relatively high amounts for marine mammals. For example, a maximum concentration of 250 ng/kg OCDD in one of these dugongs exceeds all published estimates for this congener in blubber of cetaceans of which I am aware, including over 50 individuals in seven species of odontocetes from widely scattered areas of the world's oceans (Kannan *et al.*, 1989; Buckland *et al.*, 1990; Norstrom *et al.*, 1990; Jarman *et al.*, 1996; van Scheppingen *et al.*, 1996). Only congeners with at least 2,3,7,8-chlorination were found in these dugongs, PCDDs predominated over PCDFs, and OCDD was highest. Octa- and hepta-chlorinated compounds predominated over tetra- and penta-forms. Dugongs appear to be unlike other marine mammals, where the total toxic equivalents (TEQs) are attributed primarily to certain classes of PCBs rather than PCDDs/PCDFs. The sources and possible effects of these dibenzodioxins and dibenzofurans on dugongs are not fully understood. Possible sources could include higher exposure than in other marine mammals because of feeding close to sediments, relatively higher inputs of OCDDs and hepta-chlorinated dibenzodioxins (hepta-CDDs) from adjacent agricultural regions, biotransformation from precursors during hindgut digestion, or unique metabolic degradation capacities (Haynes *et al.*, 1999b). Recent analysis of PCDD and PCDF concentrations in sediments and seagrass samples from dugong habitat in the Great Barrier Reef supports the hypothesis that high exposure from contaminated sediments is the likely explanation (McLachlan *et al.*, 2001).

Organochlorine pesticides and metabolites reported from fat of dugongs include *p,p*-DDE and dieldrin, at concentrations ranging up to 8.0 and 2.1 µg/g, respectively (Table 11.1). Several other organochlorines have been reported as not detectable in fat or other tissues of most dugongs sampled thus far, including *p,p*-DDD, *p,p*-DDT, aldrin, HCB, α-benzene hexachloride (BHC), β-BHC, γ-BHC, heptachlor and heptachlor epoxide, and various other components of chlordane and toxaphene (Table 11.1; Miyazaki *et al.*, 1979; Haynes *et al.*, 1999a; Vetter *et al.*, 2001). In addition to organochlorines, Haynes *et al.* (1999a; David Haynes, 27 September 2001, personal communication) examined fat samples of three dugongs for a range of 2- to 6-ring PAHs, but found none (detection limit 2.6 ng/g). Vetter *et al.* (2001) reported the presence of polybrominated compounds in tissues of dugongs from coastal north-eastern Queensland, but concentrations appeared to be about two orders of magnitude less than in small cetaceans from the same region. Blubber of these dugongs was also reported to contain a naturally occurring persistent organochlorine, 'Q1' ($C_9H_3C_{17}N_2$), most likely a heptachlorobipyrrole (Vetter *et al.*, 2001).

Metals have been reported in various tissues of about 57 dugongs (Table 11.2). In the most extensive study, Denton *et al.* (1980) reported on cadmium, chromium, cobalt, copper, iron, lead, manganese, nickel, silver and zinc in muscle, liver and kidneys of dugongs collected from northern Queensland in 1974–1978. They also analyzed elements in seagrasses used as forage. The focus of the study was based on nutritional considerations, rather than environmental pollution. Negligible amounts of nickel, lead and chromium were found in dugong tissues. Findings included numerous correlations among elemental concentrations in various tissues, correlations with element concentrations in tissues and age, as well as differences by sex. Concentrations of cadmium in kidneys were positively correlated with age, consistent with findings in other mammals, and marine mammals in particular (O'Shea, 1999). Maximum concentrations of 309 µg/g (dry weight) cadmium were reported in a 30-year-old dugong from Mornington Island (Denton *et al.*, 1980). Other elements that occurred in exceptionally high concentrations were iron, zinc, copper, cadmium, cobalt and silver in livers. Manganese and copper decreased with age in livers and kidneys, whereas zinc increased; iron increased with age in muscle and liver, silver increased in kidneys and cobalt increased in livers; and no relationships were observed with age and silver in livers, iron and cobalt in kidneys, and zinc in muscle (Denton *et al.*, 1980). Maximum concentrations of iron, zinc, copper, cobalt and silver in livers were much higher than in other marine mammals. It was hypothesized that unusual aspects of the metal status of dugongs were related to dietary imbalances of elements in their seagrass forage. High natural levels of elements such as iron were found in seagrasses from the study area, whereas copper was low. Extensive hemosiderin pigment granulation was observed in histological examination of livers, particularly in older specimens. No pathological changes were observed. It was suggested that with high dietary iron, and relatively low dietary phosphates and copper, dugongs store excess iron in the liver as hemosiderin. These authors also suggested that depigmented areas of skin may be associated with dietary deficiencies of copper. The many intriguing relationships of elements in tissues of dugongs merits further study, particularly from a nutritional standpoint.

Analyses of mercury in dugong tissues have been limited to muscle tissues of two individuals from Sulawesi Island, Indonesia, and liver, muscle and kidney of one and liver of a second from northern Queensland (Table 11.2). Concentrations were very low. Analyses for methylmercury were only carried out in muscle of the dugongs from Indonesia. One individual had 0.004 µg/g (wet weight) methylmercury and 0.005 µg/g total mercury; the other had no detectable methylmercury and 0.002 µg/g total mercury (Miyazaki *et al.*, 1979). Arsenic has also been reported at low concentrations in tissues of a few dugongs (Table 11.2). Concentrations of aluminum, arsenic, cadmium, chromium, cobalt, copper, iron, mercury, manganese, nickel, lead, selenium and zinc were determined in liver, muscle and fat of three dugongs from the Great Barrier Reef region of Australia (Table 11.2; Haynes *et al.*, 1999a). Concentrations of elements in liver and muscle of these three dugongs were generally consistent with the findings of others and were largely unremarkable. Results of elemental analyses in fat were uniformly low (fat tissue is not usually an important depot for potentially toxic elements) and are not tabulated in this review.

There have been no verified reports of die-offs of dugongs unequivocally attributed to specific toxic substances. St. Aubin and Lounsbury (1990) summarize reports by others of 53 dugongs found dead in 1983 in the Persian Gulf. Numerous other marine vertebrates were also found dead at this time, which coincided with major oil pollution events from drilling well platforms damaged during the Iran–Iraq war. Preen (1988) reported the occurrence of at least 37 dugong carcasses on the shores of Saudi Arabia and Bahrain during the Nowruz oil spill from 1983 to 1985, but no carcasses were examined (Preen, 1988). Unlike manatees, few dugongs have been held captive. To my knowledge, none have been used to document clinically normal values for biomarkers or other toxicologic data in the scientific literature. One report (Oke, 1967) attributed the death of a captive dugong to exposure to copper sulfate added to its holding tank as an algicide, but did not provide further information to substantiate this hypothesis.

Manatees

There are three species of manatees, which inhabit large rivers, estuaries or coastal areas of the tropical and subtropical Atlantic. To my knowledge, there is no toxicological information available on West African (*Trichechus senegalensis*) or Amazonian manatees (*T. inunguis*), but a limited amount of data exist for the West Indian manatee (*T. manatus*). PCBs in the Florida manatee subspecies (*Trichechus manatus latirostris*) are very low in comparison with those in cetaceans and pinnipeds. Ames and Van Vleet (1996) detected no PCBs (Aroclor standard) in 11 blubber, 19 kidney and 15 liver samples (at 0.1 ng/g detection limits) of Florida manatees collected in 1990–3. In samples collected about a decade earlier (1977–81), O'Shea *et al.* (1984) reported PCBs (Aroclor standard) in 13 of 26 blubber samples (at 0.1 µg/g detection limits), ranging between 0.5 and 4.6 µg/g in positive samples (Table 11.3); all of the 13 with PCBs were from carcasses recovered from relatively urbanized areas of Florida. PCBs were not detected in four blubber samples collected from south-western Florida manatees in 1982 (O'Shea

Table 11.3 Summary of toxic organic contaminant residue surveys in tissues of Florida manatees (wet weight basis)

Contaminant	Tissue	n	Concentration mean (range)	Reference and comments
ΣPCBs	Blubber	26	0.7 (ND–4.6) μg/g	A
	Blubber	4	ND	B
	Blubber	1	<1.0 μg/g	C
	Blubber	11	ND	D
	Liver	15	ND	D
	Liver	1	<1.0 μg/g	C
	Kidney	19	ND	D
	Brain, Muscle, Mammary	1	<1.0 μg/g	C
OC Pesticides				
ΣDDT	Blubber	26	0.04 (ND–0.28) μg/g	A
p,p'-DDE	Blubber	4	0.01 (ND–0.25) μg/g	B
	Blubber	1	<1.0 μg/g	C
	Blubber	11	ND	D
	Brain, Liver, Muscle	1	<1.0 μg/g	C
	Mammary	1	ND	C
o,p'-DDT	Blubber	11	ND	D
	Liver	15	0.03 (ND–0.36) μg/g	D
o,p'-DDD	Liver	15	0.04 (ND–0.67) μg/g	D
	Kidney	19	0.04 (ND–0.67) μg/g	D
	Blubber	11	ND	D
Dieldrin	Blubber	26	0.04 (ND–0.36) μg/g	A
	Blubber	11	ND	D
	Blubber, Mammary, Muscle	1	ND	C
	Liver	1	<1.0 μg/g	C
	Brain	1	<1.0 μg/g	C

ND, not detected, entered into computations as 0.
Summaries of other organochlorines investigated but not detected appear in the text of this chapter.
References and comments: A, throughout Florida, 1977–81 (O'Shea *et al.*, 1984); B, south-western Florida, 1982 (O'Shea *et al.*, 1991); C, north-eastern Florida, 1974 (Forrester *et al.*, 1975); D, throughout Florida, 1990–3 (Ames and Van Vleet, 1996).

et al., 1991). Individual PCB congener determinations have not been made in Florida manatee tissues, nor have determinations of PCDDs/PCDFs or PAHs.

Organochlorine pesticides and metabolites have been analyzed in various tissues of up to 50 individual Florida manatees. Concentrations are also very low in comparison with some other marine mammals. Positive analytical results (Table 11.3) include detection of dieldrin (at 0.1 μg/g detection limits) in just 4 of 26 blubber samples collected in 1977–81, none of four collected in 1982, and in none of 11 blubber, 19 kidney and 11 liver samples collected in 1990–3. ΣDDT was reported at 0.5–4.6 μg/g in only 5 of the 26 blubber samples from 1977 to 1981, and as *p,p'*-DDE in 2 of 4 sampled in 1982 (Table 11.3). More recent samples from 1990–3 reported *o,p'*-DDD in 1 of 15 livers (at 0.67 μg/g) and 1 of 19 kidneys (at 0.67 μg/g), and *o,p'*-DDT in 2 of 15 livers (at 0.087 and 0.356 μg/g), but no DDT metabolites in 11 blubber samples (at 0.1 ng/g detection limits). No *p,p'*-isomers were detected in these recent samples, which is unusual in that in most marine mammals these isomers are much more prevalent and at higher concentrations than the *o,p'*-isomers (O'Shea, 1999). Trace amounts of HCB were found in 1 of 11 blubber and 1 of 15 liver samples, and traces of γ-BHC were found in 1 of 15 liver and 4 of 15 kidney samples from manatee carcasses found from 1990 to 1993. It is noteworthy to also summarize what was not detected in these studies. O'Shea *et al.* (1984, 1991) failed to detect (at 0.1 μg/g detection limits) the following organochlorines in any of 30 blubber samples: heptachlor epoxide, oxychlordane, *cis*-chlordane, *trans*-nonachlor, *cis*-nonachlor, endrin, HCB, mirex or toxaphene. Ames and Van Vleet (1996) failed to detect (at 0.1 ng/g detection limits) the following organochlorines in any liver, kidney, or blubber samples from up to 19 individuals: aldrin, α-BHC, β-BHC, γ-BHC, *cis*-chlordane, *trans*-chlordane, *p,p'*-DDE, *p,p'*-DDD, *p,p'*-DDT, *o,p'*-DDE, dieldrin, endrin, heptachlor, heptachlor epoxide, mirex, *trans*-nonachlor, simazine or toxaphene. Clifton and Wright (1995) examined a series of Florida manatee livers for 34 organic compounds, but found no detectable residues.

The occurrence of potentially toxic elements has been reported for tissues of Florida manatees, and for the Antillean manatee subspecies (*Trichechus manatus manatus*) from the Bay of Chetumal, Mexico (Table 11.4). Concentrations reported have not been suggestive of toxicity. Clifton and Wright (1995) also recently examined a series of Florida manatee livers for 28 elements, but detected only aluminum, cadmium and copper in both fresh and moderately decomposed samples. Elements investigated in Florida manatees collected during the late 1970s and early 1980s were cadmium, copper, iron, lead, mercury and selenium. Mercury was found at very low concentrations (maximum of 0.5 μg/g dry weight basis) in livers of 27 Florida manatees, and was not detected in muscle (O'Shea *et al.*, 1984, 1991). Some species of cetaceans and pinnipeds, in contrast, have mercury concentrations of hundreds and even thousands of μg/g in livers, without apparent harm because of detoxification mechanisms that include a positive correlation with selenium (summarized in O'Shea, 1999). Selenium concentrations in Florida manatee tissues are uniformly low (Table 11.4). Methylmercury investigations have not been reported in manatees.

Lead concentrations in livers and kidneys of Florida manatees were higher and more variable than mercury or selenium, and maxima were typical of upper estimates for other marine mammals (summarized in O'Shea, 1999). An unusually high concentration of 128 μg/g (wet weight) lead in bone of a manatee from Chetumal Bay, Mexico is amongst the highest lead

Table 11.4 Summary of surveys of West Indian manatee tissues for toxic elements (µg/g dry weight basis except where noted)

Subspecies	Tissue	Element	n	Concentration mean (range)	Reference and comments
Trichechus	Kidney	Cd	38	25.7 (ND–190)	A
manatus		Cd	8	– (1.2–22.4)	B; wet weight basis
latirostris		Pb	20	5.2 (3.3–7.1)	A
	Liver	Cu	54	175 (4.4–1200)	A
		Cu	8	42.8 (21.8–56.8)	B
		Fe	35	1920 (460–8200)	A
		Hg	19	0.01 (ND–0.2)	A
		Hg	8	0.2 (ND–0.54)	B
		Pb	19	2.7 (1.8–4.4)	A
		Pb	8	1.7 (0.44–5.1)	B
		Se	19	0.42 (ND–1.1)	A
	Muscle	Hg	27	ND	A
	Brain	Hg	2	ND–0.11	B
		Pb	2	ND–0.5	B
Trichechus	Bone	Hg	19	0.1–3.2	C
manatus	Bone	14 elements	19	See text	C
manatus	Blood	4 elements	9	See text	C

ND, not detected, entered in computations as zero.
Dashes indicate data unavailable in original report.
References and comments: A, throughout Florida, 1977–81 (O'Shea *et al.*, 1984); B, southwestern Florida, 1982 (O'Shea *et al.*, 1991); C, Chetumal Bay, Mexico, wet weight basis (Austrebertha *et al.*, 1999).

concentrations reported in marine mammals (Austrebertha *et al.*, 1999). Cadmium concentrations in kidneys showed high variability in Florida manatees, reaching as high as 109 µg/g dry weight. Much of this variability was positively correlated with body length, a surrogate for age (O'Shea *et al.*, 1984). Age-related increases in cadmium in kidneys of other marine mammals, to even higher concentrations, are well known, and cadmium toxicity has not been observed in these animals, presumably because of an ameliorative effect of metallothionein proteins (O'Shea, 1999). Evidence of lead or cadmium poisoning has not been reported in Florida manatees. However, high concentrations of lead and cadmium can be reached in tissues of aquatic plants growing in contaminated environments (Mayes *et al.*, 1977; Cooley and Martin, 1979; Welsh and Denny, 1980). Because of this, future studies of manatee carcasses should continue to determine concentrations of these elements and examine tissues for associated pathology.

One interesting finding regarding potentially toxic elements in Florida manatees surrounded exposure to copper during the late 1970s and early 1980s. Copper-based herbicides were used to control nuisance aquatic plants in some parts of the state, and were particularly heavily used in one important manatee refuge. In most mammals, copper typically decreases with age and is highest in neonates. This was also found in Florida manatees, but even after statistical adjustment for age-related effects, manatee carcasses in areas with high copper herbicide use had significantly greater concentrations of copper in livers (O'Shea et al., 1984). Concentrations as high as 1200 µg/g (dry weight basis) were observed. No evidence of an influence of dietary iron on copper concentrations was found. Investigators in this study were unable to uncover field evidence of copper poisoning in manatees, and did not claim that toxic impacts had actually occurred. However, they noted that prudent aquatic weed management should reduce the use of copper as an herbicide in areas relied on by manatees for feeding. They based this recommendation on the findings that some of these manatees had very high copper levels in livers, comparable to some other mammalian species in which toxicity had been demonstrated, and that treated aquatic plants can concentrate copper to amounts known to be toxic in feeding studies of other species of mammals (O'Shea et al., 1984).

Bones from 19 and blood from nine individual manatees from Quintana Roo, Mexico, were analyzed for up to 14 elements (Austrebertha et al., 1999). In bone, highest variability was found in copper, iron, mercury and potassium, and lowest in calcium, cadmium, chromium, lead, magnesium, nickel and zinc. Element concentrations varied according to bone type, and numerous inter-element correlations were found. In comparison with values in the literature on bone from other species of marine mammals, manatee bone from Quintana Roo was relatively high in cadmium, copper, lead, manganese and mercury.

There have been reports of multiple deaths of manatees tied to red-tide outbreaks of the dinoflagellate *Karenia brevis* on the Gulf coast of Florida (Layne, 1965; O'Shea et al., 1991; Bossart et al., 1998). Until the most recent and largest mass mortality event, in which at least 149 manatees were found dead along the south-western coast of Florida in March–April 1996, solid toxicological findings implicating the responsible agent were elusive, and conclusions about causative agents relied on circumstantial and epidemiological evidence. However, Bossart et al. (1998) were able to uncover direct evidence based on the 1996 event, and could retrospectively assign similar characteristics based on tissues saved from an earlier event in 1982. They detected the presence of brevetoxin in manatee tissues with a synaptosomal assay, and utilized immunohistochemical staining with a polyclonal primary antibody to brevetoxin to demonstrate that the presence of the agent produced by this *Karenia brevis* red-tide was correlated with histopathologic lesions. Results suggested that both acute and chronic exposure responses may occur. Inhalation of aerosolized toxins was the likely mode of exposure, a hypothesis which was supported by findings of extensive lesions in the upper respiratory tract (Bossart et al., 1998). Two pathways of toxicosis are probably involved with brevetoxin exposure: a neurotoxic mode, and a hemotoxic mode, involving hemolytic anemia and hemosiderosis in multiple tissues following more chronic exposure. There is also potential for a fatal toxic shock syndrome through the release of inflammatory mediators, consistent with immunostaining results and general lesions observed in the manatees. The immunohistochemical findings also supported theoretical explanations for the molecular and cellular modes of action of these polyether marine toxins (Bossart *et al.*, 1998).

Several studies of captive and wild manatees have reported clinical ranges for various hematological and blood chemistry parameters that can be considered biomarkers of potential use as reference values for future toxicological studies (e.g. Medway *et al.*, 1982a, b; O'Shea *et al.*, 1985; Converse *et al.*, 1994; Walsh and Bossart, 1999). Although no controlled toxicological experiments have been carried out on manatees, they are generally easier to maintain in captivity than dugongs, and dozens are held at various facilities around the world. Phillips (1964) reported that 2–3 captive manatees died at Miami Seaquarium in the 1950s after the area adjacent to the tank had been sprayed with DDT and chlordane, and suggested that airborne oil droplets containing these organochlorines had reached the water surface and were subsequently absorbed, causing death. However, no further documentation to substantiate this suggestion is available. Although not directly related to chemical toxicology, Florida manatees are known to ingest marine debris and to suffer injuries and death as a consequence (Beck and Barros, 1991). No large oil spills have occurred in Florida or elsewhere accompanied by reports of toxic impacts on manatees. It is possible that manatees could ingest tar balls or fresh petroleum, and could encounter spilled oil, resulting in cutaneous and mucous membrane exposure, but these possibilities and subsequent toxicological effects are limited to the realm of speculation (St. Aubin and Lounsbury, 1990).

Conclusions

Studies undertaken thus far suggest that anthropogenic chemicals have not reached high concentrations in sirenians in comparison with other marine mammals. This is likely due in part to their low position in marine food webs. No evidence has been uncovered that suggests direct toxic effects of environmental pollutants on sirenians. However, this does not mean that future research on this topic should be of low priority. Not only are these species of conservation concern, but they also present unique comparative models for the science of toxicology. The number of studies directly addressing the topic of toxicology in sirenians has been very limited, particularly considering the extensive distribution of these taxa. No data are available concerning aspects of toxicology in two of the four species. Much of the kind of information that has emerged from studies of other marine mammals does not exist for any sirenians, including activities of P450 enzymes or the presence of other biomarkers. The determination of specific congeners and complex metabolites of various organochlorines in sirenians also suffers from inadequate sampling. Conservation implications remain important because many of the areas where sirenians currently exist are in developing tropical countries, where pollution controls may be weak, and where some organochlorines and other chemicals banned in developed nations continue to be produced and applied. Some sirenian populations occupy coastal nearshore areas and rivers, where agricultural and industrial contamination may be very high (e.g. Mou Sue *et al.*, 1990). Because manatees and dugongs spend a considerable time in contact with sediments where certain contaminants can accumulate, their exposures may contrast strongly with those of pelagic marine mammals. Certain contaminants may be more prevalent in sediments, such as the butyltins (Tanabe, 1999) (not yet investigated in sirenians) or the dioxins (found in interesting patterns in three dugongs). Furthermore, aquatic vegetation can accumulate unusual concentrations of some potentially toxic elements (e.g. Denton *et al.*, 1980; O'Shea *et al.*, 1984) not typically encountered by other marine mammals. The patterns of contamination seen with some organochlorines in sirenians appears to be unique in marine mammals, albeit with low amounts of compounds present but at atypical relative abundances of certain metabolites and congeners in comparison with cetaceans or pinnipeds. This may imply unique metabolic degradation systems in sirenians, perhaps tied to natural selection for biochemical resistance to plant defensive compounds during their evolutionary history. The possible existence of such pathways are worthy subjects for further toxicological study. The presence of a reasonably large number of manatees in captivity could allow such studies if designed in a non-harmful way. The large number of manatees recovered dead in Florida each year in a systematic carcass recovery program (Ackerman *et al.*, 1995) could also provide considerable numbers of samples for study. The development of unique immunohistochemical techniques to study brevetoxins in manatees in response to a die-off in 1996 (Bossart *et al.*, 1998) is a prime example of the potential applications of studies of sirenians to more general advances in the science of toxicology.

References

Ackerman, B.B., Wright, S.D., Bonde, R.K., Odell, D.K. and Banowetz, D.J. 1995. Trends and patterns in mortality of manatees in Florida, 1974–1992. In *Population biology of the Florida manatee*. (O'Shea, T.J., Ackerman, B.B. and Percival, H.F., editors), pp. 223–258. National Biological Service, Information and Technology Report 1. Washington, DC.

Ames, A.L. and Van Vleet, E.S. 1996. Organochlorine residues in the Florida manatee, *Trichechus manatus latirostris*. *Marine Pollution Bulletin* 32: 374–377.

Austrebertha, R.-M., Morales-Vela, B., Cuspinera Mercadillo, M.E. and Rosiles Martines, R. 1999. Metals in bone and blood of manatees (*Trichechus manatus manatus*) from Chetumal Bay, Quintana Roo, Mexico. Unpublished manuscript, El Colegio de la Frontera Sur, Chetumal, Quintana Roo, Mexico.

Beck, C.A. and Barros, N.B. 1991. The impact of debris on the Florida manatee. *Marine Pollution Bulletin* 22: 508–510.

Bossart, G.D., Baden, D.G., Ewing, R.Y., Roberts, B. and Wright, S.D. 1998. Brevetoxicosis in manatees (*Trichechus manatus latirostris*) from the 1996 epizootic: gross, histologic, and immunohistochemical features. *Toxicologic Pathology* 26: 276–282.

Buckland, S.J., Hannah, D.J., Taucher, J.A., Slooten, E. and Dawson, S. 1990. Polychlorinated dibenzo-*p*-dioxins and dibenzofurans in New Zealand's Hector's dolphin. *Chemosphere* 20: 1035–1042.

Clifton, K.B. and Wright, S.D. 1995. Reliability of toxicological assays conducted on decomposed manatee tissues. *Eleventh Biennial Conference on the Biology of Marine Mammals, Orlando, FL*, Abstract.

Converse, L.J., Fernandes, P.J., Macwilliams, P.S. and Bossart, G.D. 1994. Hematology, serum chemistry, and morphometric reference values for Antillean manatees (*Trichechus manatus manatus*). *Journal of Zoo and Wildlife Medicine* 25: 423–431.

Cooley, T.N. and Martin, D.F. 1979. Cadmium in naturally-occurring water hyacinths. *Chemosphere* 2: 75–78.

Denton, G.R.W. and Breck, W.G. 1981. Mercury in tropical marine organisms from North Queensland. *Marine Pollution Bulletin* 12: 116–121.

Denton, G.R.W., Marsh, H., Heinsohn, G.E. and Burdon-Jones, C. 1980. The unusual metal status of the dugong (*Dugong dugon*). *Marine Biology* 57: 201–219.

Dight, I.J. and Gladstone, W. 1993. *Trace metal concentrations in sediments and selected marine biota as indicator organisms and food items in the diet of Torres Strait islanders and coastal Papuans*. Torres Strait baseline study: pilot study final report June 1993. Research Publication No. 29. Great Barrier Reef Marine Park Authority, Townsville.

Forrester, D.J., White, F.H., Woodard, J.C. and Thompson, N.P. 1975. Intussusception in a Florida manatee. *Journal of Wildlife Diseases* 11: 566–568.

Gallivan, G.J. and Best, R.C. 1980. Metabolism and respiration of the Amazonian manatee (*Trichechus inunguis*). *Physiological Zoology* 53: 245–253.

Gladstone, W. 1996. *Trace metals in sediments, indicator organisms and the traditional seafoods of the Torres Strait*. Great Barrier Reef Marine Park Authority, Townsville.

Haynes, D., Müller, J., McLachlan, M. and Carter, S. 1999a. Pollutant concentrations in Great Barrier Reef dugongs (*Dugong dugon*). Unpublished manuscript, Great Barrier Reef Marine Park Authority, Townsville, Australia.

Haynes, D., Müller, J.F. and McLachlan, M.S. 1999b. Polychlorinated dibenzo-*p*-dioxins and dibenzofurans in Great Barrier Reef (Australia) dugongs (*Dugong dugon*). *Chemosphere* 38: 255–262.

Irvine, A.B. 1983. Manatee metabolism and its influence on distribution in Florida. *Biological Conservation* 25: 315–334.

Jarman, W.W., Norstrom, R.J., Muir, D.C.G., Rosenberg, B., Simon, M. and Baird, R.W. 1996. Levels of organochlorine compounds, including PCDDs and PCDFs, in the blubber of cetaceans from the west coast of North America. *Marine Pollution Bulletin* 32: 426–436.

Kannan, N., Tanabe, S., Ono, M. and Tatsukawa, R. 1989. Critical evaluation of polychlorinated biphenyl toxicity in terrestrial and marine mammals: increasing impact of non-ortho and mono-ortho coplanar polychlorinated biphenyls from land to ocean. *Archives of Environmental Contamination and Toxicology* 18: 850–857.

Layne, J.N. 1965. Observations on marine mammals in Florida waters. *Bulletin of the Florida State Museum* 9: 131–181.

Marsh, H. 1989. Mass stranding of dugongs by a tropical cyclone in northern Australia. *Marine Mammal Science* 5: 78–84.

Marsh, H. and Lefebvre, L.W. 1994. Sirenian status and conservation efforts. *Aquatic Mammals* 20: 155–170.

Marsili, L. and Focardi, S. 1996. Organochlorine levels in subcutaneous blubber biopsies of fin whales (*Balaenoptera physalus*) and striped dolphins (*Stenella coeruleoalba*) from the Mediterranean Sea. *Environmental Pollution* 91: 1–9.

Mayes, R.A., McIntosh, A.W. and Anderson, V.L. 1977. Uptake of cadmium and lead by a rooted aquatic macrophyte (*Elodea canadensis*). *Ecology* 58: 1176–1180.

McLachlan, M.S., Haynes, D. and Müller, J.F. 2001. PCDDs in the water/sediment–seagrass–dugong (*Dugong dugon*) food chain on the Great Barrier Reef (Australia). *Environmental Pollution* 113: 129–134.

Medway, W., Bruss, M.L., Bengtson, J.L. and Black, D.J. 1982a. Blood chemistry of the West Indian manatee (*Trichechus manatus*). *Journal of Wildlife Diseases* 18: 229–234.

Medway, W., Rathbun, G.B. and Black, D.J. 1982b. Hematology of the West Indian Manatee (*Trichechus manatus*). *Veterinary Clinical Pathology* 11: 11–15.

Miyazaki, N., Itano, K., Fukushima, M., Kawai, S.I. and Honda, K. 1979. Metals and organochlorine compounds in the muscle of dugong from Sulawesi Island. *Scientific Reports of the Whales Research Institute* (Japan) 31: 125–128.

Mou Sue, L.L., Chen, D.H., Bonde, R.K. and O'Shea, T.J. 1990. Distribution and status of manatees (*Trichechus manatus*) in Panama. *Marine Mammal Science* 6: 234–241.

Norstrom, R.J., Simon, M. and Muir, D.C.G. 1990. Polychlorinated dibenzo-*p*-dioxins and dibenzofurans in marine mammals in the Canadian north. *Environmental Pollution* 66: 1–19.

Oke, V.R. 1967. A brief note on the dugong *Dugong dugon* at Cairns Oceanarium. *International Zoo Yearbook* 7: 220–221.

O'Shea, T.J. 1999. Environmental contaminants and marine mammals. In: *Biology of marine mammals* (Reynolds, J.E. III and Rommel S.A., editors), pp. 485–564. Smithsonian Institution Press, Washington, DC.

O'Shea, T.J., Moore, J.F. and Kochman, H.I. 1984. Contaminant concentrations in manatees in Florida. *Journal of Wildlife Management* 48: 741–748.

O'Shea, T.J., Rathbun, G.B., Asper, E.D. and Searles, S. 1985. Tolerance of West Indian manatees to capture and handling. *Biological Conservation* 33: 335–349.

O'Shea, T.J., Rathbun, G.B., Bonde, R.K., Buergelt, C.D. and Odell, D.K. 1991. An epizootic of Florida manatees associated with a dinoflagellate bloom. *Marine Mammal Science* 7: 165–179.

O'Shea, T.J., Ackerman, B.B. and Percival, H.F. (editors) 1995. *Population biology of the Florida manatee*. National Biological Service, Information and Technology Report 1, Washington, DC.

Parry, D.L. and Munksgaard, N.C. 1992. *Heavy metal baseline data for sediment, seawater, and biota, Bing Bong, Gulf of Carpentaria*. Unpublished report, University of Northern Territory.

Parry, D.L. and Munksgaard, N.C. 1993. *Heavy metal baseline data for sediment, seawater, and biota, Bing Bong, Gulf of Carpentaria*. Unpublished report, University of Northern Territory.

Phillips, C. 1964. *The Captive Sea*. Chilton, Philadelphia.

Preen, A. 1988. *The status and conservation of the Dugong in the Arabian region.* Final Report, Meteorology and Environmental Protection Administration, Saudia Arabia. Vols. 1–2.

Rathbun, G.B. 1984. Sirenians. In *Orders and families of recent Mammals of the world* (Anderson, S. and Jones, J.K. Jr, editors), pp. 537–547. John Wiley & Sons, New York.

St. Aubin, D.J. and Lounsbury, V. 1990. Oil effects on manatees: evaluating the risks. In *Sea mammals and oil: Confronting the risks* (Geraci, J.R. and St. Aubin, D.J., editors), pp. 241–251. Academic Press, San Diego, CA.

Tanabe, S. 1999. Butyltin contamination in marine mammals—a review. *Marine Pollution Bulletin* 39: 62–72.

Tanabe, S., Madhusree, B., Öztürk, A.A., Tatsukawa, R., Miyazaki, N., Özdamar, E., Aral, O., Samsun, O. and Öztürk, B. 1997. Isomer-specific analysis of polychlorinated biphenyls in harbour porpoise (*Phocoena phocoena*) from the Black Sea. *Marine Pollution Bulletin* 34: 712–720.

van Scheppingen, W.B., Verhoeven, A.J.I.M., Mulder, P., Addink, M.J. and Smeenk, C. 1996. Polychlorinated biphenyls, dibenzo-*p*-dioxins, and dibenzofurans in harbor porpoises (*Phocoena phocoena*) stranded on the Dutch coast between 1990 and 1993. *Archives of Environmental Contamination and Toxicology* 30: 492–502.

Vetter, W., Scholz, E., Gaus, C., Muller, J.F. and Haynes, D. 2001. Anthropogenic and natural organohalogen compounds in blubber of dolphins and dugongs (*Dugong dugon*) from northeastern Australia. *Archives of Environmental Contamination and Toxicology* 41: 221–231.

Walsh, M.T. and Bossart, G.D. 1999. Manatee medicine. In *Zoo and wild animal medicine: current therapy* (Fowler, M.E. and Miller, R.E., editors), 4th edition, pp. 507–516. W.B. Saunders Co., Philadelphia, PA.

Welsh, R.P.H. and Denny, P. 1980. The uptake of lead and copper by submerged aquatic macrophytes in two English lakes. *Journal of Ecology* 68: 443–455.

Whitehead, P.J.P. 1977. The former southern distribution of New World manatees (*Trichechus* spp.). *Biological Journal of the Linnean Society* 9: 165–189.

Wolkers, J., Burkow, I.C., Lydersen, C., Dahle, S., Monshouwer, M. and Witkamp, R.F. 1998. Congener specific PCB and polychlorinated camphene (toxaphene) levels in Svalbard ringed seals (*Phoca hispida*) in relation to sex, age, condition and cytochrome P450 enzyme activity. *Science of the Total Environment* 216: 1–11.

Part III

Cetaceans

12 Cetaceans and contaminants

Theo Colborn and Michael J. Smolen

Introduction

This chapter builds on an earlier epidemiological analysis of the literature on cetaceans that focused on the health status of individuals and populations in relation to concentrations of organochlorine contaminants in the animals (Colborn and Smolen, 1996). It includes recent literature on cetaceans and incorporates information about other chlorinated and non-chlorinated contaminants that are now becoming recognized as hazardous to living organisms. It also takes into consideration the growing knowledge about the physical and chemical condition of the cetaceans' environment and focuses on those contaminants, newly reported or increasing in background concentrations, in marine systems. When available, information from laboratory studies about the health effects of the contaminants is presented. In closing, this chapter relates this information to how contaminant-driven, phenotypic shifts among a sizable proportion of individuals in a population might lead to population destabilization.

There are still no directed, long-term studies on the status of the health of large whales and their populations relative to their exposure to contaminants. Instead, the synthesis that follows includes information about patterns of damage and the mechanisms of action of chemicals that can lead to the effects reported in other cetaceans, pinnipeds, other wildlife that thrive in marine systems, laboratory animals and humans exposed to the same chemicals. Parallel results from other species are presented as insight into possible health effects in cetaceans. By applying what is known about the concentrations of chemicals in individual animals, and what is known about the impacts of these chemicals on cell, tissue and organ function in other vertebrates, an attempt is made to provide an estimation of the potential risks to large cetaceans. This approach is based on the shared and highly conservative biochemical pathways and processes among vertebrate taxa, with an emphasis on animals' endocrine systems.

Demonstrating a relationship between an adverse health effect and an environmental contaminant in humans poses severe challenges because there are no unexposed populations for comparison and because, for ethical reasons, experiments cannot be undertaken on humans. Because so little is known about cetaceans compared with humans, any attempt to provide unequivocal evidence of a causal relationship between the health of a cetacean and a specific chemical would be impossible unless the exposure were extremely high and large-scale mortality was evident. Because this chapter addresses insidious health hazards from exposure to environmentally relevant concentrations of a number of widely distributed anthropogenic chemicals, inference and the weight of evidence are used to interpret the findings.

Background

Since the mid-1940s, large numbers of industrial chemicals and pesticides have been released in the environment, which have changed the chemical composition of the biosphere. Today, the most persistent, large-volume chemicals of the past 75 years have become ubiquitous, their final resting place, the large seas and oceans (Mackay *et al.*, 1992; Iwata *et al.*, 1993). As these man-made chemicals continue to build up in the marine environment, they pose a threat to top predators such as large odontocete and mysticete whales (Tanabe *et al.*, 1994). Even though organic chemical analytical techniques over the past two decades have become increasingly more sensitive, the high cost of this analysis has limited their use. As a result, chemists tracked only a small group of compounds in the organochlorine family that were fat soluble and accumulated in animal tissue. Recently, however, scientists have discovered that other large-volume chlorinated and non-chlorinated compounds (e.g. brominated and chlorinated fire retardants, phenolic detergents, fluorinated compounds and plastics components) that were overlooked in the past also have the potential to interfere with the development and function of all animals, including cetaceans. Like the organochlorine chemicals, these newly discovered toxicants were considered to be benign at the time they were first released, and, consequently, are now widely dispersed in the marine environment.

Persistent organochlorine contaminants

Evidence of the pervasiveness and persistence of several industrial and agricultural organochlorine chemicals led to their restricted use, or bans, in a number of developed countries in the early 1970s. This included polychlorinated biphenyls (PCBs), 1,1,1-trichloro-2,2′-bis-*p*-chlorophenyl-ethane (DDT), dieldrin, chlordane, camphenes (toxaphene), lindane (hexachlorocyclohexane or HCH), hexachlorobenzene (HCB), and mirex. As a result, concentrations of the above chemicals declined in European and North American waters, as well as in animal tissues, between the late 1970s and mid-1980s (Addison and Zinck, 1986; Gregor *et al.*, 1995). Since the late 1980s, concentrations have remained relatively constant in these regions, with the exception of DDT, which appears to be declining once again in some regions (Aono *et al.*,

1998). However, in other regions of the world, and especially in marine systems, PCBs appear to be increasing (Gregor *et al.*, 1995). For example, in an extensive survey of organochlorine contaminants in minke whales from the Antarctic (*Balaenoptera bonaerensis*) and the North Pacific (*Balaenoptera acutorostrata*), Aono *et al.* (1998) found that PCB and HCB concentrations increased significantly in the Antarctic whales sampled during 1984–5 and 1992–3. There was little change in the concentrations of the other organochlorine chemicals since 1985, which is similar to the pattern in the developed regions of the northern hemisphere. PCBs also increased in North Pacific whale samples between 1987 and 1994 (Aono *et al.*, 1998). Concentrations of organochlorine chemicals in tissues of the Antarctic minke whales were lower than in the North Pacific minke whales, perhaps because the Antarctic whales feed lower in the food web (krill) than the northern whales. Also, the animals in the 1990s survey were older and therefore exposed longer, which may add bias to the results. Aono *et al.* (1998) suggest that their results may reflect the continuous input of PCBs into the system and that perhaps some equilibrium of PCBs is taking place in the North Pacific.

A number of the organochlorine chemicals found in cetacean tissues have the potential to interfere with an animal's ability to develop and behave normally, to reproduce, or to cope with stress and infectious diseases (Colborn *et al.*, 1993). Today, no young whale is born without some of these chemicals in its body, and in some cases, at relatively high concentrations. The chemicals accumulate in the mother over her lifetime and are transferred to the offspring from her blood during gestation and from her milk during lactation (Béland, 1996a). For example, a female beluga (*Delphinapterus leucas*) in the St. Lawrence River estuary delivers 10 ppm PCBs to her calf in her milk fat (Béland, 1996a). This is ten times higher than the concentration in the breast milk fat of human mothers living at lower latitudes in the industrialized world (Jensen and Slorach, 1991). It is also ten times higher than the concentration at which neurodevelopmental deficits are measurable in human infants (Jacobson and Jacobson, 1996). It is acknowledged here that analytical protocols for determining total PCB concentrations in an individual can vary among and within laboratories over time. Tissue selection varies as well, making comparisons difficult at the individual level.

Unequivocal evidence from humans and laboratory animals reveals that the unborn and nursing offspring are far more sensitive to chemical perturbation than their adult counterparts (Bern, 1992). Prenatal or lactational exposure to chemicals that interfere with the natural hormones that control development can affect the offspring's ability to reach its fullest potential and impair its reproductive success (McLachlan *et al.*, 1981). In most cases, such prenatal effects are permanent and may not be recognizable until long after birth (Gray, 1992), if they are recognized at all. Such shifts in functionality imposed indirectly by synthetic chemicals from one generation to the next make it difficult to determine with any accuracy the health status of an individual animal or its population.

In light of the growing knowledge about these transgenerational effects, concern has broadened beyond acute toxicity and gross and obvious birth defects when determining the hazards of exposure to toxic chemicals. It now includes the invisible, but equally devastating, damage to an individual's immune, endocrine, nervous and reproductive systems, and the developing brain that undermine the individual's ability to function at its fullest potential and to reproduce. For example, cetaceans have evolved with unique auditory and visual faculties that allow them to live in the ocean environment. They need the fullest complement of their special sensory and behavioral systems to insure their reproductive success and survival. Biologists may never be able to evaluate hearing integrity or any of the other vital functions in whales that assure their survival. However, they should not dismiss concerns that the performance of a whale's vital systems has been altered. Consequently, biologists cannot assume that a population of whales is stable when they observe breeding-age adult whales accompanied by what appear to be healthy young. Only continued monitoring of survival of individual whales and their offspring and long-term population recruitment studies will provide the necessary assessment to determine whether the offspring have the functional capacity to survive and reproduce (Rolland *et al.*, 1995, 1997; Colborn *et al.*, 1998).

Health status of large cetaceans

Not much is reported in the scientific literature about the health status of individual larger toothed whales, and less about baleen whales. This is also the case for the amount of data on contamination in the larger whales. The whales' pelagic environment, secretive lifestyle and migratory behavior limit human observation and contribute to this lack of information.

Contaminants in large cetaceans

Because the concentration of a synthetic chemical in cetacean tissue alone provides little insight about its toxicity, scientists must rely on studies involving other species of wildlife, laboratory animals and humans to estimate the toxicity of the chemical. As specific mechanisms of action of contaminants at the molecular level are described for other species, considerations about similar processes can then be applied to cetaceans. For example, a slight change or shift from normal in a biochemical process or physiological function, as a result of exposure, may not be visible or immediately appear harmful. On the other hand, it has now been demonstrated that such a change at the molecular or cellular level can set off cascades of events that eventually lead to changes in an individual's phenotype, although the individual's life expectancy may not be shortened. These phenotypic differences might include structural, physiological and behavioral shifts from the norm. Widespread, insidious changes of this nature among individuals can undermine their potential and the stability of a population, thereby increasing its risk of extirpation (Rolland *et al.*, 1997; Colborn *et al.*, 1998).

Dioxin toxicity factors and equivalents

An assay that measures a compound's ability to bind to the aryl hydrocarbon (Ah) receptor and activate the cytochrome P450 system in rat liver hepatoma cells (H4IIE) is frequently used by wildlife biologists to link quantitative chemical data in animal tissue with a measurable biochemical response (Cook *et al.*, 1997). The assay provides additional evidence of the presence of certain compounds in a tissue sample, as well as a measure of a compound's potential toxicity. Of the chemicals that bind to the Ah receptor, 2,3,7,8-TCDD (dioxin) has the greatest potency. The non-, mono-, or di-*ortho* chlorine-substituted PCB congeners and some furan congeners have planar configurations similar to dioxin that facilitate their binding to the receptor as well. Upon binding to the receptor, they elicit the same receptor-mediated changes in *in vitro* cell assays, but with less potency. Because the planar organochlorine chemical, dioxin, activates the strongest response in living cells, it has been used as a standard to establish toxicity equivalence factors (TEFs) for other chemicals that bind to the Ah receptor. Chemicals that induce the Ah enzyme have been associated with immune suppression (Ross *et al.*, 1996a), dermal disorders, birth defects and cancer (Safe, 1990). They have also been associated with neurodevelopmental and neurobehavioral deficits in wildlife and human offspring as the result of prenatal exposure (Brouwer *et al.*, 1995), interfering with production, metabolism and excretion of estrogenic and androgenic hormones (Keys *et al.*, 1985; Sanderson *et al.*, 1997), and interfering with normal sexual development and behavior (Peterson *et al.*, 1992).

There are a number of instances where application of Ah enzyme assays have been employed to determine potential adverse health effects in cetaceans (Gauthier *et al.*, 1997; Weisbrod *et al.*, 2000). For example, by multiplying standardized dioxin TEFs adopted by the World Health Organization (WHO) for 12 PCB congeners (77, 81, 105, 114, 118, 123, 126, 156, 157, 167, 169 and 189) with the amount of each of these congeners in a biological sample and summing them, the total of the activity of PCBs quantified in samples can be reported in terms of dioxin toxicity equivalents (TEQs) (van Leeuwen, 1997) (Table 12.1). Tissue samples can also be introduced into the Ah *in vitro* assay system to measure their Ah activity directly. It should be pointed out that in almost every instance when the results of this direct technique were compared with the results from the TEF approach, the summed TEQ values were higher, suggesting that some antagonistic activity takes place in the real world (Jones *et al.*, 1996). The WHO has standardized TEFs for mammals, fish and birds, and found little difference in induction activity among laboratory animals and wild mammals. Because there are not enough data to provide TEFs specifically for cetaceans, the WHO TEF values for humans and mammals were used in this estimation.

Table 12.1 Example of computation of dioxin toxicity equivalents (TEQs) for No. 3 porpoise of van Scheppingen *et al.* (1996) in Tables 12.4 and 12.5

	PCB congener	77	105	118	126	156	169	Total TEQs
A	Dioxin TEFs	0.0001	0.0001	0.0001	0.01	0.0005	0.01	
B	Conc. (ppb)	1.18	375	1250	0.30	96	0.38	
C	A × B	1.18×10^{-4}	0.0375	0.125	3×10^{-3}	0.048	3.8×10^{-3}	
	C × 1000 (ppt)	0.118	37.5	125	3	48	3.8	217.42

Computed minimum TEQs = 217.42 ppt based on measurements of 6 of 12 possible PCB congeners, using the formula Σ TEQ = Σ (TEF × Conc. PCB congener ppb × 1000).

An interim method for estimating baleen whale risk from exposure to dioxin-like PCBs

Specific PCB congener data for baleen whale tissue are rare. However, an estimation of the hazard of exposure to dioxin and dioxin-like PCBs, based on TEQs, posed to baleen whales is possible using the data from Gauthier *et al.* (1997) and Weisbrod *et al.* (2000). Gauthier *et al.* (1997) report concentrations of 19 PCB congeners for 49 individuals from four species of baleen whales: northwestern Atlantic minke whales (*Balaenoptera acurostrata*), fin whales (*Balaenoptera physalis*), blue whales (*Balaenoptera musculus*), and humpback whales (*Megaptera novaeangliae*). Of the congeners recognized by the WHO as Ah enzyme inducers, only the values for PCB 105 and 118 were reported. Under ideal circumstances, values for all of the WHO congeners with known Ah activity would be available for the baleen whales. We propose that when planar congener data are missing, an approximation of TEQs for baleen whales can be made by using congener patterns found in other cetaceans. Deriving such estimates will not provide exact values but could provide valuable insight for determining what the risk might be to baleen whales from exposure to planar PCB congeners.

van Scheppingen *et al.* (1996) report concentrations of six planar PCB congeners (PCB 77, 105, 118, 126, 156 and 169) in harbor porpoises (*Phocoena phocoena*), two of which (105 and 118) are reported in the whales by Gauthier *et al.* (1997). They also report concentrations of other PCB congeners, seven of which are also reported by Gauthier *et al.* (1997) (Table 12.2). A comparison of the proportion of the nine common congeners in each taxon reveals that their profiles are quite similar (Table 12.3). It is then possible to estimate the four other planar PCB congeners in the whales, based on their proportions in the porpoises (Table 12.4). TEQs for the baleen whales can then be inferred based on the two planar PCB congeners measured by Gauthier *et al.* (1997) or based on the approximation including the four additional planar PCB congeners in the porpoises (see TEQs estimated in Table 12.4).

A similar comparison is also possible using data available for right whales (*Balaena glacialis*) as reported by Weisbrod *et al.* (2000). In this analysis,

Table 12.2 Percent total weight of nine PCB congeners quantified in each study and common to four species of baleen whales ($n = 49$) and harbor porpoises ($n = 26$, $n = 10$)

	PCB congeners (percent total weight)								
	105	118	138	153	170	180	194	195	209
Harbor porpoise (*n*=26)	1.56	5.69	28.01	46.18	6.26	10.28	1.38	0.48	0.12
Harbor porpoise (*n*=10)	1.14	4.87	27.62	47.78	6.04	10.28	1.37	0.49	0.12
Baleen whale (*n*=49)	0.06	16.68	25.59	37.48	4.55	13.35	1.50	0.61	0.17

Table 12.3 Proportion by weight of PCB 77, 126, 156 and 169 in harbor porpoises based on the sum of the nine congeners common to those found in 49 baleen whales

	PCB congeners (proportion)			
	77	126	156	169
Harbor porpoise	0.000070	0.000025	0.003276	0.000085

Table 12.4 Estimated dioxin toxicity equivalents (TEQs) (pg/g) for six PCB planar congeners in four species of baleen whales reported in Gauthier *et al.* (1997)

	#77* [0.0001]	#126* [0.01]	#105** [0.0001]	#118** [0.0001]	#156* [0.0005]	#169* [0.01]	∑PCB** [ppm]	TEQs [total]
Minke whale no.								
9138	0.01	4.39	0	33.1	2.85	1.48	3.55	41.8
9139	0.01	3.66	0	14.3	2.38	1.23	2.96	21.6
9140	0.00	1.62	0	13.6	1.05	0.55	1.58	16.8
9141	0.01	2.24	0	9.3	1.45	0.76	1.80	13.8
9142	0.01	2.85	0	21.5	1.85	0.96	2.60	27.2
9143	0.01	4.83	0	34.0	3.14	1.63	3.46	43.6
9144	0.01	2.20	0	10.9	1.43	0.74	2.23	15.3
9145	0.02	6.29	0	49.4	4.09	2.12	4.78	61.9
9146	0.03	9.43	0	63.3	6.13	3.18	7.02	82.1
9147	0.01	3.53	0	14.1	2.30	1.19	2.93	21.1
9148	0.01	1.80	0	13.0	1.17	0.61	1.52	16.6
9149	0.00	1.34	0	8.7	0.87	0.45	1.22	11.4
9150	0.01	3.09	0	19.0	2.01	1.04	2.27	25.1
9152	0.01	2.36	0	19.7	1.53	0.80	1.91	24.4
9155	0.01	1.82	0	8.4	1.18	0.61	1.21	12.0
9157	0.01	2.85	0	1.2	1.85	0.96	1.96	6.9
9158	0.01	3.80	0	26.5	2.47	1.28	2.50	34.1
9159	0.00	1.18	0	8.7	0.77	0.40	1.19	11.0
9162	0.00	1.34	0	5.6	0.87	0.45	1.30	8.3
9163	0.01	5.20	0	28.3	3.38	1.76	3.41	38.7
91114	0.02	8.46	0	33.1	5.50	2.86	5.25	49.9

Table 12.4 (cont'd)

	#77* [0.0001]	#126* [0.01]	#105** [0.0001]	#118** [0.0001]	#156* [0.0005]	#169* [0.01]	ΣPCB** [ppm]	TEQs [total]
Fin whale								
no.								
9107	0.00	1.32	0	8.9	0.86	0.44	1.07	11.5
9116	0.01	3.29	0	23.8	2.14	1.11	2.49	30.3
9117	0.00	0.29	0	1.1	0.19	0.10	0.23	1.7
9118	0.00	1.55	0	7.8	1.01	0.52	1.39	10.9
9119	0.01	4.45	3.3	32.4	2.90	1.50	3.43	44.6
9123	0.01	2.27	0	11.7	1.48	0.77	1.70	16.2
9131	0.01	3.01	0	21.1	1.96	1.02	2.47	27.1
9132	0.00	1.13	0	7.5	0.73	0.38	0.91	9.7
9133	0.00	1.46	0	9.9	0.95	0.49	1.09	12.8
9134	0.00	0.43	0	2.9	0.28	0.15	0.32	3.8
9135	0.04	14.71	0	102.1	9.56	4.97	10.22	131.4
9153	0.03	12.28	0	87.2	7.98	4.14	8.36	111.6
9156	0.01	2.31	0	12.2	1.50	0.78	1.63	16.8
9160	0.01	2.53	0	17.1	1.64	0.85	2.05	22.1
91113	0.01	4.38	0	25.7	2.85	1.48	2.76	34.4
Blue								
whale no.								
9114	0.02	6.33	0	48.7	4.11	2.14	4.02	61.3
91109	0.00	1.75	0	14.3	1.14	0.59	1.58	17.8
9201	0.00	1.05	0.6	6.0	0.68	0.35	0.79	8.7
9202	0.03	9.33	0	70.4	6.06	3.15	6.07	89.0
9204	0.00	1.11	0	6.9	0.72	0.38	0.85	9.1
9222	0.01	3.48	0	30.9	2.26	1.17	2.70	37.8
Humpback								
whale no.								
9113	0.01	3.30	0	24.6	2.14	1.11	3.88	31.2
9122	0.00	1.51	0	10.7	0.98	0.51	1.18	13.7
9127	0.01	1.85	0	20.1	1.20	0.63	1.40	23.8
9128	0.01	2.73	0	26.5	1.77	0.92	2.79	31.9
9130	0.02	8.20	0	58.3	5.33	2.77	5.02	74.6
9151	0.01	2.55	0	16.3	1.66	0.86	1.60	21.4
9164	0.01	2.42	0	19.7	1.57	0.82	2.71	24.5

* Estimated concentration; ** measured concentration.
Congeners 105 and 118 were measured by the authors while congeners 77, 126, 156, and 169 are estimates based on the proportion of these congeners in harbor porpoises (van Scheppingen et al., 1996). Sum of measured PCBs is reported. TEQs are computed as the sum of measured and estimated congeners. See text for description of the analysis. Brackets include the TEFs recognized by the WHO.

eight PCB congeners common to the harbor porpoise from van Scheppingen et al. (1996) were used to estimate the congeners 77, 126, 169 and 156 in right whales. These four estimated values were added to those measured in the study (Table 12.5).

The estimates of the dioxin TEQs in these North Atlantic baleen whales should be considered conservative. They are based on only six of the 12 dioxin-like PCB congeners recognized by the WHO. They do not take into

Table 12.5 Estimated total TEQs (ng TEQ/kg) in baleen whales using six PCB congeners

PCB congener:	105	118	77	126	156	169	
	Measured		*Estimated*				*TEQs*
Porpoise	27.55	94.87	0.14	4.89	31.98	16.32	175.75
Minke whale	0.00	20.75	0.01	0.35	2.29	1.17	24.56
Fin whale	0.22	24.76	0.01	0.36	2.29	1.22	28.96
Blue whale	0.10	29.53	0.01	0.38	2.46	1.26	33.74
Humpback whale	0.00	25.17	0.01	0.32	2.09	1.06	28.65
Right whale 94	6.90	23.60	0.01	0.32	2.10	1.07	34.00
Right whale 95	6.90	33.40	0.01	0.48	3.17	1.62	45.59
Right whale 96	2.90	9.00	0.00	0.12	0.76	0.39	13.16
Right whale 'Jan'	5.10	93.30	0.08	2.70	17.67	9.02	127.87

consideration the additional toxicity of any dioxin and furan congeners that might be lodged in the whales' tissues that were not quantified in the chemical analyses. The estimated TEQs for the individual baleen whales reported by Gauthier *et al.* (1997) range from 1.7 to 131.4 ng TEQ/kg with a mean of 30.88. The estimated TEQs for four groups of right whales ranged from 13.16 to 127.87 ng TEQ/kg. The unusually high measurement for the sample from the right whale 'Jan' skews the mean; consequently the median is used for an estimate of the right whale's TEQ. The median TEQ for the four samples is 39.79 ng TEQ/kg.

The estimated TEQs are ranked as right > blue > fin > humpback > minke for the five species of baleen whales (Table 12.5). The four congeners that were estimated based on proportions in harbor porpoises do not contribute appreciable TEQs. Congener 118, which was measured by the investigators, accounted for a majority of the computed TEQ value in the whales when estimated in this manner.

Effects associated with this range of exposure vary among vertebrates. Semi-confined seals fed Atlantic herring had 61.8 ng TEQ/kg (ppt) (based on Ahlborg, 1994) in their blubber after 104 days, representing what would be a low-dose cohort. In contrast, seals fed more highly contaminated Baltic herring had 208.7 ng TEQ/kg (ppt) (Ross *et al.*, 1996a). PCBs contributed 93 per cent of the TEQs in the seals, and the balance was from dioxins and furans. Similarly, three dioxin-like PCBs contributed 95 per cent of the toxicity, and the balance from dioxins and furans, in a Great Lakes study looking at a troubled population of Forster's terns (*Sterna forsteri*) (Kubiak *et al.*, 1989). In this study, reduced parental care, wasting, birth defects, reduced hatchability, and reduced fledgling success in the population were associated with the level of dioxin-like enzyme activity in the terns' tissues (Gilbertson *et al.*, 1991). However, in a North Pacific study of albatrosses, furans (30 per cent) and dioxins contributed a much larger percentage of

the TEQs (Jones *et al.*, 1996). In this study, eggs of the black-footed albatrosses (*Diomedea nigripes*) that feed only on the North Pacific Ocean had
123.4 pg/g TEQs computed using Ahlborg (1994) TEFs. The black-footed
albatrossess had higher embryo mortality and lower egg hatchability than
Laysan albatrosses (*Diomedea immutabilis*) with 48.4–51.7 TEQs in their
eggs. Not all the toxic congeners listed by Ahlborg (1994) were analyzed in
this study; thus this estimate may be low.

This discovery of the elevated contribution of furans and dioxins to the
total TEQs in the North Pacific raises concerns for the baleen whales because
of the recent discovery that 5 pg/g TCDD TEQs in lake trout eggs is a
threshold level for lake trout survival (Rolland *et al.*, 1997). In this study,
researchers developed fish-specific TEFs for estimating the hazards to early
life-stages of freshwater fish. Core drills now confirm that dioxin exceeded
this level in the Great Lakes in the 1940s and 1950s and undoubtedly contributed to the extirpation of lake trout and herring in this freshwater system.

Three North Pacific killer whales (*Orcinus orca*), a pregnant and a non-
pregnant female and a male, had 480, 380 and 300 ng/kg (ppt) total furans
in their blubber, respectively (Ono *et al.*, 1987). No dioxins were found
in these animals, similar to the findings with beluga whales. However, this
suggests that TEQs contributed by furans should always be considered when
estimating exposure of baleen whales to known developmental contaminants.
Furans were not measured in the North Atlantic baleen whales. The baleen
TEQs in this study fall in the range of human exposure (Table 12.6). The
implications for the baleen whales will be discussed in greater detail later.

Organotin compounds

Like concentrations of organochlorine compounds in the aquatic environment that responded to regulation in the early 1970s, concentrations of
organotin compounds in some regions of the industrialized world are
showing signs of decline. However, reports continue to demonstrate that the
compounds are more widely dispersed than realized. Organotin compounds
have been used since the 1960s in paints to prevent barnacle, mussel, and
tubeworm build-up on ships and stationary structures such as piers, drill-
rigs, etc. They are used as slimicides on wood products, sisal ropes, leather
and jute; as helminthides by the poultry industry; and as rodenticides. Tin
compounds are also used as stabilizers and disinfectants in the production
of plastic products, such as polyvinyl chloride (PVC), and in the production
of polishes, floor waxes and laundry products. They are also found in sludge
from municipal waste treatment plants, which is used in fertilizer and thus
widely dispersed on land (World Health Organization, 1980, 1990).

About 10 years after organotins were introduced, reports of gastropod
population declines and extirpations associated with tributyltin (TBT) commenced (Oehlmann *et al.*, 1996; Nicholson and Evans, 1997). When it was
discovered that TBT at extremely low concentrations (2.5 ng/l or ppt) in sea

Table 12.6 Dioxin TEQs (pg/g) and effects reported for wildlife and humans

Species	TCDD-TEQ (ppt)	Effects	Reference
Wood duck (*Aix sponsa*)	5[a]	Egg hatchability	White and Seginak (1994)
Wood duck (*Aix sponsa*)	>20–50[a]	Reduced number of fledglings*	White and Seginak (1994)
Forster's tern (*Sterna forsteri*)	2175[b] (1983)	Egg hatchability,* wasting,* reduced number of fledglings*	Kubiak *et al.* (1989)
Forster's tern (*Sterna forsteri*)	913[b] (1988)	Wasting onset day 17	Kubiak *et al.* (1989)
Double-crested cormorant (*Phalacrocorax auritus*)	85[c]	Egg mortality	Tillett *et al.* (1992)
Harbor seal (*Phoca vitulina*)	61.8[d]	No measured effects	Ross *et al.* (1995)
Harbor seal (*Phoca vitulina*)	208.7[d]	Immune suppression	Ross *et al.* (1995)
Lake trout (*Salvelinus namaycush*)	5.0[e]	Survival threshold	Rolland *et al.* (1997)
Human (*Homo sapiens*)	30.75[f]	Lower maternal plasma[g] T_3 and T_4; higher infant TSH	Koopman-Esseboom *et al.* (1994)

Superscript letters represent methods used to determine TEQs:
[a] TCDD-TEQ furans and dioxins (pg/g);
[b] dioxin and seven congeners of PCB TCDD-EQ (pg/g) H4IIE assay;
[c] direct in vitro H4IIE assay (pg/g);
[d] lipid adjusted serum, based on dioxin, furans, PCBs;
[e] from laboratory and field studies;
[f] TEQs derived from five dioxin, three furan and 14 PCB congeners.
[g] Triiodothyronine (T_3), thyroxine (T_4), thyroid stimulating hormone (TSH).
* Significant difference, $P < 0.05$.

water caused female whelks (*Buccinum undatum*) to grow penises, restrictions were put on its use on small ships and boats by several developed nations. However, its use on large ships was not banned, and in many parts of the world TBT is still used on small vessels (Bryan *et al.*, 1987). Three years after the European Union banned the use of TBT, it was still present at 36.2 ng/l and total TBT (including metabolites) at 119.3 ng/l ($n = 5$) in a natural sheltered harbor on the west coast of Sweden (Stuer-Lauridsen and Dahl, 1995). There are few reports of improved gastropod recruitment in harbors and coastal areas where TBT was banned (Tester *et al.*, 1996). Despite restrictions on its use, like the organochlorines, TBT is becoming ubiquitous in the global environment, even measurable in tissue of marine mammals from mid-oceans (Iwata *et al.*, 1997).

Larval and early stage female gastropods are especially sensitive to low concentrations of TBT. They develop visible male external genitalia, a condition called imposex, that leads to poor reproductive success and sterility (Gibbs et al., 1987). Adult females exposed to TBT are not affected. This phenomenon, where an effect is only expressed in individuals exposed very early in development, is typical of chemicals that interfere with the endocrine system (Colborn and Clement, 1992). TBT inhibits human placental aromatase activity in vitro through competitive inhibition of aromatase cytochrome P450 activity (Heidrich et al., 1999). Aromatase plays a critical role in the conversion of testosterone to estradiol and maintaining the ratio of female hormones to male hormones during sexual differentiation in embryonic development.

Organotin compounds accumulate readily in animals at the top of the marine food web (Kannan and Falandysz, 1997) and, as a result, seafood holds higher concentrations of organotin compounds than other sources of protein for human consumption (Kannan et al., 1995). The highest burdens of TBT are reported in tissue from animals in coastal areas and inland seas surrounding developed nations (Tanabe et al., 1998). TBT accumulates more in the liver than the other organs and fat, much like mercury. Iwata et al. (1997) measured TBT in a number of odontocetes from the North Pacific and Asian coastal waters of Japan. Mean concentrations in the livers ranged from 210 ng/g (ppb) wet weight (ww) in Baird's beaked whales (Berardius bairdii) (n = 3) to 2500 ng/g ww (ppb) in killer whales (n = 3) [see Tanabe et al. (1998) for a list of large whales and their TBT residue levels].

When comparing the profiles of TBT metabolites among a number of marine species, Iwata et al. (1997) discovered that cetaceans are not as efficient in metabolizing and excreting TBT as pinnipeds. Steller's sea lions (Eumatopias jubatus) accumulate more TBT in their hair than any other body tissue: 1500 ng/g ww (ppb or 1.5 ppm). About 26 per cent of the total sea lion TBT body burden is eliminated from the body through shedding hair (Kim et al., 1996a). By comparison, Risso's dolphins (Grampus griseus) carry as much as 6 ppm TBT in their liver and blubber. This may reflect the dolphins' lack of hair as an excretory pathway and may account for their vulnerability to infections (Kim et al., 1996b).

TBT was measured in bottlenose dolphins (Tursiops truncatus) stranded along the south-eastern US Atlantic and Gulf coasts between 1989 and 1994. The concentrations in the dolphins were the highest recorded to date, suggesting that TBT may have been involved in the episodes (Kannan et al., 1997). This is the first study that suggested a relationship between organotins and cetacean health.

The rodent immune system appears to be the most sensitive to TBT and its metabolites during the earliest stages of development (prenatal and early postnatal) when compared with adults (Smialowicz et al., 1989). Immune suppression induced by long-term exposure to TBT caused reduced resistance to infections by the parasite, Trichinella spiralis, and the bacterium,

Listeria monocytogenes (Vos *et al.*, 1990). Ingestion of di- or tri-*n*-butyltin dichloride led to a dose-dependent increase in thymic atrophy and resultant suppression of T-cell responses in rats (Snoeij *et al.*, 1988). In this study, di-*n*-butyltin was the most potent. Maternal exposure to dibutyltin chloride in rats also caused cleft jaw and lip, exencephaly, club foot and other malformations (Ema *et al.*, 1995).

DeLong and Rice (1997) report that 1.0 mg/kg (ppm) TBT potentiates 20-fold the Ah enzyme activity of 0.01 mg/kg (ppm) PCB 126 in mice treated for 7 days. This effect was not seen at higher doses with either both chemicals together or with each chemical alone. These levels are within the range of both PCB 126 and TBT found in fin fish and shellfish. In this study, high doses of TBT alone *inhibited* Ah enzyme activity (DeLong and Rice, 1997). The discovery of two common marine contaminants, TBT and PCB 126, acting synergistically and producing an opposite effect than when one chemical is administered alone at a higher dose reveals the complexity of dealing with the vast numbers of chemicals in the marine environment. This is only one combination of chemicals out of an enormous number of combinations that are collecting in marine sediments and the water column. In addition, as contaminants such as the vast mixtures of tin compounds travel from one environmental compartment to another, they are susceptible to many conditions that change their chemical and biochemical characteristics through complex degradation and metabolic dynamics.

Mercury

Evidence from atmospheric and oceanic studies indicates that mercury concentrations in certain regions of the global environment are increasing (Swain *et al.*, 1992; Arctic Monitoring and Assessment Programme, 1997). The Arctic Monitoring and Assessment Program (AMAP) reported that mercury was accumulating in the environment in north-west Arctic Canada three times faster in the 1990s than in the 1970s. A threefold increase was also found in the Baltic Sea, but in a shorter time span, between 1980 and 1993. According to AMAP, both cadmium and mercury are building up in wildlife and human tissue in the Arctic to concentrations where they have health implications. Dietz *et al.* (1996) found high levels of mercury (and cadmium) in marine mammals and seabirds in Greenland, in some instances exceeding Danish and FAO/WHO food standard limits. They provide a comprehensive overview of mercury contamination in Greenland waters from mollusks to fish to birds to the minke whale (sampled in 1980), beluga (in 1985), and narwhal (*Monodon monoceros*) (in 1985) (Dietz *et al.*, 1996). The narwhal was the most highly contaminated. Riget and Dietz (2000) and Johansen *et al.* (2000) provide evidence from the mid-1990s of continuing high concentrations of mercury in Greenland marine species.

The US National Oceanographic and Atmospheric Administration's (NOAA) Coastal Sediment Database reveals that mercury and cadmium were the most frequent metals to appear above its '5× high' contamination concentrations in 3878 coastal sites off the US (Daskalakis and O'Connor, 1995). These values are five times higher than what the NOAA National Status and Trends Program has established as high levels, based on their Biological Effects Database for Sediments, using data on estuarine and marine sediment toxicity. The '5× high' sites, in most instances, were located near coastal urban areas (Daskalakis and O'Connor, 1995).

Metallic mercury vaporizes readily at low temperatures and, riding on hydrogen ions, can stay airborne for over 40 days. Depending on where it settles, it can bind with local anions and form inorganic salts or, under conditions of low oxygen in the presence of microbes, can be converted to the organic form, or revaporize and travel once more on prevailing winds.

Mercury is often reported as total mercury in environmental samples, which includes metallic mercury, the inorganic salts and the organic form. The organic form, methylmercury, is fat-soluble and therefore more readily assimilated than metallic mercury when ingested. It moves through cell membranes and builds up in animal tissue. It is a developmental neurotoxicant, readily crossing the placenta and sequestering in the fetal brain. Anaerobic bacteria in sediments convert metallic mercury to methylmercury, which is rapidly picked up by bottom-dwelling organisms. Via biomagnification, methylmercury is conserved in animal tissue as it moves up the food web to top predators in the system. In contrast, almost all metallic mercury passes through the body unabsorbed. The total mercury burden in animal tissue generally increases with age (Drescher *et al.*, 1977; Dietz *et al.*, 1996). In fish, it can either become predominantly methylmercury with age, or the reverse, where the inorganic mercury fraction increases with age, depending where the fish are found in the water column (Holsbeek *et al.*, 1997).

It appears that the anoxic conditions in the pelagic marine ecosystem facilitate conversion of inorganic mercury to methylmercury. Some animals using the deep oceans and seas carry higher burdens of mercury than coastal animals. Seabirds show higher levels of mercury than coastal birds and thus are excellent sentinels for mercury contamination in oceans (Furness and Camphuysen, 1997; Monteiro and Furness, 1997). Burger and Gochfeld (1995) report that among avian species they have studied, pelagic species accumulate the highest mercury concentrations. Similarly, small cetaceans carry higher concentrations of the organochlorine chemicals than terrestrial animals; fish from the deep-sea carry higher concentrations of organochlorines than some coastal fish (Ballschmiter *et al.*, 1997); and storm petrels (*Bulweria bulwerii* and *Oceanodroma castro*) (pelagic seabirds) carry more organic mercury than coastal birds (Monteiro and Furness, 1997). In a scenario such as this, animals at the top of the food web are most vulnerable.

Mercury is often found in highest concentrations in the liver of marine mammals when compared with other organs (reviewed by Augier *et al.*, 1993). Harp seals (*Pagophilus groenlandicus*) fed fish containing 0.25 mg/kg (ppm) methylmercury experienced a decline in appetite and weight loss in

60 days and lethargy in 90 days. Seals exposed to 25 mg/kg (ppm) showed weight loss in 3 days and died between days 20 and 30. Following the low-dose regimen (0.25 mg/kg), concentrations of mercury in livers ranged from 64 ppm at 60 days to 82.5 ppm at 90 days, and following a 25 mg/kg regimen, the concentration in liver was 126 ppm at death (Tessaro *et al.*, 1977). For a comprehensive review of the health effects of mercury on wildlife, see Wolfe *et al.* (1998).

Recent findings of a Canadian research team concerning the effects of mercury on the ability of beluga whale splenocytes to respond to immune challenges raises questions about the hazards posed to marine mammals in a system where mercury concentrations are increasing (De Guise *et al.*, 1996b). It appears that the immune system may be more sensitive to mercury exposure than the nervous system. Using splenocytes from freshly collected beluga whales in the Canadian Arctic, De Guise *et al.* (1996b) demonstrated that $HgCl_2$ [2 µg/g (ppm) Hg] significantly increased the number of dead thymocytes with a Con-A challenge and reduced the proliferative response of splenocytes and thymocytes compared with controls. Most important, methylmercury [0.2 µg/g (ppm) MeHg] affected splenocytes at a dose 10 times lower than that of mercury chloride (De Guise *et al.*, 1996a). Among 30 dead adult St. Lawrence beluga whales aged 13–31 years, concentrations of mercury in livers ranged from 2.71 to 97.66 µg/g (ppm) MeHg, and in five juveniles aged 0–3.5 years, they ranged from 0.07 to 1.84 µg/g (ppm). Mercury concentrations in livers of some of the older whales exceeded 64 µg/g (ppm), the level at which the seals lost their appetites and became lethargic in the Augier *et al.* (1993) study.

Emerging concerns

PCBs have generated more concern about the health hazards of synthetic chemicals to cetaceans than any other class of chemicals. They are mentioned here because of the growing knowledge about their biotransformation, accumulation, metabolism and toxicity in humans and laboratory and marine animals (Troisi *et al.*, 1998; Letcher *et al.*, 2000b) that heretofore were not understood. Methylsulfone and hydroxylated metabolites of PCBs are demonstrated endocrine disruptors that interfere with the thyroid system and estrogen receptors. For a complete review see Letcher *et al.* (2000a).

Plastics and plastics components

Over the past 10 years, a great deal of attention has been directed toward the safety of a suite of large-volume chemicals that are used in the production of plastics. There are no reports of plastic debris building up in the gastrointestinal tracts of cetaceans, unlike the smaller vertebrates (Goldberg, 1997). However, an immense volume of plastics is accumulating in aquatic and marine systems, and their components are leaching into the water in which the whales live.

Plastic products have become a delivery system not only for the additives that give them their unique characteristics but also for the basic molecules of the polymers that become unattached upon use and disposal. Plastics were intentionally designed to be durable, some with projected half-lives of over 1000 years. In many cases, the chemicals used in plastic production are just as, or more, persistent than the industrial and agricultural organochlorine chemicals that were regulated in the 1970s. In the US, this industry has grown at the rate of 6–12 per cent per year since the mid-1940s, with annual pro- duction of plastics in the US reaching 38.5 million metric tons (0.15 metric tons per person) in 1996 (Society of the Plastics Industry, 1997). Globally, 122 million metric tons were produced in 1996, and it is estimated that plastic production in the developing world will grow at the rate of 40 per cent per year in the near future (Society of the Plastics Industry, 1997). In some instances, additives contribute as much as 50 per cent or more to the final weight of a plastic product.

Bulky plastic products as well as the invisible compounds can pose a problem to marine animals. For example, adult albatrosses pick up plastic debris from the surface of the North Pacific Ocean as they forage and take the material back to their chicks on Midway Island, where they regurgitate the plastic into the chicks as they are fed (Auman *et al.*, 1997). The plastic litter-polluted region off eastern Indonesia supports lower densities of di- atoms and different macrofauna than litter-free regions (Uneputty and Evans, 1997). An Australian study revealed that the new generation soft scrub cleansers contain granulated polyethylene, polypropylene or polystyrene (Gregory, 1996). These minute particles accumulate at the air–water inter- face and form a film that potentially interferes with the production of algae, which is food for microscopic filter feeders at the base of the food web (Goldberg, 1997). The energy lost as the filter feeders graze plastic must be considered when determining the status of the marine food web and the well being of all the animals in the oceans.

Up until a decade ago, plastics were generally considered biologically inert. Consequently, no plastic additives or monomers have been monitored systematically in tissues of whales or in any other animals, including humans. However, it is now recognized that some plastic monomers and additives have deleterious effects on the endocrine and immune systems and can inter- fere with development and reproduction (Fowles *et al.*, 1994; Jobling *et al.*, 1995; Sharpe *et al.*, 1995). Several compounds have been discovered in the tissues of other marine mammals, birds and fish (Peakall, 1975; Takei and Sawa, 1995; Jones *et al.*, 1996).

A commonly found biologically active plastic monomer, bisphenol A (BPA), is used to make polycarbonates that have many modifications and applications. For instance, BPA is used in the linings of food cans, five- gallon water demijohns, high-impact glass, sporting goods and recreational equipment (e.g. boats, ships, waterskis, goggles, etc.), and was used to make 4 billion compact discs in 2000. It is among the plastic products found floating on the surface of the oceans (Blight and Burger, 1997). In the laboratory, male offspring of mice fed BPA during gestation (2 or 20 ng/kg) experience permanently heavier prostates, and in adulthood have reduced sperm production (vom Saal *et al.*, 1998). The dose used in this study was 25 000 times lower than the 50 mg/kg approved for daily human consumption by the US Food and Drug Administration. BPA is an estrogen mimic, about 1000 to 10 000 times weaker than free estradiol *in vitro*, the most potent female hormone. However, if BPA is present during murine embryonic development *in vivo* (0.23 ppb), it enhances the rate of pre- and postnatal development (Takai *et al.*, 2001) and can cause latent effects, such as heavier prostates in males and early onset of puberty in females (2.4 ppb) (Howdeshell *et al.*, 1999).

Phthalates are a large and complex class of chemicals used to make plastics flexible. They are found in all water samples (including tap water) and in marine water around the world (Wams, 1987). Large amounts have already been released into the environment. Among the phthalates, 3.9 million metric tons of terephthalic acid and dimethylterephthalate were produced in the US in 1995 (Chem Expo, 1998). As early as the late 1970s, the combined atmospheric contribution of di-*n*-butyl phthalate (DBP) and di-2-ethylhexyl phthalate (DEHP) to the Great Lakes ranged from 74.7 metric tons per year in Lake Ontario to 32 tons in Lake Superior (Giam *et al.*, 1978; Eisenreich *et al.*, 1981). DEHP was found in the air over the Gulf of Mexico at 0.4 ng/m^3 and the North Atlantic at 2.9 ng/m^3. It was found in air samples at 500 m over Chiba Prefecture, Japan, at levels from 3 to 5300 µg/m^3 (Watanabe, 2001). Phthalic acid esters were found in the sea-surface microlayer in harbors in the North Sea (16 µg/l or 16 ppb) and Baltic Sea (not detected (n.d.), 9, and 25 µg/l or 25 ppb) in 1985 (Kocan *et al.*, 1987). DEHP was found in fish from the Gulf of Mexico at 4.5 ng/g (ppb) (Giam *et al.*, 1978). DBP was found at 800 ppb ww in flamingo testes ($n = 35$) and 250 ppb ww in Hawaiian monk seal (*Monachus schauinslandi*) livers ($n = 20$), though none was detected in the fat of the monk seal (Takei and Sawa, 1995).

Male offspring of rats receiving 10–1000 µg/l (ppb) of DBP in their water during gestation experienced degeneration of the testes, hyperplasia of the urogenital tract and reduced sperm production (Sharpe *et al.*, 1995). The concentrations used in this study were within the range of ambient human consumption. DBP is used extensively in vinyl tiles and other vinyl products. Male rats exposed perinatally to DBP and DEHP had reduced anogenital distances, retained nipples, epididymal agenesis, undescended testes and hypospadias (Gray, 1998; Gray *et al.*, 1999b). The authors did not find a no-effect level for these phthalates. Follow-up studies revealed that DBP, DEHP and di-isononyl phthalate (DNP) inhibit fetal testosterone production rather than competitively binding to the androgen receptor (Gray *et al.*, 1999a; Parks *et al.*, 1999).

Detergents

Alkyl phenol ethoxylates are used as surfactants and antioxidants in plastic products, inert ingredients in pesticides, commercial detergents in large volumes, and household detergents in lesser volumes. Their estrogenic activity was discovered when nonylphenol leached from modified polystyrene laboratory equipment and caused stored breast cancer cells to proliferate (Soto *et al.*, 1991). Almost simultaneously, a team of British scientists discovered that alkyl phenolic ethoxylates and their breakdown products in rivers in the UK were potentially causing male fish to produce unusually high quantities of vitellogenin (Jobling and Sumpter, 1993). Vitellogenin is a protein produced by the liver in egg-laying females in response to surges of the female hormone, estradiol. The vitellogenin is picked up by the blood and delivered to the developing egg where it is cleaved and sequestered in the yolk. This class of chemicals has also been shown to cause a reduction in testicular size in male fish and rats, and reduced sperm production in male rats (White *et al.*, 1994). The majority of the field work on the alkyl phenols has been in freshwater streams in the UK, other countries in Europe, and most recently the United States (Goodbred *et al.*, 1997). The ethoxylates do not degrade readily and bioconcentrate in fat in marine mussels (*Mytilus edulis*) up to 216 600 times; shrimp (*Crangon crangon*) to 7500 times; and three-spined stickleback (*Gasterosteus aculeatus*) to 17 800 times (Ekelund *et al.*, 1990).

Fire and smoke retardants

Fire and smoke retardants often contribute as much as 15–40 per cent of the final weight of plastics used for electrical and packaging material (Society of the Plastics Industry, 1997). Tetrabromobisphenol A and chlorinated and polybrominated diphenyl ethers (PBDEs) are used for this purpose. Estimated use in 1999 was 121 400 tonnes of tetrabromobisphenol A and 67 125 tonnes of PBDEs (Renner, 2000). Total PBDEs were found in sperm whale blubber at 100 µg/kg (de Boer *et al.*, 1998). Chlorinated diphenyl ether was discovered at 6 ppb in the blood of an albatross that feeds only on the surface of the North Pacific Ocean (Jones *et al.*, 1996). From 1972 to 1997, PBDEs in human breast milk doubled every 5 years in Sweden (Noren and Meironyte, 2000). Pentabromodiphenyl ether suppressed circulating total thyroxine (T_4) levels at all single doses tested, even the lowest dose (0.8 mg/kg or 800 ppb), and with a chronic dose (0.25 mg/kg or 250 ppb) for 2 weeks. It also suppressed antibody response to an immune challenge (sheep red blood cell, SRBC) only at the highest chronic dose tested (1000 mg/kg) in mice (Fowles *et al.*, 1994).

Since 1989, when Walker *et al.* (1989) first report finding *tris*(4-chlorophenyl)methane (TCPMe) and *tris*(4-chlorophenyl)methanol (TCPMOH) in biological samples, interest has grown in the pattern and significance of these chemicals in marine mammals. Most reports have focused on dolphins, porpoises, and beluga and killer whales from the Pacific Ocean, North Sea and coastal Florida, United States (Jarman *et al.*, 1992; Watanabe *et al.*, 1999, 2000; Minh *et al.*, 2000a; Lebeuf *et al.*, 2001). There are also reports of either or both of these chemicals in pinnipeds, polar bears (*Ursus maritimus*), peregrine falcons (*Falco peregrinus*) and herring gulls (*Larus argentatus*) (Walker *et al.*, 1989; Jarman *et al.*, 1992; Watanabe *et al.*, 1999). The pattern of exposure in nine species of cetaceans indicates that occurrence is widespread in both coastal and offshore regions of the oceans and associated with industrialized areas of the mainland. TCPMOH and TCPMe are also widely available in terrestrial systems (Minh *et al.*, 2000b). Both TCPMOH and its proposed precursor TCPMe are lipophilic and bioaccumulate at rates similar to DDT and PCBs. They accumulate preferentially in polar bear livers (6.8 ppm), compared with 0.1 ppm in kidney (Jarman *et al.*, 1992). The origin of these chemicals is unclear, and it is proposed that they are co-contaminants with pesticides, such as DDT and dicofol, and with synthetic polymers and dyes used in commerce.

There are no data on concentrations of TCPMOH or TCPMe in baleen whales. These chemicals should be monitored in whales that feed in shallow waters along industrialized coastal regions, such as the coast of California to the Puget Sound, US, where elevated concentrations of TCPMOH and TCPMe are found. Whales foraging off eastern Canada, near the confluence of the St. Lawrence River, should be monitored as well, because elevated concentrations of TCPMOH and TCPMe have been discovered in the blubber of beluga whales and seals in this region (Jarman *et al.*, 1992; Minh *et al.*, 2000a).

The health effects of TCPMe and TCPMOH are unclear. Preliminary *in vivo* and *in vitro* studies indicate that TCPMOH and TCPMe induce hepatic enzymes and appear to have antiandrogenic effects. Exposure of MFM-223 human breast cancer cells to 1.8 ppm TCPMOH caused cell proliferation. The androgen receptor binding affinity of TCPMOH is as high as DDE, the breakdown product of DDT (Foster *et al.*, 1999) (see more about the antiandrogenic effects of DDE below). Letcher *et al.* (1999) have demonstrated that both TCPMOH and TCPMe reduce aromatase enzyme activity *in vitro*.

Other chemicals

The polychloro-*n*-alkanes (sPCAs) (paraffins) and perfluorooctane sulfonates (water and stain repellents) were analyzed in cetacean tissue only within the very recent past, although the compounds have been used widely in commerce for decades (Tomy *et al.*, 2000; Giesy and Kannan, 2001). They bear watching as their health effects come under more scrutiny. Both classes of chemicals are highly mobile and persistent, found in beluga whales from Greenland to the MacKenzie Delta, Canada (Tomy *et al.*, 2000), in polar bears in Alaska, and in dolphins in the Mediterranean Sea and Ganges River, India (Giesy and Kannan, 2001).

Indirect concerns

Not only do synthetic chemicals pose a direct health threat to cetaceans, they may pose indirect threats that would be far more difficult to identify and quantify. What follows is a discussion concerning their possible impact on cetaceans via direct impacts on carbon fixation.

Arctic haze

In the 1950s, it was noticed that air pollution was building up on an annual basis each winter over the Arctic. Throughout the years this visible collection of organic and inorganic material has become known as the Arctic haze (Barrie *et al.*, 1992). It becomes very pronounced from January through April and generally hangs over the Arctic in an area the size of Africa. It extends over all of Greenland, northern Canada, the northernmost regions of Alaska, northern Russia, eastern Europe, and northern Finland, Sweden and Norway. The Arctic Ocean is completely covered. The sulfur in the haze over Canada and Greenland can be traced back to Europe (Lowenthal *et al.*, 1997; Nriagu *et al.*, 1991). The very fine sulfur particles in the stagnant air carry persistent organochlorine industrial chemicals and pesticides, metals (including mercury), and other industrial pollutants that this chapter does not address, such as polycyclic aromatic hydrocarbons and radio-isotopes (Barrie *et al.*, 1992). Winter air currents drive this pollution toward eastern North America, the Aleutians or southern Greenland. The particles in the haze eventually precipitate in the spring months on to the ice cover and open ocean, at the same time as photosynthesis is at its peak (Barrie *et al.*, 1992).

The air–water interface, surface microlayer and sea fog

Large cetaceans spend a great deal of their time at the air–water interface. Whether they use benthic food sources (amphipods/krill) or sources nearer the ocean surface, they must come to the surface to breathe. A growing number of reports on the quality of the air at the water interface of the oceans suggest that animals using the surface can come in contact with a wide array of contaminants (Hardy, 1987; Fellin *et al.*, 1996). Traces of widely used industrial and agricultural chemicals are detectable even in the most remote ocean spans. The highest concentrations are found at the air–water interface, in the top 100 µm of the sea-surface microlayer and in sea fog (Kocan *et al.*, 1987). What this contributes to cetaceans in terms of direct and inhalation exposure is unknown. However, primary production at the air–water interface is critical for the marine food web and cetaceans' energy needs. Anything that interferes with the natural process of carbon fixing, and therefore reduces algal production, can have an indirect effect on cetaceans.

The photosynthesis that takes place at or below the surface of the ocean provides food for the larval stages of invertebrates and fishes at the base of the food web. When daylight returns in the Arctic, intense primary production starts under the sea ice or along the ice edge as the ice recedes (Alexander, 1995). Because of limited nutrients in certain regions of Arctic waters, this intense productivity may only take place over a period of 2 weeks. Where sea ice exists, ice algae attached to the underside of the ice bloom first. However, as the ice melts, diatoms also burst into production at its edge. These blooms provide most of the primary production and energy for the year in some unique Arctic regions. The majority of the algae and diatoms drop to the sea bottom where they support invertebrates and the larval and early life stages of other bottom species (Alexander, 1995). These, in turn, support larger amphipods and fishes that support eider ducks (*Somateria* spp.), flatfish, bearded seals (*Erignathus barbatus*), walruses (*Odobenus rosmarus*), and whales. Bioturbation caused by these large animals feeding on the bottom keeps the energy moving in the system throughout the year (Alexander, 1995). Gray whales (*Eschrichtius robustus*), the largest bottom feeders, use the Bering and Chukchi Seas for six summer months and contribute to, and take advantage of, this energy cycle. They return to southern waters off Baja California and Mexico, where they fast for the winter months, throughout the calving and breeding seasons.

Certain geographic regions are more productive than others (Alexander, 1995). Algal blooms generally continue northward as the ice melts throughout the north-west Bering Sea toward the Bering Strait and Chukchi Sea, where the highest rate of primary production occurs in the Arctic as water flows north-westward into the Arctic Ocean (Barrie *et al.*, 1992). The eastern Bering Sea is much less productive. For example, the Bering Strait produces as much as 400 g carbon (C)/m^2/year, while the Arctic Ocean basin produces as little as 4 g C/m^2/year. In contrast, the Greenland Sea has continual blooms as the ice edge moves northward. Here the warmer North Atlantic waters come in contact with the receding ice edge, thus allowing for constant production over a longer time period. Water temperature is not a deterrent to photosynthesis. Polynyas, areas of ocean that do not freeze, hold the coldest open water and have the highest productivity. Production in a big polynya in the Northwest Water off the north-east Greenland coast supports walruses (Alexander, 1995).

Contaminants flowing northward in Russian rivers become frozen amidst the sediments in the shallow Siberian Arctic Sea shelves. These entrained contaminants are exported 1000 km by currents, across the Arctic Ocean toward the Greenland and Barents Seas (Pfirman *et al.*, 1995). As this ice melts in the spring it releases the contaminants into the water column. This source, plus the Arctic haze and surface microlayer, provide a continual source of contaminants to the ocean water in the spring. Contaminants landing on the ice can collect in brine channels and penetrate through the ice, or sit on the surface and flush off in high concentrations at the ice edge in the spring. Because contaminants have a tendency to cling to organic material, the algae serve as a carrier for the contaminants into the food web at its base when both are at their highest concentration (Alexander, 1995).

Hardy and Crecelius (1981) suggested that the rate of deposition of atmospheric particulate matter under certain conditions might be great enough to interfere with primary production at the sea surface microlayer. This physical impediment, in addition to the biologically active contaminants that are delivered to surface microlayer, raises further questions. For instance, herbicides that have been designed to prevent primary production could impede algal development if they reached the Arctic region. Along with mercury and the persistent organochlorine chemicals, a number of currently used pesticides have now been discovered in sea ice and fog (Chernyak *et al.*, 1996). Chlorpyrifos, trifluralin, metolachlor, chlorothanil, terbufos and endosulfan were detected in the fog at concentrations several times higher than in the surrounding water or ice. Atrazine, a herbicide, was discovered in melted ice at 0.400 ng/l (ppb) and chlorpyrifos at 0.170 ng/l (ppb). Atrazine was discovered in the Lake Superior water column at 3 ng/l (ppb) and in Lake Erie from 40 to 110 ng/l (ppb), adjacent to agricultural areas where the herbicide is applied (Schottler and Eisenreich, 1994). It was found at 120–5800 ng/l (ppb) in the coastal waters of the German Bight (Bester *et al.*, 1995). These unexpected findings in the Arctic reflect the fact that atrazine was the most widely used herbicide in the US and heavily used in Europe, and chlorpyrifos the most widely used insecticide in the US. Not only is atrazine found in groundwater and surface waters throughout the US (Schottler *et al.*, 1994), it has also been found around the world (Bintein and Devillers, 1996). This also demonstrates how pesticides and other atmospherically delivered chemicals can build up on the ice surface over the winter months when low temperatures and lack of sunlight retard their degradation.

Weight of evidence

Because of the almost impossible opportunity to work with live cetaceans, alternatives are needed to determine the hazards posed to cetaceans by the growing list of xenobiotics that can undermine their development and function. What follows is a discussion of how the chemicals that have been found in the oceans and in cetacean tissues could affect their health.

The endocrine system

The endocrine system functions through the use of chemical messengers called hormones that are produced by various glands and the brain. They have control over the developing embryo from conception to birth. In free-ranging bottlenose dolphins, the mean circulating blood level of free triiodothyronine (fT_3) was 1.38 pg/ml (ppt), and of free thyroxine (fT_4),

13.5 ng/dl (ppb) (St. Aubin *et al.*, 1996). Gray seals (*Halichoerus grypus*) had a mean of 1.3 pmol/l fT_3 and 25.7 pmol/l fT_4 (Hall *et al.*, 1998). These concentrations of thyroid hormones, as well as the proteins that carry the hormones in the blood, are comparable to those in humans (Porterfield and Stein, 1994). Aldosterone, an adrenal hormone, circulates at 1.16 pg/ml (ppt) in bottlenose dolphins (St. Aubin *et al.*, 1996). The female hormone, estradiol, operates in the range of 1/10 of a trillionth of a gram (<0.1 ppt) of free hormone in rats and humans (vom Saal *et al.*, 1992). Free hormones are not bound to blood components and, therefore, are accessible to bind to their respective receptors in organ tissue and initiate a biochemical response. In some instances, contaminants circulating in humans and wildlife are as much as a thousandfold or higher than the natural hormones that control development. Even if a synthetic chemical were biologically weaker, it would have a competitive advantage over the more potent hormone for receptor sites in target organ tissue. For example, DDE competes with testosterone for the androgen receptor and is only one-tenth as potent as flutamide, a pharmaceutical that interferes with testosterone at the cellular level to reduce testosterone circulation in men with prostate cancer (Kelce *et al.*, 1995). DDE is found in concentrations in marine animals at or above concentrations at which development of the male reproductive tract is impaired in laboratory animals (see more about DDE below).

Neuroendocrine effects

In a US study that involved only healthy human infants and mothers, infants whose mothers had more than 1 ppm of PCBs in their blood lipid during gestation had measurable neurological deficits at birth (Jacobson and Jacobson, 1996). The mothers were selected for this study based on whether or not they ate fish from Lake Michigan prior to their pregnancies. The average PCB concentrations in human lipids in the industrialized world hover around 1 ppm. The neurological deficits, measured as reduced memory and attention span, became significant across the study population when the mother's blood lipid reached 1.25 ppm PCBs. At age 11, the affected children's IQ scores averaged 6.2 points below normal, and the most severely affected were more than twice as likely to be 2 years behind their peers in reading. The children's memory, verbal comprehension and attention were affected (Jacobson and Jacobson, 1996). In this study, 17 children were removed at age 4 because they were intractable and refused to be tested. They behaved as predicted from studies in which rats had been fed fish from Lake Ontario (Daly, 1991, 1993). Their mothers also had the highest PCB lipid concentrations in the study. At age 11, another child was removed from the study because the subject's IQ was below 70, which is in the range of mental retardation. With the outliers removed, 11 per cent of the children in the study were affected. Leaving in the outliers, 20 per cent of the study population was affected. The implications of this research are important from a population perspective. In a carefully selected cohort of healthy infants based on traditional health parameters, approximately 1 in every 5 infants in the study suffered measurable neurological impairment as the result of exposure to background levels of PCBs. This study was later followed up with a new cohort of healthy mothers who ate fish from Lake Ontario prior to and during their pregnancies. Up to age 5, thus far, the same developmental problems were found (Lonky *et al.*, 1996).

In a Netherlands study of 78 healthy mother/infant pairs, the mean breast milk dioxin TEQ was 32 pg/g (ppt) fat (Koopman-Esseboom *et al.*, 1994). The mean concentration of total PCB and dioxin TEQs in mother's milk was 74.86 pg TEQ/g fat. They found significant correlations between (1) higher dioxin TEQs, (2) higher total PCB and dioxin TEQs, and (3) higher planar and non-planar PCB concentrations and lower total T_3 concentrations in the mothers' plasma during the last month of pregnancy. The breast-fed infants were divided into two groups, based on their exposure to dioxin TEQs, below and above the median at 30.75 pg/g fat. Two weeks after birth, total T_4 was significantly lower and TSH significantly higher in the 'high' exposure group, although none of the hormone levels in the infants exceeded the normal range. In a follow-up study, higher prenatal exposure to PCBs produced measurable neurodevelopmental delays at 3 months (Koopman-Esseboom *et al.*, 1996). A sizeable segment of the baleen whales reported herein have an estimated dioxin TEQ above 30.75 pg/g for PCBs alone, excluding all exposure to dioxins, furans and other dioxin-like compounds.

Thyroid hormone imbalances among fishes and birds in the Great Lakes region have been reported consistently since the early 1970s, and the problem continues today (see Leatherland, 1998 for review). An extensive literature describes how PCBs and specific PCB congeners interfere with thyroid hormone production and transport, peripheral metabolism and deiodination of T_4 to T_3, and interfere with retinol (vitamin A) levels (see Brouwer *et al.*, 1998). Vitamin A is transported on the same carrier protein, transthyretin, as thyroid hormone and, through dimerization with thyroid hormone, plays a critical role in thyroid hormone function in development (Rolland, 2000). A positive correlation between total TEQs and free and bound retinol is reported by Simms *et al.* (2000) in recently weaned and fasted harbor seal pups, suggesting a depletion of liver stores of retinol.

The medical community has recognized for many years that hearing loss in humans and laboratory animals is commonly the result of reduced exposure to thyroid hormones during development (Brucker-Davis *et al.*, 1996). The most relevant research for cetaceans demonstrated that exposure to PCBs during development leads to decreased auditory responses to low frequencies in laboratory rat pups (Goldey *et al.*, 1995; Goldey, 1996; Herr *et al.*, 1996). In these studies, researchers discovered a dose-dependent reduction in circulating T_4 and reduced auditory thresholds in the rat. They administered the commercial PCB product Aroclor 1254 (A1254) at 1, 4 or

8 mg/kg/day from day 6 of gestation to postnatal day 21, and measured brainstem auditory evoked responses in the offspring at 1 year of age with simulated filtered clicks at 1, 4, 16 and 32 kHz. They discovered that A1254 permanently decreased auditory response capability in the lowest kHz ranges administered (1 and 4 kHz). The authors suggest that the site of damage is in the cochlea and/or the auditory nerve (Goldey *et al.*, 1995). Interestingly, in the mid-1970s, a series of papers from Canada described hearing loss and middle-ear infections in Inuit children (Shephard and Itoh, 1976) from a population whose concentrations of PCBs in blood are sevenfold higher than those found in individuals at lower latitudes. These levels are within the range found in cetaceans. Whether the findings in rats extrapolate to cetaceans remains to be determined. However, the implications of these discoveries for cetaceans, which communicate with low-frequency sounds, should not be dismissed.

Impaired sexual development

Recent necropsies of highly contaminated beluga whales from the St. Lawrence River revealed that the prostate gland in all adult males had atrophied (De Guise *et al.*, 1994a). Other accessory glands of the male reproductive tract were either absent or atrophied, which the authors suggested might be the result of exposure to contaminants. The authors' description of the condition of the prostatic tissue in the whales is similar to a condition induced in laboratory mice exposed prenatally to high levels of the female hormone, estradiol (vom Saal *et al.*, 1997). Recently, in addition to the discovery of a first true hermaphrodite beluga, with two separate ovaries and two separate testes (De Guise *et al.*, 1991) – a very rare event – a second animal in the same St. Lawrence River population has been described as a pseudohermaphrodite (Sylvain De Guise and Michel Fournier, 1997, personal communication). Females in this population show little ovarian cyclical activity, and the number of pregnant animals in the population is significantly low compared to Arctic populations (Béland *et al.*, 1992, 1993).

The list of proven environmental, anti-testosterone chemicals continues to grow following the 1995 discovery that DDE is a testosterone antagonist (Kelce *et al.*, 1995). The number of mechanisms leading to abnormal male sexual differentiation continues to grow as well. It also appears that each identified chemical has its own unique mechanisms expressed differently in the phenotype (Gray *et al.*, 1999b). For instance, dihydro-testosterone (DHT) is critical for the urethra and penis to enlarge and lengthen together. However, if 5α-reductase, which converts testosterone to DHT, is inhibited at the onset of sexual differentiation, development of the penis will be impaired, resulting in hypospadias. In this condition, the urethra does not open at the end of the penis but is anywhere along the shaft or in the perineum. DDE, DEHP, DBP and DNP, and some fungicides, herbicides and insecticides have induced hypospadias in laboratory rats (Gray *et al.*, 1999b). In other cases, the synthetic chemicals interfere with the production of testosterone by the fetus. [See Baskin *et al.* (2001) for an in-depth discussion.]

Reproductive impairment

The North Atlantic right whale, protected since 1937, is showing no sign of population recovery. Its growth rate is 33 per cent of that among the South Atlantic species (*Eubalaena australis*), and the females are not as reproductively successful as the southern hemisphere population. Researchers found that 13 adult females had not calved for 11 years (Brown *et al.*, 1994). Inbreeding and food deprivation have been suggested as underlying problems in this case. However, in light of the growing knowledge about the latent effects of the contaminants in the marine environment, the role of contaminants as an additional stressor is now being considered.

It is estimated that the sperm whale (*Physeter macrocephalus*) population around the Galapagos Islands has declined at a rate of approximately 20 per cent a year between 1985 and 1995 (Whitehead *et al.*, 1997). Those studying the whales closely suggest that some of this depletion is due to emigration of male whales into whale populations along the mainland of Colombia and Ecuador that were depleted of males as a result of years of heavy exploitation of the large males in the coastal population. Adult female sperm whales will not accept younger males in the absence of larger, older males, and this could be contributing to the decrease in their numbers. This is a plausible hypothesis and should not be ruled out. However, the possibility that contaminants may be partially involved in this loss should not be ruled out either. For example, the sperm whales' feeding habits restrict them to the near coastal shelves in the more contaminated latitudes, both north and south of the equator. In concert with increased industrialization along the Pacific coastline of South America, agricultural activity has also increased significantly. Ecuador's and Chile's coastal banana plantations and vegetable farms use exceptionally high volumes of fungicides and nematocides. Vast shrimp farms along the Pacific coastlines of Colombia, Ecuador and Chile are dependent upon frequent applications of pesticides. The shrimp industry has been damaged by Taura Syndrome, a non-infectious disease in shrimp, the onset of which coincided with a heavy increase in the use of fungicides in the region.[1] Three fungicides in particular, propiconazole, tridemorph and benomyl are used to control a fungus, black sigatoka, on bananas. Propiconazole interferes with fat metabolism and the production of steroid hormones, and questions have already been raised about the role it is playing in Taura Syndrome, which prevents normal development of the earliest life stages of the shrimp.

---

[1] *Shrimp News International*, March/April 1994. This news item closed with the following recommendation from 17 scientists: '. . . that the Ecuadorian government call an immediate moratorium on the use of sterol biosynthesis inhibiting fungicides in the Guayas River Basin until their possible role in causing Taura Syndrome is resolved'.

The immune system

In the past decade, more and more studies have been directed toward the role of the immune system in maintaining the well being of marine mammals. Many of the studies were conducted because of the widespread viral outbreaks among populations of small cetaceans and pinnipeds that commenced in 1987 (de Swart *et al.*, 1996). In the 1987 episode in north-western Europe, approximately 20 000 European harbor and gray seals died. A newly identified and characterized morbillivirus (phocine distemper virus, or PDV) was assigned as primary cause of the event. However, there lingered concerns that the dioxin-like PCBs, dioxins and furans, many of which are immunotoxicants, contributed to the mass mortality. In order to investigate this, two groups of 11 captive harbor seals were fed herring from either the relatively uncontaminated Atlantic Ocean or the contaminated Baltic Sea for 30 months. Seals fed Baltic herring had significantly diminished T-cell function *in vitro* and *in vivo* and reduced natural killer cell function, both of which are crucial to anti-virus defenses in vertebrates. Comparison of concentrations of contaminant residues in blubber of these animals with those in the literature led the authors to suggest that many free-ranging harbor seal populations may be at risk to immunotoxicity (Ross *et al.*, 1996b; see Chapter 21 in this volume). Another study using free-ranging bottlenose dolphins found an association with the amount of PCBs and DDT isomers in their tissue and suppression of their response to immune challenges (Lahvis *et al.*, 1995). A third study revealed that those animals in a stricken population of striped dolphins (*Stenella coeruleoalba*) carrying the highest concentrations of persistent organochlorine compounds were more likely to die from a viral infection than those dolphins with a lower body burden (Borrell *et al.*, 1996). An ongoing, long-term, directed research project on the immune status of the St. Lawrence estuary population of beluga whales strengthens these findings. Using concentrations of toxic trace metals found in the highly exposed St. Lawrence beluga whale livers, scientists inhibited the proliferation of white blood cells in fresh blood from healthy Arctic beluga whales (De Guise *et al.*, 1996b). The endangered status of this population complicates research that requires catching the whales for close examination or for collecting blood samples. The team's only source of information about the population comes from very close surveillance of the approximately 500 animals still existing (where there were over 5000 at the turn of the century) and what they can learn from the dead animals (Béland, 1996b). They are also studying more pristine beluga whales from the Canadian Arctic and using them for comparison (see p. 304).

The St. Lawrence population of whales suffers various severe infections caused by otherwise mildly pathogenic bacteria, a signal of immune suppression (De Guise *et al.*, 1996a). Some animals had a rare condition, abscesses in the thyroid, and one animal had a thyroid adenoma (benign). They had unusually high incidences of nodules and cysts in their adrenal glands, similar to conditions associated with DDT exposure. Pathological examination of 24 carcasses revealed excessive degenerative, infectious, hyperplastic or necrotic lesions, including pneumonia, peridontitis, stomach ulcers and mastitis (De Guise *et al.*, 1995). The St. Lawrence beluga whales also carry more species of parasitic worms than any other population examined, with the exception of beluga whales from the White, Kara and Barents Seas, other contaminated marine systems. Belugas from the St. Lawrence have excessively high concentrations of DDT, PCBs, mirex, metals and other contaminants in their bodies (Muir *et al.*, 1996) (see also p. 305).

Cancer is often described as a disease resulting from a weakened immune system. Until recently, the St. Lawrence River beluga whale population was the only marine mammal population reported to be plagued with tumors. Before 1994, only 75 tumors had been reported in cetaceans. Of those, 28 were from the necropsied St. Lawrence belugas (or 40 per cent of the animals examined). The belugas had assorted types of tumors and some animals had more than one type (De Guise *et al.*, 1994b). Recently, veterinarians examining stranded live sea lions off the Pacific Coast of California between 1979 and 1994 reported 66 cases of transitional cell carcinomas, a cancer of the urogenital tract, out of 370 sub-adult and adult animals examined (Gulland *et al.*, 1996). Over a period of 24 months, five dolphins taken off the coast of Florida, US, were discovered with immunoblastic lymphoma (Bossart *et al.*, 1997). This is the first time this malignancy has ever been diagnosed in dolphins. The mammals included three adult female bottlenose dolphins, one adult female Atlantic spotted dolphin (*Stenella frontalis*), a year-old female pantropical spotted dolphin (*Stenella attenuata*), and one adult that had a 2-year history of anorexia and lethargy while confined in a marine mammal facility.

In the Netherlands study of healthy human mothers and infants mentioned above, changes in the infants' T-cell subpopulations increased in a dose response manner as dioxin TEQs increased (Weisglas-Kuperus *et al.*, 1995). Although CD4 helper cells did not increase, there was a significant increase in CD8 (cytotoxic) cells. The authors suggest '. . . that background levels of PCB/dioxin [measured as dioxin TEQs] exposure influences the human fetal and neonatal immune system'.

Conclusions

Even after significant reductions of organochlorine chemicals in the environment in the late 1970s, there is a pattern of widespread contamination of PCBs and associated organochlorine chemicals at background levels that are still high enough to cause harm to wildlife and humans. The most persistent PCBs are slowly moving toward the poles in air and water currents, and in some regions of the world are still increasing. It is now evident that cetaceans carry a suite of synthetic organochlorine chemicals, methylmercury, organotins and other industrial chemicals in concentrations at or above those where damage has been reported in other wildlife, humans and laboratory animals.

Keeping the organochlorine chemical experience in mind, it is imperative to look at new classes of chemicals that have penetrated the remotest ecosystems, the impacts of which on development and reproductive success are just coming to light. Like humans, the baleen and toothed whales carry background concentrations of organochlorine contaminants, such as PCBs, dioxins and furans, that vary considerably among and within populations. This variation among whales depends upon their geographic coordinates, species, trophic position, sex and age. Every newborn whale will have been exposed from conception to birth and throughout lactation to these chemicals. We expect that a sizeable proportion of calves in each whale population will have been perinatally exposed to concentrations of PCBs, dioxins and furans at which measurable neurological impairment is reported in human offspring. We base this on the estimated biochemical activity of the contaminants in several populations of baleen whales, using dioxin TEFs established by the WHO. The outcome of prenatal exposure to these chemicals in humans is just beginning to be understood. Could similar exposure in a whale impair its awareness of its surroundings, or its capacity to respond to seasonal stimuli or its ability to cope with stress? There is, however, enough evidence to consider seriously whether similar exposure can lead to hearing loss that could undermine a whale's sense of direction or interfere with its communication among individuals and pods. It is imperative to keep in mind that if exposure to PCBs took place during adulthood, auditory damage would not occur. The damage occurs while the auditory system is developing, during instances of thyroid hormone deprivation. An effect such as this can only be initiated during prenatal or early postnatal exposure. Allowing for a delay time of 20 years for contaminants to slowly build up in the marine system, the first generation of whale offspring with hearing impairment, or any other impairment, as the result of *in utero* exposure would not have been seen until the 1970s.

In addition to the global background exposure of organochlorine chemicals, the baleen and toothed whales are also exposed to the expanding global distribution of mercury and tin compounds. In the case of each metal, the body burden of marine animals varies significantly and is determined by the animals' geographic range. Exposure depends a great deal on regional natural sources and anthropogenic activity in the case of mercury, and anthropogenic activity alone in the case of tin. However, concentrations of mercury in livers in a sizable number of whales were above those at which adverse immune effects are reported in other animals. Concentrations exceeded those at which the immune system was adversely affected in dolphins and in an *in vitro* assay where the white blood cells from healthy beluga whales became immunologically suppressed. The synergy between environmentally relevant concentrations of TBT and PCB 126 that leads to a twenty-fold increase in Ah enzyme activity could enhance considerably the dioxin-like effect of PCB 126 *in vivo* and raises the prospect that other compounds with dioxin-like activity could be potentiated as well. This effect would never be detected using only dioxin TEQs.

New information about the developmental effects of plastic additives and monomers found in the whales' environment makes estimates of exposure more complicated. As global patination with toxic chemicals continues, the toxicity mixtures, including additive and synergistic effects, must be considered in all hazard assessments of cetacean health. Samples of new and archived baleen and toothed whale tissues should be analyzed not only for organochlorine chemicals and metals but also for the material leaching from plastics. Many of these chemicals were manufactured before most of the organochlorine chemicals and have been building up in the environment since the First World War. Northern and southern hemisphere prevailing winds are delivering the volatile plastic additives, as well as herbicides, insecticides, fungicides and nematocides, toward the poles. Biologists should watch vigilantly over the fragile polar systems for signals from their impacts on both wildlife and primary production.

A logical strategy now is to focus on the embryo, fetus, newborns and gravid females in whale populations. More information is needed about where the females spend their time feeding during gestation and lactation. More insight is needed about both male and female metabolic systems, especially in those species where the females fast throughout their calves' early months, when contaminants stored in the mothers' tissue may be mobilized. Whale experts should be working closely with bear specialists who are attempting to learn more about the metabolism of fasting. More data on the positions of various cetacean species in the marine trophic system is needed. Most important, there is the need for a fully equipped expert team of marine biologists trained to collect fresh blood from free-ranging whales, in order to address many of these questions. It will take international and intergovernmental agency cooperation to address these questions.

Because of the widespread perturbation of the oceans' chemistry over the past 75 years, systematic surveillance of large cetaceans and contaminants must broaden to include plastics additives and components, other industrial chemicals such as those mentioned above, contemporary-use pesticides, and metals such as mercury and tributyltin that are still accumulating in the marine ecosystem. Marine biologists should incorporate into their studies the rapidly growing knowledge about low-dose effects of environmental contaminants that interfere with the endocrine system and development. They should think in terms of insidious, generational effects, and watch cetacean populations carefully for changes in successive age-classes that might include alterations in feeding habits, social grouping, geographic distribution, migration behavior, seasonal activity, susceptibility to disease, reproductive success and population age structure. Changes in these parameters are already reported in the literature about several populations of large cetaceans. It is time to consider the role of contaminants in these findings.

References

Addison, R.F. and Zinck, M.E. 1986. PCBs have declined more than DDT-group residues in Arctic ringed seals (*Phoca hispida*) between 1972 and 1981. *Environmental Science and Technology* 20(3): 253–256.

Ahlborg, U.G. 1994. Toxic equivalency factors for dioxin-like PCBs. *Chemosphere* 28: 1049–1067.

Alexander, V. 1995. The influence of the structure and function of the marine food web on the dynamics of contaminants in Arctic Ocean ecosystems. *Science of the Total Environment* 160–161: 593–603.

Aono, S., Tanabe, S., Fujise, Y., Kato, H. and Tatsukawa, R. 1998. Persistent organochlorines in minke whale (*Balaenoptera acutorostrata*) and their prey species from the Antarctic and the North Pacific. *Environmental Pollution* 98(1): 81–89.

Arctic Monitoring and Assessment Programme. 1997. *Arctic pollution issues: a state of the Arctic environment report*. Oslo: Arctic Monitoring and Assessment Programme.

Augier, H., Park, W. and Ronneau, C. 1993. Mercury contamination of the striped dolphin *Stenella coeruleoalba Meyen* from the French Mediterranean coasts. *Marine Pollution Bulletin* 26(6): 306–311.

Auman, H.J., Ludwig, J.P., Giesy, J.P. and Colborn, T. 1997. Plastic ingestion by Laysan albatross chicks on Sand Island, Midway Atoll, in 1994 and 1995. In: *Albatross Biology and Conservation* (Robinson, G. and Gales, R., eds), pp. 239–244. Chipping Norton, NSW: Surrey Beatty & Sons.

Ballschmiter, K.H., Froescheis, O., Jarman, W.M. and Caillet, G. 1997. Contamination of the deep-sea. *Marine Pollution Bulletin* 14(5): 288–289.

Barrie, L.A., Gregor, D., Hargrave, B., Lake, R., Muir, D., Shearer, R., Tracey, B. and Bidleman, T. 1992. Arctic contaminants: sources, occurrence and pathways. *Science of the Total Environment* 122: 1–74.

Baskin, L.S., Himes, K. and Colborn, T. 2001. Hypospadias and endocrine disruption: is there a connection? *Environmental Health Perspectives* 109(11): 1175–1183.

Béland, P. 1996a. The beluga whales of the St. Lawrence River. *Scientific American* 274(5): 74–81.

Béland, P. 1996b. *Beluga: A Farewell to Whales*. New York, NY: Lyons and Burford.

Béland, P., De Guise, S. and Plante, R. 1992. Toxicology and pathology of St. Lawrence marine mammals. *Final Report: World Wildlife Fund's Wildlife Toxicology Fund Research Grant, 1988–1991*. St. Lawrence National Institute of Ecotoxicology, Québec, Canada.

Béland, P., De Guise, S., Girard, C., Lagace, A., Martineau, D., Michaud, R., Muir, D.C.G., Norstrom, R.J., Pelletier, E., Ray, S. and Shugart, L.R. 1993. Toxic compounds and health and reproductive effects in St. Lawrence beluga whales. *Journal of Great Lakes Research* 19(4): 766–775.

Bern, H. 1992. The fragile fetus. In: *Chemically-Induced Alterations in Sexual and Functional Development: The Wildlife/Human Connection* (Colborn, T. and Clement, C. eds), pp. 9–15. Princeton, NJ: Princeton Scientific Publishing.

Bester, K., Hühnerfuss, H., Brockmann, U. and Rick, H.J. 1995. Biological effects of triazine herbicide contamination on marine phytoplankton. *Archives of Environmental Contamination and Toxicology* 29(3): 277–283.

Bintein, S. and Devillers, J. 1996. Evaluating the environmental fate of atrazine in France. *Chemosphere* 32(12): 2441–2456.

Blight, L.K. and Burger, A.E. 1997. Occurrence of plastic particles in seabirds from the eastern North Pacific. *Marine Pollution Bulletin* 34(5): 323–325.

Borrell, A., Aguilar, A., Corsolini, S. and Focardi, S. 1996. Evaluation of toxicity and sex-related variation of PCB levels in Mediterranean striped dolphins affected by an epizootic. *Chemosphere* 32(12): 2359–2369.

Bossart, G.D., Ewing, R., Herron, A.J., Cray, C., Mase, B., Decker, S.J., Alexander, J.W. and Altman, N.H. 1997. Immunoblastic malignant lymphoma in dolphins: histologic, ultrastructural, and immunohistochemical features. *Journal of Veterinary Diagnostic Investigation* 9(4): 454–458.

Brouwer, A., Ahlborg, U.G., Van den Berg, M., Birnbaum, L.S., Boersma, E.R., Bosveld, B., Denison, M.S., Gray, L.E. Jr, Hagmar, L., Holene, E., Huisman, M., Jacobson, S.W., Jacobson, J.L., Koopman-Esseboom, C., Koppe, J.G., Kulig, B.M., Morse, D.C., Muckle, G., Peterson, R.E., Sauer, P.J.J., Seegal, R.F., Smits-van Prooije, A.E., Touwen, B.C.L., Weisglas-Kuperus, N. and Winneke, G. 1995. Functional aspects of developmental toxicity of polyhalogenated aromatic hydrocarbons in experimental animals and human infants. *European Journal of Pharmacology* 293(1): 1–40.

Brouwer, A., Morse, D.C., Lans, M.C., Schuur, A.G., Murk, A.J., Klasson-Wehler, E., Bergman, Å. and Visser, T.J. 1998. Interactions of persistent environmental organohalogens with the thyroid hormone system: mechanisms and possible consequences for animal and human health. *Toxicology and Industrial Health* 14(1–2): 59–84.

Brown, M.W., Kraus, S.D., Gaskin, D.E. and White, B.N. 1994. Sexual composition and analysis of reproductive females in the North Atlantic right whale, *Eubalaena glacialis*, population. *Marine Mammal Science* 10(3): 353–365.

Brucker-Davis, F., Skarulis, M.C., Pikus, A., Ishizawar, D., Mastroianni, M., Koby, M. and Weintraub, B.D. 1996. Prevalence and mechanisms of hearing loss in patients with resistance to thyroid hormone. *Journal of Clinical Endocrinology and Metabolism* 81(8): 2768–2772.

Bryan, G.W., Gibbs, P.E., Burt, G.R. and Hummerstone, L.G. 1987. The effects of tributyltin (TBT) accumulation on adult dog-whelks, *Nucella lapillus*: long-term field and laboratory experiments. *Journal of the Marine Biology Association* 67: 525–544.

Burger, J. and Gochfeld, M. 1995. Biomonitoring of heavy metals in the Pacific basin using avian feathers. *Environmental Toxicology and Chemistry* 14(7): 1233–1239.

Chem Expo. 1998. *PTA/DMT: Chem Expo Industry News Chemical Profile*. Available: http://www.chemexpo.com/news/newsframe.cfm?framebody=/news/profile.cfm.

Chernyak, S.M., Rice, C.P. and McConnell, L.L. 1996. Evidence of currently-used pesticides in air, ice, fog, seawater and surface microlayer in the Bering and Chukchi Seas. *Marine Pollution Bulletin* 32(5): 410–419.

Colborn, T. and Clement, C., eds. 1992. *Chemically-Induced Alterations in Sexual and Functional Development: The Wildlife/Human Connection*. Princeton, NJ: Princeton Scientific Publishing.

Colborn, T. and Smolen, M. 1996. Epidemiological analysis of persistent organochlorine contaminants in cetaceans. *Reviews of Environmental Contamination and Toxicology* 146: 91–172.

Colborn, T., vom Saal, F. and Soto, A. 1993. Developmental effects of endocrine-disrupting chemicals in wildlife and humans. *Environmental Health Perspectives* 101(5): 378–384.

Colborn, T., vom Saal, F.S. and Short, P., eds. 1998. Environmental endocrine-disrupting chemicals: neural, endocrine, and behavioral effects. *Toxicology and Industrial Health* 14(1–2).

Cook, P.M., Zabel, E.W. and Peterson, R.E. 1997. The TCDD toxicity equivalence approach for characterizing risks for early life-stage mortality in trout. In: *Chemically Induced Alterations in Functional Development and Reproduction of Fishes: Proceedings from a Session at the 1995 Wingspread Conference* (Rolland, R., Gilbertson, M. and Peterson, R.E., eds), pp. 9–22. Pensacola, Fla: Society of Environmental Toxicology and Chemistry.

Daly, H.B. 1991. Reward reductions found more aversive by rats fed environmentally contaminated salmon. *Neurotoxicology and Teratology* 13(4): 449–453.

Daly, H.B. 1993. Laboratory rat experiments show consumption of Lake Ontario salmon causes behavioral changes: support for wildlife and human research results. *Journal of Great Lakes Research* 19(4): 784–788.

Daskalakis, K.D. and O'Connor, T.P. 1995. Distribution of chemical concentrations in US coastal and estuarine sediment. *Marine Environmental Research* 40(4): 381–398.

de Boer, J., Wester, P.G., Klamer, H.J.C., Lewis, W.E. and Boon, J.P. 1998. Do flame retardants threaten ocean life? *Nature* 394(6688): 28–29.

De Guise, S., Lagace, A. and Béland, P. 1991. True hermaphroditism in a St. Lawrence beluga whale (*Delphinatperus leucas*). *Journal of Wildlife Diseases* 30(2): 287–290.

De Guise, S., Bisaillon, A., Segun, B. and Lagace, A. 1994a. The anatomy of the male genital system of the beluga whale, *Delphinapterus leucas*, with special reference to the penis. *Anatomia, Histologia, Embryologia* 23(3): 207–216.

De Guise, S., Lagace, A. and Béland, P. 1994b. Tumors in St. Lawrence beluga whales (*Delphinapterus leucas*). *Veterinary Pathology* 31(4): 444–449.

De Guise, S., Lagace, A., Béland, P., Girard, C. and Higgins, R. 1995. Non-neoplastic lesions in beluga whales (*Delphinapterus leucas*) and other marine mammals from the St. Lawrence estuary. *Journal of Comparative Pathology* 112(3): 257–271.

De Guise, S., Bernier, J., Dufresne, M.M., Martineau, D., Béland, P. and Fournier, M. 1996a. Immune functions in beluga whales (*Delphinapterus leucas*): evaluation of mitogen-induced blastic transformation of lymphocytes from peripheral blood, spleen and thymus. *Veterinary Immunology and Immunopathology* 50: 117–126.

De Guise, S., Bernier, J., Martineau, D., Béland, P. and Fournier, M. 1996b. Effects of *in vitro* exposure of beluga whale splenocytes and thymocytes to heavy metals. *Environmental Toxicology and Chemistry* 15(8): 1357–1364.

DeLong, G.T. and Rice, C.D. 1997. Tributyltin potentiates 3,3′,4,4′,5-pentachlorobiphenyl-induced cytochrome P4501A-related activity. *Journal of Toxicology and Environmental Health* 51: 131–148.

de Swart, R.L., Ross, P.R., Vos, J.G. and Osterhaus, A.D.M.E. 1996. Impaired immunity in harbour seals (*Phoca vitulina*) exposed to bioaccumulated environmental contaminants: review of a long-term feeding study. *Environmental Health Perspectives* 104(Suppl 4): 823–828.

Dietz, R., Riget, F. and Johansen, P. 1996. Lead, cadmium, mercury and selenium in Greenland marine animals. *Science of the Total Environment* 186(1–2): 67–93.

Drescher, H., Harma, U. and Huschenbeth, E. 1977. Organochlorines and heavy metals in the harbour seal *Phoca vitulina* from the German North Sea coast. *Marine Biology* 141: 99–106.

Eisenreich, S.J., Looney, B.B. and Thornton, J.D. 1981. Airborne organic contaminants in the Great Lakes ecosystem. *Environmental Science and Technology* 15(1): 30–38.

Ekelund, R., Bergman, A., Granmo, A. and Berggren, M. 1990. Bioaccumulation of 4-nonylphenol in marine animals: a re-evaluation. *Environmental Pollution* 64: 107–120.

Ema, M., Kurasaka, R., Amano, H. and Ogawa, Y. 1995. Comparative developmental toxicity of butyltin trichloride, dibutyltin dichloride and tributyltin chloride in rats. *Journal of Applied Toxicology* 14(4): 297–302.

Fellin, P., Barrie, L.A., Dougherty, D., Toom, D., Muir, D., Grift, N., Lockhart, L. and Billeck, B. 1996. Air monitoring in the Arctic: results for selected persistent organic pollutants for 1992. *Environmental Toxicology and Chemistry* 15(3): 253–261.

Foster, W.G., Desaulnier, D., Leingartner, K., Wade, M.G., Poon, R. and Chu, I. 1999. Reproductive effects of *tris*(4-chlorophenyl)methanol in the rat. *Chemosphere* 39: 709–724.

Fowles, J.R., Fairbrother, A., Baecher-Steppan, L. and Kerkvliet, N.I. 1994. Immunologic and endocrine effects of the flame-retardant pentabromodiphenyl ether (DE-71) in C57BL/6J mice. *Toxicology* 86(1–2): 49–61.

Furness, R.W. and Camphuysen, K.C.J. 1997. Seabirds as monitors of the marine environment. *ICES Journal of Marine Science* 54: 726–737.

Gauthier, J.M., Metcalf, C.D. and Sears, R. 1997. Chlorinated organic contaminants in blubber biopsies from northwestern Atlantic Balaenopterid whales summering in the Gulf of St. Lawrence. *Marine Environmental Research* 44(2): 201–223.

Giam, C.S., Chan, H.S., Neff, G.S. and Atlas, E.L. 1978. Phthalate ester plasticizers: a new class of marine pollutant. *Science* 199(4327): 419–421.

Gibbs, P.E., Bryan, G.W., Pascoe, P.L. and Burt, G.R. 1987. The use of the dog-whelk, *Nucella lapillus*, as an indicator of tributyltin (TBT) contamination. *Journal of the Marine Biology Association UK* 67: 507–523.

Giesy, J.P. and Kannan, K. 2001. Global distribution of perfluorooctane sulfonate in wildlife. *Environmental Science and Technology* 35: 1339–1342.

Gilbertson, M., Kubiak, T.J., Ludwig, J. and Fox, G. 1991. Great Lakes embryo mortality, edema, and deformities syndrome (GLEMEDS) in colonial fish-eating birds: similarity to chick-edema disease. *Journal of Toxicology and Environmental Health* 33(4): 455–520.

Goldberg, E.D. 1997. Plasticizing the seafloor: an overview. *Environmental Technology* 18: 195–201.

Goldey, E.S. 1996. Chemical disruption of auditory system development with a closer look at the effects of polychlorinated biphenyls. *Neurobehavioral Teratology Society Abstracts* 18: 330.

Goldey, E.S., Kehn, L.S., Rehnberg, G.L. and Crofton, K.M. 1995. Effects of developmental hypothyroidism on auditory and motor function in the rat. *Toxicology and Applied Pharmacology* 135: 67–76.

Goodbred, S.L., Gilliom, R.J., Gross, T.S., Denslow, N.P., Bryant, W.L. and Schoeb, T.R. 1997. *Reconnaissance of 17β-estradiol, 11-ketotestosterone, vitellogenin, and gonadal histopathology in common carp of United States streams: potential for contaminant-induced endocrine disruption.* Sacramento, Calif: US Geological Survey Open-File Report 96–627.

Gray, L.E. Jr 1992. Chemical-induced alterations of sexual differentiation: a review of effects in humans and rodents. In: *Chemically Induced Alterations in Sexual and Functional Development: The Wildlife/Human Connection* (Colborn, T. and Clement, C., eds), pp. 203–230. Princeton, NJ: Princeton Scientific Publishing.

Gray, L.E. Jr 1998. Xenoendocrine disrupters: laboratory studies on male reproductive effects. *Toxicology Letters* 102–103: 331–335.

Gray, L.E. Jr, Price, M., Lambright, C., Wolf, C., Hotchkiss, A., Parks, L. and Ostby, J. 1999a. Environmental antiandrogens: the malformation pattern varies with the mechanism of antiandrogenic action. *Biology of Reproduction* 60(Suppl 1): 201.

Gray, L.E. Jr, Wolf, C., Lambright, C., Mann, P., Price, M., Cooper, R.L. and Ostby, J. 1999b. Administration of potentially antiandrogenic pesticides (procymidone, linuron, iprodione, chlozolinate, *p,p'*-DDE, and ketoconazole) and toxic substances (dibutyl- and diethylhexyl phthalate, PCB 169, and ethane dimethane sulphonate) during sexual differentiation produces diverse profiles of reproductive malformations in the male rat. *Toxicology and Industrial Health* 15(1–2): 94–118.

Gregor, D.J., Peters, A.J., Teixeira, C., Jones, N. and Spencer, C. 1995. The historical residue trend of PCBs in the Agassiz Cap, Ellesmere Island, Canada. *Science of the Total Environment* 160–161: 117–126.

Gregory, M.R. 1996. Plastic 'scrubbers' in hand cleansers: a further (and minor) source for marine pollution identified. *Marine Pollution Bulletin* 32(12): 867–871.

Gulland, F.M.D., Trupkiewicz, J.G., Spraker, T.R. and Lowenstine, L.J. 1996. Metastatic carcinoma of probable transitional cell origin in 66 free-living California sea lions (*Zalophus californianus*), 1979 to 1994. *Journal of Wildlife Diseases* 32(2): 250–258.

Hall, A.J., Green, N.J.L., Jones, K.C., Pomeroy, P.P. and Harwood, J. 1998. Thyroid hormones as biomarkers in grey seals. *Marine Pollution Bulletin* 36(6): 424–428.

Hardy, J.T. 1987. Anthropogenic alteration of the sea surface. *Marine Environmental Research* 23: 223–225.

Hardy, J.T. and Crecelius, E.A. 1981. Is atmospheric particulate matter inhibiting marine primary productivity? *Environmental Science and Technology* 15(8): 1103–1105.

Heidrich, D., Steckelbroeck, S., Bidlingmaier, F. and Klingmuller, D. 1999. Effect of tributyltinchloride (TBT) on human aromatase activity. In: *Proceedings of the 81st Annual Meeting of the Endocrine Society*. Bethesda, Md.: The Endocrine Society.

Herr, D.W., Goldey, E.S. and Crofton, K.M. 1996. Developmental exposure to Aroclor 1254 produces low-frequency alterations in adult rat brainstem auditory evoked responses. *Fundamental and Applied Toxicology* 33(1): 120–128.

Holsbeek, L., Das, H.K. and Joiris, C.R. 1997. Mercury speciation and accumulation in Bangladesh freshwater and anadromous fish. *Science of the Total Environment* 198: 201–210.

Howdeshell, K.L., Hotchkiss, A.K., Thayer, K.A., Vandenbergh, J.G. and vom Saal, F.S. 1999. Exposure to bisphenol A advances puberty. *Nature* 401(6755): 763–764.

Iwata, H., Tanabe, S., Sakai, N. and Tatsukawa, R. 1993. Distribution of persistent organochlorines in the oceanic air and surface seawater and the role of ocean on their global transport and fate. *Environmental Science and Technology* 27: 1080–1098.

Iwata, H., Tanabe, S., Mizuno, T. and Tatsukawa, R. 1997. Bioaccumulation of butyltin compounds in marine mammals: the specific distribution and composition. *Applied Organometallic Chemistry* 11: 257–264.

Jacobson, J.L. and Jacobson, S.W. 1996. Intellectual impairment in children exposed to polychlorinated biphenyls *in utero*. *New England Journal of Medicine* 335(11): 783–789.

Jarman, W.M., Simon, M., Norstrom, R.J., Burns, S.A., Bacon, C.A., Simoneit, B.R.T. and Risebrough, R.W. 1992. Global distribution of *tris*(4-chlorophenyl)methanol in high trophic level birds and mammals. *Environmental Science and Technology* 26: 1770–1774.

Jensen, A.A. and Slorach, S.A. 1991. *Chemical Contaminants in Human Milk*. Boston, Mass: CRC Press.

Jobling, S. and Sumpter, J.P. 1993. Detergent components in sewage effluent are weakly oestrogenic to fish: an *in vitro* study using rainbow trout (*Oncorhynchus mykiss*) hepatocytes. *Aquatic Toxicology* 27: 361–372.

Jobling, S., Reynolds, T., White, R., Parker, M.G. and Sumpter, J.P. 1995. A variety of environmentally persistent chemicals, including some phthalate plasticizers, are weakly estrogenic. *Environmental Health Perspectives* 103(6): 582–587.

Johansen, P., Pars, T. and Bjerregaard, P. 2000. Lead, cadmium, mercury and selenium intake by Greenlanders from local marine food. *Science of the Total Environment* 245(1–3): 187–194.

Jones, P.D., Hannah, D.J., Buckland, S.J., Day, P.J., Leathem, S.V., Porter, L.J., Auman, H.J., Sanderson, J.T., Summer, C., Ludwig, J.P., Colborn, T. and Giesy, J.P. 1996. Persistent synthetic chlorinated hydrocarbons in albatross tissue samples from Midway Atoll. *Environmental Toxicology and Chemistry* 15(10): 1793–1800.

Kannan, K. and Falandysz, J. 1997. Butyltin residues in sediment, fish-eating birds, harbour porpoise and human tissues from the Polish coast of the Baltic Sea. *Marine Pollution Bulletin* 34(3): 203–207.

Kannan, K., Tanabe, S. and Tatsukawa, R. 1995. Occurrences of butyltin residues in certain foodstuffs. *Bulletin of Environmental Contamination and Toxicology* 55: 510–516.

Kannan, K., Senthilkumar, K., Lohanathan, B.G., Takahashi, S., Odell, D.K. and Tanabe, S. 1997. Elevated accumulation of tributyltin and its breakdown products in bottlenose dolphins (*Tursiops truncatus*) found stranded along the U.S. Atlantic and Gulf Coasts. *Environmental Science and Technology* 31(1): 296–301.

Kelce, W.R., Stone, C., Laws, S., Gray, L.E. Jr., Kemppainen, J.A. and Wilson, E.M. 1995. Persistent DDT metabolite *p,p'*-DDE is a potent androgen receptor antagonist. *Nature* 375(6532): 581–585.

Keys, B., Hlavinka, M., Mason, G. and Safe, S. 1985. Modulation of rat hepatic microsomal testosterone hydroxylases by 2,3,7,8-tetrachlorodibenzo-*p*-dioxin and related toxic isostereomers. *Canadian Journal of Physiology and Pharmacology* 63(12): 1537–1542.

Kim, G.B., Lee, J.S., Tanabe, S., Iwata, H., Tatsukawa, R. and Shimazaki, K. 1996a. Specific accumulation and distribution of butyltin compounds in various organs and tissues of the Steller sea lion (*Eumetopias jubatus*): comparison with organochlorine accumulation pattern. *Marine Pollution Bulletin* 32(7): 558–563.

Kim, G.B., Tanabe, S., Iwakiri, R., Tatsukawa, R., Amano, M., Miyazaki, N. and Tanaka, H. 1996b. Accumulation of butyltin compounds in Risso's dolphin

(*Grampus griseus*) from the Pacific Coast of Japan: comparison with organochlorine residue pattern. *Environmental Science and Technology* 30: 2620–2625.

Kocan, R.M., von Westernhagen, H., Landolt, M.L. and Furstenberg, G. 1987. Toxicity of sea-surface microlayer: effects of hexane extract on Baltic herring (*Clupea harengus*) and Atlantic cod (*Gadus morhua*) embryos. *Marine Environmental Research* 23: 291–305.

Koopman-Esseboom, C., Morse, D.C., Weisglas-Kuperus, N., Lutkeschipholt, I.J., van der Paauw, C.G., Tuinstra, L.G.M.T., Brouwer, A. and Sauer, P.J.J. 1994. Effects of dioxins and polychlorinated biphenyls on thyroid hormone status of pregnant women and their infants. *Pediatric Research* 36(4): 468–473.

Koopman-Esseboom, C., Weisglas-Kuperus, N., de Ridder, M.A.J., van der Paauw, C.G., Tuinstra, L.G.M.T. and Sauer, P.J.J. 1996. Effects of polychlorinated biphenyl/dioxin exposure and feeding type on infants' mental and psychomotor development. *Pediatrics* 97(5): 700–706.

Kubiak, T.J., Harris, H.J., Smith, L.M., Schwartz, T.R., Stalling, D.L., Trick, J.A., Sileo, L., Doucherty, D.E. and Erdman, T.C. 1989. Microcontaminants and reproductive impairment of the Forster's tern on Green Bay, Lake Michigan, 1983. *Archives of Environmental Contamination and Toxicology* 18(5): 706–727.

Lahvis, G.P., Wells, R.E., Kuehl, D.W., Stewart, J.L., Rhinehart, H.L. and Via, C.S. 1995. Decreased *in vitro* lymphocyte responses in free-ranging bottlenose dolphins (*Tursiops truncatus*) are associated with increased whole blood concentrations of polychlorinated biphenyls (PCBs) and *o,p'*-DDT, *p,p'*-DDE, and *o,p'*-DDE. *Environmental Health Perspectives* 103(Suppl 4): 67–72.

Leatherland, J.F. 1998. Changes in thyroid hormone economy following consumption of environmentally contaminated Great Lakes fish. *Toxicology and Industrial Health* 14(1–2): 41–58.

Lebeuf, M., Bernt, K.E., Trottier, S., Noel, M., Hammill, M.O. and Measures, L. 2001. *Tris*(4-chlorophenyl)methane and *tris*(4-chlorophenyl)methanol in marine mammals from the Estuary and Gulf of St. Lawrence. *Environmental Pollution* 111(1): 29–43.

Letcher, R.J., van Holsteijn, I., Drenth, H.J., Norstrom, R.J., Bergman, A., Safe, S., Pieters, R. and van den Berg, M. 1999. Cytotoxicity and aromatase (CYP19) activity modulation by organochlorines in human placental JEG-3 and JAR choriocarcinoma cells. *Toxicology and Applied Pharmacology* 160(1): 10–20.

Letcher, R.J., Klasson-Wehler, E. and Bergman, A. 2000a. Methyl sulfone and hydroxylated metabolites of polychlorinated biphenyls. *The Handbook of Environmental Chemistry* 3: 315–359.

Letcher, R.J., Norstrom, R.J., Muir, D.C.G., Sandau, C.D., Koczanski, K., Nichaud, R., De Guise, S. and Béland, P. 2000b. Methylsulfone polychlorinated biphenyl and 2,2-bis(chlorophenyl)-1,1-dichloroethylene metabolites in beluga whales (*Delphinapterus leucas*) from the St. Lawrence River estuary and western Hudson Bay, Canada. *Environmental Toxicology and Chemistry* 19(5): 1378–1388.

Lonky, E., Reihman, J., Darvill, T., Mather, J. and Daly, H. 1996. Neonatal behavioral assessment scale performance in humans influenced by maternal consumption of environmentally contaminated Lake Ontario fish. *Journal of Great Lakes Research* 22(2): 198–212.

Lowenthal, D.H., Borys, R.D. and Mosher, B.W. 1997. Sources of pollution aerosol at Dye 3, Greenland. *Atmospheric Environment* 31(22): 3707–3717.

Mackay, D., Paterson, S. and Shiu, W.Y. 1992. Generic models for evaluating regional fate of chemicals. *Chemosphere* 24: 695–717.

McLachlan, J.A., Newbold, R.R., Korach, K.S., Lamb, J.C. and Suzuki, Y. 1981. Transplacental toxicology: prenatal factors influencing postnatal fertility. In: *Developmental Toxicology* (Kimmel, C.A. and Buelke-Sam, J., eds), pp. 213–232. New York: Raven Press.

Minh, T.B., Watanabe, M., Tanabe, S., Miyazaki, N., Jefferson, T.A., Prudente, M.S., Subramanian, A. and Karuppiah, S. 2000a. Widespread contamination by *tris*(4-chlorophenyl)methane and *tris*(4-chlorophenyl)methanol in cetaceans from the North Pacific and Asian coastal waters. *Environmental Pollution* 110(3): 459–468.

Minh, T.B., Watanabe, M., Tanabe, S., Yamada, T., Hata, J. and Watanabe, S. 2000b. Occurrence of *tris*(4-chlorophenyl)methane, *tris* (4-chlorophenyl) methanol, and some other persistent organochlorines in Japanese human adipose tissue. *Environmental Health Perspectives* 108(7): 599–603.

Monteiro, L.R. and Furness, R.W. 1997. Accelerated increase in mercury contamination in north Atlantic mesopelagic food chains as indicated by the time series of seabird feathers. *Environmental Toxicology and Chemistry* 16(12): 2489–2493.

Muir, D.C.G., Ford, C.A., Rosenbeng, B., Norstrom, R.S., Simon, M. and Béland, P. 1996. Persistent organochlorines in beluga whales (*Delphinapterus leucas*) from the St. Lawrence River Estuary. I. Concentrations and patterns of specific PCBs, chlorinated pesticides and polychlorinated dibenzo-*p*-dioxins and dibenzofurans. *Environmental Pollution* 93(2): 219–234.

Nicholson, G.J. and Evans, S.M. 1997. Anthropogenic impacts on the stocks of the common whelk *Buccinum undatum* (L.). *Marine Environmental Research* 44(3): 305–314.

Noren, K. and Meironyte, D. 2000. Certain organochlorine and organobromine contaminants in Swedish human milk in perspective of past 20–30 years. *Chemosphere* 40(9–11): 1111–1123.

Nriagu, J.O., Cokjer, R.D. and Barrie, L.A. 1991. Origin of sulphur in Canadian Arctic haze from isotope measurements. *Nature* 349: 142–145.

Oehlmann, J., Fioroni, P., Stroben, E. and Markert, B. 1996. Tributyltin (TBT) effects on *Ocinebrina aciculata* (Gastropoda: Muricidae): imposex development, sterilization, sex change and population decline. *Science of the Total Environment* 188: 205–223.

Ono, M., Kannan, N., Wakimoto, T. and Tatsukawa, R. 1987. Dibenzofurans a greater global pollutant than dioxins? Evidence from analyses of open ocean killer whale. *Marine Pollution Bulletin* 18(12): 641–643.

Parks, L.G., Ostby, J.S., Lambright, C.R., Abbott, B.D., Gray, L.E. Jr 1999. Perinatal butyl benzyl phthalate (BBP) and bis(2-ethylhexyl)phthalate (DEHP) exposure induce antiandrogenic effects in Sprague–Dawley (SD) rats. *Biology of Reproduction* 60(Suppl 1): 153.

Peakall, D.B. 1975. Phthalate esters: occurrence and biological effects. *Residue Reviews* 54: 1–41.

Peterson, R.E., Moore, R.W., Mably, T.A., Bjerke, D.L. and Goy, R.W. 1992. Male reproductive system ontogeny: effects of perinatal exposure to 2,3,7,8-tetrachlorodibenzo-*p*-dioxin. In: *Chemically-Induced Alterations in Sexual and Functional Development: The Wildlife/Human Connection* (Colborn, T. and Clement, C., eds). Princeton, NJ: Princeton Scientific Publishing.

Pfirman, S.L., Eicken, H., Bauch, D. and Weeks, W.F. 1995. The potential transport of pollutants by Arctic Sea ice. *Science of the Total Environment* 159: 129–146.

Porterfield, S. and Stein, S.A. 1994. Thyroid hormones and neurological development: update 1994. *Endocrine Reviews* 3(1): 537–563.

Renner, R. 2000. Increasing levels of flame retardants found in North American environment. *Environmental Science and Technology* 34(21): 452A–453A.

Riget, F. and Dietz, R. 2000. Temporal trends of cadmium and mercury in Greenland marine biota. *Science of the Total Environment* 245(1–3): 49–60.

Rolland, R. 2000. A review of chemically-induced alterations in thyroid and vitamin A status from field studies of wildlife and fish. *Journal of Wildlife Diseases* 36(4): 615–635.

Rolland, R., Gilbertson, M. and Colborn, T., eds. 1995. Environmentally induced alterations in development: a focus on wildlife. *Environmental Health Perspectives* 103(Suppl 4).

Rolland, R.M., Gilbertson, M., Peterson, R.E., eds. 1997. *Chemically Induced Alterations in Functional Development and Reproduction of Fishes: Proceedings from a Session at the 1995 Wingspread Conference.* Pensacola, Fla: Society of Environmental Toxicology and Chemistry.

Ross, P.S., de Swart, R.L., Reijnders, P.J.H., Lovern, N.V., Vos, J.G. and Osterhaus, A.D.M.E. 1995. Contaminant-related suppression of delayed-type hypersensitivity and antibody responses in harbor seals fed herring from the Baltic Sea. *Environmental Health Perspectives* 103(2): 162–167.

Ross, P.S., de Swart, R., Addison, R., van Loveren, H., Vos, J. and Osterhaus, A. 1996a. Contaminant-induced immunotoxicity in harbour seals: wildlife at risk? *Toxicology* 112(2): 157–169.

Ross, P.S., de Swart, R.L., Timmerman, H.H., Reijnders, P.J.H., Vos, J.G., van Loveren, H. and Osterhaus, A.D.M.E. 1996b. Suppression of natural killer cell activity in harbour seals (*Phoca vitulina*) fed Baltic Sea herring. *Aquatic Toxicology* 34(1): 71–84.

Safe, S. 1990. Polychlorinated biphenyls (PCBs), dibenzo-*p*-dioxins (PCDDs), dibenzofurans (PCDFs) and related compounds: environmental and mechanistic considerations which support the development of toxic equivalent factors (TEFs). *Critical Reviews in Toxicology* 21(1): 51–88.

Sanderson, J.T., Janz, D.M., Bellward, G.D. and Giesy, J.P. 1997. Effects of embryonic and adult exposure to 2,3,7,8-tetrachlorodibenzo-*p*-dioxin on hepatic microsomal testosterone hydroxylase activities in great blue herons (*Ardea herodias*). *Environmental Toxicology and Chemistry* 16(6): 1304–1310.

Schottler, S.P. and Eisenreich, S.J. 1994. Herbicides in the Great Lakes. *Environmental Science and Technology* 28: 2228–2232.

Schottler, S.P., Eisenreich, S.J. and Capel, P.D. 1994. Atrazine, alachlor, and cyanazine in a large agricultural river system. *Environmental Science and Technology* 28: 1079–1089.

Sharpe, R.M., Fisher, J.S., Millar, M.M., Jobling, S. and Sumpter, J.P. 1995. Gestational and lactational exposure of rats to xenoestrogens results in reduced testicular size and sperm production. *Environmental Health Perspectives* 103(12): 1136–1143.

Shephard, R.J. and Itoh, S., eds. 1976. *Circumpolar Health. Proceedings of the Third International Symposium, Yellowknife, NWT.* Toronto: University of Toronto Press.

Simms, W., Jeffries, S., Ikonomou, M. and Ross, P. S. 2000. Contaminant-related disruption of vitamin A dynamics in free-ranging harbor seal (*Phoca vitulina*) pups from British Columbia, Canada, and Washington State, USA. *Environmental Toxicology and Chemistry* 19(11): 2844–2849.

Smialowicz, R.J., Riddle, M.M., Rogers, R.R., Luebke, R.W. and Copeland, C.B. 1989. Immunotoxicity of tributyltin oxide in rats exposed as adults or pre-weanlings. *Toxicology* 57(1): 97–111.

Snoeij, N.J., Penninks, A.H. and Seinen, W. 1988. Dibutyltin and tributyltin compounds induce thymus atrophy in rats due to selective action on thymic lymphoblasts. *International Journal of Immunopharmacology* 10(7): 891–899.

Society of the Plastics Industry 1997. *Facts and figures of the U.S. plastics industry: 1996 Edition*. Washington, DC: The Society of the Plastics Industry.

Soto, A.M., Justice, H., Wray, J.W. and Sonnenschein, C. 1991. *p*-Nonyl-phenol: an estrogenic xenobiotic released from 'modified' polystyrene. *Environmental Health Perspectives* 92: 167–173.

St. Aubin, D.J., Ridgway, S.H., Wells, R.S. and Rhinehart, H. 1996. Dolphin thyroid and adrenal hormones: circulating levels in wild and semidomesticated *Tursiops truncatus*, and influence of sex, age, and season. *Marine Mammal Science* 12(1): 1–13.

Stuer-Lauridsen, F. and Dahl, B. 1995. Source of organotin at a marine water/ sediment interface: a field study. *Chemosphere* 30(5): 831–845.

Swain, E.B., Engstrom, D.R., Brigham, M.E., Henning, T.A. and Brezonik, P.L. 1992. Increasing rates of atmospheric mercury deposition in midcontinental North America. *Science* 257: 784–787.

Takai, Y., Tsutsumi, O., Ikezuki, Y., Kamei, Y., Osuga, Y., Yano, T. and Taketan, Y. 2001. Preimplantation exposure to bisphenol A advances postnatal development. *Reproductive Toxicology* 15(1): 71–74.

Takei, G.H. and Sawa, T.R. 1995. The distribution and characterization of organochloride residues in animal tissues: an environmental and toxicological assessment. Report for the Department of Natural Resources, Hawaii.

Tanabe, S., Iwata, H. and Tatsukawa, R. 1994. Global contamination by persistent organochlorines and their ecotoxicological impact on marine mammals. *Science of the Total Environment* 154: 163–177.

Tanabe, S., Prudente, M., Mizuno, T., Hasegawa, J., Iwata, H. and Miyazaki, N. 1998. Butyltin contamination in marine mammals from North Pacific and Asian coastal waters. *Environmental Science and Technology* 32(2): 193–198.

Tessaro, R.K., Uthe, J.F., Freeman, H.C. and Frank, R. 1977. Methylmercury poisoning in the harp seal. *Science of the Total Environment* 8: 1–11.

Tester, M., Ellis, D.V. and Thompson, A.J. 1996. Neogastropod imposex for monitoring recovery from marine TBT contamination. *Environmental Toxicology and Chemistry* 15(4): 560–567.

Tillett, D.E., Ankley, G.T., Giesy, J.P., Ludwig, J.P., Kurita-Matsuba, H., Weseloh, D.V., Ross, P.S., Bishop, C.A., Sileo, L., Stromborg, K.L., Larson, J. and Kubiak, T.J. 1992. Polychlorinated biphenyl residues and egg mortality in double-crested cormorants from the Great Lakes. *Environmental Toxicology and Chemistry* 11: 1281–1288.

Tomy, G.T., Muir, D.C.G., Stern, G.A. and Westmore, J.B. 2000. Levels of C_{10}–C_{13} polychloro-*n*-alkanes in marine mammals from the Arctic and the St. Lawrence River estuary. *Environmental Science and Technology* 34(9): 1615–1619.

Troisi, G.M., Haraguchi, K., Simmonds, M.P. and Mason, C.F. 1998. Methyl sulphone metabolites of polychlorinated biphenyls (PCBs) in cetaceans from the Irish and the Aegean seas. *Archives of Environmental Contamination and Toxicology* 35(1): 121–128.

Uneputty, P. and Evans, S.M. 1997. The impact of plastic debris on the tidal flats in Ambon Bay (eastern Indonesia). *Marine Environmental Research* 44(3): 233–242.

van Leeuwen, F.X.R. 1997. WHO toxic equivalency factors (TEFs) for dioxin-like compounds for humans and wildlife. In: *Dioxin '97: The 17th International Symposium on Chlorinated Dioxins and Related Compounds*. Washington, DC: US Environmental Protection Agency, National Center for Environmental Assessment.

van Scheppingen, W.B., Verhoeven, A.J.I.M., Mulder, P., Addink, M.J. and Smeenk, C. 1996. Polychlorinated biphenyls, dibenzo-*p*-dioxins, and dibenzofurans in harbor porpoises (*Phocoena phocoena*) stranded on the Dutch coast between 1990 and 1993. *Archives of Environmental Contamination and Toxicology* 30(4): 492–502.

vom Saal, F.S., Montano, M.M. and Wang, M.H. 1992. Sexual differentiation in mammals. In: *Chemically-Induced Alterations in Sexual and Functional Development: The Wildlife/Human Connection* (Colborn, T. and Clement, C., eds), pp. 17–84. Princeton, NJ: Princeton Scientific Publishing.

vom Saal, F.S., Timms, B.G., Montano, M.M., Palanza, P., Thayer, K.A., Nagel, S.C., Dhar, M.D., Ganjam, V.K., Parmigiani, S. and Welshons, W.V. 1997. Prostate enlargement in mice due to fetal exposure to low doses of estradiol or diethylstilbestrol and opposite effects at high doses. *Proceedings of the National Academy of Sciences U.S.A.* 94(5): 2056–2061.

vom Saal, F.S., Cooke, P.S., Buchanan, D.L., Palanza, P., Thayer, K.A., Nagel, S.C., Parmigiani, S. and Welshons, W.V. 1998. A physiologically based approach to the study of bisphenol A and other estrogenic chemicals on the size of reproductive organs, daily sperm production and behavior. *Toxicology and Industrial Health* 14(1–2): 239–260.

Vos, J.G., De Klerk, A., Krajnc, E.I., van Loveren, H. and Rozing, J. 1990. Immunotoxicity of bis(tri-*n*-butyltin)oxide in the rat: effects on thymus-dependent immunity and on nonspecific resistance following long-term exposure in young versus aged rats. *Toxicology and Applied Pharmacology* 105(1): 144–155.

Walker, W., Risebrough, R.W., Jarman, W.M., Lappe, B.W., Lappe, J.A. and Tefft, J.A. 1989. Identification of *tris*(4-chlorophenyl)methanol in blubber of harbour seals from Puget Sound. *Chemosphere* 18: 1799–1804.

Wams, T.J. 1987. Diethylhexylphthalate as an environmental contaminant: a review. *Science of the Total Environment* 66: 1–16.

Watanabe, M., Tanabe, S., Miyazaki, N., Petrov, E.A. and Jarman, W.M. 1999. Contamination of *tris*(4-chlopopheny) methane and *tris*(4-chlorophenyl) methanol in marine mammals from Russia and Japan: body distribution bioaccumulation and contamination status. *Marine Pollution Bulletin* 39(1–12): 393–398.

Watanabe, M., Kannan, K., Takahashi, A., Loganathan, B.G., Odell, D.K., Tanabe, S. and Giesy, J.P. 2000. Polychlorinated biphenyls, organochlorine pesticides, *tris*(4-chlorophenyl)methane, and *tris*(4-chlorophenyl)methanol in livers of small cetaceans stranded along Florida coastal waters, USA. *Environmental Toxicology and Chemistry* 19(6): 1566–1574.

Watanabe, T. 2001. Determination of dialkyl phthalates in high altitude atmosphere for validation of sampling method using a helicopter. *Bulletin of Environmental Contamination and Toxicology* 66: 456–463.

Weisbrod, A.V., Shea, D., Moore, M.J. and Stegeman, J.J. 2000. Organochlorine exposure and bioaccumulation in the endangered Northwest Atlantic right whale (*Eubalena glacialis*) population. *Environmental Toxicology and Chemistry* 19(3): 654–666.

Weisglas-Kuperus, N., Sas, T.C.J., Koopman-Esseboom, C., van der Zwan, C.W., de Ridder, M.A.J., Beishuizen, A., Hooijkaas, H. and Sauer, P.J.J. 1995. Immunologic effects of background prenatal and postnatal exposure to dioxins and polychlorinated biphenyls in Dutch infants. *Pediatric Research* 38(3): 404–410.

White, D.H. and Seginak, J.T. 1994. Dioxins and furans linked to reproductive impairment in wood ducks. *Journal of Wildlife Management* 58: 100–106.

White, R., Jobling, S., Hoare, S.A., Sumpter, J.P. and Parker, M.G. 1994. Environmentally persistent alkylphenolic compounds are estrogenic. *Endocrinology* 135(1): 175–182.

Whitehead, H., Christal, J. and Dufault, S. 1997. Past and distant whaling and the rapid decline of sperm whales off the Galapagos Islands. *Conservation Biology* 11(6): 1387–1396.

Wolfe, M.F., Schwarzbach, S. and Sulaiman, R.A. 1998. Effects of mercury on wildlife: a comprehensive review. *Environmental Toxicology and Chemistry* 17(2): 146–169.

World Health Organization. 1980. *Environmental Health Criteria 15: Tin and Organotin Compounds: A Preliminary Review.* Geneva: United Nations Environment Programme and the World Health Organization.

World Health Organization. 1990. *Environmental Health Criteria 116: Tributyltin Compounds.* Geneva: United Nations Environment Programme and the World Health Organization.

13 Pathology of cetaceans.
A case study: Beluga from the St. Lawrence estuary

Daniel Martineau, Igor Mikaelian, Jean-Martin Lapointe, Philippe Labelle and Robert Higgins

Introduction

The St. Lawrence River estuary, Quebec, Canada (Figure 13.1), is the major effluent of one of the most industrialized regions of the world. It is inhabited by the southernmost population of beluga (*Delphinapterus leucas*), a population unique by its accessibility to investigation, and by its geographical isolation from the Arctic habitat of other populations of beluga. The St. Lawrence River estuary (SLE) beluga population has dwindled from an

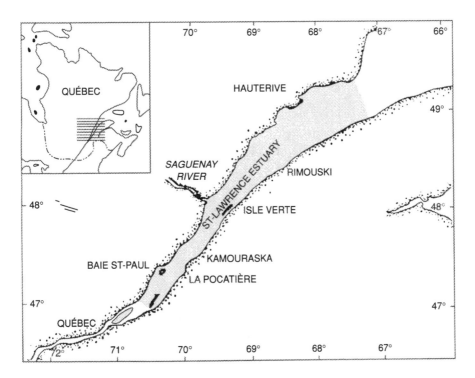

Figure 13.1 The Saint Lawrence estuary. Beluga habitat (shadowed area); eastern Canada, Quebec province (inset).

estimated 5000 to the current estimated 600–700 animals, largely because of the hunting pressure that continued until 1979 (Reeves, 1984; Sergeant, 1986; Lesage and Kingsley, 1998). Because of this dramatic decline, SLE beluga received the status of endangered species by the Canadian Government in 1980 (Cook and Muir, 1984); yet, there are no solid data supporting population recovery. To explain this apparent failure to recover, a study was initiated in 1982 to carry out systematic post-mortem examination of dead SLE beluga that drift ashore and to determine tissue levels of chemical contaminants.

Live and dead SLE beluga have been found to be heavily contaminated by agricultural and industrial contaminants, such as polychlorinated biphenyls (PCBs), DDT, and their metabolites, heavy metals and polycyclic aromatic hydrocarbons (PAH) (Martineau *et al.*, 1987, 1988; Wagemann *et al.*, 1990; Muir *et al.*, 1996; Letcher *et al.*, 2000a).

The present chapter provides an overview of the results of necropsies conducted over 16 years (1983–1998). Primary causes of death and significant lesions are identified. The rate of cancer in this population is estimated and compared to that of humans and animals.

Sample representativeness

This study has been carried out for well over a decade. Considering the beluga life span (35–40 years), a sizeable proportion of the population died (246/650 or 38 per cent) and has been examined over this period (Brodie, 1982). We assume that all carcasses have equal chances of being recovered and examined whatever the cause of death, for the following reasons. These whales live in a restricted range, as shown by thorough surveys from airplanes and boats. All carcasses have been found within that range or downstream, as a result of drift (Michaud *et al.*, 1990; Michaud, 1993). No criteria other than reasonable preservation and carcass accessibility were used to determine whether a given carcass would be examined in the post-mortem room. In conclusion, whereas some deaths may occasionally escape our attention, our sampling is most likely representative of the population in terms of causes of mortality.

Mortality

Between 1983 and 1998, 246 beluga have been found dead, drifting or stranded along the shoreline of the St. Lawrence estuary, Quebec, Canada. (Since the study was initiated in September 1982, the latter year is not included in the present review, in order to report only complete years.) One hundred and eighty-two carcasses were aged by counting dentine layers, and 120 carcasses were examined in the post-mortem room of the Faculté de Médecine Vétérinaire (College of Veterinary Medicine) of Université de Montréal, in St. Hyacinthe, PQ (FMV) (Table 13.1; Figure 13.2). Of the 120 whales examined, 94 were considered reasonably well preserved (78 per cent),

Table 13.1 Annual number of stranded beluga from the St. Lawrence estuary (1983–98). Five females were pregnant, and near-term fetuses are not taken into account

Year	Necropsied	Stranded/drifting, not necropsied	Total	Fetuses
83	7	8	15	0
84	9	3	12	1
85	4	9	13	1
86	6	4	10	1
87	3	8	11	0
88	11	10	21	0
89	8	14	22	0
90	6	11	17	1
91	6	10	16	0
92	7	8	15	0
93	9	12	21	1
94	7	7	14	0
95	12	5	17	0
96	9	6	15	0
97	7	6	13	0
98	9	5	14	0
Total	120	126	246	5

based on the firm consistency of the liver. Seventy-four of the relatively fresh 94 beluga (79 per cent) were adult (over 7 years old). To interpret these data, the following observations must be considered. First, 51 per cent (or 126/246) of reported stranded animals have not been necropsied, and thus the cause of death and lesions affecting these animals remain unknown. The number of animals stranded dead during winter is also unknown. In the spring, summer and fall, carcasses with terminal diseases are often found after several days of rough weather, suggesting that the number of strandings occurring during winter is at least the same as that reported during the rest of the year, because of the harsh weather conditions prevailing in that season. Young calves (age = 0) are difficult to find in the wild because of their small size and blue–gray color. Therefore, calf mortality is probably underestimated.

Most beluga stranded dead were between 16 and 35 years old, with a mortality peak between 21 and 25 years. Beluga that died of hunting in Northwest Alaska (NA) and SLE beluga found stranded dead have been used to infer the age-structures of the respective populations following accepted procedures in field population biology that are appropriate to each case. An earlier study on age structure was updated using the mortality records of SLE beluga from 1983 to 2001 (Béland *et al.*, 1988) (Figure 13.2b). The modeled age structures of SLE and NA standing populations are similar except for the presence of older animals in the Alaska population, implying that SLE beluga with cancer are not particularly old. In addition, the model does not support that the SLE population includes a large proportion of

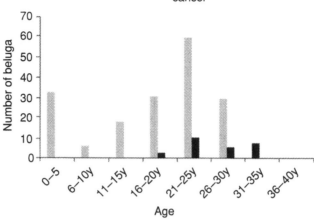

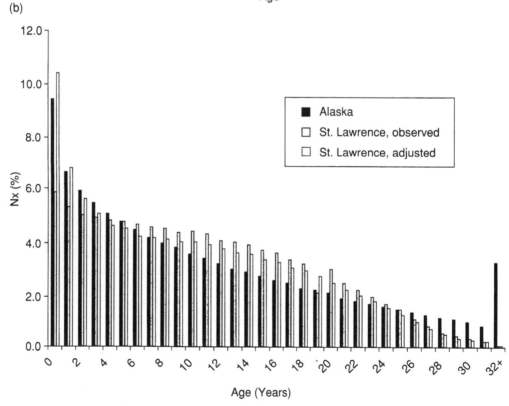

Figure 13.2 a) Age structure of SLE beluga found dead on the shoreline (1983–1997). b) Age structures of beluga populations from Alaska (as in (Burns and Seaman, 1986), with juvenile mortality adjusted) and from the SLE with [*adjusted*] or without [*observed*] modifying the record to account for under-representation of juvenile mortality (updated from Béland *et al.* 1988, using mortality records of SLE beluga from 1983 to 2001). N(x): proportion of individuals in each age class.

Table 13.2 Primary causes of death in stranded beluga (*Delphinapterus leucas*) from the St. Lawrence estuary (1983–98)

St. Lawrence beluga whales, Primary causes of death (1983–98)	*Number of whales (%)*
Infectious (bacteria (11), virus (2), and protozoa (2))	15 (16)
Parasites (lungs: 16; digestive: 4)	20 (21.3)
Cancer (terminal)	17 (18)
Not determined	23 (24.6)
Neonatal mortality, dystocia	6 (6.4)
Dissecting aneurysm, pulmonary trunk	3 (3.2)
Other[1]	5 (5.3)
Motor boat (wound due to propeller)	2 (2.1)
Trauma (probably from boat collision)	3 (3.2)
Total	94 (100)

[1] Intestinal volvulus (8315), congenital malformation (9203), severe centrolobular hepatic fibrosis (9702), severe centrolobular necrotizing hepatitis (9206), diffuse alveolar damage (8813).

animals in the age groups with elevated cancer rates (16–30 years) (Figure 13.2).

The age distribution of animals with cancer was the same as that of animals dying of causes other than cancer (data not shown), and cancer did not affect the oldest animals (Figure 13.2). These observations suggest that other factors are involved in the development of cancer in that population in addition to aging.

Primary causes of death

Overall, the three major primary causes of death were parasites (21.3 per cent), cancer (18 per cent) and infectious agents (bacterial, viral or protozoal) (16 per cent) (Table 13.2). The major cause of death in adult beluga (*n* = 74) was cancer (17 animals with terminal cancers, excluding three ovarian cancers, or 23 per cent).

In wild animal populations, cancer has not been reported as a major cause of mortality in adults, with the possible exception of virus-induced liver cancer in woodchucks (*Marmota monax*) and leukemia in the feral mouse. This finding stands in sharp contrast with the causes of mortality in terrestrial mammals where major causes of mortality are trauma and starvation and where neoplasia is rare (2 per cent) (Aguirre *et al.*, 1999). It also stands in sharp contrast with the causes of mortality in other cetacean populations where cancer has not been reported, and where major causes of death are, excluding recent viral epizootics, entanglement in fishing gear, pneumonia and/or parasitism, abandonment and starvation of neonates (Stroud and Roffe, 1979; Howard *et al.*, 1983a; Kirkwood *et al.*, 1997).

Parasites

Pulmonary infection by *Halocercus monoceris* and gastrointestinal infection with *Anisakis simplex* contributed to the death of 16 and 4 whales, respectively. A beluga had severe intestinal diphyllobothriosis. Two animals had a perforated ulcer of the first gastric compartment with numerous intralesional adult *A. simplex*, and a fourth beluga had a peritonitis most likely elicited by the presence of *A. simplex* in the abdominal cavity. In that beluga, no perforated gastric ulcers were found, but there were multiple chronic gastric ulcers suggestive of previous perforation.

The widespread pulmonary infestation by nematode larvae surrounded by dense inflammatory infiltrate has been interpreted by us (Table 13.2) and others as the primary cause of death in marine mammals with such lesions. The primary role played by parasites in the etiology of these lesions might be re-examined, however, considering the profound effects of organochlorines (OCs) on specific pulmonary cell types. Mammalian lungs contain xenobiotic-metabolizing enzymes similar to those found in the liver. In contrast to liver enzymes, those found in the lung are concentrated in specific cell types, such as the bronchiolar Clara cells. In mammals, several PCB metabolites, and particularly methylsulphone-PCBs (MeSO$_2$-PCBs), accumulate in the lung and airways epithelium, where they induce, particularly in Clara cells, cytochrome P450-dependent monooxygenases (Stripp *et al.*, 1996; Fouchecourt *et al.*, 1998). In turn, the degradation of many xenobiotics by P450-dependent monooxygenases generates harmful reactive metabolites (Smith and Brian, 1991).

Bacteria

Bacterial infections were considered to be the primary cause of death in 11 SLE animals (11/94 or 11.7 per cent) (Tables 13.2 and 13.3). The bacterial species involved are known as opportunistic, and many are commonly present in organically polluted waters (McDermott and Mylotte, 1984; Jaruratanasirikul and Kalnauwakul, 1991; Noga, 1996; Chen *et al.*, 1997). (Opportunistic bacteria are normally found in the environment and/or in their hosts, have low pathogenicity and take advantage of compromised host defenses to cause diseases.)

In several other SLE beluga, where other primary causes of death were determined (cancer, severe parasitic pneumonia, perforated gastric ulcer and cutaneous wounds), the bacteria that were isolated were considered secondary invaders (data not shown). Importantly, bacteria such as *Salmonella* spp., *Mannheimia (Pasteurella) haemolytica, Erysipelothrix rhusiopathiae, Corynebacterium* spp. and *Clostridium* spp., reported as primary causes of mortality in other cetaceans, either captive or free-ranging, were not isolated from SLE beluga (reviewed by Howard *et al.*, 1983b; Dunn, 1990).

Bacteriological studies have been conducted on 13 beluga captured in the Canadian Arctic, of which seven were subsequently kept in captivity in three

Table 13.3 Bacterial species cultured from 11 beluga carcasses from the Saint Lawrence estuary (SLE) and implicated as primary causes of death in these animals (1983–98) (number of animals). Comparison with bacterial species cultured from blowhole and/or anal swabs of 13 healthy Arctic beluga either released immediately or kept in captivity

Bacteria	Lesions found in SLE beluga (number of cases)	Bacteria isolated from blowhole in 13 Churchill River beluga released immediately (Buck et al., 1989)	Bacteria isolated from blowhole and anal swabs of seven Churchill River beluga kept in captivity in three US aquariums from moment of capture (day 0) through day 945 (Buck et al., 1989)
Aeromonas hydrophila	Acute lymphadenitis with lymphoid necrosis (1)	7/13	5/7
Edwardsiella tarda	Glomerular thrombosis, alveolar proteinosis (lungs) (1); perirenal abscess and septicemia (1); pulmonary abscess (75% of left lung) (1)	0/13	0/7
Kingella kingae	Chronic peritonitis (1)	0/13	0/7
Morganella morganii	Cystitis (1) (De Guise et al., 1995); enteritis (1) (De Guise et al., 1995)	0/13	2/7
Nocardia spp.	Multisystemic pyogranulomas (1) (Martineau et al., 1988)	0/13	0/7
Plesiomonas shigelloides	Omphalophlebitis (1)	0/13	1/7*
Shewanella putrefaciens	Peritonitis (1) (De Guise et al., 1995)	0/13	4/7
Vibrio fluvialis	Fibrinopurulent valvular endocarditis and bacteremia (1)	0/13	5/7

* *Plesiomonas shigelloides* isolated only after 945 days of captivity.

US aquaria (Table 13.3) (Buck *et al.*, 1989). Of the eight bacterial species considered as primary causes of death in SLE beluga, only one was also isolated from healthy Arctic beluga that were sampled during capture (Buck *et al.*, 1989), and four species were isolated from two apparently healthy Arctic beluga, only after these whales were kept in an aquarium for several days to several years. This finding is in accordance with the low pathogenicity of these bacteria and/or their presence in organically loaded waters. The association of these bacteria with SLE beluga mortality, their presence in healthy captive beluga and their known low pathogenicity in healthy hosts suggest that SLE beluga are immunocompromised (Table 13.3).

Edwardsiella tarda was considered as the primary cause of death of three SLE beluga, and was also isolated from five diseased non-cancerous SLE beluga, and from one SLE beluga with intestinal cancer (data not shown) (Table 13.3). In contrast, this bacterium was not isolated from apparently healthy Arctic beluga, captive or not (Buck *et al.*, 1989). Elsewhere, marine mammals from which *E. tarda* has been isolated were debilitated by conditions such as fractures, dystocia, perforated colonic ulcer, fishhooks in the stomach, gunshot wounds or morbilliviral infections (Coles *et al.*, 1978; Stroud and Roffe, 1979; Geraci, 1989). Morbilliviruses are known to be immunosuppressive in their human and animal hosts (McCullough *et al.*, 1974; Oldstone *et al.*, 1999). Considered together, these observations suggest that SLE beluga are immunocompromised.

Shewanella putrefaciens was isolated from a SLE beluga with peritonitis (Table 13.3) and from two SLE beluga with verminous pneumonia (data not shown). *Plesiomonas shigelloides* was isolated from a SLE beluga with omphalophlebitis (Table 13.3) and from two beluga with intestinal cancer (data not shown). *Morganella morganii* was recovered from a SLE beluga with cystitis and from another with enteritis (Table 13.3) and from a female with fatal dystocia (data not shown). *Vibrio fluvialis* was recovered from a SLE beluga with fibrinopurulent valvular endocarditis (Table 13.3). These bacteria were recovered from healthy Arctic beluga only after extended periods in an aquarium, reflecting the different bacteriological flora of the new enclosed environment (Table 13.3) (for instance, *Plesiomonas* spp. was recovered from a single Arctic beluga, only after 945 days of captivity; Buck *et al.*, 1989). The higher bacterial load of the St. Lawrence estuary, the warmer water temperature, and/or the immunocompromised status of SLE beluga may explain these differences.

Viruses

Two animals (8504, 9406) had microscopic lesions of non-suppurative encephalitis, most consistent with a viral etiology. A third animal (8409) had a multifocal necrotizing dermatitis, where necrotic keratinocytes contained intranuclear inclusion bodies containing clusters of herpesvirus-like viral particles (Martineau *et al.*, 1988). The latter finding, along with the occurrence of worldwide morbilliviral epizootics in marine mammals, prompted us to determine the seroprevalence of selected viruses within the SLE population. Forty six per cent of SLE beluga ($n = 13$) cross-reacted serologically with bovine herpesvirus 1, and all animals tested were seronegative for morbilliviruses ($n = 13$). Thus the SLE population is probably vulnerable to a large-scale epizootic with a cetacean morbillivirus (Mikaelian *et al.*, 1999b).

Cancer

Cancer rate

Post-mortem examinations of SLE belugas carried out from 1983 to 1998 revealed a high rate of cancer. Twenty cancers, of which 17 were terminal (cancers that led to death), were found in 94 well-preserved carcasses examined at necropsy. Thus, 18 per cent of deaths (or 17/94) were caused by cancer and 27 per cent (20/74) of adult animals were affected by cancer. In contrast, no tumors were observed in four carcasses of SLE Atlantic white-sided dolphins (*Lagenorhyncus acutus*), two carcasses of harbor porpoises (*Phocoena phocoena*) and 15 carcasses of three species of seals (*Phoca vitulina*, *Pagophilus groenlandicus* and *Halichoerus grypus*) from the same waters, nor in five Arctic beluga examined by a veterinary pathologist during the same period, using the same post-mortem examination protocol (De Guise *et al.*, 1994b).

In the Western world, cancer causes 23 per cent of all deaths in humans, a percentage comparable to that seen in SLE beluga (18 per cent) (Cotran *et al.*, 1994). To our knowledge, this is the first time that a population of free-ranging mammals (terrestrial or aquatic) is comparable to humans in this regard. The cancer rate in stranded SLE belugas is also much higher than that observed in other cetaceans. Only 30 other cases of cancer have been reported worldwide in captive and wild cetaceans (Tables 13.4–13.6). Including the 20 cancers reported in SLE beluga, the 50 cancers observed in cetaceans affected two species of mysticetes and 12 species of odontocetes.

Cancers seen in SLE beluga also differ widely in type from those seen in other cetaceans (Tables 13.5 and 13.6). Gastrointestinal epithelial cancers were the most frequent cancers seen in SLE beluga, whereas hemopoietic cancers are the most frequent types of cancer observed in other cetaceans (Tables 13.4–13.6). [Only one of the 30 cancers (3.3 per cent) seen in other cetaceans was a gastrointestinal epithelial cancer.] Furthermore, no mammary gland cancers had been reported previously in cetaceans, while three of these cancers have been reported in SLE beluga (Mikaelian *et al.*, 1999a).

A single cancer was found in over 1800 other cetaceans examined, and tumors were not found in approximately 50 belugas examined in the Canadian Arctic [David J. St. Aubin, personal communication, cited by De Guise *et al.* (1994b)]. Arctic data may not be fully representative, however, because

Table 13.4 Cancers reported in stranded beluga from the St. Lawrence estuary (1983–98)

Organ	Cancer type	Identification	Age	Sex	References	AFIP accession number
Intestine	Adenocarcinoma	8907	29.5+	M	(De Guise et al., 1994b)	2 462 295–3
Intestine	Adenocarcinoma	8908	20.5+	M	(De Guise et al., 1994b)	2 462 247–4
Intestine	Adenocarcinoma	9302	25+	M	(Martineau et al., 1995)	2 461 200
Intestine	Adenocarcinoma	9402	27.5+	M	(Martineau et al., 1995)	2 464 226–6
Intestine	Adenocarcinoma	9407	27+	F	(Martineau et al., 1999)	2 508 083–900
Intestine	Adenocarcinoma	9605	23+	F	This report	2 573 961–6
Intestine	Adenocarcinoma	9809	18+	M	This report	
Skin	Squamous cell carcinoma	Same animal (9809)				
Stomach	Adenocarcinoma	8804	21+	F	(De Guise et al., 1994b)	2 456 949–3
Stomach	Adenocarcinoma	9401	27.5+	M	(Martineau et al., 1999)	2 508 095–300
Salivary gland	Adenocarcinoma	8606	24.5++	M	(Girard et al., 1991)	2 457 053–3
Mammary gland	Adenocarcinoma	8809	22+	F	(De Guise et al., 1994b)	2 456 952–7
(Liver)	Poorly differentiated carcinoma, probably a metastasis[1]	Same animal (8809)				
Mammary gland	Adenocarcinoma	9804	20.5	F	(Mikaelian et al., 1999a)	267 466 900
Mammary gland	Adenocarcinoma	9803	26+	F	(Mikaelian et al., 1999a)	267 486 100
Uterus	Adenocarcinoma	9502	26+	F	(Lair et al., 1998)	2 573 966–5
All organs	Metastatic carcinoma	9609	23++	M	This report	2 656 322–100
Ovary	Granulosa cell tumor	8502	24.5	F	(Martineau et al., 1988)	2 519 612
Ovary	Granulosa cell tumor	8813	21+	F	(De Guise et al., 1994b)	2 462 292–0
Ovary	Dysgerminoma[2]	8906	25+	F	(De Guise et al., 1994b)	2 462 229–2
Thymus	Lymphosarcoma	9001	18.5++	M	(De Guise et al., 1994b)	2 519 747
Urinary bladder	Transitional cell carcinoma	8318	16.5	M	(Martineau et al., 1985)	

[1] Was originally classified as a primary liver cancer (beluga no. 5 in De Guise *et al.*, 1994b). Reclassified as metastasis after consultation with the Armed Forces Institute of Pathology (AFIP).
[2] Was originally classified as Granulosa cell tumor (beluga no. 16 in DeGuise *et al.*, 1994b). Reclassified as dysgerminoma after consultation with AFIP.

Table 13.5 Cancers reported in cetaceans other than St. Lawrence Estuary beluga

Species	Organ	Cancer	Age	Sex	Sources	Reference
Bottlenose dolphin	Liver, lungs	Reticuloendotheliosis	U	F	U	Ridgway, pers. comm. (Landy, 1980)
Bottlenose dolphin (3), -	Multisystemic	Immunoblastic malignant lymphoma	Adults	F	Florida	Bossart *et al.* (1997)
Atlantic spotted dolphin (1), -	Multisystemic	Immunoblastic malignant lymphoma	Adult	F	Florida	
Pantropical spotted dolphin (1)	Multisystemic	Immunoblastic malignant lymphoma	Adult	F	Florida	
Bottlenose dolphin	Blood	Myelogenous leukemia	Adult	F	Gulf coast of Texas	DF Cowan, pers. comm.
Bottlenose dolphin	Liver	Cholangiocarcinoma	Adult	F	Gulf coast of Texas	DF Cowan, pers. comm.
Bottlenose dolphin	Spleen, lymph nodes	Lymphosarcoma	Adult	F	U	Taylor and Greenwood, pers. comm. (Landy, 1980)
Pacific white-sided dolphin	Spleen, lymph nodes	Lymphosarcoma	Adult	M	U	Howard *et al.* (1983b)
Pacific white-sided dolphin	Spleen, lymph nodes, liver, kidney	Eosinophilic leukemia	Adult	M	U	Howard *et al.* (1983b)
Bottlenose dolphin	Pancreas	Carcinoma	Adult	M	U	Taylor and Greenwood, pers. comm. (Landy, 1980)
Shortfinned pilot whale	Ovary	Granulosa cell tumor	Adult	F	U	Benirschke and Marsh (1984)
Harbor porpoise	Unknown	Adenocarcinoma	Adult	F	British waters	Baker and Martin (1992)
Harbor porpoise	Stomach	Adenocarcinoma	Adult	F	Northern Wadden Sea	Breuer *et al.* (1989)
Amazon River dolphin	Lung	Squamous cell carcinoma	Adult	F	U	Geraci *et al.* (1987)
Blue whale	Ovary	Granulosa cell tumor	Adult	F	Antarctic	Rewell and Willis (1949)
Fin whale	Ovary	Granulosa cell tumor	Adult	F	Antarctic	Rewell and Willis (1949)
Fin whale	Ovary	Granulosa cell tumor	Adult	F	Antarctic	Rewell and Willis (1949)
Fin whale	Ovary	Carcinoma[1]	Adult	F	Antarctic	Stolk (1950)

[1] Reclassified as granulosa cell tumor by Geraci *et al.* (1987).
U, Unknown.
Scientific names: bottlenose dolphin (*Tursiops truncatus*); Atlantic spotted dolphin (*Stenella frontalis*); pantropical spotted dolphin (*Stenella attenuata*); Pacific white-sided dolphin (*Lagenorhynchus obliquidens*); shortfinned pilot whale (*Globicephala macrorhynchus*); harbor porpoise (*Phocoena phocoena*); Amazon River dolphin (*Inia geoffrensis*); blue whale (*Balaenoptera musculus*); fin whale (*Balaenoptera physalus*).

Table 13.6 Cetaceans affected by cancer listed in the marine mammal database, AFIP (updated June 1998)

Species	Status	Organ	Cancer type	Age	Sex	Sources	%[1]	Accession number
Beluga	Captive	Brain	Carcinoma	19 yr	M	Arctic	7.1% (14)	2 034 441
Bottlenose dolphin	Wild	U[2]	Tubulopapillary adenocarcinoma	Adult	M	Gulf of Mexico	0.3% (384)	2 304 654
Bottlenose dolphin	Wild	Kidney	Renal cell carcinoma	Adult	F	Atlantic Ocean (South Carolina)	0.3% (384)	2 445 679
Common dolphin	Wild	Multiple	Anaplastic carcinoma (primary site undetermined)	Adult	M	California coast		2 529 715
Common dolphin	Wild	Multiple	Solid carcinoma (primary site undermined)	Adult	M	New Jersey coast		
Fin whale	Wild	Kidney	Lymphosarcoma	Adult	F	U	20% (5)	1 470 245
Killer whale	Captive	Liver, lymph node, spleen	Reticuloendotheliosis	Adult	F	U	5.3% (19)	1 626 236
Killer whale	Captive	Lymph node	Hodgkin's disease-like[3]	Adult	M	Iceland	5.3% (19)	2 337 420
Pygmy sperm whale	U	Liver	Cholangiocarcinoma	U	U	U	1.7% (57)	1 777 514
Atlantic spotted dolphin	Wild	Testis, lymph nodes, adrenal glands	Malignant seminoma, pheochromocytoma	Adult	M	Gulf of Mexico	16.7% (6)	2 428 264

[1] (Number of animals affected by cancers of a given species) ÷ (the total number of animals for a given species examined and listed in the Marine Mammal Database of the AFIP) × 100.

[2] U, Unknown.

[3] Yonezawa *et al.* (1989) (Yonezawa *et al.*, 1989).

AFIP others: brain (1), kidney (1), testis and adrenals (1 animal), ND (3).

Scientific names: beluga (*Delphinapterus leucas*); bottlenose dolphin (*Tursiops truncatus*); common dolphin (*Delphinus delphinus*); fin whale (*Balaenoptera physalus*); killer whale (*Orcinus orca*); pygmy sperm whale (*Kogia breviceps*); Atlantic spotted dolphin (*Stenella fontalis*).

the ages of Arctic beluga were unknown and these were randomly selected live animals and not stranded animals. A single benign tumor was observed in 55 slaughtered long-finned pilot whales (*Globicephala melas*) in New-foundland, and only two benign tumors (0.1 per cent) were reported in 2000 mysticetes (baleen whales) hunted in South Africa (Cowan, 1966; Uys and Best, 1966).

The present study can best be compared to studies on cetaceans singly stranded rather than to studies carried out on cetaceans shot, or caught singly or collectively in fishing gear. None of these studies have shown a cancer rate comparable to that of SLE beluga. Neither Stroud and Roffe (1979), nor Howard *et al.* (1983b) reported neoplasia from 21 stranded cetaceans and 65 stranded common dolphins (*Delphinus delphis*), respect-ively. Among 90 bottlenose dolphins (*Tursiops truncatus*) stranded along the Gulf coast of Texas from 1991 to 1998, two cancers (or 2 per cent), a myelogenous leukemia (a hemopoietic cancer) and a bile duct carcinoma were found (Daniel F. Cowan, 23 Sept 2002, personal communication). No cancers were found in 28 harbor porpoises stranded on the British coast (of which three were killed by net entanglement) (Kuiken *et al.*, 1993). Three cancers were found during the post-mortem examination of 422 odontocetes (or 0.7 per cent) from British waters (Kirkwood *et al.*, 1997). A viral etiology was probably associated with a cluster of five lymphomas in bottlenose dolphins on the US east coast (Bossart *et al.*, 1997).

Cancer epidemiology in SLE beluga

The annual rate of cancer in the SLE beluga population was estimated as the number of new cases of cancers over a given time (1 year). The estimated annual rate (AR) was calculated as an annual rate per 100 000 animals (as in Dorn and Priester, 1987); the actual number of stranded SLE beluga with cancer examined in the necropsy room was divided by 16 years (1983–98), by the estimated number of SLE beluga, and the result was multiplied by 100 000 animals. A recent index estimate of 650 SLE beluga was used (Lesage and Kingsley, 1998).

$$AR = \frac{\text{SLB with cancer}}{t} \times \frac{100\,000\,\text{SLB}}{\text{Estimated current population}} \qquad (1)$$

where AR is the estimated annual rate of cancer in SLE beluga, and t is the study period (1983–98).

$$AR = \frac{20\,\text{SLB}}{16\,\text{years}} \times \frac{100\,000\,\text{SLB}}{650\,\text{SLB}} = 192\,\text{SLB with cancer per year.}$$

Stranded carcasses are very rarely reported in winter (January to March) because of the ice cover, difficult access to most of the shoreline, inclement

weather and the presence of few human observers on the shoreline. To estimate the number of dead animals that strand over a complete year (12 months),
it was assumed that the frequency of death during the winter months is equal
to that of other seasons, although it is probably higher because of the inclement weather, and thus a correction factor of 12 months/9 months (1.33) was
used. Assuming that all carcasses have an equal chance of being seen and
collected whatever the cause of death, we estimated the minimum number of
SLE beluga with cancer over the past 16 years (1983–98) (EMC) as follows:

$$\text{EMC} = \frac{\text{SLB with cancer}}{94\,\text{SLB}} \times \text{dead SLB} \times 1.33 \tag{2}$$

where EMC is the estimated minimum number of SLE beluga with cancer
per year, and dead SLB is the total number of beluga reported dead and/or
examined during the study period (1983–98).

$$\text{EMC} = \frac{20\,\text{SLB}}{94\,\text{SLB}} \times 246\ \text{dead SLB} \times 1.33 = 70\ \text{SLB with cancer.}$$

The adjusted estimated annual rate (AAR) of cancer for a complete year (12
months) is:

$$\text{AAR} = \frac{\text{EMC}}{t} = \frac{100\,000\,\text{SLB}}{650\,\text{SLB}} \tag{3}$$

where AAR is the adjusted annual rate of cancer cases, and t, the study
period (1983–98).

$$\text{AAR} = \frac{70\,\text{SLB}}{16\ \text{years}} \times \frac{100\,000\,\text{SLB}}{650\,\text{SLB}} = 673\ \text{SLB with cancer.}$$

In veterinary and human epidemiology, the number of individuals at
risk must be known precisely in order to determine disease prevalence.
This requirement explains why few epidemiological cancer studies have been
carried out in wild mammal populations, which are notoriously ill defined
and/or widespread. SLE beluga stand as an exception in this regard. This
population is reasonably well characterized, geographically isolated, restricted
to a relatively small area and has been the object of numerous censuses,
often carried out using different techniques (Michaud *et al.*, 1990; Kingsley
and Hammill, 1991; Michaud, 1993; Lesage and Kingsley, 1998). All censuses
have provided similar results and thus, the population at risk, the denominator used to calculate the AR (Dorn and Priester, 1987), is reasonably well
defined. The AR and the AAR in SLE beluga were compared to those of
domestic animals and humans (Table 13.7). The AR of cancer, of epithelial
cancer of the proximal intestine, of stomach cancer, of epithelial cancer of

Table 13.7 Frequency of cancer in St. Lawrence estuary beluga over 16 years (1983–98) compared to that of man and domestic animals. Estimated annual rate (AR) of cancer per 100 000 animals (AR)

	Total cancer AR (AAR)	Epithelial cancer of small intestine[4] AR (AAR)	Epithelial cancer of stomach[5] AR (AAR)	Epithelial cancer of the gastrointestinal tract[1] AR (AAR)	Mammary gland cancer[6] AR (AAR)
Beluga	192 (673)[2]	67 (230)[3]	19.2 (66.9)	86 (301)	29 (100)
Human (Anonymous, 1991)	475.6 (Gaudette et al., 1998)	0.8 (Anonymous, 1991)	14.7 (Gaudette et al., 1998)	55.6 (Anonymous, 1991)	99.5 (Gaudette et al., 1998)
Cattle	75	0.1	0	0.2	0.1
Dog	133 (Priester and McKay, 1980); 507 (MacVean et al., 1978)	0.5	0.8	2.6	11.2 (Priester and McKay, 1980); 133.8 (MacVean et al., 1978)
Cat	91 (Priester and McKay, 1980); 412 (MacVean et al., 1978)	1.8 (Priester and McKay, 1980)	0	2.2	4.3 (Priester and McKay, 1980); 37.3 (MacVean et al., 1978)
Horse	41	0	0.5	0.6	0.2
Sheep (Georgsson and Vigfusson, 1973)	0.03	Up to 2000 in some regions	ND	ND	ND

[1] Gastrointestinal tract: glandular stomach, proximal (small), large intestine, intestine not otherwise specified (NOS), and rectum and anus combined.
[2] ( ): AAR: adjusted AR, accounting for mortality in winter months and for total mortality.
Cattle, dog, cat, horse: Priester and McKay (1980).
[3] Seven cancers of the small intestine (most distal intestinal adenocarcinoma considered as in small intestine).
[4] Sum of adenocarcinoma, squamous cell carcinomas, and carcinomas not otherwise specified (NOS) under small intestine in Priester and McKay (1980).
[5] Sum of gastric carcinomas NOS and adenocarcinomas.
[6] Sum of mammary gland adenocarcinomas, carcinomas NOS, mucinous and squamous cell carcinomas.

the gastrointestinal tracts and of mammary glands was generally higher in SLE beluga than in domestic animals seen in veterinary hospitals. The AAR was higher in SLE beluga than in all other species including humans for all types of cancer, except for mammary gland cancer, where the AAR in SLE beluga was the same as the rate seen in humans, and was comparable to the rate seen in dogs examined in veterinary hospitals (Table 13.7).

Paradoxically, the collection of epidemiological data from the SLE population is more similar to that of humans than to that of domestic animals, because the denominator used in human studies, like the denominator used to determine the cancer rate in SLE belugas, is derived from periodical census, whereas the data obtained from domestic animals originates from veterinary hospitals. Because the population of domestic animals seen in veterinary hospitals comprises many (if not mostly) sick animals, the epidemiological data from domestic animals is expected to contain a higher number of animals with cancer than the general population of domestic animals (Priester and McKay, 1980).

Free-ranging animals generally have a shorter life span than captive animals because of predation, harsh environmental conditions, and malnutrition. Older animals are more numerous in the pet animal population than in free-ranging animals because of the absence of adverse conditions and because of curative and preventive improvements in veterinary medicine (Dorn and Priester, 1987). Because the risk of developing cancer increases with age, cancer rates in pet, zoo and aquarium animals are expected to be higher than in free-ranging mammals (Fowler, 1987).

Considered together, these observations indicate that cancer rates in domestic animals, as shown in Table 13.7, are clearly overestimated. Yet, for all cancer types, the AAR in SLE beluga is much higher than that observed in cattle, horse and sheep examined in veterinary hospitals, and is higher than the rate observed in dogs and cats examined in veterinary hospitals and in humans.

Possible etiological factors

Thirty per cent (6/20) of the cancers affecting SLE beluga originated from the intestine close to the stomach, while a seventh intestinal cancer was closer to the anus (Martineau *et al.*, 1995) (Table 13.4). Cancer of the proximal intestine is rare in all animal species, including humans, but is frequent in certain bovine and ovine populations in certain parts of the world where it has been etiologically associated with the ingestion of herbicides such as 2,4-dichlorophenoxyacetic acid (2,4-D) (Cordes and Shortridge, 1971; Georgsson and Vigfusson, 1973; Jarrett *et al.*, 1978; Newell *et al.*, 1984; Campo *et al.*, 1994). In cattle, small intestinal cancers result from an interaction between exogenous carcinogens and viruses. Bovine papillomavirus type 4 causes papillomas in the bovine upper digestive tract. In cattle infected with that virus and fed with bracken fern (which contains powerful carcinogens), papillomas become malignant and are accompanied by intestinal adenomas and adenocarcinomas (Campo *et al.*, 1994). Thus, co-carcinogens are necessary to cause small intestinal carcinomas in cattle. A similar interplay might be at work in SLE beluga because gastric papillomatosis has been observed in a significant number of carcasses, and particles consistent with papillomaviruses have been observed in papillomas (Martineau *et al.*, 1988; De Guise *et al.*, 1994a). Thus, a virus might contribute to the etiology of proximal intestinal cancers in SLE beluga.

Carcinogens are present in the environment of SLE belugas and are likely ingested by these animals. Tissular PAH concentrations found in blue mussels (*Mytilus edulis*) collected from Tadoussac and Baie Sainte-Catherine (at the mouth of the Saguenay River in the SLE) were approximately three times higher than concentrations found in mussels collected elsewhere in the Saint Lawrence estuary and gulf (Cossa *et al.*, 1983). In addition, the tissue benzo[*a*]pyrene (B[a]P) concentrations of blue mussels were 200 times higher after their transplantation into the Saguenay River than before (Picard-Bérubé and Cossa, 1983). The sediments of the Saguenay River, which is a part of the SLE beluga habitat, contain 500–4500 ppb of total PAH (dry weight), and these compounds originate from aluminum factories located upstream (Martel *et al.*, 1986). Invertebrates living in sediments contaminated by PAH accumulate PAH, in contrast to vertebrates (Ferguson and Chandler, 1998). At least in summer, SLE beluga feed in significant amounts on bottom invertebrates (Vladykov, 1946). In addition, field observations suggest that these whales dig into sediments (Dalcourt *et al.*, 1992). Benzopyrene DNA adducts have been detected in the brain tissue of stranded SLE beluga, by acid hydrolysis of DNA followed by HPLC and fluorescence detection, and were not detected in the brain tissue of Arctic beluga (Martineau *et al.*, 1988; Shugart *et al.*, 1990). However, in another study, analyzing stranded SLE beluga liver using ^{32}P post-labeling, no difference was observed between Arctic and SLE beluga (Ray *et al.*, 1991). These apparently conflicting results are not surprising given the different methods and tissues that were used in the two studies. ^{32}P post-labeling is highly sensitive but not very specific – it detects any aromatic or non-aromatic/bulky compound (either exogenous or endogenous) bound to DNA – whereas acid hydrolysis of DNA followed by HPLC and fluorescence detection is highly specific for B[a]P diolepoxide, the ultimate carcinogenic form of B[a]P. Moreover, the metabolism and the capacity of neurons and hepatocytes for DNA replication and repair are very different.

Considered together, these observations suggest that SLE beluga ingest PAH present in benthic invertebrates, which may contribute to the elevated rate of digestive tract cancers seen in this population (Martineau *et al.*, 1988, 1994; De Guise *et al.*, 1994b). This relationship between intestinal adenocarcinoma and PAH is supported by the recent demonstration that in mice, chronic ingestion of coal-tar mixtures (which contain benzopyrene) causes small intestinal adenocarcinomas, in addition to forestomach papilloma and stomach carcinoma (Culp *et al.*, 1998). In humans, there is an epidemiological relation between the ingestion of smoked food (which contains benzopyrene) and small intestinal cancer (Chow *et al.*, 1993).

Cytochromes P450 (CYP) present in the small intestinal epithelium are the first and major molecules implicated in the biotransformation and subsequent detoxification or toxification of ingested xenobiotics, and high levels of intestinal P450 have been related with gastrointestinal cancer (Kaminsky and Fasco, 1991). In the rat small intestine, the highest CYP concentrations occur in the duodenum, and the most abundant CYP is CYP1A1, known to activate PAH into carcinogenic metabolites. It is the most abundant inducible form and its inducibility decreases dramatically from the duodenum to the ileum (Zhang *et al.*, 1996, 1997).

From the tissular profiles of PCB congeners found in cetaceans, it has been concluded that these animals have high levels of CYP1A and low levels of CYP2B. Compared to Arctic beluga, SLE beluga have elevated levels of both types of enzymes, probably because CYP are induced by exposure to high levels of PCBs (reviewed in Muir *et al.*, 1996). Considered together, the above observations suggest that intestinal CYP1A levels are elevated in SLE beluga, and might be responsible for the formation of intestinal cancer by the activation of ingested PAHs into carcinogenic compounds. Accordingly, we plan to determine whether CYP1A (because it is involved in the activation of PAHs) is highly expressed in the proximal intestine of SLE beluga.

A relation between ingestion of carcinogens and cancer in wildlife is not without precedent. In bottom-dwelling fish, labial papilloma and liver cancer are strongly associated with chemical contamination of sediments (Harshbarger, 1997). Lake whitefish (*Coregonus clupeaformis*) are the only salmonids that feed on benthic fauna. Lake whitefish inhabiting the St. Lawrence estuary have tissue concentrations of organochlorinated compounds (OC) and of heavy metals 3 to 5 times higher than those of sympatric fish species (including non-salmonid bottom-dwelling species), and these high concentrations coincide with a high prevalence of liver cancer (Mikaelian *et al.*, 1998, 2002). Thus beluga and lake whitefish, two aquatic vertebrate species that widely diverge taxonomically, might be affected by cancer because both feed on the bottom, an unusual feature within their respective taxonomic groups.

Few odontocetes feed on benthic invertebrates: the Amazon River dolphin the franciscana (*Pontoporia blainvillei*), the susu (*Platanista gangetica*) and the Irrawaddy dolphin (*Orcaella brevirostris*) (Ridgway and Harrison, 1989). Because these species generally inhabit rivers, and some populations live in contaminated habitats, high rates of cancers of the digestive epithelial lining might also be found in these species.

Mammary gland cancers were the cause of death of 7.7 per cent (*n* = 39) of adult female beluga examined from 1983 to 1998 (Mikaelian *et al.*, 1999a). Mammary gland adenocarcinomas have not been reported in other marine mammals, and are rare in herbivores, including cattle, which are phylogenetically close to cetaceans (Moulton, 1990; Buntjer *et al.*, 1997) (Table 13.7). Only isolated cases have been reported in other wildlife (Gillete *et al.*, 1978; Veatch and Carpenter, 1993). In contrast, these tumors are common in humans, domestic carnivores and rodents. Mammary gland cancers have been etiologically related with viruses only in rodents [retroviral sequences have recently been found in human cancerous breast (Pogo *et al.*, 1999)].

Environmental pollutants have been related etiologically with human breast cancer, and two mechanisms for chemical breast carcinogenesis have been proposed. The first one, tumor promotion, has been supported by the fact that most breast cancers are estrogen-dependent (Russo and Russo, 1998). Many OCs and some of their metabolites have estrogenic activities (Dees *et al.*, 1997). Concentrations of DDE and PCBs in tissues have sometimes been found to be higher in women with breast cancer than in unaffected women but this finding has not been consistent (Blackwood *et al.*, 1998; Güttes *et al.*, 1998). In SLE beluga, strong expression of estrogen receptors was detected in two mammary gland cancers, suggesting that these tumors are estrogen responsive, and that xeno-estrogens could be associated with their progression (Mikaelian *et al.*, 1999a).

The second proposed mechanism implicates tumor initiation by PAHs. It is supported by the detection of higher levels of aromatic DNA adducts in breast tissues from women with breast cancer compared to women unaffected by breast cancer (Perera *et al.*, 1995). SLE beluga tissues are contaminated with high levels of OC and contain DNA adducts, which indicates exposure to PAHs (Martineau *et al.*, 1987, 1988). Therefore, both mechanisms of carcinogenesis could be implicated in the occurrence of mammary tumors in this population. Alternatively, or concurrently, these tumors might be causally related to the extended hormonal stimulation associated with the long pregnancy and lactation of cetaceans (Geraci *et al.*, 1987).

Aluminum workers in the Saguenay-Lac Saint-Jean area (Fjord-du-Saguenay census division), which comprises a part of the beluga habitat (Figure 13.1), are affected by a high prevalence of lung and urinary bladder cancer, which has been epidemiologically associated with exposure to coal-tar volatile components (Armstrong *et al.*, 1994; Tremblay *et al.*, 1995). Other types of human cancer, including those of the gastrointestinal tract (gastric cancers in particular) are also more prevalent in that area than in the rest of Quebec, Canada (Table 13.8). Elsewhere, gastric cancers have been epidemiologically related with working in aluminum plants where carbon anodes are prebaked (Bye *et al.*, 1998). It has recently been shown that the chronic ingestion of coal-tar mixture causes gastric cancers in rodents (Culp *et al.*, 1998). Breast cancers are also more prevalent in the Saguenay-Lac Saint-Jean area when compared to the rest of Canada and Quebec (Table 13.8). Together, these observations suggest that a human population and a population of long-lived mammals may be affected by specific types of cancer, because they share the same habitat and are exposed to the same environmental contaminants.

Table 13.8 Incidence of cancer for Quebec census division 94 (Fjord-du-Saguenay) for the period 1986–94 based on *Fichier des Tumeurs* du Québec using the Canada 1991 census population[1] and Quebec province as reference region

Site	Sex	ASR	Obs.	SIR	CI 95%
Stomach	M	27.1	146	1.47**	1.24–1.73
	F	11.5	80	1.40**	1.11–1.74
Small intestine	M	2.3	13	1.76	0.94–3.00
	F	1.3	9	1.34	0.61–2.53
Colon, rectum	M	75.2**	41.2	1.14**	1.03–1.26
	F	46.6	331	0.99	0.89–1.10
Pancreas	M	14.1	77	1.07	0.84–1.34
	F	12.1	85	1.36**	1.09–1.69
Liver	M	6.2	34	1.11	0.77–1.55
	F	3.3	24	1.61*	1.03–2.39
Breast	M	0.7		<1.00	
	F	103.7**	790	1.13**	1.06–1.22
Lungs, bronchi, trachea	M	157.1**	881	1.30**	1.21–1.39
	F	40.9*	309	1.12*	1.00–1.26
Pleura	M	3.2	21	1.89*	1.17–2.88
	F	0.5		<1.00	

[1] A search was carried out within the Canadian provincial and territorial cancer registries. Statistics Canada provided data on a subset of the variables to Health Canada, and this file was used to generate the table of observed and expected cases. The expected number of cases was calculated by multiplying the age-specific Census Division (CD) population times the age-specific Canadian incidence rate, that is, the reference rates are for Canada as a whole. A description of the observed to expected ratio, also known as the Standardized Incidence Ratio (SIR) can be found in Breslow and Day (1987). The 95% CI was based on the approximation by Ury and Wiggins (1985). Stomach cancer and cancer of the small intestine are from International Classification of Diseases (ICD), 9th edition, Codes 151 and 152, respectively. However, about 5% of cancers occurring in the stomach should be classified as lymphomas and should not be included in the ICD code 151. Nevertheless, a small number of cases in CDs other than the study CD (Quebec CD 94) were listed as histology lymphoma with ICD-9 151. This will have a negligible impact on the reference rates.
Reference region: Quebec; ASR: age-standardized rate; Obs: number of observed cases; SIR: standardized incidence ratio (observed to expected ratio); CI: confidence interval. Absence of '*': non-significantly elevated ASR or SIR; *: significantly elevated ASR or SIR ($P < 0.05$); **: significantly elevated ASR or SIR ($P < 0.01$).

Other types of cancer are also more frequent in people living in that region (namely colon, rectum and pleura cancers in men, and pancreas and liver cancers in women), and thus a definitive etiological association with pollutants will require molecular epidemiological evidence. Several chemical carcinogens leave a signature on the host genome by causing mutations at specific sites in genes involved in cell proliferation, such as *p53* and *ras*. The finding of the same signature in tumors of SLE beluga and humans would strongly support the etiological role of contaminants in carcinogenesis (Perera and Dickey, 1997; Perera, 1998).

In people, genetic susceptibility to cancer takes two forms, hereditary cancer syndromes (HCS), such as familial adenomatous polyposis, and population susceptibility, where an ensemble of individuals has an increased risk of cancer (but not as high as in HCS). Because inbreeding has led to some degree of genetic homogeneity in SLE beluga, the possibility of a HCS within the SLE beluga population has to be considered (Patenaude *et al.*, 1994). In HCS, the types of cancer involved are usually rare in the general population. When these are not rare, multiple primary tumors develop in affected individuals. In addition, HCS affect multiple – and most often young – members of the same family (Fearon, 1997; Perera, 1997). There were no significant differences between the age of SLE beluga affected by intestinal cancer and that of beluga affected by other types of cancer ($P = 0.35$), and beluga with cancer were not younger than beluga that died of other causes (Table 13.4, Figure 13.2). In addition, SLE beluga are affected by single (as opposed to multiple) intestinal tumors. Finally, no other genetically homogeneous free-ranging or captive wildlife populations have been found to be affected by high rates of cancer (O'Brien, 1994; Munson *et al.*, 1999). An apparent exception is the highly inbred black-footed ferret (*Mustela nigripes*) where a high prevalence of cancers has been observed (Lair, 1998). However, all black-footed ferrets affected by cancer have been kept in captivity. Captivity greatly extends the life span of these animals, from 4 years in the wild to 7–9 years in captivity. Because tumors develop only in female ferrets older than 5 years, and because most tumors in males develop in animals older than 5 years, captivity clearly plays a major role in the etiology of these tumors, by extending the life span of the affected animals. In addition, these animals may have been exposed to carcinogenic compounds in captivity (Lair, 1998).

In population susceptibility to cancer, an ensemble of individuals has an increased risk of developing cancer because these individuals have a specific and common genetic feature due to normal polymorphism. This feature most often influences the metabolism of carcinogenic xenobiotics. In contrast to HCS, the types of cancer involved in population susceptibility are relatively frequent in the general population. An example of population susceptibility is represented by smokers who have a highly inducible form of CYP1A1 that makes them more susceptible to lung cancer (Perera, 1997). Similarly, some SLE beluga might have highly induced CYP1A1 in the proximal intestinal epithelium, rendering cells susceptible to mutagenesis by DNA-damaging metabolites generated from specific xenobiotics such as benzopyrene.

There is no evidence that cancer is frequent in beluga as a species. A single case of cancer is listed among the 17 beluga whales listed in the Marine Mammal Database (Tables 13.5 and 13.6). The few cancers that are significantly prevalent in wild mammals have a viral etiology [with the possible exception of metastatic carcinomas in California sea lions (*Zalophus californianus*) (Gulland *et al.*, 1996; Lipscomb *et al.*, 2000)]. These cancers are hepatocellular carcinomas of woodchucks (WHC) due to a species-specific hepadnavirus (Snyder *et al.*, 1982), and lymphomas of wild rodents due to a retrovirus (Teich, 1984). Both virus and carcinogenic contaminants have been suspected to cause a high prevalence of metastatic carcinomas in California sea lions (Gulland *et al.*, 1996; Lipscomb, 1998; Lipscomb *et al.*, 2000).

Two lines of evidence are not consistent with the hypothesis that the high rate of cancer is due to an aging population. First, SLE beluga affected with cancer did not reach the maximum life span reached by Arctic beluga, and secondly, SLE beluga with cancer showed the same age distribution as beluga dead of other causes (Figure 13.2).

Future studies

Because cancer is an ultimate but rare consequence of chemical mutagenesis, the epidemiological association of xenobiotics with carcinogenesis requires the examination of large numbers of animals. To demonstrate the role of xenobiotics in carcinogenesis in SLE beluga, convincing statistics would require much larger numbers of whales and/or the follow-up of SLE beluga for many more decades. The observation of high prevalences of cancer in other populations of marine mammals similarly exposed to carcinogens would strengthen an etiological relationship with chemical carcinogenesis.

Three new cases of cancer in SLE beluga

The carcass of an adult female beluga (9605) was found at Saint-Joseph-de-la-Rive, Quebec, on July 23, 1996. Two meters distally to the pylorus, a 5 mm-diameter transmural circular orifice was present in the duodenal wall. The intestinal wall bordering the defect was diffusely and moderately thickened (1.5 cm) over a 15 cm-diameter circular region, roughly centered on the defect. No adhesions or signs of inflammation were visible. Upon section, the thickened intestinal wall was homogeneously firm and white, and consisted microscopically of randomly distributed irregularly shaped, often cystic, tubuloacinar glandular structures separated by a moderately abundant and poorly cellular fibrous stroma (Figure 13.3). Rare solid nests of epithelial cells were present in venules and dilated lymphatic ducts. About half of the glandular structures were cystic (up to 4.5 mm-diameter), filled with mucus [as shown by a positive periodic acid–Schiff (PAS) reaction], and were lined by single to multiple layers of disorderly arranged, variably sized epithelial cells showing moderate anisokaryosis and loss of polarity. Accordingly, the lesion was designated adenocarcinoma of the proximal intestine.

The carcass of an adult male beluga (9609) was found at Saint-Ulric de Matane, Quebec, on November 23, 1996. The abdominal cavity contained approximately 15 liters of dark red aqueous fluid. Microscopically, nodules and nests of poorly differentiated tumor cells were observed in multiple organs. In the liver, numerous 0.5–2 cm-diameter, densely cellular nodules were randomly distributed, and were traversed by delicate irregular strands

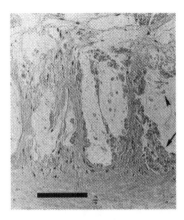

Figure 13.3 Proximal intestine of an adult beluga famale (9605) from the St. Lawrence estuary. Intestinal adenocarcinoma. Gland-like structures present in the muscularis layer are lined by epithelial cells (arrow) and are dilated by an abundant mucus material containing scattered single tumor epithelial cells (arrowhead). Bar = 100 μm. Hematoxylin–phloxine–saffron.

of fibrovascular stroma that separated disorderly arranged large tumor cells, having central, round to oval nuclei (Figure 13.4). The polygonal to flattened tumor cells had a moderately abundant cytoplasm and showed marked anisokaryosis and anisocytosis. Monstrous nuclei were occasionally seen. Moderate numbers of mitoses were seen (1–2 per 400× field).

In the lungs, variable numbers of small blood vessels contained clusters of tumor cells similar to those seen in the liver. Numerous tumor cells were present in the cortex and in the subcapsular sinuses of the mediastinal lymph nodes. In the pituitary gland, tumor cells formed emboli and 0.25 mm-diameter solid metastatic nodules. A 2 mm-diameter metastatic nodule was also observed in the testis, where mitotic tumor cells were numerous (6 mitoses/400 × field). The condition was diagnosed as a poorly differentiated metastatic malignant cancer, of probable epithelial origin.

The carcass of an adult male beluga (9809) was recovered on December 3, 1998, between Ste Flavie and Ste Luce, and examined on December 4. Ten centimeters distally to the pylorus, an ulcer was found in the duodenal ampulla, at the center of a 15 cm-diameter circular region where the duodenal wall was moderately thickened (0.5–1 cm) by tumoral glandular structures similar to those found in 9605. Tubuloacinar structures that often formed cavities up to 5 cm in diameter, filled with mucus, were distributed randomly throughout all layers of the intestinal wall (Figure 13.5). These structures were separated by abundant, fibrous, poorly cellular stroma, where small solid clusters of tumor cells, and occasionally single tumor cells, were present. In the liver, multiple solid metastatic nodules, up to 2 mm in diameter, were composed of clusters of less differentiated tumor cells that did not form

(a)

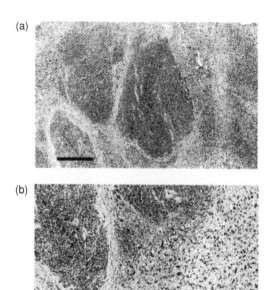

(b)

Figure 13.4 Liver of a beluga (9609) from the St. Lawrence estuary with multiple metastasis of poorly differentiated (probable epithelial) cancer. (a) Numerous, densely cellular nodules varying in shape and size are unencapsulated and randomly distributed. Bar = 0.5 mm. (b) Nodules are composed of large polygonal tumor cells showing marked anisocytosis and anisokaryosis. Note the better preservation of tumor cells compared to surrounding normal hepatocytes. Bar = 200 μm. Masson trichrome.

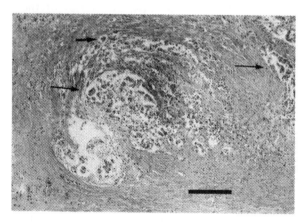

Figure 13.5 Proximal intestine of an adult beluga from the St. Lawrence estuary (9809). Intestinal adenocarcinoma. Ill-formed glandular structures (long arrows), and single tumor cells (short arrow) are present in the muscularis layer. The tumoral glandular structures are surrounded by concentric layers of connective tissue. (Desquamation of tumor cells is due to post-mortem autolysis). Bar = 200 μm. Hematoxylin–phloxine–saffron.

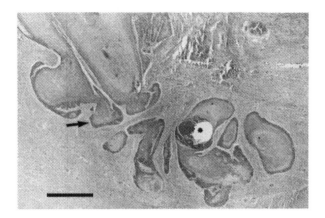

Figure 13.6 Skin of an adult beluga from the St. Lawrence estuary (9809). Spinocellular carcinoma. The dermis is invaded by irregular branching cords (arrow) and nests composed of well-differentiated keratinocyte-like cells. At the center of nests, tumor cells undergo parakeratosis (asterisk). Bar = 0.5 mm. Hematoxylin–phloxine–saffron.

glandular structures. This tumor was diagnosed as metastatic adenocarcinoma of the proximal intestine.

The same animal was also affected by a skin tumor. Ventrally, at the base of the tail, the skin showed a verrucous, slightly elevated (1–3 cm) area, 9 cm in diameter, with an ulcerated center. Upon section, digitiform extensions (3–5 mm in diameter) radiated from the ulcer into the subjacent dermis. These were composed of thick branching cords of keratinocytes surrounded by concentric layers of pale connective tissue (Figure 13.6). Several irregularly sized nests of keratinocytes were present between the branching cords and were composed of concentric layers of cells that showed gradual keratinization toward the center while they retained their pyknotic nuclei, mimicking the normal keratinization pattern of normal cetacean skin. The tumoral structures and the dermal blood vessels were surrounded by a moderately dense population of mononuclear cells. According to the above features, this tumor was termed a squamous cell carcinoma.

Hematopoietic and lymphopoietic organs

Lymphoid organs of SLE beluga often show marked lymphoid depletion. It is not possible, however, to determine whether this is due to chronic diseases or to high levels of immunosuppressive contaminants. Except for lymphomas, lymphoid lesions have seldom been described in cetaceans (Tables 13.5 and 13.6). Involution and cystic transformation of the thymus have been observed in bottlenose dolphins and in harbor porpoises (Cowan, 1994; Wunschmann *et al.*, 1999). These lesions have not been reported in other populations of marine mammals and their etiology remains undetermined.

SLE beluga are commonly affected by multisystemic infections involving lymph nodes with agents that have generally been associated with immune suppression in humans and animals. For instance, opportunistic bacteria such as *Edwardsiella tarda*, *Nocardia* spp., ciliated flagellates associated with suppurative bronchopneumonias, protozoa such as *Toxoplasma gondii* (Mikaelian *et al.*, 2000a), and herpesvirus have all been found in SLE beluga (Martineau *et al.*, 1988; De Guise *et al.*, 1995).

In other cetaceans, the finding of generalized infections with organisms not normally causing disease in immunocompetent animals has been taken as circumstantial evidence of immunosuppression induced by immunosuppressive pollutants such as PCBs (Inskeep *et al.*, 1990; Safe, 1994). Deficits in immune function are difficult to evaluate directly in free-ranging cetaceans, owing largely to the difficulty of obtaining, and rapidly processing, adequate samples. A logical approach to show that the immune functions of a given population are impaired would be to compare this population to a control population unexposed to pollutants. Many factors make such a comparison difficult: the inaccessibility of some populations, which introduce variables in the time required to collect and process samples; the stress of capture, which triggers cortisol release; and genetic differences.

Two types of indirect approaches allow these drawbacks to be avoided. The first consists of measuring a 'pollutant dose–response' effect. This has been achieved in dolphins, where increased concentrations of blood PCBs and DDT were shown to be inversely correlated with lymphocyte responses (Lahvis *et al.*, 1995). The second approach consists of measuring *in vitro* the response of immune cells from a presumably 'normal' population to pollutants added in concentrations identical or similar to those found in the tissues of contaminated animals from the same species. Thus, after obtaining baseline data on immunological parameters from Arctic beluga, despite considerable logistic problems, it has been possible to evaluate *in vitro* the effects of certain pollutants on beluga immune cells. The proliferative response of beluga lymphocytes to mitogens and their spontaneous proliferation are both impaired *in vitro* by exposure to concentrations of *p,p'*-DDT and PCB 138 similar to those found in tissues of SLE beluga (PCB 138 is one of the most abundant PCB congener present in SLE beluga tissues) (De Guise *et al.*, 1998).

More indirect evidence of OC-immunosuppressive effects in cetaceans is the observation of significantly higher tissue PCB concentrations in striped dolphins affected by the 1990–92 morbillivirus epizootic in the Mediterranean Sea, compared to concentrations observed in previous and later years. These observations led to the conclusion that PCBs may have impaired the dolphins' immune response to the viral infection (Aguilar and Borrell, 1994). More recently, a similar association between morbilliviral infection and high OC tissue levels has been observed in common dolphins (*Delphinus delphis ponticus*) from the Black Sea (Birkun *et al.*, 1999). Finally, other marine immunotoxic pollutants and their metabolites are found in cetacean tissue, such as butyltin compounds and dieldrin. Their negative impact remains to be determined (Yang *et al.*, 1998; Saint-Louis *et al.*, 1999).

Liver

SLE beluga are commonly affected by hepatic lesions. Hepatic lesions were seen in 12 of 94 animals (13 per cent) and consisted variably of portal lymphocytic infiltration, bile duct proliferation, hepatocellular degeneration and portal fibrosis, sometimes extending into the lobular parenchyma (Martineau *et al.*, 1988; De Guise *et al.*, 1995; Igor Mikaelian, personal observation). Similar lesions (chronic portal fibrosis and portal lymphocytic infiltration) are commonly found in striped dolphins (*Stenella coeruleoalba*) and aging bottlenose dolphins. The etiology of these lesions has been thought to be nutritional or toxic in some cases, while in others, the etiology is clearly parasitic (Bossart, 1984). Pigments are also commonly observed in the hepatocytes of adult SLE beluga, but their composition has not been determined (Mikaelian, unpublished observations).

Viruses are frequent causes of hepatitis in humans and in animals. In cetaceans, a single case of hepatitis has been related to a viral etiology (Bossart *et al.*, 1990).

Mercury is biomagnified in the marine food web. As a result, high hepatic levels of mercury have been detected consistently in cetaceans worldwide, and recent evidence supports the hypothesis that mercury intoxication contributes to the etiology of some of the hepatic lesions seen in cetaceans (Tilbury *et al.*, 1997; Parsons, 1998; Meador *et al.*, 1999). In bottlenose dolphins, lipofuscin-like pigments are observed in the portal connective tissue and hepatocytes. In four of nine dolphins with hepatic lipofuscin-like pigments, the pigments were accompanied with hepatic lesions (lipidosis, necrosis and inflammation), whereas only one of nine dolphins without lipofuscin-like pigments had hepatic lesions. The pigment was shown to be mainly composed of mercury by X-ray spectroscopy, and mercury concentration was significantly higher in livers with lesions (Rawson *et al.*, 1993). Because SLE beluga have similar hepatic mercury levels and lesions, further investigation on the role played by mercury contamination in the hepatic lesions seen in SLE beluga is warranted (Wagemann *et al.*, 1990). SLE beluga livers also contain other hepatotoxic pollutants (PCBs, butyltin compounds), the role of which should be further evaluated (Yang *et al.*, 1998).

Gastrointestinal tract

Periodontitis is common in SLE beluga (De Guise *et al.*, 1995). The gingiva and the teeth are separated by a gap filled with necrotic debris and the periodontal ligament is destroyed, often leading to tooth loss. In the most severely affected animals, only a few loosely attached teeth remain in the mouth. Less than 60 per cent of stranded SLE beluga have the number of teeth corresponding to the number of alveoli (28–36) (Béland *et al.*, 1993).

The impact of dental health on the survival of wildlife has been evaluated in several species. In Weddell seal (*Leptonychotes weddellii*), Darajani baboons (*Papio cynocephalus*) and feral cats, the life span of adults seems to be shortened by dental disease and tooth loss (Robinson, 1979a, b; Verstraete *et al.*, 1996a, b).

Periodontitis and maxillary and mandibular osteomyelitis are common in gray seals from the Baltic Sea, and have been attributed to environmental contamination by OCs. The prevalence of this condition increased throughout the twentieth century, along with an increase in PCB and DDT levels in Baltic biota, which suggests an etiological relationship with OC contamination (Olsson *et al.*, 1994). In humans and rats, periodontitis has been associated with chronic intoxication with PCB (Hashiguchi *et al.*, 1991, 1993). Considered together, these observations suggest that periodontitis could be related to OC chronic intoxication in SLE beluga.

Ulceration of the upper digestive tract is frequent in SLE beluga. Twelve per cent of SLE beluga had esophageal ulcers. Esophageal ulcers are frequent in stranded cetacean carcasses and have been associated with debilitation (Sweeney and Ridgway, 1975). Twenty six per cent of SLE beluga had ulcers of one of the first two gastric compartments (*n* = 24) and the first compartment was most often affected (15 beluga) (Mikaelian and Martineau, unpublished observations, 1999). Ulcers found in the first compartment were associated with nematodes, but not the ulcers found in the second compartment (Martineau *et al.*, 1988; De Guise *et al.*, 1995). Two animals had a perforated ulcer of the first compartment, and the site of perforation contained numerous adult *Anisakis simplex* (Martineau *et al.*, 1988; Mikaelian and Martineau, unpublished observation, 1999).

Gastric ulcers are common in other species of cetaceans. They are most often associated with gastric nematodiasis, and rarely result in gastric perforation (Geraci and Gertsmann, 1966; Sweeney and Ridgway, 1975; Cowan *et al.*, 1986; Baker, 1992; Babin *et al.*, 1994). In one study, however, gastric ulcers were rarely associated with nematodes (Abollo *et al.*, 1998). A possible etiological role of PCBs in gastric ulceration has been raised, because these compounds have been shown to cause gastric ulcers in laboratory and domestic animals (Hansen *et al.*, 1976; van den Berg *et al.*, 1988). Many other factors could be implicated, however, including stress and infections by ulcerogenic bacteria such as *Helicobacter* spp.

Studies comparing the prevalence of gastric ulceration in populations of cetaceans from waters contaminated to different degrees are needed to assess the potential role of contaminants in the etiology of these lesions.

Endocrine lesions

In recent years, major concerns have been raised about the presence of endocrine-disrupting chemicals in the environment and their effects on human and wildlife health (Leatherland, 1998; Kavlock, 1999). SLE beluga are contaminated by high levels of OC compounds, many of which are known to be endocrine-disruptive (Martineau *et al.*, 1987; Muir *et al.*, 1990; Brouwer *et al.*, 1998). Total PCB concentrations in blubber levels are about 30 times higher than those of Hudson Bay beluga, and mirex concentrations are 50–100 times higher (Muir *et al.*, 1996). Lesions suggestive of endocrine disruption in the SLE population include proliferative and degenerative changes of the adrenal gland, proliferative lesions of the thyroid gland, a case of true hermaphroditism (De Guise *et al.*, 1994e), a case of male pseudohermaphroditism (Martineau, unpublished data), three mammary carcinomas, a uterine adenocarcinoma, and a high prevalence of fibroleiomyomas of the female tubular genitalia (Lair *et al.*, 1998; Mikaelian *et al.*, 1999a, 2000b).

Thyroid

There are growing concerns that man-made environmental contaminants could result in the disruption of thyroid homeostasis in aquatic animals and humans. Circumstantial evidence supports a link between certain pollutants and hyperplastic thyroid lesions in wild fish and birds (Leatherland, 1998). This link has been experimentally supported in rodents fed PCBs and dioxins (Collins *et al.*, 1977; Capen, 1992; Ness *et al.*, 1993).

Thyroid lesions have rarely been reported in cetaceans. Thyroid abnormalities were reported in 4 of 55 long-finned pilot whales (7.2 per cent) collected through hunting (Cowan, 1966). The macro/microscopic appearance and size of thyroid glands showed high inter-individual variability. The four abnormally enlarged thyroid glands were grossly multinodular and had microscopic changes consistent with simple macrofollicular colloid goiter, as seen in humans (Cowan, 1966). Two of 108 cetaceans stranded on the California coast had abnormal thyroids (1.8 per cent). A northern right-whale dolphin (*Lissodelphis borealis*) had cystic changes with diffuse lymphocytic infiltration. The cystic changes were attributed to the accompanying inflammation. A second northern right-whale dolphin had nodular hyperplasia of parafollicular cells (C cells) (Howard, 1983).

From 1996 to 1998, the thyroid glands of all beluga whales from the SLE recovered by our group were sectioned in 5-mm thick slices, and all macroscopic lesions were examined microscopically. The criteria used for the classification of thyroid lesions were in accordance with those used in humans and animals (Ashley, 1978; Benjamin *et al.*, 1996). Briefly, foci consisting of proliferating follicular cells and that were not encapsulated were classified as nodules of follicular hyperplasia. Foci that were completely encapsulated were classified as adenoma.

This study indicated that eight of nine beluga whales older than 16 years had multiple nodules of follicular hyperplasia and follicular adenomas, and follicular cysts of the thyroid (Mikaelian, unpublished data). Nodules of follicular hyperplasia and adenomas both appeared as white nodular masses that markedly distorted the thyroid. Some of these masses contained cystic to polycystic areas filled with thick viscous colloid. Nodules of follicular hyperplasia were small (generally smaller than 2 mm in diameter). Adenomas were larger (generally larger than 2 mm and up to 3 cm in diameter).

Both types of lesions generally consisted of variably sized follicles lined by a cuboidal to columnar epithelium that formed numerous papillary projections. The epithelial cells had a moderately abundant acidophilic and delicately granular cytoplasm. Their nucleus was basal and round with a coarsely granular chromatin. These follicular structures were separated by variable proportions of solid regions composed of closely packed polygonal to fusiform cells. These solid areas multifocally underwent variable degrees of cystic change. The center of the largest nodules was sometimes necrotic and hemorrhagic. Invasion of the capsule and the adjacent parenchyma was occasionally present. However, these lesions were not classified as carcinomas because vascular invasion and distant metastases were not observed.

Follicular cysts were found in 8 of 12 SLE adult beluga. These lesions measured up to 40 mm in diameter and were filled with an amber-colored viscous fluid. Histologically, follicular cysts were filled with colloid and were lined by a flattened epithelium. Cholesterol clefts and thyreoliths were found in the largest lesions. The surrounding connective tissue was often mildly thickened by collagen fibers. Cowan described similar lesions in long-finned pilot whales and categorized them as colloid goiters (Cowan, 1966). Follicular cysts are common in humans, where they represent 25 per cent of cystic thyroid nodules, and in rats where they are considered as a degenerative process related to aging (Rosen *et al.*, 1986; Takaoka *et al.*, 1995).

Recognized causes of thyroid hyperplasia and neoplasia in man and domestic animals include lymphocytic thyroiditis, iodine deficiency, iodine excess, radiations and exposure to dioxins (Capen, 1990, 1992). In marine mammals, environmental contamination by OCs is strongly suspected to be a leading cause of thyroid disruption, as shown by an experimental study where harbor seals fed fish contaminated with PCBs had a significant reduction in total and free thyroxine (T_4) compared with seals fed less-contaminated fish (Brouwer *et al.*, 1989). In harbor porpoises contaminated with PCBs, thyroid fibrosis was observed, and the extent of fibrosis was inversely proportional to PCB concentrations in five of eight porpoises (Walcott, 1999). The evaluation of the role played by environmental contaminants in the etiology of thyroid lesions in SLE beluga will necessitate the examination of thyroid glands from age-matched whales from a relatively uncontaminated population.

Adrenal glands

Degenerative and proliferative changes have been observed in the adrenal cortex and medulla of SLE and western Hudson Bay beluga, and the severity of these lesions increases with age in both populations (De Guise *et al.*, 1995; Lair *et al.*, 1997). The younger age of control beluga sampled from Hudson Bay precluded a comparison of lesion severity and prevalence between age-matched groups.

Adrenocortical cysts are rare in all species except, according to existing reports, in beluga and Atlantic white-sided dolphins (Geraci and St.Aubin, 1979; Cartee *et al.*, 1995). In the latter species, these lesions were attributed to sinusoidal blockage or hypersecretion, and were considered associated with stress related and reproductive functions, because 100 per cent of females and only 20 per cent of males were affected. Adrenal cysts have been also described in a common dolphin (Cartee *et al.*, 1995). Adrenal glands of various odontocetes (including *Phocoena*, *Tursiops*, *Delphinapterus* and *Globicephala*) were examined in detail by several authors through the 1940s to the late 1960s, and no lesions have been reported (Cowan, 1966; Harrison, 1969). Consequently, the rarity of adrenal lesions in cetaceans other than beluga and white-sided dolphins is probably not an artifact, and the occurrence of adrenal cysts in the latter two species may be a recent event.

Several sources of evidence collected over the past decade suggest that OC metabolites may cause the cysts seen in the adrenal glands of Atlantic white-sided dolphins and in SLE beluga. First, some OC metabolites, $MeSO_2OC$, are adrenocorticolytic in rodents, and some of these compounds, such as $3\text{-}MeSO_2\text{-}4,4'\text{-}DDE$, have been shown to compete with glucocorticoid receptors and to inhibit glucocorticoid synthesis, which is an important function of the adrenal cortex (Durham and Brouwer, 1990; Brandt *et al.*, 1992; Johansson *et al.*, 1998; Letcher *et al.*, 2000a, b).

Secondly, in gray and harbor seals from the Baltic Sea, adrenocortical hyperplasia has been attributed to contamination with PCB and DDT, based on epidemiological data (Olsson, 1994; Olsson *et al.*, 1994; Chapter 19 in this volume). In Baltic gray seals, $3\text{-}MeSO_2$ and the PCB levels were highest in females with symptoms of adrenocortical hyperplasia (Haraguchi *et al.*, 1992). This sex distribution is reminiscent of that of the adrenal cysts seen in Atlantic white-sided dolphins.

Thirdly, SLE beluga and Atlantic white-sided dolphins are contaminated with high amounts of PCBs and other OCs (Table 13.9) (Martineau *et al.*, 1987; Muir *et al.*, 1996; McKenzie *et al.*, 1997; Troisi *et al.*, 1998). High concentrations of $MeSO_2$-PCB and $MeSO_2$-DDE have been detected in blubber of SLE beluga and are the highest among cetaceans, including Hudson Bay beluga (the concentrations found in SLE beluga are also higher than those found in humans exposed to PCB during the Yusho industrial accident) (Letcher *et al.*, 2000a).

SLE beluga and white-sided dolphins share two biological features. They have similar ability to form methylsulfones from PCBs and both species have a long life span. The $\Sigma PCB : \Sigma MeSO_2$ ratio determined in a white-sided Atlantic dolphin is similar to that measured in beluga blubber, which suggests that white-sided dolphins have the same ability as beluga to metabolize PCB through methylsulfone intermediates (Troisi *et al.*, 1998). Thus both species have been exposed to high levels of adrenotoxic OC metabolites for decades. The sex bias shown by the lesions in female

Table 13.9 PCBs, DDT, DDE and MeSO$_2$ metabolite levels in blubber of odontocetes in the northern hemisphere

Species	Location	Sex	n	Mean concentration (μg/g) ± SD (range)			References	MeSO$_2$-PCB (μg/g)	Tot MeSO$_2$-PCB/Tot. PCB	References
				Tot PCB	Tot DDT	DDE				
Beluga	Saint Lawrence estuary	M	15	78.9 (8.3–412)	81.1 (3.3–389)	47 (2.1–249)	Muir et al. (1996)	0.2 (0.02–1.01)	0.05 (0.01–0.21)	Letcher et al. (2000a)
	Hudson Bay (east and west)	F	21	29.6 (8.8–83)	17.5 (3.6–80)	10.4 (1.7–52.3)	Muir et al. (1990, 1996); Norstrom et al. (1990)	0.159 (0.12–0.18)	0.01 (<0.01–0.04)	
		M	12	2.7 ± 0.9	2.4 ± 1.0				0.03 (0.02–0.05)	
		F	3–5	nd	nd	nd		nd	–	Troisi et al. (1998)
Harbor porpoise	Great Britain	M		44.5 ± 18.8	0.1 ± 0.06	9.66 ± 4.27	Kuiken et al. (1993)		0.09	
	Saint Lawrence Gulf	F	9–12	22.4 ± 11.74	1.88 ± 2.79	4.88 ± 3.5	Westgate (1995) in Muir et al. (1996)	0.58		
	Saint Lawrence Gulf	M	13	10 ± 3.5	6.2 ± 2.1	nd		nd	nd	
Bottlenose dolphin	Gulf of Mexico	M	4	86 (72–100)	0.9 (0.234–1.9)	23 (15–40)	Kuehl and Haebler (1995)	nd	nd	
	US central/south Atlantic coast	F	6	7.2 (1.5–18)	0.009 (nd–0.015)	3.7 (0.6–14)	Kuehl et al. (1991) in Muir et al. (1996)	nd	nd	
	US central/south Atlantic coast	M, F	14	81.0 ± 66.0	15.0 ± 23	nd		nd		
Whitebeaked dolphin	Saint Lawrence Gulf	M	9	34.0 ± 22	43.0 ± 27.0	nd	Muir et al. (1988)	nd	nd	
Atlantic white-sided dolphin	Irish Sea		17	(0.773–63.4)	(0.160–54.6)	nd	McKenzie et al. (1997)	nd	nd	
Atlantic white-sided dolphin	Irish Sea	M	1	15.48	nd	nd	Troisi et al. (1998)	0.31	0.02	Troisi et al. (1998)
Atlantic white-sided dolphin	North-western Atlantic	M	2	57.9 ± 1.75	nd	16.0 ± 7.07	Kuehl et al. (1991); McKenzie et al. (1997)			

Scientific names: beluga (*Delphinapterus leucas*); harbor porpoise (*Phocoena phocoena*); bottlenose dolphin (*Tursiops truncatus*); whitebeaked dolphin (*Lagenorhyncus albirostris*); Atlantic white-sided dolphin (*Lagenorhyncus acutus*).
nd, Not determined.

white-sided dolphins and gray seals may be due to several factors. The mobilization of OCs from lipids during lactation may induce an episode of 'acute' OC toxicity by redistribution to target organs (Martineau *et al.*, 1987; Polischuk *et al.*, 1995; Letcher *et al.*, 2000a).

In SLE male beluga, 3-MeSO$_2$-4,4'-DDE levels do not increase with age, despite a significant increase in 4,4'-DDE tissue levels with age. This discrepancy has been explained by either the presence of alternative metabolic pathways for DDE and/or secondary metabolism, or by the sequestration of 3-MeSO$_2$-4,4'-DDE in adrenal glands (Letcher *et al.*, 2000a). The latter mechanism implies that 3-MeSO$_2$-4,4'-DDE concentrations increase in the adrenal cortex tissue with age, which would increase adrenal toxicity and lesion severity in males. Because there was no statistically significant difference in the severity of adrenocortical lesions between male and female beluga, it is possible that 3-MeSO$_2$-4,4'-DDE reaches threshold levels in males and females, or alternatively, that 3-MeSO$_2$-4,4'-DDE is not the major adrenotoxic OC metabolite in beluga (Lair *et al.*, 1997).

There are some apparent problems with the hypothesis that MeSO$_2$-OCs cause adrenal lesions in marine mammals. First, there seems to be a contradiction between the adrenocortical hyperplasia epidemiologically associated with MeSO$_2$-DDE in seals, and the adrenocortical degeneration induced by these compounds in laboratory animals, and possibly in SLE beluga (Brandt *et al.*, 1992; Jönsson *et al.*, 1993). A possible explanation for the lesions seen in SLE beluga is that OC metabolite-mediated degeneration of the adrenal cortex alternates with ACTH-mediated regeneration, because in mammals, the destruction of the adrenal cortex and/or the interference with glucocorticoid synthesis triggers the feedback control of the hypothalamo-pituitary–adrenocortical axis. Decreased glucocorticoid levels due to adrenocortical destruction cause increased production of ACTH and of other peptides by the pituitary. These pituitary proteins lead to hypertrophy (increased cellular size) and hyperplasia (increased cell numbers) of the adrenal cortex in order to re-establish serum glucocorticoid levels (Rijnberk, 1996). Interestingly, adrenocortical insufficiency has been reported in fish inhabiting contaminated segments of the St. Lawrence river and has been attributed to OC contamination (Hontela *et al.*, 1992). Thus, it is possible that adrenal lesions affect taxonomically divergent species because of environmental exposure to the same adrenotoxic lipophilic compounds.

Another explanation to reconcile the different nature of adrenal lesions seen in laboratory animals and in seals is that, as for most drugs and/or toxic compounds, the pathologic effects of low OC doses ingested over decades probably differ from those of large single experimental doses. SLE beluga, white-sided dolphins, harbor porpoises and Baltic gray seals (like SLE beluga) are exposed to complex cocktails of OC compounds which generate different metabolites. In turn, these metabolites alter the distribution of each other (van Birgelen *et al.*, 1996). The combined pathologic effects of these complex mixtures probably do not match the effects of single compounds or metabolites that are used in toxicological studies. For instance, the difference in tissue levels of MeSO$_2$-PCBs congeners and PCB congeners detected in Hudson Bay and SLE beluga, is due most likely to the greater induction of CYP enzymes in the latter population (Norstrom *et al.*, 1992; Muir *et al.*, 1996; Letcher *et al.*, 2000a). In addition, the effects of toxic xenobiotics vary according to species, sex, genetic background, age and the developmental stage at which experimental animals are first exposed (Jönsson *et al.*, 1995).

Secondly, harbor porpoises are also contaminated with OC and show adrenocortical hyperplasia. There is no relationship, however, between OC tissue concentrations and the severity of hyperplasia (Kuiken *et al.*, 1993). According to some authors, harbor porpoises may stand apart from other cetaceans, and most closely match pinnipeds, by their high capacity for PCB methylsulfone formation, and by high CYP2B-dependent ethoxyresorufin-*O*-deethylase (EROD) activity (Boon *et al.*, 1994; Reijnders and de Ruiter-Dijkman, 1995; Troisi *et al.*, 1998). Thus, it is possible that adrenocortical hyperplasia, opposed to adrenocortical degeneration, occurs in harbor porpoises and gray seals because both species share common pathways for metabolizing PCBs. It will be important, however, to verify this hypothesis by analyzing tissues from additional harbor porpoises.

A third issue that needs to be resolved is the role of aging in the development of adrenocortical lesions. In the original study on Baltic gray seals, there was a positive relationship between the severity of adrenocortical hyperplasia and the age of affected seals (Bergman and Olsson, 1986; Haraguchi *et al.*, 1992; Olsson *et al.*, 1994). Because OCs and their metabolites generally accumulate in tissues with increasing age (with some exceptions such as 3-MeSO$_2$-4,4'-DDE in male SLE beluga), it might be difficult to distinguish between the role of OC concentrations and degenerative changes due to aging (Letcher *et al.*, 2000a).

FUTURE STUDIES

The above observations warrant the examination of adrenal glands of other cetacean populations in order to detect morphological evidence of OC adrenotoxicity. If OC concentrations are sufficient to cause adrenotoxicity, a 'dose–response' might be observed, that is, OC concentrations should be proportional to the severity and prevalence of the adrenal lesions. In the most contaminated populations, adrenal lesions are expected to affect males and females evenly, as in SLE beluga whales, whereas in less-contaminated populations, females are expected to be most affected, as in white-sided dolphins.

Adrenal cysts can be detected and measured by echography in marine mammals (Cartee *et al.*, 1995). In future studies, the size and shape of adrenal glands (and of potential adrenal cysts) could be measured directly by echography and paralleled with serum cortisol levels, in order to determine whether chronic OC intoxication induces adrenocortical hyperplasia and/or cysts, and/or impairs adrenal function. Animals experimentally or naturally contaminated with adrenotoxic compounds for long periods are also expected to show biochemical changes specific for Addison's disease: elevated ACTH levels, and low or normal glucocorticoid levels. An ACTH-stimulation test of animals chronically intoxicated with OC and/or their adrenotoxic metabolites will not trigger an increase in serum cortisol levels, as it should do normally. The pituitary gland of contaminated animals is expected to show lesions, either of hyperplasia and hypertrophy (the glands should be heavier than normal and hypercellular), or atrophy (the glands should be lighter than normal and hypocellular) if the glucocorticoid-deficient state has been chronic. The pituitary changes would be best evaluated by weighing the gland, and by microscopic examination using immunohistochemistry to evaluate the number and size of corticotrophs (the pituitary cells producing ACTH).

In mink (*Mustela vison*), acute intoxication with DDD causes vacuolation of adrenocortical cells. Cellular vacuolation, which seems to lead to the development of cysts, is also seen in the adrenal cortex of beluga (Jönsson et al., 1993; Lair et al., 1997). Longer experimental intoxication with OC of laboratory rodents and mink would determine whether adrenocortical cell vacuolation progresses into cyst formation.

Considering that a DDT metabolite, *o,p'*-DDD, has long been used medically to destroy hyperplastic and tumoral adrenal cortex, it is paradoxical that only recently have the adrenal accumulation and adrenotoxicity of PCB and DDT metabolites been suspected of causing adrenal lesions in marine mammals (Hart et al., 1973; Durham and Brouwer, 1990; Lair et al., 1997). The emaciation of presumably starved marine mammals (as determined from blubber thickness) has sometimes been considered an artifact that hampers the analysis of data pertaining to OC contamination. Considering the above observations, future studies should address the possibility that OC intoxication plays a primary role in the starvation and/or emaciation of marine mammals. To help resolve these issues, carcasses of cetaceans submitted to analysis for OC contamination should also be examined for the presence of lesions in all organs, particularly in endocrine glands.

Because Hudson Bay beluga are also contaminated with OCs, contamination with OCs could play a role in the etiology of the adrenal lesions observed in these animals.

Conclusions

A population of approximately 650 beluga inhabits a short segment of the St. Lawrence estuary. This body of water drains one of the most industrialized areas of the world. Over 16 years (1983–98), we have examined 120 (or 49 per cent) of 246 beluga carcasses reported stranded in the St. Lawrence estuary (SLE). Of the 120 carcasses examined, 78 per cent were considered preserved adequately for diagnostic purposes. The major primary causes of death were respiratory and gastrointestinal infections with parasites (21.3 per cent), cancer (18 per cent) and bacterial, viral and protozoal infections (16 per cent). Other causes of death were dystocia and separation from mother (6.4 per cent), dissecting aneurysm of the pulmonary trunk (3.2 per cent), trauma of undetermined cause (3.2 per cent) and wound due to small motor boat propellers (2.1 per cent).

Including three ovarian tumors, cancer was observed in 27 per cent of examined adult animals found dead, a percentage similar to that found in humans. Only 30 other cases of cancer have been reported worldwide in captive and wild cetaceans. If the estimated annual rate of cancer (AR) is calculated, and assuming a stable population of 650 animals over the past 16 years, the resulting AR of 192/100 000 animals is much higher than that reported for any other population of cetacean. The resulting AR is similar to that of humans, and to that of hospitalized cats and cattle. The AR of cancer of the proximal intestine, a minimum figure of 67/100 000 animals, is much higher than that observed in domestic animals and humans, except in sheep in certain parts of the world, where environmental contaminants are believed to be involved in the etiology of this condition.

Considered together, the following factors suggest that contamination with polycyclic aromatic hydrocarbons (PAH), alone or along with a viral infection, could be etiologically related with the high rates of proximal intestinal cancer observed in this population:

(1)   The recent experimental demonstration in laboratory animals that chronic ingestion of coal tar and small intestinal cancer are etiologically related.
(2)   The presence of PAH in the sediments of the SLE beluga habitat.
(3)   SLE beluga feed significantly on benthic invertebrates.
(4)   The presence of benzo[a]pyrene DNA adducts in SLE beluga tissues and their absence from Arctic beluga tissues, as determined by HPLC and fluorescence detection.

Acknowledgments

Incidence data were provided to Health Canada from the Canadian Cancer registry at Statistics Canada. The cooperation of the provincial and territorial cancer registries which supply the data to Statistics Canada is gratefully acknowledged. We thank Drs S. De Guise, S. Lair, R.J. Letcher, L. Measures and R. Norstrom for helpful discussions. We thank R. Plante and C. Guimont (Filmar) for recovering and transporting carcasses to our facility over the past 15 years, all students who helped post-mortem examinations, and our colleagues at the FMV for their consistent support.

We acknowledge the help of Centre Océanographique de Rimouski for logistic support, and the financial support of World Wildlife Fund Canada, Alcan, Fisheries and Oceans Canada, Fondation de la Faune du Québec, Société des Parcs du Québec, and NSRC.

References

Abollo, E., Lopez, A., Gestal, C., Benavente, P. and Pascual, S. 1998. Long-term recording of gastric ulcers in cetaceans stranded on the Galician (NW Spain) coast. *Diseases of Aquatic Organisms* 32: 71–73.

Aguilar, A. and Borrell, A. 1994. Abnormally high polychlorinated biphenyl levels in striped dolphins (*Stenella coeruleoalba*) affected by the 1990–1992 Mediterranean epizootic. *Science of the Total Environment* 154: 237–247.

Aguirre, A.A., Brojer, C. and Morner, T. 1999. Descriptive epidemiology of roe deer mortality in Sweden. *Journal of Wildlife Diseases* 35: 753–762.

Anonymous 1991. *Cancer in Canada. 1985, 1986.* Vol. 3(2). Health Canada.

Armstrong, B., Tremblay, C., Baris, D. and Theriault, G. 1994. Lung cancer mortality and polynuclear aromatic hydrocarbons: a case-cohort study of aluminum production workers in Arvida, Quebec, Canada. *American Journal of Epidemiology* 139: 250–262.

Ashley, D.J.B. 1978. Tumours of the endocrine glands. In: Ashley, D.J.B. (ed.). *Evans' histopathological appearances of tumours*, pp. 221–237. Churchill Livingstone, Edinburgh.

Babin, P., Raga, J.A. and Duguy, R. 1994. Ulcères parasitaires gastriques chez les cétacés odontocètes échoués sur les côtes de France. *Point Vétérinaire* 26: 77–81.

Baker, J.R. 1992. Causes of mortality and parasites and incidental lesions in dolphins and whales from British waters. *Veterinary Record* 130: 569–572.

Baker, J.R. and Martin, A.R. 1992. Causes of mortality and parasites and incidental lesions in harbour porpoises (*Phocoena phocoena*) from British waters. *Veterinary Record* 130: 554–558.

Béland, P., Vezina, A. and Martineau, D. 1988. Potential for growth of the St. Lawrence (Québec, Canada) beluga whale (*Delphinaterus leucas*) population based on modelling. *International Council for the Exploration of the Sea (ICES). Journal of Marine Science* 45: 22–32.

Benirschke, K. and Marsh, H. 1984. Anatomic and pathologic observations of female reproductive organs in the short-finned pilot whale, *Globicephala macrorhynchus*. *Reports of the International Whaling Commission* (Special issue) 6: 451–454.

Benjamin, S.A., Stephens, L.C., Hamilton, B.F., Saunders, W.J., Lee, A.C., Angleton, G.M. and Mallinckrodt, C.H. 1996. Associations between lymphocytic thyroiditis, hypothyroidism, and thyroid neoplasia in beagles. *Veterinary Pathology* 33: 486–494.

Bergman, A. and Olsson, M. 1986. Pathology of Baltic grey seal and ringed seal with special reference to adrenocortical hyperplasia: is environmental pollution the cause of a widely distributed disease syndrome? Proceedings from the Symposium on the Seals in the Baltic and Eurasian Lakes, Savonlinna, 5–8 June 1984. *Finnish Game Research* 44: 47–62.

Béland, P., De Guise, S., Girard, C., Lagacé, A., Martineau, D., Michaud, R., Muir, D.C.G., Norstrom, R.J., Pelletier, E., Ray, S. and Shugart, L.R. 1993. Toxic compounds and health and reproductive effects in St. Lawrence Beluga whales. *Journal of Great Lakes Research* 19: 766–775.

Birkun, A. Jr, Kuiken, T., Krivokhizhin, S., Haines, D.M., Osterhaus, A.D., van de Bildt, M.W., Joiris, C.R. and Siebert, U. 1999. Epizootic of morbilliviral disease in common dolphins (*Delphinus delphis ponticus*) from the Black sea. *Veterinary Record* 144: 85–92.

Blackwood, A., Wolff, M., Rundle, A., Estabrook, A., Schnabel, F., Mooney, L.A., Rivera, M., Channing, K.M. and Perera, F. 1998. Organochlorine compounds (DDE and PCB) in plasma and breast cyst fluid of women with benign breast disease. *Cancer epidemiology, biomarkers and prevention* 7: 570–583.

Boon, J.P., Oostingh, I., van der Meer, J. and Hillebrand, M.T. 1994. A model for the bioaccumulation of chlorobiphenyl congeners in marine mammals. *European Journal of Pharmacology* 270: 237–251.

Bossart, G.D. 1984. Suspected acquired immunodeficiency in an Atlantic bottlenosed dolphin with chronic-active hepatitis and lobomycosis. *Journal of the American Veterinary Medical Association* 185: 1413–1414.

Bossart, G.D., Brawner, T.A., Cabal, C., Kuhns, M., Eimstad, E.A., Caron, J., Trimm, M. and Bradley, P. 1990. Hepatitis B-like infection in a Pacific white-sided dolphin (*Lagenorhynchus obliquidens*). *Journal of the American Veterinary Medical Association* 196: 127–130.

Bossart, G.D., Ewing, R., Herron, A.J., Cray, C., Mase, B., Decker, S.J., Alexander, J.W. and Altman, N.H. 1997. Immunoblastic malignant lymphoma in dolphins: histologic, ultrastructural and immunohistochemical features. *Journal of Veterinary Diagnostic Investigation* 9: 454–458.

Brandt, I., Jönsson, C.-J. and Lund, B.-O. 1992. Comparative studies on adrenolytic DDT metabolites. *AMBIO* 8: 602–605.

Breslow, N. and Day, N. 1987. *Statistical methods in cancer research.* International agency for research on cancer (IARC) Sci. Publ. No. 822.

Breuer, E.M., Krebs, B.H. and Hofmeister, R.J. 1989. Metastasizing adenocarcinoma of the stomach in a harbor porpoise. *Diseases of Aquatic Organisms* 7: 159–163.

Brodie, P.F. 1982. The beluga (*Delphinapterus leucas*); growth at age based on a captive specimen and a discussion of factors affecting natural mortality estimates. *Reports of the International Whaling Commission* 32: 445–447.

Brouwer, A., Morse, D.C., Lans, M.C., Schuur, A.G., Murk, A.J., Klasson-Wehler, E., Bergman, A. and Visser, T.J. 1998. Interactions of persistent environmental organohalogens with the thyroid hormone system: mechanisms and possible consequences for animal and human health. *Toxicology and Industrial Health* 14: 59–84.

Brouwer, A., Reijnders, P.J.H. and Koeman, J.H. 1989. Polychlorinated biphenyl (PCB)-contaminated fish induces vitamin A and thyroid hormone deficiency in the common seal (*Phoca vitulina*). *Aquatic Toxicology* 15: 99–106.

Buck, J.D., Shepard, L.L., Bubucis, P.M., Spotte, S., McClave, K. and Cook, R.A. 1989. Microbiological characteristics of white whale (*Delphinapterus leucas*) from capture through extended captivity. *Canadian Journal of Fisheries and Aquatic Sciences* 46: 1914–1921.

Buntjer, J.B., Hoff, I.A. and Lenstra, J.A. 1997. Artiodactyl interspersed DNA repeats in cetacean genomes. *Journal of Molecular Evolution* 45: 66–69.

Burns, J.J. and Seaman, G.A. 1986. Investigations of belukha whales in coastal waters of Western and Northern Alaska. II. Biology and Ecology. Contract NA 81 RAC 00049.

Bye, T., Romundstad, P.R., Ronneberg, A. and Hilt, B. 1998. Health survey of former workers in a Norwegian coke plant – part 2 – cancer incidence and cause specific mortality. *Occupational and Environmental Medicine* 55: 622–626.

Campo, M.S., O'Neil, B.W., Barron, R.J. and Jarrett, W.F. 1994. Experimental reproduction of the papilloma–carcinoma complex of the alimentary canal in cattle. *Carcinogenesis* 15: 1597–1601.

Capen, C.C. 1990. Tumors of the endocrine glands. In: Moulton, J.E. (ed.). *Tumors in domestic animals*, pp. 553–639. University of California Press, Berkeley.

Capen, C.C. 1992. Pathophysiology of chemical injury of the thyroid gland. *Toxicology Letters* 64–65: Spec-8.

Cartee, R.E., Tarpley, R., Mahoney, K., Ridgway, S. H. and Johnson, P.L. 1995. A case of cystic adrenal disease in a common dolphin (*Delphinus delphis*). *Journal of Zoo and Wildlife Medicine* 26: 293–297.

Chen, Y.S., Liu, Y.C., Yen, M.Y., Wang, J.H., Wann, S.R. and Cheng, D.L. 1997. Skin and soft-tissue manifestations of *Shewanella putrefaciens* infection. *Clinical Infectious Diseases* 25: 225–229.

Chow, W.H., Linet, M.S., McLaughlin, J.K., Hsing, A.W., Cochien, H.T. and Blot, W.J. 1993. Risk factors for small intestine cancer. *Cancer Causes and Control* 4: 163–169.

Coles, B.M., Stroud, R.K. and Sheggeby, S. 1978. Isolation of *Edwardsiella tarda* from three Oregon sea mammals. *Journal of Wildlife Diseases* 14: 339–341.

Collins, W.T., Capen, C.C., Kasza, L., Carter, C. and Dailey, R.E. 1977. Effects of polychlorinated biphenyls (PCB) on the thyroid gland of rats. *American Journal of Pathology* 89: 119–136.

Cook, F.R. and Muir, D. 1984. The Committee on the Status of Endangered Wildlife in Canada (COSEWIC): history and progress. *Canadian Field Naturalist* 98: 63–70.

Cordes, D.O. and Shortridge, E.H. 1971. Neoplasms of sheep: a survey of 256 cases recorded at Ruakura Animal Health Laboratory. *New Zealand Veterinary Journal* 19: 55–64.

Cossa, D., Picard-Bérubé, M. and Gouygou, J.-P. 1983. Polynuclear aromatic hydrocarbons in mussels from the estuary and northwestern gulf of St. Lawrence, Canada. *Bulletin of Environmental Contamination and Toxicology* 31: 41–47.

Cotran, R.S., Kumar, V., Robbins, S.L. and Schoen, F.J. 1994. Neoplasia. In: Cotran, R.S., Robbins, S.L. and Kumar, V. (eds). *Pathologic basis of disease*, pp. 241–304. W.B. Saunders, Philadelphia.

Cowan, D. 1966. Pathology of the pilot whale (*Globicephala melaena*). *Archives of Pathology* 82: 178–189.

Cowan, D. 1994. Involution and cystic transformation of the thymus in the bottlenose dolphin, *Tursiops truncatus. Veterinary Pathology* 31: 648–653.

Cowan, D., Walker, W.A. and Brownell, J. 1986. Pathology of small cetaceans stranded along southern California beaches. In: Bryden, M.M. and Harrison, R. (eds). *Research on Dolphins*, pp. 323–323. Oxford University Press, London.

Culp, S.J., Gaylor, D.W., Sheldon, W.G., Goldstein, L.S. and Beland, F.A. 1998. A comparison of the tumors induced by coal tar and benzo[a]pyrene in a 2-year bioassay. *Carcinogenesis* 19: 117–124.

Dalcourt, M.F., Béland, P., Pelletier, E. and Vigneault, Y. 1992. Caractérisation des communautés benthiques et étude des contaminants dans des aires fréquentées par le béluga du Saint-Laurent. *Rapport technique canadien des sciences halieutiques et aquatiques*. 1845. Fisheries and Oceans Canada.

Dees, C., Askari, M., Foster, J.S., Ahamed, S. and Wimalasena, J. 1997. DDT mimicks estradiol stimulation of breast cancer cells to enter the cell cycle. *Molecular Carcinogenesis* 18: 114.

De Guise, S., Lagacé, A. and Béland, P. 1994a. Gastric papillomas in eight St. Lawrence beluga whales (*Delphinapterus leucas*). *Journal of Veterinary Diagnostic Investigation* 6: 385–388.

De Guise, S., Lagacé, A. and Béland, P. 1994b. Tumors in St. Lawrence beluga whales (*Delphinapterus leucas*). *Veterinary Pathology* 31: 444–449.

De Guise, S., Lagacé, A. and Béland, P. 1994c. True hermaphroditism in a St. Lawrence beluga whale (*Delphinapterus leucas*). *Journal of Wildlife Diseases* 30(2): 287–290.

De Guise, S., Lagacé, A., Béland, P., Girard, C. and Higgins, R. 1995. Non-neoplastic lesions in beluga whales (*Delphinapterus leucas*) and other marine mammals from the St Lawrence Estuary. *Journal of Comparative Pathology* 112: 257–271.

De Guise, S., Martineau, D., Béland, P. and Fournier, M. 1998. Effects of *in vitro* exposure of beluga whale leukocytes to selected organochlorines. *Journal of Toxicology and Environmental Health* 55: 479–493.

Dorn, C.R. and Priester, W.A. 1987. Epidemiology. In: Theilen, G.H. and Madewell, B.R. (eds). *Veterinary cancer medicine*, pp. 27–52. Lea & Febiger, Philadelphia.

Dunn, J.L. 1990. Bacterial and mycotic diseases of cetaceans and pinnipeds. In: Dierauf, L.A. (ed.). *Handbook of marine mammal medicine: health, diseases and rehabilitation*, pp. 73–87. CRC Press, Boca Raton.

Durham, S.K. and Brouwer, A. 1990. 3,4,3',4'-tetrachlorobiphenyl distribution and induced effects in the rat adrenal gland. Localization in the zona fasciculata. *Laboratory Investigation* 62: 232–239.

Fearon, E.R. 1997. Human cancer syndromes: clues to the origin and nature of cancer. *Science* 278: 1043–1050.

Ferguson, P.L. and Chandler, G.T. 1998. A laboratory and field comparison of sediment polycyclic aromatic hydrocarbon bioaccumulation by the cosmopolitan estuarine polychaete *Streblospio benedicti* (Webster). *Marine Environmental Research* 45: 387–401.

Fouchecourt, M.O., Berny, P. and Riviere, J.L. 1998. Bioavailability of PCBs to male laboratory rats maintained on litters of contaminated soils: PCB burden and induction of alkoxyresorufin O-dealkylase activities in liver and lung. *Archives of Environmental Contamination and Toxicology* 35: 680–687.

Fowler, M.E. 1987. Zoo animals and wildlife. In: Theilen, G.H. and Madewell, B.R. (eds). *Veterinary cancer medicine*, pp. 649–662. Lea & Febiger, Philadelphia.

Gaudette, L.A., Altmayer, C.A., Wysocki, M. and Gao, R.-N. 1998. Cancer incidence and mortality across Canada. Statistics Canada. Catalogue 82–00310.

Georgsson, G. and Vigfusson, H. 1973. Carcinoma of the small intestine of sheep in Iceland. A pathological and epizootiological study. *Acta Veterinaria Scandinavica* 14: 392–409.

Geraci, J.R. 1989. Clinical Investigation of the 1987–88 mass mortality of bottlenose dolphins along the US central and South Atlantic coast. Final Report to National Marine Fisheries Service and US Navy Office of Naval Research and Marine Mammal Commission, Washington DC.

Geraci, J.R. and Gertsmann, K.E. 1966. Relationship of dietary histamine to gastric ulcers in the dolphin. *Journal of the American Veterinary Medical Association* 149: 884–890.

Geraci, J.R. and St. Aubin, D.J. 1979. *Biology of Marine Mammals. Insights through Strandings. Prepared for: Marine Mammal Commission*, MMC-77/13.

Geraci, J.R., Palmer, N.C. and St. Aubin, D.J. 1987. Tumors in cetaceans: analysis and new findings. *Canadian Journal of Fisheries and Aquatic Sciences* 44: 1289–1300.

Gillete, D.M., Acland, H.M. and Klein, L. 1978. Ductular mammary carcinoma in a lioness. *Journal of the American Veterinary Medical Association* 173: 1099–1102.

Girard, C., Lagacé, A., Higgins, R. and Béland, P. 1991. Adenocarcinoma of the salivary gland in a beluga whale (*Delphinapterus leucas*). *Journal of Veterinary Diagnostic Investigation* 3: 264–265.

Gulland, F.M., Trupkiewicz, J.G., Spraker, T.R. and Lowenstine, L.J. 1996. Metastatic carcinoma of probable transitional cell origin in 66 free-living California sea lions (*Zalophus californianus*), 1979 to 1994. *Journal of Wildlife Diseases* 32: 250–258.

Güttes, S., Failing, K., Neumann, K., Kleinstein, J., Georgii, S. and Brunn, H. 1998. Chlororganic pesticides and polychlorinated biphenyls in breast tissue of woman with benign and malignant breast disease. *Archives of Environmental Contamination and Toxicology* 35: 140–147.

Hansen, L.G., Wilson, D.W. and Byerly, C.S. 1976. Effects on growing swine and sheep of two polychlorinated biphenyls. *American Journal of Veterinary Research* 37: 1021–1024.

Haraguchi, K., Athanasiadou, M., Bergman, A., Hovander, L. and Jensen, S. 1992. PCB and methylsulfones in selected groups of seals from Swedish waters. *AMBIO* 21: 546–549.

Harrison, R.J. 1969. Endocrine organs: hypophysis, thyroid, and adrenal. In: Andersen, H.T. (ed.). *The biology of marine mammals*, pp. 349–390. Academic Press, New York.

Harshbarger, J.C. 1997. Invertebrate and cold-blooded vertebrate oncology. In: Rossi, L., Richardson, R. and Harshbarger, J.C. (eds). *Spontaneous animal tumors: a survey*, pp. 41–53. Press Point, Milano, Italy.

Hart, M.M., Reagan, R.L. and Adamson, R.H. 1973. The effect of isomers of DDD on the ACTH-induced steroid output, histology and ultrastructure of the dog adrenal cortex. *Toxicology and Applied Pharmacology* 24: 101–113.

Hashiguchi, I., Akamine, A., Miyatake, S., Anan, H., Maeda, K., Aono, M., Fukuyama, H. and Okumura, H. 1991. [Immunohistological and histopathological study of the effect of PCB on the periodontal tissue]. *Fukuoka Igaku Zasshi* 82: 256–261.

Hashiguchi, I., Akamine, A., Toriya, Y., Maeda, K., Aono, M., Fukuyama, H. and Okumura, H. 1993. [Immunohistochemical study on the distribution of CGRP-containing nerve fibers in the PCB poisoned rats with experimental periodontitis]. *Fukuoka Igaku Zasshi* 84: 232–235.

Hontela, A., Rasmussen, J.B., Audet, C. and Chevalier, G. 1992. Impaired cortisol stress response in fish from environments polluted by PAHs, PCBs, and mercury. *Archives of Environmental Contamination and Toxicology* 22: 278–283.

Howard, E.B. 1983. Miscellaneous diseases. In: Howard, E.B. (ed.). *Pathobiology of Marine Mammal Diseases*, pp. 163–233. CRC Press, Boca Raton.

Howard, E.B., Britt, J.O. Jr, Matsumoto, G.K., Itahara, R. and Nagano, C.N. 1983a. Bacterial diseases. In: Howard, E.B. (ed.). *Pathobiology of marine mammal diseases*, pp. 68–118. CRC Press, Boca Raton.

Howard, E.B., Britt, J.O. Jr and Simpson, J.G. 1983b. Neoplasms in marine mammals. In: Howard, E.B. (ed.). *Pathobiology of marine mammal diseases*, pp. 95–162. CRC Press, Boca Raton, FL.

Inskeep, W., Gardiner, C.H., Harris, R.K., Dubey, J.P. and Goldston, R.T. 1990. Toxoplasmosis in Atlantic bottle-nosed dolphins (*Tursiops truncatus*). *Journal of Wildlife Diseases* 26: 377–382.

Jarrett, W.H.F., McNeil, P.E., Grimshaw, W.T.R., Selman, I.E. and McIntyre, W.I.M. 1978. High incidence area of cattle cancer with a possible interaction between an environmental carcinogen and a papillomavirus. *Nature* 274: 215–217.

Jaruratanasirikul, S. and Kalnauwakul, S. 1991. *Edwardsiella tarda*: a causative agent in human infections. *Southeast Asian Journal of Tropical Medicine and Public Health* 22: 30–34.

Johansson, M., Larsson, C., Bergman, A. and Lund, B. O. 1998. Structure–activity relationship for inhibition of CYP11B1-dependent glucocorticoid synthesis in Y1 cells by aryl methyl sulfones. *Pharmacology and Toxicology* 83: 225–230.

Jönsson, C.-J., Lund, B.-O. and Brandt, I. 1993. Adrenocorticolytic DDT-metabolites: studies in mink, *Mustela vison* and otter, *Lutra lutra*. *Ecotoxicology* 2: 41–53.

Jönsson, C-J., Rodriguez-Martinez, H. and Brandt, I. 1995. Transplacental toxicity of 3-methylsulphonyl-DDE in the developing adrenal cortex in mice. *Reproductive Toxicology* 9: 257–264.

Kaminsky, L.S. and Fasco, M.J. 1991. Small intestinal cytochromes P450. *Critical Reviews in Toxicology* 21: 407–422.

Kavlock, R.J. 1999. Overview of endocrine disruptor research activity in the United States. *Chemosphere* 39: 1227–1236.

Kingsley, M.C.S. and Hammill, M.O. 1991. Photographic census surveys of the St. Lawrence population. 1988 and 1990. *Rapport technique canadien des sciences halieutiques et aquatiques* 1776: 1–19.

Kirkwood, J.K., Bennett, P.M., Jepson, P.D., Kuiken, T., Simpson, V.R. and Baker, J.R. 1997. Entanglement in fishing gear and other causes of death in cetaceans stranded on the coasts of England and Wales. *Veterinary Record* 141: 94–98.

Kuehl, D.W. and Haebler, R. 1995. Organochlorine, organobromine, metal, and selenium residues in bottlenose dolphins (*Tursiops truncatus*) collected during an unusual mortality event in the Gulf of Mexico, 1990. *Archives of Environmental Contamination and Toxicology* 28: 494–499.

Kuehl, D.W., Haebler, R. and Potter, C. 1991. Chemical residues in dolphins from the US Atlantic coast including Atlantic bottlenose obtained during the 1987–88 mass mortality. *Chemosphere* 22: 1071–1084.

Kuiken, T., Höfle, U., Benett, P.M., Allchin, C.R., Kirkwood, J.K., Baker, J.R., Appleby, E.C., Lockyer, C.H., Walton, M.J. and Sheldrick, M.C. 1993. Adrenocortical hyperplasia, disease and chlorinated hydrocarbons in the harbour porpoise (*Phocoena phocoena*). *Marine Pollution Bulletin* 26: 440–446.

Lahvis, G.P., Wells, R.S., Kuehl, D.W., Stewart, J.L., Rhinehart, H.L. and Via, C.S. 1995. Decreased lymphocyte responses in free-ranging bottlenose dolphins (*Tursiops truncatus*) are associated with increased concentrations of PCBs and DDT in peripheral blood. *Environmental Health Perspective* 103: 67–72.

Lair, S. 1998. Epidemiology and pathology of neoplasia in the captive population of the black-footed ferret (*Mustela nigripes*). D.V.Sc., Department of Pathobiology, University of Guelph. Guelph, Ontario, Canada.

Lair, S., Béland, P., De Guise, S. and Martineau, D. 1997. Adrenal hyperplastic and degenerative changes in beluga whales. *Journal of Wildlife Diseases* 33: 430–437.

Lair, S., De Guise, S. and Martineau, D. 1998. Uterine adenocarcinoma with abdominal carcinomatosis in a beluga whale. *Journal of Wildlife Diseases* 34: 373–376.

Landy, R.B. 1980. A review of neoplasia in marine mammals (Pinnepedia and Cetacea). In: Montali, R.J. and Migaki, G. (eds). *The comparative pathology of zoo animals*, pp. 579–591. Smithsonian Institution Press, Washington.

Leatherland, J.F. 1998. Changes in thyroid hormone economy following consumption of environmentally contaminated Great Lakes fish. *Toxicology and Industrial Health* 14: 41–57.

Lesage, V. and Kingsley, M.C.S. 1998. Updated status of the St. Lawrence River population of the beluga, *Delphinapterus leucas*. *Canadian Field-Naturalist* 112: 98–113.

Letcher, R.J., Norstrom, R.J., Muir, D.C.G., Sandau, K., Koczanski, R., Michaud, R., De Guise, S. and Béland, P. 2000a. Methylsulfone PCB and DDE metabolites in beluga whale (*Delphinapterus leucas*) from the St. Lawrence River estuary and western Hudson Bay. *Environmental Toxicology and Chemistry* 19(5): 1378–1388.

Letcher, R.J., Klasson-Wehler, E. and Bergman, A. 2000b. Methyl sulfone and hydroxylated metabolites of polychlorinated biphenyls. In: Paasivirta, J. (ed.). *The handbook of environmental chemistry. New types of persistent halogenated compounds*, pp. 317–359. Springer-Verlag, Berlin.

Lipscomb, T.P. 1998. Metastatic carcinoma of California Sea Lions: Evidence of genital origin and association with a gamma-herpesvirus infection. Abstract. American College of Veterinary Pathologists, 49th Annual meeting, 1998. *Veterinary Pathology* 35: 421.

Lipscomb, T.P., Scott, D.P., Garber, R.L., Krafft, A.E., Tsai, M.M., Lichy, J.H., Taubenberger, J.K., Schulman, F.Y. and Gulland, F.M. 2000. Common metastatic carcinoma of California sea lions (*Zalophus californianus*): evidence of genital origin and association with novel gammaherpesvirus. *Veterinary Pathology* 37: 609–617.

MacVean, D.W., Monlux, A.W., Anderson, P.S. Jr, Silberg, S.L. and Roszel, J.F. 1978. Frequency of canine and feline tumors in a defined population. *Veterinary Pathology* 15: 700–715.

Martel, L.M., Gagnon, J., Massé, R., Leclerc, A. and Tremblay, L. 1986. Polycyclic aromatic hydrocarbons in sediments from the Saguenay fjord, Canada. *Bulletin of Environmental Contamination and Toxicology* 37: 133–140.

Martineau, D., Lagacé, A., Massé, R., Morin, M. and Béland, P. 1985. Transitional cell carcinoma of the urinary bladder in a beluga whale (*Delphinapterus leucas*). *Canadian Veterinary Journal* 26: 297–302.

Martineau, D., Béland, P., Desjardins, C. and Lagacé, A. 1987. Levels of organochlorine chemicals in tissues of beluga whales (*Delphinapterus leucas*) from the St. Lawrence Estuary, Quebec, Canada. *Archives of Environmental Contamination and Toxicology* 16: 137–147.

Martineau, D., Lagacé, A., Béland, P., Higgins, R., Armstrong, D. and Shugart, L. 1988. Pathology of stranded beluga whales (*Delphinapterus leucas*) from the St. Lawrence Estuary, Quebec. *Journal of Comparative Pathology* 98: 287–311.

Martineau, D., De Guise, S., Fournier, M., Shugart, L., Girard, C., Lagace, A. and Beland, P. 1994. Pathology and toxicology of beluga whales from the St. Lawrence Estuary, Quebec, Canada. Past, present and future. *Science of the Total Environment* 154: 201–215.

Martineau, D., Lair, S., De Guise, S. and Béland, P. 1995. Intestinal adenocarcinomas in two beluga whales (*Delphinapterus leucas*) from the estuary of the St. Lawrence River. *Canadian Veterinary Journal* 36: 563–565.

Martineau, D., Lair, S., De Guise, S., Lipscomb, T.P. and Béland, P. 1999. Cancer in beluga whales from the St. Lawrence Estuary, Quebec, Canada: A potential biomarker of environmental contamination. *Journal of Cetacean Research Management* Special Issue 1: 249–265.

McCullough, B., Krakowka, S. and Koestner, A. 1974. Experimental canine distemper virus-induced lymphoid depletion. *American Journal of Pathology* 74: 155–170.

McDermott, C. and Mylotte, J.M. 1984. *Morganella morganii*: epidemiology of bacteremic disease. *Infection Control* 5: 131–137.

McKenzie, C., Rogan, E., Reid, R.J. and Wells, D.E. 1997. Concentrations and pattern of organic contaminants in Atlantic white-sided dolphins (*Lagenorhynchus acutus*) from Irish and Scottish coastal waters. *Environmental Pollution* 98: 15–27.

Meador, J.P., Ernest, D., Hohn, A.A., Tilbury, K., Gorzelany, J., Worthy, G. and Stein, J.E. 1999. Comparison of elements in bottlenose dolphins stranded on the beaches of Texas and Florida in the Gulf of Mexico over a one-year period. *Archives of Environmental Contamination and Toxicology* 36: 87–98.

Michaud, R. 1993. Distribution estivale du beluga du Saint-Laurent; synthèse 1986 à 1992. Rapport technique canadien des sciences halieutiques et aquatiques 1906.

Michaud, R., Vézina, A., Rondeau, N. and Vigneault, Y. 1990. Distribution annuelle et caractérisation préliminaire des habitats du béluga. Rapport technique canadien des sciences halieutiques et aquatiques 1757.

Mikaelian, I., de Lafontaine, Y., Ménard, C., Harshbarger, J. and Martineau, D. 1998. Tumors and contaminants in Lake Whitefish (*Coregonus clupeaformis*) from the St. Lawrence River, Quebec, Canada. SETAC. *25th Annual Aquatic Toxicology Workshop, Quebec, Canada*.

Mikaelian, I., Labelle, P. and Martineau, D. 1999a. Metastatic mammary adenocarcinomas in two beluga whales (*Delphinapterus leucas*) from the St. Lawrence Estuary, Quebec, Canada. *Veterinary Record* 145: 738–739.

Mikaelian, I., Tremblay, M.P., Montpetit, C., Tessaro, S.V., Cho, H.J., House, C., Measures, L. and Martineau, D. 1999b. Seroprevalence of selected viral infections in a population of beluga whales (*Delphinapterus leucas*) in Canada. *Veterinary Record* 144: 50–51.

Mikaelian, I., Boisclair, J., Dubey, J.P., Kennedy, S. and Martineau, D. 2000a. Toxoplasmosis in beluga whales (*Delphinapterus leucas*) from the St. Lawrence estuary: Two case reports and a serological survey. *Journal of Comparative Pathology* 122: 73–76.

Mikaelian, I., Labelle, P., Dore, M. and Martineau, D. 2000b. Fibroleiomyomas of the tubular genitalia in female beluga whales. *Journal of Veterinary Diagnostic Investigation* 12: 371–374.

Mikaelian, I., de Lafontaine, Y., Harshbarger, J.C., Lee, L.L.J. and Martineau, D. 2002. Health of lake whitefish (*Coregonus clupeaformis*) with elevated tissue levels of environmental contaminants. *Environmental Toxicology and Chemistry* 21(3): 532–541.

Moulton, J.E. 1990. Tumors of the mammary gland. In: Moulton, J.E. (ed.). *Tumors in domestic animals*, pp. 518–552. University of California Press, Berkeley.

Muir, D.C.G., Wagemann, R., Grift, N.P., Norstrom, R.J., Simon, M. and Lien, J. 1988. Organochlorine and heavy metal contaminants in whitebeaked dolphin (*Lagenrhynchus albirostris*) and Pilot Whale (*Globicephala melaena*) from the coast of Newfoundland, Canada. *Archives of Environmental Contamination and Toxicology* 17: 613–629.

Muir, D.C.G., Ford, C.A., Stewart, R.E.A., Smith, T.G., Addison, R.F., Zinck, M.E. and Béland, P. 1990. Organochlorine contaminants in belugas, *Delphinapterus leucas*, from Canadian waters. *Canadian Bulletin of Fisheries and Aquatic Sciences* 224: 165–190.

Muir, D.C.G., Ford, C.A., Rosenberg, B., Norstrom, R.J., Simon, M. and Béland, P. 1996. Persistent organochlorines in beluga whales (*Delphinapterus leucas*) from the St. Lawrence River Estuary. 1. Concentrations and patterns of specific PCBs, chlorinated pesticides and polychlorinated dibenzo-p-dioxins and dibenzofurans. *Environmental Pollution* 93: 219–234.

Munson, L., Nesbit, J.W., Meltzer, D.G., Colly, L.P., Bolton, L. and Kriek, N.P. 1999. Diseases of captive cheetahs (*Acinonyx jubatus jubatus*) in South Africa: a 20-year retrospective survey. *Journal of Zoo and Wildlife Medicine* 30: 342–347.

Ness, D.K., Schantz, S.L., Moshtaghian, J. and Hansen, L.G. 1993. Effects of perinatal exposure to specific PCB congeners on thyroid hormone concentrations and thyroid histology in the rat. *Toxicology Letters* 68: 311–323.

Newell, K.W., Ross, A.D. and Renner, R.M. 1984. Phenoxy and picolinic acid herbicides and small-intestinal adenocarcinoma in sheep. *Lancet* 2: 1301–1305.

Noga, E.J. 1996. In: Duncan, L.L. (ed.). *Fish disease. Diagnosis and treatment.* Mosby, St. Louis, Missouri.

Norstrom, R.J., Simon, M. and Muir, D.C.G. 1990. Chlorinated dioxins and furans in marine mammals from the Canadian Arctic. *Environmental Pollution* 66: 1–20.

Norstrom, R.J., Muir, D.C.G., Ford, C.A., Simon, M., MacDonald, C.R. and Béland, P. 1992. Indications of P450 monooxygenase activities in beluga (*Delphinapterus leucas*) and narwhal (*Monodon monoceros*) from patterns of PCB, PCDD and PCDF accumulation. *Marine Environmental Research* 34: 267–272.

O'Brien, S.J. 1994. A role for molecular genetics in biological conservation. *Proceedings of the National Academy of Sciences of the United States of America* 191: 5748–5755.

Oldstone, M.B., Lewicki, H., Thomas, D., Tishon, A., Dales, S., Patterson, J., Manchester, M., Homann, D., Naniche, D. and Holz, A. 1999. Measles virus infection in a transgenic model: virus-induced immunosuppression and central nervous system disease. *Cell* 98: 629–640.

Olsson, M. 1994. Effects of persistent organic pollutants on biota in the Baltic Sea. *Archives of Toxicology, Supplement* 16: 43–52.

Olsson, M., Karlsson, B. and Ahnland, E. 1994. Diseases and environmental contaminants in seals from the Baltic and the Swedish west coast. *Science of the Total Environment* 154: 217–227.

Parsons, E.C. 1998. Trace metal pollution in Hong Kong: implications for the health of Hong Kong's Indo-Pacific hump-backed dolphins (*Sousa chinensis*). *Science of the Total Environment* 214: 175–184.

Patenaude, N.J., Quinn, J.S., Béland, P., Kingsley, M. and White, B.N. 1994. Genetic variation of the St. Lawrence beluga whale population assessed by DNA fingerprinting. *Molecular Ecology* 3: 375–381.

Perera, F.P. 1997. Environment and cancer: who are susceptible? *Science* 278: 1068–1073.

Perera, F.P. 1998. Molecular epidemiology of environmental carcinogenesis. *Recent Results in Cancer Research* 154: 39–46.

Perera, F.P. and Dickey, C. 1997. Molecular epidemiology and occupational health. *Annals of the New York Academy of Sciences* 837: 353–359.

Perera, F.P., Estabrook, A., Hewer, A., Channing, K., Rundle, A., Mooney, L.A., Whyatt, R. and Phillips, G.N. 1995. Carcinogen-DNA adducts in human breast tissue. *Cancer epidemiology, biomarkers and prevention* 4: 233–238.

Picard-Bérubé, M. and Cossa, D.P. 1983. Teneurs en benzo 3,4 pyrène chez *Mytilus edulis* L. de l'estuaire et du Golfe du Saint-Laurent. *Marine Environment Research* 10: 63–71.

Pogo, B.G., Melana, S.M., Holland, J.F., Mandeli, J.F., Pilotti, S., Casalini, P. and Menard, S. 1999. Sequences homologous to the mouse mammary tumor virus *env* gene in human breast carcinoma correlate with overexpression of laminin receptor. *Clinical Cancer Research* 5: 2108–2111.

Polischuk, S.C., Letcher, R.J., Norstrom, R.J. and Ramsay, M.A. 1995. Preliminary results of fasting on the kinetics of organochlorines in polar bears (*Ursus maritimus*). *Science of the Total Environment* 160–161: 465–472.

Priester, W.A. and McKay, F.W. 1980. *The occurrence of tumors in domestic animals.* US Department of Health and Human Services, Public Health Service, National Institutes of Health, National Cancer Institute, Bethesda, MD.

Rawson, A.J., Patton, G.W., Hofmann, S., Pietra, G.G. and Johns, L. 1993. Liver abnormalities associated with chronic mercury accumulation in stranded Atlantic bottlenose dolphins. *Ecotoxicology and Environmental Safety* 25: 41–47.

Ray, S., Dunn, B.P., Payne, J.F., Fancey, L., Helbig, R. and Béland, P. 1991. Aromatic DNA-carcinogen adducts in beluga whales from the Canadian Arctic and Gulf of St. Lawrence. *Marine Pollution Bulletin* 22: 392–396.

Reeves, R.R. 1984. Catch history and initial population of white whales (*Delphinapterus leucas*) in the River and Gulf of the St. Lawrence Eastern Canada. *Canadian Field-Naturalist* 111: 63–121.

Reijnders, P.J.H. and de Ruiter-Dijkman, E.M. 1995. Toxicological and epidemiological significance of pollutants in marine mammals. In: Blix, A.S., Walloe, L. and Ultang, O. (eds). *Whales, seals, fish and man*, pp. 575–587. Elsevier Press, Tromsø.

Rewell, R.E. and Willis, R.A. 1949. Some tumours found in whales. *Journal of Pathology and Bacteriology* 61: 454–456.

Ridgway, S.H. and Harrison, R. (eds) 1989. *Handbook of marine mammals*, pp. 1–101. Academic Press Inc., San Diego.

Rijnberk, A. 1996. Adrenals. In: Rijnberk, A. (ed.). *Clinical endocrinology of dogs and cats*, pp. 61–94. Kluwer Academic Publishers, London.

Robinson, P.T. 1979a. A literature review of dental pathology and aging by dental means in nondomestic animals. Part I. *Journal of Zoo Animal Medicine* 10: 57–65.

Robinson, P.T. 1979b. A literature review of dental pathology and aging by dental means in nondomestic animals. Part II. *Journal of Zoo Animal Medicine* 10: 81–91.

Rosen, I.B., Provias, J.P. and Walfish, P.G. 1986. Pathologic nature of cystic thyroid nodules selected for surgery by needle aspiration biopsy. *Surgery* 100: 606–613.

Russo, I.H. and Russo, J. 1998. Role of hormones in mammary cancer initiation and progression. *Journal of mammary gland biology and neoplasia* 3: 49–61.

Safe, S.H. 1994. Polychlorinated biphenyls (PCBs): environmental impact, biochemical and toxic responses, and implications for risk assessment. *Critical Reviews in Toxicology* 24: 87–149.

Saint-Louis, R., Pelletier, E., Doidge, B., Leclair, D., Mikaelian, I. and Martineau, D. 1999. Hepatic butyltin concentrations in beluga whales (*Delphinapterus leucas*) from the St. Lawrence Estuary and Northern Quebec, Canada. *Applied Organometallic Chemistry* 14: 218–226.

Sergeant, D.E. 1986. Present status of white whales (*Delphinapterus leucas*) in the St. Lawrence Estuary. *Canadian Field-Naturalist* 113: 61–81.

Shugart, L.R., Martineau, D. and Béland, P. 1990. Detection and quantitation of benzo(a)pyrene adducts in brain and liver tissue of beluga whales (*Delphinapterus leucas*). In: Prescott, J. and Gauquelin, M. (eds). *Proceedings, International Forum for the future of the Beluga. Tadoussac, Québec, Canada*, pp. 219–223. Presses de l'Université du Québec.

Smith, B.R. and Brian, W.R. 1991. The role of metabolism in chemical-induced pulmonary toxicity. *Toxicologic Pathology* 19: 470–481.

Snyder, R.L., Tyler, G. and Summers, J. 1982. Chronic hepatitis and hepatocellular carcinoma associated with woodchuck hepatitis virus. *American Journal of Pathology* 107: 422–425.

Stolk, A. 1950. Tumours in whales. *Amsterdam Naturalist* 1: 28–33.

Stripp, B.R., Lund, J., Mango, G.W., Doyen, K.C., Johnston, C., Hultenby, K., Nord, M. and Whitsett, J.A. 1996. Clara cell secretory protein: a determinant of PCB bioaccumulation in mammals. *American Journal of Physiology* 271: 656–664.

Stroud, R.K. and Roffe, T.J. 1979. Causes of death in marine mammals stranded along the Oregon coast. *Journal of Wildlife Diseases* 15: 91–97.

Sweeney, J.C. and Ridgway, S.H. 1975. Common diseases of small cetaceans. *Journal of the American Veterinary Association* 167: 533–540.

Takaoka, M., Teranishi, M., Furukawa, T., Manabe, S. and Goto, N. 1995. Age-related changes in thyroid lesions and function in F344/DuCrj rats. *Experimental Animals* 44: 57–62.

Teich, N. 1984. Taxonomy of retroviruses. In: Weiss, R., Teich, N., Varmus, H. and Coffin, J. (eds). *RNA tumor viruses*, pp. 25–209. Cold Spring Harbor Laboratory, Cold Spring Harbor, NY.

Tilbury, K.L., Stein, J.E., Meador, J.P., Krone, C.A. and Chan, S.L. 1997. Chemical contaminants in harbor porpoise (*Phocoena phocoena*) from the north Atlantic coast: tissue concentrations and intra- and inter-organ distribution. *Chemosphere* 34: 2159–2181.

Tremblay, C., Armstrong, B., Theriault, G. and Brodeur, J. 1995. Estimation of risk of developing bladder cancer among workers exposed to coal tar pitch volatiles in the primary aluminum industry. *American Journal of Industrial Medicine* 27: 335–348.

Troisi, G.M., Haraguchi, K., Simmonds, M.P. and Mason, C.F. 1998. Methyl sulphone metabolites of polychlorinated biphenyls (PCBs) in cetaceans from the Irish and the Aegean Seas. *Archives of Environmental Contamination and Toxicology* 35: 121–128.

Ury, H.K. and Wiggins, A.D. 1985. Another shortcut method for calculating the confidence interval of a poisson variable (or of a standardized mortality ratio). *American Journal of Epidemiology* 122: 197–198.

Uys, C.J. and Best, P.B. 1966. Pathology of lesions observed in whales flensed at Saldanha Bay, South Africa. *Journal of Comparative Pathology* 76: 407–412.

van Birgelen, A.P., Ross, D.G., DeVito, M.J. and Birnbaum, L.S. 1996. Interactive effects between 2,3,7,8-tetrachlorodibenzo-p-dioxin and 2,2',4,4',5,5'-hexachlorobiphenyl in female B6C3F1 mice: tissue distribution and tissue-specific enzyme induction. *Fundamental and Applied Toxicology* 34: 118–131.

van den Berg, K.J., Zurcher, C., Brouwer, A. and van Bekkum, D.W. 1988. Chronic toxicity of 3,4,3',4'-tetrachlorobiphenyl in the marmoset monkey (*Callithrix jacchus*). *Toxicology* 48: 209–224.

380   *D. Martineau* et al.

Veatch, J.K. and Carpenter, J.W. 1993. Metastatic adenocarcinoma of the mammary gland in a Père David's deer. *Journal of Veterinary Diagnostic Investigation* 5: 639–640.

Verstraete, F.J., van Aarde, R.J., Nieuwoudt, B.A., Mauer, E. and Kass, P.H. 1996a. The dental pathology of feral cats on Marion Island, part I: congenital, developmental and traumatic abnormalities. *Journal of Comparative Pathology* 115: 265–282.

Verstraete, F.J., van Aarde, R.J., Nieuwoudt, B.A., Mauer, E. and Kass, P.H. 1996b. The dental pathology of feral cats on Marion Island, part II: periodontitis, external odontoclastic resorption lesions and mandibular thickening. *Journal of Comparative Pathology* 115: 283–297.

Vladykov, V.D. 1946. Études sur les mammifères aquatiques IV. Nourriture du marsouin blanc ou béluga (*Delphinapterus leucas*) du fleuve Saint-Laurent. *Contribution du Département Pêcheries du Québec* 17: 1–155.

Wagemann, R., Stewart, R.E.A., Béland, P. and Desjardins, C. 1990. Heavy metals and selenium in tissues of beluga whales, Delphinapterus leucas, from the Canadian Arctic and the St. Lawrence estuary. In: Smith, T.G., St. Aubin, D.J. and Geraci, J.R. (eds). *Advances in research on the beluga whale*, Delphinapterus leucas, pp. 191–206. Canadian Bulletin of Fisheries and Aquatic Sciences 224, Ottawa.

Walcott, H.E. 1999. An evaluation of the thyroids and body burden of harbour porpoises (*Phocoena phocoena*) exposed to chlorinated biphenyls in British waters; qualitative morphopathology, histomorphometry and bioanalysis. 50th Annual Meeting. The American College of Veterinary Pathologists. Chicago, Ill, Nov. 14–19, 1999. *Veterinary Pathology* 36: 505.

Westgate, A.J. 1995. Concentrations and geographic variation of organochlorine contaminants in blubber of harbour porpoises (*Phocoena phocoena*) from the Canadian western North Atlantic. M.Sc. thesis, University of Guelph, Guelph, Canada.

Wunschmann, A., Siebert, U. and Frese, K. 1999. Thymic cysts in harbor porpoises (*Phocoena phocoena*) from the German North Sea, Baltic Sea, and waters of Greenland. *Veterinary Pathology* 36: 391–396.

Yang, F., Chau, Y.K. and Maguire, R.J. 1998. Occurrence of butyltin compounds in beluga whales (*Delphinapterus leucas*). *Applied Organometallic Chemistry* 12: 651–656.

Yonezawa, M., Nakamine, H., Tanaka, T. and Miyaji, T. 1989. Hodgkin's disease in a killer whale (*Orcinus orca*). *Journal of Comparative Pathology* 100: 203–207.

Zhang, Q.Y., Wikoff, J., Dunbar, D. and Kaminsky, L. 1996. Characterization of rat small intestinal cytochrome P450 composition and inducibility. *Drug metabolism and disposition: the biological fate of chemicals* 24: 322–328.

Zhang, Q.Y., Wikoff, J., Dunbar, D., Fasco, M. and Kaminsky, L. 1997. Regulation of cytochrome P4501A1 expression in rat small intestine. *Drug metabolism and disposition: the biological fate of chemicals* 25: 21–26.

14 Immune status of St. Lawrence estuary beluga whales

*Pauline Brousseau, Sylvain De Guise,
Isabelle Voccia, Sylvia Ruby and Michel Fournier*

Introduction

The St. Lawrence River estuary (SLE), downstream from the Great Lakes, is home to the southernmost population of beluga whales (*Delphinapterus leucas*). Hunted for food and as a pest until the 1950s, the population has dwindled from an estimated 5000 to approximately 500 animals in the 1970s and has been listed as an endangered population (Reeves and Mitchell, 1984; Michaud, 1993). The population has not recovered over the past 30 years and has been under scrutiny for a decade. Post-mortem examination of carcasses retrieved from the shores of the St. Lawrence estuary since 1982 has shown a high prevalence of infectious, degenerative or necrotic lesions, often associated with mildly pathogenic organisms or neoplasms (Martineau *et al.*, 1988; DeGuise *et al.*, 1994, 1995a). The frequency, diversity and severity of lesions described in this population were considerably higher than those found in marine mammals elsewhere. Worldwide, a total of 75 tumors have been reported in cetaceans, of which 28 (37 per cent) have come from the small SLE population of beluga whales. Consequently, a link was suggested between toxic contaminants in the SLE food web and these observations. Because SLE beluga whales are generally restricted to the St. Lawrence estuary, they are exposed to sediments, and prey on invertebrates and fish, contaminated from industrial and agricultural discharges originating from the Great Lakes, the St. Lawrence River and its tributaries. In addition, the relatively long life span of the beluga whale, its diet and the year-long residency in the St. Lawrence, result in a greater accumulation of contaminants compared to other marine mammals such as seals. Higher concentrations of organohalogens, benzo[a]pyrene (B[a]P), polychlorinated biphenyls (PCBs), dichlorophenyl trichloroethane (DDT), mirex, mercury and lead are found in tissues of SLE belugas compared to Arctic belugas (Martineau *et al.*, 1987; Muir *et al.*, 1990; Wagemann *et al.*, 1990). Among these chemicals, many were demonstrated to have adverse effects on different aspects of the normal physiology of laboratory animals and wildlife, including effects on the reproductive system. The immune system also may be a potential target. There is ample evidence that organohalogens (Loose *et al.*, 1977; Thomas and Hinsdill, 1978; Imanishi *et al.*, 1980; Exon *et al.*, 1985; Davis and Safe, 1988; Smialovicz *et al.*, 1989; Tryphonas *et al.*, 1989; Kerkvliet *et al.*, 1990), pesticides (see review by Voccia *et al.*, 1999) and metals (see review by Bernier *et al.* 1995) have detrimental effects on the immune system of humans and animals, decreasing resistance to viral, bacterial and parasitic infections as well as inducing neoplasms. Because little was known about the immunology of marine mammals, research was undertaken to develop assays to evaluate immune functions in belugas, to further evaluate the possible effects of environmental contaminants on their immune systems.

The immune system of beluga whales

Lymphoid organs

The information gathered on the immune system of marine mammals is scarce and comes from isolated cases without long-term follow-up. Romano *et al.* (1992) examined several organs of the immune system, including spleen, thymus, lymph nodes, pharyngeal tonsils and gut-associated lymphoid tissue. They concluded that the morphology of lymphoid organs in beluga whales is quite similar to that of other species of mammals with a few differences, such as the absence of follicles in the white pulp of the spleen. Overall, the organs of the immune system of cetaceans seem to appear and function quite similarly to those of terrestrial mammals.

Lymphoid cells

With the use of murine, bovine and human monoclonal antibodies, morphologically undistinguishable lymphocyte subpopulations were recognized and classified (DeGuise, 1997a) using flow cytometry. A scattergram of the flow cytometric profile of peripheral blood leukocytes from a beluga whale from Churchill, Hudson Bay, shows that the different populations of cells are rather easily discerned on the basis of cell size (forward side scatter, FSC) and complexity (side scatter, SSC) (Figure 14.1). Lymphocytes are rather small cells with low complexity, as are slightly larger monocytes, whereas granulocytes are clearly more complex or granular.

The proportion of B cells evaluated as the percentage of surface IgM-positive cells (3–12 per cent) is rather small in beluga peripheral blood mononuclear cells (PBMC) compared to other species of domestic animals (>15 per cent) (Table 14.1). About 98.8 per cent of beluga were labeled by an anti-bovine CD2-PBMC (Table 14.1), usually a T-cell surface marker (DeGuise *et al.*, 1997a). While an anti-TCR (γδ) labeled 31 per cent of beluga PBMC. This is similar to the proportion found in ruminants (Tizard, 1992), in which this form of TCR (T-cell antigen receptor) predominates (Hein and Mackay, 1991), as opposed to humans and mice, in which TCR (γδ) is

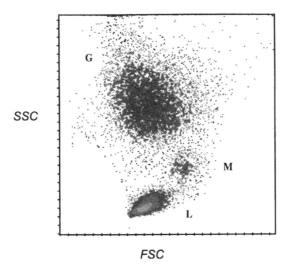

Figure 14.1 Scattergram of the flow cytometric profile of a wild beluga from Churchill, showing peripheral blood leukocytes according to their size (FSC) and complexity (SSC). Lymphocytes (L) are rather small cells with low complexity, as are the slightly larger and more complex monocytes (M), whereas granulocytes (G) are clearly more complex or granular.

Table 14.1 Cross-reactivity of different monoclonal antibodies (MoA) against beluga peripheral blood mononuclear cells (PBMC). Different subclasses of beluga PBMC were evaluated by bovine, human and mouse monoclonal antibodies. Results are expressed as the percentage of PBMC positive for that marker

Marker	*CD2* *CH61A* *(bovine)*	*CD4(Ho)* *SIM-4* *(human)*	*TCR (γδ)* *CACT148A* *(bovine)*	*BS IgM* *(bovine)*
Mean	98.8	29.8	31	6.21
SD	2	6.9	28.8	3.01
N	9	8	10	8

For the belugas tested, nine were from the arctic and three were captive (two from San Diego Zoological Park, San Diego and one from the John G. Shedd Aquarium, Chicago).

present on only 1–3 per cent of peripheral blood lymphocytes (Tizard, 1992). Because antibodies cross-reacting with total T cells and CD8 were not found, the proportion of T cells in beluga peripheral blood could not be compared to that of other species, nor could B : T and CD4 : CD8 ratios (DeGuise *et al.*, 1997a). The immunoprecipitation of surface proteins confirmed the specificity of the antibodies cross-reacting with beluga whale lymphocytes.

The major histocompatibility complex (MHC)

One of the trademarks of the immune system is its capability to differentiate between 'self' and 'non-self'. The term 'non-self' covers anything that is detectably different from an animal's own constituents. Genes of the MHC code for two types of cell-surface molecules found on all nucleated cells (Class I) or some immunological cells only (Class II). The interactions between MHC molecules and immunocompetent cells are vital. In a study conducted by DeGuise *et al.* (1997a), the authors tested five anti-MHC class I and six anti-MHC class II bovine monoclonal antibodies against beluga whale PBMC (Table 14.2). Their results showed that the MHC class I were expressed on virtually 100 per cent of the lymphocytes, as in other species. The MHC class II were expressed on the majority of lymphocytes. This corresponds to similar findings in bottlenose dolphins (*Tursiops truncatus*), swine, horses, cats and dogs, but differs from humans, where it is expressed only on B cells or activated T cells, and from mice in which MHC class II are expressed mainly by B lymphocytes, macrophages, dendritic cells and endothelial cells. In order to confirm the specificity of these bovine mono-clonal antibodies (MoAbs) in beluga, the molecular weight of the surface proteins immunoprecipitated by these antibodies were compared to those of other species. The anti-bovine MHC-I immunoprecipitated a 46 kDa as well as a 14 kDa protein in beluga. The 46 kDa protein corresponds to the molecular weight of the MHC-I α-chain in human (44 kDa) and mouse (47 kDa), whereas the 14 kDa protein corresponds to the molecular weight of β$_2$-microglobulin or the β-chain of MHC class I (12 kDa), which would correspond to the MHC-I α-chain in beluga as in other species (Abbas *et al.*, 1991; DeGuise, 1997a). The anti-bovine MHC class II antibody

Table 14.2 Cross-reactivity of different monoclonal antibodies (MoAb) against molecules of the major histocompatibility complex of class I and II (MHC-I and MHC-II) on beluga whale PBMC

	Identification of antibody	*Origin**	*Cross-reactivity*
MHC class I	H58A	Bovine	+
	PT85A	Bovine	+
	H1A	Bovine	+
	H6A	Bovine	+
	H11A	Bovine	+
MHC class II	H34A	Bovine	+
	H42A	Bovine	+
	TH16B	Bovine	−
	TH21A	Bovine	+
	TH81A5	Bovine	+
	TH14B	Bovine	+

* The bovine monoclonal antibodies were donated by VMRD (Pullman, Washington).

immunoprecipitated a 31 kDa protein in beluga, the molecular weight of which corresponds to that of a protein simultaneously immunoprecipated in bovine (33 kDa) and in humans (32–34 kDa for the α-chain and 29–32 kDa for the β-chain; Abbas *et al.*, 1991; DeGuise, 1997a). Furthermore, Bernier *et al.* (2000) showed that a stimulation of beluga PBMC with the lectin phytohemagglutinin (PHA) increases the intensity of the expression of the molecule at the cell surface, a phenomenon shared with other mammals (Rideout *et al.*, 1992).

Restriction analysis of DNA indicates that SLE beluga are genetically more homogeneous than Canadian Arctic beluga (Patenaude and White, 1994). The loss of genetic diversity at the MHC locus may be a feature of the SLE beluga population because they suffered a drastic reduction in population size over a short period of time. This phenomenon was described in the cheetah (*Acinonyx jubatus*) population which underwent a severe contraction (population bottleneck) leading to genetically identical individuals which manifested a lack of rejection of skin transplants through the loss of genetic diversity at the MHC locus (O'Brien *et al.*, 1983; Yuhki and O'Brien, 1990). Also, probably due to this monoclonality, cheetahs are homogeneously susceptible to some strains of coronaviruses (coronaviruses cause feline infectious peritonitis) not virulent in other felids.

Non-specific immune response

The non-specific immune response is crucial for effective protection of the organism against bacteria or viral challenges and neoplastic cells. Phagocytosis and natural killer (NK) cell activity are the two main cellular mechanisms of the non-specific immune response. Because the level of opportunistic infections and tumors are high in belugas living in polluted waters, the need to understand these mechanisms is obvious.

Phagocytosis

Neutrophils represent the first line of defense of the immune system against invading agents, especially bacteria. Ingestion of foreign material through the process of phagocytosis and its destruction through a series of biochemical events, known as antigen processing and the respiratory burst, are the major functions of neutrophils (Roitt *et al.*, 1989).

Using fluorescent microspheres (diameter 1 μm), phagocytosis was shown, by flow cytometry, to peak after 18 h in neutrophils (Figure 14.2). This period of time is relatively long compared to other mammals where the maximum is reached within approximately 1 h (DeGuise *et al.*, 1995b). Furthermore, average phagocytosis in female belugas is significantly higher than that of male belugas ($P < 0.01$). The capacity of neutrophils to undergo a respiratory burst was measured by the production of H_2O_2 following stimulation with

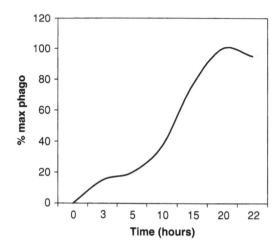

Figure 14.2 Time kinetics for phagocytosis by neutrophils in captive beluga from the John G. Shedd Aquarium, Chicago. Results are expressed as the percentage of maximum phagocytosis reached (% max phago). Phagocytosis increased for up to 18–20 hours, and then decreased after 22 hours.

phorbol myristate acetate (PMA). PMA is an activator of cells that binds directly to protein kinase C, allowing the direct stimulation of the respiratory burst. Using the probe 2′,7′-dichloro-fluorescein diacetate (DCFDA) to measure the production of H_2O_2, the respiratory burst was shown to peak after 60 minutes by flow cytometry (DeGuise *et al.*, 1997b). Contrary to phagocytosis, female belugas have a lower respiratory burst than males, which may be compensated by higher phagocytosis. DeGuise *et al.* (1997b) observed rather large variability in both functions, but variation was smaller when animals were grouped by sex.

Natural killer cell activity

Natural killer (NK) cells are also an important cellular component of the non-specific immune response. They represent the third major class of lymphocytes, but they do not express markers for either T or B lymphocytes. Initially called null cell population, these large granular lymphocytes express other surface markers, such as CD16 and CD56, in humans (O'Shea and Ortaldo, 1992). NK cell activity is directed mainly against tumor and virus-infected cells. NK cell activity is not restricted by the major histocompatibility complex (MHC), nor does it need previous sensitization. DeGuise *et al.* (1995) demonstrated an NK activity in beluga peripheral blood leukocytes with the use of two established NK-sensitive tumor cell lines, the murine

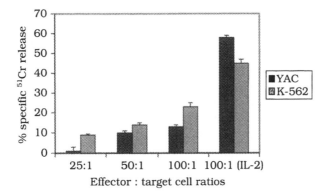

Figure 14.3 Specific ^{51}Cr release in a beluga as an evaluation of NK activity after an 18-hour incubation of PBMC with ^{51}Cr-labeled YAC lymphoma and human erythroleukemia tumor K-562 cell lines at different effector (NK cell) : target cell (tumor cell) ratios, with or without the cytokine IL-2.

YAC lymphoma, and the human K-562 erythroleukemia (Figure 14.3). They found that beluga NK cell activity was enhanced in the presence of recombinant human interleukin-2 (rh-IL-2), showing a cross-reactivity with the human cytokine IL-2 in belugas. The pattern of increased cytotoxicity with increasing effector cell (NK cell) : target cell (tumor cell) ratios and the IL-2 responsiveness of the NK activity suggest that the effector cells, responsible for the NK activity in belugas, have similar properties to those of natural killer cells for other species (Henney *et al.*, 1981; Ross *et al.*, 1995). The authors observed no significant differences in NK activity between males and females or between adult and immature belugas.

Specific immune response

Adaptive immunity is based on the capability of lymphocytes to respond specifically to non-self material, or antigens, leading to the elimination of the antigen and retaining memory of that specific antigen. Memory enables the organism to respond more rapidly and effectively to a second stimulus by the same antigen. The two responses are the humoral response (mediated by antibodies) and the T-cell-mediated immune response with the production of cytolytic T lymphocytes (CTL).

Humoral response

Following a complex series of cell-to-cell interactions (the description of which is beyond the scope of this report), resting B cells are activated to secrete antibodies of different classes. In the primary immune response, the antigen-specific antibody is initially of the IgM class of immunoglobulins, which is then replaced by IgG immunoglobulins. In marine mammals, different isotypes of immunoglobulins have been identified, although the rate at which they appear following antigenic stimulation is currently unknown (Kennedy-Stoskopf, 1990). Indeed, in the northern fur seal (*Callorhinus ursinus*), immunoglobulins of the isotype G, M, and A were identified and purified by gel filtration chromatography and were found to be structurally similar to human immunoglobulins. However, IgE has not been found in any marine mammals so far. However, this may be due to its relatively low level in serum and insensitivity of assays (Britt and Howard, 1983). Although qualitative and quantitative determination of beluga whale immunoglobulins could be performed, the active response is presently impossible to measure in beluga, because of the antigenic challenges that have to be performed.

Cell-mediated response

T lymphocytes are the main class of mononuclear cells implicated in specific cellular immunity. The cells involved in this response are cytotoxic T lymphocytes (CTL). To become efficient effector cells, the pre-cytotoxic T lymphocytes have to be activated following a classical antigenic stimulation, and must differentiate into CTLs. Following maturation, the CTLs develop the capability to kill target cells by the mediation of a lytic mechanism. Assays to measure a complete cell-mediated response, such as the mixed lymphocyte reaction, are difficult to validate in mammals due to the lack of inbred animals. However, the basic capability of CTLs to proliferate upon antigenic stimulation could be studied by the use of a mitogenic assay. This assay could be also extended for B lymphocytes. De Guise *et al.* (1996) validated a quantitative assay to measure lymphoblastic transformation in PBMC, splenocytes and thymocytes from belugas. For each source of leukocytes, the authors tested different mitogens: concanavalin A (Con-A), PHA, lipopolysaccharide (LPS) and pokeweed mitogen (PWM), for which the optimal concentrations were determined (Figures 14.4, 14.5 and 14.6). The results show that Con-A and PHA were the strongest mitogens in PBMC and thymocytes, whereas LPS was the strongest mitogen for splenocytes. The stimulation index for Con-A and PHA was higher for thymocytes than for PBMC, and was the lowest in splenocytes, while LPS was a stronger mitogen for splenocytes than thymocytes. Furthermore, the minimal stimulation by LPS with concomitant strong PHA and Con-A stimulation in the thymus, which contain only T cells, and the low Con-A and PHA but strong LPS stimulation observed in the spleen, which contains a high proportion of B cells, tend to confirm the specificity of Con-A and PHA for beluga T lymphocytes and LPS for B lymphocytes, as in other species.

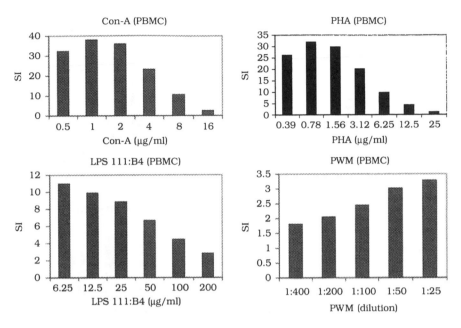

Figure 14.4 Stimulation of captive (John G. Shedd Aquarium, Chicago) beluga PBMC cultured with various doses of mitogens (Con-A, concanavalin A; PHA, phytohemagglutinin; LPS, lipopolysaccharide; PWM, pokeweed mitogen). SI, stimulation index.

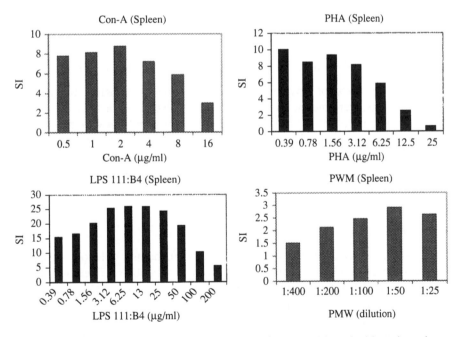

Figure 14.5 Stimulation of wild Arctic beluga splenocytes cultured with various doses of mitogens (Con-A, concanavalin A; PHA, phytohemagglutinin; LPS, lipopolysaccharide; PWM, pokeweed mitogen). SI, stimulation index.

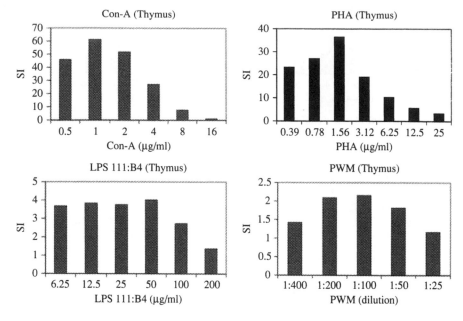

Figure 14.6 Stimulation of wild Arctic beluga thymocytes cultured with various doses of mitogens (Con-A, concanavalin A; PHA, phytohemagglutinin; LPS, lipopolysaccharide; PWM, pokeweed mitogen). SI, stimulation index.

Immunotoxicity of contaminants found in beluga whale blubber

High concentrations of environmental contaminants, most importantly organohalogens, were measured in the tissues of beluga whales (Muir *et al.*, 1990). In mammals, the suppression of immune function following exposure to organochlorines or metals is well documented (Daum *et al.*, 1993; Bernier *et al.*, 1995; Thomas, 1995). In an attempt to understand the possible immunotoxic potential of these chemicals in belugas, several studies were conducted *in vitro* as well as *in vivo*.

Effects of in vitro *exposure to metals*

In a study conducted by DeGuise *et al.* (1995c), beluga splenocytes and thymocytes were collected from two free-ranging belugas killed by Inuit hunters for food consumption in the North West Territories, Canada. Cell suspensions were exposed to 10^{-5}, 10^{-6}, 10^{-7} M of mercuric chloride ($HgCl_2$) or cadmium chloride ($CdCl_2$), as well as to 10^{-5}, 10^{-6}, 10^{-7} M of lead chloride ($PbCl_2$). The lymphocyte proliferation was then evaluated on thymocytes (Figure 14.7) and on splenocytes (Figure 14.8). In order to study the impact of possible cytotoxic effects of metals on the proliferative response of

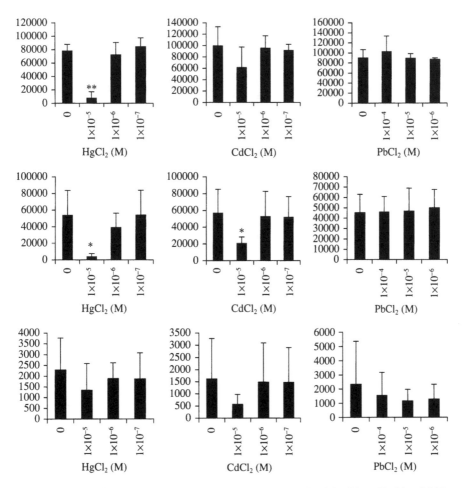

Figure 14.7 Proliferation of beluga thymocytes cultured with either HgCl$_2$, CdCl$_2$ or PbCl$_2$. The cells were either stimulated with Con-A (first row), with PHA (second row) or not stimulated (third row). The results are expressed in dpm (disintegrations per minute). * Significantly different from unstimulated control ($P < 0.05$); ** highly significantly different from unstimulated control ($P < 0.01$).

lymphocytes, cell viability was measured on cells exposed to the same experimental conditions as those for the lymphocyte proliferation assay. For the splenocytes and the thymocytes, none of the concentrations tested significantly affected cell viability. The proliferative response of beluga whale thymocytes was significantly reduced at the highest concentration of HgCl$_2$ when stimulated or not with Con-A and at the highest concentration of HgCl$_2$ and CdCl$_2$ when stimulated with PHA. However, the proliferative responses of splenocytes with Con-A and PHA were significantly reduced at the highest

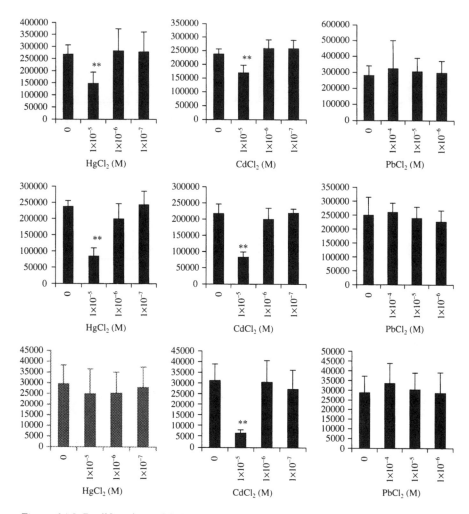

Figure 14.8 Proliferation of beluga splenocytes cultured with either HgCl$_2$, CdCl$_2$ or PbCl$_2$. The cells were either stimulated with Con-A (first row), with PHA (second row) or not stimulated (third row). The results are expressed in dpm (disintegrations per minute). * Significantly different from unstimulated control ($P < 0.05$); ** highly significantly different from unstimulated control ($P < 0.01$).

concentration of HgCl$_2$ and CdCl$_2$ (10^{-5} M). Moreover by doing some experiments with methylmercury, it was found that the latter induced an immunotoxic effect on beluga splenocytes at concentrations 10 times lower than for mercuric chloride (Figure 14.9). These results confirm what has been found in other vertebrates (Nakatsuru *et al.*, 1985; Shenker *et al.*, 1992), and suggest that mercury, as well as cadmium, is immunotoxic on splenocytes and thymocytes of beluga following *in vitro* exposure.

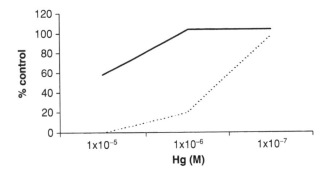

Figure 14.9 Proliferation of Con-A-stimulated beluga splenocytes exposed to either methylmercury (dotted line) or mercuric chloride (solid line), expressed as the percentage of the response of cells cultured without mercury. The proliferation of cells exposed to methylmercury was affected at concentrations tenfold lower than that of cells exposed to mercuric chloride.

Important information was provided by Béland *et al.* (1992) who measured, in the liver of 30 adult SLE belugas (13–31 years old), levels of mercury that were higher than the concentration of 10^{-5} M mercuric chloride (which represents 2.01 ppm of Hg) found to significantly suppress the proliferation of splenocytes and thymocytes *in vitro*. The concentration of 10^{-5} M $CdCl_2$ (which represents 1.11 ppm Cd), found to alter the proliferation of beluga splenocytes, was one order of magnitude lower than those measured in tissues of SLE belugas (Béland *et al.*, 1992), but within the range of those found in Arctic belugas naturally exposed to heavy metals.

Effects of in vitro *exposure to organochlorines*

A study was designed to evaluate the potential immunotoxic effects of organochlorines following *in vitro* exposure. Two parameters of the immune response, namely phagocytosis by peripheral blood neutrophils and the lymphoproliferative response of splenocytes, were evaluated following exposure to selected organochlorines. Contaminants tested were: the three most abundant PCB congeners measured in SLE beluga tissues (138, 153 and 180); the PCB coplanar congener 169, present at low concentrations but well known for its high toxicity; *p,p'*-DDT and its metabolite *p,p'*-DDE; and tetrachlorodibenzo-dioxin (TCDD) (DeGuise *et al.*, 1995). For both parameters, cells were exposed (90 minutes for phagocytosis or 66 hours for proliferation) to 5, 10, 15, 20 or 25 ppm PCB congeners, or to 10, 25, 50, 75 or 100 ppm of *p,p'*-DDT or *p,p'*-DDE, or to 1, 5, 10 or 15 pg/ml of TCDD.

Phagocytosis of beluga leukocytes exposed to various organochlorines was not significantly affected. In contrast, the proliferative response of PHA-stimulated splenocytes was significantly ($P < 0.05$) reduced following an

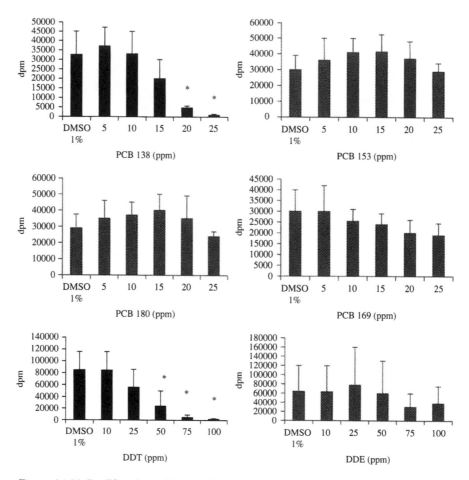

Figure 14.10 Proliferation of PHA-stimulated beluga splenocytes, exposed to one of four PCB congeners or one of two DDT metabolites, cultured in the presence of [³H]thymidine. Results are expressed in disintegrations per minute (dpm). * Significantly different from control ($P < 0.05$).

exposure to 20 and 25 ppm of PCB 138 and to 75 and 100 ppm of *p,p*′-DDT (Figure 14.10). Furthermore, the spontaneous proliferation (without mitogen) of splenocytes was significantly inhibited by PCB 138 at concentrations of 20 and 25 ppm, as well as by PCB 180 at a concentration of 25 ppm. When splenocytes were exposed to a mixture of 5 ppm of PCB congeners 138, 153 and 180, the capacity of splenocytes to proliferate was significantly reduced, whereas no effects were observed in cells exposed to each PCB individually at concentrations lower than 20 ppm (Figure 14.11). However, when PCB 169 was added to the mixture, none of the concentrations of TCDD tested had an effect on the proliferative response of PHA-stimulated splenocytes.

(A)

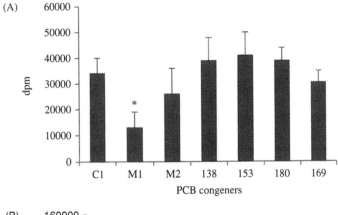

(B)

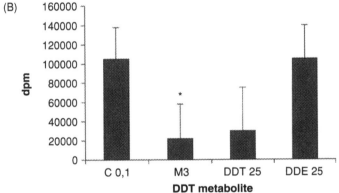

Figure 14.11 Proliferation of PHA-stimulated beluga splenocytes exposed to: (A) 5 ppm of either one or four PCB congeners, or 5 ppm of each of three (138, 153 and 180; M1) or four (138, 153, 180 and 169; M2) of these congeners mixed together (total PCB concentrations of 15 ppm and 20 ppm, respectively); (B) 25 ppm of either *p,p'*-DDE or *p,p'*-DDT, or 25 ppm of each mixed together (M3; total DDT-related compound concentration of 50 ppm); cultured in the presence of [^{3}H]thymidine. Results are expressed in disintegrations per minute (dpm). * Significantly different from control (C 1 or C 0,1) ($P < 0.05$).

Of all the organochlorines tested, PCB congener 138 was the most immunotoxic. This observation is quite significant since this congener is the most abundant one measured in SLE beluga tissues (Muir *et al.*, 1990). Furthermore, immunotoxic levels of individual organochlorine compounds or their mixtures, as reported here, are within the ranges measured in the blubber of SLE belugas. Whether immunocompetent cells are exposed to these concentrations is not clear, but these results support the hypothesis that the high concentrations of organochlorines found in the tissues of beluga whales could significantly impair immune function in these animals.

Effects on the immune response of the Fisher rat fed contaminated blubber

A study was undertaken to assess the immunotoxic potential of the complex mixtures of organohalogens accumulated in SLE belugas upon environmental exposure. From the literature, there is ample evidence that organohalogens have detrimental effects on the immune system of humans and animals (Smialowicz *et al.*, 1989; Tryphonas *et al.*, 1989; Kerkvliet *et al.*, 1990), resulting in a higher sensitivity in experimental animals to a wide array of infectious agents (Loose *et al.*, 1978; Thomas and Hinsdill, 1978; Imanishi *et al.*, 1980).

For this experiment, 60 Fisher rats were assigned to four groups of 15 and fed a diet containing 5 per cent lipids. The control group (series 100) received corn oil. The second group (series 200) received lipids from the blubber of a relatively uncontaminated Arctic Hudson Bay beluga (Table 14.3). The third group (series 300) was fed lipids made of blubber of Arctic beluga and

Table 14.3 Organochlorine contaminants in the blubber of two beluga from the St. Lawrence estuary and from western Hudson Bay, $N = 2$

	St. Lawrence, male	*Hudson Bay, male*
Contaminant		
(µg/g wet weight)		
s-PCB	64.9	3.02
s-DDT	50.0	3.13
s-CHL	10.2	2.33
CHB	25.7	5.10
s-HCH	0.29	0.24
s-CBz	1.03	0.66
PCB homolog (-chloro)		
groups (µg/g wet weight)		
Tri-	0.27	0.02
Tetra-	6.77	0.65
Penta-	12.72	0.95
Hexa-	28.10	0.89
Hepta-	13.80	0.49
Octa-	3.01	0.11
Nona-/deca-	0.26	0.016
Non-ortho PCB congeners		
(ng/g wet weight)		
PCB 37	3.43*	ND
PCB 81	3.5	10.88
PCB 77	192.5	19.58
PCB 126	16.33	52.48
PCB 169	7.45	132.48
PCB 189	79.18	ND

s, Sums of; PCB, polychlorobiphenyls; DDT, six DDT-, DDE- and DDD-related compounds; CHL, chlordane-related compounds; CHB, toxaphene; HCH, hexachlorocyclohexanes; CBz, chlorobenzene.

* Values represent an average of two analyses; ND, below the level of detection.

Table 14.4 Concentrations of major organochlorines contaminants in four diets used in rat feeding study (values in µg/g, wet weight)

Diet group	s-PCB	s-DDT	s-CHL	CHB	s-HCH	s-CBz	Total
100	0	0	0	0	0	0	0
200	0.15	0.16	0.12	0.26	0.01	0.03	0.57
300	1.70	1.33	0.31	0.77	0.01	0.04	4.17
400	3.25	2.50	0.51	1.29	0.01	0.05	7.61

s, Sums of; PCB, polychlorobiphenyls; DDT, six DDT- and DDE-related compounds; CHL, chlordane-related compounds; CHB, toxaphene; HCH, hexachlorocyclohexanes; CBz, chlorobenzene.
Diets: 100 = corn oil; 200 = Arctic beluga fat; 300 = Arctic and St. Lawrence beluga fat; 400 = SLE beluga fat

blubber of SLE beluga, in a 50 : 50 ratio, and the last group (series 400) received lipids from the blubber of a heavily contaminated SLE beluga. The rats were given a 30 g portion per day and the concentrations of s-PCBs and major organohalogen pesticides in the daily diet are given in Table 14.4. Following a 2-month feeding period, immune cells were collected from the blood, the peritoneum and the spleen to assess immune functions: phagocytosis, oxidative burst, lymphoblastic transformation, natural killer cell activity, immunophenotyping and plaque-forming cell assay. For this last assay, satellite series of 40 rats subdivided into four groups, as above, were included because the animals have to be challenge with an antigen (sheep red blood cell, SRBC). Results from this study indicate that the immune response of rats fed a diet whose lipid source (5 per cent) comes from the blubber of a beluga highly contaminated with organohalogen was not different from the control group, except for the humoral response, in which a significant stimulation was observed with rats fed on whale blubbers. Quite interestingly, the amplitude of the stimulation was greater in the group fed with low contaminated Arctic blubber. In a comparable study, Ross *et al.* (1995) fed rats a diet containing 33 per cent Baltic Sea herring (7.135 ppm s-PCBs on average) resulting in a total PCB level in the diet lower than in this study and with no observable impact on the immune response. However, seals fed a similar diet of Baltic Sea herring (*Clupea harengus*) exhibited marked reductions in immune function (De Swart *et al.*, 1996; Ross *et al.*, 1996b) demonstrating that marine mammals can be much more sensitive than rats to toxic effects of organohalogens.

Effects on the immune response and reproduction of C57Bll6 mice fed contaminated blubber

A similar study was undertaken in the mouse model. In this experiment, 105 female mice were assigned to seven groups of 15 specimens each, which were fed a diet containing 5 per cent lipids. In the first and second groups, the control groups, animals were fed on regular chow in which lipids consisted of beef tallow or corn oil, respectively. Groups 3 and 4 were fed a diet in which lipids were replaced by either blubber from a slightly contaminated Arctic beluga or blubber from a heavily contaminated St. Lawrence beluga (the same blubbers as for the rat study). Groups 5–7 were fed on mixtures of Arctic and St. Lawrence beluga blubbers in the respective following ratios: 25–75 per cent; 50–50; 75–25 per cent. The mice were fed for a period of 3 months and then sacrificed to harvest blood, peritoneal macrophages, spleen and gonads. The immunological tests used were phagocytosis, lymphoblastic transformation, immunophenotyping, natural killer cell activity and humoral response. For the reproductive system, ovaries were preserved in Bouin's solution, embedded in wax, sectioned, stained with hematoxylin and eosin, mounted on Permount and observed under a light microscope.

No significant differences in body weight gain or spleen and thymus weight were observed in exposed mice. Out of the six endpoints monitored, humoral response and phagocytosis were significantly suppressed in mice fed on blubber, irrespective of its organohalogen content (Fournier *et al.*, 2000).

The results of the present study and our previous rat study are also consistent with the observation that mice are more susceptible to organohalogen-induced immunotoxicity than rats (Smialowicz *et al.*, 1994).

For the reproductive system, the results show important ovarian and oviduct histological differences between the mice fed with beluga blubber when compared to the control groups. Indeed, ovulation took place in all the groups. However, there was an increase in the life span of the corpora lutea in all five groups fed with beluga blubber, but more importantly in groups 4 to 7 (Ruby *et al.*, 2002). The endocrine disruptor characteristics of some of the xenobiotics present in the blubber might explain this phenomenon. There was also a loss of zona pellucida, a high rate of atrasia and necrosis of follicles in all treated mice.

Effects on reproduction in mink fed contaminated blubber

In a reproductive toxicology study, Murphy *et al.* (1995) tested the effect of toxic chemicals, present in the blubber of SLE beluga, on mink (*Mustela vison*) reproduction. In experimental groups, animals were exposed daily to close to 5 mg of organohalogens, including 2.23 mg of PCBs. Sixty mink were randomly assigned to 6 groups of 10 animals, each fed for 30 days either on a diet where the lipid portion was replaced by beluga fat (experimental) or a diet where the lipid portion was replaced by beef tallow (control). The first two groups were used for early gestation studies (pre-implantation embryos), the second two groups were used for early post-implantation studies, and the third two groups were used for late gestational studies.

Results show that animals from experimental groups all lost weight and three mink (one from the second group and two from the third group) died during the period of treatment. The results are presented in Table 14.5. The

Table 14.5 Embryos and fetuses from mink fed beluga fat for 30 days prior to early gestation (February 21 to March 22), early post-implantation pregnancy (March 13 to April 11), or late gestation (March 22 to April 21)

Treatment Group No.	Number surviving	Number with embryos	Total embryos per mink (SEM)	Live; mean (SEM)	Dead; mean (SEM)
Beluga fat, early gestation	10	9	6.1 (0.9)	6.1 (0.9)	0
Control, early gestation	10	9	6.0 (1.3)	5.9 (1.4)	0.1 (0.01)
Beluga fat, early post-implantation	9	8	6.4 (0.9)*	6.4 (0.9)*	0
Control, early post-implantation	10	9	9.0 (1.2)	8.7 (1.1)	0.2 (0.01)
Beluga fat, late pregnancy	8	7	9.1 (1.5)	4.0 (1.5)*	4.2 (1.2)
Control, late pregnancy	10	8	8.8 (1.3)	8.8 (1.3)	0

* Significantly less than corresponding control at $P < 0.05$. SEM, Standard error of the mean.

absence of difference between numbers of live pre-implantation embryos indicates no apparent effect of the beluga fat on the processes of mating, ovulation or embryonic development to the blastocyst stage. The difference between numbers of early post-implantation embryos suggests that the beluga fat reduced the numbers of embryos that implanted, in the absence of an effect on the viability of embryos that successfully attached and invaded the uterus. The principal lesion was in post-implantation survival, more than half of the embryos being dead in late gestation.

These results are consistent with the view that chemicals present in beluga fat interfered with post-implantation gestation in mink. A previous study by Kihlström *et al.* (1992) reported that mink fed PCB fractions showed a dose-dependent post-implantation mortality and a weight loss in females. Furthermore, a study by Reijnders (1986) demonstated that seals fed polluted fish showed reduced pup production when compared to seals fed much less polluted fish. Although the association between reproductive failure and pollution is circumstantial for St. Lawrence beluga whales, the number of studies putting forward the reproductive toxicity of organohalogens is increasing substantially.

Conclusions

There is an increasing body of evidence showing that exposure to mixtures of persistent toxic substances, mainly through the food web, may result in detrimental effects on the health of several organisms, including marine mammals. Indeed, the resistance of beluga whales might be significantly impaired by persistant contaminants, either directly (target system: immune system) or indirectly (target system: endocrine system), that prevent the immune system from providing adequate defense. In order to tackle this problem efficiently, several assays were developed to study the immune system of beluga whales. The establishment of optimal conditions and their validation required a major commitment from our team. Nevertheless, this essential work gives us the opportunity to look at various aspects of the immune system of belugas and to study the effects of xenobiotics on their immune system in more depth. Indeed, these assays could be used to understand disease processes in cetaceans and potentially improve medical management as well as the health status of these animals.

References

Abbas, A.K., Lichtman, A.H. and Pober, J.S. (eds.) 1991. The major histocompatibility complex. In: *Cellular and Molecular Immunology*. WB Saunders, Montreal, Québec, pp. 98–114.

Béland, P., De Guise, S. and Plante, R. 1992. *Toxicology and pathology of St-Lawrence marine mammals*. Report World Wildlife Fund, Washington, DC.

Bernier, J., Fournier, M., Brousseau, P., Krzystyniak, K. and Tryphonas, H. 1995. Immunotoxicity of heavy metals pertinent to Great Lakes contamination: relevance of published data to risk assessment. *Environ. Health Persp.* 103(supp. 9): 23–34.

Bernier, J., DeGuise, S., Beaudet, M., Martineau, D., Béland, P. and Fournier, M. 2000 Immune functions in beluga whales (*Delphinapterus leucas*): T and B lymphocytes separation, and mechanisms of activation of T cells. *Dev. Comp. Immunol.* 24: 653–662.

Britt, J.O. and Howard, E.B. 1983. The hemapoietic system. In: Howard, E.B. (ed.) *Pathobiology of marine mammal diseases*, Vol. 2. CRC Press, Boca Raton, FL, pp. 65–78.

Daum, J.R., Shepherd, D.M. and Noelle, R.J. 1993. Immunotoxicity of cadmium and mercury on B lymphocytes. I. Effects on lymphocyte function. *Int. J. Immunopharmacol.* 15: 383–394.

Davis, D. and Safe, S. 1988. Immunosuppressive activities of polychlorinated dibenzofuran congeners: quantitative structure–activity relationship and interactive effects. *Toxicol. Appl. Pharmacol.* 94: 141–149.

DeGuise, S. 1995. Evaluation immunotoxicologique des bélugas du Saint-Laurent. Thèse présentée comme exigence partielle du doctorat en Sciences de l'Environnement. Université du Québec à Montréal.

DeGuise, S., Lagacé, A. and Béland, P. 1994. Tumors in St-Lawrence beluga whales (*Delphinapterus leucas*). *Vet. Pathol.* 31: 444–449.

DeGuise, S., Lagacé, A, Béland, P., Girard, C. and Higgins, R. 1995a. Non-neoplastic lesions in beluga whales (*Delphinapterus leucas*) and other marine mammals from the St-Lawrence estuary. *J. Comp. Pathol.* 112: 257–371.

DeGuise, S., Flipo, D., Boehm, J., Martineau, D., Béland, P. and Fournier, M. 1995b. Immune functions in beluga whales (*Delphinapterus leucas*): Evaluation of phagocytosis and respiratory burst with peripheral blood using flow cytometry. *Vet. Immunol. Immunopathol.* 47: 351–362.

DeGuise, S., Bernier, J., Dufresne, M.M., Martineau, D., Béland, P. and Fournier, M. 1996. Immune functions in beluga whales (*Delphinapterus leucas*): evaluation of mitogen-induced blastic transformation of lymphocytes from peripheral blood, spleen and thymus. *Vet. Immunol. Immunopathol,* 50, 117–126.

DeGuise, S., Bernier, J., Martineau, D., Béland, P. and Fournier, M. 1997a. Phenotyping of peripheral blood lymphocytes using monoclonal antibodies. *Dev. Comp. Immunol.* 21: 425–433.

DeGuise, S., Ross, P.S., Osterhaus, A.D.M.E., Martineau, D., Beland, P., Fournier, M. 1997b. Immune function in beluga whales (*Delphinapterus leucas*): Evaluation of natural killer (NK) cell activity. *Vet. Immun. Immunopat.* 58: 345–354.

Exon, J.H., Talcott, P.A. and Koller, L.D. 1985. Effect of lead, polychlorinated biphenyls and cyclophosphamide in rat natural killer cells, interleukin 2, and antibody synthesis. *Fund. Appl. Toxicol.* 5: 158–164.

Fournier, M., Dégas, V., Colborn, T., Omara. F.O., Denizeau, F., Potworowski, E.F. and Brousseau, P. 2000. Immunosuppression in mice fed on diets containing beluga whale blubber from the St. Lawrence Estuary and the Arctic populations. *Tox. Letters.* 112: 311–317.

Hein, W.R. and Mackay, C.R. 1991. Prominence of γδ T cells in the ruminant immune system. *Immunol. Today* 12: 30–34.

Henney, C.S., Kuribayashi, K., Kern, D.E. and Gillis, S. 1981. Interleukin-2 augments natural killer cell activity. *Nature* 291: 335–338.

Imanishi, J., Nomura, H., Matsubara, M., Kita, M., Won, S.-J., Mizutani, T. and Kishida, T. 1980. Effect of polychlorinated biphenyl on viral infections in mice. *Infect. Immun.* 29: 275–277.

Kennedy-Stoskopf, S. 1990. Immunology of marine mammals. In: Dierauf, L.A. (ed.) *Handbook of marine mammal medicine: health, disease, and rehabilitation.* CRC Press, Boca Raton, FL.

Kerkvliet, N.I., Baecher-Steppan, L., Smith, B.B., Youngberg, J.A., Henderson, M.C. and Buhler, D.R. 1990. Role of the Ah locus in suppression of cytotoxic T lymphocyte activity by halogenated aromatic hydrocarbons (PCBs and TCDD): structure–activity relationships and effects in C57B1/6 mice congenic at the Ah locus. *Fund. Appl. Toxicol.* 14: 532–541.

Kihlström, J.E., Olsson, M., Jensen, S., Johansson, A., Ahlbom, J. and Bergman, A. 1992. Effects of PCB and different fractions of PCB on the reproduction of the mink (*Mustela vison*). *Ambio* 21: 563–569.

Loose, L.D., Pittman, K.A., Benitz, K-F. and Silkworth, J.B. 1977. Polychlorinated biphenyl and hexachlorobenzene induced humoral immunosuppression. *J. Reticuloendothel. Soc.* 22: 253–271.

Loose, L.D., Pittman, K.A., Benitz, K-F., Silkworth, J.B., Mueller, W. and Coulston, F. 1978. Environmental chemical-induced immune dysfunction. *Ecotoxicol. Environ. Safety.* 2: 173–198.

Martineau, D., Béland, P., Déjardins, C. and Lagacé, A. 1987. Levels of organochlorine chemicals in tissues of beluga whales (*Delphinapterus leucas*) from the St-Lawrence estuary, Québec, Canada. *Arch. Environ. Contamin. Toxicol.* 16: 137–147.

Martineau, D., Lagacé, A., Béland, P., Higgins, R., Armstrong, D. and Shugart, L.R. 1988. Pathology of stranded beluga whales (*Delphinapterus leucas*) from the St-Lawrence estuary, Québec, Canada. *J. Comp. Pathol.* 98: 287–311.

Michaud, R. 1993. Distribution estivale du béluga du St-Laurent; synthèse de 1986 à 1992. *Can. Tech. Rpt. Fish. Aquat. Sci.* 1906.

Muir, D.C.G., Ford, C.A., Stewart, R.E.A., Smith, T.G., Addison, R.F., Zinck, M.E. and Béland, P. 1990. Organochlorine contaminants in belugas, *Delphinapterus leucas*, from Canadian waters. *Can. Bull. Fish Aquat. Sci.* 224: 165–190.

Murphy, B.D., Muir, D.C.G., Nostrom, R., Martineau, D., Béland, P. and DeGuise, S. 1995. Reproductive toxicity of beluga blubber in the mink model. In: Béland, P., DeGuise, S., Fournier, M. and Martineau, D. (eds) *Markers of organohalogen toxicity in St-Lawrence beluga whales* (Delphinapterus leucas). Final report, year 2. TOXEN, UQAM & St-Lawrence National Institute of Ecotoxicology.

Nakatsuru, S., Oohashi, J., Nozaki, H., Nakada, S. and Imura, N. 1985. Effects of mercurials on lymphocyte functions *in vitro*. *Toxicology* 36: 297–305.

O'Brien, S.J., Goldman, D., Merril, C.R. and Rush, M. 1983. The cheetah is depauperate in genetic variation. *Science* 221: 459–461.

O'Shea, J. and Ortaldo, J.R. 1992. The biology of natural killer cells: insights into the molecular basis of function. In: Lewis, C.E. and McGee, J.O.D. (eds) *The natural immune system: The natural killer cell*. IRL Press at Oxford University Press, New York, pp. 2–40.

Patenaude, N. and White, B. 1994. Reduced genetic variability of the St-Lawrence beluga whale population as assessed by DNA fingerprinting. *Mol. Ecol.* 3: 375–381.

Reeves, R.R. and Mitchell, E. 1984. Catch history and initial population of white whales (*Delphinapterus leucas*) in the river and gulf of St-Lawrence, eastern Canada. *Naturaliste-Can. (rev. Ecolo. Syst.)* 111: 63–121.

Reijnders, P.J. 1986. Reproductive failure in common seals feeding on fish from polluted coastal waters. *Nature* 324(6096): 456–457.

Rideout, B.A., Moore, P.F. and Pedersen, N.C. 1992. Persistent upregulation of MHC class II antigen expression on T-lymphocytes from cats experimentally infected with feline immunodeficiency virus. *Vet. Immunol. Immunopathol.* 35: 71–81.

Roitt, I., Brostoff, J. and Male, D. (eds) 1989. *Immunologie fondamentale et appliquée*. MEDSI, Paris.

Romano, T.A., Ridgway, S.H. and Quaranta, V. 1992. MHC class II molecules and immunoglobulins on peripheral blood lymphocytes of the bottlenose dolphin, *Tursiops truncatus*. *J. Exp. Zool.* 263: 96–104.

Ross, P.S., DeSwart, R.L., Timmerman, H.H., Vedder, L.J., VanLoveren, H., Vos, J.G., Reijnders, P.J.H. and Osterhaus, A.D.M.E. 1995 Suppression of natural killer cell activity in harbour seals (*Phoca vitulina*) fed Baltic Sea herring. *Aquatic Toxicol* 34: 71–84.

Ross, P.S., DeSwart, R.L., Addison, R.F., Van Loveren, H., Vos, J.G. and Osterhaus, A.D.M.E. 1996b. Contaminant-induced immunotoxicity in harbour seals: wildlife at risk? *Toxicology* 112: 157–169.

Ruby, S., Tavera Mendoza, L., Brousseau, P., Dégas, V. and Fournier, M. 2002. Reproductive system impairments in mice fed on diets containing beluya whale blubber from the St. Lawrence Estuary and the Arctic populations. *Journal of Toxicology and Environmental Health*, in press.

Shenker, B.J., Rooney, C., Vitale, L. and Shapiro, I.M. 1992. Immunotoxic effects of mercuric compounds on human lymphocytes and monocytes. I. Suppression of T-cell activation. *Immunopharmacol. Immunotoxicol.* 14: 539–553.

Smialovicz, R.J., Andrews, J.E., Riddle, M.M., Rodgers, R.R., Loebke, R.W. and Copeland, C.B. 1989. Evaluation of immunotoxicity of low level PCB exposure in the rat. *Toxicology* 56: 197–211.

Smialowicz, R.J., Riddle, M.M., Williams, W.C. and Diliberto, J.J. 1994. Effects of 2,3,7,8-tetrachloro-dibenzo-p-dioxin (TCDD) on humoral immunity and lymphocyte subpopulations: differences between mice and rats. *Toxicol. Appl. Pharmacol.* 124: 248–256.

Thomas, P.T. 1995. Pesticide-induced immunotoxicity: are Great Lakes residents at risk? *Environ. Health Persp.* 103(supp. 9): 55–61.

Thomas, P.T. and Hinsdill, R.D. 1978. Effect of polychlorinated biphenyls on the immune response of rhesus monkeys and mice. *Toxicol. Appl. Pharmacol.* 44: 41–51.

Tizard, I. 1992. Lymphocytes. In: *Veterinary immunology, an introduction*, 4th edn. WB Saunders, Montreal, Québec, pp. 72–84.

Tryphonas, H., Hayward, S., O'Grady, L., Loo, J.C.K., Arnold, D.L., Bryce, F. and Zawidzka, Z.Z. 1989. Immuno-toxicity studies of PCB (Aroclor 1254) in the adult rhesus (*Macaca mulatta*) monkey – preliminary report. *Int. J. Immunopharmacol.* 11: 199–206.

Wagemann, R., Stewart, R.E.A., Béland, P. and Lagacé, A. 1990. Heavy metals and selenium in tissues of beluga whales, *Delphinapterus leucas*, from the Canadian arctic and the St.-Lawrence estuary. *Can. Bull. Fish Aquat. Sci.* 224: 191–206.

Yuhki, N. and O'Brien, S.J. 1990. DNA variation of the mammalian major histocompatibility complex reflects genomic diversity and population history. *Proc. Natl Acad. Sci.* 87: 836–840.

15 Evaluation of genotoxic effects of environmental contaminants in cells of marine mammals, with particular emphasis on beluga whales

J. M. Gauthier, H. Dubeau and E. Rassart

Introduction

Marine mammal populations, such as the St. Lawrence beluga whales (*Delphinapterus leucas*), the Mediterranean Sea striped dolphins (*Stenella coeruleoalba*), the US Atlantic coast and Gulf of Mexico bottlenose dolphins (*Tursiops truncatus*) and the Baltic gray seals (*Halichoerus grypus*), show high concentrations of environmental contaminants which have been associated with disease in these populations (Baker, 1984; Kuehl *et al.*, 1991; Leonzio *et al.*, 1992; Olsson *et al.*, 1994; Kuehl and Haebler, 1995; Borrell *et al.*, 1996; Jenssen, 1996; Munson *et al.*, 1998). The St. Lawrence beluga whales form a small endangered population (Committee on the Status of Endangered Wildlife in Canada, COSEWIC) that inhabits the St. Lawrence estuary and the Saguenay Fjord (Michaud, 1993). Both these regions are contaminated with environmental contaminants including polycyclic aromatic hydrocarbons (PAHs), organochlorine compounds (OCs) and mercury, which originate from atmospheric transport, a variety of local industries, such as aluminum refineries and chloro-alkali plants, municipal effluents, the Great Lakes, St. Lawrence tributaries, and soil erosion (Martel *et al.*, 1986; Dalcourt *et al.*, 1992; Gobeil and Cossa, 1993; Pham *et al.*, 1993; Gearing *et al.*, 1994; Quémarais *et al.*, 1994). Necropsies of dead stranded whales have shown a high prevalence of tumors, non-neoplastic lesions and opportunistic diseases (DeGuise *et al.*, 1994, 1995; Martineau *et al.*, 1994, Chapter 13 in this volume). Cancer has been diagnosed as the principal cause of death in 18 per cent of 73 examined carcasses and the annual crude cancer rate was estimated to be 233/100 000; higher than in many human and domestic animal populations (DeGuise *et al.*, 1994, 1995; Martineau *et al.*, 1999 and more recent data in this monograph). Population modeling, field studies and necropsy observations of female reproductive organs have suggested a lower reproductive rate in this population than in Arctic populations (Béland *et al.*, 1988, 1993; Michaud, 1993). High concentrations of polychlorinated biphenyls (PCBs), OC compounds and mercury have been found in tissues of the St. Lawrence beluga whales (Muir *et al.*, 1990, 1996). On the contrary, Arctic beluga whales have low concentrations of environmental contaminants (Muir *et al.*, 1990). Benzo[*a*]pyrene (B[a]P)-DNA adducts have been detected by high pressure liquid chromatography (HPLC) fluorescence analysis in St. Lawrence beluga whales, but not in Arctic beluga whales (Martineau *et al.*, 1994; Mathieu *et al.*, 1997). However, B[a]P-DNA adducts were detected in Arctic beluga whales using competitive enzyme-linked immunosorbent assay (ELISA) with a fluorescent endpoint (Mathieu *et al.*, 1997). It remains to be determined whether this immunochemical assay will detect higher levels of B[a]P-DNA adducts in St. Lawrence beluga whales. Moreover, no gross evidence of cancer has been found in approximately 50 Arctic beluga whale carcasses sampled for routine biological purposes (David J. St. Aubin, 30 October 1997, personal communication). Factors such as tumor induction, tumor promotion and immunosuppression induced by environmental contaminants have been proposed to explain the elevated cancer prevalence in the St. Lawrence population (DeGuise *et al.*, 1994, 1995). Environmental contamination may also be involved in the high prevalence of cancer in California sea lions (*Zalophus californianus*) (Gulland *et al.*, 1996).

DNA damage plays an important role in the development of cancer and can be a risk factor for teratogenesis and other diseases (Hemminki and Vineis, 1985; Oshimura and Barett, 1986; Evans, 1990; Hagmar *et al.*, 1994; Tucker and Preston, 1996; Duffaud *et al.*, 1997). Several environmental contaminants, such as PAHs, certain OCs and mercury, found in marine mammals (Wagemann and Muir, 1984; Hellou *et al.*, 1990; Wagemann *et al.*, 1990; Kuehl *et al.*, 1991; Leonzio *et al.*, 1992; Martineau *et al.* 1994; Olsson *et al.* 1994; Borrell *et al.*, 1996; Jenssen, 1996) have been shown to cause DNA damage at the gene, strand and/or chromosome level in human and animal cells (Palmer *et al.*, 1972; Majumbar *et al.*, 1976; Hooper *et al.*, 1979; Cantoni and Costa, 1983; IARC, 1983, 1993; Sobti *et al.*, 1983; Oshimura and Barrett, 1986; Safe, 1989; Sargent *et al.*, 1989; Steinel *et al.*, 1990; Bhunya and Jena, 1992; Meisner *et al.*, 1992; Perera *et al.*, 1992; Sakar *et al.*, 1993; DeFlora *et al.*, 1994; Franchi *et al.*, 1994; Warshawsky *et al.*, 1995; Ogura *et al.*, 1996; Canonero *et al.*, 1997). PAH compounds, methylmercury (MeHg), DDT, toxaphene and, to a lesser extent, chlordane are considered carcinogens in experimental animals (IARC, 1987, 1993). Carcinogenicity to humans has been evaluated by the International Agency for Research on Cancer (IARC) (1987, 1993) to be probable for many multi-ringed (>3 rings) PAH compounds (group 2A) and possible for MeHg, toxaphene and DDT (group 2B). Environmental mixtures containing PAH compounds, such as those resulting from aluminum production, are also considered carcinogenic (IARC, 1984).

Several techniques can be used to detect DNA damage. The mammalian cytokinesis-block micronuclei (MN) assay is a well-known cytogenetic technique used to assess chromosome damage induced by environmental contaminants (Fenech, 1993). MN are chromosomal fragments or whole chromosomes that are not incorporated into daughter nuclei during mitosis because of chromosomal breakage or aneuploidy caused by dysfunction of the mitotic apparatus, respectively (Fenech, 1993). Cytochalasin-B is added during culture of cells to prevent cytokinesis and produces binucleated (BN) cells which can be scored easily and accurately for MN following one cell cycle (Fenech and Morley, 1985). The single-cell microgel electrophoresis (Comet) assay can be used to analyze single- and double-strand breaks and alkali-labile sites in DNA, as well as excision repair of these lesions (Singh *et al.*, 1988). Cells are embedded in agarose gel on a microscopic slide, lysed, electrophoresed and stained with a fluorescent dye (Singh *et al.*, 1988). The charged DNA is pulled by the electric current from the nucleus, causing greater migration of relaxed and broken DNA fragments. The subsequent images of the cells have the appearance of 'comets'. The extent of DNA strand breakage can be analyzed by measuring DNA migration distance (length of the comet tail) and tail moment, which takes into account both tail length and tail fluorescence intensity. The unscheduled DNA synthesis (UDS) assay analyzes DNA repair synthesis after excision of damaged DNA induced by genotoxic agents (Maddle *et al.*, 1994). The incorporation of [^{3}H]thymidine into DNA is used as a measure of resynthesis of the excised region.

Most cancers are of epithelial origin, including those observed in St. Lawrence beluga whales and other marine mammals (Cairns, 1975; Martineau *et al.*, 1999). Epithelial cells are difficult to obtain from tissues of wild marine mammals. Peripheral blood lymphocytes are considered the best non-invasive source of cells to analyze DNA damage because 80 per cent of these cells circulate throughout all organs and tissues and thus provide an estimate of average whole-body exposure to genotoxic compounds (Tucker and Preston, 1996). Although blood lymphocytes are often used to test genotoxic effects of environmental contaminants in humans (Evans, 1984; Tucker and Preston, 1996), they can be difficult to obtain from marine mammals and are not always available in sufficient amounts for *in vitro* cytogenetic studies in these animals. Fibroblasts are the major cell type of conjunctive tissue of organs and body tissues, and are thus omnipresent in the organism (Barlovatz-Meimon and Martelly, 1988). Fibroblasts have been widely used in genotoxicity testing studies using the MN assay and other cytogenetic endpoints. Mice injected i.p. with MeHg *in vivo* have shown increases in chromosome aberrations (CAs) in skin fibroblasts (Gilbert *et al.*, 1983), demonstrating that skin fibroblasts can also be used as a surrogate cell type for estimation of target tissue exposure. Skin samples can be obtained from stranded marine mammals or by biopsy on free-ranging animals. Skin fibroblasts can be easily isolated from small skin samples and proliferate for several generations in culture to yield great amounts of cells (Barlovatz-Meimon and Martelly, 1988).

Very few studies exist on the genotoxic effects of environmental contaminants in marine mammal cells. The objectives of this chapter are:

1. To review present studies on the evaluation of *in vitro* genotoxicity of environmental contaminants in marine mammal cells and their possible ecotoxicological relevance in wild populations. These studies comprise the assessment of the genotoxic effects of mercury, OC and PAH compounds in skin fibroblasts of a beluga whale using the MN assay, of mercury in blood lymphocytes of a bottlenose dolphin using the Comet assay, and of B[a]P in fetal bottlenose dolphin epithelial kidney cells using Comet and UDS assays.
2. To discuss the factors influencing the ecotoxicological relevance of the *in vitro* genotoxicity findings in marine mammal cells.
3. To present future research needs in this emerging field of marine mammal genotoxicology.

Review of present studies on the genotoxicity evaluation of environmental contaminants in marine mammal cells

This section provides a brief introduction of the genotoxicity of PAH, mercury and OC compounds as reported in human and terrestrial mammal studies, followed by a review of present studies on the genotoxicity evaluation of these compounds in marine mammals and their possible ecotoxicological relevance in wild populations.

PAH compounds

Many of the multi-ringed PAH compounds are potent mutagens and chromosome-damaging agents (IARC, 1983). PAHs are metabolically activated to highly reactive DNA-binding electrophiles responsible for tumor induction and other toxic effects of these compounds (Pelkonen and Nebert, 1982; Harvey, 1991). B[a]P is a model PAH compound which has been widely used as an index of PAH contamination and as a positive control in *in vitro* and *in vivo* genotoxicity studies. Positive results have been obtained *in vitro* and *in vivo* for B[a]P in CA, MN and sister chromatid exchange (SCE) induction assays (IARC, 1983; He and Baker, 1991; Warshawsky *et al.*, 1995). Other multicyclic PAHs, such as benzo[*a*]anthracene (BA), dibenzo[*a,h*]anthracene (DBA), benzo[*b*]fluoranthrene (BF) and chrysene, are also mutagenic in the AMES test and positive in the MN, CA and/or SCE *in vitro* and *in vivo* assays in human and other terrestrial mammals (IARC, 1983; He and Baker, 1991; Reddy *et al.*, 1991; Crofton-Sleigh *et al.*, 1993; Winker *et al.*, 1995). Statistically significant correlations have been found between frequency of CA in human lymphocytes and PAH air pollution (Perera *et al.*, 1992).

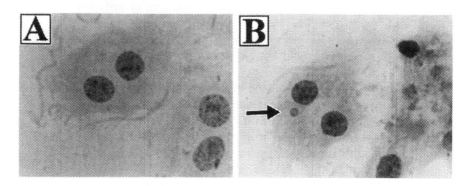

Figure 15.1 Typical binucleated beluga whale fibroblastoid cell without (A) and with (B) a micronucleus (MN). The MN is indicated by the arrow.

MN assay

Gauthier *et al.* (2002) used the MN assay to analyze the genotoxic effects of B[a]P and a mixture of PAH compounds in the presence of external metabolic factor (S9 rat liver fraction) in fibroblasts isolated from a biopsy skin sample taken from an west Hudson Bay Arctic beluga whale collected by subsistence Inuvialuit hunting. Typical BN beluga skin fibroblasts with and without a MN are shown in Figure 15.1. A range of experimental concentrations for PAH compounds were chosen according to those found in sediments of the St. Lawrence estuary and Saguenay Fjord, and according to preliminary MN induction experiments. Constituents of the PAH mixture (PAH-M) were selected for their prevalence in sediments of areas of the St. Lawrence estuary with high beluga frequentation, and/or their known genotoxic and carcinogenic activities (IARC, 1987; Harvey, 1991; Dalcourt *et al.*, 1992; Michaud, 1993). The PAH-M constitutes about 50 per cent of the mixture of 16 PAHs identified in sediments in these areas (Dalcourt *et al.*, 1992). The PAH-M was prepared according to proportions found in these sediments and is composed of 17 per cent B[a]P, 10 per cent BA, 10 per cent DBA, 42 per cent BF and 21 per cent chrysene. Although rat S9 fraction was used as the external metabolic factor, B[a]P hydroxylase (AHH) activity in beluga whales falls within the range of other odontocetes, which have comparable activities to those of rats (Watanabe *et al.*, 1989; White *et al.*, 1997).

Both B[a]P and the PAH-M induced a highly significant ($P > 0.001$) concentration–response increase in micronucleated cell (MNC) frequency (number of MNCs per 1000 BN cells). Statistically significant increases in MNCs were found for 0.5, 2 and 5 µg/ml B[a]P and 5 and 20 µg/ml of the PAH-M when compared to control cultures (Table 15.1). Concentrations of 0.5, 2 and 5 µg/ml B[a]P, respectively, induced a 2.2-, 2.7- and 5.5-fold mean increase in MNCs. Treatment with B[a]P alone was more potent in inducing

Table 15.1 Micronucleated cell (MNC) induction by B[a]P and the PAH-M in beluga whale skin fibroblasts in presence of S9 mix. Data are presented as mean ± standard deviation about the mean (SD) for total MNCs per 1000 BNs cells. DMSO was used as the negative control for B[a]P and the PAH-M

Chemical and concentration ($\mu g/ml$)	Mean ± SD	P value[a]
DMSO		
1%	9.0 ± 2.6	–
B[a]P		
0.02	11.7 ± 1.5	0.205
0.5	20.0 ± 1.2	0.007
2	24.3 ± 1.2	0.001
5	49.3 ± 6.7	0.001
PAH-M		
0.02	14.3 ± 2.1	0.052
0.1	14.7 ± 3.1	0.072
5	30.3 ± 2.5	0.001
20	53.3 ± 9.9	0.002

[a] Probability values are based on a one-way ANOVA comparing the total number of MNCs observed per 1000 BN cells for each treatment with the respective control.

MN than the PAH-M (Table 15.1), suggesting antagonistic effects of PAH compounds in this mixture. Most studies have shown that mixtures of various PAHs, including those used in this study, are less potent in inducing DNA adduct formation, mutations and MN than B[a]P alone and this is probably due to competitive inhibition of metabolic pathways involved in formation of reactive metabolites (Haugen and Peak, 1983; Springer *et al.*, 1989; Reddy *et al.*, 1991). According to concentrations tested in the study, sensitivity of beluga whale fibroblasts to the genotoxic effects of B[a]P treatment in the presence of S9 mix appeared to be similar to that of Chinese hamster fibroblasts, lower than rat skin primary fibroblasts, and greater than that of human blood lymphocytes (Ellard *et al.*, 1991; Vian et al, 1993; Vienneau *et al.*, 1995).

B[a]P-DNA adducts have been found in tissues of the St. Lawrence beluga whales (Martineau *et al.*, 1994). Stomach content analyses and field studies suggest that these whales dig in bottom sediments to feed on invertebrates (Vladykov, 1946; Robert Michaud, 2 October 1998, personal communication). PAHs have been identified as the most likely contaminants causing positive genotoxicity results in bottom sediment samples (e.g. Marvin *et al.*, 1993; Gagné *et al.*, 1995; Papoulias and Buckler, 1996). Moreover, a significant relationship was found between PAH concentrations in sediments of the Saguenay Fjord and genotoxicity of bivalve tissue extracts of this region (White *et al.*, 1997). These findings suggests that St. Lawrence beluga whales are exposed to sediment-bound PAHs and to the genotoxic effects of these compounds. Concentrations of ΣPAH (sum of 13–16 unsubstituted PAHs of pyrogenic origin) in sediments of areas of the St. Lawrence estuary and the

Saguenay Fjord frequented by the beluga whales range between 0.07–10 and 0.5–30 µg/g (dry weight), respectively (Martel *et al.*, 1986; Dalcourt *et al.*, 1992; Gearing *et al.*, 1994). Because B[a]P accounts for about 7 per cent of ΣPAH in these regions (Martel *et al.*, 1986; Dalcourt *et al.*, 1992), concentrations of B[a]P may be estimated to vary between 0.005 and 2.1 µg/g dry weight. Consequently, the authors concluded that although results of this *in vitro* study do not imply that PAH compounds are involved in the etiology of cancer in St. Lawrence beluga whales, concentrations of 0.5 and 2 µg/ml B[a]P and 5 and 20 µg/ml of the PAH-M that induced MN formation in beluga fibroblasts were within the range of concentrations of PAH combustion products found in sediments of certain regions of the habitat of these whales.

Comet and UDS assays

Both the Comet and UDS assays were used by Carvan *et al.* (1995) to analyze the *in vitro* DNA-damaging effects of B[a]P and excision repair of B[a]P-induced DNA-adducts in kidney epithelial cells isolated from a fetal bottlenose dolphin in the presence and absence of 2,3,7,8-tetrachlorodibenzo-*p*-dioxin (TCDD). TCDD is a potent inducer of cytochrome P450 CYP1A1, which is responsible for the metabolism of B[a]P to highly reactive DNA-binding electrophiles (Harvey, 1991; Schrenk, 1998). Both migration distance of DNA fragments (comet tail lengths) and [³H]thymidine incorporation as a measure of UDS were measured as a function of hours following B[a]P treatment. These assays showed that B[a]P (0.025–2.5 µg/ml) induced low levels of DNA fragment migration and/or excision repair in the absence of TCDD (Carvan *et al.*, 1995). However, when cells were pretreated with 3.32 ng/ml TCDD, comet tail lengths and UDS increased as a function of time following B[a]P treatment, peaked at 2 hours and decreased as a function of excision repair of B[a]P-DNA adducts (Carvan *et al.*, 1995). Differences between non-TCDD- and TCDD-pretreated cultures were significant ($P < 0.05$–0.005) for both endpoints. The ability of fetal bottlenose dolphin kidney epithelial cells to excise B[a]P-adducts was also analyzed by quantifying removal of incorporated [³H]B[a]P and was found to be similar to results obtained for the Comet and UDS assays. The authors concluded that these data indicate that TCDD exposure enhances the conversion of B[a]P to a reactive metabolite and subsequent B[a]P-DNA adduct formation in fetal bottlenose dolphin kidney epithelial cells.

Bottlenose dolphins feed on benthic fish and invertebrates (Wells and Scott, 1998) and could therefore be exposed to the genotoxic effects of sediment-bound PAH compounds when feeding on these organisms. To our knowledge, B[a]P-DNA adducts and TCDD have not been analyzed in dolphins. However, high concentrations of PAHs are found in sediments of certain areas inhabited by dolphins with high environmental contaminant loads (Kuehl *et al.*, 1991; Narbonne *et al.*, 1991; O'Connor, 1991; Kannan *et al.*, 1993; Borrell *et al.*, 1996). Moreover, non- and mono-*ortho*-substituted PCBs have similar modes of action and toxic effects to those of TCDD (Safe, 1993). Because high concentrations and elevated toxic equivalent (TEQ) values have been found for non- and mono-*ortho*-PCBs in certain dolphin populations, these PCB congeners may represent a greater toxicological risk than TCDD in these dolphins (Kuehl *et al.*, 1991; Kannan *et al.*, 1993; Borrell *et al.*, 1996). These PCB congeners could potentially contribute to increased metabolism of PAH compounds, leading to increased genotoxic potential of PAH exposure in dolphins.

Mercury compounds

Although mercury compounds are not mutagenic in bacterial systems, both inorganic and organic forms of mercury have been shown to induce DNA breakage at the chromosome and DNA strand level, and aneuploidy in cells of humans, domestic animals and wildlife (DeFlora *et al.*, 1994). Methylmercury chloride (CH_3HgCl_2) (MeHg) and mercury chloride ($HgCl_2$) (Hg), especially MeHg, cause c-mitosis induced aneuploidy through binding to sulfhydryl groups of tubulin proteins of the mitotic apparatus (Oshimura and Barrett, 1986). Mechanisms proposed to explain the genotoxic effects of mercury include direct binding to nitrogen atoms of thymidine bases by Hg^{2+} and CH_3Hg^{2+} cations, production of free radicals, and the depletion of glutathione reserves and inhibition of DNA repair enzymes by binding to sulfhydryl groups of proteins (Eichhorn, 1981; Morimoto *et al.*, 1982; Cantoni and Costa, 1983; Cantoni *et al.*, 1984; Snyder *et al.*, 1988). The indirect effects of lipid peroxidation, increased condensation of chromosomes and inhibition of RNA polymerase I (in the case of Hg) may also be involved in the induction of aneuploïdy (Andersen *et al.*, 1983; Verschaeve *et al.*, 1985; Önfelt, 1986). Hg and MeHg are positive in the CA, SCE, MN and Comet *in vitro* tests in human and other terrestrial mammalian cells (Morimoto *et al.*, 1982; Betti *et al.*, 1992, 1993; DeFlora *et al.*, 1994; Ogura *et al.*, 1996). Although *in vivo* studies have yielded contradictory results, some studies have shown induction of chromosome damage in experimental animals exposed to mercury experimentally and humans exposed to mercury through accidental, occupational or alimentary sources (Skerfving *et al.*, 1974; Miller *et al.*, 1979; DeFlora *et al.*, 1994; Franchi *et al.*, 1994). Significant correlations have been found between frequency of CAs or MN and total mercury concentrations between 0.01 and 1.1 µg/g wet weight in blood of humans that consume large amounts of fish and seafood (Skerfving *et al.*, 1974; Franchi *et al.*, 1994).

MN assay

Gauthier *et al.* (1998) analyzed the genotoxic effects of Hg and MeHg using the MN assay in fibroblasts isolated from a biopsy skin sample taken from

Table 15.2 Micronucleated cell (MNC) induction by Hg and MeHg in beluga whale skin fibroblasts. Data are presented as mean ± SD for total MNCs per 1000 BNs cells. Positive control is mitomycin C (MMC). Distilled water was used as the negative control for MMC and Hg, and DMSO as the negative control for MeHg

Chemical and concentration ($\mu g/ml$)	Mean ± SD	P value[a]
H$_2$O	8.0 ± 2.7	–
DMSO		
1%	7.3 ± 2.1	0.733[b]
MMC		
0.1	15.6 ± 1.0	0.010
1	49.1 ± 6.1	<0.001
2	45.1 ± 2.1	<0.001
Hg		
0.05	5.2 ± 0.7	0.163
0.5	13.5 ± 2.1	0.049
5	22.2 ± 4.8	0.011
20	28.6 ± 6.7	0.008
MeHg		
0.05	15.8 ± 1.9	0.006
0.5	18.6 ± 3.2	0.007
2	26.7 ± 3.2	0.001

[a,b] Probability values are based on a one-way ANOVA comparing the total number of MNCs observed per 1000 BN cells for each treatment with the respective control ([a]), and between DMSO and distilled water control ([b]).

a west Hudson Bay Arctic beluga whale collected by subsistence Inuvialuit hunting. The range of experimental concentrations for mercury compounds was chosen according to those found in St. Lawrence beluga whales and according to preliminary cell-killing concentration–response experiments. Both Hg and MeHg induced a highly significant ($P \leq 0.001$) concentration–response increase of MNCs. Statistically significant increases in MNCs were observed for 0.5, 5 and 20 $\mu g/ml$ Hg and 0.05, 0.5 and 2 $\mu g/ml$ MeHg when compared to control cultures (Table 15.2). Concentrations of 0.5, 5 and 20 $\mu g/ml$ Hg, respectively, induced a two-, three- and fourfold increase of MNCs. Treatment with MeHg was an order of magnitude more potent in inducing MN than Hg (Table 15.2). Sensitivity of beluga whale fibroblasts to genotoxic activity of Hg and MeHg appeared to be greater than that of Chinese hamsters fibroblasts and human lymphocytes, as induction of MN and CAs was reported for higher concentrations in these cells (Betti *et al.*, 1992; Yamada *et al.*, 1993; Ogura *et al.*, 1996).

The mean total mercury concentration on a wet weight basis in the liver of St. Lawrence beluga whales is 34 $\mu g/g$ (range = 0.4–202 $\mu g/g$) (Wagemann *et al.*, 1990). Using total mercury concentrations on a wet weight basis in liver (63 $\mu g/g$, range = 0.2–218 $\mu g/g$) and blood (0.4 $\mu g/g$, range = 0.008–1.5 $\mu g/g$) analyzed in pantropical spotted dolphins (*Stenella attenuata*) of the Pacific Ocean (André *et al.*, 1990), the authors estimated the concentrations that may be present in blood of certain St. Lawrence beluga whales to be about 0.016–1.4 µg/g. Mercury in blood of mammals, including striped dolphins, exists almost entirely as MeHg (Itano *et al.*, 1984; Vahter *et al.*, 1994). Hence, the authors concluded that although these *in vitro* results do not imply that mercury compounds are involved in the etiology of cancer in St. Lawrence beluga whales, significant increases in MN frequency were found at low concentrations of MeHg (0.05 and 0.5 µg/ml MeHg) which are believed to be comparable to concentrations present in blood of certain whales of this population.

Comet assay

Betti and Nigro (1996) used the Comet assay to evaluate the *in vitro* genotoxic effects of methylmercury in peripheral blood lymphocytes of a captive bottlenose dolphin. The extent of DNA strand breakage was analyzed by measuring DNA migration distance. Significant increases in DNA strand breakage were observed for all tested concentrations of MeHg (1–8 µg/ml) ($P < 0.0001$). An identical protocol was used to compare DNA strand breakage potency of MeHg in rat and human lymphocytes (Betti *et al.*, 1993; Betti and Nigro, 1996). Although the response of dolphin lymphocytes was similar to that of rat and human lymphocytes at 1 µg/ml MeHg, it was statistically lower ($P < 0.01$) at 2 µg/ml and higher concentrations.

Total mercury concentrations in liver and blood of pantropical spotted dolphins of the Pacific Ocean range between 0.2 and 218 µg/g and 0.008 and 1.5 µg/g wet weight, respectively (André *et al.*, 1990). Moreover, concentrations in livers of Mediterranean striped and bottlenose dolphins, respectively, reach concentrations of about 4400 and 13 150 µg/g dry weight (André *et al.*, 1990; Leonzio *et al.*, 1992), corresponding to 1180 and 3520 µg/g wet weight when the conversion factor of 3.738 proposed by Wagemann *et al.* (1990) is used. Accordingly, the authors concluded that concentrations of MeHg that induced single-strand breakage in DNA of bottlenose dolphin lymphocytes are likely to be within the range of those naturally occurring in the blood of wild dolphins.

Organochlorine compounds

Toxaphene is an electrophilic DNA-reactive compound and is the only OC compound found in beluga whales that is considered a mutagenic and genotoxic agent (Hooper *et al.*, 1979; Sobti *et al.*, 1983; Steinel *et al.*, 1990; Ashby and Tennant, 1991). Boon *et al.* (1998) proposed that toxaphene–DNA adducts are formed by interaction between the C_8 or C_9 carbons of toxaphene and the amino group of DNA bases through a biomolecular nucleophilic substitution (S_N2) reaction (Arnaud, 1983). Toxaphene is positive in the AMES and SCE tests in the absence of external metabolic factor

414   *J.M. Gauthier* et al.

(Hooper *et al.*, 1979; Sobti *et al.*, 1983; Steinel *et al.*, 1990; Ashby and Tennant, 1991). Higher frequencies of CAs have been observed in humans exposed occupationally to toxaphene (Samosh, 1974). Certain authors have suggested that OCs, such as DDT, hexachlorobenzene and PCBs, could act as aneugens through inactivation of the mitotic spindle (Al-Sabti and Metcalfe, 1995; Parry *et al.*, 1996; Canonero *et al.*, 1997). Certain studies have shown that chlordane is positive in the UDS test and that certain PCBs and their metabolites, which have a similar structure as DDT, could bind to DNA and cause single-strand lesions in this molecule (Ahmed *et al.*, 1977; Safe, 1989). Consequently, it is also possible that DDT and chlordane compounds could act at the DNA level (Hooper *et al.*, 1979). However, negative results have been obtained for *p,p'*-DDT and its metabolite *p,p'*-DDE in most bacterial mutagenicity tests, and chlordane is considered a weak mutagen in these tests (Kelly-Garvet and Legator, 1973; Mamber *et al.*, 1984; WHO, 1984; Reifferscheild and Heil, 1996). Although conflicting results have been obtained for genotoxicity of *p,p'*-DDT and chlordane, it has been shown that these compounds are capable of inducing CAs and SCEs *in vitro* and *in vivo* in certain human and terrestrial mammal cell systems (Palmer *et al.*, 1972; Larsen and Jalal, 1974; Lessa *et al.*, 1976; Sobti *et al.*, 1983; Sakar *et al.*, 1993).

MN assay

Gauthier *et al.* (1999a, b) used the MN assay to test the genotoxic effects of toxaphene, chlordane and *p,p'*-DDT with and without external metabolic factor S9 in fibroblasts isolated from a biopsy skin sample taken from a west Hudson Bay Arctic beluga whale collected by subsistence Inuvialuit hunting. Ranges of experimental concentrations were chosen according to those found in St. Lawrence beluga whales and according to preliminary cell-killing concentration–response experiments. In the absence of external metabolic factor S9, toxaphene, chlordane and *p,p'*-DDT induced significant ($P < 0.05$) concentration–response increases of MNCs. Statistically significant increases in MNCs, ranging from 1.7- to fivefold when compared to control cultures, were observed for 0.05, 0.5, 5 and 10 µg/ml toxaphene, for 2, 5 and 10 µg/ml chlordane and for 10 and 15 µg/ml *p,p'*-DDT (Table 15.3). According to concentrations used in this study, sensitivity of beluga whale fibroblasts to the genotoxic effects of *p,p'*-DDT treatment in the absence of S9 mix appears to be greater than that of Chinese hamster fibroblasts and lower than that of rat kangaroo cells (Palmer *et al.*, 1972; Mahr and Miltenburger, 1976).

The authors also showed that presence of S9 mix greatly reduced or abolished the genotoxic effects of OC pesticides in beluga whale fibroblasts (Table 15.3). No significant relationships between MNC frequency and concentration were observed for any of the tested OC pesticides in the presence of S9 mix. The MN induction potency of chlordane and *p,p'*-DDT was

Table 15.3 Micronucleated cell (MNC) induction by toxophene, chlordane and *p,p'*-DDT in beluga whale skin fibroblasts in absence and presence of S9 mix. Data are presented as mean ± SD for total MNCs per 1000 BNs cells. Positive controls are mitomycin C (MMC) and BP for experiments conducted in absence and presence of S9 mix, respectively. Distilled water was used as the negative control for MMC, and DMSO as the negative control for B[a]P and OC compounds

Chemical and concentration (*µg/ml*)	Without S9		With S9	
	Mean ± SD	*P value*[a]	*Mean ± SD*	*P value*[a]
H₂O	8.0 ± 2.7	–		
DMSO				
1%	7.3 ± 2.1	0.733[b]	9.0 ± 2.6	–
MMC/BP				
0.1/0.5	15.6 ± 1.0	0.010	20.0 ± 0.6	0.007
1/2	49.1 ± 6.1	<0.001	24.3 ± 1.2	0.001
2/5	45.1 ± 2.1	<0.001	49.3 ± 6.7	0.001
Toxaphene				
0.05	16.0 ± 2.0	0.007	12.7 ± 0.6	0.079
0.5	23.7 ± 3.5	0.002	21.7 ± 3.8	0.009
5	23.0 ± 1.0	<0.001	15.3 ± 1.5	0.023
10	34.3 ± 1.2	<0.001	13.0 ± 2.6	0.138
Chlordane				
2	9.0 ± 0.0	0.238	13.0 ± 4.4	0.246
5	22.0 ± 4.4	0.006	12.7 ± 3.5	0.222
10	19.7 ± 3.5	0.006	11.7 ± 2.1	0.242
p,p'-DDT				
5	11.3 ± 1.5	0.055	10.3 ± 1.5	0.492
10	25.3 ± 5.8	0.007	7.7 ± 2.5	0.561
15	36.7 ± 1.5	<0.001	11.0 ± 1.0	0.288

[a,b] Probability values are based on a one-way ANOVA comparing the total number of MNCs observed per 1000 BN cells for each treatment with the respective control ([a]), and between DMSO and distilled water control ([b]).

completely eliminated in the presence of S9 mix at concentrations tested in the study (Table 15.3). Contrary to this study, SCE induction in human lymphoid cells by chlordane was greater in the presence of S9 mix than in non-activated cultures (Sobti *et al.*, 1983). However, addition of S9 mix eliminated the induction of UDS by chlordane in human fibroblasts (Ahmed *et al.*, 1977). *p,p'*-DDT is principally metabolized to intermediate metabolite *p,p'*-DDE and to final metabolite *p,p'*-DDA by mixed function oxidases (Matsumura, 1985). Contradictory data have been reported on the differential chromosomal breakage potency of *p,p'*-DDT compounds (Palmer *et al.*, 1972; Kelly-Garvet and Legator, 1973; Mahr and Miltenburger, 1976). However, for concentrations similar to that used in this study (10 µg/ml), *p,p'*-DDE and *p,p'*-DDA were shown to be less potent in inducing chromosome aberrations than *p,p'*-DDT in rat kangaroo cells (Palmer *et al.*, 1972). This may explain the loss of *p,p'*-DDT-induced MN formation in beluga whale fibroblasts in the presence of S9 mix. All tested concentrations of toxaphene were more potent in inducing MN without S9 mix than in the presence of this metabolic factor ($P < 0.001$ to $P = 0.05$), with the exception of the 0.5 μg/ml treatment, which induced similar MNC frequencies in both test systems ($P = 0.539$). Decreases in toxaphene-induced SCEs have also been reported in human lymphoid cells in the presence of S9 mix (Steinel *et al.*, 1990). These results indicate detoxification of toxaphene by the mixed function oxidases of the rat liver S9 fraction and/or interactions with components of the S9 mix (Steinel *et al.*, 1990; Saleh, 1991). It has been proposed that the formation of DNA adducts are prevented by hydroxylation of the -CH$_2$Cl group at the C$_8$ or C$_9$ position of toxaphene by the P450 enzymatic system present in rat S9 fraction (Boon *et al.*, 1998).

The lowest concentrations of toxaphene, chlordane and *p,p'*-DDT which induced significant increases in MN without S9 mix in beluga fibroblasts of this study were 0.05, 5 and 10 μg/ml, respectively. OC pesticides have not been analyzed in blood of beluga whales. Because OCs in blood and other tissues reach an equilibrium, and that, at equilibrium, concentrations in tissues are principally governed by their lipid content (Clark *et al.*, 1987; Boon *et al.*, 1994), the authors estimated approximate concentrations by adjusting for the differential lipid content of blood and blubber of beluga whales. Although other factors can play a role in OC tissue distribution, partitioning principally according to lipid content is in accordance with blubber/blood concentration ratios on a lipid weight basis reported for dolphins (0.6–1.8 for OC compounds and 1.0 for ΣDDT) (Reddy *et al.*, 1998). Mean blubber lipid content of St. Lawrence beluga whales is about 85 per cent (Muir *et al.*, 1996), which is within the range found in other odontocetes (Tanabe *et al.*, 1981; Boon *et al.*, 1994). Blood lipid content in odontocetes is about 0.1 per cent (range = 0.05–0.32%) (Tanabe *et al.*, 1981; Boon *et al.*, 1994). Concentrations of toxaphene, chlordane and *p,p'*-DDT in St. Lawrence beluga whales ($n = 36$) range respectively between 2–46, 1–28 and 3–389 μg/g lipid in blubber (Muir *et al.*, 1996), corresponding to approximately 0.002–0.05, 0.001–0.03 and 0.003–0.4 μg/g wet weight in blood when 0.1 per cent blood lipid content is used. According to these concentrations, only the effects of 0.05 μg/ml toxaphene could be within the range of concentrations found in the circulating blood of some of these whales.

The authors cautioned, however, that toxaphene-induced effects were diminished in the presence of rat S9 mix, which may indicate that the *in vivo* genotoxic hazard of this compound is low at concentrations found in St. Lawrence beluga whales. Boon *et al.* (1998) evaluated the *in vitro* biotransformation capacity of rat S9 fraction and marine mammal microsomes towards toxaphene in terms of genotoxic potential, using the Microtox assay. Both the addition of rat S9 fraction and microsomes of harbor seals (*Phoca vitulina*) decreased the genotoxic response of toxaphene in the Microtox assay, but this was not observed for whitebeaked dolphin (*Lagenorhynchus albirostris*) and sperm whale (*Physeter macrocephalus*) microsomes. These results indicate that microsomes of seals are capable of metabolizing toxaphene to the less genotoxic hydroxylated metabolites, and that marine mammals with low ability to metabolize toxaphene, such as cetaceans, may be the most affected by the genotoxic and carcinogenic properties of toxaphene (Boon *et al.*, 1998). Accordingly, it is possible that *in vitro* results obtained with rat S9 fraction are not relevant with beluga microsome capability to metabolize toxaphene, and that toxaphene may thus represent a genetic hazard to these whales.

Factors influencing the ecotoxicological relevance of *in vitro* genotoxicity testing results in marine mammals

DNA damage plays an important role in the development of cancer and can be a risk factor for teratogenesis and other diseases (Hemminki and Vineis, 1985; Oshimura and Barett, 1986; Evans, 1990; Hagmar *et al.*, 1994; Tucker and Preston, 1996; Duffaud *et al.*, 1997). In beluga whale skin fibroblasts, significant increases in MN frequency were found at concentrations of PAH compounds, MeHg and toxaphene that are believed to be comparable to concentrations of MeHg and toxaphene present in the blood of certain St. Lawrence beluga whales and PAH combustion products found in sediments of the habitat of these whales. In bottlenose dolphin peripheral blood lymphocytes, concentrations of MeHg that induced significant increases in DNA-strand breakage were found to be within the range of those naturally occurring in the blood of wild dolphins. In fetal bottlenose dolphin kidney epithelial cells, significant increases of B[a]P-induced DNA damaging effects and excision repair were induced by the potent CYP1A1 inducer TCDD, which may indicate increased genotoxic exposure potential of PAH compounds in dolphins with high concentrations of TCDD-like PCB congeners.

It is important to realize that *in vitro* induction of DNA damage by MeHg, toxaphene and PAH compounds in cells of marine mammals does not signify that these compounds are involved in the etiology of cancer or other diseases in cetaceans such as the St. Lawrence beluga whales. Indeed, the ecotoxicological consequences of these findings are unknown and real effects are probably more complex because:

(1) cells of only one individual was used in MN, Comet and UDS assays and responses could vary between individuals;
(2) only five PAHs were used to represent the PAH mixture present in the St. Lawrence beluga whale habitat;
(3) synergetic, additive or antagonistic interactions between environmental compounds of highly complex mixtures present in the environment could modify their DNA-damaging potential;
(4) the extent of metabolic detoxification of PAH compounds and toxaphene *in vivo* in marine mammals is unknown;
(5) it has been showed in dolphins that most MeHg is demethylated and stored principally in liver, and to a lesser extent in other tissues, as

insoluble granules of Hg-selenide (tiemannite) or Hg-selenoproteins, which prevent the production of toxic effects of mercury (Martoja and Berry, 1980; Palmisano *et al.*, 1995);

(6) DNA repair and apoptosis of damaged cells can modify the genotoxic effects of chemical compounds;

(7) there is no recognized association between genotoxicity and carcino-genicity of MeHg and toxaphene; and

(8) *in vitro* systems are usually more sensitive than *in vivo* systems.

Although *in vitro* effects are expected to be greater than *in vivo* effects, significant correlations have been found between the frequency of MN and/or CAs in human blood lymphocytes and concentrations of mercury in blood between 0.01 and 1.1 μg/g (fresh weight) and exposure to environmental PAH air pollution (Skerfving *et al.*, 1974; Perera *et al.*, 1992; Franchi *et al.*, 1994). Moreover, considering the carcinogenicity in experimental animals and established genotoxicity of PAH compounds, MeHg and toxaphene, together with the high environmental contaminant loads and long life span of cetaceans such as beluga whales and bottlenose dolphins (~30 years), exposure to these compounds could result in accumulation of DNA lesions. Accumulation of DNA lesions has been implicated in increased risk of mutation fixation and cancer initiation (Poirier and Béland, 1992). Therefore, it cannot be excluded that these compounds may pose a long-term genetic hazard to marine mammals and could act as possible participants in the carcinogenesis process in St. Lawrence beluga whales exposed to high concentrations of these compounds.

Conclusions and future research needs

In the reviewed studies, the MN, Comet and UDS assays were applied with success to evaluate the genotoxicity potential of environmental contaminants in beluga whale and bottlenose dolphin cells. To enable further interpretation of these *in vitro* results, data are necessary on exposure and tissue distribution of OC compounds, and both inorganic and methylated forms of mercury, in Arctic and St. Lawrence beluga whales, and of B[a]P-DNA adducts and TCDD in wild populations of bottlenose dolphins. Data on the capability of beluga whale and bottlenose dolphin microsomes to biotransform environmental compounds are also needed. Future *in vitro* studies are required on the analysis of the genotoxic potential of other suspected or known genotoxic environmental compounds and mixtures in cells of beluga whales and other marine mammals, using different genotoxic endpoints, such as the MN, CA, SCE and Comet assays. These assays could be used complementarily because they analyze different types of DNA damage, which can have dissimilar sensitivities to environmental contaminants (Carrano and Natarajan, 1988; Fenech, 1993; Tucker and Preston, 1996). Evaluation of the genotoxic effects of environmental mixtures in marine mammal cells would enable further investigations of possible synergistic, additive and/or antagonistic genotoxic effects of these mixtures found in marine mammals. Future MN studies should include techniques using an anti-kinetochore antibody (immunofluorescence technique) which discriminates between MN-containing chromosome fragments and whole chromosomes (Hennig *et al.*, 1988), to elucidate the mechanisms of MN formation in marine mammal cells.

Marine mammal genotoxicology is an emerging field. These *in vitro* studies represent the initial steps for the evaluation of *in vivo* genotoxic effects of environmental pollutants in marine mammals. Hongell (1996) analyzed the frequency of CAs and SCEs in lymphocytes of gray seal pups and adult ringed seals (*Pusa hispida*) from the Baltic Sea and concluded that the presence of CAs in gray seals, particularly the high frequency found in certain animals, indicates that these seals have been exposed to genotoxins in their marine environment. However, screening of changes in genotoxic activity within a population or between different populations requires baseline (spontaneous) values to be established in populations exposed to relatively low concentrations of putative genotoxic environmental contamination. Gauthier *et al.* (2002) used the MN, SCE and CA assays to analyze baseline levels of DNA damage in blood lymphocytes of individuals of the relatively healthy and lightly contaminated Arctic beluga whale, Sarasota Bay, FL, bottlenose dolphin and north-western Atlantic gray and harp (*Pagophilus groenlandicus*) seal populations. In the future, analysis of MN, SCEs, CAs and DNA-strand breakage could be used as biomarkers of genotoxicity to compare DNA damage between relatively unexposed and highly exposed populations of marine mammals of the same species, such as Arctic and St. Lawrence beluga whales. Analysis of OC compounds and heavy metals could be used as a general contamination index and as a basis for analysis of a putative association between chemical contamination and genotoxic effects in marine mammals. Genotoxic biomarkers could act, together with measurement of other health parameters and analysis of environmental contamination, as a basis for a multiple response assessment in a non-destructive approach to predict the impact of pollution in marine mammals.

References

Ahmed, F.E., Hart, R.W. and Lewis, N.J. 1977. Pesticide induced DNA damage and its repair in cultured human cells. *Mutation Research* 42: 161–174.

Al-Sabti, K. and Metcalfe, C.D. 1995. Fish micronuclei for assessing genotoxicity in water. *Mutation Research* 343: 121–135.

Andersen, O., Rønne, M. and Nordberg, G.F. 1983. Effects of inorganic metal salts on chromosome length in human chromosomes. *Hereditas* 98: 65–70.

André, J.M., Ribeyre, F. and Boudou, A. 1990. Mercury contamination levels and distribution in tissues and organs of Delphinids (*Stenella attenuata*) from the eastern tropical Pacific, in relation to biological and ecological factors. *Marine Environmental Research* 30: 43–72.

Arnaud, P. 1983. *Cours de Chimie Organique*. 13th edition. Gauthier-Villars, Bordas, Paris.

Ashby, J. and Tennant, R.W. 1991. Definitive relationships among chemical structure, carcinogenicity and mutagenicity for 301 chemicals tested by the US NTP. *Mutation Research* 257: 229–306.

Baker, J.R. 1984. Mortality and morbidity in grey seal pups (*Halichoerus grypus*). Studies on its causes, effects of environment, the nature and sources of infectious agents and the immunological status of pups. *Journal of Zoology (London)* 203: 23–48.

Barlovatz-Meimon, G. and Martelly, I. 1988. Culture de fibroblastes. In *Culture de Cellules Animales: Methodologie d'Application*. Adolphe, M. and Barlovatz-Meimon, G. (Eds). Éditions INSERM, Paris, pp. 141–167.

Béland, P., Vézina, A. and Martineau, D. 1988. Potential for growth of the St. Lawrence population (Québec, Canada) beluga whale (*Delphinapterus leucas*) population based on modelling. *Journal du Conseil International de l'Exploration de la Mer* 45: 22–32.

Béland, P., DeGuise, S., Girard, C., Lagacé, A., Martineau, D., Michaud, R., Muir, D.C.G., Norstrom, R.J., Pelletier, É., Ray, S. and Shugart, L.R. 1993. Toxic compounds and health and reproductive effects in St. Lawrence beluga whales. *Journal of Great Lakes Research* 19: 766–775.

Betti, C. and Nigro, M. 1996. The Comet assay for the evaluation of the genetic hazard of pollutants in cetaceans: preliminary results on the genotoxic effects of methyl-mercury on the bottlenosed dolphin (*Tursiops truncatus*) lymphocytes in vitro. *Marine Pollution Bulletin* 32: 545–548.

Betti, C., Davini, T. and Barale, R. 1992. Genotoxic activity of methylmercury chloride and dimethyl mercury in human lymphocytes. *Mutation Research* 281: 255–260.

Betti, C., Barale, R. and Pool-Zobel, B.L. 1993. Comparative studies on cytotoxic and genotoxic effects of two organic mercury compounds in lymphocytes and gastric mucosa cells of Sprague-Dawley rats. *Environmental and Molecular Mutagenesis* 22: 172–180.

Bhunya, S.P. and Jena, G.B. 1992. Genotoxic potential of the organochlorine insecticide lindane (gamma-BHC): an in vivo study in chicks. *Mutation Research* 272: 175–181.

Boon, J.P., Oostingh, I., van der Meer, J. and Hillebrand, T.J. 1994. A model for the accumulation of chlorobiphenyl congeners in marine mammals. *Europe Journal of Pharmacology* 270: 237–251.

Boon, J.P., Sleiderink, H.M., Helle, M.S., Dekker, M., van Schanke, A., Roex, E., Hillebrand, M., Theo, J., Klamer, H.J.C., Govers, B., Pastor, D., Morse, D., Wester, P.G. and de Boer, J. 1998. The use of microsomal in vitro assay to study phase I biotransformation of chlorobornanes (toxaphene) in marine mammals and birds. Possible consequences of biotransformation for bioaccumulation and genotoxicity. *Comparative Biochemistry and Physiology Part C* 121: 385–403.

Borrell, A., Aguilar, A., Corsolini, S. and Focardi, S. 1996. Evaluation of toxicity and sex-related variation of PCB levels in Mediterranean striped dolphins affected by an epizootic. *Chemosphere* 32: 2359–2369.

Cairns, J. 1975. Mutation selection and the natural history of cancer. *Nature* 255: 197–200.

Canonero, R., Brambilla Campart, G., Mattioli, F., Robbiano, L. and Martelli, A. 1997. Testing of *p*-dichlorobenzene and hexachlorobenzene for their ability to induce DNA damage and micronucleus formation in primary cultures of rat and human hepatocytes. *Mutagenesis* 12: 35–39.

Cantoni, O. and Costa, M. 1983. Correlations of DNA strand breaks and their repair with cell survival following acute exposure to mercury(II) and X-rays. *Molecular Pharmacology* 24: 84–89.

Cantoni, O., Christie, N.T., Swann, A., Drath, D.B. and Costa, M. 1984. Mechanism of HgCl$_2$ cytotoxicity in cultured mammalian cells. *Molecular Pharmacology* 26: 360–368.

Carrano, A.V. and Natarajan, A.T. 1988. Considerations for population monitoring using cytogenetic techniques. *Mutation Research* 204: 379–406.

Carvan, M.J. III, Flood, L.P., Campbell, B.D. and Busbee, D.L. 1995. Effects of benzo(a)pyrene and tetrachlorodibenzo(p)dioxin on fetal dolphin kidney cells: inhibition of proliferation and initiation of DNA damage. *Chemosphere* 30: 187–198.

Clark, T.P., Norstrom, R.J., Fox, G.A. and Won, H.T. 1987. Dynamics of organochlorine compounds in herring gulls (*Larus argentus*): II. A two-compartment model and data for ten compounds. *Environmental Contamination and Toxicology* 6: 547–559.

Crofton-Sleigh, C., Doherty, A., Ellard, S., Parry, E.M. and Venitt, S. 1993. Micronucleus assays using cytochalasin-blocked MCL-5 cells, a proprietary human cell line expressing five human cytochromes P-450 and microsomal epoxide hydroxylase. *Mutagenesis* 8: 363–372.

Dalcourt, M.-F., Béland, P., Pelletier, É. and Vigneault, Y. 1992. Caractérisation des communautés benthiques et étude des contaminants dans les aires fréquentées par le béluga du St. Laurent. *Rapport Technique Canadien des Sciences Halieutiques et Aquatiques* 1845.

DeFlora, S., Bennicelli, C. and Bagnasco, M. 1994. Genotoxicity of mercury compounds. A review. *Mutation Research* 317: 57–79.

DeGuise, S., Lagacé, A. and Béland, P. 1994. Tumors in St. Lawrence beluga whales (*Delphinapterus leucas*). *Veterinary Pathology* 31: 444–449.

DeGuise, S., Martineau, D., Béland, P. and Fournier, M. 1995. Possible mechanisms of action of environmental contaminants on St. Lawrence beluga whales (*Delphinapterus leucas*). *Environmental Health Perspectives* 103 (Suppl. 4): 73–77.

Duffaud, F., Osière, T., Villani, P., Pelissier, A.L., Volot, F., Favre, R. and Botta, A. 1997. Comparisons between micronucleated lymphocyte rates observed in healthy subjects and cancer patients. *Mutagenesis* 12: 227–231.

Eichhorn, G.L. 1981. The effect of metal ions on the structure and function of nucleic acids. In *Advances in Inorganic Biochemistry*, Vol. 3. Eichhorn, G.L. and Marzilli, G. (Eds). Elsevier/North-Holland, Amsterdam, pp. 1–46.

Ellard, S., Mohammed, Y., Dogra, S., Wölfel, C., Doehmer, J. and Parry, J.M. 1991. The use of genetically engineered V79 Chinese hamster cultures expressing rat liver *CYP1A1*, *1A2* and *2B1* cDNAs in micronucleus assays. *Mutagenesis* 6: 461–470.

Evans, H.J. 1984. Human peripheral blood lymphocytes for the analysis of chromosome aberrations in mutagen tests. In *Handbook of Mutagenicity Test Procedures*. 2nd edition. Kilbey, B.J., Legator, M., Nichols, W. and Ramel, C. (Eds). Elsevier Science Publishers, Amsterdam, pp. 405–427.

Evans, H.J. 1990. Cytogenetics. In *Mutation in the Environment. Part B*. Mendelsohn, M.L. and Albertini, R.J. (Eds). Wiley-Liss, New Yark, pp. 301–323.

Fenech, M. 1993. The cytokinesis-block micronucleus technique: a detailed description of the method and its application to genotoxic studies in human populations. *Mutation Research* 285: 35–44.

Fenech, M. and Morley, A.A. 1985. Measurement of micronuclei in lymphocytes. *Mutation Research* 147: 29–36.

Franchi, E., Loprieno, G., Ballardin, M., Ptrozzi, L. and Migliore, L. 1994. Cytogenetic monitoring of fishermen with environmental mercury exposure. *Mutation Research* 320: 23–29.

Gagné, F., Trottier, S., Blaise, C., Sproull, J. and Ernst, B. 1995. Genotoxicity of sediment extracts obtained in the vicinity of a creosote-treated wharf to rainbow trout hepatocytes. *Toxicology Letters* 78: 175–182.

Gauthier, J.M., Dubeau, H. and Rassart, É. 1998. Mercury induced micronuclei in skin fibroblasts of beluga whales. *Environmental Toxicology and Chemistry* 17: 2487–2493.

Gauthier, J.M., Dubeau, H. and Rassart, É. 1999a. Induction of micronuclei in vitro by organochlorine compounds in beluga whale skin fibroblasts. *Mutation Research* 439: 87–95.

Gauthier, J.M., Dubeau, H., Rassart, É., Jarman, W.M. and Wells, R.S. 1999b. Biomarkers of DNA damage in marine mammals. *Mutation Research* 444: 427–439.

Gauthier, J.M., Dubeau, H. and Rassart, É. 2002. Micronuclei induction by polycyclic aromatic hydrocarbons in isolated beluga whale skin fibroblasts. In *Molecular and Cell Biology of Marine Mammals*. Pfeiffer, C. (Eds). Krieger Publishing, Malabar, FL.

Gearing, J.N., Gearing, P.J., Noël, M. and Smith, J.N. 1994. Polycyclic aromatic hydrocarbons in sediment of the St. Lawrence estuary. Proceedings of the 20th Annual Workshop on Aquatic Toxicology, October 17–21, 1993, Québec. *Canadian Technical Report of Fisheries and Aquatic Sciences* 1989: 58–64.

Gilbert, M.M., Sprecher, J., Chang, L.W. and Meisner, L.F. 1983. Protective effect of vitamin E on genotoxicity of methylmercury. *Journal of Toxicology and Environmental Health* 12: 767–773.

Gobeil, C. and Cossa, D. 1993. Mercury in sediments and sediment pore water in the Laurentian trough. *Canadian Journal of Aquatic Science* 50: 1794–1800.

Gulland, F.M.D., Trupkiewicz, J.G., Spraker, T.R. and Lowenstine, L.J. 1996. Metastatic carcinoma of probable transitional cell origin in 66 free-living California sea lions (*Zalophus californianus*), 1979 to 1994. *Journal of Wildlife Disease* 32: 250–258.

Hagmar, L., Brogger, A., Hansteen, I.L., Heim, S., Hogsted, B., Knudsen, L., Lambert, B., Linnainmaa, K., Mitelman, F., Nordenson, I., Reuterwall, C., Salomaa, S., Skerfving, S. and Sorsa, M. 1994. Cancer risk in humans predicted by increased levels of chromosomal aberrations in lymphocytes: Nordic study group on the health risk of chromosome damage. *Cancer Research* 54: 2919–2922.

Harvey, R.G. 1991. *Polycyclic Aromatic Hydrocarbons, Chemistry and Carcinogenicity*. Cambridge University Press, Cambridge.

Haugen, D.A. and Peak, M.J. 1983. Mixtures of polycyclic aromatic hydrocarbons inhibit mutagenesis in the Salmonella/microsome assay by inhibition of metabolic activation. *Mutation Research* 116: 257–269.

He, S.L. and Baker, R. 1991. Micronuclei in mouse skin cells following in vivo exposure to benzo[*a*]pyrene, 7,12-dimethylbenz[*a*]anthracene, chrysene, pyrene and urethane. *Environmental Molecular Mutagenesis* 17: 163–168.

Hellou, J., Stenson, G., NI, I.-H. and Payne, J.F. 1990. Polycyclic aromatic hydrocarbons in muscle tissue of marine mammals from the northwest Atlantic. *Marine Pollution Bulletin* 21: 469–473.

Hemminki, K. and Vineis, P. 1985. Extrapolation of the evidence on teratogenicity of chemicals between humans and experimental animals: chemicals other than drugs. *Teratogenesis, Carcinogenesis and Mutagenesis* 5: 251–318.

Hennig, U.G.G., Rudd, N.L. and Hoar, D.I. 1988. Kinetochore immunofluorescence in micronuclei: a rapid method for the in situ detection of aneuploidy and chromosome breakage in human fibroblasts. *Mutation Research* 203: 405–414.

Hongell, K. 1996. Chromosome survey of seals in the Baltic Sea in 1988–1992. *Archives of Environmental Contamination Toxicology* 31: 399–403.

Hooper, N.K., Ames, B.N., Saleh, M.A. and Casida, J.E. 1979. Toxaphene, a complex mixture of polychloroterpenes and a major insecticide, is mutagenic. *Science* 205: 591–592.

IARC (International Agency for Research on Cancer). 1983. *Polycyclic aromatic hydrocarbons, Part 1, Chemical, Environmental and Experimental Data*, Vol. 32. IARC, Lyon, pp. 39–91.

IARC 1984. *Polycyclic aromatic hydrocarbons, Part 3, Industrial exposures in aluminium production, coal gasification, coke production, and iron and steel founding*, Vol. 34. IARC, Lyon, pp. 37–64.

IARC 1987. *Monographs on the Evaluation of Carcinogenic Risk to Humans*. World Health Organization (WHO). Suppl. 7. WHO, Lyon.

IARC 1993. Mercury and mercury compounds. In *IARC Monograph on the Evaluation of the Carcinogenic Risk of Chemicals to Humans*, Vol. 58. IARC, Lyon, pp. 239–345.

Itano, K., Kawai, S., Miyazaki, N., Tatsukawa, R. and Fujiyama, T. 1984. Body burden and distribution of mercury and selenium in striped dolphins. *Agricultural and Biological Chemistry* 48: 1117–1121.

Jenssen, B.M. 1996. An overview of exposure to, and effects of, petroleum oil and organochlorine pollution in grey seals (*Halichoerus grypus*). *Science of the Total Environment* 186: 109–118.

Kannan, K., Tanabe, S., Borrell, A., Aguilar, A., Focardi, S. and Tatsukawa, R. 1993. Isomer-specific analysis and toxic evaluation of polychlorinated biphenyls in striped dolphins affected by an epizootic in the western Mediterranean sea. *Archives of Environmental Contamination and Toxicology* 25: 227–233.

Kelly-Garvet, F. and Legator, M.S. 1973. Cytogenetic and mutagenic effects of DDT and DDE in a Chinese hamster cell line. *Mutation Research* 17: 223–229.

Kuehl, D.W. and Haebler, R. 1995. Organochlorine, organobromine, metal, and selenium residues in bottlenose dolphins (*Tursiops truncatus*) collected during an unusual mortality event in the Gulf of Mexico, 1990. *Archives of Environmental Contamination and Toxicology* 28: 494–499.

Kuehl, D.W., Haebler, R. and Potter, C. 1991. Chemical residues in dolphins from the U.S. Atlantic coast including Atlantic bottlenose obtained during the 1987/88 mass mortality. *Chemosphere* 22: 1071–1084.

Larsen, K.D. and Jalal, S.M. 1974. DDT induced chromosome mutations in mice – Further testing. *Canadian Journal of Genetics and Cytology* 16: 491–497.

Leonzio, C., Focardi, S. and Fossi, C. 1992. Heavy metals and selenium in stranded dolphins of the Northern Tyrrhenian Sea (NW Mediterranean). *Science of the Total Environment* 119: 74–77.

424   *J.M. Gauthier* et al.

Lessa, J.M.M., Berçak, W., Nazareth Rabello, M., Pereika, C.A.B. and Ungaro, M.T. 1976. Cytogenetic study of DDT on human lymphocytes in vitro. *Mutation Research* 40: 131–138.

Maddle, S., Dean, S.W., Andrae, U., Brambilla, G., Burlinson, B., Doolittle, D.J., Furihata, C., Hertner, T., McQueen, C.A. and Mori, H. 1994. Recommendations for the performance of UDS tests in vitro and in vivo. *Mutation Research* 312: 263–285.

Mahr, U. and Miltenburger, H.G. 1976. The effect of insecticides on Chinese hamster cell cultures. *Mutation Research* 40: 107–118.

Majumdar, S.K., Kopelman, H.A. and Schnitman, M.J. 1976. Dieldrin-induced chromosome damage in mouse bone-marrow and WI-38 human lung cells. *Journal of Heredity* 67: 303–307.

Mamber, S.W., Bryson, V. and Katz, S.E. 1984. Evaluation of the *Escherichia coli* K12 inductest for detection of potential chemical carcinogens. *Mutation Research* 130: 141–151.

Martel, L., Gagnon, M.J., Massé, R., Leclerc, A. and Tremblay, L. 1986. Polycyclic aromatic hydrocarbons in sediments from the Saguenay Fjord, Canada. *Bulletin of Environmental Contamination and Toxicology* 37: 133–140.

Martineau, D., DeGuise, S., Fournier, M., Shugart, L., Girard, C., Lagacé, A. and Béland, P. 1994. Pathology and toxicology of beluga whales from the St. Lawrence estuary, Québec, Canada. Past, present and future. *Science of the Total Environment* 154: 201–215.

Martineau, D., Lair, S., De Guise, S., Lipscomb, T. and Béland, P. 1999. Cancer in beluga whales from the St. Lawrence Estuary, Quebec, Canada: A potential biomarker of environmental contamination. *Journal of Cetacean Research and Management* (Special Issue 1): 249–265 (previously 'Reports of the International Whaling Commission').

Martoja, R. and Berry, J.-P. 1980. Identification of tiemannite as a probable product of demethylation of mercury by selenium in cetaceans. A complement to the scheme of the biological cycle of mercury. *Vie Milieu* 30: 7–10.

Marvin, C.H., Allan, L., McCarry, B.E. and Bryant, D.W. 1993. Chemico/biological investigations of contaminated sediments in western lake Ontario. *Environmental Molecular Mutagenesis* 22: 61–70.

Mathieu, A., Payne, J.F., Fancey, L.L., Santella, R.M. and Young, T.L. 1997. Polycyclic aromatic hydrocarbon-DNA adducts in beluga whales from the Arctic. *Journal of Toxicology and Environmental Health* 51: 1–4.

Matsumura, F. 1985. *Toxicology of Insecticides.* 2nd edition. Plenum Press, New Yark, pp. 233–237.

Meisner, L.F., Roloff, B., Sargent, L. and Pitot, H. 1992. Interactive cytogenetic effects on rat bone-marrow due to chronic ingestion of 2,5,2′,5′ and 3,4,3′,4′ PCBs. *Mutation Research* 283: 179–183.

Michaud, R. 1993. Distribution estivale du béluga du St. Laurent: synthèse 1986 à 1992. *Rapport Technique Canadien des Sciences Halieutiques et Aquatiques* 1906.

Miller, C.T., Zawidzka, Z., Nagy, E. and Charbonneau, S.M. 1979. Indicators of genetic toxicity in leucocytes and granulocytic precursors after chronic methymercury ingestion by cats. *Bulletin of Environmental Contamination and Toxicology* 21: 296–303.

Morimoto, K., Iijima, S. and Koizumi, A. 1982. Selenite prevents induction of sister-chromatid exchanges by methyl mercury and mercuric chloride in human whole-blood cultures. *Mutation Research* 102: 183–192.

Muir, D.C.G., Ford, C.A., Stewart, R.E.A., Smith, T.G., Addison, R.F., Zinck, M.E. and Béland, P. 1990. Organochlorine contaminants in belugas, *Delphinapterus leucas*, from Canadian waters. *Canadian Bulletin of Fisheries and Aquatic Sciences* 224: 165–190.

Muir, D.C.G., Ford, C.A., Rosenberg, B., Norstrom, R.J., Simon, M. and Béland, P. 1996. Persistent organochlorines in beluga whales (*Delphinapterus leucas*) from the St. Lawrence river estuary. I. Concentrations and patterns of specific PCBs, chlorinated pesticides and polychlorinated dibenzo-p-dioxins and -dibenzofurans. *Environmental Pollution* 93: 219–234.

Munson, L., Calzada, N., Kennedy, S. and Sorenson, T.B. 1998. Lutenized ovarian cysts in Mediterranean striped dolphins. *Journal of Wildlife Disease* 34: 656–660.

Narbonne, J.F., Garrigues, P., Ribera, D., Raoux, C., Mathieu, A., Lemaire, P., Salaun, J.P. and Lafaurie, M. 1991. Mixed-function oxygenase enzymes as tools for pollution monitoring: field studies on the French coast of the Mediterranean Sea. *Comparative Biochemistry and Physiology C* 100: 37–42.

O'Connor, T.P. 1991. Concentrations of organic contaminants in mollusks and sediments at NOAA National Status and Trend sites in the coastal and estuarine United States. *Environmental Health Perspectives* 90: 69–73.

Ogura, H., Takeuchi, T. and Morimoto, K. 1996. A comparison of the 8-hydroxydeoxyguanosine, chromosome aberrations and micronucleus techniques for the assessment of the genotoxicity of mercury compounds in human blood lymphocytes. *Mutation Research* 340: 175–182.

Olsson, M., Karlsson, B. and Ahnland, E. 1994. Diseases and environmental contaminants in seals from the Baltic and Swedish west coast. *Science of the Total Environment* 154: 217–227.

Önfelt, A. 1986. Mechanistic aspects on chemical induction of spindle disturbances and abnormal chromosome numbers. *Mutation Research* 168: 249–300.

Oshimura, M. and Barrett, C. 1986. Chemically induced aneuploidy in mammalian cells: mechanisms and biological significance in cancer. *Environmental Mutagenesis* 8: 129–159.

Palmer, K.A., Green, S. and Legator, M.S. 1972. Cytogenetic effects of DDT and derivates of DDT in a cultured mammalian cell line. *Toxicology and Applied Pharmacology* 22: 355–364.

Palmisano, F., Cardellichio, N. and Zambonin, P.G. 1995. Speciation of mercury in dolphin liver: A two-stage mechanism for the demethylation accumulation process and role of selenium. *Marine Environmental Research* 40: 109–121.

Papoulias, D.M. and Buckler, D.R. 1996. Mutagenicity of Great Lakes sediments. *Journal of Great Lakes Research* 22: 591–601.

Parry, J.M., Parry, E.M., Bourner, R., Doherty, A., Ellard, S., O'Donovan, J., Hoebee, B., de Stoppelaar, J.M., Mohn, G.R., Önfelt, A., Renglin, A., Shultz, N., Söderpalm-Berndes, C., Jensen, K.G., Kirsch-Volders, M., Elhajouji, A., van Hummelen, P., Degrassi, F., Antoccia, A., Cimini, D., Izzo, M., Tanzarella, C., Adler, I.-D., Kliesch, U., Schriever-Schwemmer, G., Gasser, P., Crebelli, R., Cadere, A., Andreoli, A., Benigni, R., Leopardi, P., Marcon, F., Zinjo, Z., Natarajan, A.T., Beoi, J.J.W.A., Kappas, A., Voutsinas, G., Zarani, F.E., Patrinelli, A., Pachierotti, F., Tiveron, C. and Hess, P. 1996. The detection and evaluation of aneugenic chemicals. *Mutation Research* 353: 11–46.

Pelkonen, O. and Nebert, D.W. 1982. Metabolism of polycyclic aromatic hydrocarbons: Etiologic role in carcinogenesis. *Pharmacology Review* 34: 189–222.

Perera, F.P., Hemminki, K., Gryzbowska, E., Motykiewicz, G., Michalska, J., Santella, R.M., Young, T.-L., Dickey, C., Brandt-Rauf, P., DeVivo, I., Blaner, W., Tsai, W.-Y. and Chorazy, M. 1992. Molecular and genetic damage in humans from environmental pollution in Poland. *Nature* 360: 256–258.

Pham, T., Lum, K. and Lemieux, C. 1993. The occurence, distribution, and sources of DDT in the St. Lawrence river, Québec (Canada). *Chemosphere* 26: 1595–1606.

Poirier, M.C. and Béland, F.A. 1992. DNA adduct measurements and tumor incidence during chronic carcinogen exposure in animal models: Implications for DNA adduct-based human cancer risk assessments. *Chemical Research in Toxicology* 5: 749–755.

Quémarais, B., Lemieux, C. and Lum, K.R. 1994. Concentrations and sources of PCBs and organochlorine pesticides in the St. Lawrence river (Canada) and its tributaries. *Chemosphere* 29: 591–610.

Reddy, T.V., Stober, J.A., Olson, G.R. and Daniel, F.B. 1991. Induction of nuclear anomalies in the gastrointestinal tract by polycyclic aromatic hydrocarbons. *Cancer Letters* 56: 215–224.

Reddy, M., Echols, S., Finklea, B., Busbee, D., Reif, J. and Ridgway, S. 1998. PCBs and chlorinated pesticides in clinically healthy *Tursiops truncatus*: relationships between levels in blubber and blood. *Marine Pollution Bulletin* 36: 892–903.

Reifferschield, G. and Heil, J. 1996. Validation of the SOS/*umu* test using test results of 486 chemicals and comparison with the Ames test and carcinogenicity data. *Mutation Research* 369: 129–145.

Safe, S. 1989. Polychlorinated biphenyls (PCBs): mutagenicity and carcinogenicity. *Mutation Research* 220: 31–47.

Safe, S. 1993. Toxicology, structure–function relationship, and human and environmental health impacts of polychlorinated biphenyls: progress and problems. *Environmental Health Perspectives* 100: 259–268.

Sakar, D., Sharma, A. and Talukder, G. 1993. Differential protection of chlorophyllin against clastogenic effects of chromium and chlordane in mouse bone marrow. *Mutation Research* 301: 33–38.

Saleh, M.A. 1991. Toxaphene: chemistry, biochemistry, toxicity and environmental fate. *Reviews in Environmental Contamination and Toxicology* 118: 1–76.

Samosh, L.V. 1974. Chromosome aberration and character of satellite associations after accidental exposure of the humans to polychlorocamphene. *Tsitologiya i Genetika* 8: 24–27.

Sargent, L., Roloff, B. and Meisner, L. 1989. In vitro chromosome damage due to PCB interactions. *Mutation Research* 224: 79–88.

Schrenk, D. 1998. Impact of dioxin-type induction of drug-metabolizing enzymes on the metabolism of endo- and xenobiotics. *Biochemical Pharmacology* 55: 1155–1162.

Singh, N.P., McCoy, M.T., Tice, R.R. and Schneider, E.L. 1988. A simple technique for quantification of low levels of DNA damage in individual cells. *Experimental Cell Research* 175: 184–191.

Skerfving, S., Hansson, K., Mangs, C., Lindsten, J. and Ryman, N. 1974. Methylmercury-induced chromosome damage in man. *Environmental Research* 7: 83–98.

Snyder, R.D. 1988. Role of active oxygen species in metal induced DNA breakage in human diploid fibroblasts. *Mutation Research* 193: 237–246.

Sobti, R.C., Krishan, A. and Davies, J. 1983. Cytokinetic and cytogenetic effect of agricultural chemicals on human lymphoid cells in vitro. II. Organochlorine pesticides. *Archives of Toxicology* 52: 221–231.

Springer, D.L., Mann, D.B., Dankovic, D.A., Thomas, B.L., Wright, C.W. and Mahlum, D.D. 1989. Influences of complex organic mixtures of tumor-initiating activity, DNA binding and adducts of benzo[*a*]pyrene. *Carcinogenesis* 10: 131–137.

Steinel, H.H., Arlauskas, A. and Baker, R.S.U. 1990. SCE induction and cell-cycle delay by toxaphene. *Mutation Research* 230: 29–33.

Tanabe, S., Tatsukawa, R., Tanaka, H., Maruyama, K., Miyazaki, N. and Fujiyama, T. 1981. Distribution and total burdens of chlorinated hydrocarbons in bodies of striped dolphins (*Stenella coeruleoalba*). *Agricultural and Biological Chemistry* 45: 2569–2578.

Tucker, J.D. and Preston, R.J. 1996. Chromosome aberrations, micronuclei, aneuploidy, sister chromatid exchanges, and cancer risk assessment. *Mutation Research* 365: 147–159.

Vahter, M., Mottet, N.K., Friberg, L., Lind, B., Shen, D.D. and Burbacher, T. 1994. Speciation of mercury in the primate blood and brain following long-term exposure to methylmercury. *Toxicology and Applied Pharmacology* 124: 221–229.

Verschaeve, L., Kirsch-Volders, M., Hens, L. and Susanne, C. 1985. Comparative in vitro studies in mercury-exposed human lymphocytes. *Mutation Research* 157: 221–226.

Vian, L., Bichet, N. and Gouy, D. 1993. The in vitro micronucleus test on isolated human lymphocytes. *Mutation Research* 291: 93–102.

Vienneau, D.S., DeBoni, U. and Wells, P.G. 1995. Potential genoprotective role for UDP-glucoronyltransferases in chemical carcinogenesis: initiation of micronuclei by benzo[*a*]pyrene and benzo[*e*]pyrene in UDP-glucoronyltransferase-deficient cultured rat skin fibroblasts. *Cancer Research* 55: 1045–1051.

Vladykov, V.D. 1946. Etudes sur les mammifères aquatiques. IV. Nourriture du marsouin blanc ou béluga (*Delphinapterus leucas*) du fleuve St-Laurent. *Contribution au Département des Pêcheries de la Province de Québec* 14: 1–194.

Wagemann, R. and Muir, D.C.G. 1984. Heavy metals and organochlorine in marine mammals of northern waters: Overview and evaluation. *Canadian Technical Report of Fisheries and Aquatic Sciences* 1279.

Wagemann, R., Stewart, R.E.A., Béland, P. and Desjardins, C. 1990. Heavy metals and selenium in tissues of beluga whales, *Delphinapterus leucas*, from the Canadian Arctic and the St. Lawrence estuary. *Canadian Journal of Fisheries and Aquatic Sciences* 224: 191–206.

Warshawsky, D., Livingston, G.K., Fonouni-Fard, M. and LaDow, K. 1995. Induction of micronuclei and sister chromatid exchanges by polycyclic and N-heterocyclic aromatic hydrocarbons in cultured human lymphocytes. *Environmental Molecular Mutagenesis* 26: 109–118.

Watanabe, S., Shimada, T., Nakamura, S., Nishiyama, N., Tanabe, S. and Tatsukawa, R. 1989. Specific profile of liver microsomal cytochrome P-450 in dolphins and whales. *Marine Environmental Research* 27: 51–65.

Wells, R.S. and Scott, M.D. 1998. Bottlenose dolphins. In *Handbook of Marine Mammals, Volume 6, The Second Book of Dolphins and Porpoises*. Ridgway, S.H. and Harrison, R.J. (Eds). Academic Press, San Diego, CA.

White, P.A., Blaise, C. and Rasmussen, J.B. 1997. Detection of genotoxic substances in bivalve molluscs from the Saguenay Fjord (Canada), using the SOS chromotest. *Mutation Research* 392: 277–300.

WHO (World Health Organization) 1984. Chlordane. *Environmental Health Criteria* 34. WHO, Geneva.

Winker, N., Weniger, P., Klein, W., Ott, E., Kocsis, F., Shoket, B. and Korpet, K. 1995. Detection of polycyclic aromatic hydrocarbon exposure damage using different methods in laboratory animals. *Journal of Applied Toxicology* 15: 59–62.

Yamada, H., Miyahara, T., Kozuka, H., Matsuhashi, T. and Sasaki, Y.F. 1993. Potentiating effects of organomercuries on clastogen-induced chromosome aberrations in cultured Chinese hamster cells. *Mutation Research* 290: 281–291.

16 Mechanisms of aromatic hydrocarbon toxicity: implications for cetacean morbidity and mortality

Michael J. Carvan III and David L. Busbee

Introduction

Industrial accidents, agricultural run-off, atmospheric drift and volatization, leaching from toxic waste dumps, and various types of chemical dumping have introduced large quantities of xenobiotic pollutants into the oceans, which include some of our most fragile and productive ecosystems. Although the long-term effects of these pollutants on ecosystem health and the health of specific organisms are largely unknown, every level of life is potentially adversely affected by xenobiotic pollutants. Those effects may be most immediate, and perhaps most critical, in marine mammals, classic upper trophic level carnivores (Gaskin, 1982). Although dolphins and other marine mammals feeding at the top of the food chain bioconcentrate lipophilic organic pollutants in their fatty tissues, almost no data are available on the multiple adverse effects of pollutants in these animals. Two major classes of environmental hydrocarbon pollutants have been suggested as potential etiologic factors in disease susceptibility, pathology and mass mortality of marine mammals, the halogenated aromatic hydrocarbons (HAHs) and polycyclic aromatic hydrocarbons (PAH) (Tanabe *et al.*, 1987, 1989; Tanabe, 1989; Kannan *et al.*, 1989, 1993; Osterhaus *et al.*, 1995; Borell *et al.*, 1996). It is our objective to discuss aromatic hydrocarbons and their metabolic and physiological effects in cetaceans. Most of our discussions will be limited to the smaller odontocetes (primarily the Delphinidae, Phocoenidae, and Monodontidae) based on the wealth of information from this group of cetaceans, but most of our conclusions can be generalized to the other cetacean families.

Aromatic hydrocarbons

The HAHs, most of which are agricultural, industrial or combustion by-products, have been studied extensively and occur in all global areas, from the Arctic to the tropics (Safe, 1990, 1991; Tatsukawa, 1992). The HAH family is extensive. Among these compounds are the halogenated-biphenyls (PCBs and PBBs), dibenzofurans (PCDFs), and dibenzo-*p*-dioxins (PCDDs), the most toxic member of which is 2,3,7,8-tetrachlorodibenzo-*p*-dioxin (TCDD). Dichlorodiphenyl-trichloroethane,1,1,1-trichloro-2,2-*bis*[*p*-chlorophenyl]ethane (DDT) and its metabolites, although no longer used in developed countries, is a prototypical halogenated pesticide evaluated in essentially all studies of the HAHs. The most toxic HAHs tend to be highly stable and lipophilic and can be concentrated in the fatty tissues, leading to their bioaccumulation within the food chain. The PAHs constitute a group of related aromatic chemicals with pyrene, anthracene or phenanthrene nuclei, and include chemicals such as 3-methylcholanthrene (MC), benzo[*a*]pyrene (B[a]P), benzo[*a*]anthracene, dibenz[*a,h*]anthracene, and related compounds. Rapid metabolism of PAHs to highly reactive intermediates significantly reduces their bioaccumulation, but increases their reactivity with cellular macromolecules.

The HAHs and PAHs induce characteristic physiological response patterns that initiate disease in exposed mammals. This is due to their role in the regulation of gene expression and the biochemical reactivity of their metabolites. These disorders include a wasting syndrome, immunotoxicity, neurotoxicity, hepatotoxicity, reproductive dysfunction/developmental toxicity and carcinogenicity. Among the most alarming actions of HAHs and PAHs in animals is their initiation of fetotoxicity, which appears due largely to: (1) their induction of cytochrome P450s (CYPs) and other drug-metabolizing enzymes, and (2) their pattern of endocrine disruption or altered expression of endocrine-regulated genes (Poland and Knutson, 1982; Safe, 1986, 1990; Goldstein and Safe, 1989; Whitlock, 1990; Abbott *et al.*, 1994; Birnbaum, 1995; Hurst *et al.*, 1998). The degree to which these toxic phenomena are exhibited is species-specific; hence there are significant limitations in generalizing effects between species (Poland and Knutson, 1982; Safe, 1986; Kimbrough, 1987; Gonzalez, 1989).

Over two decades ago the use of pollution-related CYP induction was proposed as a potential early warning marker capable of fulfilling the requirements of 'most sensitive biological response' in monitoring exposure to a variety of organic contaminants (Payne *et al.*, 1987). Numerous studies and large volumes of data now suggest that CYP induction is only one component of the HAH/PAH biological response. Other components should include the relationships between CYP induction, altered gene expression in general, and endocrine disruption as facets of the biological indication of exposure to xenobiotic pollutants.

Either the acute exposure to PAH and HAH or the bioaccumulation of potentially harmful concentrations of HAH may lead to alterations in gene expression and cellular physiology. Bioaccumulation or bioconcentration of hydrocarbons may lead to physiological changes, even though environmental concentrations of the chemicals are very low. Bioaccumulated HAHs are generally sequestered in the relatively nonmetabolic fatty tissues of the body, where they have been presumed, probably incorrectly, to be inert. These stored contaminants are typically redistributed when an animal is mobilizing its lipid stores for use as an energy source, which can result in significantly elevated concentrations in serum and/or milk (Korytko *et al.*, 1999). Ideally, an exposed animal will be protected from pollutants by enzyme systems that metabolize chemicals to less toxic forms that can be excreted.

Metabolism of aromatic hydrocarbons

Two major enzyme groups, the Phase I, or oxidative enzymes, and the Phase II, or conjugation enzymes, carry out biotransformation of hydrocarbons. Enzymes of these systems metabolize chemicals to more hydrophilic forms that can be eliminated. The Phase I enzymes include cytochrome P450s, epoxide hydrolase, amine oxidase, peroxidases, various esterases and amidases, and aldehyde, alcohol and ketone oxidation–reduction systems. Of the Phase I enzymes, the rather extensive family of cytochrome P450s (CYPs by gene nomenclature convention), with more than 700 individual isoforms from 67 families identified in a variety of animals (Omiecinski *et al.*, 1999; *http://drnelson.utmem.edu/CytochromeP450.html*), are the most relevant to HAH and PAH metabolism. CYP-mediated metabolism of HAH and PAH generates electrophilic intermediates, which can be conjugated by Phase II enzymes. Epoxide hydrolase converts the electrophilic and potentially damaging arene oxides and epoxides formed as CYP-generated metabolites by hydrating the reactive intermediates. The Phase II enzymes include glutathione *S*-transferases, glucuronosyl transferases, sulfo-transferases, methyl trans-ferases, *N*-acetyl transferases, amino acid conjugation enzymes and rhodanese. Glutathione *S*-transferases are the Phase II enzymes most commonly asso-ciated with PAH conjugation, detoxication and elimination. Expression of both the inducible Phase I and II enzymes occurs when cells are exposed to suitable concentrations of hydrocarbons that interact as ligands with the Ah receptor (AHR). The ligand-bound AHR dimerizes with the AHR nuclear transporter (ARNT) and the complex binds to the xenobiotic response elements (XRE) that regulate transcription of hydrocarbon-inducible genes. Typically, inducing agents are also substrates for the Phase I enzymes.

Most of the lipophilic hydrocarbons to which an animal is exposed are metabolized and excreted, or bioconcentrate in fatty tissues. Formation of reactive intermediates is a normal part of the overall process of hydrocarbon metabolism. Reactive hydrocarbon metabolites are generally detoxified pro-vided there is a balance between the rates of formation and detoxication. When the balance is disturbed, electrophilic HAH and PAH metabolites may not be immediately conjugated and excreted, and can cause cellular injury due to their interaction with essential macromolecules (Sipes and Gandolfi, 1991; Cavalieri and Rogan, 1992) (Figure 16.1). The rates of forma-tion and detoxication of reactive intermediates can be altered by inducing CYP enzymes (leading to excess production of reactive intermediates), or by diminishing the capacity for Phase II detoxication. Exposure to high concentrations of HAHs or PAHs can reduce the rate of reactive metabolite

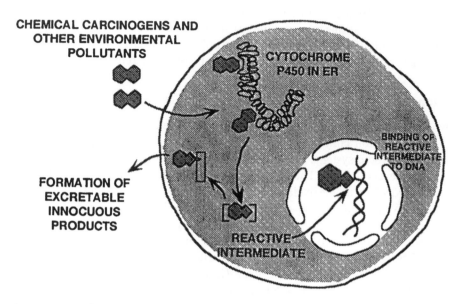

Figure 16.1 Scheme for the metabolic activation of HAH and PAH by cytochrome
P450s in the smooth endoplasmic reticulum (ER) in a typical mammalian cell to form reactive intermediates that bind to nucleophilic centers
in DNA. (Redrawn after Friedberg, 1985.)

removal by saturating the major biotransformation pathways that produce
nonreactive metabolites, or by depleting the cellular conjugation systems
that remove reactive metabolites. If this occurs, minor metabolic pathways,
which produce reactive metabolites, may play a major role in the generation
of toxic intermediates. Under conditions where detoxication pathways are
compromised, normally nontoxic doses of a xenobiotic can result in cellular
injury.

Cytochromes P450

The microsomal mixed-function monooxygenase enzymes, cytochrome P450s,
are also essential for the metabolism of a variety of normal substrates, such
as steroid hormones and fatty acids. Xenobiotic induction of a variety of
CYP isoforms, and of other enzyme systems, may not only lead to increased
rates of lipid peroxidation and free-radical generation, but may also alter
hormone metabolism and disrupt endocrine regulation of physiological systems (Luster *et al.*, 1988; Abbot and Birnbaum, 1990; Bookstaff *et al.*, 1990;
Goldstein *et al.*, 1990; Harris *et al.*, 1990; Nebert *et al.*, 1990). The HAH
and PAH may act as either agonists or antagonists for AHR-regulated gene
expression. Further, exposure to hydrocarbons that are AHR-interactive may
also result in a number of changes that are not metabolism-dependent. These
include alteration of normal receptor-mediated biochemical pathways and the

disruption of normal calcium-mediated homeostatic mechanisms (Nicotera *et al.*, 1989; Nebert *et al.*, 1990). Exposure may also lead to increased formation of macromolecular adducts and may result in altered biochemical and signal transduction pathways and in generalized oxidative stress on the cell (Dalton *et al.*, 1998). These molecular abnormalities are correlated at the tissue level with cell death, necrosis, malignancy, immune system dysfunction, decreased fecundity, fetotoxy or developmental abnormalities.

The CYP enzymes are membrane-bound terminal monooxygenases in the mixed-function oxygenase (MFO) system of the smooth endoplasmic reticulum. The CYP enzymes are spatially associated with NADPH cytochrome P450 reductase (P450 reductase). Individually, the CYPs are single-chain polypeptides, ranging in molecular weight from 45 to 60 kDa, and containing a noncovalently bound protoporphyrin ring moiety. The spectral properties of the CYPs are due to a thiolate ligand involving cysteine, the fifth ligand of the heme iron (Black and Coon, 1988). There are several CYP molecules associated with each molecule of NADPH cytochrome P450 reductase (Figure 16.2). The mixed-function monooxygenase electron transfer system consists of CYP, P450 reductase, cytochrome b_5 and cytochrome b_5 reductase, and is capable of metabolizing a variety of substrates with differing degrees of specificity and efficiency. In CYP-mediated metabolism,

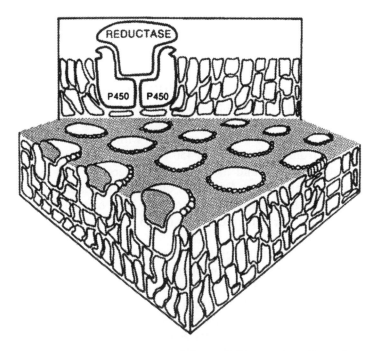

Figure 16.2 Schematic demonstration of the spatial interaction of cytochrome P450, cytochrome P450 reductase and the lipid bilayer. Note the low ratio of P450 reductase to P450. (Redrawn after Nebert *et al.*, 1981.)

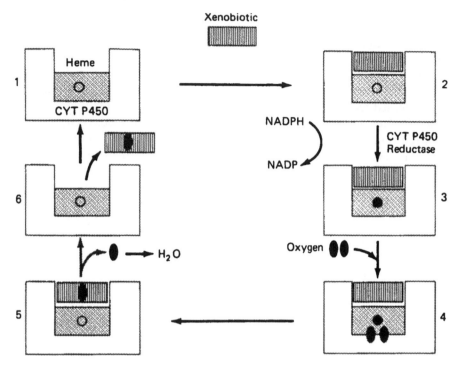

Figure 16.3 Cytochrome P450 electron transport and oxidation of a xenobiotic (Sipes and Gandolfi, 1991; reproduced with permission from Pergamon Press).

the substrate combines with oxidized CYP (Fe^{3+}) to form a substrate–CYP complex (Figure 16.3). This complex acts as an electron acceptor for an NADPH-derived electron (via P450 reductase) reducing the CYP heme moiety to Fe^{2+}. The reduced complex combines with molecular oxygen and another electron from NADPH (via P450 reductase, or in some instances NADH via cytochrome b_5 and its reductase), generating an unstable and highly reactive oxygen species. One atom of the oxygen is introduced into the substrate, while the other is reduced to water. The oxygenated substrate then dissociates from the complex, and oxidized CYP is regenerated (Sipes and Gandolfi, 1991).

Cytochrome P450s are, in general, not highly substrate-specific. This is a function of the high degree of homology between CYPs from the same families (Nebert and McKinnon, 1994; Nelson *et al.*, 1996) dictating varying degrees of affinity of CYP isoforms for a wide variety of related substrates. These substrates include endobiotic compounds such as steroid hormones, vitamins, prostaglandins, ketones, fatty acids and bile acids, and xenobiotic compounds such as drugs, pesticides, HAH and PAH. A large number of different CYP genes, and/or the corresponding cDNAs, each encoding proteins with different substrate affinities, have been sequenced from humans, rodents, rabbits, chickens and fish, and have been classified into families based on sequence similarities (Nelson *et al.*, 1996; Omiecinski *et al.*, 1999; *http://drnelson.utmem.edu/CytochromeP450.html*). These genes are thought to have arisen by duplication and divergent evolution from a single bacterial gene. It has been suggested that habitation of land and the subsequent consumption of terrestrial plants selected for the extensive diversification of CYP genes in animals in response to the quantity of toxic flavones in terrestrial plants relative to those from aquatic sources (Nelson and Strobel, 1987). This evolutionary process increased the variety of substrates that animals can metabolize. If CYP isozymes were unavailable to facilitate metabolism, and allow the excretion of hydrocarbons, toxic levels would eventually be reached in susceptible tissues through continued exposure. Alternatively, if the specific isozymes are present, toxic levels of the parent compound may not be reached because the chemical is metabolized and excreted.

CYP1A1 (cytochrome P450 gene family 1, subfamily A, gene 1), is the specific CYP that is most commonly associated with HAH and PAH metabolism. This is one of several related CYP enzymes which, as a group, were previously named aryl hydrocarbon hydroxylase, AHH. AHH activity has been highly conserved in mammals and can be induced in a multitude of organisms, ranging from fungi to insects, fish, birds and mammals.

The biotransformation of aromatic hydrocarbon contaminants in vertebrates is due largely to two major families of CYP, CYP1A and CYP2B. Hydrocarbons may be inducers (agonists), substrates and/or inhibitors (antagonists) of CYP induction, generally increasing their own metabolism by inducing the expression of genes, resulting in increased amounts of CYP within exposed cells. In the past, CYP inducers have been categorized as either 3-methylcholanthrene-type (MC-type), phenobarbital-type (PB-type), or mixed-type. The MC-type compounds include PAHs, PCDDs, PCDFs and non-*ortho*-substituted PCBs, and induce the expression and activity of CYP1A and CYP1B isozymes. The PB-type inducers, which increase the expression of CYP2B isozymes, include DDT and *ortho*-substituted PCBs. These compound classes are referred to as CYP1A- and CYP2B-type inducers to more accurately reflect their induction activities.

Hydrocarbon induction of increased CYP activity is not associated with the activation of existing cellular proteins; rather, it requires *de novo* protein synthesis resulting from increased gene expression (Okey, 1990). Some compounds, however, appear to enhance the stability of CYP mRNAs and proteins (Gonzalez, 1989) regulating the magnitude of induction. The induction of CYP1A1 via the AHR has been studied in great detail. The AHR is a multimeric protein complex, which binds TCDD, B[a]P, MC, and other structurally related HAHs and PAHs with high affinity (Okey and Vella, 1982; Manchester *et al.*, 1987). It has long been hypothesized that AHR plays a key role in mediating the diverse spectrum of sex-, strain-, age- and species-specific responses elicited by AHR-binding ligands (Poland and Knutsen, 1982; Whitlock, 1987). Analysis of AHR 'knockout' mice, in which

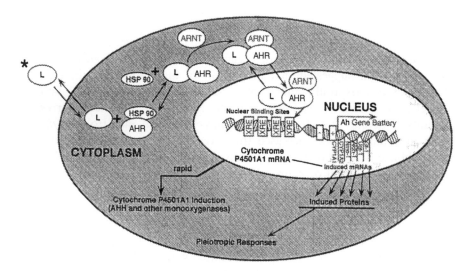

Figure 16.4 Proposed mechanism of induction of CYP1A1 gene expression by HAH and PAH. Hydrophobic ligands enter the cell where they interact with the cytosolic AHR. After ligand binding, the AHR dissociates from the HSP 90 (shown here as a dimer although the actual number of subunits is unknown; Landers and Bunce, 1991; Whitelaw *et al.*, 1993) and combines with ARNT. The ligand–AHR–ARNT complex translocates into the nucleus and binds to XREs upstream of HAH/PAH-responsive genes of the 'Ah gene battery', thereby inducing transcription and subsequent translation of Ah battery genes, which include Phase I and II biotransformation components. Also shown are an inhibitory element and a proximal transcription promotor element, which further regulate gene expression. (Modified and reproduced with permission from S. Safe.)

the *AHR* gene has been rendered non-functional by homologous recombination, has confirmed the role of the AHR as a primary mediator of HAH and PAH toxicity (Fernandez-Salguero *et al.*, 1995). It is presumed that HAH and PAH enter the cell by passive diffusion due to their lipophilicity and bind to the unoccupied cytosolic AHR (Figure 16.4), which is associated with the 90 kDa heat-shock protein (HSP 90) and possibly other factors (Landers and Bunce, 1991; Carver *et al.*, 1998; Meyer *et al.*, 1998). Ligand-induced transformation of the unbound cytosolic AHR to a lower molecular weight heterodimer with increased DNA binding affinity involves the release of HSP 90 and binding of the ARNT protein (Whitelaw *et al.*, 1993). The transformed AHR–ligand complex possesses the structural requirements not only for translocation to the nucleus, but also for association with DNA, binding to the major groove of the xenobiotic response element (XRE). There are at least several XREs upstream from AHR-regulated genes (Nebert *et al.*, 1990; Lusska *et al.*, 1993; Carvan *et al.*, 1999), where AHR-ligand binding induces a change in chromatin structure (Durrin and Whitlock,

1989; Elferink and Whitlock, 1990). The XREs are followed by a down-stream inhibitory element and a proximal transcription promotor element flanking the AHR-regulated genes. These include CYP1 genes, in addition to glutathione *S*-transferase and at least three other Phase II enzymes, each of which is regulated independently. Binding of appropriate regulatory proteins to the proximal element is necessary but not sufficient for induction of the transcription of CYP1A1 (Nebert *et al.*, 1990). Several theories have arisen concerning the regulation of CYP1A1 with regard to tissue and species differences. These include altered DNA methylation as well as tissue- and species-specific regulatory factors that remain unknown.

Prior to the significant increases in understanding of molecular biology in the past two decades, there was minimal understanding of the mechanisms of CYP induction in general. Even now, the induction of CYP1A1, the most heavily researched of the CYP enzymes and the model system for investigating CYP induction, is incompletely understood. The mechanism(s) by which CYP2B enzymes are induced by aromatic hydrocarbons has been poorly characterized. Although the existence of a receptor protein for CYP2B-type inducers has been proposed, the early events of CYP2B induction remain unknown. The diverse group of chemicals that induce CYP2B have no apparent common structural features that might predict a good fit with a receptor (Figure 16.5). In addition, a much higher inducing dose is required for CYP2B inducers. TCDD effectively induces CYP1A expression in the low nmol/kg range whereas PB is effective as a CYP2B inducer in the high μmol/kg dose range (Okey, 1990). This, in conjunction with the identification of PB-responsive enhancer sequences (Honkakoski *et al.*, 1998), suggests low binding affinity of PB to a receptor; however, the molecular mechanisms of CYP2B induction remain unclear.

The affinity with which most chemical ligands bind the AHR *in vitro* correlates well with the potency of those compounds as CYP1A inducers *in vivo*. Within each group of halogenated hydrocarbons, the congeners that are the most potent inducers of CYP1A also exhibit the highest binding affinity for AHR (Safe *et al.*, 1985; Safe, 1986, 1988). The AHR-binding HAH congeners have in common a highly planar molecular structure, such as non-*ortho*-substituted PCBs (Poland and Knutson, 1982; Safe, 1986) (Figures 16.5, 16.6). This structure–activity relationship is common to each class of HAH, with some discrepancies in comparisons across classes of chemicals. TCDD is about 3×10^4 times more potent than MC as an inducer of CYP1A1 activity in rat liver *in vivo* (Poland and Glover, 1974), but MC binds cytosolic AHR from rat liver *in vitro* with an affinity nearly equal to that of TCDD (Okey and Vella, 1982). This departure from the simple receptor affinity induction relationship may be due, in part, to the biological half-life of the respective compounds (Okey, 1990). TCDD is poorly metabolized whereas MC is readily metabolized by CYP1A1. Thus, the MC is biotransformed and excreted while TCDD remains to continue CYP induction.

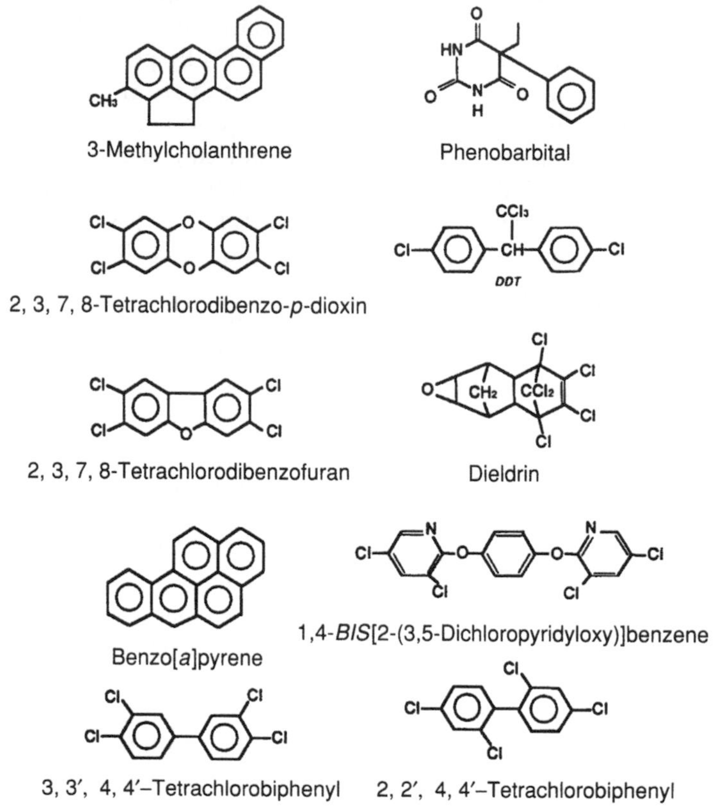

3-Methylcholanthrene

Phenobarbital

2, 3, 7, 8-Tetrachlorodibenzo-*p*-dioxin

DDT

2, 3, 7, 8-Tetrachlorodibenzofuran

Dieldrin

Benzo[*a*]pyrene

1,4-*BIS*[2-(3,5-Dichloropyridyloxy)]benzene

3, 3′, 4, 4′–Tetrachlorobiphenyl   2, 2′, 4, 4′–Tetrachlorobiphenyl

Figure 16.5 Structures of selected P450 inducers. The compounds shown on the left represent known CYP1A inducers, whereas the column on the right depicts known CYP2B inducers.

Aromatic hydrocarbons, both PAHs and HAHs, vary in the degree to which they are metabolized by CYPs, and the differences in rates of metabolism are an important element of their toxicity. For HAH, the degree of halogen substitution is directly related to both stability and toxicity. Those congeners, which are more highly halogenated, are typically the most toxic and the most lipophilic, and have low rates of metabolism by CYPs (Goldstein and Safe, 1989; Safe, 1989). The PCB congeners with fewer chlorine substitutions are more water soluble and may be readily metabolized *in vivo* and *in vitro* by CYP2B enzymes (Shimada *et al.*, 1981). For PAH, the capacity for metabolic conversion to electrophilic metabolites dictates cellular toxicity. PAHs are generally readily metabolized if the appropriate CYPs are induced. Their metabolism to reactive electrophilic intermediates, and subsequent covalent binding to nucleophilic sites on macromolecules, are dependent on the presence of one or more key reaction sites within the molecular structure (Cavalieri and Rogan, 1992). Sites in cellular macromolecules where

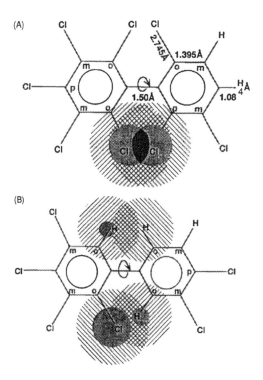

Figure 16.6 Structural features of PCB congeners influencing P450 induction and
enzymatic metabolism. Areas where the principal enzymatic reaction
occurs are given by broken lines. Bond lengths are taken from McKinney
et al. (1983). For atoms in the *ortho* position, the outer circle represents
the area within the van der Waals' radius of an atom, indicating the
maximum distance for any possible influence of an atom (H = 1.45 Å; Cl
= 1.90 Å; Huheey, 1975); the inner circle represents the part of this area
which is also within the single-bond covalent radius. In (A), the overlap-
ping covalent radii for two *ortho*-Cl show that a planar configuration is highly improbable when three or four *ortho*-Cl are present. In
(B), nonoverlapping covalent radii for *ortho*-Cl and *ortho*-H show that a
planar configuration causes a much lower energy barrier when chlorine
atoms do not oppose each other. (Redrawn after Boon *et al.*, 1992.)

electrophilic compounds bind are typically electron-dense carbon, nitrogen,
oxygen or sulfur groups (Williams and Weisburger, 1991). In proteins, the
major reactive sites for electrophilic radicals have tryptophan, tyrosine,
methionine and perhaps histidine residues. In DNA, the highly nucleophilic
sites (Figure 16.7) include the N-7 and exocyclic N-2 of guanine and the N-
3 of adenine, with the affinity for specific nucleophilic sites varying with the
type of reactive compound. For example, the electrophilic form of aflatoxin
B_1 reacts with the N-7 of guanine, whereas the 7,8-dihydrodiol-9,10-epoxy
metabolite of B[a]P preferentially reacts with the exocyclic N-2 of guanine
(Conney, 1982; Weinstein, 1988; Wogan, 1989).

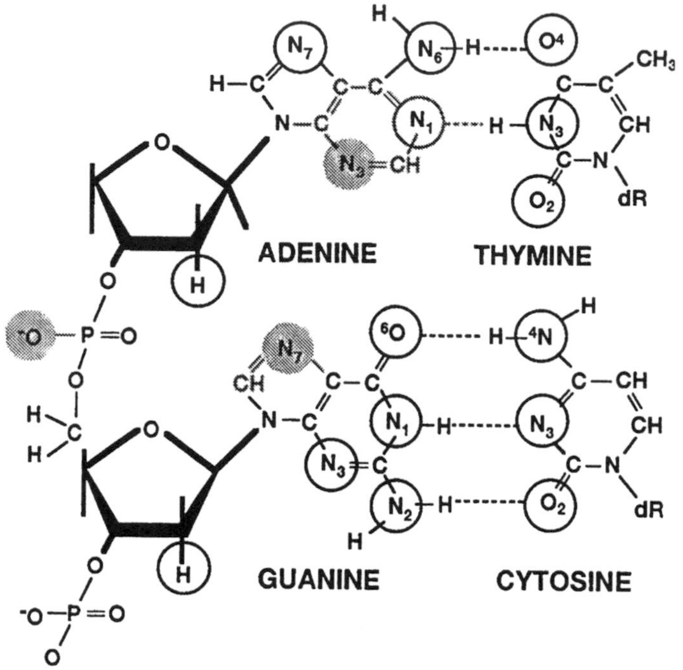

Figure 16.7 Nucleophilic centers in DNA that are the most reactive to electrophiles. In general, the ring nitrogens of the bases are more nucleophilic, hence reactive, than the ring oxygens, with the N-7 of guanine and N-3 of adenine being the most reactive. (Friedberg, 1985; reproduced with permission from W.H. Freeman and Company.)

CYPs in odontocete cetaceans

Observations of the effects of HAHs and PAHs on odontocete cetaceans are limited to studies of a few populations, such as the beluga whales (*Delphinapterus leucas*) in the St. Lawrence River, and there are very few experimental data on the potential cytotoxic or genotoxic effects of chemical contaminants on cetaceans (Carvan *et al.*, 1995). Although there have been few direct studies of dolphin CYPs, existing data suggest that CYPs are induced and active in delphinids, phocoenids and monodonts. Individual belugas of the St. Lawrence River exhibit high concentrations of HAH residues in the tissues (Massé *et al.*, 1986; Martineau *et al.*, 1987). Beluga whales have been shown to express the Ah receptor, CYP1A1 protein and CYP1A1 enzyme activities (Hahn *et al.*, 1994; White *et al.*, 1994). *In vitro* data of Carvan *et al.* (1994, 1995) indicate that cells of a dolphin epithelial line express the Ah receptor, are induced by TCDD, and metabolize benzo[*a*]pyrene with the subsequent formation of B[a]P–DNA adducts. The probability for CYP1A-associated metabolism of HAH and PAH contaminants to form damaging electrophilic metabolites in beluga whales has been suggested by several investigations. For instance, carcinomas are rarely seen in cetaceans, but Martineau *et al.* (1985) reported a transitional cell carcinoma in the urinary bladder of a beluga from the St. Lawrence River. In a later report they showed that this population has high levels of HAH residues, a high incidence of neoplasia and elevated levels of multisystemic and viral lesions (Martineau *et al.*, 1988). DNA adducts have been reported in liver tissues of belugas from the St. Lawrence River and Arctic (Ray *et al.*, 1991; Mathieu *et al.*, 1997), and the levels of B[a]P–DNA adducts in beluga brain tissues were shown to be similar to those of both terrestrial and aquatic laboratory animals exposed to carcinogenic levels of B[a]P (Varanasi *et al.*, 1981; Shugart and Kao, 1985; Stowers and Anderson, 1985).

One of the most comprehensive studies of delphinid CYPs to date is perhaps that of Watanabe *et al.* (1989) which analyzed CYP levels and activity in shortfinned pilot whales (*Globicephala macrorhynchus*). This study also included limited data on CYP levels and activity in striped dolphins (*Stenella coeruleoalba*) and killer whales (*Orcinus orca*). No differences between immature and mature (fetal activities were generally lower), or male and female pilot whales were found, and all microsomal CYP activity levels reported were lower than those found in uninduced rats. In pilot whales there were correlations between CYP levels and selected enzyme activities, including AHH, 7-ethoxyresorufin-*O*-deethylation (EROD) and activation of selected promutagens. AHH and EROD activities in pilot whales were inhibited by polyclonal antibodies to rat CYP1A1 and 1A2, suggesting that cetacean and rodent CYPs have structural homologies.

Fossi *et al.* (1992) provide one of the very few published studies comparing levels of presumed CYP activity and HAH residues in dolphins. They reported significant CYP1A-like activity in homogenized skin biopsies obtained from free-ranging striped dolphins. Their data on benzo[*a*]pyrene monooxygenase activity (BPMO, expressed as arbitrary units of fluorescence/ h/g tissue) showed considerable variation when plotted against total HAH (mg/kg dry weight) (Figure 16.8). Biopsy tissues from male fin whales (*Balaenoptera physalis*) show similar variation with HAH levels, accounting for only about 50 per cent of the variation in BPMO (Marsili *et al.*, 1998). Although these data do not demonstrate strong correlations between HAH contaminants and CYP activity, they reveal the utility of such analyses. Serum HAH concentrations are more pharmacologically relevant than residue concentrations in blubber. An animal that is mobilizing its lipid stores is also mobilizing and/or releasing stored HAH into the serum, and should present a higher BPMO activity index even though its blubber HAH stores may be significantly decreased. Unfortunately, serum HAH cannot be measured in tissue obtained by dart biopsy, the method used by Fossi *et al.* (1992) and Marsili *et al.* (1998). It may be possible in the future to account for CYP activity variation in skin biopsies using molecular markers that would indicate hormonal status, stress, lipid mobilization or other potentially confounding factors.

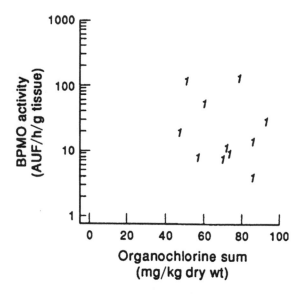

Figure 16.8 Plot of total organochlorines with BPMO activity in skin biopsies of *Stenella coeruleoalba*. (Fossi *et al.*, 1992; reproduced with permission from Pergamon Press.)

A number of independent investigators have published data on the capacity of cetacean CYPs to metabolize HAH, based on the ratios of specific HAH concentrations in tissues (Tanabe *et al.*, 1988; Duinker *et al.*, 1989; Boon *et al.*, 1992). They concluded that CYP1A-mediated metabolism of HAH was lower in small cetaceans than in terrestrial mammals (other than mink, *Mustela vison*), and that CYP2B-mediated metabolism was either very low or absent in cetaceans. Norstrom *et al.* (1992) examined possible CYP activity in beluga whales and narwhal (*Monodon monoceros*). Their studies were based on the ratios of specific HAH isomers which are generally metabolized by CYP1A or CYP2B in other mammals, and suggested that CYP2B-type activities were significant but low relative to other fish-eating birds and mammals. They also suggested that beluga whales and narwhal efficiently metabolize planar TCDD, TCDF, and PCB congeners. These data contrast with data from other mammals in which TCDD and coplanar PCB congeners are poorly metabolized, a factor contributing to their toxicity (Safe, 1986; Okey, 1990).

White *et al.* (2000) has provided the strongest evidence to date that CYP activity in cetacean tissue has the potential to shape HAH residue ratios. Long-finned pilot whale (*Globicephala melas*) and beluga hepatic microsomes were able to metabolize two tetrachlorobiphenyl (TCB) congeners, and metabolism of 3,3′,4,4′-TCB by beluga microsomes strongly correlated with both immunoreactive CYP1A content and EROD activity. 3,3′,4,4′-TCB metabolism was substantially lower in pilot whale samples, compared to those from beluga. Several reasons for this difference were addressed, such as autolysis of the pilot whale liver samples, resulting in inactivation of CYP, or possible spatial differences in CYP expression within the liver that could influence measured CYP activities if samples were taken from different hepatic regions (White *et al.*, 2000).

Several studies (Tanabe *et al.*, 1987; Kannan *et al.*, 1989, 1993; Boon *et al.*, 1992; Borell *et al.*, 1996; Minh *et al.*, 2000; Busbee *et al.*, unpublished data) have reported TCDD toxic equivalency quotients (TEQs) for various HAH isomers found in cetaceans, in order to establish the potential hazard these residues pose to this group. The concept of TEQs is based on the capacity of various HAH isomers to induce CYP1A activities (AHH and EROD) relative to those induced by 2,3,7,8-TCDD *in vitro*. Safe *et al.* (1989) established toxic equivalency factors (TEFs) for several HAH isomers in a rat hepatoma cell line, H4-IIE, and demonstrated that a direct quantitative correlation exists between the *in vitro* dose–response effects (AHH and EROD induction) and *in vivo* toxic effects (weight loss, thymic atrophy, teratogenicity) of TCDD and TCDD-like HAH isomers. The equation for determining TEQs from TEFs is: TEQ = (actual concentration of a given residue) × TEF (its capacity for EROD and AHH induction relative to TCDD).

EROD and AHH induction by HAH is species-specific, and therefore the extent of induction by TCDD and other HAH isomers must be established in cetaceans before any quantitative determination of TEFs can be made. Any HAH isomer ratio correlations with TEF data involving cetaceans must currently be considered inconclusive for several reasons. First, there are insufficient data on dietary input levels of the various HAHs. In most cases, the food habits of cetaceans are so poorly understood that the prey species have not yet been completely identified in order to be sampled for accurate determination of dietary HAH input. Second, there are little or no data on the *in vitro* metabolism of individual HAH congeners by purified dolphin CYPs in reconstituted microsomal systems, or by isolated native dolphin microsomes containing CYPs. If cetaceans have the capacity to metabolize TCDD and/or TCDD-like HAH isomers, as suggested by Norstrom *et al.* (1992), these isomers should be much less toxic in cetaceans compared to rodents, which have a very limited capacity for TCDD and TCDD-like HAH metabolism. Thirdly, there are essentially no data on inducibility of CYP2B enzymes in dolphins. The threshold for induction of CYPs and their activity is variable, depending on the inducing compound and the animal species. Perhaps the serum threshold of HAH and PAH residues that induce CYP2B enzymes and are metabolized by them have not been reached in the dolphins studied to date.

The validity of extrapolating investigative findings from laboratory animals to marine mammals, and between marine mammal species, must be established. Hazard and risk assessment of complex HAH mixtures have utilized a TEF approach, and have been used to estimate risk associated with many toxicologic outcomes, including immunotoxicity, hepatotoxicity and carcinogenicity (Andersen and Barton, 1998; Vos *et al.*, 1997–98). The toxicity of HAH congeners is species-specific, and thus TEFs must be as well because of the dependence of TEFs on receptor–ligand interactions. Several studies have demonstrated correlations between TEFs derived by *in vitro* and *in vivo* methods (Safe, 1998a), which have been used to develop toxicokinetic models (Kreuzer *et al.*, 1997) and estimate risk even to low levels of exposure (Andersen and Barton, 1998). Species-specific TEFs for marine mammals can be developed using *in vitro* tools such as chemically responsive cell lines and recombinant proteins, which can also contribute to the determination and quantitation of complex interactions such as antagonism and synergism. Such tools would allow the analysis of metabolic enzyme activity, the binding of xenobiotics to cellular receptors, and pertinent gene and protein structures. Metabolic enzyme activity differences account for 90–95 per cent of the inter-species variability in the response to xenobiotics, with the remainder being accounted for in the binding of chemicals to cellular receptors which control the levels of these enzymes (Nebert and Weber, 1990).

Interaction with steroid receptors: Endocrine disruption

Serum PAH and HAH, which induce a variety of CYP types, have also been shown to alter hormone metabolism, serum hormone levels and hormone-regulated gene expression by means of interaction with steroid receptors in cells (see also Reijnders, Chapter 3 in this volume). The superfamily of steroid hormone receptors includes cellular proteins that bind estrogens, androgens, glucocorticoids, progestins, vitamin D, retinoic acid and thyroid hormone. These receptors exert exquisite regulation of gene expression, maintaining physiological function essential for good health. Steroids bind their respective receptors, the receptor–ligand interacts with steroid response elements, and the complex of response element plus receptor–ligand acts to recruit transcriptional elements essential for gene expression (Fuller, 1991; O'Malley and Tsai, 1992; Truss and Beato, 1993; Lieberman, 1997). Steroid regulation of gene expression may be due to agonistic or antagonistic activities of the respective receptor-interactive ligand. Animals, either terrestrial or aquatic, may be chronically exposed to a variety of contaminant chemicals that bind one or more of the steroid receptors. A large body of data exists on acutely toxic effects of substituted polycyclic and/or aromatic hydrocarbons, including herbicides, pesticides and related organochlorines. However, there are few data on the physiological effects of chronic exposure to sub-acute levels of the variety of endocrine-disruptive chemicals, including plasticizers, detergents, pesticides, herbicides, AH, PAH and HAH, that bind steroid receptors and are potentially endocrine interactive in animals. In addition, no rapid, practical, readily quantifiable and inexpensive methods exist to screen routinely water and soil samples for the presence of chemical contaminants in order to identify sources of chronic exposure to endocrine disruptive agents. TEF/TEQ data are currently being generated for estrogenic compounds (Safe, 1998b), an approach which could be extended to other chemical classes that act through specific receptor–ligand interations, and which could facilitate development of practical methods for analysis of water and soil samples for xenoestrogens.

Widespread distribution of terrestrial and aquatic contaminants with steroid mimetic activities, and the tendency of those chemicals that are lipid soluble to bioaccumulate into fat reserves, raise particular concern for marine mammals. A direct cause and effect relationship was noted a number of years ago between exposure to a variety of chemicals and the initiation of altered physiological responses in humans and laboratory animals (Jackson and Halbert, 1974; Constable and Hatch, 1985; Lawrence *et al.*, 1985; Kashimoto and Miyata, 1987). These responses include weight loss, thymic atrophy and impaired immune function, hepatotoxicity, fetal toxicity, developmental disorders, loss of reproductive efficiency and increased incidence of cancer (Poland and Knutson, 1982; Safe, 1990), and are now known to be directly related to endocrine-disruptive activities of these chemicals. These responses differ dramatically between individual humans and between inbred strains of animals, dependent at least in part on endocrine receptors and endocrine-regulated cell function, and on the specific capacity for enzyme induction (Santostefano *et al.*, 1998).

Endocrine disruption in dolphins

Although acute administration of endocrine-disruptive chemicals has begun to be well investigated (Colborn, 1995; Ekena *et al.*, 1997; El-Sabeawy *et al.*, 1998; Faqi *et al.*, 1998), very little definitive data are available to predict how humans and other mammals might react to chronic low-level exposure. It is also unclear whether the bioaccumulation of lipid-soluble contaminants into blubber reserves may increase the pharmacologically effective dose of endocrine-disruptive chemicals to which a cetacean may be exposed. Subramanian *et al.* (1987) reported a negative correlation between reduced serum testosterone and increased concentrations of DDE in blubber of Dall's porpoise (*Phocoenoides dalli*), but were unable to find a statistically significant correlation between the phenomena and PCB concentrations. Moreover, their report included no data on the pathophysiological consequences of reduced testosterone or of increased HAH concentrations in tissues. A study of laboratory rats (Johnson *et al.*, 1992) suggests that animals exposed to HAH present significantly lower serum testosterone levels without changes in sperm production, so the reduced serum testosterone reported by Subramanian *et al.* (1987) in Dall's porpoise may not have significant reproductive consequences. The relationships between residues of chemicals in blubber, toxic effects of the chemicals, and reproductive efficiency or offspring survival remain a concern for marine mammals.

Data have long indicated that stressed mammals metabolize fat reserves and release bioaccumulated chemicals, resulting in elevated serum levels of chemicals that may expose cells in adult animals to toxic levels of these compounds (Korytko *et al.*, 1999). These chemicals may also cross the placental barrier and affect fetal cells. A major component of this concern is directed toward the offloading of large quantities of bioaccumulated lipophilic chemicals in milk of lactating females (Ridgway and Reddy, 1995) resulting in nursing young that have extremely high concentrations of chemicals relative to the actual environmental levels to which they are exposed (Tanabe *et al.*, 1982).

Suggestions for future research

A series of research principles was suggested by Reijnders (1988), which should be developed in order to establish a predictive concept for the effects of pollutants on marine mammals. These are applicable to odontocete cetaceans and include the establishment of baseline data on existing pollutant residues in tissues and reference values for blood chemistry and hormone profiles; development of an understanding of the toxicokinetics of organic pollutants; and determination of the physiological responses to pollutants. Based on these broad categories of data, recognizable parameters could be developed to serve as biomarkers of exposure and of physiological change caused by organic pollutants. The biomedical approach to such a complex problem must follow the lead of ongoing studies designed to determine the consequences of human HAH and PAH exposure, because (at least in the US) the general guidelines for collection of human and cetacean tissues are very similar.

Whereas a large volume of data exists on concentrations of organic pollutants in cetaceans, the numbers of investigators, the differing species from which samples have been derived, and the variety of analytical methods utilized in data collection renders any baseline determinations incomplete at this time. Data must be collected from a significant number of individuals of the same species, including adequate samples from both sexes across a range of ages in a given geographical area, to allow for proper statistical treatment. Samples must include animals of both sexes and many age classes to provide meaningful data, because of the continued accumulation of residues as animals age and the differential accumulation of HAH residues between sexes. Lactating females accumulate lower levels of HAH residues with time because they offload lipophilic compounds during milk formation. Appropriate emphasis must be placed on the physiological state of the animal at the time of collection, the foraging ecology, sampling over an appropriate time period, and standardization of collection and analytical techniques. The physiological state of animals is important, because the best data will come from live individuals in which the residue concentrations in blubber and serum have been determined and for which any confounding parameters have been identified (such as lactation, illness, or reduced caloric intake resulting in lipid metabolism in blubber). Foraging ecology of cetaceans is particularly important for comparison of samples from different species or populations of dolphins. For example, dietary HAH and PAH exposure will vary significantly between pelagic teuthivores and coastal piscivores. Samples collected over a period of a few years can be considered within the same data set if one can assume that input variables have not changed significantly. Data collected over many years or decades may show temporal trends rendering them inappropriate for establishing baseline values, but may be quite valuable in establishing the changing status of contaminants in a specific ecosystem. Standardization of collection and analytical techniques to allow interlaboratory comparisons of data will be valuable to ensure proper quantitation of residues. The standardization of analytical methods was significantly enhanced by establishment of the marine mammal tissue bank by the US National Marine Fisheries Service, which makes blubber samples with known residue concentrations available for use as analytical quality control standards.

Captive dolphins and samples from live-capture and release studies are invaluable resources that can be used to establish reference values for normal blood chemistry and hormone profiles (see also Reddy and Ridgway, Chapter 5 in this volume). As mentioned earlier, the toxic effects of HAH and PAH are often the result of alterations in hormone metabolism and in the disruption of receptor-mediated biochemical pathways. Standardization in the analysis of blood samples to provide a minimum data set, and generation of an accessible database, are important for the establishment of reference values of the highest quality. Captive dolphins and live-capture and release programs can provide samples for blood reference values, but baseline data on toxicokinetics and residue-initiated physiological changes can only be established using tissues from animals that are theoretically healthy, either live animals or very fresh carcasses with direct human-related causes of death (ship strikes, nets, fishing lines and hooks, etc.). These samples are rare, inconsistently available, and difficult to obtain. Several laboratories are currently developing the technology to utilize cellular and molecular techniques and *in vitro* model systems to examine contaminant metabolism and biotransformation, cellular physiological response to contaminant insult, and toxicokinetics of contaminant HAH and PAH. The tools for those studies are, however, limited at this time. Cetacean-specific molecular probes and monoclonal antibodies must be developed for maximal utilization of *in vitro* modeling systems. When the molecular tools become available, *in vitro* models will allow mechanistic investigations of the effects of HAH and PAH contaminants on dolphin cells and tissues, and will provide the necessary information required to predict mortality and morbidity in dolphins resulting from environmental pollutants.

Conclusions

HAH and PAH residues have been found in the tissues of a variety of terrestrial, marine and aquatic animals. Dolphins and other marine mammals sampled in virtually all oceans, including some waters generally considered to be uncontaminated, have also been found to have alarmingly high, but not acutely toxic, HAH residues in tissues (Clausen *et al.*, 1974; Gaskin, 1982; Cockcroft *et al.*, 1989, 1990; Buckland *et al.*, 1990). Although the physiological consequences of chronic cellular exposure to both low environmental levels and significantly higher bioaccumulated levels of potentially toxic HAH in dolphins are unknown, elevated concentrations of PCB residues in tissues are proposed to have occurred concurrently with epizootic deaths of dolphins. In the western Mediterranean epizootic event, PCB concentrations as high as 3000 µg/g were detected in blubber (Aguilar and Borrell, 1994). Although no cause and effect relationship has been drawn between PCB contamination and toxic effects, such as immune dysfunction, in cetaceans, the relationships certainly have been shown in other mammals (Kashimoto *et al.*, 1981; Chen *et al.*, 1985; Chen and Hsu, 1987; Kashimoto and Miyata, 1987; Gobin and Phillips, 1991; Neubert *et al.*, 1991; Tryphonas *et al.*, 1991a, b; Hardin *et al.*, 1992). Enormous volumes of data collected from almost all of the oceans show HAH residues persistent in tissues of dolphins (reviewed in Gaskin, 1982; Martin *et al.*, 1987; Cockcroft *et al.*, 1989, 1990; Kannan *et al.*, 1989; Borrell and Aguilar, 1990; Buckland *et al.*, 1990; Loganathan *et al.*, 1990). Data from these studies typically do not address the physiological consequences of chemical residues, and neither support nor refute the proposed association between chemical residues and chronic toxicity in dolphins. It is well established, however, that HAH and other aromatic hydrocarbon residue levels are found in dolphins, that significant CYP activity levels have been detected in dolphin tissues (Geraci and St. Aubin, 1982; Geraci, 1990; Fossi *et al.*, 1992), that CYPs metabolize aromatic hydrocarbons to electrophilic forms capable of causing DNA, RNA and protein adducts (Poland *et al.*, 1979; Whitlock, 1987; Goldstein and Safe, 1989; Carvan *et al.*, 1995), and that DNA–hydrocarbon adducts have been reported for cetaceans from both pristine and polluted waters (Ray *et al.*, 1991; Mathieu *et al.*, 1997). It is well established in other mammalian systems that both DNA and protein adducts decrease DNA synthesis required for both DNA repair and mitosis in rapidly dividing cells, and that unrepaired DNA adducts are associated with a variety of adverse health effects (Busbee *et al.*, 1984; Brown and Romano, 1991; Hardin *et al.*, 1992). We can confidently say that DNA adducts in cetacean cells decrease DNA synthesis, based on studies in other organisms; however, to assume adverse health effects in cetaceans as a result of the measured adduct levels would be far too speculative.

Changes in gene expression which occur in cells of animals exposed to chemical pollutants are typically initiated by receptor-mediated interactions, resulting in induction or inhibition of transcription, and are known to be associated with at least:

(1) induction of cytochrome P450s resulting in increased biotransformation, activation and/or elimination of HAH, PAH and other aromatic hydrocarbons, and in the formation of electrophilic metabolites that are capable of interacting with nucleic acids and proteins to cause adducts;
(2) decreased capacity of cells to repair damaged DNA;
(3) altered metabolism of endogenous steroids, resulting in changed physiological states, including reproductive dysfunction; and
(4) systemic interactions that decrease immune responses to viruses, parasites and bacteria.

These interacting phenomena may result in an aggregate chronic state of physical debilitation, leading to reproductive failure and/or death of the organism, even though concentrations of the inducing chemicals may not have reached acutely toxic levels. In order to address questions regarding the effects of pollutants on cetaceans, we must develop a comprehensive strategy to investigate the mechanisms of aromatic hydrocarbon toxicity in a variety of cetacean systems and the impact of aromatic hydrocarbons on species-specific cetacean physiology. Such a comprehensive examination must include a variety of investigations at the relatively simple cellular and molecular level, and also involve thorough analyses at the much more complex individual and population levels in several species. Without such analyses, we can only continue to speculate as to the effects of these potentially harmful ubiquitous compounds on cetaceans.

References

Abbot, B.D. and Birnbaum, L.S. 1990. Effects of TCDD on embryonic uteric epithelial EGF receptor expression and cell proliferation. *Teratology* 41: 71–84.

Abbott, B.D., Perdew, G.H., Buckalew, A.R. and Birnbaum, L.S. 1994. Interactive regulation of Ah and glucocorticoid receptors in the synergistic induction of cleft palate by TCDD and hydrocortisone. *Toxicology and Applied Pharmacology* 128: 138–150.

Aguilar, A. and Borrell, A. 1994. Abnormally high polychlorinated biphenyl levels in striped dolphins (*Stenella coeruleoalba*) affected by the 1990–1992 Mediterranean epizootic. *Science of the Total Environment* 154: 237–247.

Andersen, M.E. and Barton, H.A. 1998. The use of biochemical and molecular parameters to estimate dose–response relationships at low levels of exposure. *Environmental Health Perspectives* 106(suppl 1): 349–355.

Birnbaum, L.S. 1995. Developmental effects of dioxins. *Environmental Health Perspectives* 103(suppl 7): 89–94.

Black, S.D. and Coon, M.J. 1988. P-450 cytochromes: structure and function. *Advances in Enzymology* 60: 35–87.

Bookstaff, R.C., Kame, F., Moore, R.W., Bjerke, D.L. and Petersen, R.E. 1990. Altered regulation of pituitary gonadotropin-releasing hormone (GnRH) receptor number and pituitary responsiveness to GnRH in 2,3,7,8-tetrachlorodibenzo-*p*-dioxin-treated male rats. *Toxicology and Applied Pharmacology* 105: 78–92.

Boon, J.P. *et al.* 1992. The toxicokinetics of PCBs in marine mammals with special reference to possible interactions of individual congeners with the cytochrome P450-dependent monooxygenase system – an overview. Pages 119–159 *in* Walker, C.H. and Livingstone, D.R., editors. *Persistent pollutants in marine ecosystems.* Pergamon Press, Oxford.

Borrell, A. and Aguilar, A. 1990. Loss of organochlorine compounds in the tissues of a decomposing stranded dolphin. *Bulletin of Environmental Contamination and Toxicology* 45: 46–53.

Borrell, A., Aguilar, A., Corsolini, S. and Focardi, S. 1996. Evaluation of toxicity and sex-related variation of PCB levels in Mediterranean striped dolphins affected by an epizootic. *Chemosphere* 32: 2359–2369.

Brown, W.C. and Romano, L.J. 1991. Effects of benzo[*a*]pyrene-DNA adducts on a reconstituted replication system. *Biochemistry* 30: 1342–1350.

Buckland, S.J., Hannah, D.J., Taucher, J.A., Slooten, E. and Dawson, S. 1990. Polychlorinated dibenzo-*p*-dioxins and dibenzofurans in New Zealand's Hector's dolphin. *Chemosphere* 20: 1035–1042.

Busbee, D.L., Joe, C.O., Norman, J.O. and Rankin, P.W. 1984. Inhibition of DNA synthesis by an electrophilic metabolite of benzo(*a*)pyrene. *Proceedings of the National Academy of Science, USA* 81: 5300–5304.

Carvan, M.J. III, Santostefano, M., Safe, S. and Busbee, D.L. 1994. Characterization of a bottlenose dolphin (*Tursiops truncatus*) kidney epithelial cell line. *Marine Mammal Science* 10: 52–69.

Carvan, M.J., Flood, L.P., Campbell, B.D. and Busbee, D.L. 1995. Effects of benzo(*a*)pyrene on CDK cells: Inhibition of proliferation and initiation of DNA damage. *Chemosphere* 30: 187–198.

Carvan, M.J. III, Ponomareva, L.V., Solis, W.A., Matlib, R.S., Puga, A. and Nebert, D.W. 1999. Trout *CYP1A3* gene: recognition of fish DNA motifs by mouse regulatory proteins. *Marine Biotechnology* 1: 155–166.

Carver L.A., LaPres, J.J., Jain, S., Dunham, E.E. and Bradfield, C.A. 1998. Characterization of the Ah receptor-associated protein, ARA9. *Journal of Biological Chemistry* 273: 33580–33587.

Cavalieri, E.L. and Rogan, E.G. 1992. The approach to understanding aromatic hydrocarbon carcinogenesis. The central role of radical cations in metabolic activation. *Pharmacology and Therapeutics* 55: 183–199.

Chen, P.H. and Hsu, S.-T. 1987. PCB poisoning from toxic rice-bran oil in Taiwan. Pages 27–38 *in* Waid, J.S., editor. *PCBs and the environment*, volume 3. CRC Press, Boca Raton, Florida.

Chen, P.H., Wong, C.K., Rappe, C. and Nygren, M. 1985. Polychlorinated biphenyls, dibenzofurans and quarterphenyls in toxic rice-bran oil and in the blood and tissues of patients with PCB poisoning (YU-Cheng) in Taiwan. *Environmental Health Perspectives* 59: 59–65.

Clausen, B., Braestrup, L. and Berg, O. 1974. The content of polychlorinated hydrocarbons in arctic mammals. *Bulletin of Environmental Contamination and Toxicology* 12: 529–534.

Cockcroft, V.G., DeKock, A.C., Lord, D.A. and Ross, G.J.B. 1989. Organochlorines in bottlenose dolphins *Tursiops truncatus* from the east coast of South Africa. *South African Journal of Marine Science* 8: 207–217.

Cockcroft, V.G., DeKock, A.C., Ross, G.J.B. and Lord, D.A. 1990. Organochlorines in common dolphins caught in shark nets during the Natal 'sardine run'. *South African Journal of Zoology* 25: 144–148.

Colborn, T. 1995. Environmental estrogens: Health implications for humans and wildlife. Estrogens in the Environment, III: Global Health Implications. *Environmental Health Perspectives* 103(suppl 7): 135–136.

Conney, A.H. 1982. Induction of microsomal enzymes by foreign chemicals and carcinogenesis by polycyclic aromatic hydrocarbons. *Cancer Research* 42: 4875–4917.

Constable, J.D. and Hatch, M.C. 1985. Reproductive effects of herbicide exposure in Vietnam: recent studies by the Veitnamese and others. *Teratogenesis, Carcinogenesis, and Mutagenesis* 5: 231–250.

Dalton, T.P., Shertzer, H.G. and Puga, A. 1998. Regulation of gene expression by reactive oxygen. *Annual Reviews in Pharmacology and Toxicology* 39: 67–101.

Duinker, J.C., Hillebrand, M.T.J., Zeinstra, T. and Boon, J.P. 1989. Individual chlorinated biphenyls and pesticides in tissues of some cetacean species from the North Sea and the Atlantic Ocean; tissue distribution and biotransformation. *Aquatic Mammals* 15: 95–124.

Durrin, L.K. and Whitlock, J.P. Jr 1989. 2,3,7,8-tetrachlorodi-benzo-*p*-dioxin-inducible aryl hydrocarbon receptor-mediated change in CYP1A1 chromatin structure occurs independently of transcription. *Molecular and Cellular Biology* 9: 5733–5737.

Ekena, K., Weiss, K.E., Katzenellenbogen, J.A. and Katzenellenbogen, B.S. 1997. Different residues of the human estrogen receptor are involved in the recognition of structurally diverse estrogens and antiestrogens. *Journal of Biological Chemistry* 272: 5069–5075.

Elferink, C.J. and Whitlock, J.P. Jr 1990. 2,3,7,8-Tetrachlorodi-benzo-*p*-dioxin-inducible, Ah receptor-mediated bending of enhancer DNA. *Journal of Biological Chemistry* 265: 5718–5721.

El-Sabeawy, F., Wang, S., Overstreet, J., Miller, M., Lasley, B. and Enan, E. 1998. Treatment of rats during pubertal development with 2,3,7,8-tetrachlorodibenzo-p-dioxin alters both signaling kinase activities and epidermal growth factor receptor binding in the testis and the motility and acrosomal reaction of sperm. *Toxicology and Applied Pharmacology* 150: 427–442.

Faqi, A.S., Dalsenter, P.R., Merker, H.-J. and Chahoud, I. 1998. Reproductive toxicity and tissue concentrations of low doses of 2,3,7,8-tetrachlorodibenzo-*p*-dioxin in male offspring rats exposed throughout pregnancy and lactation. *Toxicology and Applied Pharmacology* 150: 393–401.

Fernandez-Salguero, P., Pineau, T., Hilbert, D.M., McPhail, T., Lee, S.S., Kimura, S., Nebert, D.W., Rudikoff, S., Ward, J.M. and Gonzalez, F.J. 1995. Immune system impairment and hepatic fibrosis in mice lacking the dioxin-binding Ah receptor. *Science* 268: 722–726.

Fossi, M.C., Marsili, L., Leonzio, C., Notarbartolo di Sciara, G., Zanardelli, M. and Focardi, S. 1992. The use of non-destructive biomarker in Mediterranean cetaceans:

preliminary data on MFO activity in skin biopsies. *Marine Pollution Bulletin* 24: 459–461.

Friedberg, E.C. 1985. *DNA repair.* W.H. Freeman and Company, New York.

Fuller, P. 1991. The steroid receptor superfamily: mechanisms of diversity. *FASEB Journal* 5: 3092–3099.

Gaskin, D.E. 1982. *The ecology of whales and dolphins.* Heinemann Press, London.

Geraci, J.R. 1990. Physiological and toxic effects on cetaceans. Pages 167–197 *in* Geraci, J.R. and St. Aubin, D.J., editors. *Sea mammals and oil: Confronting the risks.* Academic Press, New York.

Geraci, J.R. and St. Aubin, D.J. 1982. *Study of the effects of oil on cetaceans.* Final Report US Department of Interior, Bureau of Land Management, Washington, DC.

Gobin, S.J.P. and Phillips, J.A. 1991. Immunosuppressive effects of 2-acetyl-4-tetrahydroxybutyl imidazole (THI) in the rat. *Clinical and Experimental Immunology* 85: 335–340.

Goldstein, J.A. and Safe, S.H. 1989. Mechanism of action and structure–activity relationships for the chlorinated dibenzo-*p*-dioxins and related compounds. Pages 239–293 *in* Kimbrough, R.D., editor. *Halogenated biphenyls, terphenyls, naphthalenes, dibenzodioxins and related products.* Elsevier, Amsterdam.

Goldstein, J.A., Lin, F.H., Stohs, S.J., Graham, M., Clarke, G., Birnbaum, L. and Lucier, G. 1990. The effects of TCDD on receptors for epidermal growth factor, glucocorticoid, and estrogen in Ah-responsive and -nonresponsive congenic mice and the effects of TCDD on estradiol metabolism in a liver tumor promotion model in female rats. *Progress in Clinical and Biological Research* 331: 187–202.

Gonzalez, F.J. 1989. The molecular biology of cytochrome P450s. *Pharmacological Reviews* 40: 243–288.

Hahn, M.E., Poland, A., Glover, E. and Stegeman, J.J. 1994. Photoaffinity labeling of the Ah receptor: phylogenetic survey of diverse vertebrate and invertebrate species. *Archives of Biochemistry and Biophysics* 310: 218–228.

Hardin, J.A., Hinoshita, F. and Sherr, D.H. 1992. Mechanisms by which benzo[*a*]pyrene, an environmental carcinogen, suppresses B cell lymphopoiesis. *Toxicology and Applied Pharmacology* 117: 155–164.

Harris, M., Zacharewski, T. and Safe, S. 1990. Effects of 2,3,7,8-tetrachlorodibenzo-*p*-dioxin and related compounds of the occupied nuclear estrogen receptor in MCF-7 human breast cancer cells. *Cancer Research* 50: 3579–3584.

Honkakoski, P., Moore, R., Washburn, K.A. and Negishi, M. 1998. Activation by diverse xenochemicals of the 51-base pair phenobarbital-responsive enhancer module in the CYP2B10 gene. *Molecular Pharmacology* 53: 597–601.

Huheey, J.E. 1975. *Inorganic chemistry: Principles of structure and reactivity.* Harper & Row, London.

Hurst, C., Abbott, B., DeVito, M. and Birnbaum, L.S. 1998. 2,3,7,8-Tetrachlorodibenzo-*p*-dioxin in pregnant Long Evans rats: Disposition to maternal and embryo/fetal tissues. *Toxicological Sciences* 45: 129–136.

Jackson, T.F. and Halbert, F.L. 1974. A toxic syndrome associated with the feeding of polybrominated biphenyl-contaminated protein concentrate to dairy cattle. *Journal of the American Veterinary Medical Association* 165: 437–439.

Johnson, L., Dickerson, R., Safe, S.H., Nyberg, C.L., Lewis, R.P. and Welsh, T.H. Jr 1992. Reduced Leydig cell volume and function in adult rats exposed to 2,3,7,8-tetrachlorodibenzo-*p*-dioxin without a significant effect on spermatogenesis. *Toxicology* 76: 103–118.

Kannan, K., Tanabe, S., Borrell, A., Aguilar, A., Focardi, S. and Tatsukawa, R. 1993. Isomer-specific analysis and toxic evaluation of polychlorinated biphenyls in striped dolphins affected by an epizootic in the western Mediterranean Sea. *Archives of Environmental Contamination and Toxicology* 25: 227–233.

Kannan, N., Tanabe, S., Ono, M. and Tatsukawa, R. 1989. Critical evaluation of polychlorinated biphenyl toxicity in terrestrial and marine mammals: increasing impact of non-ortho and mono-ortho coplanar polychlorinated biphenyls from land to ocean. *Archives of Environmental Contamination and Toxicology* 18: 850–857.

Kashimoto, T., Miyata, H., Kunita, S., Tung, T.C., Hsu, S.T., Chang, K.J., Tang, S.Y., Ohi, G., Nakagawa, J. and Yamamoto, S. 1981. Role of polychlorinated dibenzofuran in yusho (PCB poisoning). *Archives of Environmental Health* 36: 321–326.

Kashimoto, T. and Miyata, H. 1987. Differences between Yusho and other kinds of poisoning involving only PCBs. Pages 1–26 *in* Waid, J.S., editor. *PCBs and the environment*, volume 3. CRC Press, Boca Raton, FL.

Kimbrough, R.D. 1987. Human health effects of polychlorinated biphenyls (PCBs) and polybrominated biphenyls (PBBs). *Annual Review of Pharmacology and Toxicology* 27: 87–111.

Korytko, P.J., Casey, A.C., Bush, B. and Quimby, F.W. 1999. Induction of hepatic cytochromes P450 in dogs exposed to a chronic low dose of polychlorinated biphenyls. *Toxicological Sciences* 47: 52–61.

Kreuzer, P.E., Csanady, G.A., Baur, C., Kessler, W., Papke, O., Greim, H. and Filser, J.G. 1997. 2,3,7,8-Tetrachlorodibenzo-*p*-dioxin (TCDD) and congeners in infants. A toxicokinetic model of human lifetime body burden by TCDD with special emphasis on its uptake by nutrition. *Archives of Toxicology* 71: 383–400.

Landers, J.P. and Bunce, N.J. 1991. The Ah receptor and the mechanism of dioxin toxicity. *Biochemical Journal* 276: 273–287.

Lawrence, C., Reilly, A., Quyickenton, P., Greenwald, P., Page, W. and Kuntz, A. 1985. Mortality patterns of New York State Vietnam veterans. *American Journal of Public Health* 75: 277–283.

Lieberman, B.A. 1997. The estrogen receptor activity cycle: dependence on multiple protein–protein interactions. *Critical Reviews in Eukaryotic Gene Expression* 7: 43–59.

Loganathan, B.G., Tanabe, S., Tanaka, H., Watanabe, S., Miyazaki, N., Amano, M. and Tatsukawa, R. 1990. Comparison of organochlorine residue levels in the striped dolphin from western North Pacific, 1978–79 and 1986. *Marine Pollution Bulletin* 21: 435–439.

Lusska, A., Shen, E. and Whitlock, J.P. Jr 1993. Protein–DNA interactions at a dioxin-responsive enhancer: analysis of six bona fide DNA-binding sites for the liganded Ah receptor. *Journal of Biological Chemistry* 268: 6575–6580.

Luster, M.I., Germolec, D.R., Clark, G., Wiegand, G. and Rosenthal, G.J. 1988. Selective effects of 2,3,7,8-tetrachlorodibenzo-*p*-dioxin and corticocosteroid on in vitro lymphocyte maturation. *Journal of Immunology* 140: 928–935.

Manchester, D.K., Gordon, S.K., Golas, C.L., Roberts, E.A. and Okey, A.B. 1987. Ah receptor in human placenta: stabilization by molybdate and characterization of binding of 2,3,7,8-tetrachlorodibenzo-*p*-dioxin, 3-methylcholanthrene, and benzo[*a*]pyrene. *Cancer Research* 47: 4861–4868.

Marsili, L., Fossi, M.C., Notarbartolo di Sciara, G., Zanardelli, M., Nani, B., Panigada, S. and Focardi, S. 1998. Relationship between organochlorine contaminants and mixed function oxidase activity in skin biopsy specimens of Mediterranean fin whales (*Balaenoptera physalus*). *Chemosphere* 37: 1501–1510.

Martin, A.R., Reynolds, P. and Richardson, M.G. 1987. Aspects of the biology of pilot whales (*Globicephala melaena*) in recent mass strandings on the British coast. *Journal of Zoology* 211: 11–23.

Martineau, D., Lagacé, A., Massé, R., Morin, M. and Béland, P. 1985. Transitional cell carcinoma of the urinary bladder in a beluga whale (*Delphinapterus leucas*). *Canadian Veterinary Journal* 26: 297–302.

Martineau, D., Béland, P., Desjardins, C. and Lagacé, A. 1987. Levels of organochlorine chemicals in tissues of beluga whales (*Delphinapterus leucas*) from the St. Lawrence estuary, Quebec, Canada. *Archives of Environmental Contamination and Toxicology* 16: 137–147.

Martineau, D., Lagacé, A., Béland, P., Higgins, R., Armstrong, D. and Shugart, L.R. 1988. Pathology of stranded beluga whales (*Delphinapterus leucas*) from the St. Lawrence estuary, Quebec, Canada. *Journal of Comparative Pathology* 98: 287–311.

Massé, R., Martineau, D., Tremblay, L. and Béland, P. 1986. Concentrations and chromatographic profile of DDT metabolites and polychlorobiphenyl (PCB) residues in stranded beluga whales (*Delphinapterus leucas*) from the St. Lawrence estuary, Canada. *Archives of Environmental Contamination and Toxicology* 15: 567–579.

Mathieu, A., Payne, J.F., Fancey, L.L., Santella, R.M. and Young, T.L. 1997. Polycyclic aromatic hydrocarbon–DNA adducts in beluga whales from the Arctic. *Journal of Toxicology and Environmental Health* 51: 1–4.

McKinney, J.D., Gottschalk, K.E. and Pedersen, L. 1983. The polarizability of aromatic systems. An application to polychlorinated biphenyls (PCBs), dioxins and polyaromatic hydrocarbons. *Journal of Molecular Structure* 105: 427–438.

Meyer, B.K., Pray-Grant, M.G., Vanden Heuvel, J.P. and Perdew, G.H. 1998. Hepatitis B virus X-associated protein 2 is a subunit of the unliganded aryl hydrocarbon receptor core complex and exhibits transcriptional enhancer activity. *Molecular and Cellular Biology* 18: 978–988.

Minh, T.B., Nakata, H., Watanabe, M., Tanabe, S., Miyazaki, N., Jefferson, T.A., Prudente, M. and Subramanian, A. 2000. Isomer-specific accumulation and toxic assessment of polychlorinated biphenyls, including coplanar congeners, in cetaceans from the North Pacific and Asian coastal waters. *Archives of Environmental Contamination and Toxicology* 39: 398–410.

Nebert, D.W. and McKinnon, R.A. 1994. Cytochrome P450: Evolution and functional diversity. *Progress in Liver Disease* 12P: 63–97.

Nebert, D.W. and Weber, W.W. 1990. Pharmacogenetics. Pages 469–531 in Pratt, W.B. and Taylor, P.W., editors, *Principles of drug action. The basis of pharmacology*, 3rd edition. Churchill Livingstone, New York.

Nebert, D.W., Eisen, H.J., Negishi, M., Lang, M.A., Hjelmeland, L.M. and Okey, A.B. 1981. Genetic mechanisms controlling the induction of polysubstrate monooxygenase (P-450) activities. *Annual Review of Pharmacology and Toxicology* 21: 431–462.

Nebert, D.W., Petersen, D.D. and Fornace, A.J. Jr 1990. Cellular responses to oxidative stress: the [Ah] gene battery as a paradigm. *Environmental Health Perspectives* 88: 13–25.

Nelson, D.R. and Strobel, H.W. 1987. Evolution of cytochrome P-450 proteins. *Molecular Biology and Evolution* 4: 572–593.

Nelson, D.R., Koymans, L., Kamataki, T., Stegeman, J.J., Feyereisen, R., Waxman, D.J., Waterman, M.R., Gotoh, O., Coon, M.J., Estabrook, R.W., Gunsalus, I.C. and Nebert, D.W. 1996. P450 superfamily: Update on new sequences, gene mapping, accession numbers, and nomenclature. *Pharmacogenetics* 6: 1–42.

Neubert, R., Jacob-Muller, U., Helge, H., Stahlmann, R. and Neubert, D. 1991. Polyhalogenated dibenzo-*p*-dioxins and dibenzofurans and the immune system. 2. *In vitro* effects of 2,3,7,8-tetrachlorodibenzo-*p*-dioxin (TCDD) on lymphocytes of the venous blood from man and another nonhuman primate (*Callithrix jacchus*). *Archives of Toxicology* 65: 213–219.

Nicotera, P., Thor, H. and Orrenius, S. 1989. Cytosolic free Ca^{2+} and cell killing in hepatoma-1c1c7 cells exposed to chemical anoxia. *FASEB Journal* 3: 59–64.

Norstrom, R.J., Muir, D.C.G., Ford, C.A., Simon, M., Macdonald, C.R. and Béland, P. 1992. Indications of P450 monooxygenase activities in beluga (*Delphinapterus leucas*) and narwhal (*Monodon monoceros*) from patterns of PCB, PCDD, and PCDF accumulation. *Marine Environmental Research* 34: 267–272.

Okey, A.B. 1990. Enzyme induction in the cytochrome P-450 system. *Pharmacology and Therapeutics* 45: 241–298.

Okey, A.B. and Vella, L.M. 1982. Binding of 3-methylcholanthrene and 2,3,7,8-tetrachlorodibenzo-*p*-dioxin to a common Ah receptor site in mouse and rat hepatic cytosols. *European Journal of Biochemistry* 127: 39–47.

O'Malley, B. and Tsai, M.-J. 1992. Molecular pathways of steroid receptor action. *Biology of Reproduction* 46: 163–167.

Omiecinski, C.J., Remmel, R.P. and Hosagrahara, V.P. 1999. Concise review of the cytochrome P450s and their roles in toxicology. *Toxicological Sciences* 48: 151–156.

Osterhaus, A.D., de Swart, R.L., Vos, H.W., Ross, P.S., Kenter, M.J. and Barrett, T. 1995. Morbillivirus infections of aquatic mammals: newly identified members of the genus. *Veterinary Microbiology* 44: 219–227.

Payne, J.F., Fancey, L.L, Rahimtula, A.D. and Porter, E.L. 1987. Review and perspective on the use of mixed-function oxygenase enzymes in biological monitoring. *Comparative Biochemistry and Physiology* 86C: 233–245.

Poland, A. and Glover, E. 1974. Comparison of 2,3,7,8-tetrachlorodibenzo-*p*-dioxin, a potent inducer of aryl hydrocarbon hydroxylase, with 3-methylcholanthrene. *Molecular Pharmacology* 10: 349–359.

Poland, A. and Knutson, J.C. 1982. 2,3,7,8-Tetrachlorodibenzo-*p*-dioxin and related aromatic hydrocarbons: examination of the mechanism of toxicity. *Annual Reviews in Pharmacology and Toxicology* 22: 517–554.

Poland, A., Greenlee, W.F. and Kende, A.S. 1979. Studies on the mechanism of action of the chlorinated dibenzo-*p*-dioxins and related compounds. *Annals of the New York Academy of Science* 320: 214–230.

Ray, S., Dunn, B.P., Payne, J.F., Fancey, L., Hilbig, R. and Beland, P. 1991. Aromatic DNA–carcinogen adducts in beluga whales from the Canadian arctic and Gulf of St. Lawrence. *Marine Pollution Bulletin* 22: 392–396.

Reijnders, P.J.H. 1988. Ecotoxicological perspectives in marine mammalogy: research principles and goals for a conservation policy. *Marine Mammal Science* 4: 91–102.

Ridgway, S. and Reddy, M. 1995. Residue levels of several organochlorines in *Tursiops truncatus* milk collected at varied stages of lactation. *Marine Pollution Bulletin* 30: 609–614.

Safe, S. 1986. Comparative toxicology and mechanism of action of polychlorinated dibenzo-p-dioxins and dibenzofurans. *Annual Reviews in Pharmacology and Toxicology* 26: 371–399.

Safe, S. 1988. The aryl hydrocarbon (Ah) receptor. *ISI Atlas of Science: Pharmacology* 2: 78–83.

Safe, S. 1989. Polychlorinated biphenyls (PCBs): mutagenicity and carcinogenicity. *Mutation Research* 220: 31–47.

Safe, S. 1990. Polychlorinated biphenyls (PCBs) and polybrominated biphenyls (PBBs): biochemistry, toxicology and mechanisms of action. *Toxicology* 21(1): 51–88.

Safe, S. 1991. Polychlorinated dibenzo-*p*-dioxins and related compounds: sources, environmental distribution and risk assessment. *Environmental Carcinogenesis and Ecotoxicological Reviews* C9: 261–302.

Safe, S.H. 1998a. Development validation and problems with the toxic equivalency factor approach for risk assessment of dioxins and related compounds. *Journal of Animal Science* 76: 134–141.

Safe, S.H. 1998b. Hazard and risk assessment of chemical mixtures using the toxic equivalency factor approach. *Environmental Health Perspectives* 106(suppl 4): 1051–1058.

Safe, S., Safe, L. and Mullin, M. 1985. Polychlorinated biphenyls: Congener-specific analysis of a commercial mixture and a human milk extract. *Journal of Agricultural and Food Chemistry* 33: 24–29.

Safe, S., Zacharewski, T., Safe, L., Harris, M., Yao, C. and Holcomb, M. 1989. Validation of the AHH induction bioassay for the determination of 2,3,7,8-TCDD toxic equivalents. *Chemosphere* 18: 941–946.

Santostefano, M.J., Wang, X., Richardson, V.M., Ross, D.G., DeVito, M.J. and Birnbaum, L.S. 1998. A pharmacodynamic analysis of TCDD-induced cytochrome P450 gene expression in multiple tissues: dose- and time-dependent effects. *Toxicology and Applied Pharmacology* 151: 294–310.

Shimada, T., Imai, Y. and Sato, R. 1981. Covalent binding *in vitro* of polychlorinated biphenyls to proteins by reconstituted monooxygenase system containing cytochrome P-450. *Chemico-Biological Interactions* 38: 29.

Shugart, L.R. and Kao, J. 1985. Examination of adduct formation in vivo in the mouse between benzo(*a*)pyrene and DNA of skin and hemoglobin of red blood cells. *Environmental Health Perspectives* 62: 223–226.

Sipes, G. and Gandolfi, A.J. 1991. Biotransformation of toxicants. Pages 88–126 in Amdur, M.O., Doull, J. and Klaassen, C.D., editors. *Casarett and Doull's toxicology: The basic science of poisons*. Pergamon Press, New York.

Stowers, S.J. and Anderson, M.W. 1985. Formation and persistence of benzo(*a*)pyrene metabolite-DNA adducts. *Environmental Health Perspectives* 62: 31–39.

Subramanian, A., Tanabe, S., Tatsukawa, R., Saito, S. and Miyazaki, N. 1987. Reduction in the testosterone levels by PCBs and DDE in Dall's porpoises of northwestern North Pacific. *Marine Pollution Bulletin* 18: 643–646.

Tanabe, S. 1989. A need for reevaluation of PCB toxicity. *Marine Pollution Bulletin* 20: 247–248.

Tanabe, S., Tatsukawa, R., Maruyama, K. and Miyazaki, N. 1982. Transplacental transfer of PCBs and chlorinated hydrocarbon pesticides from the pregnant striped dolphin (*Stenella coerulealba*) to her fetus. *Agricultural Biological Chemistry* 46: 1249–1254.

Tanabe, S., Kannan, N., Subramanian, A., Watanabe, S. and Tatsukawa, R. 1987. Highly toxic coplanar PCBs: occurrence, source, persistency and toxic implications to wildlife and humans. *Environmental Pollution* 47: 147–163.

Tanabe, S., Watanabe, S., Kan, H. and Tatsukawa, R. 1988. Capacity and mode of PCB metabolism in small cetaceans. *Marine Mammal Science* 4: 103–124.

Tanabe, S., Kannan, N., Ono, M. and Tatsukawa, R. 1989. Toxic threat to marine mammals: increasing toxic potential of non-ortho and mono-ortho coplanar PCBs from land to ocean. *Chemosphere* 18: 485–490.

Tatsukawa, R. 1992. Contamination of chlorinated organic substances in the ocean ecosystem. *Water Science and Technology* 25: 1–8.

Truss, M. and Beato, M. 1993. Steroid hormone receptors: interaction with deoxyribonucleic acid and transcription factors. *Endocrinology Reviews* 14: 450–479.

Tryphonas, H. *et al.* 1991a. Effects of PCB (Aroclor 1254) on non-specific immune parameters in rhesus (*Macaca mulatta*) monkeys. *International Journal of Immunopharmacology* 13: 639–648.

Tryphonas, H. *et al.* 1991b. Effect of chronic exposure of PCBs (Aroclor 1254) in specific and nonspecific immune parameters in the rhesus (*Macaca mulatta*) monkey. *Fundamental and Applied Toxicology* 16: 773–786.

Varanasi, U., Stein, J.E. and Hom, T. 1981. Covalent binding of benzo(a)pyrene to DNA in fish liver. *Biochemical and Biophysical Research Communications* 103: 780–787.

Vos, J.G., De Heer, C. and Van Loveren, H. 1997–98. Immunotoxic effects of TCDD and toxic equivalency factors. *Teratogenesis, Carcinogenesis, and Mutagenesis* 17: 275–284.

Watanabe, S. *et al.* 1989. Specific profile of liver microsomal cytochrome P-450 in dolphin and whales. *Marine Environmental Research* 27: 51–65.

Weinstein, I.B. 1988. The origins of human cancer: molecular mechanisms of carcinogenesis and their implications for cancer prevention and treatment. *Cancer Research* 48: 4135–4143.

White, R.D., Hahn, M.E., Lockhart, W.L. and Stegeman, J.J. 1994. Catalytic and immunochemical characterization of hepatic microsomal cytochromes P450 in beluga whale (*Delphinapterus leucas*). *Toxicology and Applied Pharmacology* 126: 45–57.

White, R.D., Shea, D., Schlezinger, J.J., Hahn, M.E. and Stegeman, J.J. 2000. In vitro metabolism of polychlorinated biphenyl congeners by beluga whale (*Delphinapterus leucas*) and pilot whale (*Globicephala melas*) and relationship to cytochrome P450 expression. *Comparative Biochemistry and Physiology* 126C: 267–284.

Whitelaw, M., Pongratz, I., Wilhelmsson, A., Gustafsson, J.-Å. and Poellinger, L. 1993. Ligand-dependent recruitment of the ARNT coregulator determines DNA recognition by the dioxin receptor. *Molecular and Cellular Biology* 13: 2504–2514.

Whitlock, J.P. Jr 1987. The regulation of gene expression of 2,3,7,8-tetrachlorodibenzo-*p*-dioxin. *Pharmacological Review* 39: 147–161.

Whitlock, J.P. Jr 1990. Genetic and molecular aspects of 2,3,7,8-tetrachlorodibenzo-*p*-dioxin action. *Annual Reviews in Pharmacology and Toxicology* 30: 251–277.

Williams, G.M. and Weisburger, J.H. 1991. Chemical carcinogenesis. Pages 127–200 in Amdur, M.O., Doull, J., Klaassen, C.D., editors. *Casarett and Doull's toxicology: The basic science of poisons.* Pergamon Press, New York.

Wogan, G.N. 1989. Markers of exposure to carcinogens: method for human biomonitoring. *Journal of the American College of Toxicology* 8: 871–881.

17 Ecotoxicological investigations of bottlenose dolphin (*Tursiops truncatus*) strandings: Accumulation of persistent organic chemicals and metals

*John E. Stein, Karen L. Tilbury,
James P. Meador, Jay Gorzelany,
Graham A.J. Worthy and Margaret M. Krahn*

Introduction

An unprecedented mass mortality of bottlenose dolphins (*Tursiops truncatus*) occurred along the US Atlantic coast from June 1987 until March 1988 (Geraci, 1989). During this event more than 740 dolphins stranded. This was followed by unusual mortalities of bottlenose dolphins in the Gulf of Mexico in 1990 and in 1992. In Europe, strandings of striped dolphins (*Stenella coeruleoalba*) in the Mediterranean Sea began in 1990 and continued through 1992 (Aguilar and Raga, 1993). These and other recent stranding events have raised recurring questions about whether toxic chemicals and elements (metals) may be contributing factors in cetacean mass mortality.

Strong ecological evidence is available indicating that chemical pollution in coastal areas near urban centers exerts a variety of deleterious biological effects in marine fish (Myers *et al.*, 1987; Arkoosh *et al.*, 1991; Varanasi *et al.*, 1992; Arkoosh *et al.*, 1998a, b). These findings support concerns that marine mammals, particularly coastal species, may also experience adverse physiological effects, such as reproductive and immune dysfunction (DeLong *et al.*, 1973; Duinker *et al.*, 1979; Reijnders, 1986; Ross *et al.*, 1995). Demonstrating a causal link between pollution and adverse effects is particularly difficult in marine mammals. This is due to problems inherent in obtaining sufficient numbers of samples from both healthy individuals and those exhibiting specific diseases, and because of limitations in conducting controlled experimental studies with live marine mammals. Nevertheless, both noninvasive and invasive biological markers of potential adverse effects can be collected from live or recently deceased animals, substantially improving our capability to assess contaminant effects on marine mammals. The use of biological markers, however, requires detailed information on exposure in order to determine associations between exposure and effects. Here we report data we have collected on contaminant exposure in bottlenose dolphins from analyses of 218 tissue samples from 78 animals that stranded primarily in the northern Gulf of Mexico and the north-western Atlantic off Massachusetts, USA. These include specimens collected during mass mortality events.

Halogenated aromatic hydrocarbons (HAHs) are a class of contaminants of particular concern. Odontocetes, which are often top predators in marine food webs, can accumulate high levels of certain lipophilic contaminants, such as polychlorobiphenyls (PCBs) and chlorinated pesticides (Waid, 1986). The HAHs are among the most widespread and persistent chemical contaminants in coastal environments (Varanasi *et al.*, 1992). Because of their lipophilicity and resistance to metabolism, these pollutants tend to bioaccumulate in aquatic organisms, particularly in lipid-rich tissues of marine mammals. Furthermore, HAHs are suspected of exerting immunological, reproductive and carcinogenic effects in wildlife (e.g. Safe, 1984). Of particular interest are the dioxin-like PCB congeners that have a stereochemically similar (dioxin-like) conformation to that of 2,3,7,8-tetrachlorodibenzo-*p*-dioxin. These congeners exhibit dioxin-like toxic effects mediated through binding to the aryl hydrocarbon (Ah) receptor (Birnbaum, 1985; Carvan *et al.*: Chapter 16 in this volume). Although the dioxin-like PCB congeners are present in the environment at much lower concentrations than are the non-dioxin-like congeners, the majority of PCB toxicity is thought to be contributed by the dioxin-like congeners (Safe, 1990; Ahlborg *et al.*, 1994). Although not measured in this study, metabolites of HAH compounds, such as methylsulfones, can also bioaccumulate in mammals, and certain metabolites have been linked with mammalian toxicity (Lund *et al.*, 1988). The methylsulfone of DDE, a metabolite of DDT, causes severe adrenal damage in various species (Jonsson *et al.*, 1992) and is suspected of being associated with subsequent effects on other organs in gray (*Halichoerus grypus*) and ringed (*Pusa hispida*) seals from the Baltic Sea (Olsson *et al.*, 1994). These findings suggest a putative causal relationship between DDE and PCB methylsulfones and a disease complex involving the adrenal gland in seals, and supports further investigation of the toxicology of these HAH metabolites and other metabolites such as hydroxylated PCB congeners (Bergman *et al.*, 1994 and Chapter 19 in this volume).

Determination of toxic and essential element concentrations in marine mammals is also important because of the physiological and toxicological significance of these substances and their accumulation in liver and kidney (Marcovecchio *et al.*, 1990; Law *et al.*, 1991; Meador *et al.*, 1993). For example, cadmium, mercury and lead are nonessential elements that can accumulate to elevated levels in mammals (Venugopal and Luckey, 1978; Meador *et al.*, 1999). As in many marine mammals, the highest concentrations of mercury are found in livers of cetaceans, whereas cadmium concentrations are generally greatest in kidneys (Honda *et al.*, 1983; Marcovecchio *et al.*, 1990; Meador *et al.*, 1993). Methylmercury is also accumulated by marine mammals and is more toxic than inorganic mercury (Venugopal and Luckey, 1978). The effects of high concentrations of these metals have been studied only rarely in cetaceans. Excess cadmium, for example, can result in renal dysfunction and adverse effects on reproduction, growth and bone structure in mammals (Kostial, 1986). However, interactions among some metals (e.g. mercury and selenium) can reduce the toxic potential. In addition, the effect of accumulation of nonessential metals on the homeostasis of essential metals (e.g. zinc, copper) warrants further investigation.

Methods

Field sampling

Samples of blubber, liver and kidney were collected from 78 bottlenose dolphins: 20 dolphins were a subset of approximately 350 bottlenose dolphins that stranded from February to June 1990 along the Gulf of Mexico coast from Texas to Florida (Texas, $n = 14$; Louisiana, $n = 1$; Mississippi, $n = 1$; Alabama, $n = 1$; Florida, $n = 3$), 20 were from a stranding event in March and April 1992 along the Texas coast, 20 stranded along the Texas coast between October 1991 and December 1993, 14 stranded along the Florida coast between March 1990 and September 1992, and 4 stranded on the coast of Massachusetts in December 1992 (Figure 17.1; Appendix at end of chapter). The stranding sites, including latitude and longitude, date collected, sex, length, age, condition of the animal at the time the tissues were collected and the tissues analyzed are listed in the Appendix. The numbers of animals in the tables and figures do not correspond exactly to the numbers in the Appendix, because blubber and liver samples were not available for all of the animals. The lengths of the animals ranged from 96 to 279 cm, ages ranged from 0.02 to more than 45 years, and tissues were collected from 38 females and 38 males; the sex of two dolphins was not known (Table 17.1). A subset ($n = 32$) of dolphins was assigned ages by determining growth layer groups in teeth (Fernandez and Hohn, 1998). For the remainder, age was estimated from mercury concentrations in liver (Figure 1 in Meador *et al.*, 1999) with the inverse prediction model using the equation from Figure 1 in Meador *et al.* (1999) for animals from Texas ($Y = 1.42 + 0.88 \times$ log age, with $r^2 = 0.71$). The inverse prediction model does not generate a new equation, but uses the data from the original equation to predict X by using Y, which is the 'inverse' of how it is usually done. Age estimation is particularly important, rather than relying on body length as a surrogate, because the lengths of several of the Gulf of Mexico dolphins were similar to the asymptotic length determined for bottlenose dolphins from the Atlantic coast (Mead and Potter, 1990). However, we could not determine ages of

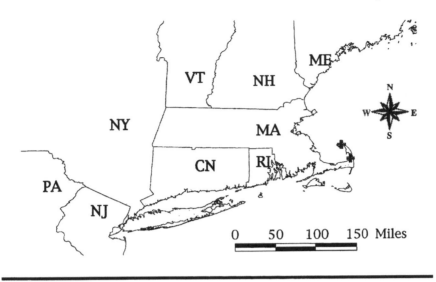

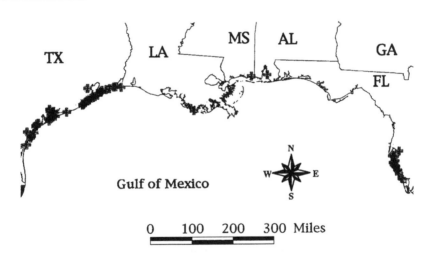

Figure 17.1 Stranding sites of bottlenose dolphins.

some animals, specifically those from the Massachusetts coast, because these animals may be from different populations than the Gulf of Mexico dolphins and the regression equation that relates ages and concentrations of mercury in livers may not apply (Table 17.1).

Analytical procedures for organochlorines

Subsamples of blubber and liver tissue were analyzed for selected individual PCB congeners, DDT and its metabolites (DDE and DDD), selected pesticides (Krahn *et al.*, 1988a, b; Sloan *et al.*, 1993), and percent lipid (Varanasi *et al.*,

Table 17.1 Life history data for bottlenose dolphins (*Tursiops truncatus*)

	Length (cm)	n	Age (years)	n
Gulf of Mexico				
Texas				
female	199 (9)[a]	28	8.6 (2.1)[b]	27
male	195 (13)	24	6.8 (2.1)	22
unknown	247 (9)	2	9	1
Florida				
female	203 (20)	8	8.5 (2.6)	8
male	200 (12)	9	5.1 (1.8)	9
Louisiana				
female	251	1	na	–
Mississippi				
male	161	1	na	–
Alabama				
male	261	1	na	–
North-west Atlantic				
Massachusetts				
female	279	1	na	–
male	250 (10)	3	na	–

[a] Mean value (standard error of the mean).
[b] Age was estimated from dental growth layers or predicted from mercury in liver (Appendix).
n = the number of animals.
na = not available. These animals may be from a different population of dolphins, especially the Massachusetts animals, and the regression equation that related age and liver concentration of mercury may not apply (Meador *et al.*, 1999).

1994) following standard methods and quality assurance (QA) protocols. Tissue (1–3 g) was macerated with sodium sulfate and methylene chloride. The methylene chloride extract was filtered through a column of silica gel and alumina, and concentrated for further clean-up; the analytical method was slightly modified for the lipid-rich tissue of marine mammals. Size exclusion chromatography with high performance liquid chromatography was used to separate lipids and other biogenic material from a fraction containing the HAHs. The HAH fraction was analyzed by capillary column gas chromatography (GC) with an electron capture detector, and peak identifications were confirmed on selected samples using GC–mass spectrometry with selected ion monitoring.

The HAHs are reported here as follows: 'Σchlordanes' refers to the sum of concentrations of *cis*-chlordane, oxychlordane, *trans*-nonachlor, heptachlor, and heptachlor epoxide; 'ΣDDTs' is the sum of concentrations of *o,p'*-DDD, *p,p'*-DDD, *o,p'*-DDE, *p,p'*-DDE, *o,p'*-DDT and *p,p'*-DDT; and 'ΣPCBs' is the sum of concentrations of chlorinated biphenyl congeners 18, 28, 44, 52, 66, 101, 105, 118, 128, 138, 153, 170, 180, 187, 195, 206 and 209, following the nomenclature of Ballschmiter and Zell (1980). All concentrations are given on a wet weight basis, unless otherwise stated.

Analytical procedures for elements

Subsamples of liver and kidney were analyzed for selected elements and per cent dry weight (Robisch and Clark, 1993). Tissue (1–2 g) was digested with 10 ml of concentrated ultra pure nitric acid for 2 h at room temperature and subsequently heated in a microwave oven in a sealed Teflon bomb. The digestate was further treated to destroy organic matter by digestion with 4 mL hydrogen peroxide and further heating in the microwave oven. The concentrations of elements were determined by atomic absorption spectro-photometry (mercury, lead) and inductively coupled argon plasma emission spectroscopy (cadmium). All concentrations of elements are given on a wet weight basis, unless otherwise stated (see Meador *et al.*, 1999 for details).

Statistical analysis

Analysis of variance (ANOVA) was used to determine whether differences in the concentrations and proportions (percentage of the total HAHs) of analytes in tissues were dependent on the sex, length, estimated age of the animal, or the stranding site. Length and estimated age were used as covariates and the data for the concentrations of analytes were log trans-formed to reduce deviations from normality (i.e. to reduce heteroscedasticity in the variances) and then analyzed using the SuperANOVA statistical pack-age (Abacus Concepts, 1989). The results of the statistical analyses were very similar whether concentrations were expressed on a wet weight or lipid weight basis. Differences were considered significant at $\alpha \leq 0.05$. Possible differences in the HAHs accumulated in blubber (i.e. patterns) between bottlenose dolphins that stranded along different areas of the Gulf of Mexico and Massachusetts were assessed by principal component analysis (PCA) using the JMP statistical package (SAS Institute, Inc. 1994). Before PCA was performed, the concentrations of individual HAHs in each blubber sample were normalized (concentrations of individual HAHs in each sample were divided by the sum of the HAHs in the sample). The use of PCA is advantageous when the concentrations of several analytes (e.g. chlorobiphenyl congeners) are evaluated statistically. The dimensionality (i.e. number of measurements) is reduced to fewer variables (i.e. principal components) without losing information, thus providing a better understanding and clearer visualization of the interrelationships among analyte patterns and the loca-tions of strandings.

Bioaccumulation of halogenated aromatic hydrocarbons

Concentrations in tissues

A wide range of HAH concentrations was found in blubber and liver of stranded bottlenose dolphins from the different regions (Table 17.2). For

Table 17.2 Concentrations of organochlorines and per cent lipid in blubber and liver samples collected from bottlenose dolphins (*Tursiops truncatus*)

	Hexachlorobenze	ΣChlordanes	ΣDDTs	ΣPCBs	Per cent lipid
Blubber					
Gulf of Mexico					
Texas (*n* = 44)	570 (180)[a] (7–6600)[b]	2800 (520) (51–17 000)	21 000 (3800) (300–110 000)	15 000 (2500) (280–72 000)	42 (2.7)[c] (8.0–80)
Louisiana (*n* = 1)	300	1500	6600	9800	23
Mississippi (*n* = 1)	590	10 000	100 000	39 000	58
Alabama (*n* = 1)	sample not available				
Florida (*n* = 17)	34 (5) (nd–78)	7500 (1400) (64–20 000)	11 000 (2600) (240–43 000)	15 000 (2300) (1100–34 000)	48 (4.3) (1.2–74)
North-west Atlantic					
Massachusetts (*n* = 4)	190 (55) (33–270)	5500 (2700) (450–13 000)	33 000 (21 000) (1500–96 000)	43 000 (23 000) (4000–110 000)	74 (3.1) (68–83)
Liver					
Gulf of Mexico					
Texas (*n* = 53)	55 (15) (nd–620)	180 (51) (nd–2000)	2700 (930) (6–36 000)	1400 (290) (17–12 000)	6.8 (1.0) (1.3–39)
Louisiana (*n* = 1)	53	68	220	600	4.8
Mississippi (*n* = 1)	55	660	12 000	3700	5.4
Alabama (*n* = 1)	89	3200	32 000	12 000	2.2
Florida (*n* = 17)	14 (11) (nd–190)	230 (55) (5–900)	350 (73) (3–1100)	730 (130) (48–2000)	3.4 (0.3) (1.2–5.6)
North-west Atlantic					
Massachusetts (*n* = 4)	16 (6) (2–31)	180 (120) (4–540)	3300 (3100) (17–12 000)	1100 (640) (56–2900)	3.3 (1.1) (1.7–6.4)

[a] Mean values (standard error of the mean) were calculated using one-half the detection limit for any analytes that were not detected (nd). The means are expressed as ng/g wet weight.
[b] The range of concentrations or range of per cent lipid.
[c] The means are expressed as per cent lipid.
n = the number of animals.

example, the concentrations of ΣDDTs in blubber of dolphins that stranded in Texas ranged from 300 to 110 000 ng/g. A wide range in the concentrations of HAHs was also detected in liver, but at much lower concentrations. This difference between tissues was largely accounted for by the differences in total lipids. The mean and standard error of the mean (SEM) of the ratio of concentrations of HAHs in blubber/liver on a wet weight basis was 26 (2.4) and on lipid basis the ratio was 1.9 (0.1).

The concentrations of HAHs in tissues reported for these dolphins are within the range of concentrations found in other small cetaceans (Muir *et al.*, 1988b; Kuehl *et al.*, 1991) and considerably higher than in larger cetaceans (O'Shea and Brownell, 1994). A major factor for this difference is that most of the larger cetaceans are mysticetes, which feed predominantly on lower trophic-level organisms, and generally do not feed near coastal urban sites, which are often more contaminated. Trophic position is known to be an important factor in the bioaccumulation of lipophilic compounds that are resistant to metabolism, such as PCBs (Muir *et al.*, 1988a; Hargrave *et al.*, 1992; Harding *et al.*, 1997). A study of fish, for example, has demonstrated that trophic position, measured by stable isotopes, was the major factor controlling accumulation of PCBs by burbot (*Lota lota*) from a subarctic lake (Kidd *et al.*, 1995). In addition to diet and trophic position, physiological differences [e.g. excretion rate (Parke, 1980)], activity of detoxifying enzymes (Walker, 1980), metabolism (Moriarty, 1984), concentration of pollutants within prey, nutritive condition, disease, age and sex and reproductive differences among species of cetaceans are also believed to contribute to the differences in the body burden of HAHs (i.e. total mass of HAHs in the entire organism). Also, body size and metabolic rate play a role in accumulation of HAHs, with smaller species of cetaceans having higher concentrations in tissues, but sometimes lower body burdens (Aguilar *et al.*, 1999). All of these complex factors together contribute to the lower tissue concentrations of HAHs in species such as gray whales (*Eschrichtius robustus*) and bowhead whales (*Balaena mysticetus*), compared to the odontoceti, such as bottlenose dolphins and harbor porpoises (*Phocoena phocoena*). For example, recent analyses (O'Hara *et al.*, 1999) of tissues from bowhead whales, which feed on invertebrates in the Arctic, show that the mean blubber concentration of ΣPCBs (180 ± 20 ng/g) in these samples is nearly 100-fold lower than concentrations in bottlenose dolphins from the Gulf of Mexico (Table 17.2).

Texas and Florida dolphins showed no significant differences in total HAH concentrations in blubber or liver after adjusting for age in the covariance analysis. There were significant differences in the concentrations of the pesticide hexachlorobenzene (HCB) and Σchlordanes in blubber and liver. The results were similar among bottlenose dolphins from Texas, Florida and Massachusetts, with length used as a covariate (length was used as a covariate instead of age because ages were not available for all of the dolphins, as explained in Table 17.1). Similar statistical evaluation for the effect of

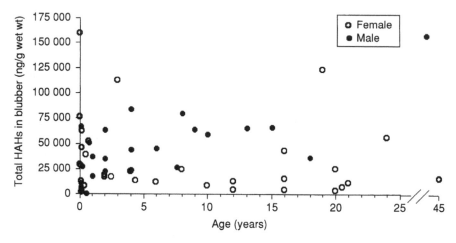

Figure 17.2 The concentrations of total halogenated aromatic hydrocarbons (HAHs) in blubber of female and male bottlenose dolphins in relation to estimated age.

sex, using length as a covariate, showed significant differences in the concentrations of total HAHs, including Σchlordanes, ΣDDTs and ΣPCBs, in blubber and liver between the female and male dolphins from the coasts of Texas, Florida and Massachusetts. The concentrations in males were approximately two to three times higher than the concentrations in females. For example, the mean concentration (SEM) of total HAHs in blubber for females (n = 37) was 29 000 (5800) ng/g and the mean concentration for males (n = 30) was 57 000 (9300) ng/g. Using only the animals for which age was estimated, there were no significant statistical correlations. Curvilinear effects between the concentrations of HAHs in blubber with estimated age in either females or males were tested and the r^2 ranged from 0.02 to 0.52. However, for dolphins of similar age, the males generally had higher concentrations of HAHs than the females that were 4 or more years old (Figure 17.2). It is also interesting to note that the largest difference between male and female dolphins was between the oldest dolphins, with the concentration in the oldest male (43 years old) 12 times greater than the concentration in the oldest female (>45 years old). This difference between female and male dolphins is not surprising. Female–male differences in the concentrations of ΣDDTs and ΣPCBs in blubber from 33 bottlenose dolphins that stranded along the Gulf of Mexico coasts were also reported by Salata *et al.* (1994). Lower concentrations of HAHs in blubber are observed in sexually mature females due to transfer of HAHs during lactation (e.g. Cockcroft *et al.*, 1989). This was shown in a recent study by Ridgway and Reddy (1995) in which milk periodically sampled from captive bottlenose dolphins had decreasing HAH concentrations during lactation. The lower concentrations of HAHs in milk were probably reflective of lower concentrations in the body

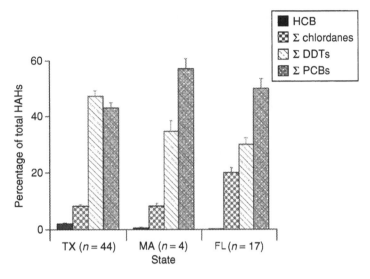

Figure 17.3 The percentages of summed (Σ) analytes (see Methods) in blubber of bottlenose dolphins (*Tursiops truncates*) of the total halogenated aromatic hydrocarbons (HAHs) reported. Texas (TX) and Florida (FL) were significantly ($\alpha \leq 0.05$) different from each other for hexachlorobenzene, Σchlordanes and ΣDDTs; Massachusetts (MA) and FL for Σchlordanes. *n* = number of animals.

after HAHs were transferred to offspring in milk, therefore reducing concentrations in the body. In addition, the concentrations of HAHs were higher in females that had never calved or had not calved for at least 26 years (Ridgway and Reddy, 1995). These findings highlight that knowledge of reproductive history and age data is critical for interpreting concentrations of lipophilic contaminants in tissues of dolphins.

Organochlorine profiles

The ΣDDTs and ΣPCBs generally accounted for the greatest proportion of the total HAHs in blubber of the stranded bottlenose dolphins (Figure 17.3). For example, the blubber of dolphins that stranded in Texas waters contained high proportions of ΣDDTs (47 per cent) and ΣPCBs (43 per cent), with a lesser contribution from the Σchlordanes (8 per cent); HCB accounted for only 2 per cent of the total HAHs. The contribution of Σchlordanes (20 per cent) to the total HAHs in stranded bottlenose dolphins from Florida was significantly higher (more than twofold) than in the dolphins stranding in Texas, whereas the proportion of ΣDDTs and HCB were significantly higher in the Texas dolphins. For the dolphins that stranded on the Massachusetts coast, the mean percentages for the Σchlordanes, ΣDDTs, and ΣPCBs in blubber were 8, 35 and 57 per cent, respectively. The percentage

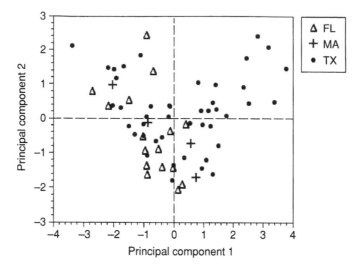

Figure 17.4 The first two principal components (see Methods) based on five chlordanes (see Methods) in blubber of bottlenose dolphins (*Tursiops truncates*) (*n* = 65) stranded on Florida (FL), Massachusetts (MA) and Texas (TX) coasts. *n* = number of animals.

of Σchlordanes in blubber for the dolphins from Florida was significantly higher than that for dolphins from Massachusetts.

There were no distinct patterns for individual DDTs and PCBs in blubber of stranded bottlenose dolphins sampled from the Texas, Florida and Massachusetts coasts (not shown) when assessed by PCA (see Methods). However, a plot of the first two principal components, which represented 75 per cent of the total variation in the normalized concentrations of individual chlordanes (*cis*-chlordane, oxychlordane, *trans*-nonachlor, heptachlor, heptachlor epoxide), showed a small distinction in the profile of these analytes in dolphins among the various sites (Figure 17.4). The dolphins that stranded in Florida (represented by triangles) are grouped mainly in the upper and lower left quadrants whereas more than 50 per cent of the animals stranded in Texas (represented by circles) are in the upper and lower right quadrants. This distinction in the profile of chlordanes between dolphins from Florida and Texas corroborates the statistical differences shown in Figure 17.3, where Σchlordanes as a per cent of the total HAHs are statistically higher in dolphins from Florida than in those from Texas and Massachusetts. These findings indicated that because there were no distinct patterns for individual DDTs and PCBs and only apparent differences in the profiles of the chlordanes in bottlenose dolphins stranding at these three sites, the variation among individual animals is greater than differences among regions. This may be due to age and reproductive status.

Relation to stranding

A preliminary study (Lahvis *et al.*, 1995) of free-ranging bottlenose dolphins from the Gulf of Mexico suggested a correlation between increasing concentrations of PCBs and DDTs in blood and altered leukoproliferative response of plasma lymphocytes. These are intriguing findings that suggest, as observed in pinnipeds (Ross *et al.*, 1996), that HAHs can have a deleterious effect on immunocompetence. This may lead to increased parasitism as well as increased susceptibility to disease. It is noteworthy that studies with wild fish have demonstrated an association between altered immunocompetence and disease susceptibility (Arkoosh *et al.*, 1998a). However, whether exposure to HAHs in cetaceans contributes significantly to unusual mass mortality events is contentious and a full review of data on both sides of this issue is beyond the scope of this paper.

Our results support some studies with small cetaceans showing elevated tissue concentrations of HAHs in stranded or diseased animals. Abnormally high PCB levels were found in striped dolphins affected by the 1990–1992 Mediterranean morbillivirus epizootic (Aguilar and Borrell, 1994). The median concentration of ΣPCBs (780 000 ng/g lipid weight) was significantly higher in dolphins that died in the epizootic than the concentrations of ΣPCBs (280 000 ng/g lipid weight) in biopsies from a putative healthy group. In some of the diseased dolphins the PCB concentrations in blubber were as high as 1 000 000 ng/g lipid weight. Additionally, a comparison among species of Delphinidae from various geographical areas (Figure 17.5) shows higher concentrations of ΣPCB congeners 138 and 153 (sums of total HAHs were not available in many of the papers reviewed and these congeners were reported most consistently) and ΣDDTs in striped dolphins, Risso's dolphins (*Grampus griseus*) and bottlenose dolphins stranded during the Mediterranean epizootic. In contrast, a recent study of an unusual stranding event of short-beaked (*Delphinus delphis*) and long-beaked (*Delphinus capensis*) common dolphins on the US west coast showed no marked difference in concentrations in blubber of PCBs between the common dolphins that stranded and apparently healthy dolphins that were incidentally taken in a fishery interaction (Susan Chivers, 1 October 1998, personal communication). Moreover, the concentrations of PCBs in blubber of the common dolphins appeared comparable to the concentrations in bottlenose dolphins reported herein, although comparisons among species should be made with caution. These findings indicate that there is little compelling evidence of a direct cause-and-effect association between HAHs and specific bottlenose dolphin stranding events. However, the elevated body burden of PCBs and other HAHs in bottlenose dolphins could potentially have deleterious effects on physiological functions, such as immunocompetence, which may lead to a decrease in disease resistance or increased parasitism in the general population, as suggested by Lahvis *et al.* (1995). Clearly, additional comparative studies of the body burden of toxic compounds in wild bottlenose

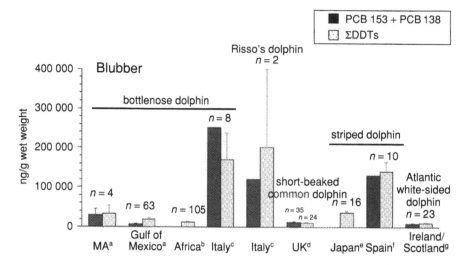

Figure 17.5 The concentrations of PCB 138 plus PCB 153 and the sum of DDTs in blubber of dolphins collected from several geographical areas. The species included are: bottlenose dolphins (*Tursiops truncatus*), Risso's dolphins (*Grampus griseus*), short-beaked common dolphins (*Delphinus delphis*), striped dolphins (*Stenella coeruleoalba*), and Atlantic white-sided dolphins (*Lagenorhynchus acutus*). [a]This study; [b]Cockcroft *et al.*, 1989; [c]Corsolini *et al.*, 1995; [d]Law *et al.*, 1994; [e]Loganathan *et al.*, 1990; [f]Kannan *et al.*, 1993; [g]McKenzie *et al.*, 1997.

dolphins from the general population and from captive animals, as well as those taken in fishery interactions, will provide important baseline data for determining the role of chemical contaminants in unusual mortality events. Live-capture studies of bottlenose dolphins are being conducted to assess relationships between health parameters and levels of contaminants in blood and blubber taken by biopsy (Larry Hansen, 4 March 1999, personal communication).

It is worth noting that in all studies relating effects to exposure (e.g. assessing the health risk posed by chemical contaminants) the value of the data will be dependent on the QA procedures employed in the analyses. The comparison of chemical contaminant data among different studies often has been difficult because of insufficient information on QA measures employed in both the sample collection and chemical analyses. The use of rigorous QA protocols will increase greatly the comparability among studies, which is important for comparisons among geographical regions, as well as mandatory in reliably assessing temporal trends in contaminant concentrations in marine species and in monitoring the effectiveness of management actions in improving the quality of our coastal ecosystems. Without high-quality data over extended time frames, the ability to identify and quantify improvements in environmental quality can be problematic. The use of rigorous performance-based QA/quality control (QC) protocols includes the use of standard reference materials appropriate to the matrix (e.g. tissue) being analyzed (Krahn *et al.*, 1997). Additionally, the value of performance-based QA/QC has been demonstrated in studies with composited whale blubber material (Schantz *et al.*, 1993, 1995; Wise *et al.*, 1993) and composited mussel tissue (Wise *et al.*, 1991).

Bioaccumulation of metals

Mercury

The mean concentrations of mercury were highest in liver compared to kidney. The range in mercury concentrations in liver was 25–350 000 ng/g. Mercury concentrations in kidney ranged from 'not detected' to 280 000 ng/g (Table 17.3). Mercury in liver and kidney of bottlenose dolphins

Table 17.3 Concentrations of selected toxic elements in liver and kidney samples collected from bottlenose dolphins (*Tursiops truncatus*)

	Cadmium	Lead	Mercury
Liver			
Gulf of Mexico			
Texas	56 (10), $n = 36^a$ (nd–190)[b]	81 (18), $n = 40$ (1–660)	42 000 (10 000), $n = 44$ (25–350 000)
Louisiana ($n = 1$)	79	54	2800
Mississippi ($n = 1$)	nd	44	1800
Alabama ($n = 1$)	31	45	11 000
Florida	84 (66), $n = 6$ (nd–410)	46 (6), $n = 6$ (27–62)	58 000 (20 000), $n = 16$ (730–320 000)
North-west Atlantic			
Massachusetts ($n = 4$)	2100 (810) (590–4100)	16 (1) (nd–21)	7500 (2300) (3000–14 000)
Kidney			
Gulf of Mexico			
Texas	370 (55), $n = 42$ (nd–1600)	39 (8), $n = 40$ (1–330)	17 000 (5700), $n = 43$ (97–210 000)
Louisiana ($n = 1$)	610	59	2100
Mississippi ($n = 1$)	19	23	970
Alabama ($n = 1$)	170	28	1700
Florida	710 (190), $n = 11$ (nd–1900)	22 (5), $n = 5$ (7–32)	50 000 (17 000), $n = 16$ (nd–280 000)
North-west Atlantic			
Massachusetts ($n = 4$)	7800 (3500) (2400–18 000)	21 (6) (nd–43)	4100 (1600) (660–6900)

[a] Mean values (standard error of the mean) were calculated using one-half the detection limits for any analytes that were not detected (nd). The means are expressed as ng/g wet weight. The mean (SEM) of dry to wet weight ratio is 0.26 (0.9) for liver and 0.22 (0.8) for kidney.
[b] Range of concentrations.
n = the number of animals.
nd = not detected.

from the Gulf of Mexico was significantly higher compared to dolphins from Massachusetts when length was used as a covariate. The trend of higher mercury concentrations (Table 17.3, Figures 17.6a, b) in the Florida animals in comparison with Texas dolphins was likely due to higher exposure. A comparison among delphinids (Figure 17.7) shows geographical differences in the concentrations of mercury in several species of dolphins. Geographical differences could be due to natural or anthropogenic sources and may result from differential concentrations in prey. The type of prey and diet can contain varying concentrations of contaminants and can vary by species of dolphins (Meador *et al.*, 1999). Geraci (1989) also reported elevated concentrations of up to 110 000 ng/g of mercury in liver of bottlenose dolphins from the 1987–88 stranding event along the US mid-Atlantic coast. Additionally, concentrations up to 89 000 ng/g were detected in a subset of 15 bottlenose dolphins that stranded along the Gulf of Mexico during the 1990 event (Kuehl and Haebler, 1995).

Mercury concentrations exceeding 100 000 ng/g are suggested by Wagemann and Muir (1984) to be associated with toxic effects in marine mammals. In the present study, concentrations of mercury in liver samples from 8 of the 64 bottlenose dolphins exceeded 100 000 ng/g. Rawson *et al.* (1993) associated liver abnormalities in Atlantic bottlenose dolphins with chronic exposure to mercury, although the correlation between age and mercury concentrations may be a confounding factor. In dolphins with abnormalities, concentrations of up to 440 000 ng/g were found in livers, whereas concentrations up to 50 000 ng/g were measured in dolphins that exhibited no liver abnormalities. The assessment of the toxicity of mercury in cetaceans is not straightforward, however, because mercury forms a 1 : 1 complex with selenium, which is generally thought to be biologically inert. As seen in many studies, there is an inverse relationship between the proportion of highly toxic methylmercury and total mercury in tissues (e.g. liver, kidney, brain). The concentration of methylmercury in long-finned pilot whales (*Globicephala melas*) decreases in a nonlinear manner with total mercury, indicating *in vivo* demethylation (Meador *et al.*, 1993). Similarly, we also found a nonlinear relationship between methylmercury and total mercury in bottlenose dolphins (Figure 17.8, Meador *et al.*, 1999). A detailed assessment of the inter-relationships between toxic and trace elements in bottlenose dolphins from the Gulf of Mexico, which also accounts for age, can be found in Meador *et al.* (1999).

Many studies have shown that odontocete cetaceans may accumulate more mercury than mysticetes. Concentrations of 9–120 ng/g have been reported in livers of stranded gray whales from the North American west coast (Varanasi *et al.*, 1993, 1994), 32–100 ng/g in bowhead whales from the Arctic (Krone *et al.*, 1999), and 61–390 ng/g in Antarctic minke whales (*Balaenoptera bonaerensis*) (Honda *et al.*, 1987). In contrast, the maximum concentrations of mercury (350 000 ng/g) in livers of bottlenose dolphins

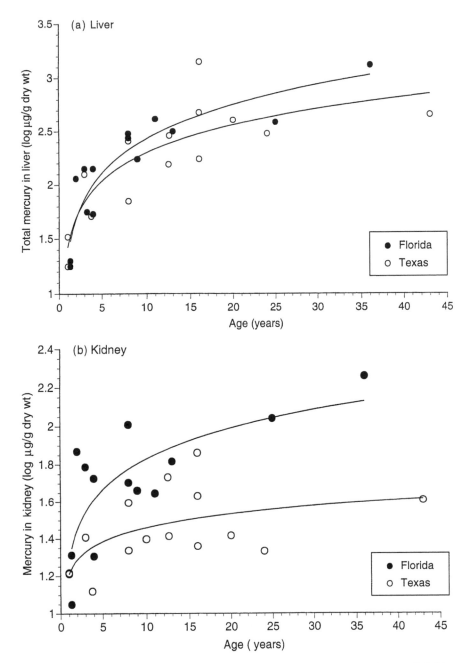

Figure 17.6 (a) The concentrations of mercury in livers of bottlenose dolphins (*Tursiops truncates*) versus age. Regressions for Florida animals: $Y = 1.36 + 1.08 \times \log$ age ($r^2 = 0.83$), Texas animals: $Y = 1.42 + 0.88 \times \log$ age ($r^2 = 0.71$). (Redrawn from Meador *et al.*, 1999.) (b) The concentrations of mercury in kidneys of bottlenose dolphins (*Tursiops truncates*) versus age. Regressions for Florida animals: $Y = 1.29 + 0.54 \times \log$ age ($r^2 = 0.56$), Texas animals: $Y = 1.20 + 0.25 \times \log$ age ($r^2 = 0.36$). (Redrawn from Meador *et al.*, 1999.)

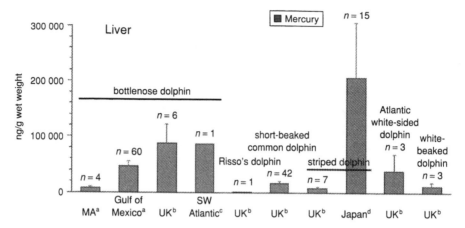

Figure 17.7 The concentrations of mercury in livers of dolphins collected from several geographical areas. The species included are: bottlenose dolphins (*Tursiops truncatus*), Risso's dolphins (*Grampus griseus*), short-beaked common dolphins (*Delphinus delphis*), striped dolphins (*Stenella coeruleoalba*), Atlantic white-sided dolphins (*Lagenorhynchus acutus*), and white-beaked dolphins (*Lagenorhynchus albirostris*). [a]This study; [b]Law *et al.*, 1994; [c]Marcovecchio *et al.*, 1990; [d]Itano and Kawai, 1981.

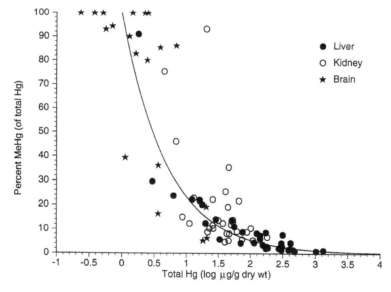

Figure 17.8 Methylmercury as a percentage of total mercury in liver, kidney and brain of bottlenose dolphins (*Tursiops truncates*). Texas and Florida sites combined. $Y = 99.6 \times e^{(-1.45X)}$ ($r^2 = 0.92$). (Redrawn from Meador *et al.*, 1999.)

found in our study are much higher than that reported in baleen whales. Mercury concentrations (170 000 ng/g) in livers of long-finned pilot whales reported by Meador *et al.* (1993) were also much higher than in baleen whales. These findings are consistent with the greater bioaccumulation of mercury in the odontoceti (bottlenose dolphins and pilot whales) that often feed at higher trophic levels, compared to mysticetes (gray, bowhead and minke whales) that feed predominantly on krill.

Cadmium

Mean concentrations of cadmium in liver and kidney were significantly higher in the Massachusetts bottlenose dolphins than in the Gulf of Mexico animals (Table 17.3). The concentrations of cadmium in liver (58 ng/g) and kidney (430 ng/g) of the Gulf animals were much lower than mean concentrations in liver (4300 ng/g) and kidney (4100 ng/g) of gray whales (Varanasi *et al.*, 1993, 1994), whereas the mean concentrations of cadmium in the liver and kidney of the Massachusetts dolphins (2100 and 7800 ng/g, respectively) were more comparable to concentrations in gray whales. In addition, the mean concentration for cadmium in liver of bowhead whales was 7800 ng/g (Krone *et al.*, 1999) with larger burdens of cadmium found in liver (9500 ng/g) and kidney (53 000 ng/g) of pilot whales (Meador *et al.*, 1993). The differences in cadmium concentrations in the Gulf dolphins compared to the Massachusetts dolphins, as well as gray, pilot or bowhead whales, may be explained by diet. The predominant prey of bottlenose dolphins that may strand along the Texas coast are teleosts (Hansen, 1992), whereas the primary prey of the Massachusetts dolphins, probably from an offshore population (Greg Early, 1 March 1994, personal communication) may include squid (Barros and Odell, 1990). Pilot whales that stranded on the Atlantic coast of the US (Meador *et al.*, 1993) also feed on squid (Sergeant, 1962), gray whales feed on benthic invertebrates (Rice and Wolman, 1971; Nerini, 1984), and bowhead whales feed mainly on planktonic crustaceans (Lowry, 1993). Previous studies have shown that invertebrates, including squid, can accumulate high concentrations of cadmium (Smith *et al.*, 1984; Finger and Smith, 1987). In addition, gray whales incidentally ingest sediment during foraging (Haley, 1986), which also could be a significant source of cadmium; many US west-coast sediments have been shown to contain more than 1000 ng/g dry weight of cadmium (Meador *et al.*, 1998). The higher cadmium concentrations in liver and kidney of the Atlantic bottlenose dolphins, and kidneys of pilot whales and gray whales, compared to bottlenose dolphins from the Gulf coasts suggest that a diet with a higher percentage of squid and other invertebrates is a major factor leading to high body burdens of this trace element.

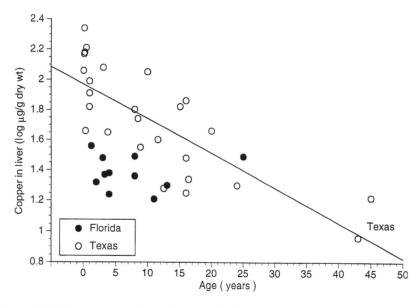

Figure 17.9 The concentrations of copper in livers of bottlenose dolphins (*Tursiops truncates*) versus age. Regression for Texas animals: $Y = 1.91 - 0.33 \times X$ ($r^2 = 0.60$). (Redrawn from Meador *et al.*, 1999.)

Lead and copper

The ranges in concentrations for lead in liver (not detected to 660 ng/g) and kidney (not detected to 330 ng/g) of the bottlenose dolphins were relatively low. These ranges were comparable to those found in liver (not detected to 250 ng/g) and kidney (14–29 ng/g) of long-finned pilot whales that stranded near Cape Cod, Massachusetts (Meador *et al.*, 1993). Additionally, Meador *et al.* (1999) reported the concentrations of several additional toxic and essential elements in liver and kidney from a subset of the animals reported herein. For example, the concentrations of copper, an essential element, in liver are higher for the dolphins from Texas than for those from Florida, even when controlled for age. Copper in the liver is known to decline normally with age; however, differences between populations were not expected (Figure 17.9). Normal physiological concentrations of copper in liver and kidney are not known for bottlenose dolphins. Altered concentrations of essential elements, such as copper, may be an indication of ill health. Several liver diseases in humans are known to cause elevated copper levels in this organ (Davis and Mertz, 1987). Conversely, deficiencies in essential elements in these organs may be an indication of animals in less than optimal health, assuming that dietary intake is sufficient (Meador *et al.*, 1999).

Accumulation with age

There were significant correlations of cadmium and mercury concentrations in both liver and kidney with age for the Gulf of Mexico dolphins, indicating that bioaccumulation occurred over time. The relationship between the concentrations of mercury in liver and the ages of the dolphins was very strong (Figure 17.6a), and less so for mercury in kidney (Figure 17.6b). A relatively constant input of mercury over time can be assumed because the association between mercury concentrations and age is linear based on a log–log plot. In long-finned pilot whales, a correlation between concentrations of mercury and lead in liver with length was observed (Meador *et al.*, 1993); for the set of pilot whales analyzed, length was a relatively accurate predictor of age (Sergeant, 1962; Kasuya *et al.*, 1988). These findings for bottlenose dolphins and pilot whales indicate that bioaccumulation is occurring over time at a fairly predictable rate, probably due to dietary input.

Conclusions

Our analyses of tissues from bottlenose dolphins that stranded revealed elevated concentrations of HAHs and trace elements that potentially could induce adverse sublethal physiological effects. However, determining whether there is a relationship between chemical contaminants and sublethal physiological effects that may have contributed to these or other mortality events is difficult to assess with certainty; all evidence to date is still circumstantial. Moreover, identifying toxicological effects of chemical contaminants is also often confounded because we sample moribund or dead animals that also exhibit multisystemic diseases. For example, in this and other mortality events in bottlenose dolphins, it is suspected that a highly virulent morbillivirus has been a major factor in the mortality of dolphins in the Gulf of Mexico since 1990, and may have also been the etiologic agent in the mass mortality on the Atlantic seaboard in 1987–88 (Lipscomb *et al.*, 1994). Some researchers contend that for some pathogens (e.g. morbillivirus), pathogenic virulence, the lack of previous exposure of the population to the pathogen, and ease of horizontal transmission are the major factors contributing to the high risk of an epizootic (Kennedy, 1999), whereas, others contend that a multifactorial etiology in viral diseases must be considered (Eis, 1989). Thus, whether chemical contaminants contribute directly to the onset or extent of a mortality event in the presence of a highly infectious agent is still debated. Nevertheless, both field studies and experimental research have linked elevated concentrations of contaminants in cetaceans and pinnipeds to diseases (Aguilar and Raga, 1993; Olsson *et al.*, 1994) and to alterations of hormone homeostasis (Subramanian *et al.*, 1987; Brouwer *et al.*, 1989; Chiba *et al.*, 2001), immune function (Lahvis *et al.*, 1995; Ross *et al.*, 1995) and reproduction (DeLong *et al.*, 1973; Reijnders, 1986). Concomitantly, several extensive studies and literature reviews conclude that chemical contaminants have little impact on cetaceans. For example, O'Shea and Brownell (1994) for baleen whales, Kuiken *et al.* (1994) for harbor porpoise, and Perrin (1990) for toxic elements in pilot whales find little evidence to support significant physiological effects from chemical contaminants. These apparently conflicting findings show that detailed ecotoxicological investigations of several case studies are necessary. Previous investigations of other species, such as lake trout and bald eagles (Bowerman *et al.*, 1995) in the US Great Lakes, have shown that eco-epidemiological studies are needed to provide substantive evidence associating biological disorders to chemical contaminant exposure. Studies of marine mammals must also include an assessment of the cumulative effects of multiple environmental factors, of which marine pollution is one, to determine the relative risk from chemical contaminants on the health of cetaceans and pinnipeds.

Overall, evidence from both the laboratory and the field suggest that toxic compounds, such as dioxin-like PCB congeners, are immunosuppressive and alter hormone homeostasis, which may lead to increased risk for parasitism, infection and altered reproduction and development in cetacean populations. Thus, to better identify the hazards posed by chemical contaminants, an improved integration of measurements of physiological parameters shown to be affected by toxic chemicals with analytical chemical measurements is necessary, especially in those populations exposed to high levels of contaminants, such as coastal odontoceti. Moreover, it is critical to collect information on apparently healthy animals through non-destructive sampling (biopsy) and to sample animals taken incidentally during fisheries. Such information is essential in defining what is physiologically normal and the threshold of exposure at which deviation from normality occurs. It has been proposed (Peakall, 1999) for marine mammals living in the open ocean that physiological normality is a reasonable and realistic goal, because a return to a pristine environment is not practical. The anticipated continued presence and input of persistent chemical contaminants to marine ecosystems support concern for adverse effects on marine mammals. It is important to assess the health risk from chemical contaminants and to determine temporal trends in exposure to known substances as well as to yet unidentified substances of concern.

Acknowledgments

We thank Dr Teri Rowles of NOAA for providing support for these studies under the Marine Mammal Health and Stranding Response Program. We also thank Dr Susan Chivers of the Southeast Fisheries Science Center for collection of samples of common dolphins. Additionally, we thank Donald Brown and his colleagues in our Environmental Chemistry program for assistance in chemical and data analyses. A special acknowledgment must be given to the late Dr Nancy Foster. Dr Foster was instrumental in establishing the

National Marine Mammal Tissue Bank and Stranding Network Program, which is now the Marine Mammal Health and Stranding Response Program, and for supporting the early phases of this investigation.

Mention of trade names is for information only and does not constitute endorsement by the US Department of Commerce.

References

Abacus Concepts. 1989. *SuperANOVA*. Abacus Concepts, Inc., Berkeley, California.

Aguilar, A. and Borrell, A. 1994. Abnormally high polychlorinated biphenyl levels in striped dolphins (*Stenella coeruleoalba*) affected by the 1990–1992 Mediterranean epizootic. *Science of the Total Environment* 154: 237–247.

Aguilar, A. and Raga, J.A. 1993. The striped dolphin epizootic in the Mediterranean Sea. *Ambio* 22: 524–528.

Aguilar, A., Borrell, A. and Pastor, T. 1999. Biological factors affecting variability of persistent pollutant levels in cetaceans. *Journal of Cetacean Research and Management (Special Issue)* 1: 83–116.

Ahlborg, U.G., Becking, G.C., Birnbaum, L.S., Brouwer, A., Derks, H.J.G.M., Feeley, M., Golor, G., Hanberg, A., Larsen, J.C., Liem, A.K.D., Safe, S.H., Schlatter, C., Wærn, F., Younes, M. and Yrjanheikki, E. 1994. Toxic equivalency factors for dioxin-like PCBs. *Chemosphere* 28: 1049–1067.

Arkoosh, M.R., Casillas, E., Clemons, E., Kagley, A.N., Olson, R., Reno, P. and Stein, J.E. 1998a. Effect of pollution on fish diseases: potential impacts on salmonid populations. *Journal of Aquatic Animal Health* 10: 182–190.

Arkoosh, M.R., Casillas, E., Huffman, P., Clemons, E., Evered, J., Stein, J.E. and Varanasi, U. 1998b. Increased susceptibility of juvenile Chinook salmon from a contaminated estuary to *Vibrio anguillarum*. *Transactions of the American Fisheries Society* 127: 360–374.

Arkoosh, M.R.E., Casillas, E., Clemons, E., McCain, B.B. and Varanasi, U. 1991. Suppression of immunological memory in juvenile chinook salmon (*Oncorhynchus tshawychta*) from an urban estuary. *Fish and Shellfish Immunology* 4: 261–278.

Ballschmiter, K. and Zell, M. 1980. Analysis of polychlorinated biphenyls with capillary gas chromatography. *Fresenius Journal of Analytical Chemistry* 302: 20–31.

Barros, N.B. and Odell, D.K. 1990. Food habits of bottlenose dolphins in the southeastern United States. In *The bottlenose dolphin*, Leatherwood, S. and Reeves, R.R., eds. San Diego: Academic Press, 309–328.

Bergman, A., Klasson-Wehler, E. and Kuroki, H. 1994. Selective retention of hydroxylated PCB metabolites in blood. *Environmental Health Perspectives* 102: 464–469.

Birnbaum, L.S. 1985. The role of structure in the disposition of halogenated aromatic xenobiotics. *Environmental Health Perspectives* 61: 11–20.

Bowerman, W.W., Giesy, J.P., Best, D.A. and Karamer, V.J. 1995. A review of factors affecting productivity of bald eagles in the Great Lakes region: implications for recovery. *Environmental Health Perspectives* 103: 51–59.

Brouwer, A., Reijnders, P.J.H. and Koeman, J.H. 1989. Polychlorinated biphenyl (PCB)-contaminated fish induces vitamin A and thyroid hormone deficiency in the common seal (*Phoca vitulina*). *Aquatic Toxicology* 15: 99–106.

480   *J.E. Stein* et al.

Chiba, I., Sakakibara, A., Goto, Y., Isono, T., Yamamoto, Y., Iwata, H., Tanabe, S., Shimazaki., K., Akahori, F., Kazusaka, A. and Fujita, S. 2001. Negative correlation between plasma thyroid hormone levels and chlorinated hydrocarbon levels accumulated in seals from the coast of Hokkaido, Japan. *Environmental Toxicology and Chemistry* 20: 1092–1097.

Cockcroft, V.G., DeKock, A.C., Lord, D.A. and Ross, G.J.B. 1989. Organochlorines in bottlenose dolphins *Tursiops truncatus* from the east coast of South Africa. *South African Journal of Marine Science* 8: 207–217.

Corsolini, S., Focardi, S., Kannan, K., Tanabe, S., Borrell, A. and Tatsukawa, R. 1995. Congener profile and toxicity assessment of polychlorinated biphenyls in dolphins, sharks, and tuna collected from Italian coastal waters. *Marine Environmental Research* 40: 33–53.

Davis, G.K. and Mertz, W. 1987. Copper. In *Trace elements in human and animal nutrition*, Mertz, W., ed., 5th edition, vol. 1. San Diego, CA: Academic Press, 301–364.

DeLong, R.L., Gilmartin, W.G. and Simpson, J.G. 1973. Premature births in California sea lions: association with high organochlorine pollutant residue levels. *Science* 181: 1168–1170.

Duinker, J.C., Hildebrand, M., Nolting, T.J. and Nolting, R.F. 1979. Organochlorines and metals in harbour seals (Dutch Wadden Sea). *Marine Pollution Bulletin* 10: 360–364.

Eis, D. 1989. Simplification in the etiology of recent seal deaths. *Ambio* 18: 144.

Fernandez, S. and Hohn, A.A. 1998. Age, growth, and calving season of bottlenose dolphins, *Tursiops truncatus*, off coastal Texas. *Fishery Bulletin* 96: 357–365.

Finger, J.M. and Smith, J.D. 1987. Molecular association of Cu, Zn, Cd, and ^{210}Po in the digestive gland of the squid *Nototodarus gouldi*. *Marine Biology* 95: 87–91.

Geraci, J.R. 1989. *Clinical investigation of the 1987–88 mass mortality of bottlenose dolphins along the US central and south Atlantic coast*. National Marine Fisheries Service and US Navy, Office of Naval Research and Marine Mammal Commission.

Haley, D. 1986. *Marine mammals of eastern north Pacific and Arctic waters*. Seattle, Washington: Pacific Search Press.

Hansen, L.J. 1992. *Report on investigation of 1990 Gulf of Mexico bottlenose dolphin strandings*. NOAA/NMFS/SEFSC.

Harding, G.C., LeBlanc, R.L., Vass, W.P., Addison, R.F., Hargrave, B.T., Pearre, S. Jr, Dupuis, A. and Brodie, P.F. 1997. Bioaccumulation of polychlorinated biphenyls (PCBs) in the marine pelagic food web, based on a seasonal study in the southern Gulf of St. Lawrence, 1976–1977. *Marine Chemistry* 56: 145–179.

Hargrave, B.T., Harding, G.C., Vass, W.P., Erickson, P.E., Fowler, B.R. and Scott, V. 1992. Organochlorine pesticides and polychlorinated biphenyls in the Arctic ocean food web. *Archives of Environmental Contamination and Toxicology* 22: 41–54.

Honda, K., Tatsukawa, R., Itano, K., Miyazaki, N. and Fujiyama, T. 1983. Heavy metal concentrations in muscle, liver and kidney tissue of striped dolphin, *Stenella coeruleoalba*, and their variations with body length, weight, age and sex. *Agricultural and Biological Chemistry* 47: 1219–1228.

Honda, K., Yamamoto, Y., Kato, H. and Tatsukawa, R. 1987. Heavy metal accumulations and their recent changes in southern minke whales, *Balaenoptera acutorostrata*. *Archives of Environmental Contamination and Toxicology* 16: 209–216.

Itano, K. and Kawai, S. 1981. Mercury and selenium levels in striped dolphins in the Pacific coast of Japan. In *Studies on the levels of organochlorine compounds and heavy metals in the marine organisms*. Okinawa, Japan, University of the Ryukyus, 73–83.

Jonsson, C.J., Lund, B.O., Bergman, A. and Brandt, I. 1992. Adrenocortical toxicity of 3-methylsulphonyl-DDE; 3: Studies in fetal and suckling mice. *Reproductive Toxicology* 6: 233–240.

Kannan, K., Tanabe, S., Borrell, A., Aguilar, A., Focardi, S. and Tatsukawa, R. 1993. Isomer-specific analysis and toxic evaluation of polychlorinated biphenyls in striped dolphins affected by an epizootic in the western Mediterranean Sea. *Archives of Environmental Contamination and Toxicology* 25: 227–233.

Kasuya, T., Sergeant, D.E. and Tanaka, K. 1988. Re-examination of life history parameters of long-finned pilot whales in the Newfoundland waters. *Scientific Reports of the Whales Research Institute* 39: 77–90.

Kennedy, S. 1999. Morbillivirus infections in marine mammals. *Journal of Cetacean Research and Management (Special Issue)* 1: 267–273.

Kidd, K.A., Schindler, D.W., Hesslein, R.H. and Muir, D.C.G. 1995. Correlation between stable nitrogen isotope ratios and concentrations of organochlorines in biota from a freshwater food web. *Science of the Total Environment* 160–161: 381–390.

Kostial, K. 1986. Cadmium. In *Trace elements in human and animal nutrition*, Mertz, W., ed., vol. 2. New York: Academic Press.

Krahn, M.M., Becker, P.R., Tilbury, K.L. and Stein, J.E. 1997. Organochlorine contaminants in blubber of four seal species: integrating biomonitoring and specimen banking. *Chemosphere* 34: 2109–2121.

Krahn, M.M., Moore, L.K., Bogar, R.G., Wigren, C.A., Chan, S-L. and Brown, D.W. 1988a. High-performance liquid chromatographic method for isolating organic contaminants from tissue and sediment extracts. *Journal of Chromatography* 437: 161–175.

Krahn, M.M., Wigren, C.A., Pearce, R.W., Moore, L.K., Bogar, R.G., MacLeod, W.D. Jr, Chan, S.-L. and Brown, D.W. 1988b. Standard Analytical Procedures of the NOAA National Analytical Facility, 1988. New HPLC Cleanup and Revised Extraction Procedures for Organic Contaminants. NOAA Technical Memorandum. NMFS F/NWC 153, 1–52.

Krone, C.A., Robisch, P.A., Tilbury, K.L., Stein, J.E., Mackey, E.A., Becker, P.R., O'Hara, T.M. and Philo, L.M. 1999. Elements in liver tissues of bowhead whales (*Balaena mysticetus*). *Marine Mammal Science* 15:123–142.

Kuehl, D.W. and Haebler, R. 1995. Organochlorine, organobromine, metal, and selenium residues in bottlenose dolphins (*Tursiops truncatus*) collected during an unusual mortality event in the Gulf of Mexico, 1990. *Archives of Environmental Contamination and Toxicology* 28: 494–499.

Kuehl, D.W., Haebler, R. and Potter, C. 1991. Chemical residues in dolphins from the US Atlantic coast including Atlantic bottlenose obtained during the 1987/88 mass mortality. *Chemosphere* 22: 1071–1084.

Kuiken, T., Bennet, P.M., Allchin, C.R., Kirkwood, J.K., Baker, J.R., Lockyer, C.H., Walton, M.J. and Sheldrick, M.C. 1994. PCBs, cause of death and body condition in harbour porpoises (*Phocoena phocoena*) from British waters. *Aquatic Toxicology* 28: 13–28.

Lahvis, G.P., Wells, R.S., Kuehl, D.W., Stewart, J.L., Rhinehart, H.L. and Via, C.S. 1995. Decreased lymphocyte responses in free-ranging bottlenose dolphins (*Tursiops truncatus*) are associated with increased concentrations of PCBs and DDT in peripheral blood. *Environmental Health Perspectives* 103(May 1995): 67–72.

Law, R.J. (compiler) 1994. *Collaborative UK Marine Mammal Project: summary of data produced 1988–1992*. Fisheries Research Technical Report No. 97. Directorate of Fisheries Research, Lowestoft.

Law, R.J., Fileman, C.F., Hopkins, A.D., Baker, J.R., Harwood, J., Jackson, D.B., Kennedy, S., Martin, A.R. and Morris, R.J. 1991. Concentrations of trace metals in the livers of marine mammals (seals, porpoises and dolphins) from waters around the British Isles. *Marine Pollution Bulletin* 22: 183–191.

Lipscomb, T.P., Schulman, F.Y., Moffett, D. and Kennedy, S. 1994. Morbilliviral disease in Atlantic bottlenose dolphins (*Tursiops truncatus*) from 1987–88 epizootic. *Journal of Wildlife Diseases* 30: 567–571.

Loganathan, B.G., Tanabe, S., Tanaka, H., Watanabe, S., Miyazaki, N., Amano, M. and Tatsukawa, R. 1990. Comparison of organochlorine residue levels in the striped dolphin from western North Pacific, 1978–79 and 1986. *Marine Pollution Bulletin* 21: 435–439.

Lowry, L.F. 1993. Foods and feeding ecology. In *The bowhead whale*, Burns, J.J., Montague, J.J. and Cowles, C.J., eds, vol. 2. Kansas: The Society for Marine Mammalogy, 201–238.

Lund, B.-O., Bergman, A. and Brandt, I. 1988. Metabolic activation and toxicity of a DDT-metabolite, 3-methylsulfonyl-DDE, in adrenal zona fasciculata in mice. *Chemico-Biological Interactions* 65: 25–40.

Marcovecchio, J.E., Moreno, V.J., Bastida, R.O., Gerpe, M.S. and Rodriguez, D.H. 1990. Tissue distribution of heavy metals in small cetaceans from the southwestern Atlantic Ocean. *Marine Pollution Bulletin* 21: 299–304.

McKenzie, C., Rogan, E., Reid, R.J. and Wells, D.E. 1997. Concentrations and patterns of organic contaminants in Atlantic white-sided dolphins (*Lagenorhynchus acutus*) from Irish and Scottish coastal waters. *Environmental Pollution* 98: 15–27.

Mead, J.G. and Potter, C.W. 1990. Natural history of bottlenose dolphins along the central Atlantic coast of the United States. In *The bottlenose dolphin*, Leatherwood, S. and Reeves, R.R., eds. San Diego: Academic Press, 165–195.

Meador, J.P., Varanasi, U., Robisch, P.A. and Chan, S.-L. 1993. Toxic metals in pilot whales (*Globicephala melaena*) from strandings in 1986 and 1990 on Cape Cod, Massachusetts. *Canadian Journal of Fisheries and Aquatic Science* 50: 2698–2706.

Meador, J.P., Robisch, P.A., Clark, R.C. Jr and Ernest, D.W. 1998. Elements in fish and sediment from the Pacific Coast of the United States: results from the National Benthic Surveillance Project. *Marine Pollution Bulletin* 37: 56–66.

Meador, J.P., Ernest, D., Hohn, A.A., Tilbury, K., Gorzelany, J., Worthy, G. and Stein, J.E. 1999. Comparisons of elements in bottlenose dolphins stranded on the beaches of Texas and Florida in the Gulf of Mexico over a one-year period. *Archives of Environmental Contamination and Toxicology* 36: 87–98.

Moriarty, F. 1984. Persistent contaminants, compartmental models and concentration along food-chains. *Ecological Bulletin* 36: 35–45.

Muir, D.C.G., Norstrom, R.J. and Simon, M. 1988a. Organochlorine contaminants in Arctic marine food chains: accumulation of specific polychlorinated biphenyls and chlordane-related compounds. *Environmental Science and Technology* 22: 1071–1079.

Muir, D.C.G., Wagemann, R., Grift, N.P., Norstrom, R.J., Simon, M. and Lien, J. 1988b. Organochlorine chemical and heavy metal contaminants in white-beaked dolphins (*Lagenorhynchus albirostris*) and pilot whales (*Globicephala melaena*) from the coast of Newfoundland, Canada. *Archives of Environmental Contamination and Toxicology* 17: 613–629.

Myers, M.S., Rhodes, L.D. and McCain, B.B. 1987. Pathologic anatomy and patterns of occurrence of hepatic neoplasms, putative preneoplastic lesions, and other idiopathic hepatic conditions in English sole (*Parophrys vetulus*) from Puget Sound, Washington. *Journal of the National Cancer Institute* 78: 333–363.

Nerini, M. 1984. A review of gray whale feeding ecology. In *The gray whale* (Eschrichtius robustus), Jones, M.L., Swartz, S.L. and Leatherwood, S., eds. New York: Academic Press, 423–448.

O'Hara, T., Krahn, M., Boyd, D., Becker, P. and Philo, L. 1999. Organochlorine contaminant levels in Eskimo harvested bowhead whales of Arctic Alaska. *Journal of Wildlife Diseases* 35: 742–752.

Olsson, M., Karlsson, B. and Ahnland, E. 1994. Diseases and environmental contaminants in seals from the Baltic and the Swedish west coast. *Science of the Total Environment* 154: 217–227.

O'Shea, T.J. and Brownell, R.L. Jr 1994. Organochlorine and metal contaminants in baleen whales: a review and evaluation of conservation implications. *Science of the Total Environment* 154: 179–200.

Parke, D.V. 1980. The metabolism and chemobiokinetics of environmental chemicals. In *The principles and methods in modern toxicology*, Galli, C.L., Murphy, S.D. and Paleotti, R., eds. Amsterdam: Elsevier, 85–105.

Peakall, D.B. 1999. Biomarkers as pollution indicators with special reference to cetaceans. *Journal of Cetacean Research and Management (Special Issue)* 1: 117–124.

Perrin, W.F. 1990. *Report*. International Whaling Commission, 40.

Rawson, A.J., Patton, G.W., Hofmann, S., Pietra, G.G. and Johns, L. 1993. Liver abnormalities associated with chronic mercury accumulation in stranded Atlantic bottlenose dolphins. *Ecotoxicology and Environmental Safety* 25: 41–47.

Reijnders, P.J.H. 1986. Reproductive failure in common seals feeding on fish from polluted waters. *Nature* 324: 456–457.

Rice, D.W. and Wolman, A.A. 1971. *The life history and ecology of the gray whale* (Eschrichtius robustus). The American Society Mammalogists. Special Publication 3, 1–142.

Ridgway, S. and Reddy, M. 1995. Residue levels of several organochlorines in *Tursiops truncatus* milk collected at varied stages of lactation. *Marine Pollution Bulletin* 30: 609–614.

Robisch, P.A. and Clark, R.C. Jr 1993. Sampling and analytical methods of the National Status and Trends Program, National Benthic Surveillance and Mussel Watch Projects 1984–1992. Volume III. Comprehensive descriptions of elemental analytical methods. Sample preparation and analyses of trace metals by atomic absorption spectrocopy. NOAA Technical Memorandum. NOS ORCA 71, 111–150.

Ross, P.S., De Swart, R.L., Reijnders, P.J.H., Van Loveren, H., Vos, J.G. and Osterhaus, A.D.M.E. 1995. Contaminant-related suppression of delayed-type hypersensitivity and antibody responses in harbor seal fed Baltic Sea herring. *Environmental Health Perspectives* 103: 162–167.

Ross, P.S., De Swart, R.L., Addison, R., Van Loveren, H., Vos, J.G. and Osterhaus, A.D.M.E. 1996. Contaminant-induced immunotoxicity in harbour seals: Wildlife at risk? *Toxicology* 112: 157–169.

Safe, S. 1984. Polychlorinated biphenyls (PCBs) and polybrominated biphenyls (PBBs): biochemistry, toxicology, and mechanism of action. *CRC Critical Reviews in Toxicology* 13: 319–395.

Safe, S. 1990. Polychlorinated biphenyls (PCBs), dibenzo-*p*-dioxins (PCDDs), dibenzofurans (PCDFs), and related compounds; environmental and mechanistic considerations which support the development of toxic equivalency factors (TEFs). *CRC Critical Reviews in Toxicology* 21: 51–88.

Salata, G.G., Wade, T.L., Sericano, J.L., Davis, J.W. and Brooks, J.M. 1994. Analysis of Gulf of Mexico bottlenose dolphins for organochlorine pesticides and PCBs. *Environmental Pollution* 88: 167–175.

SAS Institute, Inc. 1994. *JMP Statistics*. Version 3. Cary, North Carolina.

Schantz, M.M., Koster, B.J., Wise, S.A. and Becker, P.R. 1993. Determination of PCBs and chlorinated hydrocarbons in marine mammal tissues. *Science of the Total Environment* 139/140: 323–345.

Schantz, M.M., Koster, B., Oakley, L.M., Schiller, S.B. and Wise, S.A. 1995. Certification of polychlorinated biphenyl congeners and chlorinated pesticides in a whale blubber standard reference material. *Analytical Chemistry* 67: 901–910.

Sergeant, D.E. 1962. *The biology of the pilot or pothead whale* Globicephala melaena *(Traill) in Newfoundland waters*. Fisheries Research Board of Canada, Bulletin, 132.

Sloan, C.A., Adams, N.G., Pearce, R.W., Brown, D.W. and Chan, S.-L. 1993. Sampling and analytical methods of the National Status and Trends Program, National Benthic Surveillance and Mussel Watch Projects 1984–1992. Volume IV. Comprehensive descriptions of trace organic analytical methods. Northwest Fisheries Science Center Organic Analytical Procedures. NOAA Technical Memorandum. NOS ORCA 71, 53–98.

Smith, L.M., Stalling, D.L. and Johnson, J.L. 1984. Determination of part-per-trillion levels of polychlorinated dibenzofurans and dioxins in environmental samples. *Analytical Chemistry* 56: 1830–1842.

Subramanian, A., Tanabe, S., Tatsukawa, R., Saito, S. and Miyazaki, N. 1987. Reduction in the testosterone levels by PCBs and DDE in Dall's porpoises of northwestern North Pacific. *Marine Pollution Bulletin* 18: 643–646.

Varanasi, U., Stein, J.E., Reichert, W.L., Tilbury, K.L., Krahn, M.M. and Chan, S.-L. 1992. Chlorinated and aromatic hydrocarbons in bottom sediments, fish and marine mammals in US coastal waters: laboratory and field studies of metabolism and accumulation. In *Persistent pollutants in marine ecosystems*, Walker, C.H. and Livingstone, D.R., eds. New York: Pergamon Press, 83–115.

Varanasi, U., Stein, J.E., Tilbury, K.L., Meador, J.P., Sloan, C.A., Brown, D.W., Calambokidis, J. and Chan, S.-L. 1993. Chemical contaminants in gray whales

(*Eschrichtius robustus*) stranded in Alaska, Washington, and California, USA. NOAA Technical Memorandum. NMFS NWFSC 11, 1–115.

Varanasi, U., Stein, J.E., Tilbury, K.L., Meador, J.P., Sloan, C.A., Clark, R.C. Jr and Chan, S.-L. 1994. Chemical contaminants in gray whales (*Eschrichtius robustus*) stranded along the west coast of North America. *Science of the Total Environment* 145: 29–53.

Venugopal, B. and Luckey, T.D. 1978. Metal toxicity in mammals. In *Chemical toxicity of metals and metalloids*, vol. 2. New York: Plenum Press, 24–32.

Wagemann, R. and Muir, P.C.G. 1984. *Concentration of heavy metals and organochlorines in marine mammals of northern waters: overview and evaluation.* Department of Fisheries and Oceans. Canadian Technical Report of Fisheries and Aquatic Sciences, 1279.

Waid, J.S., ed. 1986. *PCBs and the environment*, vol. II. Boca Raton, Florida: CRC Press.

Walker, C.H. 1980. Species variations in some hepatic microsomal enzymes. *Progress in Drug Metabolism* 5: 113–164.

Wise, S.A., Benner, B.A. Jr, Christensen, R.G., Koster, B.J., Kurz, J., Schantz, M.M. and Zeisler, R. 1991. Preparation and analysis of a frozen mussel tissue reference material for the determination of trace organic constituents. *Environmental Science and Technology* 25: 1695–1704.

Wise, S.A., Schantz, M.M., Koster, B.J., Demiralp, R., Mackey, E.A., Greenberg, R.R., Burow, M., Ostapczuk, P. and Lillestolen, T.I. 1993. Development of frozen whale blubber and liver reference materials for the measurement of organic and inorganic contaminants. *Fresnius Journal of Analytical Chemistry* 345: 270–277.

Appendix

Bottlenose dolphin (Tursiops truncatus) specimen number, stranding site, date tissue samples collected, sex, length, animal condition, and tissues analyzed

Sample number	Stranding site	Latitude North	Longitude West	Date collected	Sex	Length (cm)	Age (years)[a]	Age (years)[b]	Condition[c]	Tissues analyzed
Gulf of Mexico										
Texas										
TX01	Kleberg County	27.20.0	97.19.8	02/12/92	Female	242	21		2 to 3	Blubber, liver, kidney
TX02	North-east of Galveston	29.27.8	94.36.2	02/13/90	Female	219		4.4	3	Blubber, liver, kidney
TX03	South-west of Galveston	29.06.0	95.07.0	03/04/90	Female	233		20.5	2	Blubber, liver, kidney
TX04	Near Freeport	29.00.7	95.12.5	03/10/90	Male	109		0.05	3	Blubber, liver, kidney
TX05	East Galveston	29.26.2	94.36.6	03/24/90	Male	106		0.07	3	Blubber, liver, kidney
TX06	West Galveston	29.13.7	94.53.7	03/28/90	Female	106		0.06	3	Blubber, liver
TX07	Near Port Arthur	29.15.8	94.50.0	03/29/90	Male	118		0.05	3	Blubber, liver, kidney
TX08	South-west of Galveston	29.11.5	94.57.3	03/29/90	Male	194		0.7	3	Blubber, liver, kidney
TX09	North-east of Galveston	29.26.4	94.39.2	03/30/90	Male	118		0.02	3	Blubber, liver, kidney
TX10	West Galveston Bay	29.30.8	95.10.8	04/25/90	Male	220		7.6	1	Blubber, liver, kidney
TX11	Galveston	na	na	06/08/90	Female	206		0.2	2	Blubber, liver, kidney
TX12	Brazoria County	28.55.8	95.18.4	02/28/92	Male	224	4		2	Blubber, liver, kidney
TX13	Brazoria County	28.58.6	95.15.1	03/02/92	Female	174	2		2	Blubber, liver, kidney
TX14	Galveston County	29.32.1	94.25.6	03/21/92	Female	255	24		2	Blubber, liver, kidney
TX15	Brazoria County	na	na	03/24/92	Male	247	43		2	Blubber, liver, kidney
TX16	Galveston County	29.07.6	95.03.2	05/01/92	Male	195		1.0	2	Blubber, liver, kidney
TX17	Brazoria County	na	na	06/26/92	Female	244	16		2	Blubber, liver, kidney
TX18	Galveston County	29.13.6	94.53.6	12/16/92	Male	226		0.2	1	Blubber, liver, kidney
TX19	Galveston County	29.28.6	94.38.7	02/15/93	Female	219		0.5	2	Blubber, liver, kidney
TX20	Galveston County	29.07.6	95.03.4	08/22/93	Female	216		2.5	2	Blubber, liver, kidney
TX21	Galveston County	29.29.7	94.35.3	12/01/93	Female	205		0.04	2	Blubber, liver, kidney
TX22	Aransas Pass	27.47.0	97.06.0	02/02/90	Female	205		0.7	2	Blubber, liver, kidney
TX23	Nueces County	27.34.8	97.13.2	10/17/91	Female	247	> 45		2	Blubber, liver, kidney
TX24	Nueces County	27.37.1	97.12.4	02/05/92	Female	230	12		2	Blubber, liver, kidney
TX25	Fulton Beach/Aransas County	28.05.6	97.01.9	03/17/92	Female	170		0.4	3-early	Blubber, liver, kidney
TX26	Goose Is. St. Pk./Aransas County	28.07.2	96.59.1	03/17/92	Male	101		0.1	3	Liver, kidney
TX27	CC Marina/Nueces County	27.46.9	97.23.2	04/04/92	Male	96		0.3	4	Liver, kidney
TX28	Nueces County	27.39.4	97.10.8	04/11/92	Male	263	13		2	Blubber, liver, kidney

ID	Location	Lat	Long	Date	Sex	Length			Condition	Tissues
TX29	Rockport/Aransas County	28.00.8	97.04.4	04/12/92	Male	98		0.5	3	Blubber, liver, kidney
TX30	Rockport/Aransas County	28.00.8	97.04.4	04/17/92	Female	115		0.2	3-early	Blubber, liver, kidney
TX31	Grace Island/Aransas County	28.07.4	96.59.1	04/29/92	Male	231	8		3	Blubber, liver, kidney
TX32	Matagorda Bay	28.37.7	95.53.9	01/21/90	Female	235		19	2	Blubber, liver, kidney
TX33	Espiritu Santo Bay/Calhoun County	28.26.0	96.25.7	03/24/92	na	240		9	3-late	Liver, kidney
TX34	Long Island/Calhoun County	28.19.3	96.37.0	03/30/92	Female	113		9	3-early (very)	Blubber, liver, kidney
TX35	Matagorda Bay/Calhoun County	28.26.6	96.24.2	03/30/92	Male	211		0.2	3-late	Liver, kidney
TX36	Espiritu Santo Bay/Calhoun County	28.26.2	96.24.8	04/11/92	Female	202		3	3-late	Liver, kidney
TX37	Espiritu Santo Bay/Calhoun County	28.24.2	96.25.2	04/11/92	Male	250	16	15	3-late	Liver, kidney
TX38	Magnolia Beach/Calhoun County	28.27.0	97.30.0	04/16/92	Female	111			4	Liver, kidney
TX39	Magnolia Beach/Calhoun County	28.39.2	96.35.8	04/11/92	Female	230	3	0.2	3-early	Blubber, liver, kidney
TX40	Sand Point Beach/Calhoun County	28.36.0	96.25.2	04/14/92	Male	177	1		4	Blubber, liver, kidney
TX41	Sand Point Beach/Calhoun County	28.35.1	96.26.8	04/14/92	na	253	na		4	Liver
TX42	Magnolia Beach/Calhoun County	28.35.1	96.35.1	04/24/92	Male	253	16		4	Liver, kidney
TX43	Espiritu Santo Bay/Calhoun County	28.24.0	96.29.0	04/21/92	Male	234	8		4	Liver, kidney
TX44	Port O'Connor/Calhoun County	28.30.0	97.33.0	04/23/92	Male	248		9	3	Blubber, liver, kidney
TX45	Espiritu Santo Bay/Calhoun County	28.19.5	96.30.8	04/23/92	Male	235	10		4	Blubber, liver, kidney
TX46	Espiritu Santo Bay/Calhoun County	28.24.8	96.27.5	04/22/92	Female	116	na		4	Blubber
TX47	Blackberry Island	28.24.0	96.28.1	09/13/92	Female	193	15	0.04	2	Blubber, liver, kidney
TX48	na	na	na	01/24/93	Male	272	na		2	Blubber, liver, kidney
TX49	Sabine Pass	29.39.0	94.07.5	02/14/90	Male	260	na		2	Blubber, liver, kidney
TX50	Sabine Pass/High Island	29.35.0	94.17.4	02/22/90	Female	251	16		2	Blubber, liver, kidney
TX51	Jefferson County	29.40.3	94.03.0	02/16/92	Female	237	20		2	Blubber, liver, kidney
TX52	Jefferson County	29.39.0	97.02.2	03/15/93	Female	236		2	2	Blubber, liver, kidney
TX53	Jefferson County	29.39.8	93.50.1	08/17/93	Female	233			2	Blubber, liver, kidney
TX54	Jefferson County	29.39.8	93.50.1	08/17/93	Female	138		0.02	2	Blubber, liver, kidney
Florida										
FL01	Tampa Bay	27.42.2	82.40.2	03/03/90	Female	137		0.2	3	Blubber, liver, kidney
FL02	Tampa Bay	27.43.8	82.44.7	05/20/90	Female	96		0.09	3	Blubber, liver, kidney
FL03	Northeast Tampa Bay	27.53.2	82.28.2	05/27/90	Male	118		0.08	3	Blubber, liver, kidney
FL04	Manatee County	27.26.2	82.40.5	09/03/91	Female	239	12		2	Blubber, liver, kidney
FL05	Charlotte County	26.56.3	82.20.1	09/06/91	Female	209	6		2	Blubber, liver, kidney
FL06	Charlotte County	26.54.4	82.20.1	09/06/91	Male	220	8		2	Blubber, liver, kidney
FL07	Anna Maria Island	27.27.6	82.41.4	09/25/91	Male	195	2		Good	Blubber, liver, kidney
FL08	Manatee County	27.29.6	82.42.6	09/26/91	Male	210	4		Good	Blubber, liver, kidney
FL09	Manatee County	27.25.3	82.39.6	10/05/91	Male	226	18		Good	Blubber, liver, kidney
FL10	Sarasota County	27.14.4	82.33.2	10/17/91	Male	176	2		Good	Blubber, liver, kidney
FL11	Manatee County	27.22.5	82.37.6	10/20/91	Male	226	6		3	Blubber, liver, kidney
FL12	Sarasota County	27.06.5	82.28.3	10/29/91	Female	214	4		Mod. good	Blubber, liver, kidney
FL13	Sarasota County	27.14.5	82.32.4	10/29/91	Male	192	2		Good	Blubber, liver, kidney

Sample number	Stranding site	Latitude North	Longitude West	Date collected	Sex	Length (cm)	Age (years)[a]	Age (years)[b]	Condition[c]	Tissues analyzed
FL14	Longboat Key	27.20.4	82.36.3	03/09/92	Female	246	16		Good	Blubber, liver, kidney
FL15	Sarasota Bay	27.23.0	82.38.0	05/13/92	Male	233	4		2	Blubber, liver, kidney
FL16	Siesta Key	27.15.6	82.33.1	06/30/92	Female	246	20		na	Blubber, liver, kidney
FL17	Sarasota County	27.20.6	82.36.4	09/10/92	Female	236	10		Good	Blubber, liver, kidney
Louisiana										
LA01	Isle of Dernieres	29.05.0	90.50.0	03/25/90	Female	251	na		3	Blubber, liver, kidney
Mississippi										
MS01	Pascagoula	30.22.9	88.33.9	03/14/90	Male	161	na		2	Blubber, liver, kidney
Alabama										
AL01	Mobile Bay	30.27.5	87.55.0	05/23/90	Male	261	na		2	Liver, kidney
North-west Atlantic										
Massachusetts										
MA01	Provincetown	na	na	12/12/92	Male	230	na		2	Blubber, liver, kidney
MA02	Provincetown	na	na	12/12/92	Female	279	na		2	Blubber, liver, kidney
MA03	Eastham	na	na	12/12/92	Male	263	na		2	Blubber, liver, kidney
MA04	Eastham	na	na	12/12/92	Male	257	na		2	Blubber, liver, kidney

[a] Age is estimated from dental growth layers (Meador et al., 1999).
[b] Age is predicted from mercury in liver (Meador et al., 1999).
[c] The following animal condition codes were used:
   1 – live stranding;
   2 – extremely fresh, no bloating, as if just died (<24 h);
   3 – early – minor bloating, skin peeling (1 d – 1 wk);
   4 – late – moderate decomposition, major bloating, skin peeling, penis may be extended in males (1 d – 1 wk);
   4 – advanced decomposition, bone exposed due to decomposition, major bloating, skin exposed, penis extended in males;
   5 – mummified carcass, no organs present.

na = not available. These animals may be from a different population of dolphins, especially the Massachusetts animals, and the regression equation that related age and liver concentration of mercury may not apply (Meador et al., 1999).

Part IV
Pinnipeds

18 Global temporal trends of organochlorines and heavy metals in pinnipeds

Peter J.H. Reijnders and Mark P. Simmonds

Introduction

Residues of organochlorines (OCs) and heavy metals in pinnipeds from different parts of the globe published in the literature have been the subject of recent extensive reviews (Wagemann and Muir *et al.*, 1999; Borrell and Reijnders, 1999; O'Shea, 1999). Assessment of general time trends in concentrations of those compounds in tissues of pinnipeds may appear theoretically possible. However, a further examination of the quality and the quantity of the available data reveals that this is a complex matter. The basic problem is that the levels found in pinnipeds are a reflection of the fate of the compound in the environment, culminating in exposure and then followed by the toxicokinetics in the animal. The biological factors affecting variability of pollutants in cetaceans, although these also hold for pinnipeds, have been elegantly described in detail elsewhere (Aguilar *et al.*, 1999); this will therefore not be discussed in this chapter.

The levels of release of specific contaminants in the environment are highly variable between compounds, due to differences in their history of production and use. In addition, their subsequent fate in the environment depends on many factors (such as their persistence and dispersal routes), which are influenced by properties of the environment. Besides these aspects, there is also a clear difference between the discharge of synthetic chemicals such as OCs and trace elements. Some of the latter chemicals have always been in the environment (Murozumi *et al.*, 1969; Weiss *et al.*, 1971). Therefore, in several cases, trace element concentrations in pinnipeds can be considered natural except in enclosed seas or riverine systems. However, concentrations of mercury, cadmium and lead in marine mammal tissues have increased considerably since the turn of the twentieth century (Wagemann *et al.*, 1990). Temporal trends for mercury and cadmium as environmental contaminants are superimposed on the geochemical trends. This is particularly clear in the remarkable case of cadmium in Greenland ice which, compared to some thousand years ago, is now more than 200 times more concentrated (Goyer, 1991).

Two other handicaps in assessing general temporal trends of contaminants in pinnipeds are the lack of longer times series and poor comparability of published data. This latter aspect is particularly manifest in analyses of OCs. The analytical techniques used by different laboratories have changed considerably over time and lead to large differences in accuracy.

We conclude that, for the above-mentioned reasons, much of the published data on organochlorines and heavy metals are not comparable. This makes it difficult to describe in general terms global trends in residues of those chemicals in marine mammals, including pinnipeds. Nevertheless, in some laboratories the methods of sampling and analyses have been carried out by the same research group, often even involving the same researchers, and on the basis of comparable internal standards. We therefore consider it scientifically justifiable to select and use these specific data sets as case studies to indicate temporal trends of contaminants in some pinnipeds.

Areas and species covered

In this chapter, three different types of areas have been selected. Relatively unpolluted areas: Greenland and Canada (with exception of the Gulf of St. Lawrence); intermediate polluted areas: north north-east Pacific; and relatively highly polluted areas: the Wadden Sea and the Baltic Sea. These areas differ, furthermore, in their distance from the sources of organic contaminants: the N.NE Pacific, Wadden Sea, Baltic and Gulf of St. Lawrence are under direct influence of discharge sources, whereas Greenland and Canada are more influenced by long-range transport of contaminants. The relationship between these circumstances and trends is particularly of interest. The species covered are harbour seal, *Phoca vitulina* (Wadden Sea); grey seal, *Halichoerus grypus* (Baltic and Canada); Arctic ringed seal, *Pusa hispida hispida* (Canada, Greenland), Baltic ringed seal, *Pusa hispida botnica* (Baltic); northern fur seal, *Callorhinus ursinus* (N.NE Pacific); harp seal, *Pagophilus groenlandicus* (Canada); and Atlantic walrus, *Odobenus rosmarus rosmarus* (Canada).

Results and discussion

Heavy metals

Information on heavy metals in pinnipeds (Table 18.1) is more fragmentary than that for OCs. It would seem that the emphasis of toxicological research shifted from heavy metals to organochlorines during the 1970s.

Canadian Arctic

Wagemann *et al.* (1996) analysed mercury in livers of Arctic ringed seals from the Canadian Arctic. After correction for the effect of age, they found that higher mean concentrations of mercury were found in samples collected in 1987–93 compared to 1972–73, indicating a positive time trend. Wagemann

Table 18.1 Time variation in heavy metal concentrations in pinnipeds from different regions

Compound	Species	Region	Period	Trend	Reference
Mercury	*P. hispida h.*	Can. Arctic	1972–73/1987–93	↑	Wagemann *et al.* (1996)
	O. rosmarus r.	Can. Arctic	1982–88	↔	Wagemann and Stewart (1994)
Mercury	*P. hispida h.*	Greenland	mid-1980s–mid-1990s	↔	Riget and Dietz (2000)
Mercury	*P. vitulina* (juv)	Wadden Sea	1975–76/1988	↔	Drescher *et al.* (1977)
	P. vitulina (adult)			↓	Harms *et al.* (1978)
					Kremer (1994)
Methylmercury	*P. vitulina* (adult)	Wadden Sea	1975–76/1988	↓	Reijnders (1980)
					Kremer (1994)
Cadmium	*O. rosmarus r.*	Can. Arctic	AD 1200–1500/1988	↓	Outridge *et al.* (1997)
Cadmium	*P. hispida h.*	Greenland	1979–85	↔	Riget and Dietz (2000)
			1985–95	↓	
Lead	*O. rosmarus r.*	Can. Arctic	AD 1200–1500/1988	↔	Outridge *et al.* (1997)
Lead	*P. vitulina*	Wadden Sea	1975–76/1988	↓	Drescher *et al.* (1977)
					Kremer (1994)

↑ increasing; ↓ decreasing; ↔ no trend.

and Stewart (1994) did not find a trend in Atlantic walrus sampled over a 6-year period (1982–88). Outridge *et al.* (1997) analysed *inter alia* lead and cadmium in teeth from Atlantic walrus dated from AD 1200–1500 and teeth from 1987–98. They concluded that there was no difference in levels of lead between the two periods. However, cadmium was significantly lower in the recent samples compared to the historical material.

Greenland

Riget and Dietz (2000) analysed mercury and cadmium in Arctic ringed seals from Greenland, sampled between the late 1970s and the mid-1990s. There were no consistent overall time trends in concentrations of both metals over these two decades. But, between the late 1970s and the mid-1980s, cadmium tended to increase, whereas it decreased from the mid-1980s to the mid-1990s. There was a notable increase in mercury in young Arctic ringed seals between the mid-1980s and mid-1990s and a concurrent decrease in cadmium. Riget and Dietz (2000) suggest that this indicated a change in feeding pattern rather than a change in anthropogenic exposure.

Wadden Sea

Kremer (1994) provides data on total mercury in harbour seals from the northern Wadden Sea (Schleswig Holstein) sampled in 1988. Compared with data provided by Drescher *et al.* (1977) and Harms *et al.* (1978) on harbour seals sampled in 1975/76, there was no difference for juvenile seals. The situation is different for adult seals. When, for example the age group of 6–8 years from 1988 is compared with the same age group from 1975/76, the level of total mercury is clearly lower in the more recent samples. The ratio of methylmercury/total mercury in the 1988 samples of adult seals is also lower compared to the 1975/76 samples from the same area analysed by Reijnders (1980). Because the uptake of mercury by seals via their prey is entirely in the form of methylmercury, this supports the finding of a negative time trend for total mercury in adult seals.

Comparison of data on lead in liver tissue of seals sampled in 1988 (Kremer, 1994) and in 1975/76 (Drescher *et al.*, 1977; Harms *et al.*, 1978) shows that the levels in the more recent samples are one order of magnitude lower. This indicates a strong negative time trend for lead in harbour seals from the northern Wadden Sea region.

Organochlorines

Data on concentrations of organochlorines (OCs) in blubber of pinnipeds are more numerous than for the previously discussed heavy metals (Table 18.2). This is most likely due to the specific features of this group of compounds, many of which are rather persistent, exhibit a global atmospheric transport

Table 18.2 Time variation in organochlorine residue levels in pinnipeds in different regions

Compound	Species	Period	Trend	Reference
Canadian Arctic				
ΣPCB	*P. hispida hispida*	1972–81	↓	Addison and Smith (1998)
ΣDDT	*P. hispida hispida*	1982–91	↔	Addison and Smith (1998)
		1972–81	↔	Addison and Smith (1998)
HCB	*P. hispida hispida*	1982–89	↔↓	Addison and Smith (1998)
HCH, chlordanes mirex, HEPOX, dieldrin	*P. hispida hispida*	1981–91	↔	Addison and Smith (1998)
Canadian north-west Atlantic				
ΣPCB, ΣDDT toxaphene, chlordanes	*P. hispida hispida*	1985–95	↔	Muir et al. (1997)
ΣPCB	*H. grypus*	1976–82	↔	Addison et al. (1984)
ΣDDT	*H. grypus*	1976–82	↔↓	Addison et al. (1984)
Gulf of St. Lawrence				
ΣPCB	*P. groenlandicus*	1971–82	↓↓	Addison et al. (1984); Beck et al. (1993)
ΣDDT	*P. groenlandicus*	1971–82	↓	Addison et al. (1984); Beck et al. (1993)
		1982–89	↔	Beck et al. (1993)
Greenland				
ΣPCB	*O. rosmarus rosmarus*	1978–88	↔	Muir et al. (2000)
ΣDDT	*O. rosmarus rosmarus*	1978–88	↔	Muir et al. (2000)
Dieldrin, α+γ-HCH, toxaphene	*O. rosmarus rosmarus*	1978–88	↔	Muir et al. (2000)
Dieldrin, α+γ-HCH, toxaphene	*O. rosmarus rosmarus* (females only)	1978–88	↑	Muir et al. (2000)
North north-west Pacific				
ΣPCB	*C. ursinus*	1971–76	↑	Tanabe et al. (1994)
		1976–end 1970s	↑↓	
		1979–88	↔	

Table 18.2 (cont'd)

Compound	Species	Period	Trend	Reference
ΣDDT	*C. ursinus*	1971–76	↑	Tanabe *et al.* (1994)
		1976–end 1970s	→	
		1979–88	→	
HCH	*C. ursinus*	1971–88	↔	Tanabe *et al.* (1994)
Dutch Wadden Sea				
ΣPCB	*P. vitulina* (juv.)	1975/1988	↔	Reijnders (1980); Reijnders and Dijkman, unpubl. data
	P. vitulina (adults)	1975/1988	→	Reijnders and Dijkman, unpubl. data: Boon *et al.*, unpubl. data
	P. vitulina (juv.)	1988/mid-1990s	→	
ΣDDT	*P. vitulina* (juv.)	1975/1988	→	Reijnders (1980); Reijnders and Dijkman, unpubl. data
	P. vitulina (adults)		→	Reijnders (1980); Reijnders and Dijkman, unpubl. data
α-HCH, dieldrin, HEPOX	*P. vitulina* (juv. + adults)	1975/1988	→	Reijnders (1980); Reijnders and Dijkman, unpubl. data
Baltic				
ΣPCB	*P. hispida hispida* (juv.)	1969/1973–1980/1988	→	Olsson *et al.* (1975); Blomkvist *et al.* (1992); Roos *et al.* (1998)
ΣDDT	*P. hispida hispida* (juv.)	1969/1973–1980/1988	→	Olsson *et al.* (1975); Blomkvist *et al.* (1992); Roos *et al.* (1998)
ΣPCB	*P. hispida hispida* (juv.)	1981–1986	↔	Stenman *et al.* (1987)
ΣDDT	*P. hispida hispida* (juv.)	1981–1986	→	Stenman *et al.* (1987)
ΣPCB	*H. grypus* (juv.)	1969/1973–1980/1988	↔	Olsson *et al.* (1975); Blomkvist *et al.* (1992)
ΣPCB	*H. grypus* (juv.)	1969–1997	→	Roos *et al.* (1998)
ΣDDT	*H. grypus* (juv.)	Early 1970s until end-1970s	→	Olsson *et al.* (1975); Blomkvist *et al.* (1992)
		1979–1988	↔	Olsson *et al.* (1975); Blomkvist *et al.* (1992); Roos *et al.* (1998)
		1969–1997	→	

↑ increasing; ↓ decreasing; ↔ no trend.

cycle, and have major effects on marine mammals attributed to them, in particularly the PCBs. It is beyond the context of this paper to discuss the properties of these compounds, but recent comprehensive overviews on the state of science on persistent organic pollutants (POPs) are provided by Vallack *et al.* (1998), Aguilar *et al.* (1999), Jones and de Voogt (1999), O'Shea *et al.* (1999) and Reijnders *et al.* (1999).

Northwest Territories (Canadian Arctic)

Addison and Smith (1998) report on OC residue concentrations in Arctic ringed seals sampled in 1972, 1981, 1989 and 1991. They state that ΣPCB concentrations (expressed as Aroclor 1254) dropped significantly between 1972 and 1981, being about one-third of their 1972 levels by 1981. Thereafter, no significant decline was observed. DDT group (DDT, DDE and DDD) concentrations did not decline between 1972 and 1981. But, after 1981, p,p'-DDE and p,p'-DDT showed declines of 50 and 80 per cent, respectively. The authors suggest that the observed decline in the DDT group is mainly caused by a reduction in supply to the environment of p,p'-DDT.

Hexachlorobenzene (HCB) concentrations fell by 40–50 per cent between 1981 and 1991. Hexachlorocyclohexanes (α- and β-HCH) and other OC pesticides – chlordanes, mirex, heptachlorepoxide (HEPOX) and dieldrin – did not change, or changed only marginally.

Lancaster and Cumberland Sound (Canadian Arctic)

Muir *et al.* (1997) did not find significant declines of ΣPCB, ΣDDT, toxaphene and chlordanes in Arctic ringed seal blubber sampled from the mid-1980s to mid-1990s.

North-west Atlantic (Canada)

Addison *et al.* (1984) analysed grey seal blubber samples from Sable Island, collected in 1976 and 1982. In contrast to trends in residue concentrations in Arctic ringed seals from the Northwest Territories, DDT-concentrations in the Sable Island samples fell by about 60 per cent between 1976 and 1982, although the ΣPCB concentrations did not change over that period.

Gulf of St. Lawrence

Addison *et al.* (1984) and later Beck *et al.* (1993) report on harp seal blubber samples collected in the Gulf of St. Lawrence in 1971, 1982 and 1988/89. The combined studies indicate that ΣPCB concentrations declined slightly between 1971 and 1982, and continued to decline after 1982. ΣDDT levels declined significantly between 1971 and 1982 (by nearly 80 per cent); however, there was no significant decrease between 1982 and 1988/89. The authors postulate cryptically that, if changing analytical methods and biological variables were taken into account, the concentrations of the DDT group may also have declined after 1982.

Greenland

Persistent organochlorines were determined by Muir *et al.* (2000) in blubber of Atlantic walruses sampled in 1978 and 1988 in the Avanersuaq (Thule) region of north-west Greenland. ΣPCB and ΣDDT concentrations did not differ significantly between the 1978 and 1988 samples. For males, this also holds true for the other OCs studied. In females, however, some PCB congeners and dieldrin, toxaphene, α-HCH and ΣHCH were significantly higher in the 1988 samples.

Northern North Pacific

Female northern fur seals were sampled between 1971 and 1988 to allow analyses of OCs in their blubber. This sampling protocol and subsequent analyses is probably one of the most useful time series to be used for assessing temporal trends in OCs in pinnipeds. In this series, biological variables have been kept to a minimum: for example, only females in the age group of 20 years and older are used, the scientists and laboratories involved have not changed; and, most important of all, samples for nearly each year over an 18-year period are available. Temporal variations in the concentrations of ΣPCB, ΣDDT and HCH have been reported by Tanabe *et al.* (1994).

The ΣPCB concentrations show an increasing trend between 1971 and 1976/77, and a strong decrease in the end of the 1970s. Between 1979 and 1988, levels were rather steady, being about 60–65 per cent of the maximum values found in the mid-1970s. The authors suggest that the PCB concentrations in these fur seals reflect a global contamination pattern and closely resemble the pattern found in European and North American coastal biota, rather than a pattern from a Japanese-derived source.

ΣDDT residue levels also increased until 1976 and declined afterwards. Contrary to PCBs, a further decrease was observed in the 1980s. The concentrations reached mostly below 10 per cent of the maximum value in 1976. ΣHCH shows a relatively smaller variation and a slowly declining trend.

Wadden Sea

Blubber samples of harbour seals from the Dutch Wadden Sea were analysed for ΣPCB, ΣDDT and DDT metabolites, α-HCH, HEPOX and dieldrin. A first set was sampled in 1975–76 and analyses published in Reijnders (1980). A second set was sampled in 1988 and subsequently analysed. Preliminary results are provided further on in this section. The 1975–76 samples were analysed for ΣPCB using a different analytical method from that used for the 1988 samples. To compare the results from both time periods, the first step was to convert the 1975–76 data obtained by the decachlorobiphenyl method (Reijnders, 1980) into results comparable to results obtained with the most widely used method to estimate concentrations at that time. PCBs were quantified by comparing the total peak area in the sample with the total peak area of an Aroclor 1254 external standard. The second step was to re-analyse stored samples, which had been analysed in the mid-1970s following the Aroclor 1254 method, with the capillary column technique presently in use, where concentrations of individual PCB congeners are measured and summed. This provided a regression equation and enabled the second conversion step, providing data comparable with the 1988 data. The comparison shows that the concentrations of ΣPCB in juvenile harbour seals (3 years and less) are not statistically different between the 1975–76 and 1988 samples, whereas in adults (4 years and older) concentrations are significantly lower ($P < 0.05$) in the 1988 samples. Average concentrations in 1988 were about 60 per cent lower compared to the mid-1970s.

ΣDDT concentrations in juveniles were lower ($P < 0.1$) in 1988 compared to 1975–76, the average concentrations being 80 per cent lower. In adults, the ΣDDT concentrations in 1988 were significantly lower ($P < 0.05$), the average in the 1988 samples being about 90 per cent lower.

Comparing the ratio of DDE/ΣDDT revealed that in juvenile and adult seals from 1975–76 the ratio was respectively 0.54 and 0.56, whereas it was respectively 0.71 and 0.79 in the 1988 samples. These data fit the observation of Aguilar (1984) of an 'ageing' of DDT pollutants in the NE Atlantic. DDT will decompose in the environment into its metabolite forms, predominantly DDE and, to a lesser extent, DDD. The relative abundance of DDE to total DDT can therefore be used to assess the chronology of DDT input. The results from the Dutch Wadden Sea indicate that such a process is still continuing and has not stopped at an equilibrium level of 0.60 as Aguilar (1984) predicted.

The α-HCH concentrations in both juvenile and adult seals are significantly lower, respectively $P < 0.1$ and $P < 0.005$, in the 1988 samples. The average in juveniles had declined by about 80 per cent, and in adults by nearly 90 per cent.

HEPOX concentrations were also lower ($P < 0.1$) in juvenile seals from 1988, and average concentrations dropped by 70 per cent. In adults, significantly lower ($P < 0.05$) concentrations were observed in the 1988 samples, the average being 80 per cent lower.

Dieldrin concentrations in the 1988 juvenile, as well as in adult seals, were significantly lower, respectively $P < 0.1$ and $P < 0.001$. The average values in juveniles showed a decline of 65 per cent and in adults 60 per cent.

The general trend for all OC concentrations is that levels in 1988 are considerably lower than in 1975–76. The only exception is ΣPCB in juvenile seals. This could be due to the small sample size, but could also have a biological reason, as will be discussed below. The decrease seen is most strongly in the pesticides group – DDT, HEPOX, HCH and dieldrin levels in 1988 being generally about 80–90 per cent lower than the mid-1970s values. By contrast, ΣPCB concentrations only fell in adults by about 60 per cent. Some additional data have been obtained on ΣPCB concentrations in juvenile seals from the mid-1990s (J. Boon, personal communication). Concentrations in those animals are, on average, 80 per cent lower than in the juveniles from 1988.

Baltic Sea

Time-trend monitoring of organochlorine pollution has been carried out in Sweden since the late 1960s. Investigations conducted under the auspices of the Contaminant Monitoring Programme of the Swedish Environmental Protection Agency provide excellent information on changes in concentrations of, for example, ΣPCB and ΣDDT in biota in Swedish waters (Olsson and Reutergård, 1986; Blomkvist *et al.*, 1992; Bignert *et al.*, 1993, 1998). This programme can serve as a model to be applied in studies aimed at obtaining a time perspective on environmental pollution. Quality assurance, both for biological sampling and analytical procedures, is of a high standard. With respect to pinnipeds, the investigations published by Olsson *et al.* (1975), Blomkvist *et al.* (1992) and Roos *et al.* (1998) have been used here to discuss time trends.

The overall conclusion is that for juvenile ringed seals the concentrations of ΣPCB and ΣDDT in blubber have considerably decreased between 1969–73 and 1980–88.The mean value in the more recent samples for ΣPCB is around 80 per cent lower, for ΣDDT about 90 per cent.

Stenman *et al.* (1987) compared PCB and DDT levels in ringed seals sampled between 1981 and 1986 in the Gulf of Finland (eastern Baltic). Their results confirmed the trend in PCBs described in other areas in the Baltic. They found a continued decrease of ΣDDT, which might point to an ongoing drop of ΣDDT levels in the entire Baltic. However, caution is needed in this interpretation, because the Gulf of Finland could be under the influence of different (direct) sources of DDT discharge than the rest of the Baltic.

In juvenile grey seals, the picture is somewhat different. Blomkvist *et al.* (1992) discussed results of analyses obtained from juvenile grey seals sampled until 1988. They concluded that PCBs showed a slight downward trend until the late 1970s, but that no significant decrease in PCBs was discernible. In contrast, ΣDDT decreased significantly from the early 1970s, the mean value being reduced by around 90 per cent. The strongest decrease was noted between the early 1970s and the late 1970s, therafter concentrations levelled off. However, Roos *et al.* (1998) analysed an additional number of samples collected between 1989 and 1997. They distinguished two age classes of young grey seals: pups (2–6 months) and juveniles (7–20 months). By combining their results obtained from more recent samples, with the data obtained by Blomkvist *et al.* (1992), they could compare concentrations in samples collected between 1969 and 1997. Their calculations showed that between 1969 and 1997, a significant decrease of ΣPCB concentrations in both pups and juveniles was found. The annual decrease was 2 per cent in pups and 4 per cent in juveniles. ΣDDT also decreased significantly in both age classes, albeit at a much higher level, with an annual decrease of 11 per cent in pups and 12 per cent in juveniles.

Concluding summary

The literature available for assessment of time trends of heavy metals in pinnipeds is scarce. There are, in fact, only two regions from which data are available. The relatively little polluted Canadian Arctic, and Greenland, and the more polluted Wadden Sea. Levels of, for example, total mercury in pinnipeds from the Wadden Sea area were, in the 1970s, tenfold higher on average than in pinnipeds from the more pristine area.

For total mercury: levels in Arctic ringed seals from the Canadian Arctic increased between 1972–73 and 1987–93, whereas no increase was found in Arctic ringed seals from Greenland between the mid-1980s and the mid-1990s. Mercury levels in Atlantic walruses from Greenland did not change between 1982 and 1988. Mercury levels, total mercury as well as methylmercury, in harbour seals from the Wadden Sea did not change in juveniles (3 years and less), but were considerably lower in adults (4 years and more) in 1988 compared to 1975–76.

Cadmium levels in Arctic ringed seals from Greenland did not change over the last two decades of the twentieth century. However, an increase was observed between 1979 and 1985, followed by a subsequent decrease between 1985 and 1995.

Lead concentrations in Atlantic walrus from pre-industrial times were not different from samples collected in 1987–88, whereas cadmium was lower in the recent samples. Lead concentrations in tissue from harbour seals of the Wadden Sea collected in 1988 were lower compared to 1975–76.

An overall conclusion is that between the mid-1970s and mid-1990s, there was either no clear trend for mercury and cadmium, or a slight increase of mercury, in pinnipeds from the Canadian Arctic and Greenland, whereas the more recent levels of mercury and lead are considerably lower in pinnipeds from the Wadden Sea.

The assessment of time trends observed in OC levels can be based on our own material and literature sources, describing changes in residue concentrations in blubber of pinnipeds occurring in several regions with differing pollution burdens.

The overall pattern for ΣPCB has two distinct periods. In the first one, lasting from the mid-1970s onwards until approximately the early 1980s, concentrations in pinnipeds generally decreased strongly, thereafter the decrease levelled off. There are two exceptions: PCBs in juvenile harbour seals in the Dutch Wadden Sea and in juvenile grey seals from the Baltic do not show a clear decrease in the early part of the period. An explanation for this (provided by Roos et al., 1998) is that young seals are unlikely to be found with high concentrations of PCBs, because mothers with high levels will not produce pups, because of the toxic effects of PCBs on reproduction (Helle et al., 1976; Reijnders, 1986). This would imply the existence of a threshold, separating non-fertile and fertile females. Another, perhaps complementary, explanation is that, irrespective of the PCB burden of the mother, only a certain amount of the PCBs in mothers can be transferred to the offspring during pregnancy and subsequent lactation. This latter phenomenon will only affect fertile animals with a relative moderate to high PCB burden. The support for this hypothesis is derived from the fact that in 1975–76 young harbour seals in the Dutch Wadden Sea had about the same ΣPCB concentrations as young harbour seals from the northernmost Wadden Sea, despite the fact that average levels in adult seals differed by at least a factor of five. The fact that between 1988 and the mid-1990s concentrations of ΣPCBs in subadult seals from the Dutch Wadden Sea have decreased by 80 per cent, indicates that the already noted decrease in concentrations in adults is now also followed by a strong reduction of ΣPCB concentrations in the younger age classes.

The pattern for DDT is similar to that for PCBs, although the decrease seems, in most pinniped species, to have levelled off at the end of the 1970s. The trends in the Canadian Arctic and Baltic ringed seal are the exceptions, because concentrations continued to decrease during the 1980s. The picture for Baltic juvenile grey seals is somewhat obscure. The concentrations decreased in the 1970s, showed no trend from 1979 until 1988, and the combined results from 1969 until 1997 showed an annual decrease of 11–12 per cent. This might indicate a renewed decrease since the 1990s. The largest difference between the trends seen in PCB and DDT relates to the magnitude of the decrease. ΣPCB has, in general, decreased in the late 1970s until the mid-1980s by, at maximum, 60 per cent, whereas ΣDDT levels, in general, dropped over that period by 80–90 per cent. This is remarkable because production and new applications of both compounds on a global scale should have discontinued midway through the 1970s. We conclude that the slower decrease of PCBs is, to a large extent, caused by a continuing dispersal into the environment, because 30 per cent of all the PCBs ever produced are still in use. Of the other 70 per cent, 30 per cent has accumulated in dump sites, sediments of lakes and coastal zones, only 1 per cent has reached the oceans, and the fate of the other nearly 40 per cent is unknown (Marquenie and Reijnders, 1979; Reijnders and de Ruiter-Dijkman, 1995). Given these conclusions, and the predictions by Tanabe (1988) and Tateya et al. (1988) on future trends of global PCB levels in marine biota, it is expected that the observed levelling off of the decrease of PCB concentrations in pinnipeds, will not be followed by a further strong reduction in the near future.

For the other OCs – HCH, HCB, dieldrin, HEPOX, chlordanes and toxaphene – fewer data are available. A tentative conclusion is that there is no clear trend discernible in most pinniped species in the regions investigated. The exception is the Dutch Wadden Sea, where concentrations of α-HCH, HEPOX and dieldrin in harbour seal blubber were significantly lower in 1988 compared to 1975–76. The other remarkable finding is that female Atlantic walruses showed an increase in dieldrin, α- and γ-HCH and toxaphene.

As stated in the introduction, lack of widescale surveys, particularly of longer time series, is a serious handicap in assessing time trends in pollutants. It caused a general lack of reference data for times where levels of given pollutants were absent or still very low. In order to avoid being confronted again in the near or mid-term future with the same problem, we emphasize the importance of starting and/or maintaining adequate monitoring programmes for novel compounds in marine mammals and their environment. There are already a few potentially important novel compounds, which are known to be very persistent, manufactured and used in large quantities, and which are being detected in tissues of marine mammals and their environment. These include organotin compounds in many cetaceans and pinnipeds (Iwata *et al.*, 1995; Law *et al.*, 1998; Tanabe *et al.*, 1998), and polybrominated biphenyls and polybrominated diphenyl ethers in cetacean and pinniped species (de Boer *et al.*, 1998).

Monitoring of these compounds and continued monitoring of the 'classical' compounds, including heavy metals, is considered highly relevant for assessment of time trends as well as the potential threats of these compounds to pinnipeds and other marine mammals.

References

Addison, R.F. and T.G. Smith 1998. Trends in organochlorine residue concentrations in ringed seal (*Phoca hispida*) from Holman, Northwest Territories, 1972–91. *Arctic* 51: 253–261.

Addison, R.F., P.F. Brodie and M.E. Zinck 1984. DDT has declined more than PCBs in Eastern Canadian seals during the 1970s. *Environm. Sci. Techn.* 18: 935–937.

Aguilar, A. 1984. Relationship of DDE/tDDT in marine mammals to the chronology of DDT input into the ecosystem. *Can. J. Fish. Aquat. Sci.* 41: 840–844.

Aguilar, A., A. Borrell and T. Pastor 1999. Biological factors affecting variability of persistent pollutant levels in cetaceans. In P.J.H. Reijnders, A. Aguilar and G.P. Donovan (eds), Chemical Pollutants and Cetaceans. *J. Cetacean Res. Manage.* (Special Issue 1): 83–116.

Beck, G.G., T.G. Smith and R.F. Addison 1993. Organochlorine residues in harp seals, *Phoca groenlandica*, from the Gulf of St. Lawrence and Hudson Strait: an evaluation of contaminant concentrations and burdens. *Can. J. Zool.* 72: 174–182.

Bignert, A., A. Göthberg, S. Jensen, K. Litzén, T. Odsjö, M. Olsson and L. Reutergårdh 1993. The need for adequate biological sampling in ecotoxicological investigations: a retrospective study of twenty years pollution monitoring. *Sci. Tot. Environm.* 128: 121–139.

Bignert, A., M. Olsson, W. Persson, S. Jensen, S. Zakrisson, K. Litzén, U. Eriksson, L. Häggberg and T. Alsberg 1998. Temporal trends of organochlorines in Northern Europe, 1967–1995. Relation to global fractionation, leakage from sediments and international measures. *Environm. Poll.* 99: 177–198.

Blomkvist, G., A. Roos, S. Jensen, A. Bignert and M. Olsson 1992. Concentration of tDDT and PCB in Seals from Swedish and Scottish Waters. *Ambio* 21: 539–545.

Boer, J. de, P.G. Wester, H.J.C. Klamer, W.E. Lewis and J.P. Boon 1998. Do flame retardants threaten ocean life? *Nature* 394: 28–29.

Borrell, A. and P.J.H. Reijnders 1999. Summary of temporal trends in pollutant levels observed in marine mammals. In P.J.H. Reijnders, A. Aguilar and G.P. Donovan (eds), Chemical Pollutants and Cetaceans. *J. Cetacean Res. Manage.* (Special Issue 1): 149–157.

Drescher, H.E., U. Harms and E. Huschenbeth 1977. Organochlorines and heavy metals in the harbour seal (*Phoca vitulina*) from the German North Sea coast. *Mar. Biol.* 41: 99–106.

Goyer, R.A. 1991. Toxic effects of metals. In M.O. Ambur, J. Doull and C.D. Klaassen (eds), *Toxicology. The basic science of poisons*. McGraw-Hill Inc., New York, 623–680.

Harms, U., H.E. Drescher and E. Huschenbeth 1978. Further data on heavy metals and organochlorines in marine mammals from German coastal waters. *Meeresforsch.* 26: 153–161.

Helle, E., M. Olsson and S. Jensen 1976. PCB levels correlated with pathological changes in seal uteri. *Ambio* 5: 261–263.

Iwata, H., S. Tanabe, T. Mizuno and R. Tatsukawa 1995. High accumulation of toxic butyltins in marine mammals from Japanese coastal waters. *Environm. Sci. Techn.* 29: 2959–2962.

Jones, K.C. and P. de Voogt 1999. Persistent organic pollutants (POPs): state of the science. *Environm. Poll.* 100: 209–221.

Kremer, H. 1994. Verteilungsmuster der Schwermetalle Blei, Cadmium und Quecksilber in Weich- und Hartgeweben mariner Säugetiere aus deutschen Küstengewässern. Schriftenr. Bundesforschungsanstalt f. Fischerei, Hamburg, No. 21.

Law, R.J., S.J. Blake, B.R. Jones and E. Rogan 1998. Organotin compounds in liver tissue of harbour porpoises (*Phocoena phocoena*) and grey seals (*Halichoerus grypus*) from the coastal waters of England and Whales. *Mar. Poll. Bull.* 36: 241–247.

Marquenie, J.M. and P.J.H. Reijnders 1989. PCBs, an increasing concern for the marine environment. ICES CM, 1989/N:12.

Muir, D., B. Braune and B. DeMarch 1997. Ecosystem uptake and effects. In J. Jensen, K. Adhare and R. Shearer (eds), *Canadian Arctic Contaminants Assessment Report*. Indian and Northern Affairs Canada, Ottawa, 183–294.

Muir, D., B. Braune, B. DeMarch, R. Nortrom, R. Wagemann, L. Lockhart, B. Hargrave, D. Bright, R. Addison, J. Payne and K. Reimer 1999. Spatial and temporal trends and effects of contaminants in the Canadian Arctic marine ecosystem: a review. *Sci. Tot. Environm.* 230: 83–144.

Muir, D.C.G., E.W. Born, K. Koczansky and G.A. Stern 2000. Temporal and spatial trends of persistent organochlorines in Greenland walrus (*Odobenus rosmarus rosmarus*). *Sci. Tot. Environm.* 245: 73–86.

Murozumi, M., T.J. Chow and C. Petterson 1969. Chemical concentrations of pollutant lead aerosols, terrestrial dusts and sea salts in Greenland and Antarctica snow strata. *Geochim. Cosmochim. Acta* 33: 1249–1294.

Olsson, M. and L. Reutergårdh 1986. DDT and PCB pollution trends in the Swedish aquatic environment. *Ambio* 15: 103–109.

Olsson, M., A.G. Johnels and R. Vaz 1975. DDT and PCB levels in seals from Swedish waters. The occurrence of aborted seal pups. In *Proceedings from the Symposium on the Seal in the Baltic, June 4–6, Lidingo, Sweden.* SNV PM 591, Swedish Environmental Protection Agency, Solna, Sweden, 43–65.

O'Shea, T.J. 1999. Environmental contaminants and marine mammals. In J.E. Reynolds III and S.A. Rommel (eds), *Biology of marine mammals.* Smithsonian Institution Press, Washington, USA, 485–564.

O'Shea, T.J., R.R. Reeves and A.K. Long 1999. Marine mammals and persistent ocean contaminants. *Proceedings of the Marine Mammal Commission Workshop, Keystone, Colorado, 12–15 October 1998.*

Outridge, P.M., R.D. Evans, R. Wagemann and R.E. Stewart 1997. Historical trends of heavy metal and stable lead isotopes in beluga (*Delphinapterus leucas*) and walrus (*Odobenus rosmarus rosmarus*) in the Canadian Arctic. *Sci. Tot. Environm.* 203: 209–219.

Reijnders, P.J.H. 1980. Organochlorine and heavy metal residues in harbour seals from the Wadden Sea and their possible effects on reproduction. *Neth. J. Sea Res.* 14: 30–65.

Reijnders, P.J.H. 1986. Reproductive failure in common seals feeding on fish from polluted coastal waters. *Nature* 324: 456–457.

Reijnders, P.J.H. and E.M. de Ruiter-Dijkman 1995. Toxicological and epidemiological significance of pollutants in marine mammals. In A.S. Blix, L. Walløe and Ø. Ulltang (eds), *Whales, seals, fish and man.* Elsevier Science, Amsterdam, 575–587.

Reijnders, P.J.H., A. Aguilar and G.P. Donovan 1999. Chemical Pollutants and Cetaceans. *J. Cetacean Res. Manage.* (Special Issue 1).

Riget, F. and R. Dietz 2000. Temporal trends of cadmium and mercury in Greenland marine biota. *Sci. Tot. Environm.* 245: 49–60.

Roos, A., A. Bergman, E. Greyerz and M. Olsson 1998. Time trend studiers on ΣDDT and PCB in juvenile grey seals (*Halichoerus grypus*), fish and guillemot eggs from the Baltic Sea. *Organohalogen Compounds* 39: 109–112.

Stenman, O., E. Helle and M. Perttilä 1987. Concentrations des organochlorés et des métaux lourds dans les jeunes phoques annelés et phoques gris de la mer Baltique en 1981–1986. ICES Symposium sur les sciences de la mer des regions arctiques et sub-arctiques. Poster 25.

Tanabe, S. 1988. PCB problems in the future: foresight from current knowledge. *Environm. Poll.* 50: 5–28.

Tanabe, S., J.K. Sung, D.Y. Choi, N. Baba, M. Kiyota, K. Yoshida and R. Tatsukawa 1994. Persistent organochlorine residues in northern fur seal from the Pacific coast of Japan since 1971. *Environm. Poll.* 85: 305–314.

Tanabe, S., M. Prudente, T. Mizuno, J. Hasegawa, H. Iwata and N. Miyazaki 1998. Butyltin contamination in marine mammals from North pacific and Asian coastal waters. *Environm. Sci. Techn.* 32: 193–198.

Tateya, S., S. Tanabe and R. Tatsukawa 1988. PCBs on the globe: possible trend of future levels in the open ocean environment. In N.W. Schmidtke (ed.), *Toxic*

Contaminants in Large Lakes. Proceedings of the World Conference on Large Lakes. May 1986. Mackinac Island, Michigan, US. Lewis Publishers Inc., 237–281.

Vallack, H.W., D.J. Bakker, I. Brandt, E. Brostrøm-Lundén, A. Brouwer, K.R. Bull, C. Cough, R. Guardans, I. Holoubek, B. Jansson, R. Koch, J. Kuylenstierna, A. Lecloux, D. Mackay, P. McCutcheon, P. Mocarelli and R.D.F. Taalman 1998. Controlling organic pollutants – what next? *Environm. Tox. Pharmacol.* 6: 143–175.

Wagemann, R. and D.C.G. Muir 1984. Concentrations of heavy metals and organochlorines in marine mammals of northern waters: overview and evaluation. *Can. Techn. Rep. Fish. Aquat. Sci.,* 1297, 97 p.

Wagemann, R. and R.E.A. Stewart 1994. Concentrations of heavy metals and selenium in tissues and some food of walrus (*Odobenus rosmarus rosmarus*) from the eastern Canadian Arctic and sub-Arctic, and associations between metals, age and gender. *Can. J. Fish. Aquat Sci.* 51: 426–436.

Wagemann, R., R.E.A. Stewart, P. Béland and C. Desjardins 1990. Heavy metals and selenium in tissues of beluga whales, *Delphinapterus leucas*, from the Canadian Arctic and the St. Lawrence estuary. *Can. Bull. Fish. Aquat. Sci.,* 224: 191–206.

Wagemann, R., S. Innes and P.R. Richard 1996. Overview and regional and temporal differences of heavy metals in Arctic whales and ringed seals in the Canadian Arctic. *Sci. Tot. Environm.* 186: 41–66.

Weiss, H.V., M. Koide and E.D. Goldberg 1971. Mercury in a Greenland icesheet: evidence of recent input by man. *Science* 174: 692–694.

19 Pathology in Baltic grey seals (*Halichoerus grypus*) in relation to environmental exposure to endocrine disruptors

A. Bergman, A. Bignert and M. Olsson

Introduction

Historically, products from seals have provided an important extra income for people inhabiting the Swedish coast (Almkvist *et al.*, 1980). Thus seals have been the subject of intense hunting. Furthermore, seals were regarded as vermin by fishermen, and bounties were paid for killing seals between 1808 and 1864, and again from 1891 to 1974. After the Second World War the number of seals was low because of this intensive hunting. Despite decreased hunting pressure after the war, the Swedish seal populations did not recover (Bergman, 1956). The populations of all three species of seals which inhabit the Baltic, the grey seal (*Halichoerus grypus*), the ringed seal (*Pusa hispida botnica*) and the harbour seal (*Phoca vitulina*), were found to further decrease rapidly in the 1960s and 1970s (Almkvist *et al.*, 1980). According to a recent report (Hårding and Härkönen, 1999), the grey seal population in the Baltic Sea decreased from 88 000–100 000 at the beginning of the twentieth century to approximately 4000 in the late 1970s. The corresponding decrease for the ringed seal in the Baltic Sea during the same period was from 190 000–220 000 to approximately 5000. The harbour seal population in the Baltic Sea at the beginning of the twentieth century was about 5000 (Hårding and Härkönen, 1999). At present, the Baltic harbour seal population is estimated at 600 (Helander and Härkönen, 1999). From the beginning of the 1980s a slight increase in seal populations has been recorded in the Baltic Sea. Based on photo-ID data the number of Baltic grey seals recently is estimated to be about 12 000 (Hiby *et al.*, 2001). Low numbers of ringed seals have been recorded in the Baltic Sea: 200–300 in the Gulf of Finland and 1400 in the Gulf of Riga (Härkönen *et al.*, 1998).

Jensen *et al.* (1969) disclosed serious DDT and PCB pollution of the Baltic. Hook and Johnels (1972) and Olsson *et al.* (1975) suspected pollution to be a factor involved in the decrease of the Baltic seal populations, and that this contamination was related to findings of aborted seal pups in the southern part of the Baltic Sea. The main reason for the decrease of the Baltic seal populations after the Second World War and through the 1970s was reproductive failure. PCB compounds were suspected to be associated with signs of interrupted pregnancies: high prevelence of uterine horn obstruction (stenoses and occlusions), which were observed in Baltic ringed and grey seals in the 1970s (Helle *et al.*, 1976a, b). During that decade, the harbour seal population in the Dutch Wadden Sea was also reported to have decreased (Reijnders, 1976), and their decline was suspected to be caused by PCBs (Reijnders, 1980). In an experimental study, female harbour seals fed contaminated fish from the Dutch Wadden Sea showed lowered reproduction compared to seals fed fish that were less contaminated (Reijnders, 1986).

Cooperative research on contaminants and seals in the Baltic Sea was initiated at the Swedish Museum of Natural History in Stockholm and at the Department of Veterinary Pathology, Swedish University of Agricultural Sciences, Uppsala in 1977. Since then seals found dead at the shoreline or found drowned in fishing gear along the Swedish coast have been the subjects of post-mortem studies. Autopsies revealed a high prevalence of uterine occlusions and stenoses in grey and ringed seals and a high prevalence of uterine tumours (leiomyomas) in grey seals (Bergman and Olsson, 1985). Like the uterine lesions, certain chronic lesions in non-reproductive organs also commonly occurred in both these species (Bergman and Olsson, 1985, 1989). Taken together, the post-mortem studies revealed a disease complex in the Baltic seals with regularly occurring specific organ lesions.

Interestingly, the phocine distemper epizootic in 1988, associated with a mass mortality in harbour seals in areas of the North Sea and on the Swedish west coast (Dietz *et al.*, 1989; see also Chapter 20 in this volume), had a limited impact upon seal populations in the Baltic Sea. Only harbour seals from the south-western part of the Baltic and a few grey seals (three old females), were found to have died from this disease (Bergman *et al.*, 1990). The present (2002) outbreak of phocine distemper in the same area shows a similarly high mortality (more than 50 per cent) in harbour seals on the Swedish west coast as in 1988 (Harding *et al.*, 2002). According to our findings in pathology, harbour seals or grey seals from the Baltic proper (north of Hanö Bight) have, to date, shown no lesions compatible with phocine distemper virus.

This chapter focuses on the character and prevalence of various lesions present in grey seals of the Baltic Sea, based on results from post-mortem investigations during 1977–96 (Bergman, 1999). Observations on grey seals are related to temporal variation of concentrations of contaminants (especially organochlorines, some of which are also known to act as endocrine disruptors) in the food of the seals. The number of ringed seals and harbour seals we examined is too low to allow a similar comparison.

Pathology in seals of the Baltic Sea

Seals were collected in the Baltic Proper and in the Gulf of Bothnia. Most were either found drowned in fishing gear or were found dead at the shoreline. Autopsy procedures included determination of body and organ weights, body length, blubber thickness (nutritional status) and tissue sampling for histology, environmental contaminant chemistry and, when indicated, for parasitology, bacteriology and virology. Semiquantitative evaluation of the degree and extent of lesions was performed by using a five-degree scale: no or non-evident change, slight, moderate, severe change or fatal lesion. Age determination was performed by examination of the annual growth pattern in cementum zones in undecalcified tooth sections (Johnston and Watt, 1981).

Results of the pathology investigations in grey seals of the Baltic Sea

Common findings during post-mortem investigations were lesions of the integument, intestine, large arteries, adrenals, kidneys, skull bones and female reproductive organs.

Integument

Lesions occurred predominantly on digits and claws. These were characterized by claw fold inflammation, softened and brittle claw horn with local or more generalized deformation of claws and claw fractures; in severe cases these were associated with wounds in the claw regions (Figure 19.1). Regional skin changes appeared as a more or less severe hypotrichosis (thin hair coat), localized in the ventral thoracic and abdominal regions with extension to medial surfaces of fore- and hind flippers, and, in one case, also

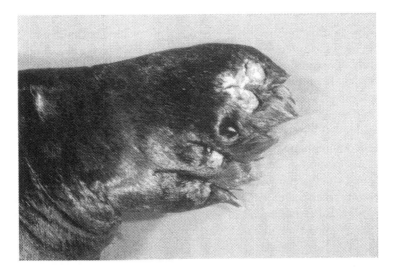

Figure 19.1 Severe claw lesions: loss of claws, wounds at claw sites. Male grey seal, 14–16 years old. Photo: Anders Bergman. (From Bergman, 1999; with permission.)

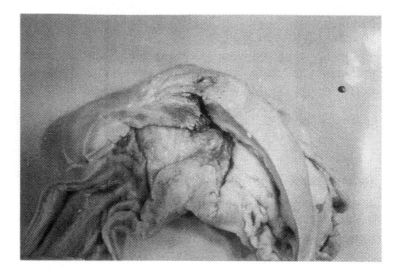

Figure 19.2 Ulcerous process in the ileocaeco-colonic region with perforation of colonic wall. Formalin-fixed material. Female grey seal, 24 years old. Photo: Anders Bergman. (From Bergman, 1999; with permission.)

around the supercilia of both eyes. Histologically, the latter case showed chloracne-like changes, including a thin epidermis, hyperkeratosis and dilation of hair follicles, focally with formations of large subepidermal cysts.

Intestine

Often-fatal intestinal ulcers were found in the ileum, caecum and colon (predominantly the anterior part). These were nearly always associated with hookworms (*Corynosoma* sp.). Lesions denoted as slight implied minor ulcerations (3–10 mm in diameter) of the mucous membrane. Lesions of a moderate degree affected larger mucosal areas. In severe lesions, the ulcerous process had reached the muscular tunic of the intestinal wall, quite often with spread to the serosal tunic, at which fibrinous and/or fibrous adherences to adjacent abdominal organs sometimes occurred. In fatal lesions there was total perforation of the intestinal wall. The perforation site was almost always found some centimetres from the head of the caecum (Figure 19.2), opposite to the ileo-colic orifice. The mucosal lesions sometimes occupied ileum, caecum and a large part of colon, with thickening (hypertrophy) of the muscular tunic of the diseased part of the intestine.

Large arteries

Vascular changes, present as arteriosclerosis, were most evident in the distal abdominal aorta and its bifurcations. Occurrence of some whitish intimal spots and streaks was classified as a mild change, whereas more extensive

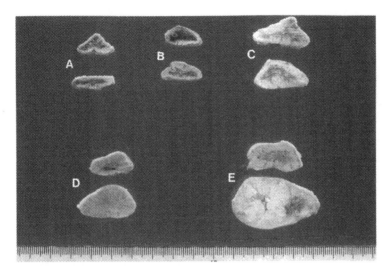

Figure 19.3 Cross-sections of adrenals from five grey seals (A–E). (A) normal relationship between cortex and medulla. (B) slight, (C) moderate and (D)
severe cortical hyperplasia. (E) Large adenoma of one gland. Formalin-
fixed material. Photo: Bengt Ekberg, National Veterinary Institute,
Upsala, Sweden. (From Bergman, 1999; with permission.)

changes with intimal granularity and roughness were classified as moderate.
In severe cases, the intima, as well as the vessel as a whole, showed prominent irregularities and loss of elasticity.

Adrenals

A common finding in the adrenals was cortical hyperplasia that appeared as
different degrees of thickening or nodular changes of the adrenal cortices
(Figure 19.3). Histologically, this thickening appeared to be due to proliferation of cells in the zona fasciculata and zona reticularis of the gland.

Female reproductive organs

Lesions observed were uterine occlusions, uterine stenoses and uterine
muscle cell tumours (leiomyomas). Stenoses and occlusions were most often
localized at about the middle part of the uterine horn. At the site of the
obstruction the uterine wall was usually thickened. In grey seals, the areas
of obstruction generally occured as stenoses (Figure 19.4) while in ringed
seals they were occlusions. Both stenosis and occlusion led to upstream
accumulation of fluid, with flattening of the mucosal folds. Uterine tumours
showed a characteristic gross appearance: they were of firm consistency and
most often localized in the wall of the uterine corpus (Figure 19.5). They
were white-brown to white-yellow on cut surfaces, which showed a whirled

Figure 19.4 Stenosis (arrow) with dilation of anterior part of uterine horn (arrow-heads). Female grey seal 40–42 years old. Photo: Bengt Ekberg, National Veterinary Institute, Upsala, Sweden. (From Bergman, 1999; with permission.)

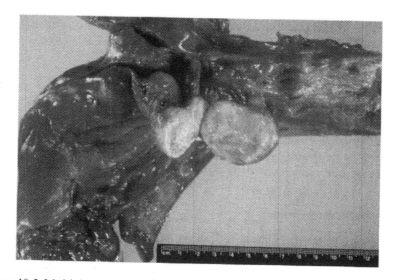

Figure 19.5 Multiple tumours (leiomyomas) in uterine wall at corpus (cut surfaces of large tumour) and at posterior part of uterine horn. Female grey seal, 35 years old. Photo: Bengt Ekberg, National Veterinary Institute, Upsala, Sweden. (From Bergman, 1999; with permission.)

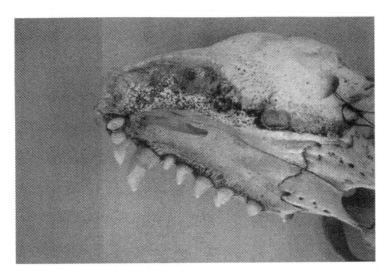

Figure 19.6 Severe erosion of masticatory bones. Female grey seal, 25 years old. Photo: Göran Frisk, Swedish Museum of Natural History.

pattern, as occurred in leiomyomas. Occasionally, large tumours showed central areas of necrosis. They were often multiple and sometimes reached 10 cm in diameter. Most were confirmed histologically as leiomyomas, although a few were diagnosed tentatively on the basis of macroscopy (Bergman, 1999).

Kidneys

Lesions in the kidneys occurred in glomeruli as well as in tubules, and were found both in grey and ringed seals. The histology displayed different degrees of diffuse glomerular capillary wall thickening and occurrence in the glomeruli of ovoid, brightly periodic acid–Schiff (PAS)-positive hyaline bodies. A quite unique change frequently occurred in the urinary tubules in the shape of islands of multilayered tubular cells, which in places obliterated the tubular lumen (Bergman *et al.*, 2001).

Skull bones

Lesions occurred most prominently in masticatory bones (Bergman *et al.*, 1992), with periodontitis and loss of alveolar bone, loss of teeth and deformations in severe cases. The changes were often found in the incisival parts of the jaws or around the canine teeth, but sometimes also occurred around premolars and molars. In severe cases there was substantial loss of bone, especially in the area of the incisors (Figure 19.6); the changes sometimes comprised all parts of the jaws, with bone defects, bone apposition and

Table 19.1 Female grey seals collected 1977–86. Number of animals in different age classes, type of examination, and causes of death

Age class years	Animals examined	Whole-body exam.	Organ sample exam.	Drowned in fishing gear	Diseased, shot	Found dead at shore	Perforating colonic ulcer	Pneumonia	Other causes	Cause not confirmed
1–3	7	7	0	6	0	1	1	0	0	0
4–15	3	3	0	2	0	1	0	0	1[a]	0
16–25	4	2	2	1	1	2	1	0	1[b]	0
≥ 26	12	10	2	5	1	6	2	0	2[b,c]	2
Total	26	22	4	14	2	10	4	0	4	2

[a] Dystocia?
[b] Chronic renal disease.
[c] Multiple serious organ lesions.

deformation. In more mild cases (early lesions) the bone loss was localized to the periodontal lamella with widening of the sockets of the teeth.

Cause of death and changes over time in prevalence of various lesions

High concentrations of DDTs and PCBs were evident in Baltic biota during the 1970s (Olsson and Reutergård, 1986). Since then, concentrations of DDTs, PCBs, HCHs, HCB and dioxins have decreased (Odsjö *et al.*, 1997; Bignert *et al.*, 1998a). The concentrations of PCBs in guillemot eggs are at present less than 15 per cent of the concentrations at the beginning of the 1970s (Bignert *et al.*, 1998a). An increase in seal populations in the Baltic Sea has been observed concurrently. Results from post-mortem investigations during 1977–96 (Bergman, 1999) were used to study tendencies of disease variables through the period of decreasing concentrations of organochlorines in biota of the Baltic Sea.

Material was collected from 159 seals (76 females and 83 males). These were grouped by sex and age as 1–3-year-olds (subadults), or as adults aged 4–15, 16–25, and ≥26 years. Of these, 115 specimens were found drowned in fishing gear, 39 were found dead at the shore and five were shot to reduce suffering in cases of severe illness. Whole-body investigations were performed on 145 seals (Tables 19.1–19.4) and investigation of organ samples on 14 (Tables 19.1, 19.3 and 19.4). Twelve of the 14 seals from which only organ samples were available were females and two were males. Four of the females were collected in the 1977–86 period and eight during 1987–96. The two males were collected in the period from 1987 to 1996.

Fatal colonic ulcer occurred in 11 cases (7 per cent). Ten of these were autopsied and one occurred in a seal which was subjected to organ sample examination only. After drowning, perforating colonic ulcer was the most common cause of death. The most extreme proportion attributable to this cause of death (3 of 6) was recorded in males, aged 16–25 years, subject to whole-body investigation between 1989 and 1996. (Table 19.4). Death from pneumonia occurred in five cases (3 per cent). These were all females collected in the recent autopsy period (Table 19.3), and included three elderly

Table 19.2 Male grey seals collected 1977–86. Number of animals in different age classes, type of examination, and cause of death

Age class years	Animals examined	Whole-body exam.	Organ sample exam.	Drowned in fishing gear
1–3	14	14	0	14
4–15	7	7	0	7
16 ≥ 26	0			
Total	21	21	0	21

Table 19.3 Female grey seals collected 1987–96. Number of animals in different age classes, type of examination, and causes of death

Age class years	Animals examined	Whole-body exam.	Organ sample exam.	Drowned in fishing gear	Diseased, shot	Found dead at shore	Perforating colonic ulcers	Pneumonia	Other causes	Cause not confirmed
1–3	13	13	0	12	0	1	1	0	0	0
4–15	17	16	1	16	0	1	0	0	0	1
16–25	5	3	2	1	0	4	0	0	2[a,b]	2
≥ 26	15	10	5	3	2	10	0	5	0	5
Total	50	42	8	32	2	16	1	5	2	8

[a] Chronic hepatitis and peritonitis.
[b] Colonic and renal lesions.

Table 19.4 Male grey seals collected 1987–96. Number of animals in different age classes, type of examination, and causes of death

Age class years	Animals examined	Whole-body exam.	Organ sample exam.	Drowned in fishing gear	Diseased, shot	Found dead at shore	Perforating colonic ulcers	Pneumonia	Other causes	Cause not confirmed
1–3	30	29	1	27	0	3	0	0	2[a,b]	1
4–15	25	25	0	20	0	5	3	0	2[a,c]	0
16–25	7	6	1	1	1	5	3	0	1[d]	1
≥26	0									
Total	62	60	2	48	1	13	6	0	5	2

[a] Septicaemia.
[b] Thymic lymphosarcoma.
[c] Abscesses of mesenteric lymph nodes.
[d] Abscesses of internal organs after illegal shotgun hunting.

individuals who died from distemper in 1988. Eleven cases (7 per cent) showed various other serious lesions, judged to be the main cause of death (Tables 19.1, 19.3 and 19.4). The cause of death was not established for 12 seals (7.5 per cent, Tables 19.1, 19.3 and 19.4) in which the investigation was limited to organ samples. Twenty-six females (Table 19.1) and 21 males (Table 19.2) were collected during 1977–86. Fifty females (Table 19.3) and 62 males (Table 19.4) were sampled during 1987–96. It should be noted, that the 12 females in the oldest age class (>25 years) of the material collected in 1977–86 (Table 19.1) were born between 1942 and 1957, whereas in the material collected in 1987–96 the 15 females of the same age class (Table 19.3) were born between 1948 and 1964. Thus, both groups of aged females were exposed during the era of the greatest pollution of the Baltic Sea with organochlorines.

To distinguish between seals exposed during the most heavily polluted time period and those less exposed in more recent times (after the 1970s, when both PCBs and DDTs decreased), the material was divided into two groups according to time of birth: seals born before 1980 (≤1979) and seals born after 1979 (≥1980). This division resulted in only seven specimens (males and females combined) in the 1–3-year age class in the early group, but 57 in the later group. This finding is in agreement with the improving trend in reproduction in the Baltic grey seal population.

Seal specimens were also categorized based on collection time, to compare the prevalences of lesions in subadults (1–3 years old) and in seals older than 15 years. Thus, seals collected in 1977–86 were compared with those collected in 1987–96. The number of 1–3-year-old seals in the earlier collection period was 21 (7 females and 14 males, Tables 19.1 and 19.2) and for the later period 43 (13 females and 30 males, Tables 19.3 and 19.4). The corresponding number of seals older than 15 years was 16 (all females) in the early collection period, and 27 (20 females and 7 males) in the later period.

We present the prevalences (percentages) of uterine occlusions/stenoses and uterine tumours, as well as the rate of pregnancy, in adult females collected in the two periods, 1977–86 and 1987–96. For determination of the rate of pregnancy, females sampled from mid-August up to the second week of February were investigated, that is the period in which all fertilized grey seal females in the Baltic Sea are expected to be pregnant (King, 1983).

Changes over time between animals grouped according to the year of birth or period of collection were tested using contingency table analysis (Chi-square test with Yates' correction). This statistical analysis was not carried out in the groups of 16–25-year-old seals because of low sample size. Only a few differences between groups turned out to be statistically significant. Despite the relatively large sample size collected during a period of 20 years and the fairly high incidence of pathological changes, the number of individuals in each group was often low when the data were partitioned by various periods of exposure, age and sex.

In non-reproductive organs, lesions in digits and claws, intestine (colonic ulcers), large arteries (arteriosclerosis) and adrenals (cortical hyperplasia)

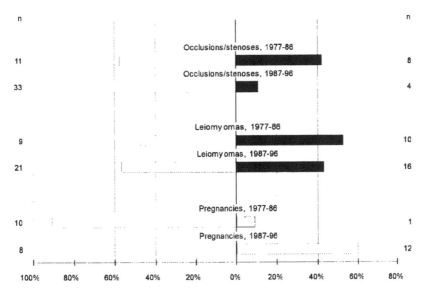

Figure 19.7 Gyaecological health of adult grey seals in two collection periods. The percentage of animals with occlusions/stenoses and leiomyomas in the uterus, as well as the percentage of pregnant animals is shown in the right part of the figure; animals lacking these conditions are shown in the left part (*n* = number of animals).

were used as disease variables. The prevalences of animals in different sex and age groups with moderate to severe degree of these lesions were recorded to compare health during the different birth and collection periods, as specified above.

Uterine occlusions/stenoses, leiomyomas and pregnancies

Improvement in gynaecological health is indicated by comparisons of prevalences of these three conditions during 1977–86 with those of 1987–96 (Figure 19.7). A decreased frequency of uterine occlusions/stenoses, from 42 to 11 per cent ($P < 0.05$), and an increased frequency of pregnancies, from 9 to 60 per cent ($P < 0.05$), were recorded in the recent group. In total, 13 females out of 31 were pregnant but only one of these was collected during 1977–86. A decreasing trend was found for uterine leiomyomas (from 53 to 43 per cent).

Digits and claws

These lesions were rather common and those of moderate to severe degree began to occur in the 4–15-year classes (Figure 19.8). Although the prevalence seemed higher in 4–15-year-old animals born 1979 and earlier, compared to animals born thereafter, this difference was not statistically significant.

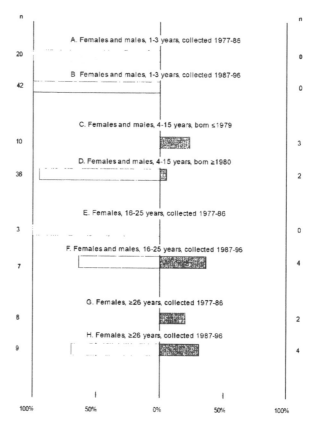

Figure 19.8 Claw lesions in female and male grey seals. The percentage of animals with no or slight lesions is shown on the left-hand side of the figure; animals with moderate or severe lesions are shown on the right. A–H denote different age classes, periods of collection or of birth (*n* = number of animals). No male aged >15 years old was autopsied during 1977–1986.

Colonic ulcers

Apparently high prevalences of colonic ulcers were found in 4–15-year-old females and males born 1980 and later, as well as in seals older than 15 years (Figure 19.9). Remarkably, the prevalence was higher in more recent groups than in the earlier ones. A significantly higher frequency ($P < 0.01$) of this lesion was found in the 1–3-year-olds collected during 1987–96 compared with the corresponding group collected during 1977–86.

Arteriosclerosis

The prevalence of this lesion was strongly related to age (Figure 19.10). Moderate to severe lesions were apparently high in elderly females and were found in 18 out of 20 animals.

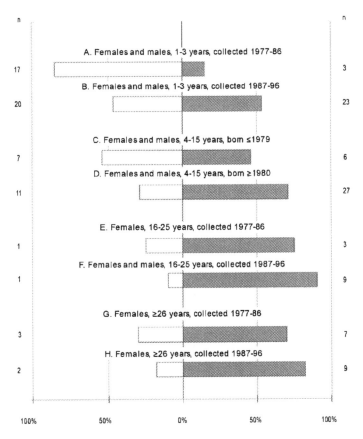

Figure 19.9 Ulcers in the colon of female and male grey seals (for explanation see the legend to Figure 19.8).

Adrenocortical hyperplasia

All seals older than 15 years showed moderate to severe hyperplastic changes (Figure 19.11). Comparison of records of this lesion in animals born 1979 and earlier with those born thereafter suggests a decline in prevalence, but this is not statistically significant.

Skull bone lesions

In an earlier investigation (Bergman *et al.*, 1992), skull bones of Baltic grey seals, aged from 6 years and collected during the two periods 1960–69 and 1971–85, showed high frequencies of lesions: >30 per cent and >50 per cent, respectively. This study also showed that the prevalence of lesions was lower in seals collected in the Baltic Sea before 1950 (around 10 per cent), and in samples collected in waters around the British Isles (<10 per cent). Recent

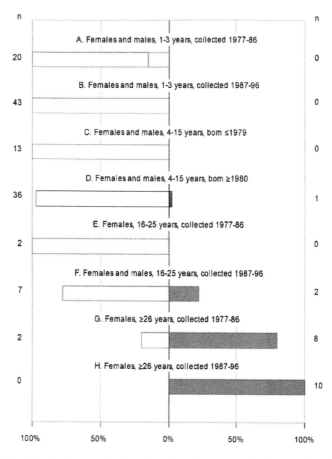

Figure 19.10 Arteriosclerosis in female and male grey seals (for explanation see the legend to Figure 19.8).

results (Bergman *et al.*, unpublished data) indicate an improvement and lowered prevalence of skull bone lesions since 1980. In the latter study grey seals born 1979 and earlier and aged 6–15 years showed a rather high frequency of skull bone lesions (28 affected out of 72: 29 per cent). Seventeen of the 28 skull bones affected showed severe lesions. In seals of the same age group but born 1980 and later, 4 out of 39 skull bones (10 per cent) showed lesions, none of which was severe.

Temporal variation of environmental contaminants in seals of the Baltic Sea and in their food

Since the end of the 1960s, the temporal variation of environmental contaminants has been studied by annual collection and analyses of defined biological matrices of the Baltic Sea (Olsson and Reutergårdh, 1986; Bignert

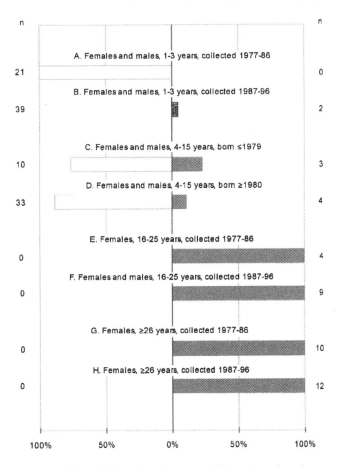

Figure 19.11 Hyperplasia of the adrenal cortex of female and male grey seals (for explanation see the legend to Figure 19.8).

et al., 1998a). The programme was extended at the end of the 1970s to include the Swedish west coast. The main matrices used are herring (*Clupea harengus*) and cod (*Gadus morrhua*), and eggs from the guillemot (*Uria aalgae*). The guillemot egg is an important matrix with many advantages (Bignert *et al.*, 1995), because this bird is non-migratory in the Baltic and preys on pelagic fish. Concentrations of all organochlorines measured have been decreasing since the 1970s. DDTs declined quite remarkably from the beginning of that decade. PCBs and dioxins also showed an initial rapid decrease, starting in the middle of the 1970s. A continuing decrease for PCB is not evident in samples from the Baltic Proper during the 1990s, whereas for dioxins, the decline seems to have stopped in the middle of the 1980s (Olsson *et al.*, 2002). Recent studies have revealed a high similarity in slopes of the decrease in DDTs and PCBs, regardless of whether marine or freshwater fish or bird eggs were used as a matrix (Bignert *et al.*, 1998a, b). The generalized change over time, regardless of species investigated, shows that contamination of the seal food chain has decreased.

Long-term monitoring of trends through time in concentrations of DDTs and PCB compounds in various matrices from the Baltic Sea and the Swedish west coast has shown decreasing concentrations of both these classes of organochlorines. During the entire study period, 1967–95, the PCB concentrations were two to three times, and DDTs concentrations three to five times, higher in the Baltic than at the Swedish west coast (Bignert *et al.*, 1998a).

Some of the seal specimens from the Baltic Sea that were examined for pathological changes have also been analysed for concentrations of heavy metals and for various chlorinated organic compounds. The aims of these analyses have been to investigate spatial exposure as well as exposure in relation to the health of the seals (Roos *et al.*, 1992).

Yearlings of subpopulations of the three species inhabiting the Swedish waters were used to study the spatial distribution of contaminants. Studies on contaminant exposure in relation to health concerned levels in harbour seal yearlings hit by the phocine distemper virus as well as levels in apparently healthy seals. Chemical analyses were conducted on three groups of adult female grey seals: group one, seals in normal nutritional condition, lacking uterine lesions and other severe pathological changes; group two, seals in normal nutritional condition but with one or several of the lesions discussed above; group three, seals in subnormal nutritional condition but having lesions similar to those seals in group two.

DDTs, PCBs and their methyl sulfone metabolites

Studies on juvenile harbour seals from the late 1980s indicate that the concentration of PCBs in specimens from the Baltic Sea was about twice as high as in the same species from the Swedish west coast (Blomkvist *et al.*, 1992). Of the three species of seals inhabiting the Baltic Sea, grey seals had the highest concentrations. Furthermore, the females belonging to the groups with various pathological changes had higher concentrations than 'healthy' ones. The concentrations found in diseased and emaciated females were often extremely high, reaching 1900 µg/g lipid weight DDTs and 5300 µg/g lipid weight PCBs in blubber.

A congener-specific analysis of PCBs that included samples from the grey seals mentioned above (Haraguchi *et al.*, 1992) revealed that specimens with high concentrations of PCBs did not biomagnify the mono-*ortho* PCB 118 (Olsson *et al.*, 1992a). Data from a study on Danish harbour seals also indicated a similar difference in biomagnification based on body burden of total PCBs (Storr-Hansen and Spliid, 1992). The findings indicate that the most severely contaminated specimens had an enzyme-induced degradation capability for some of the dioxin-like PCB congeners (Olsson *et al.*, 1992b). Studies of PCB methyl sulfones and DDE methyl sulfone revealed high concentrations in adult grey seals, especially in groups of females with pathological changes (Haraguchi *et al.*, 1992).

Polychlorinated camphenes (Toxaphene), chlordane compounds and polybrominated diphenyl ethers

Concentrations of these compounds were much lower than concentrations of DDTs and PCBs (Andersson and Wartanian, 1992). The concentrations of these compounds were generally slightly higher in seals of the Baltic Sea compared to those of the Swedish west coast. Concentrations found in harbour seals infected by the phocine distemper virus were similar to those in unaffected specimens. Only concentrations of chlordane compounds diverged, and were much higher in the groups of females showing pathologic changes.

Dioxins

Concentrations of polychlorinated dioxins and dibenzofurans were remarkably low (Bergek *et al.*, 1992). Concentrations were often slightly higher in juvenile seals than in adult seals, including the diseased females. The levels were lower in grey seals than in herring, their main food. Concentrations were similar in seals of the Baltic and the Swedish west coast. Harbour seals infected by the seal distemper virus had concentrations similar to those in unaffected ones. However, ringed seals of the Baltic Sea had higher concentrations than grey and harbour seals, but concentrations in Arctic ringed seals and Baltic ringed seals were similar (Bignert *et al.*, 1989).

Heavy metals

A study of heavy metal concentrations in kidneys and livers from seals of various Swedish waters, including the Swedish west coast, revealed no significant spatial variations or significant variation in relation to health of the seals (Frank *et al.*, 1992). Metals analysed were Al, As, Ca, Cd, Co, Cr, Cu, Fe, Hg, Mg, Mn, Ni, Pb, Se, V and Zn.

Discussion

Post-mortem investigations on grey seals of the Baltic Sea during 1977–96 showed a significant improvement of the gynaecological health (decreasing frequency of uterine occlusions/stenoses and an increasing frequency of pregnancies) and indications of an improvement in other health variables (claw lesions, leiomyomas and adrenocortical hyperplasia) in the latter part of the period. However, the prevalence of some lesions, such as intestinal ulcers, has increased (Bergman, 1999). Among ringed seals of the Baltic Sea, the prevalence of uterine occlusions is still high, ranging between 25 and 30 per cent in recent years (Mattson and Helle, 1995).

The broad analytical chemistry approach on tissues, presented above, does not provide evidence for any metal-induced disease among seals of the Baltic Sea. In addition, the chemical analytical results indicated that dioxins and coplanar PCBs, which have similar but less toxic effects (Safe, 1990), were not suspected to be the main cause of the disease complex among Baltic seals (Olsson *et al.*, 1992a). Similarly, thymus tissue has been observed macroscopically in rather old Baltic grey and ringed seals (Bergman and Olsson, 1985), as well as in ringed seals from Svalbard (Bergman, unpublished data), with no indications of pathology. Because dioxins and related compounds are known to affect the thymus severely (Vos and Luster, 1989), it is less likely that they had an impact on the health of Baltic seals.

On the other hand, studies on the methyl sulfone metabolites of both PCBs and DDTs revealed high concentrations in adult grey seals, and especially in the groups of females showing various pathological changes (Haraguchi *et al.*, 1992). Of special interest was the finding of high concentrations of DDE methyl sulfones (Haraguchi *et al.*, 1992), which, like DDD, are known to be highly adrenocorticolytic compounds and covalently bind to the adrenal cortex (Lund *et al.*, 1988; Jönsson *et al.*, 1992, 1993; Lund, 1994). Monitoring studies aiming to reveal the temporal trends of DDTs and PCBs have clearly shown that concentrations have decreased significantly in the Baltic Sea (Bignert *et al.*, 1998a). The census studies on Baltic seal populations clearly show that conditions have improved for all three species simultaneously, with decreased concentrations of these compounds (Helander and Bignert, 1992; Hårding and Härkönen, 1999; Helander *et al.*, 2001).

A hypothesis regarding the development of a disease complex in Baltic seals was presented by Bergman and Olsson (1985). The hypothesis focused on possible effects associated with the hyperplasia of the adrenal zona fasciculata and zona reticularis commonly found in these animals. If the proliferated adrenal cells are functional, the pathophysiological effect is hyperfunction. The resulting clinical condition in animals and humans is known as hyperadrenocorticism, and is characterized by increased levels of glucocorticoids in the blood. As a consequence, the patient develops symptoms and lesions due to the gluconeogenic, lipolytic, protein-catabolic, anti-inflammatory and immunosuppressive effects of the glucocorticoid hormones. Clinical symptoms and lesions frequently occurring in hyperadrenocorticism are elevated fasting plasma glucose concentrations, increased blood lipid and cholesterol concentrations, elevated blood pressure (hypertension), muscle wasting, osteoporosis, bilaterally symmetric alopecia, skin infections and poor wound healing. Alterations to the oestrous cycle and persistent anoestrus have also been reported (for further information on the syndrome, see Cotran *et al.*, 1999).

The autopsy findings on seals of the Baltic Sea indicate effects on the endocrine system with probable involvement of the hypothalamic–pituitary–gonadal and adrenal axis (Bergman and Olsson, 1985). Brandt (1975) showed strong uptake of PCBs in corpora lutea, adrenal cortex and the pituitary in mice by using whole-body autoradiography. The accumulation of certain PCB congeners in the pituitary and/or hypothalamus may lead to effects on gonadotrophin release and gonadal function. Jansen *et al.* (1993) showed an influence on pituitary function by PCBs by studying the effect of Aroclor 1242 on cultured anterior pituitary cells from immature female rats, where they found that these cells exhibited increased gonadotrophin responses.

Results on ringed seals of the Baltic Sea (Bergman *et al.*, unpublished data) indicate that occlusions and stenoses are associated with hormonal imbalance and immune suppression. This is probably due to interference of organochlorines with steroid metabolism, causing a disturbed puerperium or fetal death. Also according to previous findings in ringed seals (Helle *et al.*, 1976b; Helle, 1980) and grey seals (Bergman and Olsson, 1985) of the Baltic Sea, uterine occlusions and stenoses are related to pregnancy – some cases have shown retained fetal membranes containing fetuses, aligned to stenosed parts of the horns.

Traditionally, oestrogen has been considered the major promoter of leiomyoma growth. According to Kennedy and Miller (1993) uterine leiomyoma is a common tumour in the dog but uncommon in other domestic species. As found in our studies in grey seals, it rarely occurs in female dogs before middle age and is often associated with ovarian follicular cysts or oestrogen-secreting tumours, endometrial and mammary hyperplasia and mammary neoplasia. When female dogs are castrated early in life, leiomyomas do not develop, and, if present, they regress following castration, circumstances which further indicate an endocrine factor in their development. In humans, uterine leiomyoma is a common tumour and affects about 30 per cent of women above 35 years of age; there is strong evidence of an oestrogen dependency for their growth in humans, which is supported by evidence of regression after menopause (Crum, 1999). The high prevalence of uterine leiomyomas in Baltic grey seals is believed to be due to the influence of a factor, or factors, in the environment causing hyperoestrogenic states (Bergman, 1999).

Regarding the defects on digits and claws observed in the Baltic seals, it is of interest to mention experimental studies in which abnormalities in claws were induced by feeding PCBs to European ferrets (*Mustela putorius furo*) (Bleavins *et al.*, 1982) and American mink (*Mustela vison*) (Aulerich *et al.*, 1987).

The high concentrations of DDTs and PCBs in seals of the Baltic Sea, the many different aspects of the toxicity of DDTs and PCBs, and the character of the lesions suggesting an endocrine origin, all support the probability of a causal linkage between DDTs and PCBs and the disease complex seen in Baltic seals. However, it is difficult to distinguish between effects caused by DDTs and effects caused by PCBs. Roos *et al.* (1998) presented results on long-term monitoring of concentrations of DDTs and PCBs in annual samples of various biological matrices from the Baltic, e.g. guillemot eggs, herring, cod and yearling grey seals. The material was collected from the late 1960s to the 1990s, and showed a remarkable consistency with respect to annual decreases in concentrations of DDTs and PCBs in all matrices except yearling grey seals. The concentrations of DDTs decreased in the grey seals at a similar rate as for other matrices studied, but PCB concentrations remained more or less constant. The authors interpreted the results in the following way: the concentrations of lipophilic persistent compounds that biomagnify in yearling seals are largely a result of suckling of the lipid-rich seal milk. Thus a high concentration of a compound in a young seal implies that the dam had a high body burden of the compound. The decreasing concentrations of DDTs, but more constant concentrations of PCBs, evident in Baltic grey seals but not in their food over the 1960s–90s indicate that PCBs, in particular, have adversely affected reproduction of seals in the past. Thus, females with high concentrations of DDTs in the past were able to produce offspring, but females with high concentrations of PCBs did not. These results are in line with experimental results obtained in mink, where feeding PCBs but not DDT compounds caused reproductive failure (Kihlström *et al.*, 1976; Aulerich and Ringer, 1977; Jensen *et al.*, 1977).

The increasing prevalence of intestinal ulcers, especially in young grey seals, might indicate immunosuppression (Bergman, 1999). In recent experiments in which groups of captive harbour seals were fed herring from the Baltic Sea or Atlantic Ocean, immune suppression was observed in the seals fed the Baltic herring (De Swart *et al.*, 1994, 1995; Ross *et al.*, 1995, 1996; see also Vos *et al.*, Chapter 21 in this volume). These findings indicate that the food available to Baltic seals is still sufficiently contaminated to affect health.

Conclusions

Post-mortem investigations of grey and ringed seals from the Baltic Sea show that certain lesions are common in adults. These involve changes of female sex organs (occlusions or stenoses of the uterus in sexually mature grey and ringed seals, and uterine leiomyomas in grey seals) and changes in other organs of both sexes (lesions of claws and digits, skull bone erosions, colonic ulcers, arteriosclerosis and adrenocortical hyperplasia and adrenocortical adenomas). Our main hypothesis is that the adrenal changes are crucial, explaining the occurrence of most of the other lesions. When translated to clinical conditions, the complex of lesions is reminiscent of hyperadrenocorticism or Cushing's syndrome, which occurs as a disease entity in domestic animals and humans.

During recent years a decreasing tendency in the prevalence of certain lesions appeared in Baltic grey seals, especially those involving female reproductive organs. The prevalence of uterine obstructions decreased significantly and the rate of pregnancy increased, while a decreasing trend was found for leiomyomas. Improvements seem to be indicated for the incidence of adrenocortical hyperplasia, as well as for skull bone lesions, but these are not statistically significant. There are indications of an interrelation between these signs of decreasing severity of the disease complex in grey seals of the Baltic Sea and the decreasing levels of organochlorines, in particular PCBs, in Baltic biota over the period of investigation.

However, unfavourable tendencies remain. The finding by Finnish scientists of a continued high prevalence of uterine occlusions in Baltic ringed seals is a matter of concern. The incidence of colonic ulcers has continued to increase in grey seals of the Baltic Sea, despite the reduced concentrations of known organochlorines, indicating a continuing but unknown role for other factors. These may include contaminant-induced immunosuppression.

References

Almkvist, L., Olsson, M. and Söderberg, S. 1980. *Sälar i Sverige.* Svenska Naturskyddsföreningen, Stockholm (in Swedish).

Andersson, Ö. and Wartanian, A. 1992. Levels of polychlorinated camphenes (toxaphene), chlordane compounds and polybrominated diphenyl ethers in seals from Swedish waters. *Ambio* 21: 550–552.

Aulerich, R.J. and Ringer, R.K. 1977. Current status of PCB toxicity to mink, and effect of their reproduction. *Arch. Environ. Contam. Toxicol.* 5: 279–292.

Aulerich, R.J., Bursian, S.J., Evans, M.G., Hochstein, J.R., Koudele, K.A., Olson, B.A. and Napolitano, A.C. 1987. Toxicity of 3,4,5,3′,4′,5′-hexachlorobiphenyl to mink. *Arch. Environ. Contam. Toxicol.* 16: 53–60.

Bergek, S., Bergqvist, P.-A., Hjelt, M., Olsson, M., Rappe, C., Roos, A. and Zook, D. 1992. Concentrations of PCDDs and PCDFs in seals from Swedish waters. *Ambio* 21: 553–556.

Bergman, G. 1956. Sälbeståndet vid våra kuster. *Nordenskiöldsamfundets tidskrift* 1956: 49–65 (in Swedish).

Bergman, A. 1999. Health condition of the Baltic grey seal (*Halichoerus grypus*) during two decades. Gynaecological health improvement but increased prevalence of colonic ulcers. *APMIS* 107: 270–282.

Bergman, A. and Olsson, M. 1985. Pathology of Baltic grey seal and ringed seal females with special reference to adrenocortical hyperplasia: Is environmental pollution the cause of a widely distributed disease syndrome? *Finnish Game Research* 44: 47–62.

Bergman, A. and Olsson, M. 1989. Pathology of Baltic grey seal and ringed seal males. Report regarding animals sampled 1977–1985. In: A.V. Yablokov and M. Olsson (eds). *Influence of human activities on the Baltic ecosystem.* Proceedings of the Soviet–Swedish symposium 'Effects of Toxic substances on dynamics of seal populations', Moscow, USSR, April 14–18, 1986, pp. 74–86. Leningrad Gidrometeoizdat.

Bergman, A., Järplid, B. and Svensson, B.-M. 1990. Pathological findings indicative of distemper in European seals. *Vet. Microbiol.* 23: 331–341.

Bergman, A., Olsson, M. and Reiland, S. 1992. Skull-bone lesions in the Baltic grey seal (*Halichoerus grypus*). *Ambio* 21: 517–519.

Bergman, A., Bergstrand, A. and Bignert, A. 2001. Renal lesions in Baltic grey seals (*Halichoerus grypus*) and ringed seals (*Phoca hispida botnica*). *Ambio* 30: 397–409.

Bignert, A., Olsson, M., Bergqvist, P.-A., Bergek, S., Rappe, Ch., de Wit, C. and Janson, B. 1989. Polychlorinated dibenzo-p-dioxins (PCDD) and dibenzo-furans (PCDF) in seal blubber. *Chemosphere* 19 (1–6): 551–556.

Bignert, A., Litzén, K., Odsjö, T., Persson, W. and Reutergårdh, L. 1995. Time-related factors influence the concentrations of sDDT, PCBs and shell parameters in eggs of Baltic guillemot (*Uria aalge*), 1861–1989. *Environ. Pollut.* 89: 27–36.

Bignert, A., Olsson, M., Persson, W., Jensen, S., Zakrisson, S., Litzén, K., Eriksson, U., Häggberg, L. and Alsberg, T. 1998a. Temporal trends of organochlorines in northern Europe, 1967–1995. Relation to global fractionation, leakage from sediments and international measures. *Environ. Pollut.* 99: 177–198.

Bignert, A., Greyerz, E., Olsson, M., Roos, A., Asplund, L. and Kärsrud, A.-S. 1998b. Similar decreasing rate of OCs in both eutrophic and oligotrophic environments – a result of atmospheric degradation? Part 2. Proceedings from the 18th Symposium or Halogenated Environmental Organic Pollutants, Stockholm, Sweden, August 17–21, 1998. In: DIOXIN-98. Transport and Fate PI. N. Johansson, Å. Bergman, D. Broman, H. Håkansson, B. Jansson, E. Klasson Wehler, L. Poellinger and B. Wahlström (eds). *Organohalogen Compounds* 36: 459–462.

Bleavins, M.R., Aulerich, R.J., Ringer, R.K. and Bell, T.G. 1982. Excessive nail growth in the European ferret induced by Aroclor 1242. *Arch. Environ. Contam. Toxicol.* 11: 305–312.

Blomkvist, G., Roos, A., Jensen, S., Bignert, A. and Olsson, M. 1992. Concentrations of sDDT and PCB in seals from Swedish and Scottish waters. *Ambio* 21: 539–545.

Brandt, I. 1975. The distribution of 2,2′,3,4,4′,6′ and 2,3′,4,4′,5′,6-hexachlorobiphenyl in mice studied by whole-body autoradiography. *Toxicology* 4: 275–287.

Cotran, R.S., Kumar, V. and Collins, T. 1999. The endocrine system. In: R.S. Cotran, V. Kumar and T. Collins (eds). *Robbins Pathologic Basis of Disease*, 6th edn. W.B. Saunders Company, Philadelphia, pp. 1121–1169.

Crum, P.C. 1999. Female genital tract. In: R.S. Cotran, V. Kumar and T. Collins (eds). *Robbins Pathologic Basis of Disease*, 6th. edn. W.B. Saunders Company, Philadelphia, pp. 1035–1091.

De Swart, R.L., Ross, P.S., Vedder, L.J., Timmerman, H.H., Heisterkamp, S.H., van Loveren, H., Vos, J.G., Reijnders, P.J.H. and Osterhaus, A.D.M.E. 1994. Impairment of immune function in harbour seals (*Phoca vitulina*) feeding on fish from polluted waters. *Ambio* 23: 155–159.

De Swart, R.L., Ross, P.S., Timmerman, H.H., Vos, H.W., Reijnders, P.J.H., Vos, J.G. and Osterhaus, A.D.M.E. 1995. Impaired cellular immune response in harbour seals (*Phoca vitulina*) fed environmentally contamined herring. *Clin. Exp. Immunol.* 101: 480–486.

Dietz, R., Heide-Jörgensen, M.-P. and Härkönen, T. 1989. Mass deaths of harbour seals (*Phoca vitulina*) in Europe. *Ambio* 18: 258–264.

Frank, A., Galgan, V., Roos, A., Olsson, M., Petersson, L.R. and Bignert, A. 1992. Metal concentrations in seals from Swedish waters. *Ambio* 21: 529–538.

Haraguchi, K., Athanasiadou, M., Bergman, Å., Hovander, L. and Jensen, S. 1992. PCB and PCB methyl sulfones in selected groups of seals from Swedish waters. *Ambio* 21: 546–549.

Hårding, K.C., Härkönen, T. and Caswell, H. 2002. The European seal plague: Epidemiology and population consequences. *Ecology Letters*. (In print).

Hårding, K.C. and Härkönen, T.J. 1999. Developments of the Baltic grey seal (*Halichoerus grypus*) and ringed seal (*Phoca hispida*) populations during the 20th century. *Ambio* 28: 619–627.

Härkönen, T., Stenman, O., Jüssi, M., Jüssi, I., Sagitov, R. and Verevkin, M. 1998. *Population Size and Distribution of the Baltic Ringed Seal* (*Phoca* hispida botnica). NAMMCO Scientific Publications.

Helander, B. and Bignert, A. 1992. Harbour seal (*Phoca vitulina*) on the Swedish Baltic coast: Population trends and reproduction. *Ambio* 21: 504–510.

Helander, B. and Härkönen, T. 1999. *Säl och havsörn. Östersjö '98*, Stockholms Marina Forskningscentrum, pp. 47–52 (in Swedish).

Helander, B., Karlsson, O. and Lundberg, T. 2001. *Inventering av gråsäl vid svenska östersjökusten 2000.* Sälinformation 2001: 1, Swedish Museum of Natural History, Box 500 07, SE-104 05 Stockholm (in Swedish).

Helle, E. 1980. Lowered reproductive capacity in female ringed seals (*Pusa hispida*) in the Bothnian Bay, northern Baltic Sea, with special reference to uterine occlusions. *Annal. Zool. Fenn.* 17: 147–158.

Helle, E., Olsson, M. and Jensen, S. 1976a. DDT and PCB levels and reproduction in ringed seal from the Bothnian Bay. *Ambio* 5: 188–189.

Helle, E., Olsson, M. and Jensen, S. 1976b. PCB levels correlated with pathological changes in seal uteri. *Ambio* 5: 261–263.

Hiby, L., Lundberg, T., Karlsson, O., Watkins, J. and Helander, B. 2001. An estimate of the size of the Baltic grey seal population based on photo-id data. Report to the project 'Sälar och Fiske', Administrative Board, Västernorrland County, SE-871 86 Härnösand, Sweden.

Hook, O. and Johnels, A.G. 1972. The breeding and distribution of the grey seal (*Halichoerus grypus* Fab.) in the Baltic Sea, with observations on other seals in the area. *Proc. R. Soc. Lond. B* 182: 37–58.

Jansen, H.T., Cooke, P.S., Porcelli, J., Liu, T.C. and Hansen, L.G. 1993. Estrogenic and antiestrogenic actions of PCBs in the female rat: in vitro and in vivo studies. *Reprod. Toxicol.* 7: 237–248.

Jensen, S., Johnels, A.G., Olsson, M. and Otterlind, G. 1969. DDT and PCB in marine animals from Swedish waters. *Nature* 224: 247–250.

Jensen, S., Kihlström, J.E., Olsson, M., Lundberg, C. and Örberg, J. 1977. Effects of PCB and DDT on mink (*Mustela vison*) during the reproductive season. *Ambio* 6: 239.

Johnston, D.H. and Watt, I.D. 1981. A rapid method for sectioning undecalcified carnivore teeth for aging. In *Proceedings, Worldwide Furbearer Conference, August 3–11, 1980, Frostburg, Maryland, USA*, Vol. 1, pp. 407–422.

Jönsson, C.-J., Lund, B.O., Bergman, Å. and Brandt, I. 1992. Adrenocortical toxicity of 3-methylsulfonyl-DDE in mice; 3: Studies in fetal and suckling mice. *Reprod. Toxicol.* 6: 233–240.

Jönsson, C.-J., Lund, B.O. and Brandt, I. 1993. Adrenocorticolytic DDT-metabolites: studies in mink, *Mustela vison* and otter, *Lutra lutra. Ekotoxicology* 2: 41–53.

Kennedy, P.C. and Miller, R.B. 1993. The female genital system. In K.F.V. Jubb, P.C. Kennedy and N. Palmer (eds), *Pathology of Domestic Animals*, 4th edn, Vol. 3. Academic Press, San Diego, pp. 349–470.

Kihlström, J.E., Olsson, M. and Jensen, S. 1976. Effekter på högre djur av organiska miljögifter. I: Organiska miljögifter i vatten. *Tolfte nordiska symposiet om vatten-*

532   *A. Bergman* et al.

forskning, Visby 11–13 maj 1976. Nordforsk, Miljövårdssekretariatet. Publikation 1976: 2, pp. 567–576 (in Swedish).

King, J.E. 1983. Chapter 11: Reproduction, pp. 181–190. In: J.E. King (ed.), *Seals of the World*, 2nd edn. British Museum (Nat. Hist.), London and Oxford University Press, Oxford, Chapter 11.

Lund, B.O. 1994. In vitro adrenal bioactivation and effects on steroid metabolism of DDT, PCBs, and their metabolites in the gray seal (*Halichoerus grypus*). *Environ. Toxicol. Chem.* 13: 911–917.

Lund, B.O., Bergman, Å. and Brandt, I. 1988. Metabolic activation and toxicity of a DDT-metabolite, 3-methylsulphonyl-DDE, in the adrenal zona fasciculata in mice. *Tem. Biol. Interactions* 65: 25–40.

Mattson, M. and Helle, E. 1995. Reproductive recovery and PCBs in Baltic seal populations. *Eleventh Biennal Conference on the Biology of Marine Mammals 14–18 December 1995, Orlando, Florida, USA.* Abstracts, p. 74.

Odsjö, T., Bignert, A., Olsson, M., Asplund, L., Eriksson, U., Häggberg, L., Litzén, K., de Wit, C., Rappe, C. and Åslund K. 1997. The Swedish environmental specimen bank – application in trend monitoring of mercury and some organo-halogenated compounds. *Chemosphere* 34: 2059–2066.

Olsson, M. and Reutergårdh, L. 1986. DDT and PCB pollution trends in the Swedish aquatic environment. *Ambio* 15: 103–109.

Olsson, M., Johnels, A.G. and Vaz, R. 1975. DDT and PCB levels in seals from Swedish waters. The occurrence of aborted seal pups. In *Proceedings of the Symposium on the Seal in the Baltic, Lidingö, Sweden, June 4–6, 1974.* Report from the National Swedish Environment Protection Board, PM 591: 43–65.

Olsson, M., Andersson, Ö., Bergman, Å., Blomkvist, G., Frank, A. and Rappe, C. 1992a. Contaminants and diseases in seals from Swedish waters. *Ambio* 21: 561–562.

Olsson, M., Karlsson, B. and Ahnland, E. 1992b. Seals and seal protection: Summary and comments. *Ambio* 21: 606.

Olsson, M., Bignert, a., Aune, M., Haarich, M., Harms, U., Jansson, M., Korhonen, M., Leivuori, M., Pedersen, B., Poutanen, E.L., Roots, O., Sapota, G. and Simm, M. 2002. Contaminant concentrations in biota. In *Fourth Periodic Assessment of the State of the Marine Environment of the Baltic Sea Environment* (chapter 6.3). Helsinki Commission (in print).

Reijnders, P.J.H. 1976. The harbour seal (*Phoca vitulina*) in the Dutch Wadden Sea: size and composition. *Neth. J. Sea Res.* 10: 223–235.

Reijnders, P.J.H. 1980. Organochlorine and heavy metal residues in harbour seals from the Wadden Sea and their possible effects on reproduction. *Neth. J. Sea Res.* 14: 30–65.

Reijnders, P.J.H. 1986. Reproductive failure in common seals feeding fish from polluted coastal waters. *Nature* 324: 456–457.

Roos, A., Blomkvist, G., Jensen, S., Olsson, M., Bergman, A. and Härkönen, T. 1992. Sample selection and preparation procedures for analyses of metals and organohalogen compounds in Swedish seals. *Ambio* 21: 525–528.

Roos, A., Bergman, A., Greyerz, E. and Olsson, M. 1998. Time trend studies on ΣDDT and PCB in juvenile grey seals (*Halichoerus grypus*), fish and guillemot eggs from the Baltic Sea. Proceedings of the 18th Symposium on Halogenated Environmental Organic Pollutants, Stockholm, Sweden, August 17–21, 1998. In: DIOXIN-98. Environmental Levels II. N. Johansson, Å. Bergman, D. Broman,

H. Håkansson, B. Jansson, E. Klasson Wehler, L. Poellinger and B. Wahlström (eds). *Organohalogen Compounds* 39: 109–112.

Ross, P.S., De Swart, R.L., Reijnders, P.J.H., van Loveren, H., Vos, J.G. and Osterhaus, A.D.M.E. 1995. Contaminant-related suppression of delayed-type hypersensivity and antibody responses in harbour seals fed herring from the Baltic sea. *Environ. Health Perspect.* 103: 162–167.

Ross, P.S., De Swart, R.L., Timmerman, H.H., Reijnders, P.J.H., Vos, J.G., van Loveren, H. and Osterhaus, A.D.M.E. 1996. Suppression of natural killer cell activity in harbour seals (*Phoca vitulina*) fed Baltic sea herring. *Aquat. Toxicol.* 34: 71–84.

Safe, S. 1990. Polychlorinated biphenyls (PCBs), dibenzo-*p*-dioxins (PCDDs), dibenzofurans (PCDFs), and related compounds: Environmental and mechanistic considerations which support the development of toxic equivalency factors (TEFs). *CRC Crit. Rev. Toxicol.* 21: 51–88.

Storr-Hansen, E. and Spliid, H. 1992. Levels and patterns of toxic coplanar polychlorinated biphenyl (CB) congeners in harbour seal (*Phoca vitulina*): Geographic variability of CB levels and congener patterns. In: E. Storr-Hansen. Coplanar polychlorinated biphenyl congeners. Developmental of an analytical method and its application to samples from Danish harbour seal (*Phoca vitulina*). PhD thesis, Department of Environmental Chemistry, Ministry of the Environment, National Environment Research Institute, Paper V, pp. 139–176.

Vos, J.G. and Luster, M.I. 1989. Immune alterations. In: R.D. Kimbrough and A.A. Jensen (eds). *Halogenated Biphenyls, Terphenyls, Naphthalenes, Dibenzodioxins and Related Products*, 2nd edn. Elsevier, Amsterdam, pp. 295–322.

20 The immune system, environmental contaminants and virus-associated mass mortalities among pinnipeds

Peter S. Ross, Joseph G. Vos and Albert D.M.E. Osterhaus

Introduction

The task of the immune system appears reasonably simple: to fend off invasion by foreign pathogens or prevent the growth of malignant tissues, using a network of interacting cells and chemical messengers. In practice, however, the immune system is highly complex and its function difficult to characterize, often rendering an assessment of its 'status' or 'health' extremely challenging. Although comparative studies have found that some specific differences exist between pinnipeds and other mammals, the immune system of pinnipeds shares many basic features with other species (Cavagnolo, 1979; Ross *et al.*, 1994). Many of the functional assays used in assessing immune function in laboratory animals have been shown to be applicable to pinnipeds. Disease outbreaks among pinnipeds represent complex ecological events under natural circumstances (Hall *et al.*, 1992a; Thompson and Hall, 1993), but the presence of high concentrations of immunotoxic chemicals in the tissues of these animals represents an additional variable requiring scrutiny.

The immunotoxicity of low levels of many anthropogenic persistent organic pollutants (POPs) has long been known, with studies demonstrating that polychlorinated biphenyls (PCBs), polychlorinated dibenzo-*p*-dioxins (PCDDs), polychlorinated dibenzofurans (PCDFs) and a number of pesticides can alter normal immune function in rodents, rabbits and primates (Vos and De Roij, 1972; Koller and Thigpen, 1973; Faith and Moore, 1977; Tryphonas *et al.*, 1991). Contaminant-associated immunotoxicity in mammals is suspected of being mediated largely via the cytosolic aryl hydrocarbon receptor (AHR), with the 'dioxin-like', or 'planar', congeners of the PCB, PCDD and PCDF classes binding the most strongly, and hence having the greatest immunotoxic potential (Luster *et al.*, 1987; Safe, 1990). The lipophilicity of many of these chemicals, coupled with their persistence in the natural environment and relative resistance to metabolic degradation in

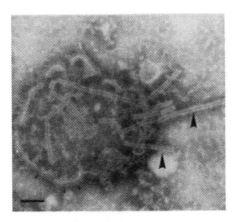

Figure 20.1 Electron micrograph of a morbillivirus, members of which include canine distemper virus (CDV), peste des petits ruminants (PPRV), rinderpest (RPV), measles virus (MV) and phocine distemper virus (PDV). The morbillivirus here (PDV) was isolated from a harbour seal, subsequently cultured on Vero cells, and the high-speed centrifuge concentrate of culture fluid was negatively stained with uranyl acetate. Note the partly disrupted virus particle with release of nucleocapsid (arrowhead). Bar represents 100 nm. (Photo courtesy J.S. Teppema, National Institute of Public Health and the Environment, Bilthoven, The Netherlands.)

different organisms, cause pinnipeds and other fish-eating wildlife to accumulate high concentrations in their fatty tissues (Peakall and Fox, 1987; Hutchinson and Simmonds, 1994; Tanabe *et al.*, 1994).

The immunotoxicity of POPs to marine mammals became a major issue in 1988, when a severe outbreak of a morbillivirus occurred among harbour (*Phoca vitulina*) and grey (*Halichoerus grypus*) seals in northern Europe. Even though a newly discovered virus, phocine distemper virus (PDV; Figure 20.1) (Cosby *et al.*, 1988; Mahy *et al.*, 1988; Osterhaus and Vedder, 1988; Osterhaus *et al.*, 1995) was found to have been introduced into an immunologically naïve population of pinnipeds (Osterhaus *et al.*, 1989a), concern about the possible contribution of immunotoxic anthropogenic chemicals was not to be allayed. The dramatic nature of the event, with heavy mortality, and the multitude of ailments with which surviving individuals washed ashore, served to attract and propagate media and scientific interest (Dietz *et al.*, 1989a). Further mortalities among pinnipeds and cetaceans in other relatively contaminated parts of the world set the stage for further speculation (Grachev *et al.*, 1989; Van Bressem *et al.*, 1991; Lipscomb *et al.*, 1994).

The immune system of pinnipeds

The immune system evolved to serve as a dynamic and functional protector of health in animals, and it is not surprising that early interest in pinniped immunology stemmed from the veterinary sciences. The reasonably conserved nature of the immune system among mammals allowed clinicians to view captive pinnipeds as they would domestic animals. Routine clinical haematology, which ably distinguished many of the basic leucocyte subpopulations, including lymphocytes, monocytes, neutrophils, basophils and eosinophils, was widely used as a diagnostic tool when captive pinnipeds suffered from viral, bacterial or parasitic infections (Engelhardt, 1979; Bossart and Dierauf, 1990; Roletto, 1993). In tandem with haematological studies, serum chemistry was often measured to assess the physiological state of organ or endocrine systems and general health (Castellini *et al.*, 1993; Roletto, 1993; De Swart *et al.*, 1995a). Although such information provides valuable quantitative details on basic aspects of health, physiology and the immune system, it does not provide any functional information on the immune system or its components.

The mammalian immune system consists of a set of lymphoid organs, circulating cells and antibodies, and soluble factors and chemical messengers. The primary lymphoid tissues consist of the leucocyte-producing bone marrow, the T-cell maturation site of the thymus, and the antigen collection sites of the spleen and the widely distributed lymph nodes. White blood cells, or leucocytes, circulate in the blood and spend time collected in sites of activity such as the spleen or lymph nodes. Leucocyte subpopulations which are non-specific consist of the anti-bacterial neutrophils, anti-parasitic eosinophils, inflammatory-associated basophils, phagocytizing and antigen-presenting macrophages and their circulating precursor monocytes, and cytotoxic virus- and tumour-killing natural killer (NK) cells. These cells are responsible for the clearance of pathogens in the absence of specific (acquired) memory, and/or in the early stages of infection. Leucocytes involved in the secondary, or specific, immune response consist of B-lymphocytes, precursors to the immunoglobulin (Ig)-secreting plasma cell, and T-lymphocytes, which are involved in the coordination of the overall immune response (T-helper cells) and the killing of virus-infected cells (cytotoxic T cells). These lymphocytes act by acquired memory, and they target only the antigen for which they possess this memory. Their response to infection by a new pathogen generally peaks 6–10 days post-exposure, several days later than the non-specific defences (Figure 20.2).

Although interspecies differences exist in leucocyte surface antigens, immunoglobulin structure, and to a lesser extent, lymphoid tissue structure, the immune systems of mammals possess fundamentally similar characteristics. Dedicated studies in pinniped immunology began in the 1970s, when an interest in comparative evolutionary immunology and ecology provided a rationale for undertaking this work. Some of these studies involved a basic histopathological examination of lymphoid tissues, while others focused on the more species-specific characterization of immunoglobulin classes in pinnipeds, including California sea lions, *Zalophus californianus* (Nash and Mach, 1971), and northern fur seals, *Callorhinus ursinus* (Cavagnolo and Vedros, 1978; Cavagnolo, 1979).

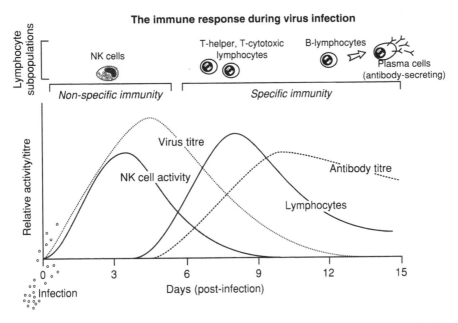

Figure 20.2 Following infection with a new virus, the non-specific arm of the immune system responds, with an increase in the activity of the natural killer (NK) cells playing an important role. These large granular lymphocytes bind to and lyse virus-infected cells in the absence of acquired memory or prior exposure, and represent an important first line of defence against new viruses. T- (helper and cytotoxic subpopulations) lymphocytes will acquire the ability to recognize portions of the virus in the days following infection, and the subsequent proliferation of like progeny will result in a highly specific and effective anti-virus response, and afford the host long-term protection against the same virus. B-lymphocytes will also acquire memory specific to the virus in the days following infection, but they develop into antibody-producing plasma cells. These lymphocyte subpopulations and immune responses have been characterized in pinnipeds using a combination of functional *in vitro* and *in vivo* tests and immunological reagents (De Swart *et al.*, 1993; Ross *et al.*, 1994, 1996a). For more information on the immune response in mammals in general, the reader is directed to (Roitt, 1990). (Adapted from Ross, 2002.)

Knowledge about pinniped immunology advanced rapidly following the 1988 PDV epizootic in northern Europe, reflecting concerns about the health of populations in this area, and the possible interaction between chemical contaminants and the immune systems of affected animals. Early studies identified methods which proved suitable for the assessment of non-specific immune function in both field (Ross *et al.*, 1993, 1994) and captive (De Swart *et al.*, 1993; Ross *et al.*, 1996a) studies of harbour seals. The methods used in these studies provide a means of evaluating the *in vitro* responses of lymphocytes using peripheral blood mononuclear cells (PBMC) purified from whole (heparinized) blood by density gradient separation, and showed that functional tests could be adapted successfully from use with other species. Additional studies involving immunizations and subsequent delayed-type hypersensitivity (DTH) tests *in vivo* (Ross *et al.*, 1995), as well as the determination of PBMC proliferative responses to antigen *in vitro* (De Swart *et al.*, 1995b) provided more systemic measures of specific immune function. These studies have contributed to the development of relatively non-invasive techniques suitable for use in pinnipeds.

The *in vitro* functional tests of immune function used in these studies suggest that the leucocyte subpopulations present in other mammals are also present in pinnipeds. Mitogens known to specifically stimulate T-lymphocytes (concanavalin A or Con-A; phytohaemagglutinin or PHA), B-lymphocytes (lipopolysaccharide or LPS), or both T- and B-lymphocytes (pokeweed mitogen or PWM) in laboratory animals were found to stimulate the same lymphocyte subpopulations in *in vitro* assays using PBMC from harbour seals (De Swart *et al.*, 1993). Natural killer cells were also characterized in harbour seals, with these lymphocytes showing properties similar to those of other mammals, including:

(1) cytotoxic activity against both tumour cells (YAC-1, P815, K562) and virus-infected cells (influenza-infected Nalm-6 cells) in *in vitro* assays;
(2) increased cytotoxic properties when co-incubated with human recombinant interleukin-2 (hrIL-2) *in vitro*; and
(3) diminished cytotoxic properties when coincubated with the NK surface protein-disrupting anti-asialo antibody (Ross *et al.*, 1996a).

Harbour seal T-lymphocytes have also been shown to be responsive to the stimulatory properties of hrIL-2, suggesting that this stimulatory cytokine is highly conserved (De Swart *et al.*, 1993). A cytokine similar to human IL-6 has been characterized in harbour and grey seals (King *et al.*, 1993a, 1996), suggesting a similar role for this pleiotropic chemical messenger in pinnipeds.

Although gross similarities have been shown to exist in the structure of the different immunoglobulin (Ig) classes, species-specific differences in binding affinities require the development of new reagents for each species studied. This has been largely circumvented in pinnipeds by the use of the IgG-binding protein A from *Staphylococcus aureus* as a reagent in the determination of both total and specific serum IgG antibody titres (Visser *et al.*, 1990; Ross *et al.*, 1993, 1994). Further characterization of harbour and grey seal immunoglobulins, and the development of monoclonal antibodies against these, are likely to provide more accurate means of assessing total and specific antibody levels and responses in some pinnipeds (Carter *et al.*, 1990a, b, 1992; King *et al.*, 1993b, 1994). Cell line-based virus neutralization assays are also used routinely in specific antibody titre determinations for studies of pinniped serology, disease status and epidemiology (Visser *et al.*, 1990; Ross *et al.*, 1992; Duignan *et al.*, 1997).

Developmental studies have demonstrated that the newborn harbour seal pup is precocious and possesses an immune system which is highly functional, contrasting with that of other newborn mammals. Highly responsive T- and B-lymphocytes compared to their mothers, and vigorous antibody responses following immunization, were observed in young, free-ranging harbour seals on Sable Island, Nova Scotia, Canada (Ross *et al.*, 1993, 1994). The lack of specific immunological memory, which is characteristic of newborn mammals, was compensated for by the lactational transfer of total IgG and pathogen-specific antibodies (in this case, specific against PDV) (Ross *et al.*, 1994). The newborn pups would therefore be passively and temporarily protected against infection by pathogens for which the mother had acquired memory. The relative immunocompetence of the newborn harbour seal was thought to have evolved as a result of the short nursing period (24 days) and limited maternal care in this species (Ross *et al.*, 1994).

A considerable amount of information now exists on the immune system of harbour seals, which have conveniently become the marine mammal of choice for immunological studies. Interest in the harbour seal stemmed partly from an earlier captive study on the reproductive toxicity of environmental contaminants using this species (Reijnders, 1986), but more importantly, as a consequence of the 1988 PDV epizootic. The harbour seal is also reasonably small (adults weigh 65–130 kg) and non-aggressive, enabling researchers to handle these animals with relative ease under both field and captive situations. Given the difficulties involved in the handling and sampling of many of the other pinnipeds, as well as cetaceans, extrapolation should allow the harbour seal to serve as a model for studies in immunology, toxicology and health assessment for other marine mammals (Ross and De Guise, 1998; Ross, 2000).

Environmental contaminants in the environment

The twentieth century represented a time of profound technological change, but the complex array of POPs that became widely distributed in the global environment reflected part of the cost of progress. The pesticide DDT was applied to many agricultural regions of the world, but was later discovered to be highly persistent in the environment, bioaccumulative in the food chain, and difficult to metabolize by wildlife (Carson, 1962; Hickey and Anderson, 1968). It was also found in animals, including pinnipeds, which inhabited regions of the world previously considered remote and pristine (Sladen *et al.*, 1966; Bowes and Jonkel, 1975), attesting to the relative ease with which such a chemical could be distributed around the world. The endocrine-disrupting effects of DDT on fish-eating birds were devastating, with eggshell thinning causing the extirpation of many species from large parts of Europe and North America (Hickey and Anderson, 1968; Wiemeyer and Porter, 1970; Lundholm, 1997).

DDT was pinpointed as a chemical which had direct effects on wildlife, but other POPs had similar physico-chemical characteristics. The industrial PCBs, produced as stabilizing and heat-resistant lubricants and transformer fluids, were also found in pinnipeds inhabiting both waters adjacent to industrial and population centres (Helle *et al.*, 1976a) and remote areas of the world (Addison and Smith, 1974; Risebrough *et al.*, 1976). In addition to the DDT group, numerous other organochlorine pesticides have been found in pinnipeds, including hexachlorocyclohexane (HCH), chlordane, hexachlorobenzene (HCB) and dieldrin (Muir *et al.*, 1992; Addison and Smith, 1998). It has become clear that persistent, lipophilic chemicals are bioaccumulating in marine food chains on a global scale, with higher trophic-level organisms being the most contaminated (Muir *et al.*, 1988; Norstrom *et al.*, 1988).

Advances in analytical chemistry have enabled the detection of PCDDs and PCDFs in pinniped blubber samples more recently (Rappe *et al.*, 1981; Norstrom *et al.*, 1990; Jarman *et al.*, 1996). Although some of the congeners of these chemicals are considered highly toxic, they did not appear to bio-accumulate to the same extent as the DDTs and PCBs, as evidenced from minimal differences in PCDD and PCDF concentrations in blubber samples between pinnipeds inhabiting industrial and remote areas (Oehme *et al.*, 1988; Bergek *et al.*, 1992). While this pattern may have partly reflected a wider atmospheric dispersal of the PCDDs and PCDFs, compared to the PCBs and DDTs, it has more recently become evident that pinnipeds are able to preferentially metabolize the former contaminants (Boon and Eijgenraam, 1988; Goksøyr *et al.*, 1992; De Swart *et al.*, 1995c; Troisi and Mason, 1997).

Whereas many studies documented the widespread contamination of the world's pinnipeds, others were slowly beginning to document a number of unusual observations in the most contaminated populations. A large number of premature births and abortions were observed in highly contaminated California sea lions in the late 1960s and early 1970s (Delong *et al.*, 1973; Gilmartin *et al.*, 1976). Low reproductive success was observed in harbour seals of the contaminated Wadden Sea in the late 1970s (Reijnders, 1980). High rates of skeletal lesions were observed in ringed (*Pusa hispida*), grey and harbour seals of the Baltic Sea from the 1950s onwards, when compared to earlier samples or to samples examined from the less contaminated North Sea (Zakharov and Yablokov, 1990; Bergman *et al.*, 1992; Mortensen *et al.*, 1992; Olsson *et al.*, 1994). A high rate of reproductive impairment was associated with pathological uterine lesions among female ringed seals of the contaminated Baltic Sea (Helle *et al.*, 1976a, b). A disease complex consisting of claw malformations, uterine lesions, uterine tumours and intestinal ulcers was tentatively associated with a high level of adrenal hyperplasia among the contaminated ringed and grey seals of the Baltic Sea in the 1970s and 1980s (Bergman and Olsson, 1985).

Although many of these studies were unable to identify contaminants con-clusively as the cause of the abnormalities, they did provide circumstantial evidence that the high levels of chemicals found in the animals' tissues were adversely affecting normal physiological function. Confounding factors which complicated the interpretation of results in some of these studies included variable age, condition and reproductive status of sampled animals, as well as the possible involvement of additional external factors in the abnormalities, such as pathogens and prey abundance. Nevertheless, the observational evidence of contaminant-related bioeffects in free-ranging pinnipeds fuelled the debate about an immunotoxic role for contaminants when the 1988 PDV outbreak struck the harbour seal population of northern Europe.

The pinniped immune system under assault from environmental contaminants

Many of the POPs to which free-ranging pinnipeds are exposed are immunotoxic to laboratory animals at low doses; these include PCBs (Vos and Van Driel-Grootenhuis, 1972), PCDDs (Vos *et al.*, 1973), DDT (Street and Sharma, 1975), butyltin compounds (Vos *et al.*, 1984, 1990; Smialowicz *et al.*, 1990), hexachlorobenzene (HCB) (Van Loveren *et al.*, 1990) and numerous other pesticides (Street and Sharma, 1975). Such studies have helped to identify the relative risk that each of these chemical classes might present to non-target species, including humans and wildlife, but exposure of laboratory animals to complex environmental mixtures have also resulted in immunotoxicity (Germolec *et al.*, 1989) and other toxicities (Sonstegard and Leatherland, 1979; Cleland *et al.*, 1987; Ross *et al.*, 1997).

Exposure to immunotoxic chemicals not only causes measurable alterations in parameters of immune function in laboratory animals, but also affects the immune system as a whole, and leads to an increased sensitivity to infection by pathogens. Delayed-type hypersensitivity (DTH) reactions, reflecting cell-mediated immune responses, were reduced in PCB-fed guinea-pigs (Vos and Van Driel-Grootenhuis, 1972). Ducks exposed to dietary PCBs had elevated dose-dependent mortality rates following challenge with duck hepatitis virus (Friend and Trainer, 1970). Following exposure to dietary PCBs, mice challenged with *Salmonella typhimurium* had elevated bacterial counts in organs, and elevated mortality rates (Thomas and Hinsdill, 1978). Exposure to low levels of 2,3,7,8-TCDD resulted in increased mortality rates after *Salmonella* challenge in mice (Thigpen *et al.*, 1975). Exposure to 2,3,7,8-TCDD administered by intraperitoneal (i.p.) injection resulted in a dose-dependent effect on survival following challenge with influenza virus in mice (House *et al.*, 1990). Exposure to low levels of 2,3,7,8-TCDD administered by i.p. injection resulted in elevated parasite loads in mice infected with *Trichinella spiralis* (Luebke *et al.*, 1994). Immunotoxicity is routinely measured as a change in one or more functional components of the immune system, but an increased sensitivity to infection clearly represents the biological significance of such a change. In wildlife, such biological alterations can be of ecological significance, because immunotoxicity might lead to changes in the ability of a population to cope with infection by new or enzootic pathogens.

In a practical sense, immunotoxicological studies can be difficult to perform, even when carefully controlled conditions exist in which the laboratory animals are kept, and possible confounding factors, such as stress, age and sex, are minimized. The complexity of the immune response makes a selection of appropriate endpoints difficult in immunotoxicological testing. In an attempt to evaluate the utility of immune-function tests in predicting susceptibility to disease, a comparative evaluation was made of both immune function and host resistance results from a database of experiments (Luster *et al.*, 1993). The authors found that over 70 per cent of the immune-function tests used were good predictors for alterations in host resistance, and that susceptibility to infection by a pathogen was related to the degree of immunosuppression and the amount of pathogen to which the study animal was exposed. Such considerations are important for immunotoxicological studies in pinnipeds, because immune-function tests are less invasive, and therefore more ethically acceptable, than host-resistance tests.

One feeding study has been carried out in captive harbour seals, in which tests of immune function were carried out. The study is described more fully elsewhere in this volume (Chapter 21), but we will summarize briefly the results that are relevant. In an attempt to assess the role of contaminants in the 1988 epizootic in Europe, two groups of 11 captive seals each were fed herring (*Clupea harengus*) from either the relatively uncontaminated Atlantic Ocean or from the contaminated Baltic Sea over the course of 2.5 years (for reviews see De Swart *et al.*, 1996; Ross *et al.*, 1996b, c). The Baltic Sea herring had up to 4–10 times higher concentrations of POPs than the Atlantic Ocean herring (De Swart *et al.*, 1995c).

Within 1 year of the start of the feeding experiment, results of *in vitro* tests of immune function using PBMC isolated from whole blood revealed significantly lower T-cell function (De Swart *et al.*, 1994) and NK cell function (Ross *et al.*, 1996a) in the seals fed the Baltic Sea herring. The specific immune responses of the seals fed Baltic herring were also affected, with lower *in vitro* proliferative responses to different antigens following a sensitizing immunization (De Swart *et al.*, 1995b), and lower *in vivo* delayed-type hypersensitivity responses (DTH) and antibody responses to ovalbumin 10 days after sensitization (Ross *et al.*, 1995). Reduced serum retinol (vitamin A) levels in the seals fed the Baltic Sea herring were consistent with the earlier reproductive toxicity study in which captive harbour seals were fed fish from the contaminated western Wadden Sea (Brouwer *et al.*, 1989), and with studies of laboratory rodents exposed to PCBs (Brouwer *et al.*, 1998).

The greater availability of specific immunological reagents, and the possibility of carrying out more invasive sampling, histopathology and host resistance assays in laboratory rodents, led researchers to carry out immunotoxicological studies of rats alongside the captive seal feeding study. In the first of these, juvenile rats were fed a supplemented diet of freeze-dried

Atlantic or Baltic herring for 4.5 months (Ross *et al.*, 1996d). Rats of the group fed herring from the Baltic Sea had diminished thymocyte numbers, lower thymus CD4 : CD8 ratios, and elevated salivary gland virus titres following challenge with rat cytomegalovirus (RCMV). Splenic lymphocyte proliferation and NK cell activity were unaffected.

A second parallel study was carried out using rats exposed perinatally to a maternally administered Atlantic or Baltic herring oil extract, or Atlantic herring oil spiked with 2,3,7,8-TCDD as a positive control (Ross *et al.*, 1997). In this study, an apparent dose-related response was observed in several parameters of thymus cellularity, virus-associated NK cell activity, splenic T-cell function, and RCMV-specific antibody (IgG) titres in the order of Atlantic > Baltic > (Atlantic + TCDD spike) rats.

These studies of laboratory rodents are significant for several reasons:

(1) they confirm that the Baltic Sea herring is immunotoxic, but also that the immunotoxic contaminants are found in the lipid fraction of the herring (this would include the PCBs, PCDDs and PCDFs);
(2) the similar, yet more pronounced, effects in the TCDD-spiked Atlantic group of rats provides evidence of a dioxin-like, or Ah-receptor-mediated, immunotoxicity in rats exposed to the complex contaminant mixture in the Baltic Sea herring;
(3) the similar pattern of immune alterations in the harbour seals and the rats exposed to the contaminants in the Baltic Sea herring provides evidence of a similar mechanism of immunotoxic action in the two species;
(4) the effects observed in the rats exposed to Baltic Sea herring contaminants suggest that the thymus may be an important target for contaminants in the Baltic seals, might explain the diminished T-lymphocyte function in these seals, and might suggest that immunosuppressed seals may be vulnerable to an increased severity of virus infection; and
(5) the high sensitivity of the perinatally exposed rats compared to the juvenile rats suggests that perinatally exposed free-ranging pinnipeds may be more vulnerable to the immunotoxic actions of environmental contaminants than the captive feeding study of older juvenile seals suggested.

The contaminant–immune system–virus link in pinniped mortalities

The deaths of approximately 20 000 harbour and several hundred grey seals in northern Europe in 1988, followed by a series of other mass mortalities among both pinnipeds (see Table 20.1) and cetaceans, fuelled a lasting debate about the possible involvement of immunotoxic pollutants. The agent responsible for the 1988 event, PDV, first affected harbour seals on Anholt Island in the Kattegat between Denmark and Sweden in mid-April of 1988, and spread to other continental populations over the subsequent months (Dietz *et al.*, 1989a). Harbour seals in the United Kingdom were the last to

Table 20.1 Outbreaks of morbillivirus-associated mass mortalities among pinnipeds

Responsible virus	Year	Location	Species	Estimated mortality	Reference
CDV	1957	Antarctica	Crabeater seal (*Lobodon carcinophaga*)	3000	Bengtson *et al.* (1991), Laws and Taylor (1957)
CDV	1987–88	Lake Baikal	Baikal seal (*Pusa sibirica*)	Several thousand	Grachev *et al.* (1989), Osterhaus *et al.* (1989b)
PDV	1988	Europe	Harbour seal (*Phoca vitulina*)	20 000	Osterhaus and Vedder (1988), Mahy *et al.* (1988)
PDV	1988	Europe	Grey seal (*Halichoerus grypus*)	400	Osterhaus and Vedder (1988), Mahy *et al.* (1988)
CDV	1997	Caspian Sea	Caspian seal (*Pusa caspica*)	Several thousand	Forsyth *et al.* (1998), Kennedy *et al.* (2000)

CDV, Canine distemper virus; PDV, phocine distemper virus.

be affected, with mortality peaking in September, October and November of the same year (Dietz *et al.*, 1989a; Hall *et al.*, 1992a). Although many disease symptoms were reported, including fever, gastrointestinal disorders, cutaneous lesions, neurological disturbances and respiratory ailments (Baker, 1992; Heide-Jörgensen *et al.*, 1992a), seals often succumbed to acute bacterial pneumonia (Osterhaus, 1988; Dietz *et al.*, 1989a). Numerous types of bacteria were isolated from victims in different areas (reviewed by Heide-Jörgensen *et al.*, 1992a).

Following the PDV epizootic, comparative serological evidence established that harbour and grey seals from the less contaminated Canadian east coast had been infected with a virus which was similar or identical to PDV prior to 1988, with no signs of unusual mortalities (Henderson *et al.*, 1992; Ross *et al.*, 1992). Additional studies demonstrating antibodies against morbilliviruses in Arctic Walrus (*Odobenus rosmarus*), hooded (*Cystophora cristata*), harp (*Pagophilus groenlandicus*) and ringed seals (Dietz *et al.*, 1989b; Markussen and Have, 1992; Duignan *et al.*, 1994, 1997), and in Antarctic crabeater seals (*Lobodon carcinophaga*) (Bengtson *et al.*, 1991), demonstrated that morbilliviruses represent widely occurring pathogens in apparently healthy, free-ranging pinniped populations, with no observations of unusual mortality. On the other hand, morbilliviruses have been known to cause serious mortality events in seronegative populations of various mammals in the absence of elevated contaminant concentrations (Norrby and Oxman, 1990; Bengtson *et al.*, 1991; Harder *et al.*, 1995). Members of the genus *Morbillivirus* include canine distemper virus (CDV), measles virus (MV), rinderpest virus (RPV), peste des petits ruminants (PPRV), as well as the recently identified dolphin morbillivirus (DMV), porpoise morbillivirus (PMV) and PDV (Osterhaus *et al.*, 1995).

Another significant pinniped die-off took place among Baikal seals (*Pusa sibirica*) in Russia's Lake Baikal in 1987–88, claiming 10 000 victims (Grachev *et al.*, 1989; Osterhaus *et al.*, 1989b). Although a morbillivirus was identified as the causative agent, the virus proved not to be PDV, but rather, the related CDV (Visser *et al.*, 1990). Later studies showed that seals inhabiting the Baikal Sea were heavily contaminated with PCBs, PCDDs and PCDFs, leading to further speculation about the role of immunotoxic chemicals in outbreaks of disease (Nakata *et al.*, 1995, 1997; Tarasova *et al.*, 1997).

Several other pinniped mortalities have been documented. Approximately 400 harbour seals died along the coast of New England in 1979, and although influenza A was isolated from some of the victims (Geraci *et al.*, 1982), an absence of contaminant information prevents a solid basis for speculation about possible immunotoxicity. Both a morbillivirus (Osterhaus *et al.*, 1998) and toxic dinoflagellates in fish (Costas and Lopez-Rodas, 1998) have been put forward to explain mortality in a population of highly endangered Mediterranean monk seals (*Monachus monachus*) along the coast of the former Spanish Sahara in 1997. Relatively low levels of contaminants in this population would tend to rule these out as a factor in the event (Borrell *et al.*, 1997). Several thousand freshwater Caspian seals (*Pusa caspica*) died in the spring of 1997 during an outbreak of CDV, but little is known about contaminant levels in this population (Forsyth *et al.*, 1998; Kennedy *et al.*, 2000).

In the aftermath of the 1988 PDV epizootic, laboratory animal-based immunotoxicological studies provided an immediate basis for speculation about the involvement of environmental contaminants. It was also widely known that the populations of seals affected had high tissue concentrations of PCBs and organochlorine pesticides, and that these had previously been associated with a number of abnormalities in Baltic and North Sea pinnipeds. In a comparison of 34 harbour seals that died during the epizootic, and 41 live, free-ranging seals sampled in early 1989, PCB levels were found to be significantly higher in seals that died than in survivors (Hall *et al.*, 1992b). Although the authors were cautious in their interpretation of these findings, because of possible differences in body condition (hence blubber thickness) between the two study groups, such an observation might be expected if contaminants were contributing to the severity of the epizootic. Interestingly, males accounted for almost 60 per cent of deaths during the epizootic in both the United Kingdom (Hall *et al.*, 1992a) and in the Baltic Sea (Heide-Jörgensen *et al.*, 1992b), which could be attributed to the timing of the infection and sex-related differences in seasonal haul-out behaviour and moulting times. Should contaminants have played a role in the event, however, this finding might also be expected, since adult males have higher concentrations of POPs than do age-matched females, and might therefore be more affected by immunotoxic contaminants. Females lose fat-soluble contaminants through reproduction and lactation, whereas males continue to accumulate contaminants throughout their lifetime (Addison and Smith, 1974; Addison and Brodie, 1977, 1987).

The subsequent captive feeding study of harbour seals provided the first evidence of contaminant-related immunotoxicity in mammalian wildlife (De Swart *et al.*, 1996; Ross *et al.*, 1996b), and parallel rodent studies provided supporting evidence about the role of dioxin-like contaminants in the immunotoxicity observed in the seals (Ross *et al.*, 1996d, 1997). Since NK cells are an important first line of defence against virus-infected cells (Welsh and Vargas-Cortes, 1992), and T cells represent a memory-based lymphocyte subpopulation which is vital to a specific anti-virus response (Finberg and Benacerraf, 1981), reductions in their performance could be expected to jeopardize the ability of seals to defend themselves against PDV.

PCBs contributed to over 90 per cent of the total toxic equivalents (TEQ) to 2,3,7,8-TCDD measured in the blubber of captive seals fed the Baltic Sea herring for 2 years; the immunotoxicity observed in the Baltic group of seals was consequently ascribed largely to this chemical class (Ross *et al.*, 1996b). Other studies have also demonstrated that PCBs present the greatest dioxin-like risk to free-ranging pinniped populations in different parts of the world (Oehme *et al.*, 1995; Ross *et al.*, 1998). Many harbour seal populations in northern Europe exceeded the accumulated blubber total PCB levels which were associated with immunotoxicity in the captive harbour seals fed Baltic herring, suggesting that immunotoxicity was likely occurring in several populations of free-ranging harbour seals in northern Europe (Ross *et al.*, 1996b).

Epi-immunotoxicology and the weight of evidence

The many epizootiological and scientific studies of the morbillivirus outbreak among harbour and grey seals in 1988, in particular, attest to the complexity of disease outbreaks, and the many factors that play a role (Thompson and Hall, 1993; Hall, 1995; Simmonds and Mayer, 1997). Given the contribution that immunotoxic contaminants are thought to have made to the event (Ross *et al.*, 1996b), one can speculate about the way in which the outcome of a virus outbreak might be affected by such contaminants. Because of the multifactorial nature of disease outbreaks, it is impossible to establish conclusively the involvement of immunotoxic chemicals. In a hypothetical model, we speculate that reduced non-specific (NK cells) and specific (T-lymphocytes) anti-viral defences will lead to a more severe expression of disease in individuals, a more rapid spread of virus in a population, and elevated mortality rates (Figure 20.3). This conceptual approach, which we describe here as 'epi-immunotoxicology', combines the known immunotoxic actions of environmental contaminants in harbour seals with the introduction of a new virus into a population, and considers the additional factors which may play a role in the outcome of an epizootic.

No single study or study design will be able to fully address the multiple aspects of the recent mass mortalities among pinnipeds. Whether a mass mortality caused by a morbillivirus represents a natural occurrence in an immunologically naïve pinniped population, or an unusual event partly due

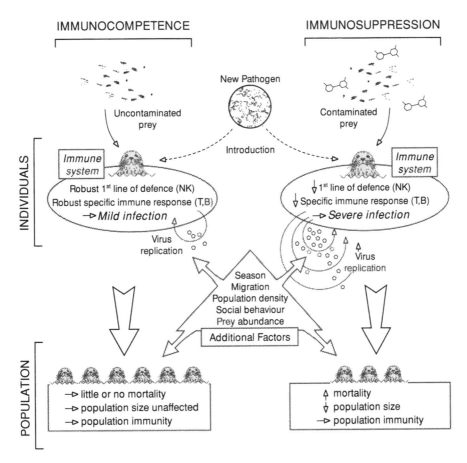

Figure 20.3 Epi-immunotoxicology: A novel virus is introduced into two immunologically naïve populations of seals. The first population lives in a remote, relatively uncontaminated environment, while the second population has been exposed to high levels of persistent, fat-soluble chemicals through nursing as young pups, and subsequently through feeding on contaminated fish. Diminished natural killer (NK) cell and T-cell function in the individuals exposed to high contaminant levels could be expected to lead to an increased susceptibility to severe disease. This, in turn, could be expected to lead to a high rate of virus reproduction within individuals, and rapid inter-individual transmission of the virus. On a population level, this may lead to elevated mortality rates and reduced population numbers. This hypothetical description of a disease outbreak influenced by immunotoxic contaminants in pinnipeds is based upon the weight of evidence which has accumulated from a number of studies of rodents, captive seals and comparative immunotoxicology (Thompson and Hall, 1993; De Swart *et al.*, 1996; Ross *et al.*, 1996b, c, 1997; Ross, 2000). (Figure adapted from Ross, 2002.)

to immunotoxic environmental contaminants, cannot be elucidated solely on the basis of field studies. While the debate about 'cause' and 'effect' in the case of the 1988 morbillivirus outbreak and other mass mortalities will likely continue (O'Shea, 2000a, b; Ross *et al.*, 2000), no one disputes the complexity of the issues (O'Shea *et al.*, 2000). However, an increasing 'weight of evidence' suggests that environmental contaminants, and PCBs in particular, contributed to the severity of the 1988 event, and are likely to continue to present a risk to free-ranging pinnipeds for decades (De Swart *et al.*, 1995d, 1996; Ross *et al.*, 1996b; Ross, 2000, 2002). Taken together, the 'weight of evidence' gathered from a number of different kinds of studies implicates contaminants in the 1988 event, and demonstrates one of the many risks faced by free-ranging pinnipeds in today's environment.

Conclusions

An outbreak of virus disease in a wildlife population is affected by a combination of factors, including condition of the individuals, migration, social behaviour, population density, specific immunity of a population to the pathogen (or lack thereof) and, of course, the presence of the virus. The immune system represents an individual animal's functional barrier to infection by viruses, bacteria and macroparasites, and consists of a complex network of interacting cells and chemical messengers. Although the immune systems of pinnipeds have not been examined extensively, studies carried out to date suggest that the structure and function of lymphoid tissues and cells, and the nature of the immune response in pinnipeds, closely resemble those of other mammals. Studies of laboratory rodents have demonstrated the unequivocal immunotoxicity of many of the persistent organic pollutants found at high concentrations in pinnipeds, including the PCBs and related compounds.

Morbillivirus-associated mass mortalities have occurred in several pinniped populations, in some cases fuelling speculation that contaminants may have predisposed them to a more severe disease event than would have normally been the case. A captive feeding study of harbour seals has demonstrated that ambient levels of contaminants in Baltic Sea herring were immunotoxic, and a set of parallel studies in laboratory rodents confirmed and extended these findings. Serological studies have since discovered that morbilliviruses routinely infect different pinniped species inhabiting relatively uncontaminated areas, with little or no associated mortality. The 'weight of evidence' now suggests that the dioxin-like PCBs may have contributed to the severity of at least one of these mass mortalities by impairing normal immune function. Delineating the relative contribution of contaminants from the many other natural and anthropogenic factors to outbreaks of infectious disease in marine mammals represents a considerable scientific challenge for the future (Nichol *et al.*, 2000; Daszak *et al.*, 2001; Ross, 2002). Temporal and spatial trends suggest that immunotoxic compounds, and PCBs in particular, will continue to present a risk to free-ranging pinnipeds in many parts of the world for decades.

References

Addison, R.F. and Brodie, P.F. 1977. Organochlorine residues in maternal blubber, milk, and pup blubber from grey seals (*Halichoerus grypus*) from Sable Island, Nova Scotia. *Journal of the Fisheries Research Board of Canada* 34: 937–941.

Addison, R.F. and Brodie, P.F. 1987. Transfer of organochlorine residues from blubber through the circulatory system in milk in the lactating grey seal, *Halichoerus grypus*. *Canadian Journal of Fisheries and Aquatic Sciences* 44: 782–786.

Addison, R.F. and Smith, T.G. 1974. Organochlorine residue levels in Arctic ringed seals: variation with age and sex. *Oikos* 25: 335–337.

Addison, R.F. and Smith, T.G. 1998. Trends in organochlorine residue concentrations in ringed seal (*Phoca hispida*) from Holman, Northwest Territories, 1972–91. *Arctic* 51: 253–261.

Baker, J.R. 1992. The pathology of phocine distemper. *Science of the Total Environment* 115: 1–7.

Bengtson, J.L., Boveng, P., Franzen, U., Have, P., Heide-Jörgensen, M.-P. and Härkönen, T. 1991. Antibodies to canine distemper virus in antarctic seals. *Marine Mammal Science* 7: 85–87.

Bergek, S., Bergqvist, P.-A., Hjelt, M., Olsson, M., Rappe, C., Roos, A. and Zook, D.R. 1992. Concentrations of PCDDs and PCDFs in seals from Swedish waters. *Ambio* 21: 553–556.

Bergman, A. and Olsson, M. 1985. Pathology of Baltic grey seal and ringed seal females with special reference to adrenocotical hyperplasia: is environmental pollution the cause of a widely distributed disease syndrome? *Finnish Game Research* 44: 47–62.

Bergman, A., Olsson, M. and Reiland, S. 1992. Skull-bone lesions in the Baltic grey seal (*Halichoerus grypus*). *Ambio* 21: 517–519.

Boon, J.P. and Eijgenraam, F. 1988. The possible role of metabolism in determining patterns of PCB congeners in species from the Dutch Wadden Sea. *Marine Environmental Research* 24: 3–8.

Borrell, A., Aguilar, A. and Pastor, T. 1997. Organochlorine pollutant levels in Mediterranean monk seals from the Western Mediterranean and the Sahara Coast. *Marine Pollution Bulletin* 34: 505–510.

Bossart, G.D. and Dierauf, L.A. (1990). Marine mammal clinical laboratory medicine. In L.A. Dierauf (Ed.), *CRC handbook of marine mammal medicine: health, disease and rehabilitation.* (pp. 1–52). Boca Raton: CRC Press.

Bowes, G.W. and Jonkel, C.J. 1975. Presence and distribution of polychlorinated biphenyls (PCBs) in arctic and subarctic marine food chains. *Journal of the Fisheries Research Board of Canada* 32: 2111–2123.

Brouwer, A., Reijnders, P.J.H. and Koeman, J.H. 1989. Polychlorinated biphenyl (PCB)-contaminated fish induces vitamin A and thyroid hormone deficiency in the common seal (*Phoca vitulina*). *Aquatic Toxicology* 15: 99–106.

Brouwer, A., Morse, D.C., Lans, M.C., Schuur, A.G., Murk, A.J., Klasson, W.E., Bergman, A. and Visser, T.J. 1998. Interactions of persistent environmental organohalogens with the thyroid hormone system: mechanisms and possible consequences for animal and human health. *Toxicology and Industrial Health* 14: 59–84.

Carson, R. 1962. *Silent spring.* Boston: Houghton-Mifflin.

Carter, S.D., Hughes, D.E., Bell, S.C., Baker, J.R. and Cornwell, H.J.C. 1990a. Immune responses of the common seal (*Phoca vitulina*) to canine distemper

550   *P.S. Ross* et al.

antigens during an outbreak of phocid distemper viral infection. *Journal of Zoology* 222: 391–398.

Carter, S.D., Hughes, D.E. and Baker, J.R. 1990b. Characterization and measurement of immunoglobulins in the grey seal (*Halichoerus grypus*). *Journal of Comparative Pathology* 102: 13–23.

Carter, S.D., Hughes, D.E., Taylor, V.J. and Bell, S.C. 1992. Immune responses in common and grey seals during the seal epizootic. *Science of the Total Environment* 115: 83–91.

Castellini, M.A., Davis, R.W., Loughlin, T.R. and Williams, T.M. 1993. Blood chemistries and body condition of steller sea lion pups at Marmot-Island, Alaska. *Marine Mammal Science* 9: 202–208.

Cavagnolo, R.Z. 1979. The immunology of marine mammals. *Developmental and Comparative Immunology* 3: 245–257.

Cavagnolo, R.Z. and Vedros, N.A. 1978. Identification and characterization of three immunoglobulin classes in the Northern Fur Seal (*Callorhinus ursinus*). *Developmental and Comparative Immunology* 2: 689–697.

Cleland, G.B., Leatherland, J.F. and Sonstegard, R.A. 1987. Toxic effects in C57B1/6 and DBA/2 mice following consumption of halogenated aromatic hydrocarbon-contaminated Great Lakes Coho salmon (*Oncorhynchus kisutch Albaum*). *Environmental Health Perspectives* 75: 153–157.

Cosby, S.L., McQuaid, S., Duffy, N., Lyons, C., Rima, B.K., Allan, G.M., McCullough, S.J., Kennedy, S., Smyth, J.A., McNeilly, F., Craig, C. and Örvell, C. 1988. Characterization of a seal morbillivirus. *Nature* 336: 115–116.

Costas, E. and Lopez-Rodas, V. 1998. Paralytic phycotoxins in monk seal mass mortality. *Veterinary Record* 142: 643–644.

Daszak, P., Cunningham, A.A. and Hyatt, A. 2001. Anthropogenic environmental change and the emergence of infectious diseases in wildlife. *Acta Tropica* 8: 103–116.

Delong, R.L., Gilmartin, W.G. and Simpson, J.G. 1973. Premature births in California sea lions: Association with high organochlorine pollutant residue levels. *Science* 181: 1168–1170.

De Swart, R.L., Kluten, R.M.G., Huizing, C.J., Vedder, L.J., Reijnders, P.J.H., Visser, I.K.G., UytdeHaag, F.G.C.M. and Osterhaus, A.D.M.E. 1993. Mitogen and antigen induced B and T cell responses of peripheral blood mononuclear cells from the harbour seal (*Phoca vitulina*). *Veterinary Immunology and Immunopathology* 37: 217–230.

De Swart, R.L., Ross, P.S., Vedder, L.J., Timmerman, H.H., Heisterkamp, S.H., Van Loveren, H., Vos, J.G., Reijnders, P.J.H. and Osterhaus, A.D.M.E. 1994. Impairment of immune function in harbor seals (*Phoca vitulina*) feeding on fish from polluted waters. *Ambio* 23: 155–159.

De Swart, R.L., Ross, P.S., Vedder, L.J., Boink, F.B.T.J., Reijnders, P.J.H., Mulder, P.G.H. and Osterhaus, A.D.M.E. 1995a. Haematology and clinical chemistry values for harbour seals (*Phoca vitulina*) fed environmentally contaminanted herring remain within normal ranges. *Canadian Journal of Zoology* 73: 2035–2043.

De Swart, R.L., Ross, P.S., Timmerman, H.H., Vos, H.W., Reijnders, P.J.H., Vos, J.G. and Osterhaus, A.D.M.E. 1995b. Impaired cellular immune response in harbour seals (*Phoca vitulina*) feeding on environmentally contaminated herring. *Clinical and Experimental Immunology* 101: 480–486.

De Swart, R.L., Ross, P.S., Timmerman, H.H., Hijman, W.C., De Ruiter, E., Liem, A.K.D., Brouwer, A., Van Loveren, H., Reijnders, P.J.H., Vos, J.G. and Osterhaus, A.D.M.E. 1995c. Short-term fasting does not aggravate immunosuppression in harbour seals (*Phoca vitulina*) with high body burdens of organochlorines. *Chemosphere* 31: 4289–4306.

De Swart, R.L., Harder, T.C., Ross, P.S., Vos, H.W. and Osterhaus, A.D.M.E. 1995d. Morbilliviruses and morbillivirus diseases of marine mammals. *Infectious Agents and Disease* 4: 125–130.

De Swart, R.L., Ross, P.S., Vos, J.G. and Osterhaus, A.D.M.E. 1996. Impaired immunity in harbour seals (*Phoca vitulina*) exposed to bioaccumulated environmental contaminants: review of a long-term study. *Environmental Health Perspectives* 104(suppl. 4): 823–828.

Dietz, R., Heide-Jörgensen, M.-P. and Härkönen, T. 1989a. Mass deaths of harbor seals (*Phoca vitulina*) in Europe. *Ambio* 18: 258–264.

Dietz, R., Ansen, C.T. and Have, P. 1989b. Clue to seal epizootic? *Nature* 338: 627–627.

Duignan, P.J., Saliki, J.T., St. Aubin, D.J., House, J.A. and Geraci, J.R. 1994. Neutralizing antibodies to phocine distemper virus in Atlantic walruses (*Odobenus rosmarus rosmarus*) from Arctic Canada. *Journal of Wildlife Diseases* 30: 90–94.

Duignan, P.J., Nielsen, O., House, C., Kovacs, K., Duffy, N., Early, G., Sadove, S.S., St. Aubin, D.J., Rima, B.K. and Geraci, J.R. 1997. Epizootiology of morbillivirus infection in harp, hooded, and ringed seals from the Canadian arctic and western Atlantic. *Journal of Wildlife Diseases* 33: 7–19.

Engelhardt, F.R. 1979. Haematology and plasma chemistry of captive pinnipeds and cetaceans. *Aquatic Mammals* 7: 11–20.

Faith, R.E. and Moore, J.A. 1977. Impairment of thymus-dependent immune functions by exposure of the developing immune system to 2,3,7,8-tetrachlorodibenzo-*p*-dioxin (TCDD). *Journal of Toxicology and Environmental Health* 3: 451–464.

Finberg, R. and Benacerraf, B. 1981. Induction, control and consequences of virus specific cytotoxic T cells. *Immunology Reviews* 58: 157–180.

Forsyth, M.A., Kennedy, S., Wilson, S., Eybatov, T. and Barrett, T. 1998. Canine distemper virus in a Caspian seal. *Veterinary Record* 143: 662–664.

Friend, M. and Trainer, D.O. 1970. Polychlorinated biphenyl: interaction with duck hepatitis virus. *Science* 170: 1314–1316.

Geraci, J.R., St. Aubin, D.J., Barker, I.K., Webster, R.G., Hinshaw, V.S., Bean, W.J., Ruhnke, H.L., Prescott, J.H., Early, G., Baker, A.S., Madoff, S. and Schooley, R.T. 1982. Mass mortality of harbor seals: pneumonia associated with Influenza A virus. *Science* 215: 1129–1131.

Germolec, D.R., Yang, R.S.H., Ackermann, M.F., Rosenthal, G.J., Boorman, G. A., Blair, P. and Luster, M.I. 1989. Toxicology studies of a chemical mixture of 25 groundwater contaminants. II. Immunosuppression in B6C3F sub(1) mice. *Fundamental and Applied Toxicology* 13: 377–387.

Gilmartin, W.G., Delong, R.L., Smith, A.W., Sweeney, J.C., De Lappe, B.W., Risebrough, R.W., Griner, L.A., Dailey, M.D. and Peakall, D.B. 1976. Premature parturition in the California sea lion. *Journal of Wildlife Diseases* 12: 104–115.

Goksøyr, A., Beyer, J., Larsen, H.E., Andersson, T. and Forlin, L. 1992. Cytochrome P450 in seals: Monooxygenase activities, immunochemical cross-reactions and responses to phenobarbital treatment. *Marine Environmental Research* 34: 113–116.

Grachev, M.A., Kumarev, V.P., Mamaev, L.V., Zorin, V.L., Baranova, L.V., Denikjna, N.N., Belikov, S.I., Petrov, E.A., Kolesnik, V.S., Kolesnik, R.S., Dorofeev, V.M., Beim, A.M., Kudelin, V.N., Nagieva, F.G. and Sidorov, V.N. 1989. Distemper virus in Baikal seals. *Nature* 338: 209.

Hall, A.J. 1995. Morbilliviruses in marine mammals. *Trends in Microbiology* 3: 4–9.

Hall, A.J., Pomeroy, P.P. and Harwood, J. 1992a. The descriptive epizootiology of phocine distemper in the UK during 1988/89. *Science of the Total Environment* 115: 31–44.

Hall, A.J., Law, R.J., Harwood, J., Ross, H.M., Kennedy, S., Allchin, C.R., Campbell, L.A. and Pomeroy, P.P. 1992b. Organochlorine levels in common seals (*Phoca vitulina*) which were victims and survivors of the 1988 phocine distemper epizootic. *Science of the Total Environment* 115: 145–162.

Harder, T.C., Kenter, M., Appel, M.J.G., Roelke-Parker, M.E., Barrett, T. and Osterhaus, A.D.M.E. 1995. Phylogenetic evidence of canine distemper virus in Serengeti's lions. *Vaccine* 13: 521–523.

Heide-Jörgensen, M.-P., Härkönen, T., Dietz, R. and Thompson, P.M. 1992a. Retrospective of the 1988 European seal epizootic. *Diseases of Aquatic Organisms* 13: 37–62.

Heide-Jörgensen, M.-P., Härkönen, T. and Aberg, P. 1992b. Long term effects of epizootic in harbor seals in the Kattegat-Skagerrak and adjacent areas. *Ambio* 21: 511–516.

Helle, E., Olsson, M. and Jensen, S. 1976a. DDT and PCB levels and reproduction in ringed seal from the Bothnian Bay. *Ambio* 5: 188–189.

Helle, E., Olsson, M. and Jensen, S. 1976b. PCB levels correlated with pathological changes in seal uteri. *Ambio* 5: 261–263.

Henderson, G., Trudgett, A., Lyons, C. and Ronald, K. 1992. Demonstration of antibodies in archival sera from Canadian seals reactive with a European isolate of phocine distemper virus. *Science of the Total Environment* 115: 93–98.

Hickey, J.J. and Anderson, D.W. 1968. Chlorinated hydrocarbons and eggshell changes in raptorial and fish-eating birds. *Science* 162: 271–273.

House, R.V., Lauer, L.D., Murray, M.J., Thomas, P.T., Ehrlich, J.P., Burleson, G.R. and Dean, J.H. 1990. Examination of immune parameters and host resistance mechanisms in B6C3F1 mice following adult exposure to 2,3,7,8-tetrachlorodibenzo-*p*-dioxin. *Journal of Toxicology and Environmental Health* 31: 203–215.

Hutchinson, J.D. and Simmonds, M.P. 1994. Organochlorine contamination in pinnipeds. *Reviews of Environmental Contamination and Toxicology* 136: 123–168.

Jarman, W.M., Nostrom, R.J., Muir, D.C.G., Rosenberg, B., Simon, M. and Baird, R.W. 1996. Levels of organochlorine compounds, including PCDDs and PCDFs, in the blubber of cetaceans from the West coast of North America. *Marine Pollution Bulletin* 32: 426–436.

Kennedy, S., Kuiken, T., Jepson, P.D., Deaville, R., Forsyth, M., Barrett, T., Van de Bildt, M.W.G., Osterhaus, A.D.M.E., Eybatov, T., Duck, C., Kydyrmanov, A., Mitrofanov, I. and Wilson, S. 2000. Mass die-off of Caspian seals caused by canine distemper virus. *Emerging Infectious Diseases* 6(6): 637–639.

King, D.P., Robinson, I., Hay, A.W.M. and Evans, S.W. 1993a. Identification and partial characterization of common seal (*Phoca vitulina*) and grey seal (*Halichoerus grypus*) interleukin-6-like activities. *Developmental and Comparative Immunology* 17: 449–458.

King, D.P., Hay, A.W.M., Robinson, I. and Evans, S.W. 1993b. The use of monoclonal antibodies specific for seal immunoglobulins in an enzyme-linked immunosorbent assay to detect canine distemper virus-specific immunoglobulin in seal plasma samples. *Journal of Immunological Methods* 160: 163–171.

King, D.P., Lowe, K.A., Hay, A.W.M. and Evans, S.W. 1994. Identification, characterisation, and measurement of immunoglobulin concentrations in grey (*Halichoerus grypus*) and common (*Phoca vitulina*) seals. *Developmental and Comparative Immunology* 18: 433–442.

King, D.P., Schrenzel, M.D., McKnight, M.L., Reidarson, T.H., Hanni, K.D., Stott, J.L. and Ferrick, D.A. 1996. Molecular cloning and sequencing of interleukin 6 cDNA fragments from the harbor seal (*Phoca vitulina*), killer whale (*Orcinus orca*), and Southern sea otter (*Enhydra lutris nereis*). *Immunogenetics* 43: 190–195.

Koller, L.D. and Thigpen, J.E. 1973. Biphenyl-exposed rabbits. *American Journal of Veterinary Research* 34: 1605–1606.

Laws, R.J. and Taylor, R.J.F. 1957. A mass dying of Crabeater seals. *Lobodon carcinophagus* (Gray). *Proceedings of the Zoological Society of London* 129: 315–324.

Lipscomb, T.P., Schulman, F.Y., Moffett, D. and Kennedy, S. 1994. Morbilliviral disease in Atlantic bottlenose dolphins (*Tursiops truncatus*) from the 1987–1988 epizootic. *Journal of Wildlife Diseases* 30: 567–571.

Luebke, R.W., Copeland, C.B., Diliberto, J.J., Akubue, P.I., Andrews, D.L., Riddle, M.M., Williams, W.C. and Birnbaum, L. 1994. Assessment of host resistance to *Trichinella spiralis* in mice following preinfection exposure to 2,3,7,8-TCDD. *Toxicology and Applied Pharmacology* 125: 7–16.

Lundholm, C.E. 1997. DDE-induced eggshell thinning in birds: effects of p,p'-DDE on the calcium and prostaglandin metabolism of the eggshell gland. *Comparative Biochemistry and Physiology B* 118C: 113–128.

Luster, M.I., Blank, J.A. and Dean, J.H. 1987. Molecular and cellular basis of chemically induced immunotoxicity. *Annual Review of Pharmacology and Toxicology* 27: 23–49.

Luster, M.I., Portier, C., Pait, D.G., Rosenthal, G.J., Germolec, D.R., Corsini, E., Blaylock, B.L., Pollock, P., Kouchi, Y., Craig, W., White, K.L., Munson, A.E. and Comment, C.E. 1993. Risk assessment in immunotoxicology: II. Relationships between immune function and host resistance tests. *Fundamental and Applied Toxicology* 21: 71–82.

Mahy, B.W.J., Barrett, T., Evans, S., Anderson, E.C. and Bostock, C.J. 1988. Characterization of a seal morbillivirus. *Nature* 336: 115–116.

Markussen, N.H. and Have, P. 1992. Phocine distemper virus infection in harp seals (*Phoca groenlandica*). *Marine Mammal Science* 8: 19–26.

Mortensen, P., Bergman, A., Bignert, A., Hansen, H.-J., Härkönen, T. and Olsson, M. 1992. Prevalence of skull lesions in harbor seals (*Phoca vitulina*) in Swedish and Danish museum collections: 1835–1988. *Ambio* 21: 520–524.

Muir, D.C.G., Norstrom, R.J. and Simon, M. 1988. Organochlorine contaminants in Arctic marine food chains: Accumulation of specific polychlorinated biphenyls and chlordane-related compounds. *Environmental Science and Technology* 22: 1071–1079.

Muir, D.C.G., Wagemann, R., Hargrave, B.T., Thomas, D.J., Peakall, D.B. and Norstrom, R.J. 1992. Arctic marine ecosystem contamination. *Science of the Total Environment* 122: 75–134.

Nakata, H., Tanabe, S., Tatsukawa, R., Amano, M., Miyazaki, N. and Petrov, E. 1995. Persistent organochlorine residues and their accumulation kinetics in Baikal seal (*Phoca sibirica*) from Lake Baikal, Russia. *Environmental Science and Technology* 29: 2877–2885.

Nakata, H., Tanabe, S., Tatsukawa, R., Amano, M., Miyazaki, N. and Petrov, E.A. 1997. Bioaccumulation profiles of polychlorinated biphenyls including coplanar congeners and possible toxicological implications in Baikal seal (*Phoca sibirica*). *Environmental Pollution* 95: 57–65.

Nash, D.R. and Mach, J.-P. 1971. Immunoglobulin classes in aquatic mammals. *Journal of Immunology* 107: 1424–1430.

Nichol, S.T., Arikawa, J. and Kawaoka, Y. 2000. Emerging viral diseases. *Proceedings of the National Academy of Sciences of the United States of America* 97(23): 12411–12412.

Norrby, E. and Oxman, M.N. (1990). Measles virus. In B.N. Fields amd D.M. Knipe (Eds.), *Virology.* (pp. 1013–1044). New York: Raven Press.

Norstrom, R.J., Simon, M., Muir, D.C.G. and Schweinsburg, R.E. 1988. Organochlorine contaminants in arctic marine food chains: identification, geographical distribution, and temporal trends in polar bears. *Environmental Science and Technology* 22: 1063–1071.

Norstrom, R.J., Simon, M. and Muir, D.C.G. 1990. Polychlorinated dibenzo-*p*-dioxins and dibenzofurans in marine mammals in the Canadian north. *Environmental Pollution* 66: 1–19.

Oehme, M., Fürst, P., Krüger, C., Meemken, H.A. and Groebel, W. 1988. Presence of polychlorinated dibenzo-*p*-dioxins, dibenzofurans and pesticides in arctic seal from Spitzbergen. *Chemosphere* 17: 1291–1300.

Oehme, M., Schlabach, M., Hummert, K., Luckas, B. and Nordoy, E.S. 1995. Determination of levels of polychlorinated dibenzo-*p*-dioxins, dibenzofurans, biphenyls and pesticides in harp seals from the Greenland Sea. *Science of the Total Environment* 162: 75–91.

Olsson, M., Karlsson, B. and Ahnland, E. 1994. Diseases and environmental contaminants in seals from the Baltic and the Swedish west coast. *Science of the Total Environment* 154: 217–227.

O'Shea, T.J. 2000a. PCBs not to blame. *Science* 288: 1965–1966.

O'Shea, T.J. 2000b. Cause of seal die-off in 1988 is still under debate. *Science* 290: 1097–1098.

O'Shea, T.J., Reeves, R.R. and Kirk Long, A. 2000. *Marine mammals and persistent ocean contaminants: Proceedings of the Marine Mammal Commission Workshop, 12–15 October 1998.* Keystone: Marine Mammal Commission.

Osterhaus, A.D.M.E. 1988. Seal death. *Nature* 334: 302–302.

Osterhaus, A.D.M.E. and Vedder, E.J. 1988. Identification of virus causing recent seal deaths. *Nature* 335: 20.

Osterhaus, A.D.M.E., Groen, J., UytdeHaag, F.G.C.M., Visser, I.K.G., Vedder, E.J., Crowther, J. and Bostock, C.J. 1989a. Morbillivirus infection in European seals before 1988. *Veterinary Record* 125: 326–326.

Osterhaus, A.D.M.E., Groen, J., UytdeHaag, F.G.C.M., Visser, I.K.G., Van de Bildt, M.W.G., Bergman, A. and Klingeborn, B. 1989b. Distemper virus in Baikal seals. *Nature* 338: 209–210.

Osterhaus, A.D.M.E., De Swart, R.L., Vos, H.W., Ross, P.S., Kenter, M.J.H. and Barrett, T. 1995. Morbillivirus infections of aquatic mammals: newly identified members of the genus. *Veterinary Microbiology* 44: 219–227.

Osterhaus, A., Van de Bildt, M.W.G., Vedder, L.J., Martina, B., Niesters, H., Vos, J.G., Van Egmond, H., Liem, D., Baumann, R., Androukaki, E., Kotomatas, S., Komnenou, A., Abou-Sidi, B., Jiddou, A.B. and Barham, M.E. 1998. Monk seal mortality: virus or toxin? *Vaccine* 16: 979–981.

Peakall, D.B. and Fox, G.A. 1987. Toxicological investigations of pollutant-related effects in Great Lakes gulls. *Environmental Health Perspectives* 71: 187–193.

Rappe, C., Buser, H.R., Stalling, D.L., Smith, L.M. and Dougherty, R.C. 1981. Identification of polychlorinated dibenzofurans in environmental samples. *Nature* 292: 524–526.

Reijnders, P.J.H. 1980. Organochlorine and heavy metal residues in harbour seals from the Wadden Sea and their possible effects on reproduction. *Netherlands Journal of Sea Research* 14: 30–65.

Reijnders, P.J.H. 1986. Reproductive failure in common seals feeding on fish from polluted coastal waters. *Nature* 324: 456–457.

Risebrough, R.W., Walker, W., Schmidt, T.T., De Lappe, B.W. and Connors, C.W. 1976. Transfer of chlorinated biphenyls to Antarctica. *Nature* 264: 738–739.

Roitt, I. 1990. *Essential immunology*. Oxford: Blackwell Scientific Publications.

Roletto, J. 1993. Hematology and serum chemistry values for clinically healthy and sick pinnipeds. *Journal of Zoo and Wildlife Medicine* 24: 145–157.

Ross, P.S. 2000. Marine mammals as sentinels in ecological risk assessment. *Human and Ecological Risk Assessment* 6: 29–46.

Ross, P.S. 2002. The role of immunotoxic environmental contaminants in facilitating the emergence of infectious diseases in marine mammals. *Human and Ecological Risk Assessment* 8: 277–292.

Ross, P.S. and De Guise, S. 1998. Environmental contaminants and marine mammal health: Research applications. *Canadian Technical Report Of Fisheries and Aquatic Sciences* 2255: 1–29.

Ross, P.S., Visser, I.K.G., Broeders, H.W.J., Van de Bildt, M.W.G., Bowen, W.D. and Osterhaus, A.D.M.E. 1992. Antibodies to phocine distemper virus in Canadian seals. *Veterinary Record* 130: 514–516.

Ross, P.S., Pohajdak, B., Bowen, W.D. and Addison, R.F. 1993. Immune function in free-ranging harbor seal (*Phoca vitulina*) mothers and their pups during lactation. *Journal of Wildlife Diseases* 29: 21–29.

Ross, P.S., De Swart, R.L., Visser, I.K.G., Murk, W., Bowen, W.D. and Osterhaus, A.D.M.E. 1994. Relative immunocompetence of the newborn harbour seal, *Phoca vitulina*. *Veterinary Immunology and Immunopathology* 42: 331–348.

Ross, P.S., De Swart, R.L., Reijnders, P.J.H., Van Loveren, H., Vos, J.G. and Osterhaus, A.D.M.E. 1995. Contaminant-related suppression of delayed-type hypersensitivity and antibody responses in harbor seals fed herring from the Baltic Sea. *Environmental Health Perspectives* 103: 162–167.

Ross, P.S., De Swart, R.L., Timmerman, H.H., Reijnders, P.J.H., Vos, J.G., Van Loveren, H. and Osterhaus, A.D.M.E. 1996a. Suppression of natural killer cell activity in harbour seals (*Phoca vitulina*) fed Baltic Sea herring. *Aquatic Toxicology* 34: 71–84.

Ross, P.S., De Swart, R.L., Addison, R.F., Van Loveren, H., Vos, J.G. and Osterhaus, A.D.M.E. 1996b. Contaminant-induced immunotoxicity in harbour seals: wildlife at risk? *Toxicology* 112: 157–169.

Ross, P.S., De Swart, R.L., Van Loveren, H., Osterhaus, A.D.M.E. and Vos, J.G. 1996c. The immunotoxicity of environmental contaminants to marine wildlife: A review. *Annual Review of Fish Diseases* 6: 151–165.

Ross, P.S., Van Loveren, H., De Swart, R.L., Van der Vliet, H., De Klerk, A., Timmerman, H.H., Van Binnendijk, R.S., Brouwer, A., Vos, J.G. and Osterhaus, A.D.M.E. 1996d. Host resistance to rat cytomegalovirus (RCMV) and immune function in adult PVG rats fed herring from the contaminated Baltic Sea. *Archives of Toxicology* 70: 661–671.

Ross, P.S., De Swart, R.L., Van der Vliet, H., Willemsen, L., De Klerk, A., Van Amerongen, G., Groen, J., Brouwer, A., Schipholt, I., Morse, D.C., Van Loveren, H., Osterhaus, A.D.M.E. and Vos, J.G. 1997. Impaired cellular immune response in rats exposed perinatally to Baltic Sea herring oil or 2,3,7,8-TCDD. *Archives of Toxicology* 17: 563–574.

Ross, P.S., Ikonomou, M.G., Ellis, G.M., Barrett-Lennard, L.G. and Addison, R.F. 1998. Elevated levels of PCBs, PCDDs and PCDFs in harbour seals (*Phoca vitulina*) and killer whales (*Orcinus orca*) inhabiting the Strait of Georgia, British Columbia, Canada [Abstract]. *Proceedings of the 1998 World Marine Mammal Science Conference*, Abstract 117.

Ross, P.S., Vos, J.G., Birnbaum, L.S. and Osterhaus, A.D.M.E. 2000. PCBs are a health risk for humans and wildlife. *Science* 289: 1878–1879.

Safe, S.H. 1990. Polychlorinated biphenyls (PCBs), dibenzo-*p*-dioxins (PCDDs), dibenzofurans (PCDFs), and related compounds: environmental and mechanistic considerations which support the development of toxic equivalency factors (TEFs). *Critical Reviews in Toxicology* 21: 51–88.

Simmonds, M.P. and Mayer, S.J. 1997. An evaluation of environmental and other factors in some recent marine mammal mortalities in Europe: implications for conservation and management. *Environmental Reviews* 5: 89–98.

Sladen, W.J.L., Menzie, C.M. and Reichel, W.L. 1966. DDT residues in Adelie penguins and a crabeater seal from Antarctica. *Nature* 210: 670–673.

Smialowicz, R.J., Riddle, M.M., Rogers, R.R., Luebke, R.W., Copeland, C.B. and Ernst, G.G. 1990. Immune alterations in rats following subacute exposure to tributyltin oxide. *Toxicology* 64: 169–178.

Sonstegard, R.A. and Leatherland, J.F. 1979. Hypothyroidism in rats fed Great Lakes Coho salmon. *Bulletin of Environmental Contamination and Toxicology* 22: 779–784.

Street, J.C. and Sharma, R.P. 1975. Alteration of induced cellular and humoral immune responses by pesticides and chemicals of environmental concern: quantitative studies of immunosuppression by DDT, aroclor 1254, carbaryl, carbofuran, and methylparathion. *Toxicology and Applied Pharmacology* 32: 587–602.

Tanabe, S., Iwata, H. and Tatsukawa, R. 1994. Global contamination by persistent organochlorines and their ecotoxicological impact on marine mammals. *Science of the Total Environment* 154: 163–177.

Tarasova, E.N., Mamontov, A.A., Mamontova, E.A., Klasmeier, J. and McLachlan, M.S. 1997. Polychlorinated dibenzo-*p*-dioxins (PCDDs) and dibenzofurans (PCDFs) in Baikal seal. *Chemosphere* 34: 2419–2427.

Thigpen, J.E., Faith, R.E., McConnell, E.E. and Moore, J.A. 1975. Increased susceptibility to bacterial infection as a sequela of exposure to 2,3,7,8-tetrachlorodibenzo-*p*-dioxin. *Infection and Immunity* 12: 1319–1324.

Thomas, P.T. and Hinsdill, R.D. 1978. Effect of polychlorinated biphenyls on the immune responses of rhesus monkeys and mice. *Toxicology and Applied Pharmacology* 44: 41–51.

Thompson, P.M. and Hall, A.J. 1993. Seals and epizootics – what factors might affect the severity of mass mortalities. *Mammal Review* 23: 149–154.

Troisi, G.M. and Mason, C.F. 1997. Cytochromes P450, P420 and mixed-function oxidases as biomarkers of polychlorinated biphenyl (PCB) exposure in harbour seals (*Phoca vitulina*). *Chemosphere* 35: 1933–1946.

Tryphonas, H., Luster, M.I., Schiffman, G., Dawson, L.-L., Hodgen, M., Germolec, D., Hayward, S., Bryce, F., Loo, J.C.K., Mandy, F. and Arnold, D.L. 1991. Effect of chronic exposure of PCB (Aroclor 1254) on specific and nonspecific immune parameters in the Rhesus (*Macaca mulatta*) monkey. *Fundamental and Applied Toxicology* 16: 773–786.

Van Bressem, M.F., Visser, I.K.G., Van de Bildt, M.W.G., Teppema, J.S., Raga, J.A. and Osterhaus, A.D.M.E. 1991. Morbillivirus infection in Mediterranean striped dolphins (*Stenella coeruleoalba*). *Veterinary Record* 129: 471–472.

Van Loveren, H., Krajnc, E.I., Rombout, P.J.A., Blommaert, F.A. and Vos, J.G. 1990. Effect of ozone, hexachlorobenzene, and bis(tri-n-butyltin)oxide on natural killer activity in the rat lung. *Toxicology and Applied Pharmacology* 102: 21–33.

Visser, I.K.G., Kumarev, V.P., Örvell, C., De Vries, P., Broeders, H.W.J., Van de Bildt, M.W.G., Groen, J., Teppema, J.S., Burger, M.C., UytdeHaag, F.G.C.M. and Osterhaus, A.D.M.E. 1990. Comparison of two morbilliviruses isolated from seals during outbreaks of distemper in North West Europe and Siberia. *Archives of Virology* 111: 149–164.

Vos, J.G. and De Roij, Th. 1972. Immunosuppressive activity of a polychlorinated biphenyl preparation on the humoral immune response in guinea pigs. *Toxicology and Applied Pharmacology* 21: 549–555.

Vos, J.G. and Van Driel-Grootenhuis, L. 1972. PCB-induced suppression of the humoral and cell-mediated immunity in guinea pigs. *Science of the Total Environment* 1: 289–302.

Vos, J.G., Moore, J.A. and Zinkl, J.G. 1973. Effect of 2,3,7,8-tetrachlorodibenzo-*p*-dioxin on the immune system of laboratory animals. *Environmental Health Perspectives* 5: 149–162.

Vos, J.G., De Klerk, A., Krajnc, E.I., Kruizinga, W., van Ommen, B. and Rozing, J. 1984. Toxicity of bis(tri-n-butyltin)oxide in the rat. II. Suppression of thymus-dependent immune responses and of parameters of nonspecific resistance after short-term exposure. *Toxicology and Applied Pharmacology* 75: 387–408.

Vos, J.G., De Klerk, A., Krajnc, E.I., Van Loveren, H. and Rozing, J. 1990. Immunotoxicity of bis(tri-n-butyltin)oxide in the rat: effects on thymus-dependent immunity and on nonspecific resistance following long-term exposure in young versus aged rats. *Toxicology and Applied Pharmacology* 105: 144–155.

Welsh, R.M. and Vargas-Cortes, M. 1992. Natural killer cells in viral infection. In C.E. Lewis and J.O. McGee (Eds.), *The natural killer cell.* (pp. 107–150). Oxford: Oxford University Press.

Wiemeyer, S.N. and Porter, R.D. 1970. DDE thins eggshells of captive American kestrels. *Nature* 227: 737–738.

Zakharov, V.M. and Yablokov, A.V. 1990. Skull asymmetry in the Baltic grey seal: effects of environmental pollution. *Ambio* 19: 266–269.

21 The effects of chemical contaminants on immune function in harbour seals: Results of a semi-field study

Joseph G. Vos, Peter S. Ross,
Rik L. de Swart, Henk van Loveren
and Albert D. M. E. Osterhaus

Introduction

Marine mammals inhabiting polluted coastal waters are known to accumulate high concentrations of environmental contaminants, and this has been related to several abnormalities, including skeletal deformations, reproductive toxicity and hormonal alterations. Recent outbreaks of previously unidentified morbilliviruses have led to mass mortalities among several species of marine mammals. In 1988, approximately 20 000 harbour seals (*Phoca vitulina*), representing up to 60 per cent of local populations, and several hundred grey seals (*Halichoerus grypus*) died in north-western Europe due to an outbreak of a previously unknown virus (Mahy *et al.*, 1988; Osterhaus and Vedder, 1988), which was subsequently named phocine distemper virus (PDV). Affected animals suffered from fever, cutaneous lesions, gastrointestinal dysfunction, nervous disorders and respiratory distress (Osterhaus and Vedder, 1989). Although the identification of a virus as the aetiological agent appeared to solve the puzzle as to the cause of the event, chemical contaminants were suspected as possible cofactors. The appearance and outcome of an epizootic reflects the sum of many interacting factors, including genetic background and diversity, social behaviour, population density, immunological memory (prior exposure), nutrition and the presence of the pathogen. Outbreaks of morbilliviruses among previously unexposed populations have been known to result in elevated mortality rates (Harder *et al.*, 1995), but the additional stress of immunotoxic chemicals may exacerbate infection.

Although many factors, both intrinsic and extrinsic, affect the mammalian immune system, chemicals arising from human activities present an additional concern. During the past 50 years, the ubiquitous contamination of the global environment has resulted in detectable levels of many classes of anthropogenic chemicals in wildlife inhabiting even remote areas (Riseborough *et al.*, 1968; Wasserman *et al.*, 1979; Tanabe *et al.*, 1983). Classes of particular biological concern include the persistent, water-insoluble polyhalogenated aromatic hydrocarbons (PHAHs), including polychlorinated biphenyls (PCBs), polychlorinated dibenzo-*p*-dioxins (PCDDs) and polychlorinated dibenzofurans (PCDFs). Having fulfilled their role as heat transfer and dielectric fluids, or having been created as undesired by-products of pesticide production, PCB production or low-temperature combustion processes, these chemicals have made their way slowly, but steadily, into soils, water and air. These compounds are chemically stable, lipophilic and accumulate readily in the aquatic food chain, where they largely resist metabolic breakdown. Environmental mixtures of the PHAHs are complex, with 209 theoretical PCB congeners, 75 PCDD congeners and 135 PCDF congeners. Bioaccumulation in the aquatic food chain results in high concentrations of many of the lipophilic PHAH contaminants in organisms occupying high trophic levels. Consistent with this, most of the biological effects of PHAH contaminants have been observed in top predators, and organisms of particular concern include the piscivorous birds, otters, seals, dolphins and whales.

Among the broad range of effects on physiological processes, the immune system has been shown to be particularly sensitive to the toxic action of many PHAHs. Earlier studies found PHAHs to be immunotoxic in various rodents (Vos and Van Driel-Grootenhuis, 1972; Vos and Moore, 1974; Vos and Luster, 1989) and monkeys (Thomas and Hinsdill, 1978). The ready bioaccumulation of PHAHs in wildlife, and the relative immunotoxicity of these compounds, suggest that PCBs, PCDDs and PCDFs pose the greatest immunotoxic threat to organisms occupying high positions in the aquatic food chain. PHAH-induced injury to the immune system has been shown to be largely mediated by the cytosolic aryl hydrocarbon (Ah) receptor found in mammalian cells, to which 2,3,7,8-tetrachlorodibenzo-*p*-dioxin (TCDD) and related 'planar' PHAHs readily bind (Silkworth and Grabstein, 1982; Silkworth *et al.*, 1986). The resulting intracellular TCDD–receptor complex induces enzyme production by binding to the dioxin regulatory element (DRE) in the nucleus (Whitlock, 1987), which, in turn, leads to biological responses which are poorly understood. The toxicities of complex contaminant mixtures can be simplified by knowledge of the toxicities of the individual PHAH congeners relative to TCDD (Safe *et al.*, 1985). The PCB, PCDD and PCDF congeners which structurally resemble TCDD are each assigned a TCDD 'toxic equivalent factor' (TEF), and additive toxic equivalents (sum TEQ) are calculated from concentrations of the various congeners measured in the mixture (Safe, 1990). This concept also holds for immunotoxic effects (Vos *et al.*, 1998). However, certain PHAH-induced toxicities have been shown to be partly or entirely non-Ah-dependent, such as vitamin A and thyroid hormone deficiencies, and neurotoxic and developmental effects (Brouwer *et al.*, 1986; Morse *et al.*, 1992). Although the immunotoxic action of PHAH chemicals is poorly understood, it is likely that multiple targets are involved. Developing leucocyte progenitors in the bone marrow (Fine *et al.*, 1988) and thymus appear to be sensitive to TCDD. Thymus atrophy, for example, is most likely related to an effect on the epithelium of this organ and resulting diminished maturation of T-cell precursors, and immunotoxicity at the level of the thymus likely has repercussions for the mature cellular immune response, with impaired functionality of mature T-lymphocytes (De Waal *et al.*, 1997). In addition, exposure to TCDD has been shown to impair B-lymphocyte function, although the concentrations required for such an effect are higher than those that lead to effects on T-cell function (Tucker *et al.*, 1986).

The developing immune system of mammals has been shown to be particularly sensitive to TCDD-induced immunotoxicity following exposure during gestation and nursing (Vos and Moore, 1974; Thomas and Hinsdill, 1979). TCDD transfer from mother to offspring in laboratory rodents appears to take place primarily via the milk, but a small degree of transplacental transfer also takes place (Nau and Bass, 1981; Takagi *et al.*, 1986). Because seals occupy high trophic levels, and are consequently exposed to relatively high concentrations of PHAH mixtures in milk, the risk for immunotoxic effects on the developing immune system may be high in animals born in contaminated environments. Host resistance to numerous pathogens has been shown to be affected by TCDD and related PHAHs. Using different infection models, rodents exposed to relatively low doses of TCDD or related PHAHs have been shown to be less resistant to bacterial, parasitic and virus infection, in many cases leading to elevated morbidity, mortality or increased loads of pathogens in immunosuppressed individuals (reviewed by Vos *et al.*, 1991).

Speculation about the possible role of a chemical contaminant-associated immunosuppression in the 1988 mass mortality in seals in north-western Europe prompted us to initiate a series of immune studies in a $2^{1}/_{2}$-year semi-field study in which harbour seals were fed herring (*Clupea harengus*) from the relatively uncontaminated Atlantic Ocean or the contaminated Baltic Sea. Our aim was a carefully controlled captive feeding study which mimicked 'real world' conditions for free-ranging populations of seals exposed to PHAHs, rather than to directly dose the animals with pure PCBs, which would have been unethical and reflected artificial and unrealistic exposure regimens. In this chapter we present an overview of the results obtained during the course of the study, compare these to results obtained in a follow-up study using laboratory rats, and extrapolate our observations to free-ranging populations of harbour seals (see reviews by De Swart *et al.*, 1996 and Ross *et al.*, 1996a, b). For details on the 'weight of evidence' that Ah-receptor-binding PCBs contributed to the severity of the 1988 mass mortality in Europe, see Chapter 20 in this volume.

An immunotoxicological feeding study in harbour seals under semi-field conditions

Toxicological aspects

During a $2^1/2$-year period, two groups of 11 harbour seals (seven females and four males in each group) were fed herring, destined for human consumption, from either the heavily polluted Baltic Sea or the relatively uncontaminated Atlantic Ocean. During this period, changes in immune function were assessed (De Swart *et al.*, 1994). The animals had been caught as weaned pups off the relatively uncontaminated north-east coast of Scotland, and were fed Atlantic Ocean herring for an adaptation period of 1 year prior to this feeding study. Analyses of organochlorine contaminants were performed on the fish fed throughout the experiment, and on seal blood and samples taken during the final stages of the study (Table 21.1). The analyses of seal diets indicated that estimated daily intakes of potential immunotoxic chemicals, including PCBs, PCDDs and PCDFs, hexachlorobenzene, dieldrin, β-hexachlorocyclohexane and DDT, were three to more than ten times higher in seals fed on Baltic herring. Estimated daily intakes of 2,3,7,8-TCDD toxic equivalents (TEQ) of seals in this group were 288 ng TEQ/day, compared to 29 ng TEQ/day for seals fed on Atlantic herring. Persistent compounds such as *p,p'*-DDT and total PCBs were biomagnified from herring to seal blood and blubber, whereas lipid-based levels of Ah-receptor-binding organochlorines (sum TEQ) were not higher in seals than in the fish which they were fed. The latter finding confirms that seals have the capacity to metabolize or excrete these planar compounds (De Wit *et al.*, 1992). PCBs were the overwhelming 'TCDD-like' chemical measured in the study seals (93 per cent of sum TEQ compared to PCDDs and PCDFs). Sum TEQ levels in seal blubber after 2 years on the respective diets were 209 ± 12 and 62 ± 4 ng TEQ/kg lipid, respectively (Ross *et al.*, 1995). Sum PCBs in the blubber of seals fed Baltic herring were between 15 and 20 mg/kg bw (Table 21.1).

Table 21.1 Organochlorine contaminants in herring and in seal blubber or pooled blood samples taken after 2 and $2^1/2$ years on the different diets, respectively (after De Swart *et al.*, 1996)

	Herring[a]		*Seal blubber*[b]		*Seal blood*[c]	
	Atlantic	*Baltic*	*Atlantic*	*Baltic*	*Atlantic*	*Baltic*
Sum PCBs[d]	875 ± 158	4398 ± 715	6884 ± 493	$16\,488 \pm 1023$	7109	15 062
Sum TEQ[e]	42 ± 4	426 ± 83	62 ± 4	209 ± 12	26	72
p,p'-DDT[d]	31 ± 3	272 ± 35	306 ± 55	2448 ± 368	89	552

[a] Mean ± SE of three batches of herring.
[b] Means ± SE of 11 seals (7 females, 4 males).
[c] Concentration in pooled blood sample of 11 seals.
[d] Levels in microgram per kilogram lipid.
[e] Levels in nanogram per kilogram of lipid.

Ex vivo/in vitro *and* in vivo *immune-function assays*

To carry out immunotoxicological studies in seals, a series of immune-function assays routinely used in immunotoxicology (Van Loveren and Vos, 1989; IPCS, 1996) were adapted for use in seals (De Swart *et al.*, 1993, 1995c; Ross *et al.*, 1995, 1996c; Chapter 20 in this volume). For assessment of different parameters of immune function, blood was sampled during the course of the feeding experiment from both groups of seals every 6–9 weeks with minimal capture stress. Samples for these assays were processed under code.

Table 21.2 Summary of immunotoxicological effects in a semi-field study of harbour seals fed herring from the Baltic Sea[a]

Immune parameter	Effect
NK cell activity	↓
T-cell mitogen response	↓
Mixed lymphocyte response	↓
Antigen-specific proliferation	↓
Delayed-type hypersensitivity	↓
Specific antibody production	-/↓
Neutrophils in circulation	↑

[a] As compared to results of the group of seals fed herring from the relatively uncontaminated Atlantic Ocean.

Results of the different immune-function assays are summarized in Table 21.2. An early indication of a contaminant-related effect on immune function was observed within 4–6 months of the start of the feeeding study, when the natural killer (NK) activity of peripheral blood mononuclear cells against YAC-1 tumour target cells proved to be reduced in the Baltic group (De Swart *et al.*, 1994; Ross *et al.*, 1996c). During the course of the study, the cytotoxic activity was consistently and significantly reduced to a level approximately 25 per cent lower than that observed in seals fed Atlantic herring. Interestingly, an apparent seasonal pattern emerged in the responses of the seals in both groups, with the NK activity in winter being approximately half of that observed during the summer months (Figure 21.1).

Mitogen-induced T-lymphocyte proliferative responses began to decline somewhat later (6–10 months) following the start of the experiment (De Swart *et al.*, 1994, 1995c). This was the first indication of impaired T-lymphocyte responses: only responses induced by the T-cell mitogens concanavalin A (Con-A), phytohaemagglutinin (PHA) and pokeweed mitogen (PWM) proved to be reduced. Responses to the B-cell mitogen, lipolysaccharide (LPS), were unaffected. While the results of these non-specific tests of immune function indicated that contaminants in the Baltic herring were immunotoxic, further evidence was provided when impaired mixed lymphocyte reactions (MLR) and antigen-specific lymphocyte proliferative responses were observed in the Baltic group (De Swart *et al.*, 1995c). Because the MLR reflects the capacity of T-lymphocytes to respond to allogeneic

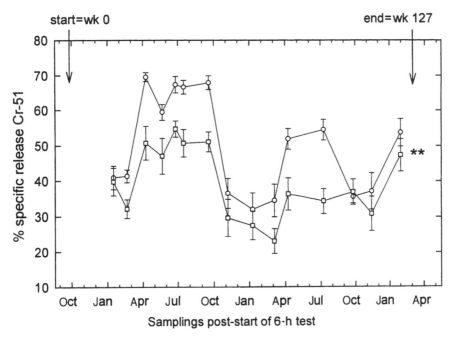

Figure 21.1 Natural killer (NK) cell activity of peripheral blood mononuclear cells is impaired in harbour seals fed herring from the contaminated Baltic Sea (squares) as compared to seals fed relatively uncontaminated herring from the Atlantic Ocean (circles). Results are mean values and SE of 11 animals per group; split plot ANOVA, ** $P < 0.01$. (Adapted from Ross *et al.*, 1996c.)

lymphocytes, it may be considered a good parameter for T-cell function. The impairment of *ex vivo/in vitro* antigen-specific proliferative responses to rabies virus and tetanus toxoid antigens confirmed that specific immune responses were impaired in the Baltic group of seals, although primary serum antibody responses following immunization of the seals with rabies virus antigen and tetanus were not reduced. Finally, impaired *in vivo* delayed-type hypersensitivity (DTH) reactions, a parameter of cellular immunity, and impaired serum antibody responses to immunization with the protein ovalbumin, a parameter of T-cell-dependent humoral immunity, provided evidence that the immune system as a whole was less capable of responding to a foreign substance in the seals fed fish from the Baltic Sea (Ross *et al.*, 1995). The DTH swelling was characterized by the infiltration of mononuclear cells in the skin and a peak in skin thickness 24 h after intradermal ovalbumin injection, as in DTH reactions observed in other animals. Significantly elevated numbers of circulating neutrophils may have reflected a compensatory increase as result of suppression of thymus-dependent immunity, as is seen in athymic animals (Vos *et al.*, 1980), an immunosuppression-related increase in the occurrence of bacterial infections in the Baltic group, or an effect of contaminants on haematopoiesis in the bone marrow (De Swart *et al.*, 1995b).

Effect of short-term fasting on immune function

In order to assess any possible additional immunotoxic risk that free-ranging seals might be exposed to during a fast-induced mobilization of lipophilic compounds, we subjected the study animals of both groups to an experimental fasting period. Following more than 2 years on their respective diets, the seals were fasted for a period of 15 days and blood sampled for a series of immune function assays (De Swart *et al.*, 1995a). The animals of both groups lost an average 11.1 kg bw, representing 16.5 per cent of their body weights. Despite a significant mobilization of PCBs, DDT and other contaminants from blubber during this period, no exacerbation of the pre-existing contaminant-related impairment in immune function could be detected. Interestingly, levels of Ah-binding PHAHs remained largely unchanged in the serum of fasting animals, which may explain the lack of immunotoxic effects observed. Alternatively, the kinetics of immunotoxicity could have led to immune-function effects at a time point subsequent to the sampling schedule. Taken together, these results suggest that short-term fasting periods, which are normal for seals, may not pose a major additional immunotoxic risk to seals with high organochlorine body burdens.

Feeding studies in rats to mimick the seal study

In order to expand our observations in seals, we carried out two long-term feeding studies in which laboratory rats (PVG strain) were either: (1) fed a diet consisting of freeze-dried herring, from the same two batches used in the seal study; or (2) orally administered oil extracted from the Atlantic or Baltic herring.

In the first of these studies, adult rats were fed the freeze-dried herring for 4 months. Rat cytomegalovirus (RCMV) titres in salivary glands were higher in the group of rats fed fish from the Baltic Sea following experimental infection, suggesting that contaminants may have affected the outcome of this infection, although no clear immunosuppression could be detected using immune-function assays (Ross *et al.*, 1996d). Because immune-function parameters were clearly impaired in seals, we concluded that the harbour seal may be more sensitive than the rat to the immunotoxic effects of the contaminants in the Baltic Sea herring. The relative insensitivity of the adult rat to the effects of TCDD or TCDD-like compounds has been shown in other studies (Vos *et al.*, 1973; Kerkvliet *et al.*, 1990; Smialowicz *et al.*, 1994), and perinatal exposure has been suggested to be a prerequisite to low-level TCDD-induced immunotoxicity in rats (Vos and Moore, 1974).

For these reasons, a second experiment was carried out in which pregnant PVG rats were dosed with the same Atlantic and Baltic Sea herring contaminant mixtures, and immune function was assessed in their pups (Ross *et al.*, 1997). In order to eliminate any possible dietary influence on immune function other than lipophilic contaminant concentrations, oil was extracted from the two herring batches and administered orally to pregnant rats on a daily basis. A positive control group was exposed to Atlantic herring oil spiked with TCDD. Exposure began on day 6 of pregnancy and continued through birth until the pups were weaned, resulting in prenatal placental transfer and postnatal lactational exposure. Rat pups exposed to the Baltic herring contaminant mixture had impaired cellular immune responses; this being most pronounced at an early age. Effects on non-specific immune-function parameters were characterized by impaired mitogen-induced T-lymphocyte proliferative responses and thymus-related effects, suggesting that developing thymocytes, or their precursors, were targeted. RCMV-directed immune responses, including virus-associated NK activity and specific antibody responses, were impaired in both the Baltic and TCDD groups, whereas RCMV-specific T-lymphocyte proliferative responses were affected in the TCDD group. Functional immune responses in the youngest rat pups of the Baltic group fell consistently between the Atlantic group and the positive control group (TCDD), but these differences became less apparent with time. At the time of necropsy, RCMV titres were similar in the salivary glands of all groups of experimentally infected rats, likely reflecting the observed recovery of immune function with time. The 24-day half-life of TCDD in rats (Rose *et al.*, 1976), would lead to rapidly diminishing contaminant burdens in the growing pups, essentially resulting in a removal of the source of immunotoxicity in the study animals. The reversibility of TCDD-induced thymus atrophy has been observed previously (Van Loveren *et al.*, 1991).

Conclusions

A virus-associated mass mortality among seals inhabiting north-western Europe generated an interest in immunotoxicology in this species. A morbillivirus was isolated from victims, but a contribution of immunotoxic contaminants to the severity of the outbreaks could not be ruled out. Fish-eating seals occupy high trophic levels in the aquatic food chain, and accumulate high concentrations of polyhalogenated aromatic hydrocarbons (PHAHs), including PCBs, PCDDs and PCDFs. Such chemicals have been found to be immunotoxic at low doses in studies of laboratory animals.

To assess whether contaminants at ambient environmental levels can affect immune function in seals, we carried out a semi-field immunotoxicological study, in which captive harbour seals were fed herring from either relatively uncontaminated sites of the Atlantic Ocean, or from the highly contaminated Baltic Sea. The seals that were fed contaminated Baltic Sea herring accumulated significantly higher body burdens of potentially immunotoxic organochlorines. Changes in immune function were monitored over a $2^1/_2$-year period by performing *ex vivo/in vitro* and *in vivo* immune-function assays, comprising non-specific as well as specific T-cell-dependent cellular and humoral immune responses. Seals from the Baltic group had impaired activity of natural killer (NK) cells and specific T-cell responses (mitogen-induced T-lymphocyte proliferation, antigen-specific response, mixed lymphocyte response, delayed-type hypersensitivity reaction and specific antibody production to ovalbumin). NK cell function and T-cell function are both crucial in vertebrates to anti-virus defences, including the clearance of morbillivirus infections.

Additional feeding studies in PVG rats using the same herring batches, and TCDD as a positive control group, suggested that an effect at the level of the thymus may be responsible for changes in cellular immunity, that virus-specific immune responses may be impaired, and that perinatal exposure to environmental contaminants represents a greater immunotoxic threat than exposure as a juvenile or adult. Together with the pattern of TCDD toxic equivalents of different PHAHs in the herring, these data indicate that current concentrations of PCBs in the aquatic food chain in northern Europe are immunotoxic to marine mammals.

Based on the combination of field, semi-field and laboratory studies carried out here and elsewhere, we conclude that dioxin-like PCBs that accumulated through the marine food chain aggravated the severity and extent of the 1988 morbillivirus-related epizootic in seals. Because PCB concentrations in free-ranging harbour seals are higher in many areas of northern Europe than in the immunosuppressed Baltic group, many populations may be at continued risk to immunotoxicity (Figure 21.2). This may result in

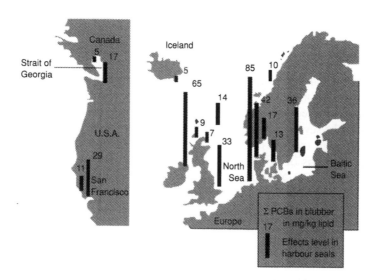

Figure 21.2 Recent mean concentrations of PCBs in blubber of harbour seals are higher in many areas of northern Europe and North America than in the immunosuppressed harbour seals of the captive feeding study (box insert at lower right). Data were compiled on the basis of sum PCBs in blubber expressed on a lipid weight basis. (For more details see review by Ross *et al.*, 1996a.)

diminished host resistance and an increased incidence and severity of infectious disease, the more so as free-ranging animals are exposed in a manner (i.e. pre- and postnatally) that ensures a more profound immunotoxicity than exposure of young adults in our semi-field study.

References

Brouwer, A., Van den Berg, K.J., Blaner, W.S., Goodman, D.S. 1986. Transthyretin (prealbumin) binding of PCBs, a model for the mechanism of interference with vitamin A and thyroid hormone metabolism. *Chemosphere* 15, 1699–1706.

De Swart, R.L., Kluten, R.M.G, Huizing, C.J., Vedder, L.J., Reijnders, P.J.H., Visser, I.K.G., UytdeHaag, F.G.C.M., Osterhaus, A.D.M.E. 1993. Mitogen and antigen induced B and T cell responses of peripheral blood mononuclear cells from the harbour seal *(Phoca vitulina)*. *Vet. Immunol. Immunopathol.* 37, 217–230.

De Swart, R.L., Ross, R.S., Vedder, L.J., Timmerman, H.H., Heisterkamp, S.H., Van Loveren, H., Vos, J.G., Reijnders, P.J.H., Osterhaus, A.D.M.E. 1994. Impairment of immune function in harbor seals *(Phoca vitulina)* feeding on fish from polluted waters. *Ambio* 23, 155–159.

De Swart, R.L., Ross, R.S., Timmerman, H.H., Hijman, W.C., De Ruiter, E., Liem, A.K.D., Brouwer, A., Van Loveren, H., Reijnders, P.J.H., Vos, J.G., Osterhaus, A.D.M.E. 1995a. Short-term fasting does not aggravate immunosuppression in harbour seals *(Phoca vitulina)* with high body burdens of organochlorines. *Chemosphere* 31, 4289–4306.

De Swart, R.L., Ross, R.S., Vedder, L.J., Boink, F.B.T.J., Reijnders, P.J.H., Mulder, P.G.H., Osterhaus, A.D.M.E. 1995b. Haematology and clinical chemistry values for harbour seals within normal ranges. *Can. J. Zool.* 73, 2035–2043.

De Swart, R.L., Ross, R.S., Timmerman, H.H., Vos, H.W., Reijnders, P.J.H., Vos, J.G., Osterhaus, A.D.M.E. 1995c. Impaired cellular immune response in harbour seals *(Phoca vitulina)* feeding on environmentally contaminated herring. *Clin. Exp. Immunol.* 101, 480–486.

De Swart, R.L., Ross, P.S., Vos, J.G., Osterhaus, A.D.M.E. 1996. Impaired immunity in harbour seals *(Phoca vitulina)* exposed to bioaccumulated environmental contaminants: review of a long-term study. *Environ. Health Perspect.* 104 (Suppl. 4), 823–828.

De Waal, E.J., Schuurman, H.J., Van loveren, H., Vos, J.G. 1997. Differential effects of 2,3,7,8-tetrachlorodibenzo-p-dioxin, bis(tri-n-butyltin)oxide and cyclospirine on thymus histophysiology. *Crit. Rev. Toxicol.* 27, 381–430.

De Wit, C., Jansson, B., Bergek, S., Hjelt, M., Rappe, C., Olsson, M., Andersson, O. 1992. Polychlorinated dibenzo-p-dioxin and polychlorinated dibenzofuran levels and patterns in fish and fish-eating wildlife in the Baltic Sea. *Chemosphere* 25, 185–188.

Fine, J.S., Gasiewicz, T.A., Silverstone, A.E. 1988. Lymphocyte stem cell alterations following perinatal exposure to 2,3,7,8-tetrachiorodibenzo-p-dioxin. *Mol. Pharmacol.* 35, 18–25.

Harder, T.C., Kenter, M., Appel, M.J.G., Roelke-Parker, M.E., Barrett, T., Osterhaus, A.D.M.E. 1995. Phylogenetic evidence of canine distemper virus in Serengeti's lions. *Vaccine* 13, 521–523.

568   *J.G. Vos* et al.

IPCS 1996. *Principles and Methods for Assessing Direct Immunotoxicity Associated with Chemical Exposure.* International Programme for Chemical Safety, Environmental Health Criteria Document No. 160, World Health Organization, Geneva.

Kerkvliet, N.I., Baecher Steppan, L., Smith, B.B., Youngberg, J.A., Henderson, M.C., Buhler, D.R. 1990. Role of the Ah locus in suppression of cytotoxic T lymphocyte activity by halogenated aromatic hydrocarbons (PCBs and TCDD): structure–activity relationships and effects in C57Bl/6 mice congenic at the Ah locus. *Fundam. Appl. Toxicol.* 14, 532–541.

Mahy, B.W.J., Barrett, T., Evans, S., Anderson, E.C., Bostock, C.J. 1988. Characterization of a seal morbillivirus. *Nature* 336, 115–116.

Morse, D.C., Koeter, H.B.W.M., Smits van Prooijen, A.E., Brouwer, A. 1992. Interference of polychlorinated biphenyls in thyroid hormone metabolism: possible neurotoxic consequences in fetal and neonatal rats. *Chemosphere* 25, 165–168.

Nau, H., Bass, R. 1981. Transfer of 2,3,7,8-tetrachlorodibenzo-p-dioxin (TCDD) to the mouse embryo and fetus. *Toxicology* 20, 299–308.

Osterhaus, A.D.M.E., Vedder, E.J. 1988. Identification of virus causing recent seal deaths. *Nature* 335, 20.

Osterhaus, A.D.M.E., Vedder, E.J. 1989. No simplification in the etiology of recent seal deaths. *Ambio* 18, 297–298.

Risebrough, R.W., Pieche, R., Peakall, D.B., Herman, S.G., Kiryen, M.N. 1968. Polychlorinated biphenyls in the global ecosystem. *Nature* 220, 1098–1102.

Rose, J.Q., Ramsey, J.C., Wentzler, T.H., Hummel, R.A., Gehring, P.J., 1976. The fate of 2,3,7,8-tetrachlorodibenzo-p-dioxin following single and repeated oral doses to the rat. *Toxicol. Appl. Pharmacol.* 36, 209–226.

Ross, R.S., De Swart, R.L., Reijnders, P.J.H., Van Loveren, H., Vos, J.G., Osterhaus, A.D.M.E. 1995. Contaminant-related suppression of delayed-type hypersensitivity and antibody responses in harbor seals fed herring from the Baltic Sea. *Environ. Health Perspect.* 103, 162–167.

Ross, R.S., De Swart, R.L., Addison, R.F., Van Loveren, H., Vos, J.G., Osterhaus, A.D.M.E. 1996a. Contaminant-induced immunotoxicity in harbour seals: wildlife at risk? *Toxicology* 112, 157–169.

Ross, P.S., De Swart, R.L., Van Loveren, H., Osterhaus, A.D.M.E., Vos, J.G. 1996b. The immunotoxicity of environmental contaminants to marine wildlife: a review. *Ann. Rev. Fish Diseases* 6, 151–165.

Ross, R.S., De Swart, R.L., Timmerinan, H.H., Reijnders, P.J.H., Vos, J.G., Van Loveren, H., Osterhaus, A.D.M.E. 1996c. Suppression of natural killer cell activity in harbour seals *(Phoca vitulina)* fed Baltic Sea herring. *Aquat. Toxicol.* 34, 71–84.

Ross, P.S., Van Loveren, H., De Swart, R.L., Van der Vliet, H., De Klerk, A., Timmerman, H.H., Van Binnenedijk, R., Brouwer, A., Vos, J.G., Osterhaus, A.D.M.E. 1996d. Host resistance to rat cytomegalovirus (RCMV) and immune function in adult PVG rats fed herring from the contaminated Baltic Sea. *Arch. Toxicol.* 70, 661–671.

Ross, P.S., De Swart, R.L., Van der Vliet, H., Willemsen, L., De Klerk, A., Van Amerongen, G., Groen, J., Brouwer, A., Morse, D.C., Van Loveren, H., Osterhaus, A.D.M.E., Vos, J.G. 1997. Impaired cellular immune response in offspring of rats exposed during pregnancy and nursing to Baltic Sea herring oil or 2,3,7,8-TCDD. *Arch. Toxicol.* 17, 563–574.

Safe, S. 1990. Polychlorinated biphenyls (PCBs), dibenzo-p-dioxins (PCDDs), dibenzofurans (PCDFs), and related compounds: environmental and mechanistic considerations which support the development of toxic equivalency factors (TEFs). *CRC Crit. Rev. Toxicol.* 21, 51–88.

Safe, S., Bandiera, S., Sawyer, T., Robertson, L., Safe, L., Parkinson, A., Thomas, R.E., Ryan, D.E., Reik, L.M., Levin, W., Denomme, M.A., Fujita, T. 1985. PCBS: structure–function relationships and mechanism of action. *Environ. Health Perspect.* 60, 47–56.

Silkworth, J.B., Grabstein, E.M. 1982. Polychlorinated biphenyl immunotoxicity: dependence on isomer planarity and the Ah aene complex. *Toxicol. Appl. Pharmacol.* 65, 109–115.

Silkworth, J.B., Antrim, L.A., Sack, G. 1986. Ah receptor mediated suppression of the antibody response in mice is primarily dependent on the Ah phenotype of lymphoid tissue. *Toxicol. Appl. Pharmacol.* 86, 380–390.

Smialowicz, R.J., Riddle, M.M., Williams, W.C., Diliberto, J.J. 1994. Effects of 2,3,7,8-tetrachlorodibenzo-p-dioxin (TCDD) on humoral immunity and lymphocyte subpopulations – differences between rats and mice. *Toxicol. Appl. Pharmacol.* 124, 248–256.

Takagi, Y., Aburada, S., Hashimoto, K., Kitaura, T. 1986. Transfer and distribution of accumulated (14-C) polychlorinated biphenyls from maternal to fetal and suckling rats. *Arch. Environ. Contam. Toxicol.* 15, 709–715.

Tanabe, S., Mori, T., Tatsukawa, R., Miyazaki, N., 1983. Global pollution of marine mammals by PCBs, DDTs and HCHs (BHCs). *Chemosphere* 12, 1269–1275.

Thomas, R.T., Hinsdill, R.D. 1978. Effect of polychlorinated biphenyls on the immune responses of rhesus monkeys and mice. *Toxicol. Appl. Pharmacol.* 44, 41–51.

Thomas, R.T., Hinsdill, R.D. 1979. The effect of perinatal exposure to tetrachlorodibenzo-p-dioxin on the immune response of young mice. *Drug Chem. Toxicol.* 2, 77–98.

Tucker, A.N., Vore, S.J., Luster, M.I. 1986. Suppression of B cell differentiation by 2,3,7,8-tetrachlorodibenzo-p-dioxin. *Mol. Pharmacol.* 29, 372–377.

Van Loveren, H., Vos, J.G. 1989. Immunotoxicological considerations: a practical approach to immunotoxicity testing in the rat. In: *Advances in applied toxicology* (A.D. Dayan and A.J. Paine, Eds.), Taylor & Francis Ltd., London, pp. 143–163.

Van Loveren, H., Schuurman, H.J., Kampinga, J., Vos, J.G. 1991. Reversibility of thymic atrophy induced by 2,3,7,8-tetrachlorodibenzo-p-dioxin (TCDD) and bis(tri-n-butyltin)oxide (TBTO). *Int. J. Immunopharmacol.* 13, 369–377.

Vos, J.G., Van Driel-Grootenhuis, L. 1972. PCB-induced suppression of the humoral and cell-mediated immunity in guinea pigs. *Sci. Total Environm.* 1, 289–302.

Vos, J.G., Luster, M.I. 1989. Immune alterations. In: *Halogenated biphenyls, terphenyls, naphthalenes, dibenzodioxins and related products* (R.D. Kimbrough and S. Jensen, Eds.), Elsevier Science Publishers, Amsterdam (Biomedical Division), Chapter 10, pp. 295–322.

Vos, J.G., Moore, J.A. 1974. Suppression of cellular immunity in rats and mice by maternal treatment with 2,3,7,8-tetrachlorodibenzo-p-dioxin. *Int. Arch. Allergy* 47, 777–794.

Vos, J.A., Moore, J.A., Zinkl, J.G. 1973. Effect of 2,3,7,8-tetrachlorodibenzo-p-dioxin on the immune system of laboratory animals. *Environm. Health Perspect.* 5, 149–162.

Vos, J.G., Kreeftenberg, J.G., Kruijt, B.C., Kruizinga, W., Steerenberg, P.A. 1980. The athymic nude rat. II. Immunological characteristics. *Clin. Immunol. Immunopathol.* 15, 229–237.

Vos, J.G., Van Loveren, H., Schuurman, H.J. 1991. Immunotoxicity of dioxin: immune function and host resistance in laboratory animals and humans. In: *Banbury Report 35: Biological basis for risk assessment of dioxins and related compounds* (M.A. Gallo, R.J. Scheuplein and K.A. van der Heijden, Eds.), Cold Spring Harbor Laboratory Press, Plainview, New York, pp. 79–93.

Vos, J.G., De Heer, C., Van Loveren, H. 1998. Immunotoxic effects of TCDD and toxic equivalent factors. *Teratogen. Carcinogen. Mutagen.* 17, 275–284.

Wasserman, M., Wasserman, D., Cucos, S., Miller, H.J. 1979. World PCBs map: storage and effects in man and his biologic environment in the 1970s. *Ann. NY Acad. Sci.* 320, 69–124.

Whitlock, J.R. 1987. The regulation of gene expression by 2,3,7,8-tetrachlorodibenzo-p-dioxin. *Pharmacol. Rev.* 39, 147–161.

22 Immunotoxicology of free-ranging pinnipeds: Approaches to study design

Peter S. Ross, Kimberlee B. Beckmen and Stéphane Pillet

Introduction: Seals, pollution and disease

The 1988 phocine distemper virus (PDV) epizootic which killed 20 000 harbour seals (*Phoca vitulina*) and several hundred grey seals (*Halichoerus grypus*) in northern Europe catapulted the issue of immunotoxicity into the public domain (Dietz *et al.*, 1989a). The toxic effects of low levels of many persistent organic pollutants (POPs) on the immune system have been clearly demonstrated in laboratory animals (Vos and Luster, 1989), and many immunotoxic contaminants were present at high levels in the blubber of pinnipeds inhabiting the areas affected by the virus (Hall *et al.*, 1992a; Hutchinson and Simmonds, 1994; Olsson, 1994). An additional mortality of 10 000 Baikal seals *(Pusa sibirica)* in 1987–88 was attributed to canine distemper virus, or CDV (Grachev *et al.*, 1989; Visser *et al.*, 1990), which, along with PDV, is a member of the genus *Morbillivirus* (Osterhaus *et al.*, 1995). Baikal seals have since been shown to be highly contaminated with POPs (Nakata *et al.*, 1997).

Although the pathogen responsible for the northern European die-off of harbour and grey seals represented a newly identified virus (Mahy *et al.*, 1988; Osterhaus *et al.*, 1988) which struck an immunologically naïve (seronegative) population (Osterhaus *et al.*, 1989), subsequent serological work has determined that members of the genus *Morbillivirus* are enzootic in many apparently unaffected, free-ranging populations of pinnipeds inhabiting less-contaminated parts of the world (Dietz *et al.*, 1989b; Ross *et al.*, 1992; Duignan *et al.*, 1997). Although no clear evidence of a contributory role of immunotoxic pollutants to the European or Baikal disease events existed, the results of laboratory animal studies, coupled with the context of a severe disease outbreak in relatively contaminated groups of free-ranging seals, provided considerable grounds for speculation.

Diseases play a role in the natural ecology of pinniped populations, with bacterial, viral and parasitic infections affecting the health and survival of susceptible individuals, and pups and juveniles in particular (Baker, 1984, 1989; Geraci and St Aubin, 1987; Hall *et al.*, 1992b; Garner *et al.*, 1997). Whereas a disease outbreak represents a complex ecological event, the immune system represents the functional barrier which is designed to control the invasion and spread of pathogens. A contaminant-related disruption of this physiological system can therefore predispose affected animals to an increased severity of disease, and lead to elevated mortality rates in a study group, and by extension, population (Loose *et al.*, 1978; Luster *et al.*, 1993; Lebrec and Burleson, 1994).

There is ample reason to examine the issue of contaminant-related immunotoxicity in pinnipeds. Most members of this group occupy high trophic levels in the aquatic environment. As primarily fish-eating animals, pinnipeds can be exposed to elevated levels of the POPs which bioaccumulate through the food chain. Despite usage restrictions which were implemented in the 1970s, chemicals such as the polychlorinated biphenyls (PCBs) and the dichlorodiphenyl-trichloroethane (DDT) group continue to be found at high concentrations in the tissues of pinniped populations inhabiting coastal waters adjacent to industrialized nations, and at moderately high concentrations even in remote regions of the planet (Hutchinson and Simmonds, 1994; Addison and Smith, 1998).

Accumulating evidence from the past three decades suggests that ambient levels of contaminants have caused a number of abnormalities, including reproductive impairment, skeletal malformations, and skin, claw and uterine lesions among the ringed (*Pusa hispida*), grey and harbour seals of northern Europe (Helle *et al.*, 1976; Reijnders, 1980; Bergman and Olsson, 1985; Olsson *et al.*, 1994), and California sea lions *(Zalophus californianus)* and northern elephant seals (*Mirounga angustirostris*) in the eastern Pacific Ocean (Delong *et al.*, 1973; Gilmartin *et al.*, 1976; Beckmen *et al.*, 1997). Along with studies of fish-eating birds, which have more definitively demonstrated an adverse effect of environmental contaminants in wildlife (Gilbertson *et al.*, 1991; Tillitt *et al.*, 1992; Fry, 1995), including immunotoxicity (Grasman *et al.*, 1996), it has become clear that high trophic level wildlife species are particularly prone to the effects of the POPs, and PCBs in particular.

Despite a number of attempts to assess the relationship between contaminants, immune function and outbreaks of infectious disease in pinnipeds and other marine mammal species, considerable challenges have prevented much progress in this subject area. These include:

(1) the dynamic nature of the vertebrate immune system, which is in a constant state of flux, with positive and negative feedback systems, and a degree of redundancy;

(2) the interactive relationship between a pathogen and a host population, which depends, in part, on poorly understood genetic, social, behavioural, seasonal, and age and sex-related factors in the population in question;

(3) a limited capacity to assess the significance of the relationship between immune function parameters measured in pinnipeds and a susceptibility to infection by a pathogen (host resistance);

(4) the small number of specialized reagents available to characterize the components of the pinniped immune system; and

(5) a limited understanding of pinniped immunology.

Overcoming these challenges is critical in immunotoxicological studies of pinnipeds. In this chapter, we will summarize some of the immunological research carried out to date on free-ranging pinnipeds, place this in the context of what we know about immunotoxicology in both laboratory animals and captive harbour seals, and present the reader with an overview of test methods and strategies available for studies of free-ranging pinnipeds. This chapter will hopefully provide a basic practical guide to those interested in carrying out immunotoxicological or immunological research in captive or free-ranging pinnipeds.

The immunotoxicity of contaminants to laboratory animals and captive pinnipeds

Many hundreds of laboratory animal studies have demonstrated the immunotoxic potential of heavy metals, organochlorine chemicals and pesticides, but the dioxin-like compounds have been found to be particularly active (Vos and De Roij, 1972; Vos *et al.*, 1973; Luster *et al.*, 1978; Tryphonas *et al.*, 1991). Some of the 419 different PCB, polychlorinated dibenzo-*p*-dioxin (PCDD) and polychlorinated dibenzofuran (PCDF) congeners share a similar planar conformation, which facilitates a toxicity mediated via the intracellular aryl hydrocarbon (Ah) receptor in mammalian tissues (Safe, 1990). Immunotoxicity of these chemicals has been shown to be largely mediated in this manner (Silkworth and Grabstein, 1982; Silkworth *et al.*, 1986), highlighting the risk that these 'dioxin-like' compounds present to wildlife in contaminated areas.

Our knowledge of contaminant-associated immunotoxicity in pinnipeds was greatly enhanced when a comprehensive captive feeding study was carried out in 1990–94. Even though the two groups, of 11 harbour seals each, were not free-ranging during the study, their respective 'contaminated' and 'relatively uncontaminated' diets of herring (*Clupea harengus*) had originated from the Baltic Sea and north Atlantic Ocean, and were destined for human consumption. The contaminant mixtures to which the seals were exposed, therefore, were representative of those to which free-ranging harbour, ringed and grey seals inhabiting the coastal waters of northern Europe were exposed. The captive design of the study enabled a rapid collection of high-quality blood samples from all animals and the implementation of concurrent immune-function tests *in vitro*, and helped to generate reproducible and comparable data. This study is described in greater detail in other chapters, so will only be mentioned briefly here. In the study, seals fed herring from the contaminated Baltic Sea had diminished natural killer cell (NK) activity *in vitro* (Ross *et al.*, 1996a) and T-lymphocyte responses *in vitro* (De Swart *et al.*, 1994; De Swart *et al.*, 1995a) and *in vivo* (Ross *et al.*, 1995), and the authors concluded that contaminant-related immunotoxicity represented a real risk to free-ranging pinnipeds and may have played a role in the 1988 PDV epizootic (reviewed by De Swart *et al.*, 1996; Ross *et al.*, 1996b, c).

Immune function in free-ranging pinnipeds

There are 33 species of pinnipeds, including 18 phocid or true seals, 14 otariid or eared seals, and the walrus (*Odobenus rosmarus*). However, immunological studies have only been carried out in a few of these, including the (phocid) harbour, grey and northern elephant seals, and the (otariid) northern fur seal (*Callorhinus ursinus*) (Cavagnolo and Vedros, 1979; De Swart *et al.*, 1993; Ross *et al.*, 1993, 1994; King *et al.*, 1994, 1998; Nielsen, 1995; Pillet *et al.*, 2000). These studies have shown that pinnipeds have lymphoid tissues, leucocyte subpopulations, immunoglobulin classes and functional immune system components which are similar to those in other vertebrates, although species-specific differences exist in leucocyte surface markers and immunoglobulin binding affinities. Although these differences limit the application of some of the tests used in other species, many of the functional tests for immunological assessment used routinely in human and veterinary studies can be, and have been, successfully adapted for use in pinnipeds.

Obtaining high-quality immune-function information from study subjects represents the greatest challenge in studies of pinniped immunotoxicology. Since inter-assay test results vary a great deal, most functional tests on a group of animals should be considered as 'relative' and carried out concurrently. If the study involves captive pinnipeds, blood samples can generally be obtained quickly and the tests carried out either the same day or the next day. In this case, care should be taken to minimize handling stress, and the blood should be taken quickly. Multiple prodding with needles and low ambient air temperatures can cause platelet activation and clotting, and lead to a low peripheral blood mononuclear cell (PBMC) yield (Ross, 1990; Ross *et al.*, 1993). Physiological stress can lead to profile shifts in circulating leucocytes (Thomson and Geraci, 1986). Captive studies using fresh PBMC and concurrent evaluation by functional assay can lead to highly reproducible results in pinnipeds, and were successfully used in the captive immunotoxicological study of harbour seals in The Netherlands.

Obtaining such reproducible and comparable results from free-ranging pinnipeds is fraught with difficulty, but is possible. Some strategies have been developed which have generated biologically relevant immune-function results from harbour seals. Immediately following the 1988 PDV epizootic in northern Europe, researchers carried out two studies of immune function in female harbour seals and their pups on Sable Island, Nova Scotia, Canada (Ross, 1990; Ross *et al.*, 1993, 1994). Mother–pup pairs were live-captured immediately following birth, marked for identification purposes, and both were sampled for blood, while the mother was also sampled for milk. They were released after sampling, and recaptured several times during the nursing period. In an electricity-supplied field laboratory, a total leucocyte count was carried out, and a blood smear made on a glass microscope slide for later Giemsa stain and differential counts of leucocyte subpopulations. Peripheral blood mononuclear cells were separated from heparinized whole blood samples by density gradient separation using a centrifuge. Peripheral blood mononuclear cells were then washed and resuspended in cryopreservation medium, transferred to cryovials, and placed into liquid nitrogen at a controlled rate. Serum and milk samples were frozen in a −20°C freezer for later total and specific antibody determinations. These field sampling procedures allowed for the collection of all the required samples, and would later enable the researchers to evaluate the functionality of PBMC in concurrent *in vitro* lymphoproliferative assays at a dedicated laboratory facility with sterile cell culture technologies and radiolabel facilities.

In addition, researchers also evaluated the ability of free-ranging harbour seal pups on Sable Island to mount a specific antibody response to an inactivated human rabies vaccine (Ross *et al.*, 1994). For this, groups of marked pups were immunized at different ages, and were recaptured for blood sampling 10 days later. Results were compared to those obtained in a parallel series of immunizations in cats and dogs of the same age as the seal pups tested. Together with the results of lymphoproliferative tests using cryopreserved PBMC, this study suggested that the newborn harbour seal pup is relatively immunocompetent compared to other mammals. These studies suggest that a cross-section of immune function tests can yield biologically and ecologically relevant information in a relatively uncontaminated population of harbour seals. Although carrying out studies in harbour seals from other study areas or in other pinniped species might be exceedingly difficult, the nature of the Sable Island harbour seal population afforded researchers a unique opportunity to recapture and sample tagged individuals on a regular basis.

The pollution–immune function link in free-ranging pinnipeds

Immunotoxicological studies in free-ranging pinnipeds are in their infancy. Two general strategies have been developed to assess immune function in free-ranging pinnipeds in the context of contaminants. Because it is considered imperative to carry out laboratory-based immune-function assays all at the same time, samples must either all be collected within hours of each other and processed within 24 hours at a cell culture laboratory, or collected on separate occasions and cryopreserved for later assessment at a laboratory. The fat-soluble POPs are quantified in blood, milk and/or blubber biopsies taken aseptically, and are normally expressed on a lipid weight basis. Because the capture of all study animals at the same time can only be carried out under exceptional circumstances for most pinnipeds, the cryopreservation of PBMC represents the simplest strategy. In addition, under isolated field conditions, distance to a dedicated immunological laboratory may necessitate cryopreservation of PBMC for storage and transportation.

Cryopreservation of PBMC in the field is being used to assess the relationship between contaminants and immune function in free-ranging northern fur seal pups in the Pribilof Islands, Alaska, USA. Commercially available PBMC preparation tubes containing acid citrate dextrose and a density gradient have made it possible to isolate and cryopreserve these cells from many individuals and at repeated time points (Blanchard and Stott, 1998). Lymphoproliferative assays are conducted simultaneously on all samples at a later date in a distant laboratory. A colorimetric enzyme-linked immunosorbent assay (ELISA) BrdU (5-bromo-2′-deoxyuridine) immunoassay kit has lowered the cost and also removed need for [³H]thymidine uptake assays. The study incorporated mitogen-induced lymphoproliferative assays to evaluate cell-mediated immune function along with other assessments of humoral immune function and health. After blood sampling, neonates were vaccinated with a benign antigen, tetanus toxoid, to stimulate antibody production and evaluate adaptive immunity. Blood was collected 4–6 weeks later in the nursing period. Tetanus antibodies were detected by ELISA (Ham-Lammé *et al.*, 2001). Changes in total immunoglobulin levels, serum haptoglobin levels and differential leucocyte counts were evaluated for indications of inflammatory reactions (such as a secondary response to opportunistic bacterial infections). Concentrations of selected PCB congeners and DDT metabolites in whole blood of pups and milk of their dams were measured to assess exposure level to these contaminants. Contaminant concentrations were determined by using rapid high-performance liquid chromatography coupled with a photodiode array detection method (Krahn *et al.*, 1994). First-born fur seal pups had the highest exposures from milk as expected but the levels of contaminants detected in pup blood were highest during the perinatal period (Beckmen *et al.*, 1999). Pups of old, multiparous dams had significantly lower peri-natal exposure. Preliminary data analysis shows that the pups in the high-exposure group had significantly lower antibody production to vaccination and increased inflammatory responses later in the nursing period. Preliminary results of the *in vitro* lymphoproliferative responses to Con-A were similar to the captive harbour seal study described previously, and showed a negative correlation between lymphocyte function and PCB-congener concentrations in whole blood. Studies are ongoing to further define the subtle effects of contaminant exposure on the developing immune system in free-ranging northern fur seals and the endangered Steller's sea lion (*Eumetopias jubatus*) in Alaska.

A study has been carried out in which harbour seal pups from three locations in the Strait of Georgia, British Columbia, Canada, were captured and temporarily housed in pools for immunotoxicological assessment. This strategy was adopted in order to minimize the negative effects of cryopreservation on PBMC viability, and to allow concurrent blood samp-lings from all study animals for immediate immune function testing. It was assumed that any contaminant-related immunotoxicity as a result of matern-ally derived contaminants would be lasting, and would be detectable in pups during their 6 weeks in captivity. Weekly and concurrent blood samplings were drawn for a series of *in vitro* tests, including natural killer (NK) cell function, and lymphoproliferative responses to mitogens (Con-A, PHA, PWM and LPS) and antigens (ovalbumin, inactivated rabies vaccine) following immunization. A blubber biopsy was taken for congener-specific analysis of PCBs, PCDDs and PCDFs. Total PCB concentrations, and total dioxin toxic equivalents (TEQ) calculated from international toxic equivalency factors (TEFs) for dioxin-like PCBs (IUPAC numbers for non-*ortho* PCBs: 77, 126, 169; mono-*ortho* PCBs: 105, 114, 118, 123, 156, 157, 167, 189; di-*ortho* PCBs 170 and 180; plus the 2,3,7,8-Cl substituted PCDDs and PCDFs) (Van Zorge *et al.*, 1989; Ahlborg *et al.*, 1994), were then compared to immune function results in a correlative approach [*note:* TEFs have since been revised by an international group (Van den Berg *et al.*, 1998)]. In a preliminary assessment of results, a negative correlation between T-lymphocyte function, as assessed by *in vitro* lymphoproliferative responses to the mitogen Con-A, and total TEQ was observed (Ross *et al.*, 1997).

Grey seal mother–pup pairs were live-captured in an effort to assess whether contaminant-related immunotoxicity might be playing a role in various non-specific infections among newborns on the Isle of May in Scotland (Hall *et al.*, 1996). The authors did not find a relationship between the cumulative lactational dose of organochlorine contaminants (as measured by 'dioxin-like' PCBs) and prevalence of infection or alterations in blood haematology or serum chemistry. The changes in haematology, serum chemistry and immune function observed in developing (nursing) pinnipeds (Ross *et al.*, 1993; Ross *et al.*, 1994; Hall *et al.*, 1996; Simms and Ross, 2000), coupled with the involvement of uncontrolled pathogens in the study animals, may have masked an effect of contaminants. None the less, such an approach provides useful ancillary information on free-ranging pinnipeds, and provides background information for future studies.

In vitro exposures to environmentally relevant concentrations of contaminants

The development of *in vitro* models for immunotoxicity testing have proven advantageous by enabling a greater understanding of the mechanisms underlying observations of immunotoxicity, particularly in humans and wildlife where ethical limitations preclude *in vivo* experimentation (Wood *et al.*, 1992). A good example of the utility of *in vitro* testing is the increasing recognition of programmed cell death (apoptosis) as a mechanism in immunotoxicity (Azzouzzi *et al.*, 1992; Corcoran *et al.*, 1994). *In vitro* exposures also provide a means of addressing mechanisms and effects under controlled conditions, something that may be lacking in field studies. This ability to perform *in vitro* exposures under carefully controlled conditions can therefore generate direct relationships between variables and effects, and can help to more directly determine the mechanisms of toxicity.

Studies of pinnipeds may expose either fresh or cryopreserved leucocytes obtained from free-ranging animals to single chemicals *in vitro* in an effort to understand certain compound-related toxicities, or to complex contaminant mixtures *in vitro* at doses which mimic those found in free-ranging populations. Advantages include the ability to examine the individual components of the complex mixtures to which pinnipeds are exposed; the ability to examine the modulatory effects of factors other than contaminants, such as sex and various hormones; and the capacity to compare the functionality of exposed to unexposed cells from the same animal (e.g. as a percentage of the latter). Such methods were used in studies that found that *in vitro* exposures to environmentally relevant concentrations of heavy metals or organochlorine compounds impaired the proliferative responses of splenocytes and thymocytes isolated from beluga whales (*Delphinapterus leucas*) (De Guise *et al.*, 1996, 1998), and that methylmercury is genotoxic to bottlenose dolphin (*Tursiops truncatus*) lymphocytes, using the Comet single-cell microgel electrophoresis assay (Betti and Nigro, 1996).

Little *in vitro* work has been carried out using pinniped cells. Using fresh peripheral blood leucocytes from harbour and grey seals live-captured in the St. Lawrence estuary (Quebec, Canada), Pillet *et al.* (2000) demonstrated that the phagocytic activity of granulocytes from adult females of both species increased following *in vitro* exposure to physiological concentrations of Zn, whereas cells from males or immature females were unaffected by the exposure (Pillet *et al.*, 2000). Phagocytosis plays a critical role in both non-specific and specific immune responses of mammals, and represents a major first line of defence against invading agents, particularly bacteria (Van Oss, 1987). While the interaction between reproductive hormones and immune function have been well established in other species (Schuurs and Verheul, 1990; Magnusson, 1991), these findings suggest a possible interaction among contaminants, reproductive hormones and immune function. *In vitro* experiments provide a means of addressing such issues under controlled conditions, something that would be impossible in the field.

In vitro exposures have also been used in order to demonstrate and characterize the induction of metallothioneins (MTHs) in peripheral blood leucocytes from grey seals (Pillet *et al.*, 2001). Metallothioneins are known to strongly modulate the toxicity of heavy metals (Klassen and Liu, 1998). More recently, these proteins have been demonstrated to modulate important immune functions (Lynes *et al.*, 1993; Crowthers *et al.*, 2000). *In vitro* exposures of grey seal peripheral blood leucocytes to Zn and Cd induced intracellular metallothioneins in grey seal PBMC (Klassen and Liu, 1998). This increase was dependent on the dose, duration of exposure, metal species and the leucocyte subpopulation in question. The relative level of intracellular metallothioneins in monocytes following *in vitro* exposure to the metal was approximately twice the level in lymphocytes. Moreover, comparative studies suggested that grey seal PBMC are less sensitive to Cd exposure than human lymphocytes (Klassen and Liu, 1998). Aside from providing us with a tool to improve the use of MTH as a biomarker of heavy metal exposure, this approach provides an additional measure of possible immunotoxicity of heavy metals in pinnipeds. *In vitro* experiments are therefore of value in assessing such factors as inter-species differences in sensitivity and gender-based differences.

Because such *in vitro* approaches enable an evaluation of contaminant exposures which are environmentally relevant, there is a distinct need for more data on concentrations of contaminants in the blood of free-ranging pinnipeds. Current approaches to contaminant measurement and reporting in pinnipeds generally rely upon concentrations in blubber, which may not accurately reflect the circulating, hence more bioavailable, contaminants in pinnipeds. In addition, there is a need for more research on the physiological factors that affect the bioavailability of contaminants, because these may modulate the toxicity of certain contaminants, e.g. metallothionein binding of heavy metals and the dynamics of organochlorine mobilization during fasting periods (Kimura, 1991; Chan *et al.*, 1993; Lynes *et al.*, 1993; De Swart *et al.*, 1995b). Further work on *in vitro* exposures of leucocytes and other cells isolated from pinnipeds may help to elucidate priority contaminants in complex environmental mixtures, and assist in the delineation of mechanisms of immunotoxic action.

Designing an immunotoxicological study of free-ranging pinnipeds

Designing an immunotoxicological study for application to a free-ranging population of pinnipeds requires that researchers consider:

(1) the practical elements involved in capturing study subjects and obtaining appropriate samples in the field, while minimizing confounding factors;
(2) the logistical details involved in the initial field processing of samples, appropriate storage of these samples, and transportation to a dedicated immunological laboratory;
(3) the selection of a suitable battery of immune-function tests which will provide an overview of the immune system; and
(4) the selection of additional tests which will complement the immune-function information and provide details on other aspects of health, disease prevalence and other possible toxicities.

A combination of access to good samples obtained from pinnipeds in the field (or pinnipeds in a captive setting), and the availability of a well-supplied cell culture laboratory, represents a fundamental backdrop for immunotoxicological studies in pinnipeds (Figure 22.1). Table 22.1 provides the reader with some practical considerations which should be taken into account in designing an immunotoxicological study in a free-ranging population of pinnipeds. Failure to address such issues adequately will likely

Field Laboratory

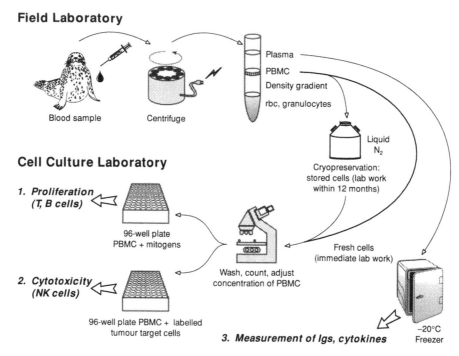

Cell Culture Laboratory

Figure 22.1 Immunological or immunotoxicological studies carried out in free-ranging pinnipeds have combined strong field programmes with access to high-quality samples and a cell culture laboratory in which assays of immune function are possible. Where distance to such a laboratory is great, or where sample collections span a number of days, cryopreservation of peripheral blood mononuclear cells (PBMC) provides a useful means of standardizing collection and subsequent assay. Plasma can be used for the measurement of immunoglobulins (Igs), cytokines and hormones. Abbreviations: rbc, red blood cells; NK cells, natural killer cells Igs, immunoglobulins; PBMC, peripheral blood mononuclear cells.

interfere with immune-function results, and diminish the interpretative power of the study.

A large number of immune-function tests can be applied to a study of pinnipeds, making a selection of relevant tests an important component of project design. Tiered series of immunological assessment tests have been designed in Europe and the US for laboratory animal studies aimed at screening for the immunotoxicity of individual chemicals. The rat has been selected as the study animal of choice in Europe, owing to its wide use in other toxicological tests. In this case, Tier I consists of immunopathology and serum immunoglobulin quantification, and Tier II consists of a comprehensive screen of cell- and humoral-mediated immunity, macrophage function, natural killer cell function and host resistance (Vos and Van Loveren, 1987).

Table 22.1 The challenges faced when obtaining samples for immunotoxicological studies of pinnipeds must be dealt with directly: many factors threaten the viability of immune-function test results. Steps must be taken which either minimize or eliminate these potentially confounding factors, or take them into account, so that a later evaluation can evaluate their possible interference in study results

Problem	Course of action
Age	Choose same-age animals (e.g. weaned pups; yearlings; juveniles)
	Choose known-age animals (e.g. tagged, tattooed)
	Remove a tooth for ageing by counting of growth layers
Sex	Choose immatures
	Choose same-sex adults
	Take reproductive cycle into account
Season	Avoid, or take into account, effects of oestrus, pupping, nursing, moulting, fasting and mating
	Carry study out in a short time within one season
Stress	Capture quickly
	Minimize handling time, handle quietly and with care
	Release immediately following sampling
Condition	Avoid non-normal study candidates (i.e. sick or stranded individuals)
	Obtain body weights and condition indices
	Obtain indication of blubber thickness
	Conduct physical exam; record scars, wounds, abrasions; eye, skin or oral lesions; umbilicus status (in pups)
	Carry out clinical haematology and serum chemistry
	Screen serum for antibodies to known pinniped pathogens (e.g. morbillivirus; leptospirosis, brucella)
Sampling times	Attempt to obtain all samples at the same time in order to allow concurrent evaluation in functional assays using fresh PBMC
	Cryopreserve PBMC (liquid N_2) in the field and store until all samples can be evaluated at the same time
Clotting of blood or PBMC	Keep sampling needles, syringes and collection tubes at room temperature
	Add anticoagulant to collection supplies if using syringes
	Add anticoagulant to dilution and washing medium during PBMC density gradient separation

Mice have been selected as the study animal of choice in the US, largely because of the availability of a large number of immunological reagents. The test battery consists of a Tier I evaluation (screen) of immunopathology, humoral- and cell-mediated immunity, and non-specific immunity. A Tier II evaluation (comprehensive), of more detailed immunopathology, humoral- and cell-mediated immunity, non-specific immunity and host resistance models, is carried out if Tier I reveals an immunotoxic effect for a chemical (Luster *et al.*, 1988). Tier I has proven consistently successful in identifying the immunotoxic properties of the more potent chemicals, whereas Tier II is required to identify weaker immunotoxicants and also provides more detailed information on the effect and its significance to the host (Luster *et al.*, 1988). Analysis of results obtained from many different chemicals has revealed a good correlation between immune function and host resistance tests, although some tests of immune function have been more predictive than others for altered host resistance (Luster *et al.*, 1993).

A tiered approach to immunological assessment can provide considerable information about the immunotoxic potential of certain individual chemicals, but the invasive nature of the methods used with laboratory rodents is not legally or ethically acceptable for application to pinnipeds under most circumstances. In addition, the nature of the question in pinnipeds is fundamentally different. Rather than trying to determine whether a one-time exposure to a certain chemical is immunotoxic, as assessed at one or more time points post-dosing, pinniped toxicologists are attempting to understand the immunotoxic effects of chronic exposures to a complex 'real world' mixture of environmental contaminants. An adaptation of these tiered approaches for use in either captive or free-ranging pinnipeds is necessary, depending on the particular circumstances of the study area and the species involved. Laboratory animal studies have identified the importance of dioxin-like effects of PCBs, PCDDs and PCDFs on the immune system, underlining the importance of including tests which assess the thymus-associated and T-lymphocyte-mediated responses. It is also important to include a cross-section of tests which are sensitive, relevant, functional and predictive for diminished host resistance.

We present here a battery of immune function tests that have been carried out successfully in harbour seals, which can provide guidance for future studies in pinniped immunotoxicology (Table 22.2). While this battery of tests is aimed at assessing immune function in free-ranging pinnipeds, it should also be noted that retinoids (e.g. vitamin A or retinol) and thyroid hormones have been proposed as 'biomarkers' of exposure to PCBs and related POPs (Brouwer *et al.*, 1989; Hall *et al.*, 1998; Woldstad and Jenssen, 1999; Rolland, 2000; Simms *et al.*, 2000; Simms and Ross, 2001). Because both contaminants and these bioactive factors interact with the immune system, their measurement should be included in any immunotoxicological study of free-ranging pinnipeds.

Pinniped immunotoxicology is a challenging field, with considerable uncertainties, confounding factors and methodological constraints which limit our capacity to establish cause-and-effect relationships, mechanisms of action, and understand the significance of results in the context of host resistance. Ultimately, however, a 'weight of evidence' approach represents a useful global means of examining the risk of an adverse effect of contaminants in a given population and identifying putative chemicals responsible for effects observed (Ross, 2000). In the absence of conclusive data which directly implicates contaminant-related immunotoxicity in the outcome of a disease event (such data are highly unlikely), information from several sources

Table 22.2 A proposed battery of tests for assessment of immune function in captive or free-ranging pinnipeds using minimally invasive sampling techniques. Most tests rely upon the collection of plasma and anticoagulated (e.g. heparinized) whole blood for subsequent density gradient isolation of peripheral blood mononuclear cells (PBMC), as indicated in the column headed '*in vitro*'. Tests for specific and systemic immune responses normally entail an immunization prior to the later collection of blood, or the carrying out of a skin test (e.g. 10–14 days post-immunization), as indicated in both the '*in vitro*' and '*in vivo*' columns below. Most tests listed have been successfully applied to harbour seals, and should be suitable for use in any pinniped species. Fresh or cryopreserved PBMC can be used in lymphoproliferative assays. In the absence of species-specific monoclonal antibodies needed for detailed quantification of immunoglobulins, protein A from *Staphylococcus aureus* can be used for estimating total and specific IgG levels

	Description of immune function test	In vitro	In vivo	*Ref.*
Haematology	Enumeration of lymphocytes, monocytes, neutrophils, basophils and eosinophils	X		1
	E-rosette test with SRBC	X		2
Non-specific immunity	Natural killer cell function (NK) using PBMC from whole blood and ^{51}Cr-labelled selected tumour target cell line (e.g. YAC-1, P815, K-562) in cytotoxicity assay	X		3
	Phagocytosis by PBMC of fluorescent bacteria or latex beads and analysis by FACS	X		4
	Cytokine receptor expression (e.g. IL-2)	X		5
Cell-mediated immunity	Lymphoproliferative responses of PBMC to T-cell-dependent mitogens (Con-A, PHA, PWM)	X		6,7
	Mixed lymphocyte reaction (MLR) using PBMC and allogeneic leucocytes or a cell line (e.g. PV1.P1)	X		8
	Lymphoproliferative responses of PBMC to specific antigens (e.g. tetanus toxoid, inactivated rabies or polio vaccine) following immunization	X	X	8
	DTH skin responses to antigen applied intradermally to flipper webbing following immunization with antigen (e.g. ovalbumin)	X	X	9
Humoral immunity	Total serum IgG estimation using protein A-binding ELISA	X		7
	If reagents available, quantification of total serum IgG, IgM and IgA levels using species-specific monoclonal antibodies	X		
	Lymphoproliferative responses of PBMC to B-cell-dependent mitogens (LPS, PWM)	X		6,8
	Specific serum antibody responses using protein A or species-specific monoclonal antibodies following immunization with antigen (ovalbumin) or inactivated vaccine (rabies, tetanus toxoid or polio)	X	X	8,9

Abbreviations: NK, natural killer; PBMC, peripheral blood mononuclear cell; ^{51}Cr, chromium-51; FACS, fluorescence-activated cell scan; Con-A, concanavalin A; PHA, phytohaemagglutinin; PWM, pokeweed mitogen; LPS, lipopolysaccharide; DTH, delayed-type hypersensitivity; Ig, immunoglobulin (G, A, M); ELISA, enzyme-linked immunosorbent assay; E-rosette, erythrocyte rosette; SRBC, sheep red blood cell.
References: 1 De Swart *et al.* (1995c); 2 Nielsen (1995); 3 Ross *et al.* (1996a); 4 De Guise *et al.* (1995); 5 DiMolfetto-Landon *et al.* (1995); 6 De Swart *et al.* (1994); 7 Ross *et al.* (1994); 8 De Swart *et al.* (1995a); 9 Ross *et al.* (1995).

should be considered together. These have been adapted from elsewhere (Ross, 2000) to include:

(1) the laboratory animal studies which demonstrate the immunotoxicity of single chemicals and/or complex mixtures *in vitro* and *in vivo*;
(2) the concentrations of immunotoxic chemicals in the tissues (blubber, blood, liver, etc.) of the pinniped population in question, and their relationship to concentrations which proved immunotoxic to captive harbour seals; and
(3) the strength of the evidence of contaminant-related toxicities in the population in question, including immunotoxicity, endocrine disruption and disease outcomes.

Conclusions

While the immunotoxicity of many of the contaminants found in pinnipeds has been established in laboratory animal studies, demonstrating their effects in free-ranging pinnipeds is considerably more challenging. The complex mixtures of many hundreds of contaminants found in the tissues of pinnipeds and the difficulties in assessing immune function under conditions which are far from controlled, represent serious obstacles to researchers (Ross *et al.*, 2000; O'Shea, 2000). The immune system of mammals is reasonably well conserved, allowing for a large degree of inter-species comparison and extrapolation, as well as the adaptation of a number of existing assays for immune function in other species to the less-studied pinnipeds. However, evolutionary adaptations to the physiological and ecological demands of a largely marine existence, together with species-specific differences in, for example, leucocyte surface markers, and immunoglobulin and cytokine structure, complicate immunotoxicological studies in pinnipeds. Basic advances in the field of pinniped immunology have arisen from the successful adaptation of a battery of veterinary diagnostic and clinical tests to studies of both captive and free-ranging pinnipeds.

In the 'real world' of highly complex contaminant mixtures, pinnipeds are exposed to elevated concentrations of fat-soluble contaminants, including the immunotoxic PCBs, PCDDs and PCDFs, the organochlorine pesticides, and certain metals. Although still in their infancy, substantial progress has been made in recent years in the areas of immunology and immunotoxicology using both captive and free-ranging pinnipeds. Current approaches to studies of free-ranging pinniped populations rely on correlations which attempt to link contaminant levels with immune-function 'responses', using fresh or cryopreserved lymphocytes. Where a re-capture or additional sampling is feasible, immunizations with peptides or inactivated vaccines can provide an indication of specific immune response. Strategies also being applied include *in vitro* exposures of lymphoid cells to environmentally relevant levels of contaminants, where different variables can be carefully controlled. Such attempts have provided insight into the immune system of pinnipeds and, together with captive studies of harbour seals and studies of laboratory animals, will help to provide a better understanding required for an assessment of the role of contaminants in virus-associated mass mortalities among pinnipeds.

The utility of pinnipeds as indicators of environmental contamination reflects their high trophic status and wide distribution. Given the persistence of many of the problematic contaminants in the environment, research in different areas of pinniped toxicology is important because it provides ecologically relevant information for conservationists, policy-makers and members of the general public.

References

Addison, R.F. and Smith, T.G. 1998. Trends in organochlorine residue concentrations in ringed seal (*Phoca hispida*) from Holman, Northwest Territories, 1972–91. *Arctic* 51: 253–261.

Ahlborg, U.G., Becking, G.C., Birnbaum, L., Brouwer, A., Derks, H.J.G.M., Feeley, M., Golor, G., Hanberg, A., Larsen, J.C., Liem, A.K.D., Safe, S.H., Schlatter, C., Waern, F., Younes, M. and Yrjanheikki, E. 1994. Toxic equivalency factors for dioxin-like PCBs. Report on a WHO-ECEH and IPCS consultation, December 1993. *Chemosphere* 28: 1049–1067.

Azzouzzi, B.E., Tsangaris, G.T., Pellegrini, O., Manuel, Y., Beneviste, J. and Thomas, Y. 1992. Cadmium induces apoptosis in a human T cell line. *Toxicology* 88: 127–139.

Baker, J.R. 1984. Mortality and morbidity in grey seal pups (*Halichoerus grypus*). Studies on its causes, effects of environment, the nature and sources of infectious agents and the immunological status of pups. *Journal of Zoology* 203: 23–48.

Baker, J.R. 1989. Natural causes of death in non-suckling grey seals (*Halichoerus grypus*). *Veterinary Record* 125: 500–503.

Beckmen, K.B., Lowenstine, L.J., Newman, J., Hill, J., Hanni, K. and Gerber, J. 1997. Clinical and pathological characterization of northern elephant seal skin disease. *Journal of Wildlife Diseases* 33: 438–449.

Beckmen, K.B., Ylitalo, G.M., Towell, R.G., Krahn, M.M., O'Hara, T.M. and Blake, J.E. 1999. Factors affecting organochlorine contaminant concentrations in milk and blood of northern fur seal (*Callorhinus ursinus*) dams and pups from St. George Island, Alaska. *Science of the Total Environment* 231: 183–200.

Bergman, A. and Olsson, M. 1985. Pathology of Baltic grey seal and ringed seal females with special reference to adrenocortical hyperplasia: is environmental pollution the cause of a widely distributed disease syndrome? *Finnish Game Research* 44: 47–62.

Betti, C. and Nigro, M. 1996. The Comet Assay for the evaluation of the genetic hazard of pollutants in cetaceans: Preliminary results on the genotoxic effects of methyl mercury on the bottlenose dolphin (*Tursiops truncatus*) lymphocytes *in vitro*. *Marine Pollution Bulletin* 32: 545–548.

Blanchard, M.T. and Stott, J.L. 1998. Marine mammal lymphocyte function: Techniques for field preparation, cryopreservation, and analysis [Abstract]. *World Marine Mammal Science Conference, Monaco, 20–25 January*.

586   *P.S. Ross* et al.

Brouwer, A., Reijnders, P.J.H. and Koeman, J.H. 1989. Polychlorinated biphenyl (PCB)-contaminated fish induces vitamin A and thyroid hormone deficiency in the common seal (*Phoca vitulina*). *Aquatic Toxicology* 15: 99–106.

Cavagnolo, R.Z. and Vedros, N.A. 1979. Serum and colostrum immunoglobulin levels in the Northern Fur Seal (*Callorhinus ursinus*). *Developmental and Comparative Immonology* 3: 139–146.

Chan, H.M., Tamura, Y., Cherian, M.G. and Goyer, R.A. 1993. Pregnancy-associated changes in plasma metallothionein concentration and renal cadmium accumulation in rats. *Proceedings of the Society for Experimental Biology and Medicine* 202: 420–427.

Corcoran, G.B., Fix, L., Jones, D.P., Moslen, M.T. and Nicotera, P. 1994. Apoptosis: molecular control point in toxicity. *Toxicology and Applied Pharmacology* 128: 169–181.

Crowthers, K.C., Kline, V., Giardina, C. and Lynes, M.A. 2000. Augmented humoral immune function in metallothionein-null mice. *Toxicology and Applied Pharmacology* 166: 161–172.

De Guise, S., Flipo, D., Boehm, J.R., Martineau, D., Béland, P. and Fournier, M. 1995. Immune functions in beluga whales (*Delphinapterus leucas*): Evaluation of phagocytosis and respiratory burst with peripheral blood leukocytes using flow cytometry. *Veterinary Immunology and Immuno-pathology* 47: 351–362.

De Guise, S., Bernier, J., Martineau, D., Béland, P. and Fournier, M. 1996. Effects of *in vitro* exposure of Beluga whale splenocytes and thymocytes to heavy metals. *Environmental Toxicology and Chemistry* 15: 1357–1364.

De Guise, S., Martineau, D., Beland, P. and Fournier, M. 1998. Effects of *in vitro* exposure of beluga whale leukocytes to selected organochlorines. *Journal of Toxicology and Environmental Health* 55: 479–493.

Delong, R.L., Gilmartin, W.G. and Simpson, J.G. 1973. Premature births in California sea lions: Association with high organochlorine pollutant residue levels. *Science* 181: 1168–1170.

De Swart, R.L., Kluten, R.M.G., Huizing, C.J., Vedder, L.J., Reijnders, P.J.H., Visser, I.K.G., UytdeHaag, F.G.C.M. and Osterhaus, A.D.M.E. 1993. Mitogen and antigen induced B and T cell responses of peripheral blood mononuclear cells from the harbour seal (*Phoca vitulina*). *Veterinary Immunology and Immunopathology* 37: 217–230.

De Swart, R.L., Ross, P.S., Vedder, L.J., Timmerman, H.H., Heisterkamp, S.H., Van Loveren, H., Vos, J.G., Reijnders, P.J.H. and Osterhaus, A.D.M.E. 1994. Impairment of immune function in harbor seals (*Phoca vitulina*) feeding on fish from polluted waters. *Ambio* 23: 155–159.

De Swart, R.L., Ross, P.S., Timmerman, H.H., Vos, H.W., Reijnders, P.J.H., Vos, J.G. and Osterhaus, A.D.M.E. 1995a. Impaired cellular immune response in harbour seals (*Phoca vitulina*) feeding on environmentally contaminated herring. *Clinical and Experimental Immunology* 101: 480–486.

De Swart, R.L., Ross, P.S., Timmerman, H.H., Hijman, W.C., De Ruiter, E., Liem, A.K.D., Brouwer, A., Van Loveren, H., Reijnders, P.J.H., Vos, J.G. and Osterhaus, A.D.M.E. 1995b. Short-term fasting does not aggravate immunosuppression in harbour seals (*Phoca vitulina*) with high body burdens of organochlorines. *Chemosphere* 31: 4289–4306.

De Swart, R.L., Ross, P.S., Vedder, L.J., Boink, F.B.T.J., Rejinders, P.J.H., Mulder, P.G.H. and Osterhaus, A.D.M.E. 1995c. Haematology and clinical chemistry values for harbor seals (*Phoca vitulina*) fed environmentally contaminated herring remains with normal ranges. *Canadian Journal of Zoology* 73: 2035–2043.

De Swart, R.L., Ross, P.S., Vos, J.G. and Osterhaus, A.D.M.E. 1996. Impaired immunity in harbour seals (*Phoca vitulina*) exposed to bioaccumulated environmental contaminants: review of a long-term study. *Environmental Health Perspectives* 104 (suppl. 4): 823–828.

Dietz, R., Heide-Jörgensen, M.-P. and Härkönen, T. 1989a. Mass deaths of harbor seals (*Phoca vitulina*) in Europe. *Ambio* 18: 258–264.

Dietz, R., Ansen, C.T. and Have, P. 1989b. Clue to seal epizootic? *Nature* 338: 627–627.

DiMolfetto-Landon, L., Erickson, K.L., Blanchard-Channell, M., Jeffries, S.J., Harvey, J.T., Jessup, D.A., Ferrick, D.A. and Stott, J.L. 1995. Blastogenesis and interleukin-2 receptor expression assays in the harbor seal (*Phoca vitulina*). *Journal of Wildlife Diseases* 31: 150–158.

Duignan, P.J., Nielsen, O., House, C., Kovacs, K., Duffy, N., Early, G., Sadove, S.S., St. Aubin, D.J., Rima, B.K. and Geraci, J.R. 1997. Epizootiology of morbillivirus infection in harp, hooded, and ringed seals from the Canadian arctic and western Atlantic. *Journal of Wildlife Diseases* 33: 7–19.

Fry, D.M. 1995. Reproductive effects in birds exposed to pesticides and industrial chemicals. *Environmental Health Perspectives* 103: 165–171.

Garner, M.M., Lambourn, D.M., Jeffries, S.J., Hall, P.B., Rhyan, J.C., Ewalt, D.R., Polzin, L.M. and Cheville, N.F. 1997. Evidence of Brucella infection in Parafilaroides lungworms in a Pacific harbor seal (*Phoca vitulina richardsi*). *Journal of Veterinary Diagnostic Investigation* 9: 298–303.

Geraci, J.R. and St. Aubin, D.J. 1987. Effects of parasites on marine mammals. *International Journal for Parasitology* 17: 407–414.

Gilbertson, M., Kubiak, T., Ludwig, J. and Fox, G.A. 1991. Great Lakes embryo mortality, edema, and deformities syndrome (GLEMEDS) in colonial fish-eating birds: similarity to chick-edema disease. *Journal of Toxicology and Environmental Health* 33: 455–520.

Gilmartin, W.G., Delong, R.L., Smith, A.W., Sweeney, J.C., De Lappe, B.W., Risebrough, R.W., Griner, L.A., Dailey, M.D. and Peakall, D.B. 1976. Premature parturition in the California sea lion. *Journal of Wildlife Diseases* 12: 104–115.

Grachev, M.A., Kumarev, V.P., Mamaev, L.V., Zorin, V.L., Baranova, L.V., Denikjna, N.N., Belikov, S.I., Petrov, E.A., Kolesnik, V.S., Kolesnik, R.S., Dorofeev, V.M., Beim, A.M., Kudelin, V.N., Nagieva, F.G. and Sidorov, V.N. 1989. Distemper virus in Baikal seals. *Nature* 338: 209.

Grasman, K.A., Fox, G.A., Scanlon, P.F. and Ludwig, J.P. 1996. Organochlorine-associated immunosuppression in prefledgling Caspian Terns and Herring gulls from the Great Lakes: An ecoepidemiological study. *Environmental Health Perspectives Supplements* 104: 829–842.

Hall, A.J., Law, R.J., Harwood, J., Ross, H.M., Kennedy, S., Allchin, C.R., Campbell, L.A. and Pomeroy, P.P. 1992a. Organochlorine levels in common seals (*Phoca vitulina*) which were victims and survivors of the 1988 phocine distemper epizootic. *Science of the Total Environment* 115: 145–162.

Hall, A.J., Pomeroy, P.P. and Harwood, J. 1992b. The descriptive epizootiology of phocine distemper in the UK during 1988/89. *Science of the Total Environment* 115: 31–44.

Hall, A., Pomeroy, P.P., Green, N., Jones, K. and Harwood, J. 1996. Infection, haematology and biochemistry in grey seal pups exposed to chlorinated biphenyls. *Marine Environmental Research* 43: 81–98.

Hall, A.J., Green, N.J.L., Jones, K.C., Pomeroy, P.P. and Harwood, J. 1998. Thyroid hormones as biomarkers in grey seals. *Marine Pollution Bulletin* 36: 424–428.

Ham-Lammé, K.D., King, D.P., Taylor, B.C., House, C., Jessup, D.A., Jeffries, S.J., Yochem, P.K., Gulland, F.M., Ferrick, D.A. and Stott, J.L. 2001. The application of rapid immunoassays for serological detection of morbillivirus exposure in free-ranging harbor seals (*Phoca vitulina*) and sea otters (*Enhydra lutris*) from the western coast of the United States. *Marine Mammal Science* 15: 601–608.

Helle, E., Olsson, M. and Jensen, S. 1976. DDT and PCB levels and reproduction in ringed seal from the Bothnian Bay. *Ambio* 5: 188–189.

Hutchinson, J.D. and Simmonds, M.P. 1994. Organochlorine contamination in pinnipeds. *Reviews of Environmental Contamination and Toxicology* 136: 123–168.

Kimura, M. 1991. Metallothioneins of monocytes and lymphocytes. *Methods in Enzymology* 205: 291–302.

King, D.P., Lowe, K.A., Hay, A.W.M. and Evans, S.W. 1994. Identification, characterisation, and measurement of immunoglobulin concentrations in grey (*Halichoerus grypus*) and common (*Phoca vitulina*) seals. *Developmental and Comparative Immonology* 18: 433–442.

King, D.P., Sanders, J.L., Nomura, C.T., Stoddard, R.A., Ortiz, C.L. and Evans, S.W. 1998. Ontogeny of humoral immunity in northern elephant seal (*Mirounga angustirostris*) neonates. *Comparative Biochemistry and Physiology B* 121: 363–368.

Klassen, C.D. and Liu, J. 1998. Induction of metallothionein as an adaptive mechanism affecting the magnitude and progression of toxicological injury. *Environmental Health Perspectives* 106: 297–300.

Krahn, M.M., Ylitalo, G.M., Buzitis, J., Sloan, C.A., Boyd, D.T., Chan, S.-L. and Varnasi, U. 1994. Screening for planar chlorobiphenyl congeners in tissues of marine biota by high-performance liquid chromatography with photodiode array detection. *Chemosphere* 29: 117–139.

Lebrec, H. and Burleson, G.R. 1994. Influenza virus host resistance models in mice and rats: utilization for immune function assessment and immunotoxicology. *Toxicology* 91: 179–188.

Loose, L.D., Silkworth, J.B., Pittman, K.A., Benitz, K.-F. and Mueller, W. 1978. Impaired host resistance to endotoxin and malaria in polychlorinated biphenyl- and hexachlorobenzene-treated mice. *Infection and Immunity* 20: 30–35.

Luster, M.I., Faith, R.E. and Clark, G. 1979. Laboratory studies on the immune effects of halogenated aromatics. *Annals New York Academy of Sciences* 320: 471–486.

Luster, M.I., Munson, A.E., Thomas, P.T., Holsapple, M.P., Fenters, J.D., White, K.L. Jr, Lauer, L.D., Germolec, D.R., Rosenthal, G.J. and Dean, J.H. 1988. Development of a testing battery to assess chemical-induced immunotoxicity: National Toxicology Program's guidelines for immunotoxicity evaluation in mice. *Fundamental and Applied Toxicology* 10: 2–19.

Luster, M.I., Portier, C., Pait, D.G., Rosenthal, G.J., Germolec, D.R., Corsini, E., Blaylock, B.L., Pollock, P., Kouchi, Y., Craig, W., White, K.L., Munson, A.E. and Comment, C.E. 1993. Risk assessment in immunotoxicology: II. Relationships between immune function and host resistance tests. *Fundamental and Applied Toxicology* 21: 71–82.

Lynes, M.A., Borghesi, L.A., Youn, J. and Olson, E.A. 1993. Immunomodulatory activities of extracellular metallothionein I. Metallothionein effects on antibody production. *Toxicology* 85: 161–177.

Magnusson, U. 1991. *In vitro* effects of prepartum concentrations of oestradiol-17β on cell-mediated immunity and phagocytosis by porcine leukocytes. *Veterinary Immunology and Immunopathology* 28: 117–126.

Mahy, B.W.J., Barrett, T., Evans, S., Anderson, E.C. and Bostock, C.J. 1988. Characterization of a seal morbillivirus. *Nature* 336: 115–116.

Nakata, H., Tanabe, S., Tatsukawa, R., Amano, M., Miyazaki, N. and Petrov, E.A. 1997. Bioaccumulation profiles of polychlorinated biphenyls including coplanar congeners and possible toxicological implications in Baikal seal (*Phoca sibirica*). *Environmental Pollution* 95: 57–65.

Nielsen, J. 1995. Immunological and hematological parameters in captive harbor seals (*Phoca vitulina*). *Marine Mammal Science* 11: 314–323.

Olsson, M. 1994. Effects of persistent organic pollutants on biota in the Baltic Sea. *Archives of Toxicology Supplement* 16: 43–52.

Olsson, M., Karlsson, B. and Ahnland, E. 1994. Diseases and environmental contaminants in seals from the Baltic and the Swedish west coast. *Science of the Total Environment* 154: 217–227.

O'Shea, T.J. 2000. PCBs not to blame. *Science* 288: 1965–1966.

Osterhaus, A.D.M.E., Groen, J., De Vries, P., UytdeHaag, F.G.C.M., Klingeborn, B. and Zarnke, R.L. 1988. Canine distemper virus in seals. *Nature* 335: 403–404.

Osterhaus, A.D.M.E., Groen, J., UytdeHaag, F.G.C.M., Visser, I.K.G., Vedder, E.J., Crowther, J. and Bostock, C.J. 1989. Morbillivirus infection in European seals before 1988. *Veterinary Record* 125: 326–326.

Osterhaus, A.D.M.E., De Swart, R.L., Vos, H.W., Ross, P.S., Kenter, M.J.H. and Barrett, T. 1995. Morbillivirus infections of aquatic mammals: newly identified members of the genus. *Veterinary Microbiology* 44: 219–227.

Pillet, S., Lesage, V., Hammill, M.O., Cyr, D., Bouquegneau, J.M. and Fournier, M. 2000. *In vitro* exposure of seal peripheral blood leukocytes to different metals reveal a sex-dependent effect of zinc on phagocytic activity. *Marine Pollution Bulletin* 40: 921–927.

Pillet, S., Cyr, D., Fournier, M. and Bouquegneau, J.M. 2001. *First characterization of metallothioneins in peripheral blood leukocytes from a marine mammal; implications in heavy metal immunotoxicity* [Abstract]. Society for Marine Mammalogy, Vancouver, Canada.

Reijnders, P.J.H. 1980. Organochlorine and heavy metal residues in harbour seals from the Wadden Sea and their possible effects on reproduction. *Netherlands Journal of Sea Research* 14: 30–65.

Rolland, R.M. 2000. A review of chemically-induced alterations in thyroid and vitamin A status from field studies of wildlife and fish. *Journal of Wildlife Diseases* 36(4): 615–635.

Ross, P.S. 1990. Immunocompetence of free-ranging harbour seal (*Phoca vitulina*) mothers and their pups over the course of lactation. MSc thesis, Dalhousie University, Halifax, Canada.

Ross, P.S. 2000. Marine mammals as sentinels in ecological risk assessment. *Human and Ecological Risk Assessment* 6: 29–46.

590   *P.S. Ross* et al.

Ross, P.S., Visser, I.K.G., Broeders, H.W.J., Van de Bildt, M.W.G., Bowen, W.D. and Osterhaus, A.D.M.E. 1992. Antibodies to phocine distemper virus in Canadian seals. *Veterinary Record* 130: 514–516.

Ross, P.S., Pohajdak, B., Bowen, W.D. and Addison, R.F. 1993. Immune function in free-ranging harbor seal (*Phoca vitulina*) mothers and their pups during lactation. *Journal of Wildlife Diseases* 29: 21–29.

Ross, P.S., De Swart, R.L., Visser, I.K.G., Murk, W., Bowen, W.D. and Osterhaus, A.D.M.E. 1994. Relative immunocompetence of the newborn harbour seal, *Phoca vitulina*. *Veterinary Immunology and Immunopathology* 42: 331–348.

Ross, P.S., De Swart, R.L., Reijnders, P.J.H., Van Loveren, H., Vos, J.G. and Osterhaus, A.D.M.E. 1995. Contaminant-related suppression of delayed-type hypersensitivity and antibody responses in harbor seals fed herring from the Baltic Sea. *Environmental Health Perspectives* 103: 162–167.

Ross, P.S., De Swart, R.L., Timmerman, H.H., Reijnders, P.J.H., Vos, J.G., Van Loveren, H. and Osterhaus, A.D.M.E. 1996a. Suppression of natural killer cell activity in harbour seals (*Phoca vitulina*) fed Baltic Sea herring. *Aquatic Toxicology* 34: 71–84.

Ross, P.S., De Swart, R.L., Addison, R.F., Van Loveren, H., Vos, J.G. and Osterhaus, A.D.M.E. 1996b. Contaminant-induced immunotoxicity in harbour seals: wildlife at risk? *Toxicology* 112: 157–169.

Ross, P.S., De Swart, R.L., Van Loveren, H., Osterhaus, A.D.M.E. and Vos, J.G. 1996c. The immunotoxicity of environmental contaminants to marine wildlife: A review. *Annual Review of Fish Diseases* 6: 151–165.

Ross, P.S., Ikonomou, M.G. and Addison, R.F. 1997. *Levels of PCBs, PCDDs and PCDFs in British Columbia harbor seals* (Phoca vitulina) *associated bioeffects* [Abstract]. Society of Environmental Toxicology and Chemistry, San Francisco, 217.

Ross, P.S., Vos, J.G., Birnbaum, L.S. and Osterhaus, A.D.M.E. 2000. PCBs are a health risk for humans and wildlife. *Science* 289: 1878–1879.

Safe, S.H. 1990. Polychlorinated biphenyls (PCBs), dibenzo-p-dioxins (PCDDs), dibenzofurans (PCDFs), and related compounds: environmental and mechanistic considerations which support the development of toxic equivalency factors (TEFs). *Critical Reviews in Toxicology* 21: 51–88.

Schuurs, A.H.W.M. and Verheul, H.A.M. 1990. Effects of gender and sex steroids on the immune response. *Journal of Steroid Biochemistry* 35: 157–172.

Silkworth, J.B. and Grabstein, E.M. 1982. Polychlorinated biphenyl immunotoxicity: dependence on isomer planarity and the *Ah* gene complex. *Toxicology and Applied Pharmacology* 65: 109–115.

Silkworth, J.B., Antrim, L.A. and Sack, G. 1986. *Ah* receptor mediated suppression of the antibody response in mice is primarily dependent on the *Ah* phenotype of lymphoid tissue. *Toxicology and Applied Pharmacology* 86: 380–390.

Simms, W. and Ross, P.S. 2000. Developmental changes in circulatory vitamin A (retinol) and its transport proteins in free-ranging harbour seal (*Phoca vitulina*) pups. *Canadian Journal of Zoology* 78(10): 1862–1868.

Simms, W. and Ross, P.S. 2001. Vitamin A physiology and its application as a biomarker of contaminant-related toxicity in marine mammals: a review. *Toxicology and Industrial Health* 16(6): 291–302.

Simms, W., Jeffries, S.J., Ikonomou, M.G. and Ross, P.S. 2000. Contaminant-related disruption of vitamin A dynamics in free-ranging harbor seal (*Phoca vitulina*) pups from British Columbia, Canada and Washington State, USA. *Environmental Toxicology and Chemistry* 19(11): 2844–2849.

Thomson, C.A. and Geraci, J.R. 1986. Cortisol, aldosterone, and leucocytes in the stress response of bottlenose dolphins, *Tursiops truncatus*. *Canadian Journal of Fisheries and Aquatic Sciences* 43: 1010–1016.

Tillitt, D.E., Ankley, G.T., Giesy, J.P., Ludwig, J.P., Kuritamatsuba, H., Weseloh, D.V., Ross, P.S., Bishop, C.A., Sileo, L., Stromberg, K.L., Larson, J. and Kubiak, T.J. 1992. Polychlorinated biphenyl residues and egg mortality in double-crested cormorants from the Great Lakes. *Environmental Toxicology and Chemistry* 11: 1281–1288.

Tryphonas, H., Luster, M.I., White, K.L. Jr, Naylor, P.H., Erdos, M.R., Burleson, G.R., Germolec, D., Hodgen, M., Hayward, S. and Arnold, D.L. 1991. Effects of PCB (Aroclor 1254) on non-specific immune parameters in rhesus (*Macaca mulatta*) monkeys. *International Society for Immunopharmacology* 13: 639–648.

Van den Berg, M., Birnbaum, L., Bosveld, A.T.C., Brunstrom, B., Cook, P., Feeley, M., Giesy, J.P., Hanberg, A., Hasegawa, R., Kennedy, S.W., Kubiak, T., Larsen, J.C., Van Leeuwen, F.X.R., Liem, A.K., Nolt, C., Peterson, R.E., Poellinger, L., Safe, S.H., Schrenk, D., Tillitt, D.E., Tysklind, M., Younes, M., Waern, F. and Zacharewski, T.R. 1998. Toxic equivalency factors (TEFs) for PCBs, PCDDs, PCDFs for humans and wildlife. *Environmental Health Perspectives* 106: 775–792.

Van Oss, C.J. 1987. Phagocytosis: An overview. *Methods in Enzymology* 132: 3–15.

Van Zorge, J.A., Van Wijnen, J.H., Theelen, R.M.C., Olie, K. and Van den Berg, M. 1989. Assessment of the toxicity of mixtures of halogenated dibenzo-p-dioxins and dibenzofurans by use of toxicity equivalency factors (TEF). *Chemosphere* 19: 1881–1895.

Visser, I.K.G., Kumarev, V.P., Örvell, C., De Vries, P., Broeders, H.W.J., Van de Bildt, M.W.G., Groen, J., Teppema, J.S., Burger, M.C., UytdeHaag, F.G.C.M. and Osterhaus, A.D.M.E. 1990. Comparison of two morbilliviruses isolated from seals during outbreaks of distemper in North West Europe and Siberia. *Archives of Virology* 111: 149–164.

Vos, J.G. and De Roij, Th. 1972. Immunosuppressive activity of a polychlorinated biphenyl preparation on the humoral immune response in guinea pigs. *Toxicology and Applied Pharmacology* 21: 549–555.

Vos, J.G. and Luster, M.I. 1989. Immune alterations. In R.D. Kimbrough and S. Jensen (eds), *Halogenated biphenyls, terphenyls, naphtalenes, dibenzodioxins and related products* (pp. 295–322). Amsterdam: Elsevier Science Publishers B.V.

Vos, J.G. and Van Loveren, H. 1987. Immunotoxicity testing in the rat. In E.J. Burger, R.G. Tardiff and J.A. Bellanti (eds), *Environmental chemical exposure and immune system integrity. Advances in Modern Environmental Toxicology*, Vol. 13 (pp. 167–180). Princeton: Princeton Scientific.

Vos, J.G., Moore, J.A. and Zinkl, J.G. 1973. Effect of 2,3,7,8-tetrachlorodibenzo-p-dioxin on the immune system of laboratory animals. *Environmental Health Perspectives* 5: 149–162.

Woldstad, S. and Jenssen, B.M. 1999. Thyroid hormones in grey seal pups (*Halichoerus grypus*). *Comparative Biochemistry and Physiology A* 122: 157–162.

Wood, S.C., Karras, J.B. and Holsapple, M.P. 1992. Integration of the human lymphocyte into immunotoxicological investigations. *Fundamental and Applied Toxicology* 18: 450–459.

Part V
Perspectives for the future

23 Conclusions and perspectives for the future

Thomas J. O'Shea, Gregory D. Bossart,
Michel Fournier and Joseph G. Vos

The emerging science of marine mammal toxicology: scope and aims

It is our hope that the chapters we have compiled in this book accurately reflect the state of the science of toxicology of marine mammals as the field crosses the bridge into the twenty-first century. This is an area of research that will most likely make its major future advances through the applications of multidisciplinary and interdisciplinary approaches that are often international in scope. This is because of the complexity of the challenges posed to the study of the effects of high numbers of chemical substances in long-lived and wide-ranging mammals of the seas and large inland waters. Indeed, the marine mammals comprise many species of diverse evolutionary origins that inhabit nearly all major aquatic reaches of the globe, occupy numerous feeding niches and trophic levels, and scale across orders of magnitude differences in body sizes, from the 4 kg marine otters (*Lontra felina*) to the blue whales (*Balaenoptera musculus*) at 150 tonnes. Superimposed on challenges to working with marine mammals, and the complexity and numbers of contaminants and biotoxins in their environment, are the ever-growing specializations of toxicology into subdisciplines.

In addition to the classical fronts of environmental chemistry and pathology (Moeller, Chapter 1; Martineau *et al.*, Chapter 13; Stein *et al.*, Chapter 17; Bergman *et al.*, Chapter 19), these subdisciplines now include immunotoxicology, toxicology of the reproductive and endocrine systems, cellular and molecular toxicology, and other areas. This is reflected in the scope of many of the chapters in this book. These include both overviews and analyses of global and temporal trends of organochlorines, inorganic pollutants and metals (e.g. O'Shea and Tanabe, Chapter 6; Das *et al.*, Chapter 7; O'Hara *et al.*, Chapter 9; Colborn and Smolen, Chapter 12; Reijnders and Simmonds, Chapter 18), specialized topics in cellular and molecular mechanisms of toxicity in marine mammals (e.g. Gauthier *et al.*, Chapter 15; Carvan and Busbee, Chapter 16), reproductive and endocrinological studies (e.g. Reijnders, Chapter 3; Gregory and Cyr, Chapter 4), and immunotoxicology (e.g. DeGuise *et al.*, Chapter 2; Brousseau *et al.*, Chapter 14; Ross *et al.*, Chapter 20). The growing reliance on multidisciplinary investigations in the science of marine mammal toxicology is also reflected in some of the diverse professional disciplines represented by contributors to this volume and, indeed, to some of its individual chapters. Some excellent case studies involving multidisciplinary approaches are reviewed in this volume, with stellar examples including recent studies of mass mortality in marine mammals due to algal toxins. These studies have combined levels of investigation ranging in scale from histological, molecular and biochemical diagnostics, through gross pathological study of stranded carcasses, to satellite imagery from space (e.g. Bossart *et al.* 1998; Scholin *et al.*, 2000; see review in Van Dolah *et al.*, Chapter 10 this volume). Recent breakthroughs in genetic character- ization and cloning of the Ah receptor in harbor seals (*Phoca vitulina*) and beluga whales (*Delphinapterus leucas*) (Jensen and Hahn, 2001; Kim and Hahn, 2002) should ensure accelerated understanding of molecular toxicology in marine mammals.

Many marine mammals are truly international resources, and this is reflected in the growing number of international collaborations that have arisen in marine mammal toxicology. We foresee that collaborations across disciplines and nations will only increase in the twenty-first century. There is a global concern for the conservation of marine mammals. In some cases this adds a sense of urgency to the need for breakthroughs in understanding the threats imposed to these species by toxic substances in the environment. The past century has seen the extinction of one species of marine mammal (the Caribbean monk seal, *Monachus tropicalis*) due to human activities, and major reductions or local elimination of populations of many others. Thus the new century has the potential to witness the complete demise of additional species that are now near the brink of extinction. However, not all prospects are bleak. Compared to 100 years ago, when attitudes about marine mammals worldwide were dominated by an exploitative philosophy, there is substantial concurrence that such extinctions must be prevented. This is now formally embodied in laws and regulations of many nations, and in meaningful international treaties and agreements. Populations have subsequently rebounded in some species of marine mammals. Current management, however, largely addresses the practices and issues of the past that were clear and obvious root causes of declines in marine mammals, such as over-hunting (for meat, oil, fur and fertilizer), intentional reductions to alleviate perceived competition with fisheries, widespread take incidental to various human activities, and habitat degradation (for overviews see Anderson, 2001; Harwood, 2001).

As documented by contributions to this volume and a very large body of other literature, threats imposed by toxic substances are now recognized as having the potential to emerge as serious negative forces on marine mammal populations. These potential negative forces could become just as powerful over the next 100 years as those threats of the twentieth century that are the current focus of management and conservation of marine mammals. Indeed, threats from toxic substances may be even more insidious and pose even greater challenges to adopting effective conservation measures. The sources of environmental contamination with anthropogenic toxic substances (and perhaps environmental degradation that may be at the roots of possible increasing threats from natural biotoxins) are pervasive but diffuse in origin. In many cases there are no single sources that can be easily eliminated. Additionally, because some marine mammal populations that may be impacted by contaminants and biotoxins occur beyond nearshore waters and are an international resource, they have the potential to suffer from 'the tragedy of the commons' (Hardin, 1968) on a grand scale. Marine mammals can be impacted by toxic substances that no single nation or human activity is alone responsible for, and many marine mammals occur in what is both the greatest 'commons' area on the surface of the planet and the ultimate sink for many toxic substances.

This situation makes it very difficult to initiate actions to stem the inputs of toxic substances into the environments of marine mammals. If efforts to overcome such inertia call on the use of solid science to guide policy, marine mammal toxicology has some formidable work cut out for it. The scientific basis to support management decisions to reduce exposure of marine mammals to some toxic substances is today perhaps more difficult to provide than the science that underlies management of many of the other human activities that more obviously impinge on marine mammal birth and death processes. The existence of mortality in marine mammals from biotoxins is an exception that is now very firmly established and the mechanisms involved are being recognized with greater precision (Van Dolah *et al.*, Chapter 10 this volume). However, the basic understanding of how human activities may or may not favor increased frequency or severity of algal blooms, for example, or what the human-influenced trajectory for future rates of occurrence of such events may be, are very uncertain. Direct mortality of marine mammals in the wild from exposure to anthropogenic chemicals through the food chain is undocumented. Evidence for effects on reproduction is growing, but limited and complicated (Reijnders, Chapter 3), as is the mounting evidence for contaminant-induced susceptibility to deaths during epizootics (Ross *et al.*, Chapter 20; Vos *et al.*, Chapter 21; Ross *et al.*, Chapter 22). The degree to which science can support links between current exposure of marine mammals to anthropogenic contaminants and negative effects at the population level is a subject of vigorous debate, even among contributors to and editors of this volume. However, all understand the likelihood of increasing contamination of the seas and the potential increasing risk this implies for marine mammals in the twenty-first century.

Perspectives for future research

Recognizing the potential for far-reaching impacts of increased exposure of marine mammal populations of the future to toxic substances, scientists working in this field have held several workshops over the past decade (e.g.

598   T.J. O'Shea et al.

Tatsukawa *et al.*, 1994; Sanderson and Gabrielson, 1996; Ross and DeGuise, 1998; Reijnders *et al.*, 1999). These workshops have, to varying degrees, assessed the major scientific issues involved in marine mammal toxicology and recommended research approaches for their resolution. Among the most recent of these was the US Marine Mammal Commission workshop held in Keystone, Colorado in 1998 (O'Shea *et al.*, 1999). This workshop was attended by 54 scientists from seven nations, representing a number of disciplines pertinent to the study of marine mammal toxicology. Indeed, 18 of the chapters in this book include authors who were participants at Keystone, and some of the chapters (DeGuise *et al.*, Chapter 2; Reijnders, Chapter 3; O'Shea and Tanabe, Chapter 6) are outgrowths of plenary presentations made at that workshop. We feel that, in addition to specific recommendations made within individual chapters in this book (e.g. Reijnders, Chapter 3; Carvan and Busbee, Chapter 16), many of the recommendations and conclusions developed by the participants at the US Marine Mammal Commission workshop also accurately reflect needed future directions for this emerging field. We therefore conclude this volume with a synopsis of the collective perspectives for the future that were synthesized at Keystone.

Recommendations for specific areas of research

Much of the focus of the US Marine Mammal Commission workshop was on the persistent organochlorine contaminants. Although some countries no longer apply or produce some of these compounds, many nations continue to manufacture, use or dispose of organochlorines that subsequently follow pathways that redistribute them to the global marine ecosystem (see, for example, O'Hara and Becker, Chapter 8). Marine mammals at the top of complex ocean food webs then become a major repository for these substances. The workshop participants therefore emphasized defining research needs based primarily on toxic substances with properties similar to the organochlorines. Participants at the Keystone workshop assembled into groups that defined the avenues for future research within four topic areas: reproduction and endocrinology; immunotoxicity, pathology and disease; risk assessment; and future trends. Findings of these groups are synopsized below. There were also a large number of common threads among most group reports. The 20 major conclusions and recommendations common across most of the four topic areas are highlighted in the subsequent section of this chapter.

Toxicology of the reproductive and endocrine systems in marine mammals

There will be several negative consequences unless there are advances in understanding in this area. These include lack of recognition of chemical-induced reproduction problems in a manner that would allow timely investigation and possible mitigation; failure to understand a full range of possible causes of declines in populations of marine mammals; failure to recognize effects of new substances before serious impacts occur; lack of validated models based on common species to apply to management of species or stocks in endangered or other critical status; and a costly lack of focus in understanding consequences of environmental change to population status of marine mammals. The 11-member Working Group on Endocrinology and Reproduction emphasized four major topics for future research (see also Reijnders, Chapter 3; Gregory and Cyr, Chapter 4, this volume).

1. To determine how organochlorines may influence reproduction at pre-implantation stages, as suggested by experiments on captive harbor seals in Europe. This work would be aimed at reducing uncertainties about which chemical agents may be responsible for disturbances observed thus far, pinpointing mechanisms leading to such effects, and understanding the overall occurrence and population significance of pre-implantation disorders.

2. To determine if and how organochlorines may be involved in causing premature births, particularly in otarrid seals. Too many confounding variables (including age, nutrition and abortion-inducing diseases) exist, in populations where this has been observed, to allow inference of cause-and-effect. Establishment of normal cycles of hormonal control of reproduction will also be necessary to make progress in establishing such inferences. Both captive and field studies are required.

3. To design and measure indices of developmental and reproductive function in marine mammals. Marine mammals are exposed to endocrine-disrupting chemicals at some level, but there is, to date, little evidence for possible effects. This may be because scientists have not been looking systematically for anatomical endpoints of such disruption in marine mammals, although effects of endocrine-disrupting chemicals have been well documented in laboratory mammals. This working group suggested that uncertainties about the existence of such effects in marine mammals can be resolved using several approaches. Endpoints that can be examined and tested in marine mammals should be defined, using controlled, longitudinal experiments on captive individuals and strategic sampling of stranded, bycaught or hunted marine mammals. These endpoints can include a wide range of morphological, histological, biochemical, endocrine, behavioral, morphometric and gravimetric assessments, described in detail in the working group report. New *in vitro* cellular and molecular assays should also be developed specifically for selected species of marine mammals, including steroid and nuclear receptors, and biomarkers of developmental exposure. Expanded sampling of multiple tissues for contaminants, and guidelines for selecting individual animals for sampling, are also recommended.

4. To use multidisciplinary approaches to the study of toxicology of reproduction in marine mammals that will allow scaling up from individual- to population-level effects.

Toxic substances and marine mammal immunology, pathology and disease

Knowledge of direct contaminant-induced pathologic effects and associated diseases in marine mammals is very limited (see also Moeller, Chapter 1; Martineau *et al.*, Chapter 13; Bergman *et al.*, Chapter 19, this volume), and immunotoxic effects of contaminants on marine mammals have also been demonstrated only in limited cases (see also DeGuise *et al.*, Chapter 2; Brousseau *et al.*, Chapter 14; Ross *et al.*, Chapter 20; Vos *et al.*, Chapter 21, this volume). Failure to increase knowledge in this area limits the ability to inform policy-makers and managers about impacts of contaminants on health status of marine mammals in relation to chemical exposure, both in terms of chronic effects and susceptibility to epizootics, possibly leading to declines or losses in populations and species. Six major areas of research needs were stressed in these related subdisciplines of marine mammal toxicology.

1. To establish linkages between exposure to contaminants, specific pathological lesions and immunosuppresive endpoints, and to improve knowledge of dose–response relationships.
2. To combine studies in this area with other subdisciplines, and in particular to apply them to understanding the health and contaminant status of otherwise well-studied (through life history research, stranded specimens, or fisheries bycatch) populations in areas of contrasting high and low contamination.
3. To determine the actual mechanisms by which toxicity to the immune system may take place in marine mammals, and thus establish the predictive power of molecular and cellular indices of chemical exposure and effect. These efforts should include *in vivo* and *in vitro* studies, and use of cell cultures, particularly for developing species-specific understanding of toxic equivalency and metabolic enzyme activity. Use of surrogate animal models will also aid in developing an understanding of mechanistic factors.
4. To employ multiple approaches in studying immunotoxicology and disease, but to ensure that these approaches are tightly coupled, include both laboratory and field components, use both wild and captive individuals, employ assays that are as complete as possible, and incorporate sample sizes large enough to ensure desired statistical power.
5. To design studies such that results will allow scaling-up of assessments of impacts of chemicals on the immune system and diseases of marine mammals from the individual to the population level.
6. To develop standards and to implement archiving of samples for long-term monitoring of disease, pathology and immune function in wild populations of marine mammals. For greater details on these suggestions, see the full report of the 13-member Working Group on Immunotoxicology, Pathology and Disease in O'Shea *et al.* (1999); for recommendations for design of immune function studies in wild pinnipeds see Ross *et al.*, Chapter 22 in this volume.

Risk assessment

The 14-member Working Group on Risk Assessment recognized that even though there were formalized, structured models that can be used to conduct risk assessments to judge possible impacts of toxic substances on organisms, such approaches have not yet been taken with marine mammals. Development and application of risk assessment procedures in marine mammal toxicology will require a variety of information, some of which may require data that are not yet readily available with suitable degrees of accuracy and precision. Needed information includes developing knowledge of exposure, defining mechanisms of threat, understanding how toxicants are distributed in populations, how toxic substances are mobilized and sequestered in the body, and what the targets of toxicity are. Risk assessment will also require understanding dose–response and cause–effect relationships, perhaps best derived through nonlethal captive experiments and experiments on meaningful surrogate species of laboratory mammals. Standard protocols for sampling and analyses should be developed and employed, including biomarkers to aid in validation of risk assessment. Risk assessment models should also incorporate margins for precaution, to insure safeguards for conservation. A number of other recommendations made by this working group were also applicable to the entire field of marine mammal toxicology, and are described further in this chapter among the 20 major conclusions and recommendations of the US Marine Mammal Commission workshop.

Future trends

The 12 scientists who participated in this working group noted that there are hundreds of lipophilic contaminants that are now in the marine environment or likely to be released to the oceans over the near future. Although marine mammals are likely to be exposed to these chemicals, their tissues are not now routinely examined for their presence. There are also classes of compounds in the marine environment that do not reach particularly high concentrations in marine mammals but, none the less, could have toxic effects. These substances have not yet been evaluated with respect to possible impacts on marine mammals. In addition, there are threats from increasing eutrophication of the nearshore environment, which may play a role in increasing exposure of marine mammals to some biotoxins. The inshore environment is also increasingly contaminated with monomers and polymers of unknown significance to marine mammals. In addition to many unknown marine contaminants, it is expected that distributions of more well-studied contaminants, such as PCBs and organochlorine pesticides, in the world's marine mammal populations will change as these contaminants continue to enter the oceans and mass balances are shifted. Even in cases where slow declines in contamination with well-known organochlorines may be expected, many will remain in the range where subtle toxic effects can be anticipated. Researchers will need to strive to develop and adhere to consistent analytical protocols, including determination of, and reporting of, specified numbers of contaminants in marine mammal samples. Participants voiced a concern that many currently known toxic substances that may affect marine mammals continue to be produced, used and disposed of in some nations, in such ways that they will continue to enter the seas in the future. The workshop stressed that much more needs to be learned about the emissions and sources of contaminants that are entering the marine environment; that specimen banking be instituted to assist in evaluation of future threats; and that chemical monitoring be more closely linked to other, more broadly based biological studies of marine mammals. The working group noted that results of environmental chemistry studies of marine mammals should be made available to managers in easily usable formats; that long-term monitoring be maintained; that international experts periodically meet to evaluate the threats of contaminants to marine mammals; and that an ongoing dialog on this topic be established and continued among international authorities.

Principal conclusions and recommendations across disciplines

Participants at the US Marine Mammal Commission workshop were concerned about the great uncertainty that remains about specific effects of contaminants in marine mammals, to what extent such effects may occur in wild marine mammals, and what impact such effects may have on their population dynamics. For most of its history, the science of marine mammal toxicology has emphasized documentation of contaminant residues in tissues (O'Shea and Tanabe, Chapter 6, this volume). However, there has been much less emphasis placed on designing and executing studies that allow interpretation of the significance of this contamination to marine mammal health and population dynamics. To achieve progress towards this goal, scientists at the US Marine Mammal Commission workshop identified 20 principal conclusions and recommendations for future directions in marine mammal toxicology that emerged in common across the various working groups. These are synopsized below as presented in the workshop report, with only minor editorial condensing. There is no priority attached to their order of presentation.

1. Integration of multiple approaches

A greatly improved understanding is needed of the linkages between specific chemical exposures (both type and amount) and endpoints of concern (e.g. impaired health, immunosuppression and reproductive disorders). Integration of laboratory, captive animal and field studies was a consistent theme of workshop deliberations. No single approach is likely to be adequate for resolving the critical uncertainties that arise in relation to contaminants and marine mammals. Thus, there is a need for multidisciplinary studies that integrate physiological, behavioral, reproductive, clinical, pathological and toxicological data, with the ultimate goal of linking immune status, health, reproduction and survival of individuals to trends observed or predicted at the population and ecosystem level. Dose–response relationships are critical, and quantification of these relationships should be a goal of all studies of individuals. Surrogate species can be particularly useful in studies of mechanisms of action.

2. Stable support for critical long-term programs, with increased emphasis on wider collaborations

Long-term monitoring and research programs provide consistent, accumulative information that is essential for understanding trends and impacts of contaminants in marine mammals. Established ongoing programs should be viewed as long-term investments that are necessary for the effective assessment and management of marine mammal populations. They should remain a high priority of sponsoring agencies and organizations and be given firm support. New efforts and enhancements to existing programs should also be implemented with assurances for long-term, stable support. In addition, the study of contaminant effects is well established in the fields of medicine and environmental chemistry. Marine mammal scientists should be encouraged to collaborate with programs in these fields.

3. Long-term interdisciplinary studies of local populations

Long-term interdisciplinary studies provide opportunities to measure contaminant exposure, monitor the health and immune responses of individuals, and relate findings to population-level trends, all in the context of knowing a great deal about the animals' life history, distribution, abundance, population dynamics and demography. Long-term studies of local populations hold out the possibility of obtaining insights on contaminant effects, grounded in observations from nature. It is important that these and similar programs continue to be supported and that long-term tracking of the health, contaminant and disease status of individuals be incorporated into the data-collection protocols. Selection of study populations should not be limited to those that are depleted or threatened or that have experienced a recent die-off. Any long-term longitudinal study offers opportunities to assess the effects of chronic pollution, and it is valuable to investigate effects during periods of population decline, stability and recovery.

4. Compilation and dissemination of information

Existing systems for compiling, interpreting and disseminating data on the production, use, physical and chemical properties, toxicology and ecological effects of persistent contaminants should be improved. It is currently estimated that there are roughly 2400 lipophilic and persistent chemicals, of which at least 390 are clearly toxic and bioaccumulative. In order to prevent long-term pollution from these largely unknown chemicals, basic information about them needs to be made widely available. This will require enhanced international cooperation, preferably within the existing framework of chemical contaminant programs, such as the Existing Chemicals Program of the Organization for Economic Co-operation and Development, the Program on Persistent Organic Pollutants of the United Nations Environment Programme, and the International Program on Chemical Safety. Such information is crucial for anticipating future problems associated with chemical contaminants in the marine environment. Contaminant categories of particular concern in relation to marine mammals include organometallic compounds and persistent, toxic, bioaccumulative compounds not on the standard persistent organic pollutants list, but that have been used or produced in large quantities, such as polychlorinated naphthalenes and polychlorinated diphenyl ethers. Other contaminants of concern that could present a risk to marine mammals include non-persistent chemicals with high bioaccumulation potential and heavy or widespread use, chemicals that contribute to coastal eutrophication, polymers and low molecular weight monomers, and a variety of chemicals that may cause endocrine disruption.

5. Monitoring environmental loads and ongoing inputs of persistent contaminants

In spite of encouraging evidence that levels of some compounds (e.g. PCBs, DDT) have declined in some areas (e.g. the North American Arctic, western Europe), production and inputs to the environment continue in areas such as the former Soviet Union (PCBs, perhaps until very recently) and many tropical countries (DDT and other organochlorine pesticides). This situation creates an ongoing need for global-scale monitoring. Monitoring contaminants in tissues of marine mammals is of direct importance for issues related specifically to this group, but more sensitive detection of contaminant trends in the marine environment in general requires monitoring of other ecosystem components. Attention needs to be paid not only to analyses of water and biological samples, but also to identifying emission sources, transport mechanisms and transport pathways, and to calculating environmental loads.

6. Establish universal protocols

Tissues are sampled and biomarkers measured from biopsies of living marine mammals and from the fresh carcasses of stranded, bycaught, or hunted animals. Universal protocols should be followed in sample collection and storage, laboratory analytical procedures and data reporting. Sample collection should include specified basic data and supporting documentation. Tissue specimens should be archived under consistent conditions in dependable, long-term programs. As analytical techniques are developed, data comparability between and among laboratories should be assured on a global scale. Laboratories should adopt a performance-based, quality-assurance approach that incorporates standard reference materials (SRMs) and inter-laboratory comparisons. Existing SRMs should be analyzed, and, if necessary, new SRMs developed for emerging chemicals. Quantitative information should be generated for as many compounds as possible in order to provide robust data sets for future use. Protocols should stipulate quantitation of a minimum number of compounds of persistent contaminants in order to enhance the comparability of data sets for purposes of risk assessment and trend analysis. Existing analytical techniques should be used to the fullest to identify all anthropogenic chemicals in tissues, and thus expand the number of existing and new chemicals known to accumulate in, and pose potential threats to, marine mammals.

7. Use formal risk assessment procedures to evaluate threats

Formal risk assessment procedures should be adapted and used to evaluate threats of contaminants to specific marine mammal populations. The Ecological Risk Assessment approach established by the US Environmental Protection Agency couples risk assessment to risk management in a process that can be iterative. This process begins with problem formulation: contaminant sources, fates and pathways are described; contaminants of concern and their modes of action are defined; and receptors of concern are identified. A risk assessment would describe specific hypotheses to be tested and measurements that can be made to evaluate endpoints of concern in the analysis phase. Risk characterization, which follows the analysis phase, uses information on exposure and effects to evaluate the risk, generally using a 'weight of evidence' approach. This approach balances the information on effects, usually obtained from mechanistic laboratory studies, against the results of captive-animal or field studies. Uncertainty about aspects of risk assessment should be clearly discussed. Discussion should include the ecological significance of the risk, how effects on individuals have been extrapolated to effects on populations, and how risks from contamination have been compared to risks from other stressors. The ideal outcome of risk characterization is identification of threshold concentrations, above which risk becomes significant and risk management is required.

8. Use of rehabilitated and captive marine mammals and associated databases

Marine mammals held in captivity that are fully rehabilitated or on display are an under-used resource for the study of contaminant impacts on physiological processes and health (see Reddy and Ridgway, Chapter 5, this volume). These animals are likely to be exposed to contaminants in routine daily rations of marine products, at least at low concentrations, and are likely to have contaminants in their bodies. Clinical measurements of blood chemistry and hematology, and detailed information on health, reproduction and survival are routinely recorded for these animals. Provided that care is taken in the design of sampling to avoid confounding effects of prior injury, disease or response to captivity, use of this resource may provide basic insight about the variability in and relationships among contaminants and basic health and physiological processes, biomarkers, reproduction and survival. Such insight could prove critical to interpretation of contaminant impacts on wild marine mammal populations.

9. *Use of surrogate animal models*

The extensive literature on mammalian toxicology provides many examples in which studies of surrogate species have been used in place of humans. A similar approach can be useful for marine mammals. Studies of surrogate models can provide insights about mechanism of action, the comparative risks presented by different chemicals, and dose–response relationships. Experimental control makes it possible to eliminate confounding factors and reduce variability within treatment groups, thus providing the statistical power necessary to detect effects. The greater availability of species-specific reagents, and the ability to carry out invasive studies, represent additional advantages of using surrogate models. The selection of an appropriate surrogate model depends on the question being posed and the endpoint being assessed. The usefulness of information generated from studies of surrogate species will depend upon assumptions and extrapolations which will require critical evaluation at all stages of the research.

10. *Controlled experimental studies to address critical questions*

The need for controlled experiments to answer critical questions about the effects of contaminants on marine mammals was generally acknowledged at this workshop (a few participants expressed opinions that use of surrogate models would suffice, or that non-lethal experiments on captive marine mammals may not be feasible). Some of the critical questions regarding impacts of even the most commonly observed persistent contaminants in marine mammals have been asked for more than 25 years, without clear answers. One reason for this chronic uncertainty is that it has been difficult to obtain funding, facilities or authorization to carry out controlled experiments on captive marine mammals. Without the highly convincing kind of evidence that only experimentation can provide, the uncertainty and debate regarding cause-and-effect relationships and necessary mitigation are bound to continue. Meanwhile, the health of individual animals and the persistence of species and populations may be in jeopardy. A widening array of 'new' chemical compounds in the immediate future could adversely affect marine mammals directly, or they could affect other components of the food webs on which marine mammals depend. Most workshop participants agreed that a two-pronged strategy should be followed in confronting this reality. First, procedures should be developed and applied globally to assess the effects of 'new' chemicals on the survival and productivity of representative ecosystem components. Such assessments should be made *before* these chemicals are mass produced. Second, when there is uncertainty about a chemical's effects, and it is determined that a delay in resolving that uncertainty could jeopardize the survival of a marine mammal species or population, it may be necessary to conduct experiments with captive marine mammals. Such experiments would involve nonlethal exposures to doses of contaminants similar to those experienced by some wild populations. The studies would need to be well justified and well designed. Design would have to include power analysis to ensure that the results would be conclusive.

11. *Understanding processes linking exposure to effects*

Understanding the linkages between contaminants and the health, immune system or reproduction of individual marine mammals at the cellular or molecular level is most likely to come from laboratory studies. Cell culture and *in vitro* techniques can contribute by providing species-specific data on topics such as toxic equivalency. In so far as it is possible, toxic equivalencies should be developed and validated using the cell lines of the marine mammal species of interest. Severe combined immune deficient (SCID) mice and other laboratory mammals can be used for the detailed, invasive types of experimentation needed to improve understanding of the biochemical processes linking exposure and effects. Data from such studies can be employed to evaluate hypotheses involving the extrapolation of *in vitro* results to whole organisms. Semi-field trials and epidemiological studies of wild populations (using biopsies of free-ranging animals, blood drawn from animals that are captured and released, and specimens salvaged from strandings, bycatch and subsistence hunts) can be used to validate the inferences based on laboratory evidence or extrapolation from model or surrogate species. A weight-of-evidence argument can be established on the basis of these different types of studies. A difficulty that will arise in evaluating the evidence for some marine mammal populations is that, under protection or well-regulated exploitation, they are in the process of recovering and experiencing phases of rapid population growth. This can mask or confound the evidence for contaminant-related effects on population dynamics and abundance, whereas such effects may be more readily discernible in depleted populations.

12. *Non-destructive sampling of biomarkers and contaminants*

The application and validation of biomarkers for marine mammals remain at an early stage of development. Biomarkers can be used to assess chemical exposure and, with further development, should, in the future, be capable of predicting effects in marine mammals. Biomarker studies should be included with other types of contaminant-related research on marine mammals. The results of such studies need to be closely linked to information on contaminant burdens and exposures, and on physiological and life-history traits. Some widespread contaminants and metabolites have low potential for bioaccumulation and are short lived, but their ingestion or inhalation by marine mammals may be a cause for concern. Biomarkers may be very appropriate to measure for assessing exposure to such compounds. Understanding the mechanisms of biomarker production is critical, but at least three key distinctions should be made: (1) biomarkers that simply demonstrate exposure versus those that show or predict effect; (2) biomarkers of persistent versus ephemeral contaminants; and (3) biomarkers for single versus multiple chemical stressors. Biomarkers for marine mammals should be sensitive, rapid, inexpensive and field-adaptable. Similarly, non-destructive sampling is often the preferred approach for collecting contaminant exposure data from marine mammals. However, samples that can be collected nondestructively are often not from the target site for the contaminants of concern. This situation is not unique to marine mammals. The same dilemma can arise in studies with humans, and the relevant literature on humans should be directly applicable. The primary issue of concern is the validation of a measurement in a tissue, such as blood, as a reliable surrogate for a target site in an internal organ. Validation studies will be necessary to increase the value of nondestructive samples as surrogates for target sites in marine mammals.

13. Expansion of sampling and monitoring programs to include histopathology, immunotoxicity and life-history information

Some contaminants to which marine mammals are exposed are known to affect the immune systems of other species. It is uncertain to what extent they may also affect the immune systems of marine mammals. There is sufficient reason for concern, however, and long-term programs for sampling marine mammals should include protocols for obtaining samples and data to support investigations of histopathology and immunotoxicology. In all cases, it is important to emphasize the need to collect associated life-history information. The latter is essential for proper interpretation of findings.

14. Selecting model species of marine mammals

It is unrealistic to expect that adequate studies will be conducted on most marine mammal species. Therefore, model species must be identified and employed in relevant studies. It was generally agreed that the most appropriate species of marine mammals to use as study subjects are those for which considerable information is already available about population dynamics, life history and physiology; that are in captivity or otherwise easily accessible; and that are already involved in related ongoing studies. Other factors to consider are the feasibility of obtaining high-quality samples in large enough numbers to ascertain sources of variation; whether the species occurs across a gradient of habitats (highly polluted versus relatively pristine); and conservation status (endangered, threatened, depleted, etc.). Among the cetaceans that come the closest to meeting these requirements are the bottlenose dolphin (*Tursiops truncatus*), harbor porpoise (*Phocoena phocoena*) and beluga whale [the 1995 International Whaling Commission workshop on contaminants and cetaceans (Reijnders *et al.*, 1999) also identified these as suitable species for study]. Captive studies of baleen whales are out of the question, but given the considerations listed above, the bowhead whale (*Balaena mysticetus*) is a suitable candidate for a model species. This is due to the availability of specimens from the subsistence hunt by Alaskan natives and the relatively solid background of life history and other information on the species. On a global scale, minke whales (*Balaenoptera acutorostrata* and *B. bonaerensis*) may serve as even more useful models, because of regional variations in ichthyophagy and availability of specimens from hunting. The harbor seal and gray seal (*Halichoerus grypus*) are the best candidates among phocid seals, whereas the California sea lion (*Zalophus californianus*) and northern sea lion (*Eumetopias jubatus*) are the prime candidates among otariids. The polar bear (*Ursus maritimus*) can be studied directly. Although sirenians were not generally seen as a high-priority group in the context of typical organochlorine contaminants, the Florida manatee (*Trichechus manatus latirostris*) is the best model sirenian because of the well-organized stranding program, number of captive individuals and ongoing life history studies. Because sirenians feed near the bottom in coastal and inland waterways and are herbivores, their exposure to contaminants may include less widely recognized chemicals that are more prevalent in such ecological contexts (see also O'Shea, Chapter 11, this volume). The mink (*Mustela vison*) is a widely used experimental animal, and it can be used as a surrogate for the closely related marine mustelids [sea otter (*Enhydra lutris*) and marine otter]. Even among mustelids, however, mink are known to be exceptionally sensitive to certain contaminants.

15. Selecting model contaminants

The 1995 workshop on cetaceans and contaminants sponsored by the International Whaling Commission (Reijnders *et al.*, 1999) identified three categories of chemicals according to how regularly they had been monitored in cetaceans. It was suggested that the criteria for selecting model compounds for priority monitoring and study should include levels of production, potency of bioaccumulation, and toxicity. The US Marine Mammal Commission workshop emphasized the need to place highest priority on chemicals with known adverse effects and/or that are most likely to bioaccumulate in marine mammals. Based on current knowledge, PCBs, DDT and metabolites, other organochlorines, butyltins, and a few trace elements are the most obvious candidates.

16. Complex mixtures

All marine mammals have body burdens of many different contaminants, in variable amounts, stored or circulating in different organ systems. Although in standard toxicological studies effects are generally related to a single chemical, marine mammals in the wild are exposed to complex mixtures rather than to single chemicals. Effects from exposure to multiple contaminants may be synergistic or antagonistic. For this reason, experimental exposures should include complex mixtures that mimic as closely as possible the types of exposure experienced by animals in nature. Care should be taken to avoid conclusions based solely on *in vitro* exposures, which may be more abrupt and thus exaggerated in their effect, compared with the gradual and perhaps lower-dose exposure experienced by the whole organism in nature. However, it should also be considered that some single substances may need to be studied in isolation in order to assist managers and policy-makers in decisions about production, use and fate of specific chemicals.

17. Dose–response relationships

Ideally, the effects of contaminants should be understood well enough to predict responses from specific doses, including likely 'safe' or no-effect levels of exposure. Analyses of the dose–response relationship must be pursued on a species-by-species, contaminant-by-contaminant basis. Experiments and sampling designs should require that exposure be controlled or at least measured, and that potentially confounding factors (e.g. age, sex, reproductive status, nutrition, etc.) be taken into account when investigating dose–response relationships.

18. Endocrine disruption

Endocrine disruption has become recognized as a potentially widespread, insidious conservation problem. Although evidence that hormonal or developmental dysfunction in marine mammals is linked to contaminant exposure has been slow to accumulate, some common contaminants of marine mammal tissues (such as several of the organochlorines) are known to affect the endocrine systems of other species. A precautionary attitude requires, at a minimum, that the potential for endocrine disruption in marine mammals be regarded as a serious possibility and be subject to aggressive research and evaluation. A weight-of-evidence approach should be used in judging whether measures should be taken to reduce exposure, out of concern about this danger. Systematic appraisal of a number of morphological and other endpoints of endocrine disruption, as described by the Working Group on Endocrinology and Reproduction (see also p. 598), and should be incorporated in routine marine mammal stranding and health evaluations. Trans-generational effects in long-lived species like marine mammals may be particularly difficult to discern without incorporating such appraisals in ongoing studies.

19. *Understanding blubber physiology and estimating total body burdens of lipophilic contaminants*

Physiological condition affects the distribution of lipophilic contaminants in the bodies of marine mammals. Further research on lipid dynamics is required to improve understanding of the processes determining this distribution, i.e. how lipid dynamics and physiological demands affect circulating levels of contaminants that ultimately reach target sites of action. Most published reports refer only to concentrations of contaminants in blubber. Concentrations in muscle or other tissues are rarely mentioned. The ability to estimate total body burden is important for determination of exposure factors in risk assessment. Estimating the total body burden of contaminants requires not only information on distribution, but also estimates of the total amount of blubber, muscle and other tissues.

20. *Statistical power in experimental designs and sampling designs*

Regardless of whether a contaminant study is to be conducted in the laboratory or the field, it is essential that hypotheses be clearly formulated and that a statistical model be developed and the appropriate sample size determined before the protocol begins. It is of no benefit to study a sample that is too small for statistically valid results. Nor is there any benefit in sampling more animals than is necessary. In the laboratory, standard statistical methods of determining significance and power can be applied. In the field, it might prove necessary to adapt the mathematical models of epidemiology to understand the impacts of contaminants on whole groups of animals. This would involve stratification by age, sex, reproductive condition, season, location and other factors.

Conclusions

The study of toxicology in marine mammals is an emerging field. Numerous subdisciplines and avenues of research are involved as the field enters the twenty-first century. Chapters included in this volume provide overviews and case studies of research in a broad spectrum of subdisciplines, areas and situations. These include environmental chemistry, biochemical and molecular research on mechanisms of action, investigations of strandings and die-offs, and impacts on the health, reproduction and immune systems of marine mammals. Geographic domains of the research as summarized in the various chapters are global, ranging from Arctic ecosystems to the tropics, but also focus on intriguing case studies from localized areas of contamination such as the Baltic Sea and the St. Lawrence River and its estuary. The conclusions of this research are not comforting. There is widespread evidence of exposure of marine mammals to anthropogenic contaminants and biotoxins. There is mounting evidence that this exposure has negative implications for marine mammal populations. With increasing contamination of the marine environment, the twenty-first century may see the influence of toxic substances emerge as a powerful negative force on marine mammal birth and death processes, perhaps comparable to those human practices that are currently the focus of management and conservation of these species.

There are formidable challenges to overcome if marine mammal toxicology is to succeed in reaching goals of providing sound science to guide future policy on pollution of the seas with toxic substances. These are due to the complexities inherent in studying multiple chemical stressors that can affect multiple functional systems in a diverse assemblage of large, mobile and often inaccessible aquatic mammals. Recognizing these challenges, scientists have developed detailed recommendations for future research directions. To meet with major success, these avenues of future research will need to be well-defined, integrated, multidisciplinary efforts that focus on key areas of uncertainty about the impacts of toxic substances on marine mammals. They must also be supported by society, with a subsequent willingness by society to then take action to rectify any negative impacts that the science may reveal.

Acknowledgments

We thank all the participants at the Keystone workshop for developing the recommendations for future research. In particular we thank R. Reeves and A. Long for their efforts in organizing and editing the original workshop report, and the following individuals who served as group leaders, subgroup leaders, rapporteurs, or steering committee members: K. Beckmen, S. De Guise, D. DeMaster, J. Geraci, L.E. Gray, J. Harwood, R. Hofman, R. Kavlock, J. Kucklick, T. Lipscomb, M. Matta, R. Mattlin, M. Moore, J. Reif, J. Reynolds, T. Rowles, L. Schwacke and R. Wells. The workshop was sponsored by the US Marine Mammal Commission, the National Marine Fisheries Service, the Environmental Protection Agency, the US Geological Survey and the National Fish and Wildlife Foundation. J. Twiss and the late N. Foster had the initial insights that provided the impetus to develop the workshop.

References

Anderson, P.K. 2001. Marine mammals in the next one hundred years: twilight for a Pleistocene megafauna? *Journal of Mammalogy* 82: 623–629.

Bossart, G.D., Baden, D.G., Ewing, R., Roberts, B. and Wright, S.D. 1998. Brevetoxicosis in manatees (*Trichechus manatus latirostris*) from the 1996 epizootic: gross, histologic, and immunohistochemical features. *Toxicologic Pathology* 26: 276–282.

Hardin, G. 1968. The tragedy of the commons. *Science* 162: 1243–1248.

Harwood, J. 2001. Marine mammals and their environment in the twenty-first century. *Journal of Mammalogy* 82: 630–640.

Jensen, B.A. and Hahn, M.E. 2001. cDNA cloning and characterization of a high affinity aryl hydrocarbon receptor in a cetacean, the beluga, *Delphinapterus leucas*. *Toxicological Sciences* 64: 41–56.

Kim, E.-Y. and Hahn, M.E. 2002. cDNA cloning and characterization of an aryl hydrocarbon receptor from the harbor seal (*Phoca vitulina*): a biomarker of dioxin susceptibility? *Aquatic Toxicology*, 58: 57–73.

O'Shea, T.J., Reeves, R.R. and Long, A.K. (eds) 1999. *Marine mammals and persistent ocean contaminants: proceedings of the marine mammal commission workshop, Keystone, Colorado, 12–15 October 1998.*

Reijnders, P.J.H., Donovan, G.P., Aguilar, A. and Bjørge, A. (eds) 1999. Report of the workshop on chemical pollution and cetaceans. *Journal of Cetacean Research and Management* (Special Issue 1): 1–42.

Ross, P.S. and DeGuise, S. 1998. Environmental contaminants and marine mammal health: research applications. *Canadian Technical Report of Fisheries and Aquatic Sciences 2255.*

Sanderson, K. and Gabrielson, G.W. (eds) 1996. Special issue: marine mammals and the marine environment. A collection of papers presented at the international conference on marine mammals and the marine environment, Lerwick, Shetland, UK, 20–21 April 1995. *Science of the Total Environment* 186: 1–179.

Scholin, C.A. *et al.* 2000. Mortality of sea lions along the central California coast linked to a toxic diatom bloom. *Nature* 403: 80–84.

Tatsukawa, R., Tanabe, S., Miyazaki, N. and Tobayama, T. (eds) 1994. Special issue: marine pollution – mammals and toxic contaminants. A collection of papers presented at the international symposium on marine pollution – mammals and toxic contaminants, Kamogawa, Japan, 6–8 February 1993. *Science of the Total Environment* 154: 107–256.

Index

Note: page numbers in italics refer to figures and tables

9 780367 395636